ENCYCLOPED   APPLICATIONS

EDITED BY G.-C. ROTA

Editorial Board

R. S. Doran, M. Ismail, T.-Y. Lam, E. Lutwak, R. Spigler

Volume 80

Lie's Structural Approach to PDE Systems

ENCYCLOPEDIA OF MATHEMATICS AND ITS APPLICATIONS

ENCYCLOPEDIA OF MATHEMATICS AND ITS APPLICATIONS

Lie's Structural Approach to PDE Systems

OLLE STORMARK

Royal Institute of Technology, Stockholm

CAMBRIDGE
UNIVERSITY PRESS

CAMBRIDGE UNIVERSITY PRESS
Cambridge, New York, Melbourne, Madrid, Cape Town,
Singapore, São Paulo, Delhi, Tokyo, Mexico City

Cambridge University Press
The Edinburgh Building, Cambridge CB2 8RU, UK

Published in the United States of America by Cambridge University Press, New York

www.cambridge.org
Information on this title: www.cambridge.org/9781107403321

First published 2000
First paperback edition 2011

A catalogue record for this publication is available from the British Library

ISBN 978-0-521-78088-9 Hardback
ISBN 978-1-107-40332-1 Paperback

Contents

Contents

Preface

Das Ziel der Wissenschaft ist einerseits, neue Tatsachen zu erobern, anderseits, bekannte unter höheren Gesichtspunkten zusammenfassen.

S. Lie

One of the principal objects of theoretical research is to find the point of view from which the subject appears in its greatest simplicity.

J.W. Gibbs

Good mathematics = down-to-earth mathematics.

A. Andreotti

Everyone knows that the theory of PDE systems is enormously rich in results—but *what about foundations?* This monograph describes one approach, but there is no claim that it is the only one, or that it is the best possible; just that there is one, and moreover one that has been around for a long time without having been as recognized as it deserves.

The study is restricted to local solvability. If a PDE system S is defined in a domain $\mathcal{D}$, and if it can be shown that S possesses local solutions at each point of $\mathcal{D}$, the question of global solvability boils down to whether it is possible to glue together local solutions in order to form global ones—in analogy with the cohomology theory for coherent analytic sheaves. But *first of all one has to know that there are local solutions.*

A major idea is to regard PDE theory from the point of view of *differential geometry*—rather than basing it on analysis, say.

ix

To illustrate this, let us first find a suitable angle from which ODE systems appear in a simple way.

Example. Look for functions $y(t)$, $z(t)$ satisfying the ODE system

$$\begin{cases} y'' = f(t, y, z, y', z'), \\ z'' = g(t, y, z, y', z'). \end{cases}$$

With $x^1 := y$, $x^2 := z$, $x^3 := y'$ and $x^4 := z'$, this goes over into the first order system

$$\begin{cases} dx^1/dt = x^3, \\ dx^2/dt = x^4, \\ dx^3/dt = f(t, x^1, x^2, x^3, x^4), \\ dx^4/dt = g(t, x^1, x^2, x^3, x^4). \end{cases}$$

Therefore the key problem in ODE theory consists in solving first order systems

$$\begin{cases} dx^1/dt = f^1(t, x^1, x^2, \ldots, x^n), \\ \ldots \\ dx^n/dt = f^n(t, x^1, x^2, \ldots, x^n). \end{cases} \tag{*}$$

The classical existence and uniqueness theorem says that for smooth f^k there is precisely one local solution $(x^1(t), \ldots, x^n(t))$ satisfying initial data $(x^1(t_0), \ldots, x^n(t_0)) = (x_0^1, \ldots, x_0^n)$.

In order to make this conceptually clearer, rewrite (*) as a pfaffian system:

$$\begin{cases} dx^1 - f^1(t, x) \, dt = 0, \\ \ldots \\ dx^n - f^n(t, x) \, dt = 0, \end{cases}$$

where $x = (x^1, \ldots, x^n)$. Setting

$$\theta^k := dx^k - f^k(t, x) \, dt \quad \text{for} \quad k = 1, \ldots, n,$$

a solution $(x^1(t), \ldots, x^n(t))$ is a function whose graph

$$\mathbb{R}_t \ni t \mapsto (t, x^1(t), \ldots, x^n(t)) \in \mathbb{R}_t \times \mathbb{R}_x^n$$

is an integral curve of the pfaffian system $\theta^1 = \cdots = \theta^n = 0$ on $\mathbb{R}_t \times \mathbb{R}_x^n$.

The dual of the latter consists of those vector fields

$$V = a\frac{\partial}{\partial t} + \sum_{k=1}^{n} a^k \frac{\partial}{\partial x^k}$$

which satisfy $0 = \theta^m(V) = a^m - a \cdot f^m(t, x)$ for $m = 1, \dots, n$. Consequently they are all multiples of the vector field

$$X = \frac{\partial}{\partial t} + \sum_{k=1}^{n} f^k(t, x) \frac{\partial}{\partial x^k}.$$

Then $(x^1(t), \dots, x^n(t))$ is a solution of (*) if and only if its graph is an integral curve of X.

Thus solving a first order ODE system is equivalent to finding the integral curves of a vector field. The latter may be done locally by introducing new coordinates making the vector field maximally simple.

The local rectification lemma. *Let*

$$X = \frac{\partial}{\partial t} + \sum_{k=1}^{n} f^k(t, x) \frac{\partial}{\partial x^k}$$

be a vector field defined near the origin of $\mathbb{R}_t \times \mathbb{R}_x^n$. *Then there is a local diffeomorphism*

$$\phi : \mathbb{R}_s \times \mathbb{R}_y^n \xrightarrow{\cong} \mathbb{R}_t \times \mathbb{R}_x^n$$

with $\phi(0) = 0$ *and* $\phi_*(\partial/\partial s) = X$.

This means that the integral curves of X are given by

$$y^k(t, x) = \text{constant} \quad \text{for } k = 1, \dots, n.$$

From a technical point of view this theorem is equivalent to the classical local existence theorem for first order ODE systems—but the advantage is that it is much more appealing to the intuition (if you agree with this, read on; otherwise stop right here).

Conclusion. *Solving an ODE system is equivalent to rectifying a vector field—which always can be done locally in the smooth category.*

Since this result is most satisfying, it is natural to ask if something similar works for general PDE systems.

Question. *Is it possible to geometrize PDE systems in an analogous way?*

The first step consists in rewriting an arbitrary PDE system as a pfaffian system, or—dually—a vector field system.

To get the idea, consider a first order ODE:

$$F(x, y, y') = 0.$$

The graph of a solution is given by

$$x \mapsto (x, y(x)).$$

Analogously the 1-graph is given by

$$x \mapsto (x, y(x), y'(x)),$$

the 2-graph by

$$x \mapsto (x, y(x), y'(x), y''(x)),$$

and so on. Let us concentrate on the 1-graph, which is a curve in a 3-dimensional space $J^1(\mathbb{R}_x, \mathbb{R}_y)$ with coordinates x, y and p, say:

$$\begin{cases} x = x, \\ y = y(x), \\ p = y'(x). \end{cases}$$

Note that on this 1-graph

$$dy - p\, dx = 0.$$

The one-form $\theta^0 = dy - p\, dx$ appearing here is called the *contact form*.

Conversely, let c be a curve in the (x, y, p)-space on which x can be used as a local coordinate, and suppose that θ^0 vanishes on c. Then c is of the form

$$x \mapsto (x, f(x), g(x))$$

with

$$0 = df - g\, dx = (f'(x) - g(x))\, dx, \quad \text{i.e.,} \quad g(x) = f'(x).$$

But this means that c is the 1-graph of the function $f(x)$.

To the ODE $F(x, y, y') = 0$ corresponds the hypersurface M in the (x, y, p)-space defined by

$$F(x, y, p) = 0.$$

Clearly solutions of $F(x, y, y') = 0$ will correspond to 1-graphs contained in M, that is, to curves in M satisfying the two requirements

(i) x can be used as a local coordinate on the curve,
(ii) $\theta^0 \equiv dy - p\, dx = 0$ vanishes on the curve.

Let us assume M to be smooth, and set

$$\theta := \theta^0{}_{|M} = (dy - p\,dx)_{|M},$$

so that θ is a one-form on M. Then the solutions of $F(x, y, y') = 0$ correspond precisely to the integral curves of θ on which $dx \neq 0$. If the latter condition is not fulfilled, the integral curve is said to represent a *generalized solution* of the ODE.

Conclusion. *The ODE $F(x, y, y') = 0$ can be regarded as a manifold M equipped with a one-form θ. The solutions of the ODE correspond to integral curves of θ.*

Let us now play the same game for a qth order PDE system S in n independent variables $x^1, \dots, x^n$ and m dependent variables $z^1, \dots, z^m$:

$$S: \quad F^a\left(x^1, \dots, x^n; z^1, \dots, z^m; \dots, \frac{\partial^k z^j}{\partial x^{i_1} \dots \partial x^{i_k}}, \dots\right) = 0,$$

with $a = 1, \dots, A$ and $k \leq q$. To this set-up corresponds the jet space $J^q(\mathbb{R}^n_x, \mathbb{R}^m_z)$ with coordinates

$$x^1, \dots, x^n, z^1, \dots, z^m, \dots, p^j_{i_1 \dots i_k}, \dots,$$

where $p^j_{i_1 \dots i_k}$ is associated to the derivative

$$\frac{\partial^k z^j}{\partial x^{i_1} \dots \partial x^{i_k}}.$$

S corresponds to the subset

$$M: \quad F^a(x^1, \dots x^n; z^1, \dots, z^m; \dots, p^k_{i_1 \dots i_k}, \dots) = 0$$

of $J^q(\mathbb{R}^n_x, \mathbb{R}^m_z)$; let us assume M to be smooth.

An n-dimensional submanifold $\mathcal{N}$ of $J^q(\mathbb{R}^n_x, \mathbb{R}^m_z)$ is a q-graph if and only if

(i) $dx^1 \wedge \cdots \wedge dx^n \neq 0$ on $\mathcal{N}$,
(ii) $\mathcal{N}$ is an integral manifold of the contact ideal ${}^q Ct^{n,m}$ generated by the one-forms

$$\begin{cases} dz^j - \sum_{i_1=1}^n p^j_{i_1}\,dx^{i_1}, \\ dp^j_{i_1} - \sum_{i_2=1}^n p^j_{i_1 i_2}\,dx^{i_2}, \\ \dots \\ dp^j_{i_1 \dots i_{q-1}} - \sum_{i_q=1}^n p^j_{i_1 i_2 \dots i_q}\,dx^{i_q}. \end{cases}$$

Let $\mathcal{P}$ denote the restriction of ${}^q Ct^{n,m}$ to M. Then *the set of solutions of S correspond precisely to the set of n-dimensional integral manifolds of $\mathcal{P}$ on which $dx^1 \wedge \cdots \wedge dx^n \neq 0$.*

Define the dual vector field system $\mathcal{V} = \mathcal{P}^\perp$ on M by

$$X \in \mathcal{V} \Longleftrightarrow \theta(X) = 0 \text{ for all } \theta \in \mathcal{P}.$$

Then

solving the PDE system S is equivalent to finding all n-dimensional integral manifolds of $\mathcal{V}$ on which $dx^1 \wedge \cdots \wedge dx^n \neq 0$.

Again, suspending the last condition gives *generalized solutions* of S.

The classical terminology for finding integral manifolds of a pfaffian or vector field system is to *integrate* the system.

Conclusion. *Solving a PDE system is equivalent to integrating a pfaffian system or its dual vector field system.*

Consider the vector field version: let $\mathcal{V} = (X_1, \ldots X_q)$ be the vector field system generated by the vector fields $X_1, \ldots, X_q$ over the ring of smooth functions on the manifold M. Shrinking M if necessary, the X_i can be assumed to be linearly independent everywhere. The *derived system* $\mathcal{V}'$ is generated by the X_i and their Lie brackets $[X_i, X_j]$; say that $\mathcal{V}' = (X_1, \ldots, X_q; Z_1, \ldots, Z_p)$. Then the *structure equations* of $\mathcal{V}$ are given by

$$[X_i, X_j] \equiv \sum_{k=1}^{p} c_{ij}^k Z_k \pmod{\mathcal{V}} \qquad \text{for } i, j = 1, \ldots r,$$

with certain *structure functions* c_{ij}^k. The latter depend on the basis of $\mathcal{V}$, and it is natural to choose one that kills as many c_{ij}^k as possible. The resulting set is called the *Lie structure* of $\mathcal{V}$, and the general claim is

the integrability properties of $\mathcal{V}$ are governed by the Lie structure!

One reason for this is Cartan's local existence theorem for integral manifolds. The proof consists of two parts: first linear algebra applied to the structure equations gives all *involutions*, and then these are specialized to *integral manifolds* by repeated applications of the Cauchy–Kowalewski theorem. Unfortunately the latter requires power series, and is not valid in the C^∞ category.

Questions:

  (i) *What is required in order to obtain integral manifolds without using power series?*

(ii) *Is there a natural intrinsic property from which more information than just existence can be derived?*

In order to tackle these it is convenient to use the classification of Drach: any PDE system is equivalent to either a first or a second order PDE system in one dependent variable.

First order systems in one dependent variable are easily understood by means of Lie's structural methods, so the real difficulties start with second order systems.

The main part of the monograph is devoted to the study of second order PDE systems in one dependent and two or three independent variables. When considering these, one is naturally led to the notion of *Monge characteristic subsystems*, which in their turn provide the following partial answer to the questions above:

> *If a vector field system admits Monge systems with enough first integrals, then it is possible to find integral manifolds without using the Cauchy–Kowalewski theorem. Moreover these Monge systems yield a lot of interesting information beyond that of pure existence—in particular as regards classifications.*

A big surprise is that looking at second order PDE systems in one dependent variable from this angle, the theory will be dominated by local Lie groups and Lie pseudogroups.

Extending these methods to the case of more than three independent variables would be quite complicated, but surely not impossible. And anyway, who would expect the general PDE theory to be simple?

Most topics treated here can be found in the classical works of Lie, Cartan and Vessiot. However, a great effort has been made to present the ideas in a *unified* and very *simple* manner. In order not to obscure the fundamental issues it has been left to the interested reader to fill in the details needed to obtain his or her own desired level of rigour.

1

Introduction and summary

This is the story of a geometric approach to the theory of PDE systems, initiated by Sophus Lie and developed by two of his disciples, Élie Cartan and Ernest Vessiot.

In chapter **2** it is explained how any decent PDE system S can be considered as a submanifold of an appropriate jet bundle. The latter is equipped with its canonical contact pfaffian system, the restriction of which to S makes S a *manifold with a pfaffian system* $\mathcal{P}$ or, dually, a *manifold with a vector field system* $\mathcal{V}$. The problem of solving the PDE system S then goes over into that of finding *integral manifolds* of $\mathcal{P}$ (or $\mathcal{V}$) of a prescribed dimension.

Of the three heroes of our tale, Lie and Vessiot favoured the vector field approch, while Cartan is the great champion of differential forms. For our purposes it is important to be able to use both approaches and to have a complete duality, so that each concept for vector field systems has a counterpart for pfaffian systems, and vice versa.

As we are only interested in local properties, S is assumed to be a small open subset of $\mathbb{R}^r$ (or $\mathbb{C}^r$), and $\mathcal{V}$ is supposed to be generated by vector fields $X_1, \ldots, X_q$, independent everywhere on S: $\mathcal{V} = (X_1, \ldots, X_q)$.

The simplest case occurs when $\mathcal{V}$ is complete with respect to Lie brackets, that is,

$$[X_i, X_j] \equiv 0 \pmod{\mathcal{V}} \quad \text{for } i, j = 1, \ldots, r.$$

According to the Frobenius theorem it is in this case possible to introduce local coordinates $x^1, \ldots, x^r$ such that $\mathcal{V} = (\partial/\partial x^1, \ldots, \partial/\partial x^q)$, whence the integral manifolds are given by

$$x^{q+1} = \text{constant}, \ldots, x^r = \text{constant}.$$

With the derived system $\mathcal{V}'$ being generated by the X_i and their Lie

1

brackets $[X_i, X_j]$, the Frobenius condition can equivalently be written as $\mathcal{V}' = \mathcal{V}$.

The general case with $\mathcal{V}' \supsetneq \mathcal{V}$ is solved in chapter **3** by means of Cartan's local existence theorem. The key idea is to first look for *maximal involutions*—with an involution being a subsystem $\mathcal{I}$ of the vector field system $\mathcal{V}$ satifying $[\mathcal{I}, \mathcal{I}] \subseteq \mathcal{V}$. Then these involutions are specialized to *complete subsystems*, i.e., subsystems $\mathcal{W}$ of $\mathcal{V}$ with $[\mathcal{W}, \mathcal{W}] \subseteq \mathcal{W}$. Thereupon the Frobenius theorem yields the wanted integral manifolds.

The step from involutions to complete subsystems is based upon a repeated application of the Cauchy–Kowalewski theorem, which unfortunately requires analyticity.

Often one wants n-dimensional integral manifolds on which $\omega^1 \wedge \cdots \wedge \omega^n \neq 0$, where the ω^i are given one-forms. If the general n-dimensional involution $\mathcal{I}_n$ satisfies $\omega^1 \wedge \cdots \wedge \omega^n|_{\mathcal{I}_n} \neq 0$, $\mathcal{V}$ is said to be involutive with respect to the ω^i. In this case Cartan's procedure yields integral manifolds of the kind wanted.

Any vector field system $\mathcal{V}$ can be *prolonged* to a vector field system $\mathcal{V}^{(1)}$ on a higher dimensional manifold, and this in turn can be prolonged to $\mathcal{V}^{(2)}$, and so on. Moreover there is a one-to-one correspondence between the integral manifolds of $\mathcal{V}$ and those of $\mathcal{V}^{(k)}$ for $k = 1, 2, 3, \ldots$.

Chapter **4** sketches the prolongation theorem of Cartan and Janet, which says that by a *finite* number of prolongations it is possible to conclude either that some $\mathcal{V}^{(m)}$ is involutive with respect to $\omega^1 \wedge \cdots \wedge \omega^n$—in which case the wanted integral manifolds are given by Cartan's existence theorem—or that $\mathcal{V}$ does not admit any integral manifold on which $\omega^1 \wedge \cdots \wedge \omega^n \neq 0$.

Drach observed that any PDE system is equivalent to either a first or a second order PDE system in one dependent variable. In chapter **5** we take a preliminary look at a single second order PDE in one dependent variable, and in particular investigate the presence of *singular* vector fields—that is, vector fields in $\mathcal{V}$ commuting modulo $\mathcal{V}$ with a greater number of vector fields than the average one does. There turn out to be either exactly two singular subsystems of $\mathcal{V}$, or none at all. If there are such, the PDE is *hyperbolic* if they are different, and *parabolic* if they coincide.

These observations give rise to the notion of *Monge characteristic subsystems* of the vector field system $\mathcal{V}$: a subsystem $\mathcal{M}$ of $\mathcal{V}$ is Monge if

(i) $\mathcal{M}$ is singular,

(ii) $\mathcal{M} \cap \mathcal{I} \neq 0$ for any maximal involution $\mathcal{I}$ of $\mathcal{V}$, and

(iii) $\mathcal{M} \cap \mathcal{W}$ is complete for any maximal complete subsystem $\mathcal{W}$ of $\mathcal{V}$.

The integral manifolds of $\mathcal{M} \cap \mathcal{W}$ are called *Monge characteristics*.

A special case is the *Cauchy charateristic subsystem* $\mathcal{C}(\mathcal{V})$ *of* $\mathcal{V}$:

$$\mathcal{C}(\mathcal{V}) := \{X \in \mathcal{V} \mid [X, \mathcal{V}] \subseteq \mathcal{V}\}.$$

$\mathcal{C}(\mathcal{V})$ is complete, and is included in any maximal complete subsystem of $\mathcal{V}$.

In chapter **6** we consider the integration of vector field systems satisfying $\dim \mathcal{V}' = \dim \mathcal{V} + 1$, which includes first order PDE systems in one dependent variable as a special case. Such a vector field system is essentially equivalent to a single pfaffian equation $\omega = 0$, which is solved by putting it into a canonical form:

$$\omega = 0 \iff dz - \sum_{i=1}^{n} p_i \, dx^i = 0.$$

The reduction procedure accomplishing this is a good demonstration of how powerful Lie's ideas are.

By Drach's classification there then remains to consider second order PDE systems in one dependent and n independent variables; the remainder of the monograph is devoted to the cases $n = 2$ and 3. The main method is

look for Monge systems and their first integrals!

At the outset there is no consideration at all of groups—but they will turn up anyway. The first example is the Lie pseudogroup of contact transformations, which consists of all local diffeomorphisms of the jet bundle $J^1(\mathbb{R}^n_x, \mathbb{R}_z)$ preserving the pfaffian equation $dz - \sum_{i=1}^{n} p_i \, dx^i = 0$. A general Lie pseudogroup is a family of local diffeomorphisms constituting the general solution of some PDE system, and being closed under composition whenever this is defined.

Chapter **7** discusses higher order contact transformations and prolongations of local diffeomorphisms to jet bundles.

In chapter **8** the general solution of the defining PDE system is supposed to depend on a *finite* number of parameters only, in which case the Lie pseudogroup is called a *local Lie group*. Acting on $\mathbb{C}^n$ and having r parameters, its elements are given by local diffeomorphisms

$$(x^1, \ldots, x^n) \mapsto (f^1(x^1, \ldots, x^n; a^1, \ldots, a^r), \ldots, f^n(x^1, \ldots, x^n; a^1, \ldots, a^r)),$$

or expressed more simply, $x \mapsto f(x;a) \equiv f_a(x)$. The group property means that whenever $f_a \circ f_b$ is defined, there is a $c = \phi(a,b)$ such that $f_a \circ f_b = f_c$. In this way there is induced a group action on the parameter space,

$$(a,b) \mapsto \phi(a,b),$$

giving this a Lie group structure in the modern sense—i.e., the parameter space acts on itself as a group.

In the action $(a,b) \mapsto \phi(a,b)$ the as can be regarded as 'variables' and the bs as 'parameters', or just the other way around. Accordingly two parameter groups arise, corresponding to right and left multiplication in modern terminology. And to these there correspond two Lie algebras of vector fields. Usually the attention is restricted to one of them, but for our applications it is essential to consider both.

Lie's original version of Lie group theory depends heavily on differential equations. In order to be able to study Lie groups and Lie algebras over more general fields than $\mathbb{C}$ and $\mathbb{R}$, 20th century mathematicians have worked hard to find purely algebraic foundations of the Lie theory. Unfortunately this has had the effect of cutting off Lie theory from differential equations—and therefore one has to struggle a bit in order to rebuild the bridge between these two disciplines. The same goes for Lie pseudogroups: what we will need here is Cartan's original version, rather than later algebraizations.

The study of hyperbolic second order PDEs requires the classification of Lie groups of dimension ≤ 3—which is derived in chapter **9**.

Cartan's local existence theorem ultimately depends on the Cauchy–Kowalewski theorem. The latter requires analyticity, and should therefore be avoided if possible. For instance, the integration of vector field systems $\mathcal{V}$ with $\dim \mathcal{V}' - \dim \mathcal{V} = 0$ or 1 is reduced to solving ODE systems only. Lie and his followers were not satisfied with general ODE systems either, but wanted to go one step further and make a reduction to what Lie called 'ausführbare Operationen'. An example is Lie's study of complete vector field systems admitting a nontrivial symmetry group, in which case he achieves a reduction to so called *Lie equations*. Originally these were defined as ODE systems corresponding to vector fields $\sum_{k=1}^{n} a^k(t) X_k(x)$, with $X_1(x), \ldots, X_n(x)$ forming a Lie algebra of vector fields. Later on Lie and Vessiot found that such systems can be characterized as ODE systems having the property that the general solution may be expressed as a certain function of a number of particular solutions.—All this is explained in chapter **10**, which also contains Vessiot's generalization to 2-dimensional Lie vector field systems.

Thus prepared we begin the discussion of second order PDEs in one dependent and two independent variables in chapter **11**. One item is the characterization of those vector field systems that arise from second order PDEs, and another is a sketch of Darboux's method for finding solutions by means of first integrals of the Monge systems.

Goursat found a remarkable classification of hyperbolic PDEs of the form

$$\frac{\partial^2 z}{\partial x \partial y} = f\left(x, y, z, \frac{\partial z}{\partial x}, \frac{\partial z}{\partial y}\right),$$

and having the property that each of the two Monge systems admits two or three functionally independent first integrals. Unfortunately he used ad hoc methods giving no clue whatsoever as to the underlying reasons.

Chapter **12** gives an account of Vessiot's theory of hyperbolic PDEs in one dependent and two independent variables, for which each of the Monge systems admits at least two independent first integrals—thus including the Goursat equations. Making use of these first integrals, the corresponding vector field systems are brought to a *finite number of canonical forms*. Remarkably the classification thus obtained ultimately depends on the classification of 2- and 3-dimensional Lie groups!

The Goursat equations are further investigated in chapter **13**.

While Vessiot's study of hyperbolic PDEs consists in a straightforward reduction to canonical forms, Cartan uses his solution of the equivalence problem in order to classify parabolic PDEs.

The equivalence problem is this: given two manifolds M_1 and M_2 of the same dimension and having local structures S_1 and S_2 respectively (in our case vector field or pfaffian systems), is it possible to find local diffeomorphisms transforming the one structure into the other? Note that if

$$\phi_k : M_1 \xrightarrow{\cong} M_2, \qquad k = 1, 2,$$

are two such diffeomorphisms for which $\phi_2^{-1} \circ \phi_1$ and $\phi_1 \circ \phi_2^{-1}$ are defined, the latter are self-equivalences of (M_1, S_1) and (M_2, S_2) respectively. The family of self-equivalences of (M_k, S_k) forms the *symmetry group* of (M_k, S_k), and it is such symmetry groups that are the original reason for the introduction of the concept of Lie pseudogroup. If (M_1, S_1) and (M_2, S_2) are locally equivalent, their symmetry groups are obviously isomorphic, but the converse need not be true—imagine for instance two different structures which are so complicated that both their symmetry groups reduce to the identity.

Cartan's idea for solving the equivalence problem consists of two steps: first determine all local diffeomorphisms making the symmetry groups isomorphic, and then, among these, determine those which in addition are local equivalences. Now in most applications the first step alone suffices, and then we are reduced to studying equivalences of Lie pseudogroups.

Chapter **14** sketches Cartan's theory of Lie psedogroups, which hence is a preparation for the equivalence problem. The latter is dealt with in chapter **15**.

After this we are ready to tackle Cartan's 'five variable paper' in chapters **16** and **17**. First it is shown that parabolic PDEs for which the (double) Monge system admits at least two independent first integrals are equivalent to systems of two second order PDEs admitting a Cauchy characteristic vector field. These systems are 2-codimensional submanifolds of the 8-dimensional jet bundle $J^2(\mathbb{C}^2, \mathbb{C})$, and so are of dimension 6. The existence of one Cauchy characteristic vector field makes it possible to reduce the integration problem to one dimension less—hence the five variables.

Chapter **18** summarizes Cartan's work on second order PDE systems in one dependent and three independent variables. The idea is to simplify the structure equations of the corresponding pfaffian equations as far as possible, and then regard systems with the same reduced structure as *structurally equivalent*—while renouncing the detailed study of local equivalence. The reduced structure equations reveal for instance that all PDE systems consisting of at least two PDEs do admit singular subsystems, and the latter are then used in order to solve the integration problem by means of the *method of Monge*. The most surprising result is that *all* systems of *two* PDEs can be solved by a reduction to ODE systems!

According to the classification of Drach there then remains to look at second order PDE systems in one dependent and more than three independent variables—but that is quite a different story.

Convention: Only being interested in local properties, we let $\mathbb{R}^n_x$ denote a generic open neighbourhood of the origin in the n-dimensional Euclidean space, with $x^1, \ldots, x^n$ as local coordinates. This neighbourhood is not fixed, but may be shrunk whenever convenient.

2

PDE systems, pfaffian systems and vector field systems

The Norwegian mathematician Sophus Lie (1842–1899) is nowadays most known for his work on Lie groups and Lie algebras. But he regarded the latter—at least initially—mainly as a tool for understanding the theory of differential equations, which for him was the most important branch of mathematics. Or as he expressed it in his review paper 'Zur allgemeinen Theorie der partiellen Differentialgleichungen beliebiger Ordnung', *Leipziger Berichte*, 1895:

> *In der ganzen modernen Mathematik ist die Theorie der Differential-gleichungen die wichtigste Disziplin.*

Before entering into any serious work on differential equations, these must be formulated intrinsically. Previous to Lie this had been done in two special cases: first order ODE systems, and first order PDE systems in one dependent variable. So let us begin by looking at these, and then consider the very simplest type of PDE systems imaginable, namely those of Frobenius type.

2.1 ODE systems, vector fields and 1-parameter groups

Let $t, x^1, \ldots, x^n$ be coordinates for $\mathbb{R}^{n+1}_{t,x} = \mathbb{R}_t \times \mathbb{R}^n_x$, and let

$$\frac{dx_i}{dt} = f^i(t; x^1, \ldots, x^n), \qquad i = 1, \ldots, n, \tag{$\mathcal{S}$}$$

be a first order ODE system, with the f^i being smooth (i.e., C^∞) functions. Solving this is equivalent to determining the integral curves on which $dt \neq 0$ of the pfaffian system

$$\theta^i := dx^i - f^i(t; x) \, dt = 0, \qquad i = 1, \ldots, n. \tag{$\mathcal{P}$}$$

7

For let C be an integral curve of $\mathcal{P}$ on which $dt \neq 0$. Then the restrictions of the x^i to C can be expressed as functions of t, $x^i = x^i(t)$, and these satisfy

$$0 = dx^i - f^i(t;x)\,dt = \left(\frac{dx^i}{dt} - f^i(t;x) \right) dt \iff \frac{dx^i}{dt} = f^i(t;x).$$

Conversely any solution of S clearly defines an integral curve of $\mathcal{P}$ on which $dt \neq 0$.

Notation. If $\theta^1, \ldots, \theta^n$ are one-forms on a certain manifold, $\mathcal{P} = (\theta^1, \ldots, \theta^n)$ denotes the module generated by the θ^i over the ring of smooth functions on the manifold. By abuse of notation $\mathcal{P}$ also denotes the corresponding system $\theta^i = 0$, $i = 1, \ldots, n$. When we are thinking of $\mathcal{P}$ in the latter form, it is called a *pfaffian system*.

By definition the vector field

$$X = a^0(t;x) \frac{\partial}{\partial t} + a^1(t;x) \frac{\partial}{\partial x^1} + \cdots + a^n(t;x) \frac{\partial}{\partial x^n}$$

is dual to $\mathcal{P} = (\theta^1, \ldots, \theta^n)$ if and only if $\theta^i(X) = 0$ for $i = 1, \ldots, n$—i.e.,

$$0 = (dx^i - f^i(t;x)\,dt) \left(a^0 \frac{\partial}{\partial t} + a^1 \frac{\partial}{\partial x^1} + \cdots + a^n \frac{\partial}{\partial x^n} \right) = a^i - a^0 \cdot f^i(t;x)$$

$$\iff a^i = a^0 \cdot f^i(t;x) \quad \text{for } i = 1, \ldots, n.$$

Hence the dual of the pfaffian system $\mathcal{P} = (\theta^1, \ldots, \theta^n)$ is the 1-dimensional vector field system $\mathcal{V} = (X_0)$ generated by

$$X_0 = \frac{\partial}{\partial t} + \sum_{k=1}^{n} f^k(t;x) \frac{\partial}{\partial x^k}.$$

Notation. If $X_1, \ldots, X_q$ are vector fields on a manifold, the module of vector fields generated by the X_i over the ring of smooth functions on the manifold is denoted by $\mathcal{V} = (X_1, \ldots, X_q)$. Such a module is called a *vector field system*.

Integrating $\mathcal{V} = (X_0)$ is the same as finding the integral curves of X_0, i.e., those curves C for which the tangent space of C at an arbitrary point $c \in C$ is spanned by the value of X_0 at the point c. Since the coefficient of $\partial/\partial t$ in X_0 equals 1, t may be used as a local coordinate on such a curve C, which accordingly can be written as $t \mapsto (t; x^1(t), \ldots, x^n(t))$. Then the tangent vectors of C have the form

$$\frac{\partial}{\partial t} + \sum_{k=1}^{n} \frac{dx^k}{dt} \frac{\partial}{\partial x^k}$$

for various t. Comparing this with X_0, we find that

$$\frac{dx^k}{dt} = f^k(t;x) \quad \text{for } k = 1,\dots,n.$$

Hence the integral curves of X_0 are in one-to-one correspondance with the solutions of S.

Conclusion. *The following are equivalent:*

- *solving $S: dx^i/dt = f^i(t;x)$,*
- *finding integral curves of $\mathcal{P} = \left(dx^1 - f^1\,dt,\dots,dx^n - f^n\,dt\right)$ on which $dt \neq 0$,*
- *integrating the vector field $X_0 = \partial/\partial t + \sum_{k=1}^n f^k(t;x)\left(\partial/\partial x^k\right)$.*

Let us consider the vector field aspect. From a geometric point of view it is natural to try to rectify X_0—that is, to find a coordinate transformation

$$\Phi : \mathbb{R}_s \times \mathbb{R}_y^n \overset{\cong}{\longrightarrow} \mathbb{R}_t \times \mathbb{R}_x^n$$

making X_0 equal to $\Phi_*(\partial/\partial s)$. Then the integral curves are given by

$$y^i = \text{constant} \quad \text{for } i = 1,\dots,n.$$

Suppose that the wanted Φ is of the form

$$\Phi : \begin{cases} t = s, \\ x^i = \xi^i(s;y^1,\dots,y^n), & i = 1,\dots,n, \end{cases}$$

with the inverse

$$\Phi^{-1} : \begin{cases} s = t, \\ y^i = \eta^i(t;x^1,\dots,x^n), & i = 1,\dots,n. \end{cases}$$

Then

$$\Phi_*\left(\frac{\partial}{\partial s}\right) = \frac{\partial}{\partial t} + \sum_{i=1}^n \frac{\partial \xi^i}{\partial s}(t;\eta^1(t;x),\dots,\eta^n(t;x))\frac{\partial}{\partial x^i},$$

so that $\Phi_*(\partial/\partial s) = X_0$ if and only if

$$\frac{\partial \xi^i}{\partial s}(t;\eta^1(t;x),\dots,\eta^n(t;x)) = f^i(t;x^1,\dots,x^n) \quad \text{for } i = 1,\dots,n,$$

or equivalently

$$\frac{\partial \xi^i}{\partial s}(s;y^1,\dots,y^n) = f^i(s;\xi^1(s;y),\dots,\xi^n(s;y)) \quad \text{for } i = 1,\dots,n.$$

In order to ensure that Φ becomes a local diffeomorphism we impose the initial conditions

$$\xi^i(0; y^1, \ldots, y^n) = y^i \quad \text{for } i = 1, \ldots, n,$$

because then

$$\frac{\partial(t; x^1, \ldots, x^n)}{\partial(s; y^1, \ldots, y^n)} = 1 \quad \text{when } t = s = 0.$$

So Φ is determined by the system

$$\frac{\partial \xi^i}{\partial s}(s; y^1, \ldots, y^n) = f^i(s; \xi^1(s; y), \ldots, \xi^n(s; y)),$$

$$\xi^i(0; y^1, \ldots, y^n) = y^i, \quad i = 1, \ldots, n.$$

If we regard s as an honest variable and the y^i as parameters, this is a parametrized ODE system. The fundamental existence and uniqueness theorem for such systems (due to Euler, Cauchy, Lipschitz, Picard, ...) then shows that there is a unique local solution Φ.

Let us state this in a slightly more general form as the **local rectification lemma for vector fields**.

Lemma 2.1.1. *Let $X = \sum_{i=0}^n a^i(x) \left(\partial/\partial x^i\right)$ be a smooth vector field defined near $x = 0$, with $a^0(0) \neq 0$. Then there is a unique local diffeomorphism*

$$\Phi : \mathbb{R}_y^{n+1} \xrightarrow{\cong} \mathbb{R}_x^{n+1}$$

such that

$$\Phi_*(\partial/\partial y^0) = X \quad \text{and} \quad x^i(0, y^1, \ldots, y^n) = y^i \quad \text{for } i = 1, \ldots, n.$$

$\square$

Notation. A *first integral* of a vector field X is a nonconstant function f satisfying $Xf = 0$.

Using this notion the rectification lemma says that the vector field $X = \sum_{i=0}^n a^i(x) \left(\partial/\partial x^i\right)$ has a uniquely determined set $\{y^1, \ldots, y^n\}$ of local first integrals, satisfying

$$y^i(0, x^1, \ldots, x^n) = x^i \quad \text{for } i = 1, \ldots, n.$$

Return to the vector field $X_0 = \partial/\partial t + \sum_{k=1}^n f^k(t; x) \left(\partial/\partial x^k\right)$. In the new coordinates $s, y^1, \ldots, y^n$,

$$X_0 = \frac{\partial}{\partial s},$$

with the dual pfaffian system $\mathcal{P} = (dy^1, \ldots, dy^n)$. The integral curves of $\mathcal{V} = (X_0)$, or equivalently of $\mathcal{P}$, are given by

$$y^1 = \text{constant}, \ldots, y^n = \text{constant}.$$

Next start with a vector field $X = \sum_{i=1}^n a^i(x) \left(\partial/\partial x^i\right)$ on $\mathbb{R}_x^n$. An integral curve C of X through a point $x_0 = (x_0^1, \ldots, x_0^n)$ is given by a mapping

$$\phi : \mathbb{R}_t \longrightarrow \mathbb{R}_x^n$$
$$t \mapsto (x^1(t), \ldots, x^n(t))$$

such that

$$\phi_* \left(\frac{d}{dt}\right) \equiv \sum_{i=1}^n \frac{dx^i}{dt} \frac{\partial}{\partial x^i} = X_{|C} \equiv \sum_{i=1}^n a^i(x^1(t), \ldots, x^n(t)) \frac{\partial}{\partial x^i},$$

and

$$x^i(0) = x_0^i \quad \text{for } i = 1, \ldots, n.$$

Thus ϕ is determined by the ODE system

$$\begin{cases} dx^i/dt = a^i(x^1(t), \ldots, x^n(t)), \\ x^i(0) = x_0^i, \end{cases} \quad \text{for } i = 1, \ldots, n,$$

which is *autonomous* in the sense that t does not appear explicitly in the right hand side.

Conversely, an autonomous system

$$\frac{dx^i}{dt} = a^i(x^1(t), \ldots, x^n(t)), \qquad i = 1, \ldots, n,$$

immediately gives rise to the pfaffian system

$$dx^i - a^i(x^1, \ldots, x^n) \, dt = 0 \quad \text{for } i = 1, \ldots, n,$$

with the dual vector field

$$X_0 = \frac{\partial}{\partial t} + \sum_{i=1}^n a^i(x^1, \ldots, x^n) \frac{\partial}{\partial x^i}.$$

Because the coefficients are independent of t, X_0 has a well-defined projection to $\mathbb{R}_x^n$:

$$X = \sum_{i=1}^n a^i(x^1, \ldots, x^n) \frac{\partial}{\partial x^i}.$$

Conclusion. *There is a one-to-one correspondence between vector fields*

$$X = \sum_{i=1}^{n} a^i(x) \frac{\partial}{\partial x^i}$$

and autonomous systems

$$\frac{dx^i}{dt} = a^i(x), \qquad i = 1, \ldots, n.$$

Once again, let

$$X = \sum_{i=1}^{n} a^i(x) \frac{\partial}{\partial x^i}$$

be a smooth vector field on $\mathbb{R}^n_x$. The integral curve $t \mapsto \xi(t; x)$ through the point $x = (x^1, \ldots, x^n)$ is the unique local solution of

$$\begin{cases} d\xi^i/dt = a^i(\xi^1, \ldots, \xi^n), \\ \xi^i(0) = x^i, \end{cases} \qquad \text{for } i = 1, \ldots, n.$$

This set-up gives rise to a 1-parameter family of local diffeomorphisms acting on $\mathbb{R}^n_x$, with t as parameter:

$$\mathbb{R}^n \longrightarrow \mathbb{R}^n$$
$$x \mapsto \xi(t; x),$$

defined at least for small t.

Examples when $n = 1$:

1. $X = d/dx$ gives the ODE system

$$\begin{cases} d\xi/dt = 1, \\ \xi(0) = x, \end{cases}$$

with the solution $\xi(t) = x + t$. If we set $b := t$ this defines the *translation* $x \mapsto x + b$.

2. $X = x\, d/dx$ gives

$$\begin{cases} d\xi/dt = \xi, \\ \xi(0) = x, \end{cases}$$

with the solution $\xi = x e^t$. If we set $a := e^t$ ($\neq 0$) the induced diffeomorphism is $x \mapsto ax$—that is, a *scaling*.

3. $X = x^2 \, d/dx$ gives

$$\begin{cases} d\xi/dt = \xi^2, \\ \xi(0) = x, \end{cases}$$

with the solution $\xi = x/(1 - tx)$, defined for $t < \frac{1}{x}$ if $x > 0$, and $t > \frac{1}{x}$ if $x < 0$. If we set $c := -t$, this defines the *inversion*

$$x \mapsto \frac{x}{cx + 1}.$$

The three transformations obtained here are special cases of the *projective group in one variable*:

$$x \mapsto \frac{ax + b}{cx + d}.$$

In the first case $a = d = 1$ and $c = 0$, in the second $b = c = 0$ and $d = 1$, and in the third $a = d = 1$ and $b = 0$.

Theorem 2.1.1. *The family of local diffeomorphisms*

$$x \mapsto \xi(t; x)$$

associated to a nonvanishing vector field

$$X = \sum_{i=1}^{n} a^i(x) \frac{\partial}{\partial x^i}$$

*constitutes a **local 1-parameter group** in the sense that*

$$\xi(s; \xi(t; x)) = \xi(s + t, x)$$

whenever t and s are such that both sides are defined.

Proof. Fix x and t, and consider the following two functions of s:

$$\begin{cases} s \mapsto \xi(s; \xi(t; x)), \\ s \mapsto \xi(s + t; x). \end{cases}$$

Both satisfy the ODE system

$$\begin{cases} d\xi/ds = a^i(\xi), \\ \xi^i(0) = \xi(t; x), \end{cases}$$

and hence they coincide by the uniqueness of local solutions to ODE systems with given initial data. $\qquad\Box$

With

$$e^{tX}(x) := \xi(t;x),$$

the local 1-parameter group induced by X is given by

$$e^{tX} : \mathbb{R}^n \to \mathbb{R}^n$$
$$x \mapsto e^{tX}(x).$$

Then

$$e^{0X}(x) = x,$$

and combining the theorem with the fact that $s + t = t + s$,

$$e^{(s+t)X}(x) = e^{sX} \circ e^{tX}(x) = e^{tX} \circ e^{sX}(x),$$

which explains why the notation e^{tX} is used.

Conversely, let

$$\xi_t : \mathbb{R}^n \longrightarrow \mathbb{R}^n$$
$$x \mapsto \xi(t;x)$$

be a local 1-parameter group. Then ξ_t defines a vector field X by the following construction.

Through each point x passes the curve $t \mapsto \xi(t;x)$ satisfying the initial condition $\xi(0;x) = x$. The tangent vector of this curve at x is given by

$$X_x = \sum_{i=1}^n a^i(x) \frac{\partial}{\partial x^i}, \quad \text{where } a^i(x) = \frac{\partial \xi^i}{\partial t}(t;x)\Big|_{t=0}.$$

Varying x we get the wanted vector field X, and it is clear that the 1-parameter group associated to this X equals the 1-parameter group we started from:

$$\xi_t = e^{tX}.$$

X is therefore called the *infinitesimal generator* of ξ_t.

Definition. Let X be a vector field defined near $x \in \mathbb{R}^n$, and let $\phi : \mathbb{R}^n \longrightarrow \mathbb{R}^n$ be a mapping. Then the push-forward vector $\phi_{x*}(X)$ is defined as follows.

Let $c : \mathbb{R}_t \longrightarrow \mathbb{R}^n$ be a curve with $c(0) = x$ and having the property that the tangent vector of c at x equals X_x. Then the curve $\phi \circ c : \mathbb{R}_t \longrightarrow \mathbb{R}^n$ satisfies $\phi \circ c(0) = \phi(x)$, and has a tangent vector at this point, which we call $\phi_{x*}(X)$. It is easily seen that this only depends on X_x and ϕ, and not on c.

Theorem 2.1.2. *Let ξ_t be a 1-parameter group with the infinitesimal generator X. Then ξ_t leaves X invariant in the sense that*

$$\xi_{t*}(X_x) = X_{\xi_t(x)} \quad \text{for all } x.$$

Proof. X_x is the tangent vector at x of the curve

$$t \mapsto \xi(t; x) = e^{tX}(x).$$

Fix a parameter value s, and consider the diffeomorphism $\xi_s = e^{sX}$. It sends x to $e^{sX}(x)$, and the curve above to

$$t \mapsto e^{sX} \circ e^{tX}(x) = e^{tX}\left(e^{sX}(x)\right).$$

But then both $X_{e^{sX}(x)}$ and $(e^{sX})_{x*}(X_x)$ are defined by the tangent vector of this curve at $e^{sX}(x)$, and so

$$\left(e^{sX}\right)_{x*}(X) = X_{e^{sX}(x)}.$$

$\square$

Example. Consider the projective group in one variable:

$$x \mapsto \frac{ax+b}{cx+d}, \quad \text{with } ad - bc \neq 0.$$

The identity transformation is given by $a = d = 1$ and $b = c = 0$. With d being nonzero close to the identity, we can divide by d and get the 3-parameter group

$$x \mapsto \frac{ax+b}{cx+1}.$$

A 1-parameter subgroup is of the form

$$x \mapsto \frac{a(t) \cdot x + b(t)}{c(t) \cdot x + 1},$$

where $a(0) = 1$ and $b(0) = c(0) = 0$. The corresponding vector field is

$$X = f(x)\frac{d}{dx},$$

where

$$f(x) = \frac{d}{dt}\left[\frac{a(t) \cdot x + b(t)}{c(t) \cdot x + 1}\right]_{|t=0} = a'(0) \cdot x + b'(0) - c'(0) \cdot x^2.$$

So

$$X = (A + Bx + Cx^2) \cdot \frac{d}{dx}$$

with $A = b'(0)$, $B = a'(0)$ and $C = -c'(0)$.

To this corresponds the autonomous differential equation

$$\frac{dx}{dt} = A + Bx + Cx^2.$$

In order to get a non-autonomous equation, we replace the constants A, B, C by functions $A(t)$, $B(t)$, $C(t)$, and obtain

$$\frac{dx}{dt} = A(t) + B(t) \cdot x + C(t) \cdot x^2$$

—that is, a Riccati equation.

2.2 First order PDE systems in one dependent variable, pfaffian equations and contact transformations

Let

$$\mathcal{S}: \quad F^i\left(x^1, \ldots, x^n; z; \frac{\partial z}{\partial x^1}, \ldots, \frac{\partial z}{\partial x^n}\right) = 0, \qquad i = 1, \ldots, m,$$

be a first order PDE system in one dependent variable z, and n independent variables x^1, ..., x^n. The key idea for finding solutions of $\mathcal{S}$ is to consider their 1-*graphs*. Let

$$z = \zeta(x^1, \ldots, x^n)$$

be a solution of $\mathcal{S}$. Then graph (ζ) is the subset of $\mathbb{R}^n_x \times \mathbb{R}_z$ defined by

$$\mathbb{R}^n_x \longrightarrow \mathbb{R}^n_x \times \mathbb{R}_z$$
$$(x^1, \ldots, x^n) \mapsto (x^1, \ldots, x^n; \zeta(x^1, \ldots, x^n)).$$

In order to define the 1-*graph* graph$^{(1)}(\zeta)$ of $\zeta(x)$, introduce n new variables p_1, ..., p_n, and let graph$^{(1)}(\zeta)$ be the subset of $\mathbb{R}^n_x \times \mathbb{R}_z \times \mathbb{R}^n_p$ given by

$$\mathbb{R}^n_x \longrightarrow \mathbb{R}^n_x \times \mathbb{R}_z \times \mathbb{R}^n_p$$
$$x \mapsto (x^1, \ldots, x^n; \zeta; \partial \zeta / x^1, \ldots, \partial \zeta / x^n).$$

To $\mathcal{S}$ we associate the following subset of $\mathbb{R}^n_x \times \mathbb{R}_z \times \mathbb{R}^n_p$, which by abuse of notation is also called $\mathcal{S}$:

$$\mathcal{S}: \quad F^i(x^1, \ldots, x^n; z; p_1, \ldots, p_n) = 0, \qquad i = 1, \ldots, m.$$

Only interested in local properties, we avoid possible singularities and assume that $\mathcal{S}$ and graph$^{(1)}(\zeta)$ are submanifolds of $\mathbb{R}^n_x \times \mathbb{R}_z \times \mathbb{R}^n_p$.

Then:

(i) ζ is a solution of $\mathcal{S} \iff$ graph$^{(1)}(\zeta) \subset \mathcal{S}$;

(ii) $dx^1 \wedge \cdots \wedge dx^n \neq 0$ on $\text{graph}^{(1)}(\zeta)$;

(iii) the restrictions of the p_i to $\text{graph}^{(1)}(\zeta)$ coincide with $\partial\zeta/x^i$, so that

$$dz - \sum_{i=1}^{n} p_i \, dx^i = 0 \quad \text{on } \text{graph}^{(1)}(\zeta).$$

Notations. $\theta^0 := dz - \sum_{i=1}^{n} p_i \, dx^i$ is called the *contact form* on $\mathbb{R}^n_x \times \mathbb{R}_z \times \mathbb{R}^n_p$, and $\theta := \theta^0{}_{|S}$ is its restriction to S. An *integral manifold* of θ is a submanifold $\mathcal{N}$ of S on which θ vanishes.

Thus, *the 1-graph of any solution of S is an n-dimensional integral manifold of θ on which $dx^1 \wedge \cdots \wedge dx^n \neq 0$.*

Suppose conversely that $\mathcal{N}$ is an n-dimensional submanifold of S on which $\theta = 0$ and $dx^1 \wedge \cdots \wedge dx^n \neq 0$. The latter property means that there are functions $\zeta, \pi_1, \dots, \pi_n$ such that

$$\begin{cases} z = \zeta(x^1, \dots, x^n), \\ p_i = \pi_i(x^1, \dots, x^n), \quad \text{for } i = 1, \dots, n \end{cases}$$

on $\mathcal{N}$. Because $\theta_{|\mathcal{N}} = 0$,

$$\sum_{i=1}^{n} \left(\frac{\partial\zeta}{\partial x^i} - \pi_i \right) dx^i = 0,$$

and therefore

$$\pi_i = \partial\zeta/\partial x^i \quad \text{for } i = 1, \dots, n.$$

But this means precisely that

$$\mathcal{N} = \text{graph}^{(1)}(\zeta).$$

Finally, $\mathcal{N} \subset S$ shows that ζ is a solution of S.

Conclusion. *There is a one-to-one correspondence between solutions of S and those n-dimensional integral manifolds of θ on which $dx^1 \wedge \cdots \wedge dx^n \neq 0$.*

Therefore it makes sense to identify the PDE system S with the structured manifold (S, θ).

Sometimes the condition $dx^1 \wedge \cdots \wedge dx^n \neq 0$ is annoying, and therefore Lie broadened the concept of solution.

Definition. A *generalized solution* of S is an n-dimensional integral manifold of θ.

So a generalized solution is not necessarily a 1-graph of a function, but it is perfectly meaningful as a geometric entity.

The main idea for finding integral manifolds of the pfaffian equation $\theta = 0$ on S—going back to Pfaff—is to write it in a simple canonical form. In fact, we will later see that it is possible to introduce local coordinates $y^1, \ldots, y^{n-m}, w, q_1, \ldots, q_{n-m}, t^1, \ldots, t^m$ for the $(2n+1-m)$-dimensional S such that

$$\theta = 0 \iff dw - \sum_{i=1}^{n-m} q_i \, dy^i = 0,$$

and it is very easy to integrate this.

A variant is to rectify S by means of a *contact transformation*.

Definitions. The *jet bundle* $J^1(\mathbb{R}^n_x, \mathbb{R}_z)$ is the Euclidean space $\mathbb{R}^n_x \times \mathbb{R}_z \times \mathbb{R}^n_p$ equipped with the pfaffian equation $\theta^0 \equiv dz - \sum_{i=1}^n p_i \, dx^i = 0$. A *contact transformation* is a local diffeomorphism ϕ of $J^1(\mathbb{R}^n_x, \mathbb{R}_z)$ which preserves the equation $\theta^0 = 0$, i.e., satisfies $\phi^*(\theta^0) = \rho \cdot \theta^0$ for some nonvanishing function ρ.

The family of contact transformations constitutes the *Lie pseudogroup of contact transformations*, denoted by ${}^1Cont^{n,1}$. The pseudogroup property is essentially this:

if $\phi_2, \phi_1 \in {}^1Cont^{n,1}$ and $\phi_2 \circ \phi_1$ is defined, then

$$(\phi_2 \circ \phi_1)^*(\theta^0) = \phi_1{}^* \circ \phi_2{}^*(\theta^0) = \phi_1{}^*(\rho_2 \cdot \theta^0)$$
$$= \phi_1{}^*(\rho_2) \cdot \phi_1{}^*(\theta^0) = \phi_1{}^*(\rho_2) \cdot \rho_1 \cdot \theta^0,$$

so $\phi_2 \circ \phi_1 \in {}^1Cont^{n,1}$ too.

Let $S \subset J^1(\mathbb{R}^n_x, \mathbb{R}_z)$ be a first order PDE system, and let

$$\phi : J^1(\mathbb{R}^n_x, \mathbb{R}_z) \xrightarrow{\ \cong\ } J^1(\mathbb{R}^n_y, \mathbb{R}_w)$$

be a contact transformation—so that

$$\phi^* \left(dw - \sum_{i=1}^n q_i \, dy^i \right) = \rho \cdot \left(dz - \sum_{i=1}^n p_i \, dx^i \right).$$

Then ϕ induces a one-to-one correspondence between the n-dimensional integral manifolds of $(dz - \sum_{i=1}^n p_i \, dx^i)_{|S}$ on S and those of $(dw - \sum_{i=1}^n q_i \, dy^i)_{|\phi(S)}$ on $\phi(S)$—i.e., between the generalized solutions of S and the generalized solutions of $\phi(S)$.

In this context Lie showed the following theorem, which will be proved in section **7.1**.

Rectification theorem for first order PDE systems in one unknown. *Let S be an involutive system of m first order PDEs in one unknown, so that S*

can be considered as an m-codimensional submanifold of $J^1(\mathbb{R}^n_x, \mathbb{R}_z)$. Then there is a contact transformation

$$\phi: J^1(\mathbb{R}^n_x, \mathbb{R}_z) \xrightarrow{\cong} J^1(\mathbb{R}^n_y, \mathbb{R}_w)$$

such that $\phi(S)$ locally is given by

$$y_1 = \cdots = y_m = 0.$$

Thus

$$\left(dw - \sum_{i=1}^{n} q_i \, dy^i\right)\Big|_{\phi(S)} = dw - \sum_{i=m+1}^{n} q_i \, dy^i,$$

making it very easy to solve the PDE system $\phi(S)$.

Conclusion. *It is easy to integrate a **single** vector field and a **single** pfaffian equation; general vector field and pfaffian systems are much harder.*

2.3 Jet bundles and contact pfaffian systems

In the preceding sections we have seen that solving a first order ODE system is the same as integrating a vector field, and that solving a first order PDE system in one unknown is equivalent to finding integral manifolds of a pfaffian equation.

Lie generalized these results to arbitrary PDE systems—see for instance his exciting book *Geometrie der Berührungstransformationen* ([Lie–Scheffers 1896]). The idea is this: a qth order PDE system S in m dependent variables $z^1, \ldots, z^m$ and n independent variables $x^1, \ldots, x^n$ is regarded as a submanifold of the jet bundle $J^q(\mathbb{R}^n_x, \mathbb{R}^m_z)$; to the latter is associated the contact pfaffian system $^qCt^{n,m}$, which by restriction induces a pfaffian system $\mathcal{P}$ on S. In this way the PDE system is considered as a *manifold with a pfaffian system*, or dually as a *manifold with a vector field system*. The solutions of the PDE system correspond to n-dimensional integral manifolds of the pfaffian (or vector field) system, on which $dx^1 \wedge \cdots \wedge dx^n \neq 0$.

The partial derivatives $\partial/\partial x^i$ in the PDE setting are replaced by the exterior differentiation operator d in the pfaffian version, and by the Lie bracket [,] in the vector field approach. Here d and [,] are *intrinsic* in the sense that for any local diffeomorphism ϕ,

$$\begin{cases} \phi^*(d\theta) = d(\phi^*(\theta)) & (\theta = \text{ differential form}), \\ \phi_*([X, Y]) = [\phi_*(X), \phi_*(Y)] & (X, Y = \text{ vector fields}). \end{cases}$$

To explain Lie's ideas, we first consider the trivial bundle

$$\mathbb{R}^n_x \times \mathbb{R}^m_z$$
$$\downarrow$$
$$\mathbb{R}^n_x$$

with $x^1, \ldots, x^n$ (the independent variables) as base coordinates, and $z^1,$ $\ldots, z^m$ (the dependent variables) as fibre coordinates. Two local sections $f = (f^1(x), \ldots, f^m(x))$ and $g = (g^1(x), \ldots, g^m(x))$ have *contact of order q at x* if

$$\frac{\partial^{k_1 + \cdots + k_n} f^i}{(\partial x^1)^{k_1} \ldots (\partial x^n)^{k_n}}(x) = \frac{\partial^{k_1 + \cdots + k_n} g^i}{(\partial x^1)^{k_1} \ldots (\partial x^n)^{k_n}}(x)$$

for $i = 1, \ldots, m$ and for all n-tuples $(k_1, \ldots, k_n)$ with $k_1 + \cdots + k_n \leq q$. It is easily seen that this notion is independent of the choice of local coordinates.

'Having contact of order q at x' is an equivalence relation. The equivalence class of a local section f at x is denoted by $j^q_x(f)$. For obvious reasons Lie called such a class a *contact element of order q at x*; for less obvious reasons it is nowadays called a *q-jet*. The set of q-jets of local sections constitutes the underlying set of the *jet bundle* $J^q(\mathbb{R}^n_x, \mathbb{R}^m_z)$ (or contact manifold in Lie's terminology). This acquires a manifold structure when equipped with the local coordinates

$$x^i, z^k, p^k_{i_1 \ldots i_a} : J^q(\mathbb{R}^n_x, \mathbb{R}^m_z) \to \mathbb{R}$$

defined by

$$\begin{cases} x^i(j^q_x(f)) := x^i & \text{for } i = 1, \ldots, n, \\ z^k(j^q_x(f)) := f^k(x) & \text{for } k = 1, \ldots, m, \\ p^k_{i_1 \ldots i_a}(j^q_x(f)) := \partial^a f^k / \partial x^{i_1} \ldots \partial x^{i_a}(x) & \text{for } a \leq q. \end{cases}$$

For instance, if $m = 1$ (in which case we write z and $p_{\ldots}$ instead of z^1 and $p^1_{\ldots}$) and $q \geq 3$,

$$p_{112}(j^q_x(f)) = \frac{\partial^3 f}{(\partial x^1)^2 \partial x^2}(x).$$

Convention. $p^k_0 := z^k.$

Note that the $p^k_{i_1 \ldots i_a}$ are symmetric with respect to the lower indices.

For $q > r \geq 0$ there are natural projections

$$\pi_r^q : J^q(\mathbb{R}^n, \mathbb{R}^m) \to J^r(\mathbb{R}^n, \mathbb{R}^m)$$
$$(x^i, z^k, p_{i_1}^k, \ldots, p_{i_1 \ldots i_q}^k) \mapsto (x^i, z^k, p_{i_1}^k, \ldots, p_{i_1, \ldots, i_r}^k).$$

The identification $J^0(\mathbb{R}^n, \mathbb{R}^m) \cong \mathbb{R}^n \times \mathbb{R}^m$ yields

$$\pi_0^q : J^q(\mathbb{R}^n, \mathbb{R}^m) \to \mathbb{R}^n \times \mathbb{R}^m$$
$$(x, z, p) \mapsto (x, z).$$

Here x is the *source* of $j_x^q(f)$, and $f(x)$ is its *target*. The *source projection* is given by

$$\alpha : J^q(\mathbb{R}^n, \mathbb{R}^m) \to \mathbb{R}^n$$
$$(x, z, p) \mapsto x,$$

and the *target projection* by

$$\beta : J^q(\mathbb{R}^n, \mathbb{R}^m) \to \mathbb{R}^m$$
$$(x, z, p) \mapsto z.$$

Let f be a local section of the bundle $\mathbb{R}^n \times \mathbb{R}^m \to \mathbb{R}^n$. Its *q-graph* graph$^{(q)}(f)$ is the submanifold of $J^q(\mathbb{R}^n, \mathbb{R}^m)$ defined by

$$j^q(f) : \mathbb{R}^n \to J^q(\mathbb{R}^n, \mathbb{R}^m)$$
$$x \mapsto j_x^q(f).$$

Clearly $x^1, \ldots, x^n$ can be used as local coordinates on graph$^{(q)}(f)$. Restricting the $p_{i_1 \ldots i_a}^k$ to such a graph and expressing them by means of the x^k, they become real derivatives:

$$p_{i_1 \ldots i_a \mid \mathrm{graph}^{(q)}(f)}^k = \frac{\partial^a f^k}{\partial x^{i_1} \ldots \partial x^{i_a}}.$$

To $J^q(\mathbb{R}^n, \mathbb{R}^m)$ is associated a canonical module of one-forms—the contact pfaffian system $^qCt^{n,m}$, generated by

$$\begin{cases} \theta_0^k := dz^k - \sum_{i_1=1}^n p_{i_1}^k \, dx^{i_1}, \\ \theta_{i_1}^k := dp_{i_1}^k - \sum_{i_2=1}^n p_{i_1 i_2}^k \, dx^{i_2}, \\ \ldots \\ \theta_{i_1 \ldots i_{q-1}}^k := dp_{i_1 \ldots i_{q-1}}^k - \sum_{i_q=1}^n p_{i_1 \ldots i_q}^k \, dx^{i_q}, \end{cases}$$

where $k = 1, \ldots, m$ and $1 \leq i_1, \ldots, i_q \leq n$.

Definition. The qth order *jet bundle* in m dependent and n independent variables is the couple $(J^q(\mathbb{R}^n, \mathbb{R}^m); {}^q Ct^{n,m})$. It is an example of a *manifold with a local structure*.

Recall that if $\theta^1, \dots, \theta^s$ are one-forms on a manifold M, $(\theta^1, \dots, \theta^s)$ denotes the pfaffian system generated by these forms over the ring of smooth functions on M.

Definition. An *integral manifold* of $(\theta^1, \dots, \theta^s)$ is a submanifold $\mathcal{N}$ of M such that

$$\theta^i{}_{|\mathcal{N}} = 0 \quad \text{for } i = 1, \dots, s.$$

Note that if $\mathcal{N}_1$ is a submanifold of such an integral manifold $\mathcal{N}$, $\mathcal{N}_1$ is automatically an integral manifold too. Therefore the most interesting integral manifolds are those of maximal dimension.

Theorem 2.3.1. *The q-graphs of* $J^q(\mathbb{R}^n, \mathbb{R}^m)$ *are exactly those n-dimensional integral manifolds of* ${}^q Ct^{n,m}$ *on which* $dx^1 \wedge \cdots \wedge dx^n \neq 0$.

Proof. The contact system ${}^q Ct^{n,m}$ is constructed so as to make any q-graph an integral manifold. Conversely, let $\mathcal{N}$ be an n-dimensional integral manifold of ${}^q Ct^{n,m}$ on which $dx^1 \wedge \cdots \wedge dx^n \neq 0$. Then $x^1, \dots, x^n$ can be used as local coordinates on $\mathcal{N}$, and the restrictions of the coordinate functions of $J^q(\mathbb{R}^n, \mathbb{R}^m)$ to $\mathcal{N}$ are functions of $(x) = (x^1, \dots, x^n)$:

$$z^k = \zeta^k(x), \ p^k_{i_1} = \pi^k_{i_1}(x), \dots, \ p^k_{i_1 \dots i_q} = \pi^k_{i_1 \dots i_q}(x).$$

Since $\mathcal{N}$ is an integral manifold of ${}^q Ct^{n,m}$,

$$d\zeta^k = \sum_{i_1=1}^n \pi^k_{i_1} \, dx^{i_1} \implies \pi^k_{i_1} = \frac{\partial \zeta^k}{\partial x^{i_1}},$$

$$\dots$$

$$d\pi^k_{i_1 \dots i_{q-1}} = \sum_{i_q=1}^n \pi^k_{i_1 \dots i_q} \, dx^{i_q} \implies \pi^k_{i_1 \dots i_q} = \frac{\partial \pi^k_{i_1 \dots i_{q-1}}}{\partial x^{i_q}},$$

and so

$$\pi^k_{i_1 i_2} = \frac{\partial \pi^k_{i_1}}{\partial x^{i_2}} = \frac{\partial^2 \zeta^k}{\partial x^{i_1} \partial x^{i_2}},$$

$$\dots$$

$$\pi^k_{i_1 \dots i_q} = \frac{\partial \pi_{i_1 \dots i_{q-1}}}{\partial x^{i_q}} = \frac{\partial^2 \pi_{i_1 \dots i_{q-2}}}{\partial x^{i_q} \partial x^{i_{q-1}}} = \cdots = \frac{\partial^q \zeta^k}{\partial x^{i_1} \dots \partial x^{i_q}},$$

i.e.,

$$\mathcal{N} = \text{graph}^q(\zeta^1, \ldots, \zeta^m).$$

□

Definition. The *generalized q-graphs* of $J^q(\mathbb{R}^n, \mathbb{R}^m)$ are the n-dimensional integral manifolds of $^qCt^{n,m}$—with no independence condition $dx^1 \wedge \cdots \wedge dx^n \neq 0$ imposed.

Next let

$$\mathcal{S}: \quad \begin{cases} F^j\left(x^i; z^k; \partial z^k/\partial x^{i_1}, \ldots, \partial^q z^k/\partial x^{i_1} \ldots \partial x^{i_q}\right) = 0 \\ \text{for } j = 1, \ldots, s, \ i = 1, \ldots, n, \ \text{and } k = 1, \ldots, m \end{cases}$$

be a qth order PDE system in m dependent and n independent variables. To $\mathcal{S}$ we associate the following subset (or submanifold locally) of $J^q(\mathbb{R}^n, \mathbb{R}^m)$, which by abuse of notation is also denoted by $\mathcal{S}$:

$$\mathcal{S} := \{(x; z; p) \in J^q(\mathbb{R}^n, \mathbb{R}^m) \mid F^j(x; z; p) = 0 \quad \text{for } j = 1, \ldots, s\}.$$

$\mathcal{S}$ inherits the pfaffian system $\mathcal{P} := {}^qCt^{n,m}|_{\mathcal{S}}$ from the contact system of $J^q(\mathbb{R}^n, \mathbb{R}^m)$.

If $\zeta = (\zeta^1(x), \ldots, \zeta^m(x))$ is a local solution of the PDE system $\mathcal{S}$,

$$\begin{cases} \text{graph}^{(q)}\zeta \text{ is an integral manifold of } {}^qCt^{n,m} \text{ (as all } q\text{-graphs)}, \\ \text{graph}^{(q)}\zeta \subset \mathcal{S} \subset J^q(\mathbb{R}^n, \mathbb{R}^m) \text{ (as a solution of } \mathcal{S}), \end{cases}$$

so that

graph$^{(q)}\zeta$ *is an n-dimensional integral manifold of* $(\mathcal{S}, \mathcal{P})$, *on which* $dx^1 \wedge \cdots \wedge dx^n \neq 0$.

Conversely any such integral manifold of $(\mathcal{S}, \mathcal{P})$ is the q-graph of a solution of the PDE system $\mathcal{S}$.

Notations.
sol$(\mathcal{S})$= the set of local solutions of the PDE system $\mathcal{S}$,
sol$^q(\mathcal{S})$ = the set of q-graphs of local solutions,
int$(\mathcal{S}, \mathcal{P}; dx)$= the set of local integral manifolds of $(\mathcal{S}, \mathcal{P})$ on which $dx^1 \wedge \cdots \wedge dx^n \neq 0$.

Using these notations we have the one-to-one correspondences

$$\text{sol}(\mathcal{S}) \longleftrightarrow \text{sol}^q(\mathcal{S}) \longleftrightarrow \text{int}(\mathcal{S}, \mathcal{P}; dx).$$

Hence solving the PDE system $\mathcal{S}$ is equivalent to determining int$(\mathcal{S}, \mathcal{P}; dx)$, and therefore it makes sense to identify the PDE system S with the structured manifold (S, P).

Alternatively, this may be viewed from a dual point of view.

If $P = (\theta^1, \ldots, \theta^s)$ is a pfaffian system on a manifold M, the dual vector field system $V = P^\perp$ is given by

$$V := \{\text{all vector fields } X \text{ on } M \text{ satisfying } \theta^i(X) = 0 \text{ for } i = 1, \ldots, s\}.$$

If we assume the θ^i to be linearly independent everywhere on M there are $q = \dim M - s$ vector fields $X_1, \ldots, X_q$ that generate V—in which case we write $V = (X_1, \ldots, X_q)$.

Conversely, if we start from a vector field system $V = (X_1, \ldots, X_q)$ on M, the dual pfaffian system $P = V^\perp$ is defined by

$$P := \{\text{all one-forms } \theta \text{ on } M \text{ satisfying } \theta(X_j) = 0 \text{ for } j = 1, \ldots, q\}.$$

Convention. Only being interested in *local and generic properties* we assume throughout that *all pfaffian and vector field systems are finitely generated*, and admit *generators that are linearly independent at each point of the domain considered*. The *dimension* equals the *number of generators*.

From this it follows that

$$(P^\perp)^\perp = P$$

for any pfaffian system P, and

$$(V^\perp)^\perp = V$$

for any vector field system V.

Let P be a pfaffian system on M, and let N be a submanifold. Then

> N is an integral manifold of $P \iff$ for each $n \in N$ the tangent space $T_n(N)$ is included in $(P^\perp)_n$.

Definition. Let V be a vector field system on a manifold M. A submanifold N is an *integral manifold* of V if and only if $T_n(N) \subseteq V_n$ for each $n \in N$.

In this way N is an integral manifold of P *(or V)* if and only if N is an integral manifold of $P^\perp$ *(or $V^\perp$)*.

2.4 The theorem of Frobenius

Let $V = (X_1, \ldots, X_q)$ be a q-dimensional vector field system on $\mathbb{R}^r$. With $[\,,\,]$ denoting the Lie bracket, the *derived system* (or *derivative*) V' of

$\mathcal{V}$ is defined as the vector field system generated by X_i and $[X_i, X_j]$ for $i, j = 1, \ldots, q$. Or a bit shorter,

$$\mathcal{V}' := \mathcal{V} + [\mathcal{V}, \mathcal{V}].$$

The second derivative is then defined as

$$\mathcal{V}'' := \mathcal{V}' + [\mathcal{V}', \mathcal{V}'],$$

and analogously for higher order derivatives.

In this section we will study the simplest case that can be imagined, namely $\mathcal{V}' = \mathcal{V}$, or

$$[X_i, X_j] \equiv 0 \pmod{\mathcal{V}} \quad \text{for all } i, j.$$

But let us first look at the concept of derived system from a dual point of view. For a pfaffian system $\mathcal{P} = (\theta^1, \ldots, \theta^s)$ on $\mathbb{R}^r$, its derived system is defined as

$$\mathcal{P}' := ((\mathcal{P}^\perp)')^\perp$$

—which requires some explanation.

Recalling that the differential $d\theta$ of a one-form θ is defined by

$$d\theta(X, Y) = X(\theta(Y)) - Y(\theta(X)) - \theta([X, Y])$$

for arbitrary vector fields X, Y, it follows that for $\theta \in \mathcal{P}$ and $X, Y \in \mathcal{P}^\perp$,

$$d\theta(X, Y) = -\theta([X, Y]).$$

Dualizing the inclusion $\mathcal{P}^\perp \subseteq (\mathcal{P}^\perp)'$ we get

$$\mathcal{P} = (\mathcal{P}^\perp)^\perp \supseteq ((\mathcal{P}^\perp)')^\perp = \mathcal{P}'.$$

Hence

$$\mathcal{P}' = \{\theta \in \mathcal{P} \mid \theta([X, Y]) = 0 \text{ for all } X, Y \in \mathcal{P}^\perp\}$$
$$= \{\theta \in \mathcal{P} \mid d\theta(X, Y) = 0 \text{ for all } X, Y \in \mathcal{P}^\perp\}.$$

Add one-forms $\pi^1, \ldots, \pi^q$ (with $q = r - s$) to the θ^i so that $\theta^1 \wedge \cdots \wedge \theta^s \wedge \pi^1 \wedge \cdots \wedge \pi^q \neq 0$, i.e., $\{\theta^1, \ldots, \theta^s, \pi^1, \ldots, \pi^q\}$ constitutes a local coframe for $\mathbb{R}^r$. Then there are smooth functions A_{ij}, B_{ik}, C_{kl} with $A_{ij} + A_{ji} = 0$ and $C_{kl} + C_{lk} = 0$ such that

$$d\theta = \frac{1}{2} \sum_{i,j=1}^{s} A_{ij}\, \theta^i \wedge \theta^j + \sum_{i=1}^{s} \sum_{k=1}^{q} B_{ik}\, \theta^i \wedge \pi^k + \frac{1}{2} \sum_{k,l=1}^{q} C_{kl}\, \pi^k \wedge \pi^l.$$

If $\theta \in \mathcal{P}'$ and $X, Y \in \mathcal{P}^{\perp}$,

$$0 = d\theta(X, Y) = \sum_{k,l=1}^{q} C_{kl}\, \pi^k(X)\, \pi^l(Y).$$

Define the vector field $\partial/\partial\pi^k$ by

$$\begin{cases} \theta^i(\partial/\partial\pi^k) = 0 & \text{for } i = 1,\ldots,s, \\ \pi^j(\partial/\partial\pi^k) = 0 & \text{for } j = 1,\ldots,k-1,k+1,\ldots,q, \\ \pi^k(\partial/\partial\pi^k) = 1, \end{cases}$$

so that $\mathcal{P}^{\perp} = (\partial/\partial\pi^1,\ldots,\partial/\partial\pi^q)$. Then

$$0 = d\theta(\partial/\partial\pi^k, \partial/\partial\pi^l) = C_{kl} \quad \text{for } k,l = 1,\ldots,q,$$

and therefore

$$d\theta \equiv 0 \quad (\text{mod } \theta^1,\ldots,\theta^s),$$

that is,

$$\theta \in \mathcal{P}' \implies d\theta \equiv 0 \quad (\text{mod } \mathcal{P}).$$

Since $\impliedby$ is clear by the above, we have reached the following result.

Conclusion. $\mathcal{P}' = \{\theta \in \mathcal{P} \mid d\theta \equiv 0 \ (\text{mod } \mathcal{P})\}$.

Definitions. The vector field system $\mathcal{V}$ is *complete* (with respect to Lie brackets) if $\mathcal{V}' = \mathcal{V}$.

The pfaffian system $\mathcal{P}$ is *complete* if $\mathcal{P}' = \mathcal{P}$.

Example. A vector field system generated by a *single* vector field X is automatically complete—simply because $[X, X] = 0$.

Dually, let $\mathcal{P} = (\theta^1,\ldots,\theta^{r-1})$ be a 1-codimensional pfaffian system on $\mathbb{R}^r$. Defining θ^r such that $\theta^1 \wedge \cdots \wedge \theta^{r-1} \wedge \theta^r \neq 0$, there are structure functions $c^i_{jk} = -c^i_{kj}$ so that

$$d\theta^i = \frac{1}{2} \sum_{j,k=1}^{r} c^i_{jk}\, \theta^j \wedge \theta^k \quad \text{for } i = 1,\ldots,r-1.$$

But since θ^r is the only one-form appearing here which does not belong to $\mathcal{P}$,

$$d\theta^i \equiv 0 \quad (\text{mod } \mathcal{P}) \quad \text{for } i = 1,\ldots,r-1,$$

that is, $\mathcal{P}$ is complete.

The fundamental theorem for complete systems is due to Frobenius. The vector field version goes as follows.

Theorem 2.4.1 (The Frobenius theorem). *Let $V = (X_1, \ldots, X_q)$ be a complete q-dimensional vector field system on $\mathbb{R}^r$. Then it is possible to find local coordinates $t^1, \ldots, t^r$ such that $V = (\partial/\partial t^1, \ldots, \partial/\partial t^q)$ locally. Hence the integral manifolds of V are given by*

$$t^{q+1} = constant, \ldots, t^r = constant.$$

Expressed a bit differently, a complete q-dimensional vector field system defines a *local foliation* with q-dimensional leaves which are integral manifolds of V.

Let V be generated by

$$X_k = \sum_{i=1}^{r} a_k^i(x) \frac{\partial}{\partial x^i} \quad \text{for } k = 1, \ldots, q.$$

Applying Gaussian elimination to the matrix $(a_k^i(x))$, and changing indices if necessary, we can find new generators

$$\hat{X}_k = \frac{\partial}{\partial x^k} + \sum_{i=q+1}^{r} b_k^i(x) \frac{\partial}{\partial x^i}, \quad k = 1, \ldots, q.$$

Definition. The generators $\hat{X}_1, \ldots, \hat{X}_q$ are said to be *resolved* with respect to $\partial/\partial x^1, \ldots, \partial/\partial x^q$. Classically a vector field system with such resolved generators is said to be in *Jacobian form*.

With $V = (\hat{X}_1, \ldots, \hat{X}_q)$ being complete there are functions $c_{ij}^k(x)$ such that

$$[\hat{X}_i, \hat{X}_j] = \sum_{k=1}^{q} c_{ij}^k \hat{X}_k \quad \text{for } i, j = 1, \ldots, q.$$

But the lefthand side does not contain any of $\partial/\partial x^1, \ldots, \partial/\partial x^q$, whence all c_{ij}^k must vanish. That is,

$$[\hat{X}_i, \hat{X}_j] = 0 \quad \text{for } i, j = 1, \ldots, q.$$

Thus prepared we are now ready to prove the Frobenius theorem.

Proof of the theorem. Applying the rectification lemma to X_1 and then Gaussian elimination as described above, it is possible to find generators $Y_1, Y_2, \ldots, Y_q$ for V of the form

$$\begin{cases} Y_1 = \partial/\partial y^1, \\ Y_k = \partial/\partial y^k + \sum_{i=q+1}^{r} c_k^i(y) \left(\partial/\partial y^i \right) & \text{for } k = 2, \ldots, q. \end{cases}$$

Then

$$0 = [Y_1, Y_k] = \sum_{i=q+1}^{r} \frac{\partial c_k^i}{\partial y^1} \frac{\partial}{\partial y^i},$$

so that the c_k^i do not depend on y^1. As a consequence the complete $(q-1)$-dimensional vector field system $(Y_2, \dots, Y_q)$ can be considered as living on $\mathbb{R}^{r-1}_{y^2,\dots,y^r}$.

For $q = 1$ Frobenius's theorem reduces to the rectification lemma. If we assume the theorem to be true for $(q-1)$-dimensional complete vector field systems there are local coordinates $t^2, \dots, t^r$ for $\mathbb{R}^{r-1}$ such that $(Y_2, \dots, Y_q) = (\partial/\partial t^2, \dots, \partial/\partial t^q)$. Setting $t^1 := y^1$ it then follows that $t^1, \dots, t^r$ are local coordinates for $\mathbb{R}^r$, and that $\mathcal{V} = (\partial/\partial t^1, \dots, \partial/\partial t^q)$. $\square$

Let $\mathcal{P} = (\theta^1, \dots, \theta^s)$ be a complete pfaffian system on $\mathbb{R}^r$. With the one-forms $\theta^1, \dots, \theta^s, \pi^1, \dots, \pi^q$ (where $q = r - s$) forming a local coframe for $\mathbb{R}^r$, $\mathcal{P}^\perp = (\partial/\partial \pi^1, \dots, \partial/\partial \pi^q)$. This is also complete since the definitions of completeness for vector field and pfaffian systems are made so that

$$\mathcal{P} \text{ is complete} \iff \mathcal{P}^\perp \text{ is complete.}$$

By the vector field version of Frobenius's theorem there are local coordinates $t^1, \dots, t^r$ such that $\mathcal{P}^\perp = (\partial/\partial t^1, \dots, \partial/\partial t^q)$. But then

$$\mathcal{P} = (\mathcal{P}^\perp)^\perp = (dt^{q+1}, \dots, dt^r).$$

Theorem 2.4.2 (Dual version of the Frobenius theorem). Let $\mathcal{P} = (\theta^1, \dots, \theta^s)$ be a complete s-dimensional pfaffian system on $\mathbb{R}^r$. Then there are local coordinates $t^1, \dots, t^r$ for $\mathbb{R}^r$ such that $\mathcal{P} = (dt^{r-s+1}, \dots, dt^r)$ locally.

For a reasonably direct proof—without using duality—see e.g. [Flanders 1989]. $\square$

Recall the formulation of the rectification lemma: if X is a nonvanishing vector field defined near the origin of $\mathbb{R}^r_x$, there is a unique local diffeomorphism

$$\Phi : \mathbb{R}^r_t \to \mathbb{R}^r_x \qquad\qquad \Phi^{-1} : \mathbb{R}^r_x \to \mathbb{R}^r_t$$
$$t \mapsto x(t), \qquad\qquad\qquad x \mapsto t(x),$$

such that $\Phi_*(\partial/\partial t^1) = X$, and

$$\begin{cases} t^1(0, x^2, \dots, x^r) = 0, \\ t^i(0, x^1, \dots, x^r) = x^i & \text{for } i = 2, \dots, r. \end{cases}$$

Using an induction argument as above, we immediately generalize this to the following **supplement to the Frobenius theorem.**

Theorem 2.4.3. *Let $\mathcal{V} = (X_1, \ldots, X_q)$ be a complete q-dimensional vector field system defined near the origin of $\mathbb{R}^r_x$. Then there is a unique local diffeomorphism*

$$\Phi : \mathbb{R}^r_t \to \mathbb{R}^r_x \qquad\qquad \Phi^{-1} : \mathbb{R}^r_x \to \mathbb{R}^r_t$$

$$t \mapsto x(t), \qquad\qquad\qquad x \mapsto t(x),$$

such that

$$\mathcal{V} = \left(\Phi_*(\partial/\partial t^1), \ldots, \Phi_*(\partial/\partial t^q) \right)$$

and

$$\begin{cases} t^i(0, \ldots, 0, x^{q+1} \ldots, x^r) = 0 & \text{for } i = 1, \ldots, q, \\ t^j(0, \ldots, 0, x^{q+1}, \ldots, x^r) = x^j & \text{for } j = q+1, \ldots, r. \end{cases}$$

Expressed differently, $\mathcal{V}$ admits a uniquely determined set of $r - q$ first integrals $t^{q+1}(x), \ldots, t^r(x)$ satisfying

$$t^j(0, \ldots, 0, x^{q+1}, \ldots, x^r) = x^j \qquad \text{for } j = q+1, \ldots, r.$$

$\square$

Earlier we have seen that solving a PDE system in n independent variables $x^1, \ldots, x^n$ is equivalent to finding n-dimensional integral manifolds of the corresponding vector field system $\mathcal{V}$ on which $dx^1 \wedge \cdots \wedge dx^n \neq 0$.

The latter problem will be approached in a 2-step process: first determine all n-dimensional complete subsystems $\mathcal{W}$ of the vector field system $\mathcal{V}$ with

$$dx^1 \wedge \cdots \wedge dx^n_{|\mathcal{W} \times \cdots \times \mathcal{W}} \neq 0,$$

and then use the Frobenius theorem in order to find the integral manifolds of each $\mathcal{W}$. The family of all integral manifolds of a complete system $\mathcal{W}$ is called a *complete integral*.

However, there is a snag to this. For instance, the solution set of a first order ODE generally consists of a 1-parameter family of solutions—constituting a complete integral—and besides this a number of singular solutions outside of the family. And the latter are lost when we consider complete integrals only.

This problem is solved by the Cartan–Kuranishi prolongation theorem (to be discussed in chapter 4): a singular integral manifold will become a member of a complete integral after having been *prolonged* far enough.

Therefore our main problem is,

given a vector field system V, find the complete subsystems of maximal dimension!

Or dually,

given a pfaffian system P, find the complete pfaffian systems $Q \supseteq P$ of smallest possible dimension!

Notation. The above is equivalently, but more briefly, expressed by *integrate V (respectively P)!*

2.5 Mayer's blowing-up method for proving the Frobenius theorem

The proof of Frobenius's theorem given in the last section involved an induction, which makes the actual construction of the new local coordinates $t^1, \ldots, t^r$ rather tedious. Fortunately there is a method due to A. Mayer—a colleague of Lie at the university of Leipzig—which reduces the proof directly to the rectification lemma for a *single* vector field.

Start with a vector field system $V = (X_1, \ldots, X_q)$ on $\mathbb{R}^r$ in Jacobian form:

$$\begin{cases} X_1 = \partial/\partial x^1 + \cdots \qquad + \sum_{i=q+1}^r a_1^i(x) \left(\partial/\partial x^i\right), \\ \ldots \\ X_q = \qquad\qquad \partial/\partial x^q + \sum_{i=q+1}^r a_q^i(x) \left(\partial/\partial x^i\right). \end{cases}$$

Mayer's idea is to use a mapping $\mu: \mathbb{R}_x^r \to \mathbb{R}_y^r$ which blows up the $(r-q)$-dimensional plane $x^1 = \cdots = x^q = 0$ into the hyperplane $y^1 = 0$. The inverse $\mu^{-1}: \mathbb{R}_y^r \to \mathbb{R}_x^r$ is given by

$$\begin{cases} x^1 = y^1, \\ x^k = y^1 \cdot y^k \qquad \text{for } k = 2, \ldots, q, \\ x^l = y^l \qquad \text{for } l = q+1, \ldots, r, \end{cases}$$

so that μ itself is defined by

$$\begin{cases} y^1 = x^1, \\ y^k = x^k/x^1 \qquad \text{for } k = 2, \ldots, q, \\ y^l = x^l \qquad \text{for } l = q+1, \ldots, r \end{cases}$$

when $x^1 \neq 0$. Then

$$\mu_*(X_1) = \frac{\partial}{\partial y^1} - \frac{1}{y^1} \sum_{k=2}^{q} y^k \frac{\partial}{\partial y^k} + \sum_{i=q+1}^{r} a_1^i(\mu^{-1}(y)) \frac{\partial}{\partial y^i}$$

and

$$\mu_*(X_k) = \frac{1}{y^1} \frac{\partial}{\partial y^k} + \sum_{i=q+1}^{r} a_k^i(\mu^{-1}(y)) \frac{\partial}{\partial y^i} \qquad \text{for } k = 2,\ldots,q.$$

To get these into Jacobian form, set

$$Y_1 := \mu_*(X_1) + \sum_{k=2}^{q} y^k \mu_*(X_k)$$

$$= \frac{\partial}{\partial y^1} + \sum_{i=q+1}^{r} \left(a_1^i(\mu^{-1}(y)) + \sum_{k=2}^{q} y^k a_k^i(\mu^{-1}(y)) \right) \frac{\partial}{\partial y^i}$$

and

$$Y_k := y^1 \mu_*(X_k) = \frac{\partial}{\partial y^k} + \sum_{i=q+1}^{r} y^1 a_k^i(\mu^{-1}(y)) \frac{\partial}{\partial y^i} \qquad \text{for } k = 2,\ldots,q.$$

The coefficients $a_k^i(x)$ are supposed to be smooth near the origin, and therefore the $a_k^i(\mu^{-1}(y))$ are smooth near the plane $y^1 = y^{q+1} = \cdots = y^r = 0$. If we set

$$b_1^i(y) := a_1^i(\mu^{-1}(y)) + \sum_{k=2}^{q} y^k a_k^i(\mu^{-1}(y))$$

and

$$b_k^i(y) := a_k^i(\mu^{-1}(y)) \qquad \text{for } k = 2,\ldots,q,$$

$\mathcal{V}$ is generated by the vector fields

$$\begin{cases} Y_1 = \partial/\partial y^1 + \sum_{i=q+1}^{r} b_1^i(y) \, (\partial/\partial y^i) & \text{and} \\ Y_k = \partial/\partial y^k + y^1 \cdot \sum_{i=q+1}^{r} b_k^i(y) \, (\partial/\partial y^i) & \text{for } k = 2,\ldots,q. \end{cases}$$

Let us return to the original x-coordinates for a moment. According to the supplement of the Frobenius theorem there is a unique set of local coordinates $t^1, \ldots, t^r$ such that $\mathcal{V} = (\partial/\partial t^1, \ldots, \partial/\partial t^q)$ and

$$\begin{cases} t^i = 0 & \text{for } i = 1,\ldots,q, \\ t^j = x^j = y^j & \text{for } j = q+1,\ldots,r, \end{cases}$$

when $x^1 = \cdots = x^q = 0 \iff y^1 = 0$.

Now if $q + 1 \le l \le r$,

$$t^l = t^l(x^1, \ldots, x^r) = t^l(y^1, y^1 y^2, \ldots, y^1 y^q, y^{q+1}, \ldots, y^r)$$

is a first integral of $\mathcal{V} = (Y_1, \ldots, Y_q)$, and in particular of Y_1, i.e.,

$$Y_1 t^l = 0 \quad \text{for } l = q + 1, \ldots, r.$$

But according to the rectification lemma there is a *unique* first integral $f^l(y)$ of Y_1 satisfying $f^l(0, y^2, \ldots, y^r) = y^l$. Hence $t^l = f^l(y)$.

Conclusion. *The functionally independent first integrals $t^{q+1}, \ldots, t^r$ of $\mathcal{V}$ with*

$$t^l(0, \ldots, 0, x^{q+1}, \ldots, x^r) = x^l \quad \text{for } l = q + 1, \ldots, r$$

*are found by applying the rectification lemma to the **single** vector field Y_1.*

Choosing $t^i := y^i$ for $i = 1, \ldots, q$ we find that

$$\frac{\partial(t^1, \ldots, t^r)}{\partial(y^1, \ldots, y^r)}(0, \ldots, 0) = 1,$$

and that

$$\mathcal{V} = \left(\frac{\partial}{\partial t^1}, \ldots, \frac{\partial}{\partial t^q} \right).$$

Example. Consider the vector field system $\mathcal{V}$ on $\mathbb{R}^4$ generated by

$$X_1 = \frac{\partial}{\partial x^1} + (x^3 + 3(x^1)^2)\frac{\partial}{\partial x^4},$$

$$X_2 = \frac{\partial}{\partial x^2} + x^2 \frac{\partial}{\partial x^4},$$

$$X_3 = \frac{\partial}{\partial x^3} + x^1 \frac{\partial}{\partial x^4}.$$

The task is to find local coordinates $t^1, \ldots, t^4$ such that $\mathcal{V} = (\partial/\partial t^1, \partial/\partial t^2, \partial/\partial t^3)$, and t^4 is a first integral of $\mathcal{V}$ satisfying $t^4(0, 0, 0, x^4) = x^4$.

As a first step we blow up the line $x^1 = x^2 = x^3 = 0$ into the hyperplane $y^1 = 0$ by

$$\mu^{-1}: \begin{cases} x^1 = y^1 \\ x^2 = y^1 y^2 \\ x^3 = y^1 y^3 \\ x^4 = y^4 \end{cases} \quad \text{and} \quad \mu: \begin{cases} y^1 = x^1 \\ y^2 = x^2/x^1 \\ y^3 = x^3/x^1 \\ y^4 = x^4 \end{cases} \quad \text{when } x^1 \ne 0.$$

Then

$$\mu_*(X_1) = \frac{\partial}{\partial y^1} - \frac{y^2}{y^1}\frac{\partial}{\partial y^2} - \frac{y^3}{y^1}\frac{\partial}{\partial y^3} + (y^1 y^3 + 3(y^1)^2)\frac{\partial}{\partial y^4},$$

$$\mu_*(X_2) = \frac{1}{y^1}\frac{\partial}{\partial y^2} + y^1 y^2 \frac{\partial}{\partial y^4},$$

$$\mu_*(X_3) = \frac{1}{y^1}\frac{\partial}{\partial y^3} + y^1 \frac{\partial}{\partial y^4}.$$

The first vector field in the corresponding Jacobian system is

$$Y_1 = \mu_*(X_1) + y^2 \mu_*(X_2) + y^3 \mu_*(X_3)$$

$$= \frac{\partial}{\partial y^1} + (2y^1 y^3 + 3(y^1)^2 + y^1 (y^2)^2)\frac{\partial}{\partial y^4}.$$

This is rectified by a coordinate transformation

$$\Phi : \mathbb{R}^4_t \longrightarrow \mathbb{R}^4_y$$

which satisfies on the one hand

$$\Phi_* \left(\frac{\partial}{\partial t^1} \right) = Y_1, \quad \text{i.e.,} \quad \begin{cases} \partial y^1 / \partial t^1 = 1, \\ \partial y^2 / \partial t^1 = \partial y^3 / \partial t^1 = 0, \\ \partial y^4 / \partial t^1 = 2y^1 y^3 + 3(y^1)^2 + y^1 (y^2)^2, \end{cases}$$

and on the other

$$y^1(0, t^2, t^3, t^4) = 0 \text{ and } y^k(0, t^2, t^3, t^4) = t^k \text{ for } k = 2, 3, 4.$$

Treating t^2, t^3 and t^4 as parameters we obtain the ODE system

$$\begin{cases} dy^1 / dt^1 = 1, \\ dy^2 / dt^1 = dy^3 / dt^1 = 0, \\ dy^4 / dt^1 = 2y^1 y^3 + 3(y^1)^2 + y^1 (y^2)^2, \end{cases}$$

with initial conditions as above. Thus

$$y^1 = t^1, \ y^2 = t^2, \ y^3 = t^3 \quad \text{and} \quad \frac{dy^4}{dy^1} = 2y^1 y^3 + 3(y^1)^2 + y^1 (y^2)^2,$$

which gives

$$y^4 = (y^1)^2 y^3 + (y^1)^3 + \frac{1}{2}(y^1 y^2)^2 + t^4.$$

If we return to the x-coordinates,

$$t^4 = x^4 - x^1 x^3 - (x^1)^3 - \frac{1}{2}(x^2)^2,$$

with $t^4(0, 0, 0, x^4) = x^4$—as required.

Finally, keeping x^1, x^2, x^3 as local coordinates but replacing x^4 by $t^4(x^1, x^2, x^3, x^4)$, we get

$$\mathcal{V} = \left(\frac{\partial}{\partial x^1}, \frac{\partial}{\partial x^2}, \frac{\partial}{\partial x^3} \right),$$

and the integral manifolds are given by

$$t^4 = \text{constant} \iff x^4 - x^1 x^3 - (x^1)^3 - \frac{1}{2}(x^2)^2 = \text{constant}.$$

3

Cartan's local existence theorem

Let $\mathcal{P} = (\theta^1, \ldots, \theta^s)$ be a pfaffian system on $\mathbb{R}^r$, and suppose that $\{\theta^1, \ldots, \theta^s, \pi^1, \ldots, \pi^q\}$ (with $q = r - s$) is a local coframe for $\mathbb{R}^r$. The *integration problem* for $\mathcal{P}$ is

find the complete pfaffian systems containing $\mathcal{P}$ that have the smallest possible dimension!

Having found such a complete system, its integral manifolds are then given by the Frobenius theorem.

The starting point for solving this problem is the *Lie structure* of $\mathcal{P}$:

$$d\theta^i \equiv \frac{1}{2} \sum_{j,k=1}^{q} \gamma_{jk}^i \pi^j \wedge \pi^k \pmod{\mathcal{P}} \quad \text{for } i = 1, \ldots, s,$$

where the structure functions γ_{jk}^i satisfy $\gamma_{jk}^i + \gamma_{kj}^i = 0$.

A tentative way of constructing a complete system of low dimension containing $\mathcal{P}$ is the following.

First use the structure equations in order to find a pfaffian system $^{(1)}\mathcal{P}$ of minimal dimension such that

$$d\theta^i \equiv 0 \pmod{^{(1)}\mathcal{P}} \quad \text{for } i = 1, \ldots, s.$$

A difficulty is that there may be several different choices of $^{(1)}\mathcal{P}$ satisfying this, and it is hard to know which will give rise to the smallest dimension in the end. But *if* there is a unique $^{(1)}\mathcal{P}$, this is clearly of interest.

Definition. If a unique $^{(1)}\mathcal{P}$ exists, it is denoted by $'\mathcal{P}$, and is called the *anti-derived system* of $\mathcal{P}$.

Example. Suppose that the structure equations of the noncomplete pfaffian system $\mathcal{P} = (\theta^1, \theta^2, \theta^3)$ are

$$\begin{cases} d\theta^1 \equiv \pi^1 \wedge \pi^2, \\ d\theta^2 \equiv \pi^1 \wedge \pi^3, \\ d\theta^3 \equiv \pi^1 \wedge (\pi^4 - \pi^5) \end{cases} \quad (\text{mod } \mathcal{P}).$$

In this case $^{(1)}\mathcal{P}$ can be nothing but $^{(1)}\mathcal{P} = (\theta^1, \theta^2, \theta^3, \pi^1)$.

After having found $^{(1)}\mathcal{P}$, use the structure equations of $^{(1)}\mathcal{P}$ in order to find a smallest pfaffian system $^{(2)}\mathcal{P} \supset {}^{(1)}\mathcal{P}$ such that

$$d^{(1)}\mathcal{P} \equiv 0 \quad (\text{mod } {}^{(2)}\mathcal{P}),$$

—and so on.

With our standard assumption that the generators of all pfaffian (and vector field) systems considered are linearly independent at each point of the domain where they are defined, this process leads after a finite number of steps to a system $^{(c)}\mathcal{P}$ satisfying

$$^{(c)}\mathcal{P} \equiv 0 \quad (\text{mod } {}^{(c)}\mathcal{P}),$$

so that $^{(c)}\mathcal{P}$ is *complete*.

However, in general it is not clear how to choose $^{(1)}\mathcal{P}$, $^{(2)}\mathcal{P}$, ... so that the resulting complete system $^{(c)}\mathcal{P} \supset \mathcal{P}$ really has the smallest possible dimension.

Cartan's solution of the integration problem is instead based on the consideration of the dual vector field system $\mathcal{V} = (X_1, \ldots, X_q)$, with $X_k = \partial/\partial\pi^k$ for $k = 1, \ldots, q$. The *integration problem* for $\mathcal{V}$ is

> *find the complete subsystems of $\mathcal{V}$ which have the largest possible dimension!*

The idea is roughly: start with a general vector field $U_1 \in \mathcal{V}$, determine the most general $U_2 \in \mathcal{V}$ such that $[U_1, U_2] = 0$, then determine the most general $U_3 \in \mathcal{V}$ satisfying $[U_1, U_3] = [U_2, U_3] = 0$, and so on. In this way there arises a chain of complete systems

$$(U_1) \subset (U_1, U_2) \subset (U_1, U_2, U_3) \subset \cdots,$$

and the last is the wanted maximal complete subsystem of $\mathcal{V}$.

To achieve this one must know the Lie structure of $\mathcal{V}$. Let $Y_k := \partial/\partial\theta^k$ for $k = 1, \ldots, s$. Then there are structure fuctions $c_{jk}^i = -c_{kj}^i$ such that

$$[X_i, X_j] \equiv \sum_{k=1}^{s} c_{ij}^k Y_k \quad (\text{mod } \mathcal{V}) \quad \text{for } i, j = 1, \ldots, q.$$

Let us compare these structure equations with those of $\mathcal{P} = (\theta^1,\ldots,\theta^s)$, remembering that $X_k = \partial/\partial\pi^k$:

$$d\theta^k \equiv \frac{1}{2}\sum_{l,m=1}^{q} \gamma^k_{lm}\,\pi^l \wedge \pi^m \quad (\text{mod } \mathcal{P}) \quad \text{for } k = 1,\ldots,s.$$

On the one hand,

$$d\theta^k(X_i, X_j) = \frac{1}{2}\sum_{l,m=1}^{q} \gamma^k_{lm}\left(\pi^l(X_i)\pi^m(X_j) - \pi^l(X_j)\pi^m(X_i)\right)$$

$$= \frac{1}{2}(\gamma^k_{ij} - \gamma^k_{ji}) = \gamma^k_{ij},$$

and on the other,

$$d\theta^k(X_i, X_j) = X_i(\theta^k(X_j)) - X_j(\theta^k(X_i)) - \theta^k([X_i, X_j])$$

$$= -\theta^k\left(\sum_{l=1}^{s} c^l_{ij}\, Y_l\right) = -c^k_{ij},$$

so that

$$c^k_{ij} = -\gamma^k_{ij}.$$

Conclusion. *The structure functions of a pfaffian system and its dual vector field system differ only in a sign.*

Recall the derived system $\mathcal{V}' = \mathcal{V} + [\mathcal{V}, \mathcal{V}]$. If $\dim \mathcal{V}' = q + p$ with $p \le s$, $\mathcal{V}'$ is of the form $(X_1,\ldots,X_q, Z_1,\ldots Z_p)$, where it can be assumed that $Z_k = \partial/\partial\theta^k$ for $k = 1,\ldots,p$. The structure equations for $\mathcal{V}$ are then

$$[X_i, X_j] \equiv \sum_{k=1}^{p} c^k_{ij}\, Z_k \quad (\text{mod } \mathcal{V}) \quad \text{for } i, j = 1,\ldots,q.$$

Cartan's key idea is to first look for *involutions*, because this only involves linear algebra.

Definition. An *involution* of the vector field system $\mathcal{V}$ is a subsystem $\mathcal{I}$ of $\mathcal{V}$ satisfying $[\mathcal{I}, \mathcal{I}] \subseteq \mathcal{V}$. In particular every complete subsystem of $\mathcal{V}$ is an involution—but not conversely.

Dually: Let $X, Y \in \mathcal{V}$ and let $\mathcal{P} = \mathcal{V}^\perp = (\theta^1,\ldots,\theta^s)$. Then

$$d\theta^i(X, Y) = -\theta^i([X, Y]) \quad \text{for } i = 1,\ldots,s,$$

so that

$$[X, Y] \in \mathcal{V} \iff d\theta^i(X, Y) = 0 \text{ for } i = 1,\ldots,s, \text{ or } d\mathcal{P}(X, Y) = 0.$$

Thus $\mathcal{I} \subseteq \mathcal{V}$ is an involution if and only if

$$d\mathcal{P}(\mathcal{I}, \mathcal{I}) = 0.$$

Cartan's 2-step process consists in:

(i) First determine the involutions of maximal dimension—which involves linear algebra and the structure functions.
(ii) Then specialize the involutions found to complete subsystems of $\mathcal{V}$—which can be done by repeated applications of the Cauchy–Kowalewski theorem.

After that the Frobenius theorem gives the wanted integral manifolds of maximal dimension.

In the next two sections we follow the presentation of Cartan's ideas given in [Vessiot 1924]; the original version is to be found in [Cartan 1901a].

3.1 Involutions and characters

Let $\mathcal{V} = (X_1, \ldots, X_q)$ be a vector field system on $\mathbb{R}^r_x$. Using Gaussian elimination and changing indices it may be assumed that

$$X_i = \frac{\partial}{\partial x^i} + \sum_{j=q+1}^{r} \xi_i^j(x) \frac{\partial}{\partial x^j} \quad \text{for } i = 1, \ldots, q.$$

Then $[X_i, X_j]$ either vanishes, or does not belong to $\mathcal{V}$. If the derivative of $\mathcal{V}$ is $\mathcal{V}' = (X_1, \ldots, X_q, Z_1, \ldots, Z_p)$, this means that the structure equations are equalities (and not just congruences (mod $\mathcal{V}$)):

$$[X_i, X_j] = \sum_{k=1}^{p} c_{ij}^k Z_k \quad \text{for } i, j = 1, \ldots, q.$$

Also let $\{\theta^1, \ldots, \theta^s, \pi^1, \ldots, \pi^q\}$ (with $s - r - q$) be a local coframe for $\mathbb{R}^r$ such that

$$X_i = \partial/\partial \pi^i \text{ for } i = 1, \ldots, q.$$

Then the dual pfaffian system of $\mathcal{V}$ is

$$\mathcal{P} = \mathcal{V}^\perp = (\theta^1, \ldots, \theta^s).$$

The *ultimate problem* consists in determining complete subsystems $\mathcal{W} = (Y_1, \ldots, Y_n)$ of $\mathcal{V}$ with $Y_i = \sum_{k=1}^{q} a_i^k(x) X_k$ being linearly independent and with n as large as possible.

Choosing the Y_i in Jacobian form it may be assumed that $[Y_i, Y_j] = 0$ for $i, j = 1, \ldots, n$. Then

$$0 = [Y_i, Y_j] = \sum_{k=1}^{q} (Y_i a_j^k - Y_j a_i^k) X_k + \sum_{k,l=1}^{q} \sum_{m=1}^{p} a_i^k a_j^l c_{kl}^m Z_m.$$

So the coefficients $a_i^k(x)$ are to satisfy

$$\sum_{k,l=1}^{q} c_{kl}^m a_i^k a_j^l = 0 \text{ for } i, j = 1, \ldots, n \text{ and } m = 1, \ldots, p, \qquad \textbf{(A)}$$

and

$$Y_i a_j^k - Y_j a_i^k = 0 \text{ for } i, j = 1, \ldots, n \text{ and } k = 1, \ldots, q. \qquad \textbf{(B)}$$

Condition **(A)** means that $[Y_i, Y_j] \equiv 0 \pmod{\mathcal{V}}$—or from a dual point of view that $d\mathcal{P}(Y_i, Y_j) = 0$. Hence

$\textbf{(A)} \Longleftrightarrow (Y_1, \ldots, Y_n)$ is an involution.

Condition **(B)** then says that the involution $(Y_1, \ldots, Y_n)$ is in fact complete.

Definition. Let $\mathcal{V}_1$ be a subsystem of the vector field system $\mathcal{V}$. A vector field $X \in \mathcal{V}$ is *in involution with* $\mathcal{V}_1$ if $[X, \mathcal{V}_1] \subseteq \mathcal{V}$. Referring to the dual pfaffian system $\mathcal{P} = \mathcal{V}^\perp$, this is equivalent to $d\mathcal{P}(X, \mathcal{V}_1) = 0$ or $X \rfloor d\mathcal{P}(\mathcal{V}_1) = 0$, where $\rfloor$ denotes interior product.

Let us now construct involutions of successively larger dimensions. The *general first order involution* is $\mathcal{I}_1 = (Y_1)$ where

$$Y_1 = \sum_{k=1}^{q} a_1^k X_k$$

with arbitrary—or *parametric*—coefficients a_1^k.
    Then let

$$Y_2 = \sum_{k=1}^{q} a_2^k X_k$$

be the most general vector field in $\mathcal{V}$ that is in involution with $\mathcal{I}_1$, i.e., satisfies

$$[Y_1, Y_2] \equiv 0 \pmod{\mathcal{V}}, \text{ or dually } Y_1 \rfloor d\mathcal{P}(Y_2) = 0.$$

Quite explicitly this means that

$$\sum_{k,l=1}^{q} c_{kl}^{m} a_1^k a_2^l = 0 \quad \text{for } m = 1, \ldots, p. \tag{1}$$

Let $\mathbf{a}_i := (a_i^1, \ldots, a_i^q)^T$—where T denotes transpose—and let $\mathrm{M}(\mathbf{a}_1)$ be the $p \times q$ matrix

$$\mathrm{M}(\mathbf{a}_1) := \begin{pmatrix} \sum_{k=1}^{q} c_{k1}^1 a_1^k & \cdots & \sum_{k=1}^{q} c_{kq}^1 a_1^k \\ \vdots & \ddots & \vdots \\ \sum_{k=1}^{q} c_{k1}^p a_1^k & \cdots & \sum_{k=1}^{q} c_{kq}^p a_1^k \end{pmatrix}.$$

Then (1) is equivalent to

$$\mathrm{M}(\mathbf{a}_1)\, \mathbf{a}_2 = \mathbf{0}. \tag{2}$$

Here $\mathrm{M}(\mathbf{a}_1)$ is regarded as a matrix over the ring of smooth functions. As such it has a certain rank.

Definitions. The *first character* of $\mathcal{V}$ (or of $\mathcal{P} = (\theta^1, \ldots, \theta^s)$) is defined by

$$s_1 := \max_{\mathbf{a}_1} \mathrm{rank}\, \mathrm{M}(\mathbf{a}_1) = \max_{Y \in \mathcal{V}} \dim (Y \rfloor d\theta^{1,}, \ldots, Y \rfloor d\theta^s).$$

$\mathbf{a}_1$ is *generic* if $\mathrm{rank}\, \mathrm{M}(\mathbf{a}_1) = s_1$. Vector fields $\sum_{k=1}^{q} a_1^k X_k$ corresponding to generic $\mathbf{a}_1$ are said to be *regular*.

Remark. The condition $\mathrm{rank}\, \mathrm{M}(\mathbf{a}_1) < s_1$ means that a certain number of determinants are to vanish. Therefore the set of regular vector fields in $\mathcal{V}$ constitutes a dense open subset of $\mathcal{V} \cong (\text{smooth functions})^q$ with respect to any reasonable topology.

Definition. The nonzero vector field $\sum_{k=1}^{q} a^k X_k$ is *singular* if $\mathrm{rank}\, \mathrm{M}(\mathbf{a}) < s_1$. If $\mathrm{rank}\, \mathrm{M}(\mathbf{a}) = 0$, it is *Cauchy characteristic*. That is, a nonzero $C \in \mathcal{V}$ is Cauchy characteristic if $[C, \mathcal{V}] \subseteq \mathcal{V}$, or equivalently $d\mathcal{P}(C, \mathcal{V}) = 0$ for $\mathcal{P} = \mathcal{V}^\perp$.

In the proof of Cartan's local existence theorem it is necessary to avoid the singular vector fields, since they eventually may give rise to overdetermined Cauchy–Kowalewski systems. However, when we are trying to solve the integration problem outside of the analytic framework, and are looking for more information about the solutions besides their existence, they are going to play a crucial role. In particular we will later discover what a blessing the Cauchy characteristic vector fields are—when existing, that is.

Let us now return to equation (2). Since

$$[Y_1, \lambda Y_1] = Y_1(\lambda) Y_1 \equiv 0 \quad (\text{mod } \mathcal{V})$$

for any smooth function λ, (2) admits at least the solution $\mathbf{a}_2 = \lambda \mathbf{a}_1$, and therefore $s_1 \leq q - 1$. If $\mathbf{a}_1$ is generic and if $s_1 < q - 1$, s_1 of the $\mathbf{a}_2$—the *principal ones*—get expressed as linear combinations of the remaining $q - s_1$ *parametric* a_2^l. Consequently there are $q - s_1$ linearly independent vector fields in involution with Y_1.

Definition. If $s_1 < q - 1$ the *general second order involution* of $\mathcal{V}$ is $\mathcal{I}_2 = (Y_1, Y_2)$, where $Y_1 = \sum a_1^k X_k$ is regular but otherwise arbitrary, and $Y_2 = \sum a_2^l X_l$ is the most general vector field in $\mathcal{V}$ in involution with Y_1. $q - s_1$ of the coefficients a_2^l are parametric—that is, can be chosen arbitrarily.

Having found $\mathcal{I}_2$ we next determine the general third order involution $\mathcal{I}_3$. To do this consider the most general vector field $Y_3 = \sum_{l=1}^q a_3^l X_l \in \mathcal{V}$ in involution with $\mathcal{I}_2$, i.e., satisfying

$$[Y_1, Y_3] \equiv 0 \quad (\text{mod } \mathcal{V}), \text{ and } [Y_2, Y_3] \equiv 0 \quad (\text{mod } \mathcal{V}), \tag{3}$$

or equivalently

$$Y_1 \rfloor d\mathcal{P}(Y_3) = Y_2 \rfloor d\mathcal{P}(Y_3) = 0 \quad \text{for} \quad \mathcal{P} = \mathcal{V}^\perp,$$

with Y_1, Y_2 as in the definition above, and thus containing $q + (q - s_1) = 2q - s_1$ parametric coefficients. Let $M(\mathbf{a}_1, \mathbf{a}_2)$ be the $2p \times q$ matrix

$$M(\mathbf{a}_1, \mathbf{a}_2) := \begin{pmatrix} \sum_{k=1}^q c_{k1}^1 a_1^k & \cdots & \sum_{k=1}^q c_{kq}^1 a_1^k \\ \vdots & \ddots & \vdots \\ \sum_{k=1}^q c_{k1}^p a_1^k & \cdots & \sum_{k=1}^q c_{kq}^p a_1^k \\ \sum_{k=1}^q c_{k1}^1 a_2^k & \cdots & \sum_{k=1}^q c_{kq}^1 a_2^k \\ \vdots & \ddots & \vdots \\ \sum_{k=1}^q c_{k1}^p a_2^k & \cdots & \sum_{k=1}^q c_{kq}^p a_2^k \end{pmatrix}.$$

Then (3) is equivalent to the following linear system for the a_3^l:

$$M(\mathbf{a}_1, \mathbf{a}_2) \mathbf{a}_3 = \mathbf{0}. \tag{4}$$

Let

$$s_1 + s_2 := \max_{\mathbf{a}_1, \mathbf{a}_2} \text{rank } M(\mathbf{a}_1, \mathbf{a}_2)$$

$$= \max_{Y_1, Y_2} \dim(Y_1 \rfloor d\theta^1, \ldots, Y_1 \rfloor d\theta^s, Y_2 \rfloor d\theta^1, \ldots, Y_2 \rfloor d\theta^s).$$

Definition. The integer s_2 appearing here is the *second character* of V (or of P). The involution (Y_1, Y_2) is *regular* if $\operatorname{rank} M(\mathbf{a}_1, \mathbf{a}_2) = s_1 + s_2$, and *singular* otherwise.

If $\mathcal{I}_2 = (Y_1, Y_2)$ is regular, (4) can be solved so as to express $s_1 + s_2$ of the coefficients a_3^l—the *principal ones*—as linear combinations of the remaining $q - s_1 - s_2$ parametric a_3^l. Consequently there are $q - s_1 - s_2$ linearly independent vector fields in V in involution with $\mathcal{I}_2$. With Y_1 and Y_2 being among these, $q - s_1 - s_2 \geq 2$. In case of equality nothing new is obtained, but otherwise we get the *general third order involution* $\mathcal{I}_3 = (Y_1, Y_2, Y_3)$ depending on $q + (q - s_1) + (q - s_1 - s_2) = 3q - 2s_1 - s_2$ arbitrary coefficients.

Having found $\mathcal{I}_3$ we next determine the general fourth order involution $\mathcal{I}_4$, whereby a regularity criterion for $\mathcal{I}_3$ is obtained, and so on. Suppose that in this process

$$q - s_1 > 1, \; q - s_1 - s_2 > 2, \ldots, \; q - s_1 - s_2 - \cdots - s_{n-1} > n - 1,$$

while

$$q - s_1 - s_2 - \cdots - s_n = n$$

for a certain integer n. Then by construction the nth order involution $\mathcal{I}_n = (Y_1, \ldots, Y_n)$ is the largest involution in V with $\mathcal{I}_{n-1} = (Y_1, \ldots, Y_{n-1})$ being regular. $\mathcal{I}_n$ depends on

$$q + (q - s_1) + \cdots + (q - s_1 - s_2 - \cdots - s_{n-1}) = nq - \sum_{k=1}^{n-1} (n-k) s_k$$

arbitrary coefficients, namely all a_1^k, $q - s_1$ of the a_2^k, $\ldots$, and finally $q - s_1 - \cdots - s_{n-1}$ of the a_n^k.

Definitions. The $\mathcal{I}_n$ found in this way is the *general n^{th} order involution* of V (or its dual pfaffian system P), and n is called the *involutive genus*; s_k is the kth *character* of V (or P) for $k = 1, \ldots, n - 1$. Being the most important of these, s_1 is sometimes simply referred to as *the character* of V (or of P).

Recall that s_1 is the rank of the system

$$[Y_1, Y] \equiv 0 \pmod{V}$$

where Y_1 is a regular vector field, while $s_1 + s_2$ is the rank of the system

$$[Y_1, Y] \equiv [Y_2, Y] \equiv 0 \pmod{V},$$

where (Y_1, Y_2) is a regular second order involution. Thus s_2 is the number of extra conditions imposed on Y from $[Y_2, Y] \equiv 0$ beyond those coming from $[Y_1, Y] \equiv 0$; hence $s_2 \leq \operatorname{rank} M(\mathbf{a}_2)$. But since Y_2 is less general than Y_1, $\operatorname{rank} M(\mathbf{a}_2) \leq \operatorname{rank} M(\mathbf{a}_1)$, and so

$$s_2 \leq \operatorname{rank} M(\mathbf{a}_2) \leq \operatorname{rank} M(\mathbf{a}_1) = s_1.$$

Analogously it follows that

$$s_1 \geq s_2 \geq \cdots \geq s_{n-1}.$$

In particular,

$$s_{p+1} = 0 \implies s_{p+2} = s_{p+3} = \cdots = s_{n-1} = 0.$$

3.2 From involutions to complete systems

Once we have found the general n^{th} order involution $\mathcal{I}_n = (Y_1, \ldots, Y_n)$ of $V = (X_1, \ldots, X_q)$ involving $nq - \sum_{k=1}^{n-1}(n-k)s_k$ parametric coefficients, the next task is to choose these coefficients so as to make $\mathcal{I}_n$ complete, i.e., so that condition **B**

$$Y_i a_j^k = Y_j a_i^k \quad \text{for } i, j = 1, \ldots, n \text{ and } k = 1, \ldots, q$$

gets satisfied. This is achieved by solving a number of first order Cauchy–Kowalewski systems. Such a system is of the form

$$\begin{cases} \partial z^j / \partial x^p = f^j \left(x^1, \ldots, x^n; z^1, \ldots, z^m; \ldots, \partial z^k / \partial x^i, \ldots \right), \\ \qquad \text{for } j, k = 1, \ldots, m, \text{ with no } \partial z^k / \partial x^p \text{ in the right hand side,} \\ z^j(x^1, \ldots, x^{p-1}, 0, x^{p+1}, \ldots, x^n) = g^j(x^1, \ldots, x^{p-1}, x^{p+1}, \ldots, x^n), \end{cases}$$

and the Cauchy–Kowalewski theorem states that it has a unique solution in the analytic category. Unfortunately Lewy's famous counterexample shows that it need not be solvable if the data are merely smooth.

The proof consists of two parts: the easy one establishes a unique formal power series solution, and the hard one uses Cauchy's majorization trick in order to demonstrate local convergence. Thus the Cauchy–Kowalewski theorem is trivially true for formal power series, nontrivially true in the real analytic realm, and false in the C^∞ category.

Note that the variable x_p plays a special role above. Because of this some preparation is necessary before it is possible to apply the Cauchy–Kowalewski theorem to our problem.

Remark. A system like

$$\begin{cases} F(x^1, x^2; z; \partial z/\partial x^1) = 0, \\ G(x^1, x^2; z; \partial z/\partial x^2) = 0 \end{cases}$$

cannot possibly be put in Cauchy–Kowalewski form. It is therefore somewhat surprising that *all* PDE sytems can be reduced to first order Cauchy–Kowalewski systems by Cartan's methods.

Let us now determine a basis for the general n^{th} order involution $\mathcal{I}_n$, allowing us to apply the Cauchy–Kowalewski theorem. Replacing the parametric coefficients a_i^k by determinate functions $\bar{a}_i^k$, $\mathcal{I}_n$ goes over into a determinate vector field system $\bar{\mathcal{I}}_n$. By Gaussian elimination and a change of indices, $\bar{\mathcal{I}}_n$ can be assumed to be generated by

$$\bar{U}_i = X_i + \sum_{j=1}^{q-n} \bar{a}_i^{n+j} X_{n+j} \quad \text{for } i = 1, \ldots, n,$$

where the $\bar{a}_i^{n+j}$ now are determinate functions. Because

$$X_i = \frac{\partial}{\partial x^i} + \sum_{j=q+1}^{n} \xi_i^j \frac{\partial}{\partial x^j} \quad \text{for } i = 1, \ldots, q,$$

the $\bar{U}_i$ are in Jacobian form.

$\bar{\mathcal{I}}_n = (\bar{U}_1, \ldots, \bar{U}_n)$ is an involution, and so are $(\bar{U}_1)$, $(\bar{U}_1, \bar{U}_2)$, $\ldots$, $(\bar{U}_1, \ldots, \bar{U}_{n-1})$—but the latter are not necessarily regular. However, the regular involutions form a dense open subset of the set of all involutions. Therefore it is possible to find a coordinate transformation $\mathbb{R}^r \longrightarrow \mathbb{R}^r$ close to the identity such that the $\bar{U}_i$ give rise to regular involutions when put in Jacobian form with respect to the new coordinates. In this way it may be assumed that the involutions $(\bar{U}_1)$, $(\bar{U}_1, \bar{U}_2)$, $\ldots$, $(\bar{U}_1, \ldots, \bar{U}_{n-1})$ are all regular.

Clearly this procedure also works for involutions $\mathcal{I}_n$ close to $\bar{\mathcal{I}}_n$, so that any such $\mathcal{I}_n$ can be written as $(U_1, \ldots, U_n)$ with

$$U_i = X_i + \sum_{j=1}^{q-n} a_i^{n+j} X_{n+j} \quad \text{for } i = 1, \ldots, n,$$

where $a_i^{n+j} \approx \bar{a}_i^{n+j}$, and (U_1), $\ldots$, $(U_1, \ldots, U_{n-1})$ all being regular.

In the preceding section there appeared nq coefficients a_i^k, and we saw that $nq - \sum_{k=1}^{n-1}(n-k)s_k$ of them were parametric. Here the total number of coefficients is $n(q - n)$ only, and therefore it is necessary to reconsider the number of parametric coefficients.

Start with

$$U_1 = X_1 + \sum_{j=1}^{q-n} a_1^{n+j} X_{n+j},$$

where the a_1^{n+j} are parametric, and let $U = \sum_{i=1}^{q} a^i X_i$ be an arbitrary vector field in $\mathcal{V}$. Expressing $X_1,\ldots,X_q$ by means of $U_1,\ldots,U_n, X_{n+1},\ldots,X_q$, we get

$$U = \sum_{i=1}^{n} a^i U_i + \sum_{j=1}^{q-n} b^{n+j} X_{n+j}.$$

The system

$$[U, U_1] \equiv 0 \quad (\mathrm{mod}\ \mathcal{V}) \tag{*}$$

is of rank s_1. Because $[U_1, U_i] \equiv 0\ (\mathrm{mod}\ V)$ for $i = 1,\ldots,n$,

$$0 \equiv [U_1, U] \equiv \sum_{j=1}^{q-n} b^{n+j} [U_1, X_{n+j}] \quad (\mathrm{mod}\ V),$$

so that $a^1,\ \ldots,\ a^n$ are left arbitrary, while s_1 of the b^{n+j} are solved as linear combinations of the remaining b^{n+j}. Consequently $a^1,\ldots,a^n$ together with $q - n - s_1$ of the b^{n+j} are parametric in the general solution of (*). Hence we are free to choose $(a^1,\ldots,a^n) = (0,1,0,\ldots,0)$, thereby obtaining just the solution we want,

$$U_2 = X_2 + \sum_{j=1}^{q-n} a_2^{n+j} X_{n+j},$$

with $q - n - s_1$ of the a_2^{n+j} being parametric.

Since the involution (U_1, U_2) is regular, the system

$$[U_1, U] \equiv [U_2, U] \equiv 0 \quad (\mathrm{mod}\ V)$$

for the coefficients of $U = \sum_{i=1}^{q} a^i X_i$ is of rank $s_1 + s_2$. As above, one first realizes that $a^1,\ldots,a^n$ and $q - n - s_1 - s_2$ of $a^{n+1},\ldots,a^q$ are parametric, and then obtains the wanted solution

$$U_3 = X_3 + \sum_{j=1}^{q-n} a_3^{n+j} X_{n+j}$$

by specializing $(a^1,\ldots,a^n)$ to $(0,0,1,0,\ldots,0)$.

This procedure eventually leads to a basis $\{U_1,\ldots,U_n\}$ for the general

n^{th} order involution $\mathcal{I}_n$ close to $\bar{\mathcal{I}}_n$, having the following parametric coefficients:

$$q - n \quad \text{(i.e. all)} \qquad \text{of} \quad a_1^{n+1}, \ldots, a_1^q,$$
$$q - n - s_1 \qquad\qquad \text{of} \quad a_2^{n+1}, \ldots, a_2^q,$$
$$\cdots$$
$$q - n - s_1 - \cdots - s_{n-1} \qquad \text{of} \quad a_n^{n+1}, \ldots, a_n^q.$$

So altogeher there are $(q-n)+(q-n-s_1)+\cdots+(q-n-s_1-\cdots-s_{n-1}) = n(q-n) - \sum_{k=1}^{n-1}(n-k)s_k$ parametric coefficients in $\mathcal{I}_n$.

Because the parametric coefficients in the general solution of the kth system

$$[U_1, U] \equiv \cdots \equiv [U_k, U] \equiv 0 \quad (\text{mod } \mathcal{V})$$

obviously are parametric for the $(k-1)^{\text{th}}$ too, it is possible to arrange the indices in such a way that the following coefficients are parametric:

$$a_1^{n+1}, \ldots, a_1^{n+1+s_1}, \ldots, a_1^{n+1+s_1+\cdots+s_{n+1}}, \ldots, a_1^q,$$
$$a_2^{n+1+s_1}, \ldots, a_2^{n+1+s_1+\cdots+s_{n-1}}, \ldots, a_2^q,$$
$$\cdots$$
$$a_n^{n+1+s_1+\cdots+s_{n-1}}, \ldots, a_n^q.$$

The transition from involutions to complete systems is made by specializing these parametric coefficients so that

$$[U_i, U_j] = 0 \quad \text{for } i, j = 1, \ldots, n.$$

We do this by an induction over the dimension. The 1-dimensional case is clear: any vector field $X \in \mathcal{V}$ gives rise to the 1-dimensional complete system (X).

Let us then look at the general second order involution $\mathcal{I}_2 = (U_1, U_2)$, where

$$\begin{cases} U_1 = X_1 + \sum_{j=1}^{q-2} a_1^{2+j} X_{2+j} = \partial/\partial x^1 + \cdots, \\ U_2 = X_2 + \sum_{j=1}^{q-2} a_2^{2+j} X_{2+j} = \partial/\partial x^2 + \cdots \end{cases}$$

with parametric coefficients $a_1^3, \ldots, a_1^q, a_2^{3+s_1}, \ldots, a_2^q$ and principal coefficients $a_2^3, \ldots, a_2^{2+s_1}$. The parametric coefficents are to be determined such that

$$U_1 a_2^{2+j} - U_2 a_1^{2+j} = 0 \quad \text{for } j = 1, \ldots, q-2.$$

This is an underdetermined system of $q-2$ first order PDEs in $(q-2)+$

$(q - 2 - s_1)$ unknowns (note that if we had used a singular involution with a smaller number of parametric coefficients, the resulting PDE system might have been overdetermined). To make it determined, the $q - 2 - s_1$ parametric a_2^{2+j} are replaced by *arbitrary functions*, whereupon the remaining principal a_2^{2+j} are expressed by means of these and the a_1^{2+j}—thus leaving $a_1^3, \ldots, a_1^q$ as $q - 2$ unknowns. Since $U_2 = \partial/\partial x^2 + \cdots$, the resulting system can be written as

$$\frac{\partial a_1^j}{\partial x^2} = A_1^j \left(x^1, \ldots, x^r; a_1^3, \ldots, a_1^q; \ldots, \frac{\partial a_1^k}{\partial x^i}, \ldots \right) \quad \text{for } j = 3, \ldots, q,$$

with no $\partial a_1^j / \partial x^2$ in the right hand side. If we add initial conditions

$$a_1^j(x^1, 0, x^3, \ldots, x^r) = g^j(x^1, x^3, \ldots, x^r) \quad \text{for } j = 3, \ldots, q,$$

the Cauchy–Kowalewski theorem yields a unique local solution provided that all data are analytic—which we assume from now on.

Conclusion. *The system* $U_1 a_2^{2+j} = U_2 a_1^{2+j}$ *for* $j = 3, \ldots, q$ *has solutions* a_1^{2+j}, a_2^{2+j} *with*

$$\begin{cases} \text{arbitrary } a_2^{3+s_1}, \ldots a_2^q, \text{ and} \\ \text{the restrictions of } a_1^3, \ldots, a_1^q \text{ to } \{x \mid x^2 = 0\} \text{ being arbitrary.} \end{cases}$$

So it seems that the general 2-dimensional complete subsystem of $\mathcal{V}$ depends on '$q - 2 - s_1$ arbitrary functions of r variables and $q - 2$ arbitrary functions of $r - 1$ variables'—an arbitrariness that will be investigated further in the next section.

When proving the existence of n-dimensional complete subsystems we will at one point need the following result.

Lemma 3.2.1 (Hadamard's lemma). *Let M be a smooth submanifold of $\mathbb{R}^n$, defined by*

$$f^1(x) = \cdots = f^k(x) = 0,$$

with $df^1 \wedge \cdots \wedge df^k \neq 0$ near M. Let $f(x)$ be a smooth function defined in a neighbourhood U of some $m \in M$, and vanishing on $U \cap M$. Then there are smooth functions $g_1(x), \ldots, g_k(x)$ such that

$$f(x) = \sum_{i=1}^{k} g_i(x) f^i(x)$$

near m.

Proof. If we introduce new local coordinates $y^1 := f^1(x), \ldots, y^k := f^k(x)$, $y^{k+1}, \ldots, y^n$ near m with $y^{k+1}(m) = \cdots = y^n(m) = 0$, the vanishing of f on $U \cap M$ means that $f(0, \ldots, 0, y^{k+1}, \ldots, y^n) = 0$ for all $(y^{k+1}, \ldots, y^n) \approx (0, \ldots, 0)$. The claim is that

$$f = \sum_{i=1}^{n} g_i(y) \, y^i.$$

This is immediate for analytic functions. In the smooth case,

$$
\begin{aligned}
f(y^1, \ldots, y^n) &= f(y^1, \ldots, y^k, y^{k+1}, \ldots, y^n) - f(0, \ldots, 0, y^{k+1}, \ldots, y^n) \\
&= \left[f(ty^1, \ldots, ty^k, y^{k+1}, \ldots, y^n) \right]_{t=0}^{1} \\
&= \int_0^1 \frac{\partial}{\partial t} f(ty^1, \ldots, ty^k, y^{k+1}, \ldots, y^n) \, dt \\
&= \int_0^1 \sum_{i=1}^{k} \frac{\partial f}{\partial y^i}(ty^1, \ldots, ty^k, y^{k+1}, \ldots, y^n) \, y^i \, dt \\
&= \sum_{i=1}^{k} y^i \int_0^1 \frac{\partial f}{\partial y^i}(ty^1, \ldots, ty^k, y^{k+1}, \ldots, y^n) \, dt.
\end{aligned}
$$

$\square$

Because of the results obtained thus far, the following hypothesis seems reasonable.

Induction hypothesis H_h for $1 \le h \le n =$ the involutive genus.

The q-dimensional vector field system $\mathcal{V}$ admits h-dimensional complete subsystems obtained from the general h^{th} order involution $\mathcal{I}_h$ by specializing its parametric coefficients in a suitable way. The data left arbitrary are

$$
\begin{cases}
a_h^{h+1+s_1+\cdots+s_{h-1}}, \ldots, a_h^q, \\
\text{the restrictions of } a_{h-1}^{h+1+s_1+\cdots+s_{h-2}}, \ldots, a_{h-1}^q \text{ to } \{x^h = 0\}, \\
\cdots \\
\text{the restrictions of } a_1^{h+1}, \ldots, a_1^q \text{ to } \{x^h = x^{h-1} = \cdots = x^2 = 0\}.
\end{cases}
$$

We have seen that H_1 and H_2 are valid. Suppose that H_{h-1} is so too, and let us prove H_h.

In order to do this, let us first recall how the general h^{th} order $\mathcal{I}_h =$

$(U_1, \ldots, U_h)$, with

$$U_i = X_i + \sum_{j=1}^{q-h} a_i^{h+j} X_{h+j} \quad \text{for } i = 1, \ldots, h$$

and

$$X_j = \frac{\partial}{\partial x^j} + \sum_{k=q+1}^{r} \xi_j^k(x) \frac{\partial}{\partial x^k} \quad \text{for } j = 1, \ldots, q,$$

is constructed.

The structure equations

$$[X_i, X_j] = \sum_{k=1}^{p} c_{ij}^k Z_k, \quad i, j = 1, \ldots, q,$$

imply that

$$[U_i, U_j] = \sum_{k=1}^{q-h} A_{ij}^{h+k} X_{h+k} + \sum_{t=1}^{p} B_{ij}^t Z_t, \quad i, j = 1, \ldots h,$$

with

$$A_{ij}^{h+k} = U_i a_j^{h+k} - U_j a_i^{h+k} \quad \text{and} \quad B_{ij}^t = c_{ij}^t + \sum_{k,l=1}^{q-h} c_{h+k,h+l}^t a_i^{h+k} a_j^{h+l};$$

note that

$$A_{ij}^{h+k} + A_{ji}^{h+k} = 0 \quad \text{and} \quad B_{ij}^t + B_{ji}^t = 0.$$

$\mathcal{I}_h$ being an involution means precisely that

$$B_{ij}^t = 0 \quad \text{for } t = 1, \ldots, p \text{ and } i, j = 1, \ldots, h.$$

After having determined the involution $\mathcal{I}_{h-1} = (U_1, \ldots, U_{h-1})$ by the conditions

$$B_{ij}^t = 0 \quad \text{for } t = 1, \ldots, p \quad \text{and} \quad i, j = 1, \ldots, h-1,$$

we find U_h (and hence also $\mathcal{I}_h$) by solving the remaining linear equations

$$B_{ih}^t = 0, \quad t = 1, \ldots, p \quad \text{and} \quad i = 1, \ldots, h-1,$$

for the coefficients $a_h^{h+1}, \ldots, a_h^q$. Now $s_1 + s_2 + \cdots + s_{h-1}$ of these equations are linearly independent; let us choose a basis

$$\{E_k(a_h^{h+j}) = 0\}_{k=1}^{s_1 + \cdots + s_{h-1}}$$

for this set of equations.

Since the vanishing of B_{ih}^t for $t = 1, \ldots, p$ and $i = 1, \ldots, h - 1$ is a consequence of the vanishing of E_k for $k = 1, \ldots, s_1 + \cdots + s_{h-1}$ and that of B_{ij}^t for $t = 1, \ldots, p$ and $i, j = 1, \ldots, h - 1$, Hadamard's lemma shows that there are functions ${}_i^t\Phi_s^{kl}(x; a_i^{h+j})$ and ${}_i^t\Psi^k(x; a_i^{h+j})$ such that

$$B_{ih}^t = \sum_{s=1}^{p} \sum_{k,l=1}^{h-1} {}_i^t\Phi_s^{kl} B_{kl}^s + \sum_{k=1}^{s_1+\cdots+s_{h-1}} {}_i^t\Psi^k E_k \tag{†}$$

for $t = 1, \ldots, p$ and $i = 1, \ldots, h - 1$.

Suppose now inductively that the vector fields $U_1, \ldots, U_{h-1}$ have been determined so that $(U_1, \ldots, U_{h-1})$ is complete. The conditions that U_h must satisfy in order for $(U_1, \ldots, U_{h-1}, U_h)$ to be complete as well are

$$\begin{cases} E_k = 0 & \text{for } k = 1, \ldots, s_1 + \cdots + s_{h-1}, \\ A_{ih}^{h+j} = 0 & \text{for } j = 1, \ldots, q - h \text{ and } i = 1, \ldots, h - 1. \end{cases}$$

Using the first set of equations we can express the principal coefficients $a_h^{h+1}, \ldots, a_h^{h+s_1+\cdots+s_{h-1}}$ by means of the parametric $a_h^{h+s_1+\cdots+s_{h-1}+1}, \ldots, a_h^q$—which are then *replaced by arbitrarily chosen functions*—and the a_i^{h+j} with $i = 1, \ldots, h - 1$ and $j = 1, \ldots, q - h$. The second set of equations amounts to

$$U_h\, a_i^{h+j} = U_i\, a_h^{h+j} \quad \text{for } i = 1, \ldots, h - 1 \text{ and } j = 1, \ldots, q - h.$$

With $U_h = \partial/\partial x^h + \cdots$, this is a first order Cauchy–Kowalewski system:

$$\frac{\partial a_i^{h+j}}{\partial x^h} = f_i^{h+j}\left(x^k; a_m^{h+s}; \frac{\partial a_m^{h+s}}{\partial x^l}\right), \tag{C–K}$$

where $i, m = 1, \ldots, h-1$, $j, s = 1, \ldots, q-h$ and $l = 1, \ldots, h-1, h+1, \ldots, r$.

In the analytic context (C–K) has a unique local solution satisfying initial conditions of the form

$$a_i^{h+j}(x^1, \ldots, x^{h-1}, 0, x^{h+1}, \ldots, x^r)$$
$$= \text{given function of } x^1, \ldots, x^{h-1}, x^{h+1}, \ldots, x^r.$$

Since

$$U_i = X_i + \sum_{j=1}^{q-h} a_i^{h+j} X_{h+j} \quad \text{for } i = 1, \ldots, h,$$

$U_1, \ldots, U_{h-1}$ do not contain $\partial/\partial x^h$. Therefore it is possible to replace x^h by 0 in these vector fields so as to get a complete $(h-1)$-dimensional vector field system on the hyperplane $\{x^h = 0\}$.

Example. Let

$$\begin{cases} A = \partial/\partial x^1 + a(x^1, x^2, x^3, x^4)\, \partial/\partial x^4, \\ B = \partial/\partial x^2 + b(x^1, x^2, x^3, x^4)\, \partial/\partial x^4. \end{cases}$$

Then $[A, B] = 0$ if and only if

$$\frac{\partial b}{\partial x^1} + a\frac{\partial b}{\partial x^4} - \frac{\partial a}{\partial x^2} - b\frac{\partial a}{\partial x^4} = 0. \qquad (\ddagger)$$

With no $\partial/\partial x^3$ present it makes sense to define the vector fields

$$\begin{cases} A^0 = \partial/\partial x^1 + a(x^1, x^2, 0, x^4)\, \partial/\partial x^4, \\ B^0 = \partial/\partial x^2 + b(x^1, x^2, 0, x^4)\, \partial/\partial x^4. \end{cases}$$

And $[A^0, B^0] = 0$ if and only if $(\ddagger)$ is true with x^3 replaced by 0. So

$$[A, B] = 0 \implies [A^0, B^0] = 0$$

—but not conversely in general.

In this way the induction hypothesis yields a complete system $(U_1, \ldots, U_{h-1})_{|x^h=0}$, with coefficients

$$a_i^{h+j}(x^1, \ldots, x^{h-1}, 0, x^{h+1}, \ldots, x^r) \text{ for } i = 1, \ldots, h-1 \text{ and } j = 1, \ldots, q-h.$$

Let us now take these functions as initial conditions for (C–K), in which $a_i^{h+j}(x^1, \ldots, x^h, \ldots, x^r)$ are regarded as unknowns for $i = 1, \ldots, h-1$ and $j = 1, \ldots, q-h$. Then these a_i^{h+j} are determined so that

$$\begin{cases} A_{ih}^{h+j} = 0 \quad \text{for } i = 1, \ldots, h-1, \ j = 1, \ldots, q-h, \\ E_k = 0 \text{ for } k = 1, \ldots, s_1 + \cdots + s_{h-1}, \\ \text{and} \\ A_{ij}^{h+l}{}_{|x^h=0} = B_{ij}^t{}_{|x^h=0} = 0 \\ \text{for } i, j = 1, \ldots, h-1, \ l = 1, \ldots, q-h \text{ and } t = 1, \ldots, p. \end{cases}$$

In order for the vector field system $(U_1, \ldots, U_h)$ found in this way to be complete there remains to show that A_{ij}^{h+l} and B_{ij}^t *vanish also outside of the hyperplane* $\{x^h = 0\}$.

By Jacobi's identity,

$$[U_h, [U_i, U_j]] = [U_i, [U_h, U_j]] - [U_j, [U_h, U_i]] \quad \text{for } i, j = 1, \ldots, h-1.$$

Because $[U_a, U_b] = \sum_{l=1}^{q-h} A_{ab}^{h+l} X_{h+l} + \sum_{t=1}^r B_{ab}^t Z_t$ for $a, b = 1, \ldots, h$ and

$A_{ih}^{h+l} = 0$ for $i = 1, \ldots, h-1$, this reduces to

$$\sum_{l=1}^{q-h} U_h(A_{ij}^{h+l}) X_{h+l} + \sum_{t=1}^{p} U_h(B_{ij}^t) Z_t + \sum_{l=1}^{q-h} A_{ij}^{h+l} [U_h, X_{h+l}] + \sum_{t=1}^{p} B_{ij}^t [U_h, Z_t]$$

$$= \sum_{t=1}^{p} \{ U_i(B_{hj}^t) - U_j(B_{hi}^t) \} Z_t + \sum_{t=1}^{p} \{ B_{hj}^t [U_i, Z_t] - B_{hi}^t [U_j, Z_t] \}.$$

The vector fields appearing here all belong to the second derivative $\mathcal{V}''$ of $\mathcal{V}$, generated by $X_1, \ldots, X_q$, $Z_1, \ldots, Z_p$ and $V_1, \ldots, V_s$, say. Expressing the brackets by means of these generators, using (†) and the fact that the E_k vanish in order to replace B_{hi}^t by linear combinations of the B_{ij}^t with $i, j = 1, \ldots, h-1$, we find that

$$\begin{cases} U_h(A_{ij}^{h+l}) = \text{linear combination of } A_{ij}^{h+l} \text{ and } B_{ij}^t, \\ U_h(B_{ij}^t) = \text{linear combination of } A_{ij}^{h+l}, B_{ij}^t \text{ and} \\ \qquad U_i(\text{linear combination of } B_{ij}^t). \end{cases}$$

This is a first order Cauchy–Kowalewski system

$$\begin{cases} \partial A_{ij}^{h+l}/\partial x^h = \cdots, \\ \partial B_{ij}^t/\partial x^h = \cdots \end{cases}$$

with the initial conditions

$$A_{ij}^{h+l}\big|_{x^h=0} = B_{ij}^t\big|_{x^h=0} = 0.$$

$A_{ij}^{h+l} \equiv B_{ij}^t \equiv 0$ is an obvious solution, and by uniqueness it is the only one. Thus $(U_1, \ldots, U_h)$ is indeed complete.

Thereby we have proved **Cartan's local existence theorem**.

Theorem 3.2.1. *If the involutive genus of the analytic vector field system $\mathcal{V}$ equals n, $\mathcal{V}$ admits n-dimensional complete subsystems $\mathcal{W}_n$.* $\square$

The general $\mathcal{W}_n$ has been constructed with the following arbitrariness:

$t_r := q - n - s_1 - s_2 - \cdots - s_{n-1}$ arbitrary functions of r variables,

$t_{r-1} := q - n - s_1 - s_2 - \cdots - s_{n-2}$ arbitrary functions of $r-1$ variables,

$\ldots$

$t_{r-n+2} := q - n - s_1$ arbitrary functions of $r - n + 2$ variables,

$t_{r-n+1} := q - n$ arbitrary functions of $r - n + 1$ variables.

However, *this result is not intrinsic*. In the next section it will be seen that if $t_m \neq 0$, while $t_{m+1} = \cdots = t_r = 0$, it does make sense to say that the general n-dimensional complete subsystem W_n depends on 't_m arbitrary functions of m variables'.

By the Frobenius theorem each W_n gives rise to a local $(r-n)$-parameter family of n-dimensional integral manifolds. On the other hand a given n-dimensional integral manifold of V may belong to several different W_n, or to none at all. Therefore the arbitrariness of the integral manifolds is not directly tied to that of the complete subsystems of V.

Example. Let $V = (X_1,\dots,X_q)$ be a vector field system on $\mathbb{R}^r$ with

$$X_i = \frac{\partial}{\partial x^i} + \sum_{k=q+1}^{r} a_i^k(x)\frac{\partial}{\partial x^k} \quad \text{for } i = 1,\dots,q.$$

Its dual pfaffian system is $\mathcal{P} = (\theta^{q+1},\dots,\theta^r)$ with

$$\theta^{q+j} = dx^{q+j} - \sum_{l=1}^{q} a_l^{q+j}(x)\,dx^l \quad \text{for } j = 1,\dots,r-q.$$

The general 1-dimensional complete sub system of V on which $dx^1 \neq 0$ is generated by

$$U_1 = X_1 + \sum_{i=2}^{q} a^i(x)\,X_i,$$

where $a^2(x),\dots,a^q(x)$ are arbitrary. So U_1 depends on $q-1$ *arbitrary functions of r variables.*

The general 1-dimensional integral manifold on which $dx^1 \neq 0$ is of the form

$$x^i = f^i(x^1) \quad \text{for } i = 2,\dots,r,$$

where f^i are solutions of the following ODE system, obtained by inserting $x^i = f^i(x^1)$ into $\theta^{q+j} = 0$:

$$\frac{df^{q+j}}{dx^1} = a_1^{q+j}(x^1,f^2(x^1),\dots,f^r(x^1))$$

$$+ \sum_{k=2}^{q} a_k^{q+j}(x^1,f^2(x^1),\dots,f^r(x^1))\frac{df^k}{dx^1} \quad \text{for } j = 1,\dots,r-q.$$

Here $f^2(x^1),\dots,f^q(x^1)$ and $f^{q+1}(0),\dots,f^r(0)$ can be chosen in an arbitrary way, whereupon $f^{q+1}(x^1),\dots,f^r(x^1)$ become uniquely determined. Thus the general 1-dimensional integral manifold depends on $q-1$ *arbitrary functions of one variable, and $r-q$ arbitrary constants.*

The conclusion is that the arbitrariness of the general complete subsystem is rather different from that of the general integral manifold of the same dimension. As regards the latter, the original proof of Cartan in [Cartan 1901a] says that the following data are arbitrary for the general n-dimensional integral manifold of an s-dimensional pfaffian system on $\mathbb{R}^r$ with the involutive genus n and the characters $s_1, \ldots, s_n$:

$$\begin{cases} s_n & \text{arbitrary functions of } n \text{ variables,} \\ s_{n-1} & \text{arbitrary functions of } n-1 \text{ variables,} \\ \ldots \\ s_1 & \text{arbitrary functions of one variable,} \\ s & \text{constants.} \end{cases}$$

Kähler's generalization of Cartan's theorem. Responding to a question posed by Goursat in his book *Leçons sur le problème de Pfaff*, Kähler showed in [Kähler 1934] that Cartan's local existence theorem can be generalized to the case where the pfaffian system is replaced by an exterior differential system $\mathcal{EDS}$, generated by

> a number of one-forms θ^i, with $i = 1, \ldots, s$, say,
>
> a number of two-forms ψ^j, including the $d\theta^i$,
>
> a number of three-forms χ^k, including the $d\psi^j$,
>
> etcetera.

Let $\mathcal{P} := (\theta^1, \ldots, \theta^s)$, and $\mathcal{V} := \mathcal{P}^\perp$. We want to find maximal complete subsystems $\mathcal{W}$ of $\mathcal{V}$ such that

$$\psi^j(X, Y) = 0 \quad \text{for any } X, Y \in \mathcal{W},$$
$$\chi^k(X, Y, Z) = 0 \quad \text{for any } X, Y, Z \in \mathcal{W},$$

and so on. Then the integral manifolds of $\mathcal{W}$ provided by the Frobenius theorem are integral manifolds of $\mathcal{EDS}$.

The route for finding such $\mathcal{W}$ is analogous to that in Cartan's theorem. Start with an arbitrary

$$Y_1 = \sum_{k=1}^{q} a_1^k X_k \quad \text{in } \mathcal{V}.$$

Then determine the most general

$$Y_2 = \sum_{k=1}^{q} a_2^k X_k \quad \text{in } \mathcal{V}$$

such that

$$\psi^j(Y_1, Y_2) = 0 \text{ for all two-forms } \psi^i \text{ in } \mathcal{EDS}.$$

As before there are certain Y_1 that allow for more Y_2s than the generic Y_1 does; these are said to be *singular*, and the others *regular*. The general 2-dimensional involution is $\mathcal{I}_2 = (Y_1, Y_2)$, with Y_1 regular and Y_2 determined as above.

After that, look for all

$$Y_3 = \sum_{k=1}^{q} a_3^k X_k \quad \text{in } \mathcal{V}$$

such that

$$\begin{cases} \psi^j(Y_1, Y_3) = 0, \\ \psi^j(Y_2, Y_3) = 0, \\ \chi^k(Y_1, Y_2, Y_3) = 0 \end{cases}$$

for Y_1, Y_2 in $\mathcal{I}_2$, and all two-forms ψ^j and all three-forms χ^k in $\mathcal{EDS}$.

Continuing in this way one will eventually find a maximal general involution $\mathcal{I}_n = (Y_1, \ldots, Y_n)$.

Finally the parametric coefficients in $\mathcal{I}_n$ are determined so that $\mathcal{I}_n$ becomes complete by using the Cauchy–Kowalewski theorem pretty much as in Cartan's theorem.

3.3 How general is the general solution?

Given a PDE system, the first question to ask is *are there any local solutions?* And if so, the second is *how many?*

This section is devoted to clarifying the arbitrariness of the general integral manifold in Cartan's local existence theorem. Cartan himself states the following concerning the result in the preceding section (see [Cartan 1945], p. 76):

> Il ne convient pas de donner un sens trop absolu à l'énoncé précédent, qui ne fait que rappeler numériquement l'ensemble des fonctions arbitraires qu'on peut se donner pour arriver à la variété intégrale à n dimensions par applications successives du théorème de Cauchy–Kowalewski. En réalité, le seul de ces entiers qui ait un sens absolu est le nombre des fonctions arbitraires du nombre maximum de variables (s_n si $s_n \neq 0$, s_{n-1} si $s_n = 0$, $s_{n-1} \neq 0$, etc.).

We will explain that statement here, following [Kuranishi 1962].

Example. In order to understand Kuranishi's idea, let $V_2 := C^\infty(\mathbb{R}^2)$, $V_1 := C^\infty(\mathbb{R})$ and $V_0 := \mathbb{R}$, where V_2 and V_1 have their usual Fréchet space topologies, and V_0 its Euclidean topology. For $f \in V_2$,

$$f(x^1, x^2) = f(x^1, x_0^2) + \int_{x_0^2}^{x^2} \frac{\partial f}{\partial x^2}(x^1, t)\, dt$$

$$= f(x_0^1, x_0^2) + \int_{x_0^1}^{x^1} \frac{\partial f}{\partial x^1}(s, x_0^2)\, ds + \int_{x_0^2}^{x^2} \frac{\partial f}{\partial x^2}(x^1, t)\, dt.$$

In this way there arises a linear homeomorphism

$$L: V_2 \xrightarrow{\cong} V_0 \oplus V_1 \oplus V_2$$

$$f(x^1, x^2) \mapsto \left(f(x_0^1, x_0^2), \frac{\partial f}{\partial x^1}(x^1, x_0^2), \frac{\partial f}{\partial x^2}(x^1, x^2) \right)$$

with inverse L^{-1} given by

$$(a, g(x^1), h(x^1, x^2)) \mapsto a + \int_{x_0^1}^{x^1} g(s)\, ds + \int_{x_0^2}^{x^2} h(x^1, t)\, dt.$$

So V_2 can be identified with $V_0 \oplus V_1 \oplus V_2$, and in particular these spaces are of the same size. Analogously we also have $V_1 \cong V_0 \oplus V_1$.

More generally, if V_i^m denotes the direct sum of m copies of V_i (where $i = 0, 1$, or 2),

$$\begin{aligned}
V_2^m &\cong V_2 \oplus V_2^{m-1} \cong (V_0 \oplus V_1 \oplus V_2) \oplus V_2^{m-1} \\
&\cong V_0 \oplus V_1 \oplus V_2^m \cong V_0 \oplus V_1 \oplus V_2 \oplus V_2^{m-1} \\
&\cong V_0 \oplus V_1 \oplus (V_0 \oplus V_1 \oplus V_2) \oplus V_2^{m-1} \\
&\cong V_0^2 \oplus V_1^2 \oplus V_2^m \cong V_0^2 \oplus (V_0 \oplus V_1) \oplus V_1 \oplus V_2^m \\
&\cong V_0^3 \oplus V_1^2 \oplus V_2^m \cong \cdots,
\end{aligned}$$

Hence for any positive integers a, b and c, $V_2^a \cong V_2^a \oplus V_1^b \oplus V_0^c$. Or to quote Cartan again: the important thing is "le nombre des fonctions de nombre maximum des variables".

In order to make an honest theorem out of this we introduce the following notations:

- A *system of characters* is an ordered set of nonnegative integers: $S = (s_0, s_1, \ldots, s_p)$ with $s_p \neq 0$.

- H_k is the vector space of convergent power series in k variables:

$$H_k := \left\{ \sum_{i_1,\ldots,i_k=0}^{\infty} a_{i_1\ldots i_k} (x^1)^{i_1} \ldots (x^k)^{i_k} \right\}.$$

- $H_k^{(l)} := \{ \sum a_{i_1\ldots i_k} (x^1)^{i_1} \ldots (x^k)^{i_k} \mid a_{i_1\ldots i_k} = 0 \text{ for } i_1 + \cdots + i_k < l \}.$
- $H_k^{s_k}$ is the direct sum of s_k copies of H_k.
- $H(S) := H_0^{s_0} \oplus H_1^{s_1} \oplus \cdots \oplus H_p^{s_p}.$
- $H^{(l)}(S) := \left(H_0^{(l)} \right)^{s_0} \oplus \left(H_1^{(l)} \right)^{s_1} \oplus \cdots \oplus \left(H_p^{(l)} \right)^{s_p}$, whence $H(S) \supset H^{(1)}(S) \supset H^{(2)}(S) \supset \cdots$.
- $H(S)$ is given the topology for which the $H^{(l)}(S)$ constitute a fundamental system of neighbourhoods of the zero element.
- If $T = (t_0, \ldots, t_q)$ is another system of characters, a linear mapping

$$L : H(S) \longrightarrow H(T)$$

is *continuous* if there is an integer c (which without loss of generality may be assumed to be nonnegative) such that

$$L(H^{(l+c)}(S)) \subseteq H^{(l)}(T)$$

for sufficiently large l.

Example. We have that $\partial/\partial x^j$ and $\int^{x^j}$ are linear continuous mappings $H_k \longrightarrow H_k$ defined on monomials by

$$\frac{\partial}{\partial x^j} \left((x^1)^{i_1} \ldots (x^k)^{i_k} \right) = i_j (x^1)^{i_1} \ldots (x^j)^{i_j-1} \ldots (x^k)^{i_k}$$

and

$$\int^{x_j} \left((x^1)^{i_1} \ldots (x^k)^{i_k} \right) = (i_j + 1)^{-1} (x^1)^{i_1} \ldots (x^j)^{i_j+1} \ldots (x^k)^{i_k}$$

respectively. We have $c = 1$ for $\partial/\partial x^i$, while c may be taken as 0 for $\int^{x^j}$. Clearly

$$\frac{\partial}{\partial x^j} \circ \int^{x^j} = \int^{x^j} \circ \frac{\partial}{\partial x^j} = \mathrm{id}_{H_k}.$$

Example. There is a linear bijection $H_1 \underset{L^{-1}}{\overset{L}{\rightleftarrows}} H_2$ defined as follows:

$$L\left(\sum_{n=0}^{\infty} a_n x^n\right) = \left(\sum_{n=0}^{\infty} a_{2n} x^n, \sum_{n=0}^{\infty} a_{2n+1} x^n\right),$$

$$L^{-1}\left(\sum_{n=0}^{\infty} b_n x^n, \sum_{n=0}^{\infty} c_n x^n\right) = \sum_{n=0}^{\infty} (b_n x^{2n} + c_n x^{2n+1}).$$

L^{-1} is continuous, but L is not. Therefore this bijection is not a homeomorphism.

Definition. The vector spaces $H(S)$ and $H(T)$ are said to be *isomorphic*, $H(S) \cong H(T)$, if there are linear continuous mappings $L : H(S) \longrightarrow H(T)$ and $K : H(T) \longrightarrow H(S)$ such that

$$K \circ L = \text{id}_{H(S)} \quad \text{and} \quad L \circ K = \text{id}_{H(T)}.$$

Theorem 3.3.1 (Kuranishi's theorem). *Let* $S = (s_0, s_1, \ldots, s_p)$ *and* $T = (t_0, t_1, \ldots, t_q)$ *be two systems of characters. Then*

$$H(S) \cong H(T) \iff (p, s_p) = (q, s_q).$$

In particular, $H(S) = H_0^{s_0} \oplus H_1^{s_1} \oplus \cdots \oplus H_p^{s_p} \cong H_p^{s_p}$.

Proof. $\Longrightarrow$: By assumption there is a bijection $L : H(S) \longrightarrow H(T)$ such that $L(H^{(l+c)}(S)) \subseteq H^{(l)}(T)$ for a certain integer c and for all l larger than some l_0. For such l, L induces a surjection

$$H(S)/H^{(l+c)}(S) \twoheadrightarrow H(T)/H^{(l)}(T)$$

between finite dimensional vector spaces. The dimension of $H_r/H_r^{(l)}$ is given by the rth order polynomial

$$P_r(l) = \binom{r+l-1}{r} = \frac{l^r}{r!} + \text{ lower powers of } l.$$

Consequently

$$\dim(H(S)/H^{(l+c)}(S)) = \sum_{r=0}^{p} s_r \, P_r(l+c),$$

$$\dim(H(T)/H^{(l)}(T)) = \sum_{r=0}^{q} t_r \, P_r(l),$$

and so

$$\sum_{r=0}^{p} s_r \, P_r(l+c) \geq \sum_{r=0}^{q} t_r \, P_r(l)$$

for $l > l_0$. The dominant term in the left hand side is $s_p \cdot \frac{l^p}{p!}$, and in the right hand side $t_q \cdot \frac{l^q}{q!}$. Therefore $p \geq q$.

The same argument applied to L^{-1} shows that $q \geq p$, so that actually $p = q$.

As we know this, another comparison of the dimensions reveals that also $t_p = s_p$.

$\Longleftarrow$: Let us first consider the special case with $s_0 = s_1 = \cdots = s_{p-1} = 0$, $s_p = 1$ and $t_0 = t_1 = \cdots = t_p = 1$, so that $S = (0, 0, \ldots, 0, 1)$ and $T = (1, 1, \ldots, 1)$. The general element in $H(S)$ is denoted by $f = (f_p)$, and the general element in $H(T)$ by $g = (g_0, g_1, \ldots, g_p)$. Define

$$L : H(S) \longrightarrow H(T)$$

by

$$(L(f))_0 := f_p(0) \quad \text{and} \quad (L(f))_r := \frac{\partial f_p}{\partial x^r}(x^1, \ldots, x^r, 0, \ldots, 0) \text{ for } r = 1, \ldots, p.$$

L is the composition of a derivation and an evaluation, and is therefore linear and continuous. Then define

$$K : H(T) \longrightarrow H(S)$$

by

$$(K(g))_p := g_0 + \int^{x^1} g_1 + \cdots + \int^{x^p} g_p;$$

it is clear that K is linear and continuous. In analogy with the example above it easily follows that

$$K \circ L = \text{id}_{H(S)} \quad \text{and} \quad L \circ K = \text{id}_{H(T)}.$$

Hence

$$H_p \cong H_0 \oplus H_1 \oplus \cdots \oplus H_p.$$

For the general case it suffices to show that

$$H_0^{s_0} \oplus \cdots \oplus H_p^{s_p} \cong H_p^{s_p},$$

and this is done by induction. If $s_0 = \cdots = s_{r-1} = 0$ and $s_r \neq 0$,

$$H(0,\ldots,0,s_r,s_{r+1},\ldots,s_p) \cong H(0,\ldots,0,s_r-1,s_{r+1},\ldots,s_p) \oplus H_r$$
$$\cong H(0,\ldots,0,s_r-1,s_{r+1},\ldots,s_p) \oplus (H_0 \oplus H_1 \oplus \cdots \oplus H_r)$$
$$\cong H(0,\ldots,0,s_r-1,s_{r+1}-1,\ldots s_p) \oplus H_{r+1} \oplus (H_0 \oplus H_1 \oplus \cdots \oplus H_r)$$
$$\cong H(0,\ldots,0,s_r-1,s_{r+1}-1,\ldots s_p) \oplus H_{r+1}$$
$$\cong H(0,\ldots,0,s_r-1,s_{r+1},\ldots,s_p).$$

Continuing in this way, all of $s_r, s_{r+1}, \ldots, s_{p-1}$ are eventually replaced by 0. $\qquad\square$

Let us now consider an s-dimensional pfaffian system $\mathcal{P}$ with the involutive genus n and the characters $s_1, s_2, \ldots, s_n$. Setting $s_0 := s$,

$$s_0 \geq s_1 \geq s_2 \geq \cdots \geq s_n.$$

Let p be the largest integer for which $s_p \neq 0$. Then $S = (s_0, \ldots, s_p)$ is a system of characters in Kuranishi's sense. Consequently the theorems of Cartan and Kuranishi give a precise meaning to the following statement:

> the general *n*-dimensional integral manifold of $\mathcal{P}$ depends on s_p arbitrary functions of p variables.

3.4 Cauchy characteristics

In order not to end up with overdetermined Cauchy–Kowalewski systems in the proof of Cartan's local existence theorem, it was necessary to avoid singular vector fields and singular involutions. However, *in the following the singular vector fields and involutions will play a decisive role.* As a preparation for this we here study the most singular vector fields that can be imagined.

Definition. Let $\mathcal{V} = (X_1, \ldots, X_q)$ be a vector field system on $\mathbb{R}^r$. A nonvanishing vector field $C \in \mathcal{V}$ is *Cauchy characteristic* if $[C, \mathcal{V}] \subseteq \mathcal{V}$.

Note that if C_1, C_2 are Cauchy characteristic and f^1, f^2 are smooth functions, then for any $X \in \mathcal{V}$

$$[f^1 C_1 + f^2 C_2, X] = f^1 [C_1, X] + f^2 [C_2, X] - Xf^1 \cdot C_1 - Xf^2 \cdot C_2 \in \mathcal{V},$$

so that $f^1 C_1 + f^2 C_2$ is Cauchy characteristic too. Hence

$$\mathcal{C}(\mathcal{V}) := \{X \in \mathcal{V} \mid X = 0 \text{ or } X \text{ is Cauchy characteristic}\}$$

is a subsystem of $\mathcal{V}$, called the *Cauchy characteristic subsystem of $\mathcal{V}$*.

Theorem 3.4.1. $C(V)$ *is complete.*

Proof. Let $C_1, C_2 \in C(V)$. By Jacobi's identity,

$$[[C_1, C_2], V] = [C_1, [C_2, V]] - [C_2, [C_1, V]] \subseteq V.$$

□

If $\dim C(V) = c$, the Frobenius theorem provides local coordinates t^1, $\ldots, t^c, y^1, \ldots, y^{r-c}$ for $\mathbb{R}^r$ such that $C(V) = (\partial/\partial t^1, \ldots, \partial/\partial t^c)$. By Gauss elimination V can then be generated by vector fields of the form

$$\begin{cases} \partial/\partial t^i & \text{for } i = 1, \ldots, c, \\ Y_j := \partial/\partial y^j + \sum_{k=1}^{r-q} a_j^{q-c+k}(t, y) \left(\partial/\partial y^{q-c+k} \right) & \text{for } j = 1, \ldots, q - c. \end{cases}$$

But with $\partial/\partial t^i$ being Cauchy characteristic,

$$[\partial/\partial t^i, Y_j] = \sum_{k=1}^{r-q} \frac{\partial a_j^{q-c+k}}{\partial t^i} \frac{\partial}{\partial y^{q-c+k}} \in V,$$

which is possible only if $\partial a_j^{q-c+k}/\partial t^i = 0$. Therefore the a_j^{q-c+k} are independent of the t-variables, and hence the Y_j may as well be regarded as vector fields living on the space $\mathbb{R}_y^{r-c}$ of y-variables. As such they generate a vector field system $V_{\text{red}} = (Y_1, \ldots, Y_{q-c})$ on $\mathbb{R}_y^{r-c}$.

Definition. The *reduced vector field system of* V is $V_{\text{red}} = (Y_1, \ldots, Y_{q-c})$, considered as a module over the smooth functions on $\mathbb{R}_y^{r-c}$. Because $y^1, \ldots, y^{r-c}$ form a fundamental set of first integrals of $C(V)$, V_{red} is said to 'live on the space of first integrals of the Cauchy characteristic subsystem'.

Let π denote the natural projection $\mathbb{R}^r = \mathbb{R}_y^{r-c} \times \mathbb{R}_t^c \longrightarrow \mathbb{R}_y^{r-c}$. If $\mathcal{M} \subset \mathbb{R}_y^{r-c}$ is an integral manifold of V_{red}, $\pi^{-1}(\mathcal{M})$ clearly is an integral manifold of V. Conversely, suppose that $\mathcal{N}$ is an integral manifold of V of *maximal dimension*. Then $\pi(\mathcal{N})$ is an integral manifold of V_{red}, and so $\pi^{-1}(\pi(\mathcal{N})) \supseteq \mathcal{N}$ is an integral manifold of V. But because of the maximality of $\mathcal{N}$ this shows that $\mathcal{N} = \pi^{-1}(\pi(\mathcal{N}))$ (locally, as always). In this way the search for maximal integral manifolds of V is reduced to the corresponding task for V_{red}.

Theorem 3.4.2. *Any maximal integral manifold $\mathcal{N}$ of V can be expressed as $\mathcal{N}_{\text{red}} \times \mathbb{R}_t^c$ locally, where $\mathcal{N}_{\text{red}}$ is a maximal integral manifold of V_{red}.* □

How does this look from a dual point of view? Let $\mathcal{P} = (\theta^1, \ldots, \theta^s)$ be a pfaffian system on $\mathbb{R}^r$, and let $\pi^1, \ldots, \pi^q$ be $q = r - s$ complementary one-forms which together with the θ^i form a local coframe for $\mathbb{R}^r$. Then the dual pfaffian system is $\mathcal{V} = (\partial/\partial\pi^1,\ldots,\partial/\partial\pi^q)$. There are structure functions c^i_{jk} with $c^i_{jk} + c^i_{kj} = 0$ such that

$$d\theta^i \equiv \frac{1}{2}\sum_{j,k=1}^{q} c^i_{jk}\,\pi^j \wedge \pi^k \quad (\mathrm{mod}\ \mathcal{P}) \quad \text{for } i = 1,\ldots,s.$$

The Cauchy characteristic subsystem of $\mathcal{V}$ is

$$\mathcal{C}(\mathcal{V}) = \{C \in \mathcal{V} \mid [C,\mathcal{V}] \subseteq \mathcal{V}\}.$$

From the pfaffian point of view $\mathcal{C}(\mathcal{V})$ consists of those vector fields C that satisfy

   (i) $\theta^i(C) = 0$ for $i = 1,\ldots,s$, or $\mathcal{P}(C) = 0$—that is, $C \in \mathcal{V}$,
   (ii) $d\theta^i(C,\mathcal{V}) = 0$ for $i = 1,\ldots,s$, or $d\mathcal{P}(C,\mathcal{V}) = 0$—because this is equivalent to $\mathcal{P}[C,\mathcal{V}] = 0$—that is, $[C,\mathcal{V}] \subseteq \mathcal{V}$.

With $\{\partial/\partial\pi^1,\ldots,\partial/\partial\pi^q\}$ being a basis for $\mathcal{V}$, the second point means that

$$(\partial/\partial\pi^j)\lrcorner\, d\theta^i(C) = 0 \text{ for } j = 1,\ldots,q \text{ and } i = 1,\ldots,s.$$

Now

$$(\partial/\partial\pi^j)\lrcorner\, d\theta^i = (\partial/\partial\pi^j)\lrcorner\left(\frac{1}{2}\sum_{k,l=1}^{q} c^i_{kl}\,\pi^k \wedge \pi^l\right) = \sum_{k=1}^{q} c^i_{jk}\,\pi^k.$$

Letting $\pi^i_j := \sum_{k=1}^{q} c^i_{jk}\,\pi^k$ we get that

$$d\theta^i \equiv \frac{1}{2}\sum_{j=1}^{q} \pi^j \wedge \pi^i_j \quad (\mathrm{mod}\ \mathcal{P}) \quad \text{for } i = 1,\ldots,s,$$

and that the dual pfaffian system of $\mathcal{C}(\mathcal{V})$ is generated by

$$\theta^i \text{ and } \pi^i_j \quad \text{for } i = 1,\ldots,s \text{ and } j = 1,\ldots,q.$$

Definition. The *Cauchy characteristic pfaffian system* $\mathcal{C}(\mathcal{P})$ of $\mathcal{P}$ is the dual pfaffian system of $\mathcal{C}(\mathcal{V})$. It is generated by the one-forms θ^i and π^i_j for $i = 1,\ldots,s$ and $j = 1,\ldots,q$.

With $\dim \mathcal{C}(\mathcal{V}) = c$ the Frobenius theorem yields local coordinates t^1, $\ldots,\ t^c,\ y^1,\ \ldots,\ y^{r-c}$ for $\mathbb{R}^r$ such that $\mathcal{C}(\mathcal{V}) = (\partial/\partial t^1,\ldots,\partial/\partial t^c)$. Then the dual pfaffian system is $\mathcal{C}(\mathcal{V}) = (dy^1,\ldots,dy^{r-c})$—which obviously is *complete*. Let us prove this without using duality, following [Cartan 1901b], section **13**.

Theorem 3.4.3. $C(\mathcal{P})$ *is complete.*

Proof. Choose complementary one-forms $\pi^1, \ldots, \pi^q$ to $\theta^1, \ldots, \theta^s$ such that the pfaffian system generated by the π^i_j equals $(\pi^1, \ldots, \pi^t)$. With $\pi^i_j = \sum_{k=1}^q c^i_{jk}\pi^k$ this means that $c^i_{jk} = -c^i_{kj}$ is zero when $j, k > t$, and that it is possible to solve the linear system

$$\sum_{k=1}^t c^i_{jk}\pi^k = \pi^i_j \quad \text{for } i = 1, \ldots, s \text{ and } j = 1, \ldots, t$$

with respect to $\pi^1, \ldots, \pi^t$. Then $d\theta^i \equiv \frac{1}{2}\sum_{j,k=1}^t c^i_{jk}\pi^j \wedge \pi^k \pmod{\mathcal{P}}$, whence

$$d\theta^i \equiv 0 \pmod{\theta^1, \ldots, \theta^s, \pi^1, \ldots, \pi^t} \quad \text{for } i = 1, \ldots, s,$$

and it remains to prove that

$$d\pi^k \equiv 0 \pmod{\theta^1, \ldots, \theta^s, \pi^1, \ldots, \pi^t} \quad \text{for } k = 1, \ldots, t.$$

Now modulo $\theta^1, \ldots, \theta^s, d\theta^s, \ldots, d\theta^s$ we have

$$0 = d^2\theta^i \equiv d\left(\frac{1}{2}\sum_{j,k=1}^t c^i_{jk}\pi^j \wedge \pi^k\right)$$

$$\equiv \frac{1}{2}\sum_{j,k=1}^t dc^i_{jk} \wedge \pi^j \wedge \pi^k + \sum_{j,k=1}^t c^i_{jk}\, d\pi^j \wedge \pi^k,$$

and so

$$\sum_{j,k=1}^t c^i_{jk}\, d\pi^j \wedge \pi^k \equiv 0 \pmod{\theta^1, \ldots, \theta^s, \pi^1 \wedge \pi^2, \ldots, \pi^{t-1} \wedge \pi^t},$$

showing that

$$\sum_{j=1}^t c^i_{jk}\, d\pi^j \equiv 0 \pmod{\theta^1, \ldots, \theta^s, \pi^1, \ldots, \pi^t}$$

for $i = 1, \ldots, s$ and $k = 1, \ldots, t$. Let $[d\pi^j]$ denote what is left from the two-form $d\pi^j$ when all terms involving at least one of the one-forms $\theta^1, \ldots, \theta^s, \pi^1, \ldots, \pi^t$ are deleted. Then the congruences above go over into the equalities

$$\sum_{j=1}^t c^i_{jk}\, [d\pi^j] = 0 \quad \text{for } i = 1, \ldots, s \text{ and } j = 1, \ldots, t,$$

or equivalently

$$\sum_{k=1}^{t} c^i_{jk} \, [d\pi^k] = 0.$$

By an earlier remark it is possible to solve the $[d\pi^k]$ from this system, whence necessarily

$$[d\pi^k] = 0,$$

that is,

$$d\pi^k \equiv 0 \quad (\mathrm{mod} \ \theta^1, \dots, \theta^s, \pi^1, \dots, \pi^t).$$

□

Remark. Recall that the integration problem for a pfaffian system $\mathcal{P} = (\theta^1, \dots, \theta^s)$ with the structure equations

$$d\theta^i \equiv \frac{1}{2} \sum_{j,k=1}^{q} c^i_{jk} \, \pi^j \wedge \pi^k = \frac{1}{2} \sum_{j=1}^{q} \pi^j \wedge \pi^i_j \quad (\mathrm{mod} \ \mathcal{P})$$

consists in finding the *minimal complete pfaffian systems containing* $\mathcal{P}$. The above theorem shows that adding all the π^i_j to $\mathcal{P}$ we do get a complete system containing $\mathcal{P}$—but in general this is not of minimal dimension.

With $\mathcal{P} = (\theta^1, \dots, \theta^s)$ and with $C(\mathcal{P})$ being a complete $(r-c)$-dimensional pfaffian system, the pfaffian version of the Frobenius theorem provides local coordinates $t^1, \dots, t^c, y^1, \dots, y^{r-c}$ such that

$$C(\mathcal{P}) = (dy^1, \dots, dy^{r-c}).$$

If $\mathcal{V} = \mathcal{P}^\perp$, $C(\mathcal{V}) = C(\mathcal{P})^\perp = (\partial/\partial t^1, \dots, \partial/\partial t^c)$. Because $\theta^i(C(\mathcal{V})) = 0$, the θ^i do not contain any dt^j. Using Gaussian elimination and a change of indices it is therefore possible to find generators for $\mathcal{P}$ of the form

$$\tilde{\theta}^i = dy^i + \sum_{k=s+1}^{r-c} b^i_k(t, y) \, dy^k \quad \text{for } i = 1, \dots, s.$$

From

$$0 = d\mathcal{P}(C(\mathcal{V}), \mathcal{V}) = (C(\mathcal{V}) \rfloor d\mathcal{P})(\mathcal{V}) \iff C(\mathcal{V}) \rfloor d\mathcal{P} \subseteq \mathcal{P},$$

it follows that

$$\frac{\partial}{\partial t^j} \rfloor d\tilde{\theta}^i = \sum_{k=s+1}^{r-c} \frac{\partial b^i_k}{\partial t^j} \, dy^k \in \mathcal{P} \quad \text{for } j = 1, \dots, c \text{ and } i = 1, \dots, s$$

—which can be true only if all $\partial b_k^i / \partial t^j$ vanish. Thus the $\tilde{\theta}^i$ can be considered as one-forms living on $\mathbb{R}_y^{r-c}$.

Definition. The *reduced pfaffian system* $\mathcal{P}_{\text{red}} = (\tilde{\theta}^1, \ldots, \tilde{\theta}^s)$ is the pfaffian system on $\mathbb{R}_y^{r-c}$ generated by $\tilde{\theta}^1, \ldots, \tilde{\theta}^s$ over the ring of smooth functions on $\mathbb{R}_y^{r-c}$.

Note that if $(\tilde{\theta}^1, \ldots, \tilde{\theta}^s)$ instead is regarded as a module over the smooth functions on $\mathbb{R}^r$, it coincides with $\mathcal{P}$. Moreover, if $\mathcal{N} \subset \mathbb{R}_y^{r-c}$ is an integral manifold of $\mathcal{P}_{\text{red}}$, $\mathcal{N} \times \mathbb{R}_t^c$ is clearly an integral manifold of $\mathcal{P}$.

If $X = \sum_{i=1}^r a_i(x) \left(\partial / \partial x^i \right)$ is a vector field on $\mathbb{R}^r$, the *integral curve* of X through a point $x_0 = (x_0^1, \ldots, x_0^r)$ is the locally defined curve

$$\xi : \mathbb{R}_t \to \mathbb{R}_x^r$$
$$t \mapsto (\xi^1(t), \ldots, \xi^r(t))$$

with $\xi^1(t), \ldots, \xi^r(t)$ forming the unique local solution of the ODE system

$$\begin{cases} d\xi^i/dt = a^i(\xi^1(t), \ldots, \xi^r(t)), \\ \xi^i(0) = x_0^i, \end{cases} \qquad \text{for } i = 1, \ldots, r.$$

For a fixed t, $e^{tX}(x_0)$ denotes the point $(\xi^1(t), \ldots, \xi^r(t))$. Letting x_0 vary we obtain a 1-parameter family of local diffeomorphisms with t as parameter, called the *flow of X*:

$$e^{tX} : \mathbb{R}^n \to \mathbb{R}^n$$
$$x \mapsto e^{tX}(x).$$

Definition. The integral curves of a Cauchy characteristic vector field are called *Cauchy characteristics*.

This set-up allows us to consider the Cauchy characteristic vector fields from a somewhat different angle, and obtain the following **theorem on Cauchy characteristics**.

Theorem 3.4.4. *Assume that $\mathcal{N}_k$ is a k-dimensional integral manifold of the vector field system $\mathcal{V}$ on $\mathbb{R}^r$, and let C be a Cauchy characteristic vector field which is transversal to $\mathcal{N}_k$. Then the image of*

$$e^C : \mathcal{N}_k \times \mathbb{R}_t \to \mathbb{R}^r$$
$$(n, t) \mapsto e^{tC}(n)$$

is a $(k+1)$-dimensional integral manifold $\mathcal{N}_{k+1}$ of $\mathcal{V}$.

Proof. We have to show that $T_m \mathcal{N}_{k+1} \subseteq \mathcal{V}_m$ for each $m \in \mathcal{N}_{k+1}$. By construction $m = e^{tC}(n)$ for some $t \in \mathbb{R}$ and some $n \in \mathcal{N}_k$. Therefore $T_m \mathcal{N}_{k+1}$ is spanned by $C_m \in \mathcal{V}_m$ and $(e^{tC})_*(\mathcal{V}_n)$. Thus it suffices to prove the following result.

Lemma 3.4.1. *The vector field system $\mathcal{V}$ is invariant under the flow generated by a vector field $X \in \mathcal{V}$, i.e.,*

$$\left(e^{tX}\right)_* : \mathcal{V} \xrightarrow{\cong} \mathcal{V} \quad \text{for } t \approx 0,$$

if and only if $X \in C(\mathcal{V})$.

Proof. Introduce local coordinates $x^1, \ldots, x^r$ such that $X = \partial/\partial x^1$, and choose vector fields

$$X_i = \frac{\partial}{\partial x^i} + \sum_{k=q+1}^{r} a_i^k(x) \frac{\partial}{\partial x^k} \quad \text{for } i = 2, \ldots, q$$

which together with X generate $\mathcal{V}$. Then e^{tX} is given by

$$(x^1, x^2, \ldots, x^r) \mapsto (x^1 + t, x^2, \ldots, x^r).$$

If $\mathcal{V}$ is preserved under $\left(e^{tX}\right)_*$,

$$\left(e^{tX}\right)_*(X_i) = \frac{\partial}{\partial x^i} + \sum_{k=q+1}^{r} a_i^k(x^1 + t, x^2, \ldots, x^r) \frac{\partial}{\partial x^k} \in \mathcal{V} \quad \text{for } i = 2, \ldots, q.$$

This is possible only if $(e^{tX})_*(X_i) = X_i$, in which case the a_i^k are independent of x^1. But then

$$[X, X_i] = 0 \quad \text{for } i = 2, \ldots, q,$$

so that X is indeed Cauchy characteristic.

Conversely, assume that $X = \partial/\partial x^1$ is Cauchy characteristic. Then

$$[X, X_i] = \sum_{k=q+1}^{r} \frac{\partial a_i^k}{\partial x^1} \frac{\partial}{\partial x^k} \in \mathcal{V} \quad \text{for } i = 2, \ldots, q,$$

which forces the $\partial a_i^k/\partial x^1$ to vanish—and so $\mathcal{V} = (X, X_2, \ldots, X_q)$ remains invariant when x^1 is replaced by $x^1 + t$. $\qquad\square$

If there are several linearly independent Cauchy characteristic vector fields which are transversal to the integral manifold $\mathcal{N}_k$, the theorem can be used inductively in order to construct integral manifolds of successively larger dimensions.

In particular, let $t^1, \ldots, t^c, y^1, \ldots, y^{r-c}$ be local coordinates such that

$C(\mathcal{V}) = (\partial/\partial t^1, \ldots, \partial/\partial t^c)$, and let $\mathcal{N}$ be an integral manifold of *maximal dimension*. Then for each point $(t_0, y_0) \in \mathcal{N}$ it follows that $\mathbb{R}_t^c \times \{y_0\} \subseteq \mathcal{N}$. So letting π denote the natural projection $\mathbb{R}_t^c \times \mathbb{R}_y^{r-c} \to \mathbb{R}_y^{r-c}$, we have

$$\mathcal{N} = \pi^{-1}(\pi(\mathcal{N})),$$

where $\pi(\mathcal{N})$ is an integral manifold of the reduced vector field system $\mathcal{V}_{\mathrm{red}}$ on $\mathbb{R}_y^{r-c}$. Hence we see one more time that each maximal integral manifold $\mathcal{N}$ of $\mathcal{V}$ is of the form

$$\mathcal{N} = \mathcal{N}_{\mathrm{red}} \times \mathbb{R}_t^c,$$

with $\mathcal{N}_{\mathrm{red}}$ being a maximal integral manifold of $\mathcal{V}_{\mathrm{red}}$.

3.5 Maximal involutions and integrable vector field systems

The biggest dimension of *regular* complete subsystems of a vector field system $\mathcal{V}$ is given by the involutive genus. But there may be larger—*singular*—ones, and how big can their dimension be?

Following [Cartan 1915] and [Hermann 1965] we in this section derive an upper bound for the dimension of complete subsystems. Vector field systems for which this is attained are said to be *integrable*, and turn out to have quite interesting properties.

Let $\mathcal{P} = (\theta^1, \ldots, \theta^s)$ be a pfaffian system on $\mathbb{R}^r$, and let $\{\pi^1, \ldots, \pi^q\}$ (where $q = r - s$) be a complementary set of one-forms in the sense that $\theta^1 \wedge \cdots \wedge \theta^s \wedge \pi^1 \wedge \cdots \wedge \pi^q \neq 0$ on $\mathbb{R}^r$. Setting $X_i := \partial/\partial \pi^i$ for $i = 1, \ldots, q$, the dual vector field system is given by $\mathcal{V} = (X_1, \ldots, X_q)$.

Recall that the characteristic vector field system of $\mathcal{V}$ is defined as

$$C(\mathcal{V}) = \{X \in \mathcal{V} \mid X \rfloor d\theta^i \equiv 0 \pmod{\mathcal{P}}, \ i = 1, \ldots, s\},$$

and that a subsystem $\mathcal{I}$ of $\mathcal{V}$ is an involution if

$$d\theta^i(X, Y) = 0 \text{ for all } X, Y \in \mathcal{I} \text{ and } i = 1, \ldots, s.$$

Replacing $d\theta^1, \ldots, d\theta^s$ by two-forms $\Omega^1, \ldots, \Omega^m$—which need not be related to the θ^i in any way—and letting $\mathcal{Q} = (\Omega^1, \ldots, \Omega^m)$ be the module of two-forms generated by $\Omega^1, \ldots, \Omega^m$ over the ring of smooth functions, the following more general concepts are introduced.

Definitions. The *characteristic vector field system* of $\mathcal{V}$ with respect to $\mathcal{Q}$ is

$$C_{\mathcal{V}}(\mathcal{Q}) := \{X \in \mathcal{V} \mid X \rfloor \Omega^i \equiv 0 \pmod{\mathcal{P}}, \text{ for } i = 1, \ldots, m\}.$$

A nonzero vector field in $\mathcal{C}_{\mathcal{V}}(\mathcal{Q})$ is said to be *characteristic with respect to* $(\mathcal{P}, \mathcal{Q})$. A subsystem $\mathcal{I}$ of $\mathcal{V}$ is an *involution with respect to* $\mathcal{Q}$ if $\Omega^i(X, Y) = 0$ for all $X, Y \in \mathcal{I}$ and for $i = 1, \ldots, m$.

Let us first study the case when $\mathcal{Q}$ is generated by a single two-form Ω, so that $\mathcal{Q} = (\Omega)$. The following notation was introduced in [Cartan 1901b].

Definition. The *genus of* Ω (mod $\mathcal{P}$) is the smallest integer g such that

$$\Omega^{\wedge(g+1)} := \Omega \wedge \cdots \wedge \Omega \text{ (with } g + 1 \text{ factors)} \equiv 0 \quad (\text{mod } \mathcal{P});$$

or equivalently the smallest integer g such that

$$\theta^1 \wedge \cdots \wedge \theta^s \wedge \Omega^{\wedge(g+1)} = 0.$$

Given an arbitrary two-form Ω there are smooth functions a_{ij} such that

$$\Omega \equiv \sum_{1 \leq i < j \leq q} a_{ij} \, \pi^i \wedge \pi^j \quad (\text{mod } \mathcal{P}).$$

The sum in the right hand side can be regarded as the restriction of Ω to $\mathcal{V} \times \mathcal{V}$:

$$\Omega|_{\mathcal{V} \times \mathcal{V}} : \mathcal{V} \times \mathcal{V} \to C^\infty(\mathbb{R}^r)$$
$$(X, Y) \mapsto \Omega(X, Y).$$

Denoting this restriction by $\Omega_{\mathcal{V}}$ we thus have

$$\Omega_{\mathcal{V}} = \sum_{1 \leq i < j \leq q} a_{ij} \, \pi^i \wedge \pi^j.$$

If $\Omega_{\mathcal{V}} \neq 0$ it may be supposed that $a_{12} \neq 0$. Then

$$\Omega_{\mathcal{V}} = a_{12} \, \pi^1 \wedge \pi^2 + a_{13} \, \pi^1 \wedge \pi^3 + \cdots + a_{1q} \, \pi^1 \wedge \pi^q$$
$$+ a_{23} \, \pi^2 \wedge \pi^3 + \cdots + a_{2q} \, \pi^2 \wedge \pi^q$$

$$\cdots$$

$$+ a_{q-1,q} \, \pi^{q-1} \wedge \pi^q$$

$$= (\pi^1 - \frac{a_{23}}{a_{12}} \, \pi^3 - \cdots - \frac{a_{2q}}{a_{12}} \, \pi^q) \wedge (a_{12} \, \pi^2 + a_{13} \, \pi^3 + \cdots + a_{1q} \, \pi^q) + \Omega',$$

where Ω' is a two-form which contains neither π^1 nor π^2. Introducing

$$\bar{\pi}^1 := \pi^1 - a_{12}^{-1} \cdot \sum_{j=3}^{q} a_{2j} \, \pi^j \quad \text{and} \quad \bar{\pi}^2 := \sum_{j=2}^{q} a_{1j} \, \pi^j$$

as new basis elements for the one-forms on $\mathbb{R}^r$ instead of π^1 and π^2, we have

$$\Omega_V = \bar{\pi}^1 \wedge \bar{\pi}^2 + \Omega'.$$

If $\Omega' \neq 0$ this procedure can obviously be repeated until one ends up with an expression of the form

$$\Omega_V = \bar{\pi}^1 \wedge \bar{\pi}^2 + \cdots + \bar{\pi}^{2g-1} \wedge \bar{\pi}^{2g}.$$

Then

$$\theta^1 \wedge \cdots \wedge \theta^s \wedge \Omega^{\wedge g} = \theta^1 \wedge \cdots \wedge \theta^s \wedge \Omega_V^{\wedge g}$$
$$= g!\, \theta^1 \wedge \cdots \wedge \theta^s \wedge \bar{\pi}^1 \wedge \bar{\pi}^2 \wedge \cdots \wedge \bar{\pi}^{2g-1} \wedge \bar{\pi}^{2g} \neq 0,$$

while

$$\theta^1 \wedge \cdots \wedge \theta^s \wedge \Omega^{\wedge(g+1)} = 0.$$

Hence the genus of Ω (mod $\mathcal{P}$) equals g.

Let

$$\rho \equiv \mathrm{rank}_V \Omega := \dim(X \rfloor \Omega_V \mid X \in V).$$

Using the canonical form above and the fact that

$$V = (\partial/\partial \bar{\pi}^1, \ldots, \partial/\partial \bar{\pi}^{2g}, \partial/\partial \pi^{2g+1}, \ldots, \partial/\partial \pi^q),$$

we get

$$(X \rfloor \Omega_V \mid X \in V) = (\bar{\pi}^1, \ldots, \bar{\pi}^{2g}),$$

and consequently

$$\rho = 2g.$$

The definition of the rank shows that a subsystem $\mathcal{I}$ of V is an involution with respect to $\mathcal{Q} = (\Omega)$ if and only if $\mathrm{rank}_{\mathcal{I}} \Omega = 0$. Let us use this fact in order to figure out how large the dimension of an involution can be. Observe first that a relation like

$$\bar{\pi}^{2g} = \sum_{i=1}^{2g-1} a_i \bar{\pi}^i$$

with smooth functions a_i makes $\sum_{i=1}^g \bar{\pi}^{2i-1} \wedge \bar{\pi}^{2i}$ equal to

$$(\bar{\pi}^1 + a_2 \bar{\pi}^{2g-1}) \wedge (\bar{\pi}^2 - a_1 \bar{\pi}^{2g-1}) + \cdots$$
$$+ (\bar{\pi}^{2g-3} + a_{2g-2} \bar{\pi}^{2g-1}) \wedge (\bar{\pi}^{2g-2} - a_{2g-3} \bar{\pi}^{2g-1}),$$

so that the genus of $\sum_{i=1}^{g} \bar{\pi}^{2i-1} \wedge \bar{\pi}^{2i}$ is diminished by one unit. Continuing in this way one realizes that it requires precisely g independent relations between $\bar{\pi}^1, \ldots, \bar{\pi}^{2g}$ in order to make the genus of $\sum_{i=1}^{g} \bar{\pi}^{2i-1} \wedge \bar{\pi}^{2i}$ vanish. As a consequence we have the following result.

Lemma 3.5.1. *The maximal dimension of involutions of* $\mathcal{P}$ *with respect to* $\mathcal{Q} = (\Omega)$ *is* $q - g = \dim \mathcal{V} - (\text{genus of } \Omega \ (\text{mod } \mathcal{P}))$. *There are several such involutions—for instance* $\mathcal{I}_1 := \{X \in \mathcal{V} \mid \bar{\pi}^{2k-1}(X) = 0, \ k = 1, \ldots, g\}$ *and* $\mathcal{I}_2 := \{X \in \mathcal{V} \mid \bar{\pi}^{2k}(X) = 0, \ k = 1, \ldots, g\}$. $\square$

Let us next study the general case with $\mathcal{Q} = (\Omega^1, \ldots, \Omega^m)$. In analogy with what was done above we set

$$\rho \equiv \text{rank}_{\mathcal{V}}\mathcal{Q} := \dim(X \rfloor \Omega^i_{\mathcal{V}} \mid X \in \mathcal{V}, \ i = 1, \ldots, m).$$

Consider for each $\lambda = (\lambda_1, \ldots, \lambda_m) \in \mathbb{R}^m$ the two-form

$$\Omega(\lambda) := \sum_{i=1}^{m} \lambda_i \Omega^i.$$

Then $\rho(\lambda) := \text{rank}_{\mathcal{V}}\Omega(\lambda)$ is a function on $\mathbb{R}^m$ taking values in the set of even integers between 0 and ρ. Clearly the maximum value of $\rho(\lambda)$ is taken on a dense open subset of $\mathbb{R}^m$, which we denote by $\mathcal{O}$. Using this the genus of $\mathcal{Q} \ (\text{mod } \mathcal{P})$ is defined as

$$g := \frac{1}{2} \max_{\lambda \in \mathbb{R}^m} \rho(\lambda)$$

or equivalently

$$g := \frac{1}{2} \rho(\lambda) \text{ for any } \lambda \in \mathcal{O}.$$

Obviously

$$2g = \max_{\lambda \in \mathbb{R}^m} \rho(\lambda) \leq \text{rank}_{\mathcal{V}}\mathcal{Q} = \rho.$$

If $\mathcal{I} \subseteq \mathcal{V}$ is an involution with respect to $\mathcal{Q}$, $\Omega^i(X, Y) = 0$ for $i = 1, \ldots, m$ and all $X, Y \in \mathcal{I}$. In particular it follows that $\Omega(\lambda)(X, Y) = 0$ for any $\lambda \in \mathbb{R}^m$. Choosing a $\lambda \in \mathcal{O}$, the previous lemma shows that

$$\dim \mathcal{I} \leq q - g.$$

Definition. The set of all $(q-g)$-dimensional involutions of $\mathcal{V}$ with respect to $\mathcal{Q}$ is denoted by $\text{inv}(\mathcal{V}, \mathcal{Q})$. $\mathcal{V}$ is said to be *involutive with respect to* $\mathcal{Q}$ if $\text{inv}(\mathcal{V}, \mathcal{Q}) \neq \emptyset$.

A $(q - g)$-dimensional involution $\mathcal{I}$ can be expressed as

$$\mathcal{I} = \{X \in \mathcal{V} \mid \alpha^i(X) = 0, \ i = 1, \ldots, g\},$$

where the one-forms $\alpha^1, \ldots, \alpha^g$ are independent. The pfaffian equations $\alpha^i = 0$ appearing here are called the *defining equations of $\mathcal{I}$*. We have previously seen that $\mathrm{inv}(\mathcal{V}, \mathcal{Q}) \neq \emptyset$ when $\mathcal{Q} = (\Omega)$, and that there are $\mathcal{I} \in \mathrm{inv}(\mathcal{V}, \mathcal{Q})$ with no common defining equations in that case. The situation is different for $\mathcal{Q} = (\Omega^1, \ldots, \Omega^m)$ with $m > 1$: it might happen that $\mathrm{inv}(\mathcal{V}, \mathcal{Q}) = \emptyset$, but it will be seen that if $\mathrm{inv}(\mathcal{V}, \mathcal{Q}) \neq \emptyset$ and $\rho > 2g$, then all $\mathcal{I} \in \mathrm{inv}(\mathcal{V}, \mathcal{Q})$ have certain common defining equations. Or, expressed differently, there is a subsystem $\mathcal{V}_0$ of $\mathcal{V}$ containing all $\mathcal{I} \in \mathrm{inv}(\mathcal{V}, \mathcal{Q})$.

Since the characteristic subsystem of $\mathcal{V}$ with respect to $\mathcal{Q}$ is defined as

$$\mathcal{C}_\mathcal{V}(\mathcal{Q}) = \{X \in \mathcal{V} \mid X \rfloor \Omega^i_\mathcal{V} = 0 \text{ for } i = 1, \ldots, m\},$$

it follows that $\dim \mathcal{C}_\mathcal{V}(\mathcal{Q}) = q - \rho$. Note that all $\mathcal{I} \in \mathrm{inv}(\mathcal{V}, \mathcal{Q})$ contain $\mathcal{C}_\mathcal{V}(\mathcal{Q})$—for otherwise it would be possible to obtain larger involutions by adding those $C \in \mathcal{C}_\mathcal{V}(\mathcal{Q})$ that are missing. Consequently, for each $\mathcal{I} \in \mathrm{inv}(\mathcal{V}, \mathcal{Q})$,

$$\mathcal{C}_\mathcal{V}(\mathcal{Q}) \subseteq \mathcal{I} \quad \subseteq \mathcal{V}$$
$$\dim: \quad q - \rho \ \leq q - g \ \leq q.$$

Analogously we have

$$\mathcal{C}_\mathcal{V}(\lambda) = \{X \in \mathcal{V} \mid X \rfloor \Omega_\mathcal{V}(\lambda) = 0\}$$

for $\lambda \in \mathbb{R}^m$. Then for $\lambda \in \mathcal{O} \subseteq \mathbb{R}^m$,

$$\dim \mathcal{C}_\mathcal{V}(\lambda) = q - 2g.$$

Clearly $\mathcal{C}_\mathcal{V}(\mathcal{Q}) \subseteq \mathcal{C}_\mathcal{V}(\lambda)$ for all λ, and therefore

$$q - \rho \leq q - 2g,$$

confirming that $\rho \geq 2g$.

Definition.

$$\mathcal{V}_0 := \left\{ X \in \mathcal{V} \mid \Omega^i(X, Y) = 0 \text{ for } i = 1, \ldots, m \text{ and for all } Y \in \bigcup_{\lambda \in \mathcal{O}} \mathcal{C}_\mathcal{V}(\lambda) \right\}.$$

Lemma 3.5.2. *Suppose that* $\mathrm{inv}(\mathcal{V}, \mathcal{Q}) \neq \emptyset$. *Then:*

(i) $\rho > 2g \Longrightarrow \mathcal{V}_0 \subset \mathcal{V}$;

(ii) $\mathcal{C}_\mathcal{V}(\lambda) \subseteq \mathcal{I} \subseteq \mathcal{V}_0$ *for each* $\lambda \in \mathcal{O}$ *and each* $\mathcal{I} \in \mathrm{inv}(\mathcal{V}, \mathcal{Q})$; *in particular* $\dim \mathcal{V}_0 \geq q - g$;

(iii) *with* $\dim \mathcal{V}_0 = q - n_0$ *for some* $n_0 \in \{0, 1, \ldots, g\}$, $\dim \mathcal{C}_{\mathcal{V}_0}(\lambda) = q - n_0 - 2(g - n_0) = q - g - (g - n_0)$ *when* $\lambda \in \mathcal{O}$.

Proof. 1. If $\rho > 2g$ it follows that $\mathcal{C}_{\mathcal{V}}(\mathcal{Q}) \subset \mathcal{C}_{\mathcal{V}}(\lambda)$ for all λ. In particular there are a $\lambda_1 \in \mathcal{O}$ and a $Y_1 \in \mathcal{C}_{\mathcal{V}}(\lambda_1)$ such that $Y_1 \notin \mathcal{C}_{\mathcal{V}}(\mathcal{Q})$, i.e., $Y_1 \rfloor \Omega_{\mathcal{V}}^k \neq 0$ for some k. But then $\mathcal{V}_0 \subseteq \{X \in \mathcal{V} \mid \Omega^k(X, Y_1) = 0\} \subset \mathcal{V}$.

2. Let $\mathcal{I} \in \text{inv}(\mathcal{V}, \mathcal{Q})$ and let $\lambda \in \mathcal{O}$. If $\{X_1, \ldots, X_{q-g}\}$ constitutes a basis for $\mathcal{I}$, extend it to a basis $\{X_1, \ldots, X_q\}$ for $\mathcal{V}$. Introduce the pfaffian systems

$$\mathcal{I} \rfloor \Omega_{\mathcal{V}}(\lambda) := (X_1 \rfloor \Omega_{\mathcal{V}}(\lambda), \ldots, X_{q-g} \rfloor \Omega_{\mathcal{V}}(\lambda))$$

and

$$\mathcal{V} \rfloor \Omega_{\mathcal{V}}(\lambda) := (X_1 \rfloor \Omega_{\mathcal{V}}(\lambda), \ldots, X_q \rfloor \Omega_{\mathcal{V}}(\lambda)).$$

Because $\Omega_{\mathcal{V}}^i(X_k, X_l) = 0$ for $k, l = 1, \ldots, q - g$, the one-forms

$$X_k \rfloor \Omega_{\mathcal{V}}(\lambda) = \sum_{i=1}^{m} \lambda_i X_k \rfloor \Omega_{\mathcal{V}}^i, \qquad k = 1, \ldots, q - g,$$

vanish on $\mathcal{I}$. The codimension of $\mathcal{I}$ in $\mathcal{V}$ being equal to g, this means that

$$\dim(\mathcal{I} \rfloor \Omega_{\mathcal{V}}(\lambda)) \leq g.$$

But since $\lambda \in \mathcal{O}$ we also have

$$\text{rank}_{\mathcal{V}} \Omega(\lambda) = \dim(\mathcal{V} \rfloor \Omega_{\mathcal{V}}(\lambda)) = 2g,$$

and therefore

$$\dim(\mathcal{I} \rfloor \Omega_{\mathcal{V}}(\lambda)) \geq 2g - g = g.$$

Hence

$$\dim(\mathcal{I} \rfloor \Omega_{\mathcal{V}}(\lambda)) = g,$$

and it follows that the g one-forms $X_{q-g+1} \rfloor \Omega_{\mathcal{V}}(\lambda), \ldots, X_q \rfloor \Omega_{\mathcal{V}}(\lambda)$ together with a g-dimensional basis for $\mathcal{I} \rfloor \Omega_{\mathcal{V}}(\lambda)$ constitute a basis for $\mathcal{V} \rfloor \Omega_{\mathcal{V}}(\lambda)$. If $X = \sum_{k=1}^{q} a^k X_k \in \mathcal{C}_{\mathcal{V}}(\lambda)$,

$$0 = X \rfloor \Omega_{\mathcal{V}}(\lambda) = \sum_{k=1}^{q} a^k (X^k \rfloor \Omega_{\mathcal{V}}(\lambda)),$$

which by above forces $a^{q-g+1}, \ldots, a^q$ to vanish. Thus

$$X = \sum_{k=1}^{q-g} a^k X_k \in \mathcal{I},$$

so that $C_\mathcal{V}(\lambda) \subseteq \mathcal{I}$. With $\mathcal{I}$ being an involution, $\Omega^i(\mathcal{I},\mathcal{I}) = 0$ for $i = 1,\dots,m$, and in particular

$$\Omega^i(\mathcal{I}, C_\mathcal{V}(\lambda)) = 0 \text{ for } i = 1,\dots,m \text{ and each } \lambda \in \mathcal{O}.$$

The definition of $\mathcal{V}_0$ then shows that $\mathcal{I} \subseteq \mathcal{V}_0$.

3. There are n_0 one-forms $\alpha^1,\dots,\alpha^{n_0}$ which are linearly independent modulo $\mathcal{P}$ such that

$$\mathcal{V}_0 = \{X \in \mathcal{V} \mid \alpha^i(X) = 0 \text{ for } i = 1,\dots,n_0\}.$$

If $\mathcal{I} \in \mathrm{inv}(\mathcal{V}, \mathcal{Q})$, then $\mathcal{I} \subseteq \mathcal{V}_0$ and $\dim \mathcal{I} = q - g$. Hence there are $g - n_0$ one-forms $\beta^1_{\mathcal{I}},\dots,\beta^{g-n_0}_{\mathcal{I}}$ such that $\alpha^1,\dots,\alpha^{n_0},\beta^1_{\mathcal{I}},\dots,\beta^{g-n_0}_{\mathcal{I}}$ are independent modulo $\mathcal{P}$, and

$$\mathcal{I} = \{X \in \mathcal{V} \mid \alpha^i(X) = 0,\ \beta^j_{\mathcal{I}}(X) = 0 \text{ for } i = 1,\dots,n_0,\ j = 1,\dots,g - n_0\}.$$

Claim. $\mathrm{rank}_{\mathcal{V}_0}\Omega(\lambda) = 2(g - n_0)$ for $\lambda \in \mathcal{O}$.

For on the one hand,

$$\mathrm{rank}_\mathcal{V}\Omega(\lambda) = 2g \implies \mathrm{rank}_{\mathcal{V}_0}\Omega(\lambda) \geq 2g - 2n_0 = 2(g - n_0),$$

and on the other the two equalities

$$\mathrm{rank}_{\mathcal{I}}\Omega(\lambda) = 0 \text{ and } \dim \mathcal{V}_0 - \dim \mathcal{I} = (q - n_0) - (q - g) = g - n_0$$

imply that

$$\mathrm{rank}_{\mathcal{V}_0}\Omega(\lambda) \leq 2(g - n_0).$$

Therefore

$$\dim C_{\mathcal{V}_0}(\lambda) = \dim\{X \in \mathcal{V}_0 \mid X \rfloor \Omega_{\mathcal{V}_0}(\lambda) = 0\} = \dim \mathcal{V}_0 - \mathrm{rank}_{\mathcal{V}_0}\Omega(\lambda)$$
$$= q - n_0 - 2(g - n_0) = q - g - (g - n_0)$$

for $\lambda \in \mathcal{O}$. $\qquad\square$

Conclusion. *Suppose that* $\mathrm{inv}(\mathcal{V}, \mathcal{Q}) \neq \emptyset$ *and* $\rho > 2q$. *Then for* $\mathcal{I} \in \mathrm{inv}(\mathcal{V}, \mathcal{Q})$ *and* $\lambda \in \mathcal{O}$

$$C_\mathcal{V}(\mathcal{Q}) \subset C_\mathcal{V}(\lambda) \subset \mathcal{I} \subseteq \mathcal{V}_0 \subset \mathcal{V}$$
$$\dim: \quad q - \rho < q - 2g < q - g \leq q - n_0 < q.$$

Since we are only interested in the maximal involutions of $\mathcal{V}$ and these are all contained in $\mathcal{V}_0$, $\mathcal{V}$ might as well be replaced by $\mathcal{V}_0$, and then

$$\mathcal{C}_{\mathcal{V}_0}(\mathcal{Q}) \subseteq \quad \mathcal{C}_{\mathcal{V}_0}(\lambda) \quad \subseteq \quad \mathcal{I} \quad \subseteq \quad \mathcal{V}_0$$

dim: $\quad q - n_0 - \rho_0 \le q - g - (g - n_0) \le q - g \le q - n_0,$

where $\rho_0 = \mathrm{rank}_{\mathcal{V}_0}(\mathcal{Q}) \ge \mathrm{rank}_{\mathcal{V}_0}\Omega(\lambda) = 2(g - n_0)$.

Let $\lambda \in \mathcal{O}$ and consider the following cases.

(1) $n_0 = g$. Then $\mathcal{I} = \mathcal{V}_0$, so that $\mathcal{V}_0$ is an involution, i.e., $\Omega^i(X, Y) = 0$ for $X, Y \in \mathcal{V}_0$ and $i = 1, \dots, m$. Accordingly $\mathcal{C}_{\mathcal{V}_0}(\mathcal{Q}) = \mathcal{V}_0$, and

$$\mathcal{C}_{\mathcal{V}_0}(\mathcal{Q}) = \mathcal{C}_{\mathcal{V}_0}(\lambda) = \mathcal{I} = \mathcal{V}_0.$$

(2) $n_0 < g$, $\rho_0 = 2(g - n_0)$. Then

$$\mathcal{C}_{\mathcal{V}_0}(\mathcal{Q}) = \quad \mathcal{C}_{\mathcal{V}_0}(\lambda) \quad \subset \quad \mathcal{I} \quad \subset \quad \mathcal{V}$$

dim: $\quad q - n_0 - \rho_0 = q - n_0 - \rho_0 < q - g < q - n_0.$

(3) $n_0 < g$, $\rho_0 > 2(g - n_0)$. Then

$$\mathcal{C}_{\mathcal{V}_0}(\mathcal{Q}) \subset \quad \mathcal{C}_{\mathcal{V}_0}(\lambda) \quad \subset \quad \mathcal{I} \quad \subset \quad \mathcal{V}_0$$

dim: $\quad q - n_0 - \rho_0 < q - g - (g - n_0) < q - g < q - n_0.$

Since in the third case $\mathcal{C}_{\mathcal{V}_0}(\mathcal{Q}) \subset \mathcal{C}_{\mathcal{V}_0}(\lambda)$, it is tempting to use the above lemma again with $\mathcal{V}$ replaced by $\mathcal{V}_0$ and q, ρ, g replaced by $q - n_0$, ρ_0 and $g - n_0$ respectively. Instead of $\mathrm{inv}(\mathcal{V}, \mathcal{Q})$ we are then to consider the set $\mathrm{inv}(\mathcal{V}_0, \mathcal{Q})$ of involutions $\mathcal{I}_0 \subseteq \mathcal{V}_0$ with respect to $\mathcal{Q}$ having the dimension

$$\dim \mathcal{I}_0 = (q - n_0) - (g - n_0) = q - g.$$

Because of the second assertion in the lemma, $\mathrm{inv}(\mathcal{V}_0, \mathcal{Q}) = \mathrm{inv}(\mathcal{V}, \mathcal{Q}) \ne \emptyset$. Therefore it is indeed possible to use the lemma again in this new situation so as to get a subsystem $\mathcal{V}_1$ of $\mathcal{V}_0$ having the property that for all $\lambda \in \mathcal{O}$ and all $\mathcal{I} \in \mathrm{inv}(\mathcal{V}_0, \mathcal{Q}) = \mathrm{inv}(\mathcal{V}, \mathcal{Q})$

$$\mathcal{C}_{\mathcal{V}_1}(\mathcal{Q}) \subseteq \quad \mathcal{C}_{\mathcal{V}_1}(\lambda) \quad \subseteq \quad \mathcal{I} \quad \subseteq \quad \mathcal{V}_1 \quad \subset \quad \mathcal{V}_0 \quad \subset \quad \mathcal{V}$$

dim: $\quad q - n_1 - \rho_1 \le q - g - (g - n_1) \le q - g \le q - n_1 \le q - n_0 \le q,$

where $n_1 := \dim \mathcal{V} - \dim \mathcal{V}_1$ and $\rho_1 := \mathrm{rank}_{\mathcal{V}_1}(\mathcal{Q})$.

This is the beginning of an obvious induction process, which can be continued as long as $\mathcal{C}_{\mathcal{V}_i}(\mathcal{Q}) \subset \mathcal{C}_{\mathcal{V}_i}(\lambda) \subset \mathcal{I}$, or equivalently $\rho_i > 2(g - n_i)$ and $n_i < g$. Since it leads to larger and larger n_i, it will end after at most g steps, and we obtain the following result.

Theorem 3.5.1. *If* inv $(V, Q) \neq \emptyset$ *there is a subsystem* $\bar{V}$ *of* V *with* $\dim \bar{V} = q - \bar{n}$ *for some* $\bar{n} \in \{0, \ldots, g\}$, *such that for all* $\mathcal{I} \in$ inv (V, Q) *and all* $\lambda \in \mathcal{O}$

$$C_{\bar{V}}(Q) \quad = \quad C_{\bar{V}}(\lambda) \quad \subseteq \quad \mathcal{I} \quad \subseteq \quad \bar{V} \quad \subseteq \quad V$$
$$\dim: \quad q - \bar{n} - 2(g - \bar{n}) = q - g - (g - \bar{n}) \leq q - g \leq q - \bar{n} \leq q.$$

$\square$

$\bar{V} = V$ if $\bar{n} = 0$; note that this always happens when $m = 1$. If $\bar{n} = g$, we have inv$(V, Q) = \{\bar{V}\}$.

$\bar{V}$ can be written as

$$\bar{V} = \{X \in V \mid \alpha^i(X) = 0, \; i = 1, \ldots, \bar{n}\},$$

where the α^i are one-forms that are linearly independent modulo $\mathcal{P}$. Then $\{\alpha^i = 0\}_{i=1}^{\bar{n}}$ are common defining equations for the involutions in inv(V, Q).

With $\mathcal{P} = (\theta^1, \ldots, \theta^s)$, we next specialize to the case when $Q = (d\theta^1, \ldots, d\theta^s)$. An involution $\mathcal{I}$ of V is then the same as an involution of V with respect to Q, i.e., $\mathcal{I}$ is a subsystem of V such that

$$d\theta^i(X, Y) = 0 \text{ for all } X, Y \in \mathcal{I} \text{ and for } i = 1, \ldots, s,$$

or equivalently $[\mathcal{I}, \mathcal{I}] \subseteq V$. Since a complete subsystem of V is a subsystem W such that $[W, W] \subseteq W \; (\subseteq V)$, any such is in particular an involution. The characteristic vector field system of V is

$$C(V) = C_V(Q) = \{X \in V \mid X \rfloor d\theta^i \equiv 0 \pmod{\mathcal{P}}, \; i = 1, \ldots, s\}.$$

For $\lambda = (\lambda_1, \ldots, \lambda_s) \in \mathbb{R}^s$ we set

$$\Omega(\lambda) := \sum_{i=1}^{s} \lambda_i \, d\theta^i.$$

The *genus of* V is by definition the same as the *genus of* Q (mod $\mathcal{P}$), i.e.,

$$g = \frac{1}{2} \max_{\lambda \in \mathbb{R}^s} \mathrm{rank}_V \Omega(\lambda).$$

This maximal value is attained on a dense open subset $\mathcal{O}$ of $\mathbb{R}^s$.

By the first lemma in this section the biggest possible dimension of the involutions of V (and a fortiori of the complete subsystems of V) equals $\dim V - (\text{genus of } V) = q - g$.

Definition. The set of $(q - g)$-dimensional complete subsystems of V is denoted by int(V). V is said to be *integrable* if int$(V) \neq \emptyset$.

Remark. Each $W \in \text{int}(\mathcal{V})$ contains $C(\mathcal{V})$, because otherwise it would be possible to obtain involutions with a larger dimension than $q - g$ by adding vector fields in $C(\mathcal{V}) \setminus W$ to W.

In the rest of this section $\mathcal{V}$ denotes an integrable vector field system with the dual pfaffian system $\mathcal{P} = (\theta^1, \ldots, \theta^s)$, and with $\mathcal{Q} = (d\theta^1, \ldots, d\theta^s)$. Then $\text{inv}(\mathcal{V}, \mathcal{Q}) \neq \emptyset$, and so the theorem above yields a subsystem $\mathcal{V}_1$ of $\mathcal{V}$ with the property that for all $W \in \text{int}(\mathcal{V}) \subseteq \text{inv}(\mathcal{V}, \mathcal{Q})$ and for all $\lambda \in \mathcal{O}$

$$C_{\mathcal{V}_1}(\mathcal{Q}) \quad = \quad C_{\mathcal{V}_1}(\lambda) \quad \subseteq \quad W \quad \subseteq \quad \mathcal{V}_1 \quad \subseteq \quad \mathcal{V}$$

dim: $\quad q - n_1 - 2(g - n_1) = q - g - (g - n_1) \leq q - g \leq q - n_1 \leq q$

for a certain $n_1 \in \{0, \ldots, g\}$. Let us assume that the complementary one-forms $\pi^1, \ldots, \pi^q$ of $\mathcal{P}$ on $\mathbb{R}^r$ have been chosen such that

$$\mathcal{V}_1 = \{X \in \mathcal{V} \mid \pi^1(X) = \cdots = \pi^{n_1}(X) = 0\}.$$

Then $\mathcal{V}_1$ is the dual vector field system of $\mathcal{P}_1 := (\theta^1, \ldots, \theta^s, \pi^1, \ldots, \pi^{n_1})$. The Cauchy characteristic vector field system of $\mathcal{V}_1$ is

$$C(\mathcal{V}_1) = \{X \in \mathcal{V}_1 \mid X \rfloor d\theta^i \equiv 0 \quad (\text{mod } \mathcal{P}), \; X \rfloor d\pi^j \equiv 0 \quad (\text{mod } \mathcal{P})$$
$$\text{for } i = 1, \ldots, s \text{ and } j = 1, \ldots, n_1\},$$

while

$$C_{\mathcal{V}_1}(\mathcal{Q}) = \{X \in \mathcal{V}_1 \mid X \rfloor d\theta^i \equiv 0 \quad (\text{mod } \mathcal{P}), \; i = 1, \ldots, s\}.$$

Therefore

$$C(\mathcal{V}_1) \subseteq C_{\mathcal{V}_1}(\mathcal{Q})$$

with equality precisely when $n_1 = 0$. So generally

$$C(\mathcal{V}_1) \subseteq C_{\mathcal{V}_1}(\mathcal{Q}) \subseteq W \subseteq \mathcal{V}_1 \subseteq \mathcal{V}. \tag{*}$$

Assume now that $n_1 > 0$, so that $C(\mathcal{V}_1) \subset C_{\mathcal{V}_1}(\mathcal{Q})$. Then we want to use the preceding theorem again, but this time with $\mathcal{V}$, $\mathcal{P}$, $\mathcal{Q}$ replaced by $\mathcal{V}_1$, $\mathcal{P}_1$ and $\mathcal{Q}_1 := (d\theta^1, \ldots, d\theta^s, d\pi^1, \ldots, d\pi^{n_1})$ respectively. To do this we have to check that the set $\text{int}(\mathcal{V}_1)$ of complete $(q_1 - g_1)$-dimensional subsystems of $\mathcal{V}_1$ is nonempty, where

$$q_1 := \dim \mathcal{V}_1 = g - n_1$$

and

$$g_1 := \frac{1}{2} \max_{(\lambda, \mu) \in \mathbb{R}^s \times \mathbb{R}^{n_1}} \text{rank}_{\mathcal{V}_1} \left(\sum_{i=1}^{s} \lambda_i \, d\theta^i + \sum_{j=1}^{n_1} \mu_j \, d\pi^j \right).$$

Set $\Omega^1 := \sum_{i=1}^{s} \lambda_i \, d\theta^i + \sum_{j=1}^{n_1} \mu_j \, d\pi^j$, and pick a $W \in \text{int}(\mathcal{V})$. Because $[W, W] \subseteq W \subseteq \mathcal{V}_1$ we have $d\theta^i(W, W) = d\pi^j(W, W) = 0$, and in particular $\Omega^1(W, W) = 0$. With $\dim \mathcal{V}_1 - \dim W = q - n_1 - (q - g) = g - n_1$ this implies that

$$\text{rank}_{\mathcal{V}_1} \Omega^1 \le 2(g - n_1),$$

and therefore $g_1 \le g - n_1$. On the other hand,

$$g_1 \ge \frac{1}{2} \max_{\lambda \in \mathbb{R}^s} \text{rank}_{\mathcal{V}_1} \left(\sum_{i=1}^{s} \lambda_i \, d\theta^i \right) = g - n_1$$

since $\dim \mathcal{C}_{\mathcal{V}_1}(\lambda) = \dim \mathcal{V}_1 - 2(g - n_1)$ when $\lambda \in \mathcal{O}$. Hence $g_1 = g - n_1$ and so $q_1 - g_1 = q - n_1 - (g - n_1) = q - g$. From this it follows that

$$\text{int}(\mathcal{V}_1) = \text{int}(\mathcal{V}) \ne \emptyset,$$

and it is indeed possible to use the theorem in this new situation involving $\mathcal{V}_1$, $\mathcal{P}_1$ and $\mathcal{Q}_1$. Using an obvious induction argument, where in each step the differentials of the n_i new one-forms occurring in $\mathcal{P}_i$ are added to the module $\mathcal{Q}_i$ of two-forms, we finally arrive at a subsystem $\tilde{\mathcal{V}}$ of $\mathcal{V}$ with the corresponding $\tilde{n}$ being equal to 0, so that there is equality between the two characteristic systems appearing in the chain of inclusions analogous to (*).

Theorem 3.5.2. *Suppose that $\mathcal{V}$ is an integrable q-dimensional vector field system of the genus g. Then the construction above yields a minimal subsystem $\tilde{\mathcal{V}}$ of $\mathcal{V}$ containing all $(q - g)$-dimensional complete subsystems of $\mathcal{V}$. With $\dim \tilde{\mathcal{V}} = q - h$ we have for each $W \in \text{int}(\mathcal{V})$*

$$\mathcal{C}(\tilde{\mathcal{V}}) \quad \subseteq \quad W \quad \subseteq \quad \tilde{\mathcal{V}} \quad \subseteq \mathcal{V}$$
$$\text{dim:} \quad q + h - 2g \; \le q - g \; \le q - h \; \le \; q.$$

$\square$

If $h = 0$,

$$\mathcal{C}(\tilde{\mathcal{V}}) \quad \subseteq \quad W \quad \subseteq \tilde{\mathcal{V}} \; = \; \mathcal{V}$$
$$\text{dim:} \quad q - 2g \; \le q - g \; \le q \; = \; q,$$

so in this case there is no simplification.

If h attains its largest possible value g,

$$\mathcal{C}(\tilde{\mathcal{V}}) \quad = \quad W \quad = \quad \tilde{\mathcal{V}} \quad \subseteq \mathcal{V}$$
$$\text{dim:} \quad q - g \; = q - g \; = q - g \; \le \; q,$$

that is, $\text{int}(\mathcal{V}) = \{\tilde{\mathcal{V}}\}$.

The moral of this theorem is that in looking for the maximal complete subsystems of a vector field system $\mathcal{V}$, it may be possible to replace $\mathcal{V}$ by a vector field system $\tilde{\mathcal{V}}$ of smaller dimension.

It may also happen that $\tilde{\mathcal{V}}$ admits Cauchy characteristic vector fields, although the original $\mathcal{V}$ does not, and then the integration problem is simplified further by considering $\tilde{\mathcal{V}}_{\text{red}}$.

4

Involutivity and the prolongation theorem

Let S be a kth order PDE system in m dependent variables $z^1, \ldots, z^m$ and n independent variables $x^1, \ldots, x^n$. S can be regarded as a submanifold of the jet bundle $J^k(\mathbb{R}^n_x, \mathbb{R}^m_z)$, with the source projection

$$S \to \mathbb{R}^n_x$$

$$(x, z, p) \mapsto x$$

being surjective (that is, locally—as always). To $J^k(\mathbb{R}^n_x, \mathbb{R}^m_z)$ belongs the contact system ${}^k Ct^{n,m}$, from which S inherits the pfaffian system

$$\mathcal{P} := {}^k Ct^{n,m}{}_{|S}.$$

The k-graph of a solution of S is then an n-dimensional submanifold $\mathcal{N}$ of S such that

$$\mathcal{P}_{|\mathcal{N}} = 0 \quad \text{and} \quad dx^1 \wedge \cdots \wedge dx^n{}_{|\mathcal{N}} \neq 0.$$

Let $\mathcal{V} := \mathcal{P}^{\perp}$ be the dual vector field system, and identify S locally with $\mathbb{R}^r$, having the coordinates $x^1, \ldots, x^n, x^{n+1}, \ldots, x^r$. In the preceding chapter we used Gaussian elimination and a *suitable change of indices* in order to find generators $X_1, \ldots, X_q$ (with $q := r - \dim \mathcal{P}$) of $\mathcal{V}$ of the form

$$X_i = \frac{\partial}{\partial x^i} + \sum_{j=1}^{r-q} \xi_i^{q+j}(x) \frac{\partial}{\partial x^{q+j}}, \quad \text{for } i = 1, \ldots, q.$$

If the involutive genus of $\mathcal{V}$ equals n, the general n^{th} order involution $(U_1, \ldots, U_n)$ can then be put in the form

$$U_i = X_i + \sum_{j=1}^{q-n} a_i^{n+j} X_j, \quad i = 1, \ldots, n,$$

by a new Gaussian elimination and change of indices.

Definition. $\mathcal{P}$ (or $\mathcal{V}$) is said to be *involutive with respect to the independent variables* if the general n^{th} order involution—with all a_1^{n+j} parametric, $q - n - s_1$ of the a_2^{n+j} parametric, and so on—can be written in this way with $x^1, \ldots, x^n$ *being local coordinates for the space of independent variables* (not necessarily the original ones though).

In this case the resulting integral manifolds will indeed be q-graphs of functions of the independent variables.

More generally, let $\mathcal{P} = (\theta^1, \ldots, \theta^s)$ be a pfaffian system on $\mathbb{R}^r$ with the involutive genus n, and let us look for n-dimensional integral manifolds of $\mathcal{P}$ on which a given decomposable n-form $\Omega^n = \varpi^1 \wedge \cdots \wedge \varpi^n$—with $\theta^1 \wedge \cdots \wedge \theta^s \wedge \Omega^n \neq 0$—is nonvanishing. Complete the θ^i and ϖ^j to a local coframe $\{\theta^1, \ldots, \theta^s; \varpi^1, \ldots, \varpi^n; \pi^1, \ldots, \pi^p\}$ for $\mathbb{R}^r$.

Definition. $\mathcal{P}$—and its dual vector field system $\mathcal{V}$—is said to be *involutive with respect to Ω^n* if there are generators $\omega^1, \ldots, \omega^n$ for the module $(\varpi^1, \ldots, \varpi^n)$ such that the general n^{th} order involution can be written in the form $(U_1, \ldots, U_n)$ with

$$U_i = \frac{\partial}{\partial \omega^i} + \sum_{j=1}^{p} a_i^j \frac{\partial}{\partial \pi^j} \quad \text{for } i = 1, \ldots, n.$$

Then $\omega^1 \wedge \cdots \wedge \omega^n \neq 0$ on the integral manifolds found by means of Cartan's local existence theorem.

In the applications it frequently happens that the given system $\mathcal{P}$ is *not* involutive, so that Cartan's theorem cannot be applied directly. The idea is then to *prolong* $\mathcal{P}$ to a pfaffian system $\mathcal{P}^{(1)}$ such that the integral manifolds of $\mathcal{P}$ are prolonged to integral manifolds of $\mathcal{P}^{(1)}$, and investigate whether $\mathcal{P}^{(1)}$ might be involutive with respect to Ω^n. If not, $\mathcal{P}^{(1)}$ is prolonged to $\mathcal{P}^{(2)} = (\mathcal{P}^{(1)})^{(1)}$, and so on.

The prolongation theorem. *After a **finite** number of prolongations one realizes*

- *either that $\mathcal{P}$ does not admit any integral manifolds on which $\Omega^n \neq 0$,*
- *or that there is a prolongation $\mathcal{P}^{(m)}$ which is involutive with respect to Ω^n.*

In the latter case Cartan's local existence theorem provides those integral manifolds which arise from the *general* n^{th} order involution. But there may be other—singular—ones too.

The more advanced prolongation theorem of Kuranishi asserts that if $\mathcal{N}$ is such a singular integral manifold it is—under certain general conditions—possible to find a prolongation $\mathcal{P}^{(m)}$ such that the corresponding prolongation $\mathcal{N}^{(m)}$ of $\mathcal{N}$ is obtained from the general n^{th} order involution of $\mathcal{P}^{(m)}$ by means of the local existence theorem. This will not be proved here, but instead we refer to [Kuranishi 1962], [Matsuda 1967], [Matsuda 1972] and [BCG³ 1991].

4.1 Independence condition and involutivity

Let $\mathcal{P} = (\theta^1, \dots, \theta^s)$ be a pfaffian system on $\mathbb{R}^r$ with the involutive genus n, let Ω^n be a decomposable n-form such that $\theta^1 \wedge \cdots \wedge \theta^s \wedge \Omega^n \neq 0$, and let $\pi^1, \dots, \pi^p$ (where $p = r - s - n$) be one-forms satisfying $\theta^1 \wedge \cdots \wedge \theta^s \wedge \Omega^n \wedge \pi^1 \wedge \cdots \wedge \pi^p \neq 0$. We want to investigate whether the general n-dimensional involution $(U_1, \dots, U_n)$ can be expressed in the form

$$U_i = \frac{\partial}{\partial \omega^i} + \sum_{\rho=1}^{p} a_i^\rho \frac{\partial}{\partial \pi^\rho},$$

where $\omega^1 \wedge \cdots \wedge \omega^n$ is a nonzero multiple of Ω^n, and

- all of the a_1^j are parametric,
- $p - s_1$ of the a_2^j are parametric,
- $\dots$
- $p - s_1 - \cdots - s_{n-1}$ of the a_n^j are parametric.

$U_1, U_2, U_3, \dots$ are obtained successively by means of the *structure equations* of $\mathcal{P}$:

$$d\theta^k \equiv \frac{1}{2} \sum_{i,j=1}^{n} c_{ij}^k \, \omega^i \wedge \omega^j + \sum_{i=1}^{n} \sum_{\rho=1}^{p} a_{i\rho}^k \, \omega^i \wedge \pi^\rho + \frac{1}{2} \sum_{\rho,\sigma=1}^{p} b_{\rho\sigma}^k \, \pi^\rho \wedge \pi^\sigma \quad (\text{mod } \mathcal{P}),$$

where $c_{ij}^k = -c_{ji}^k$ and $b_{\rho\sigma}^k = -b_{\sigma\rho}^k$.

In order to facilitate the calculations we first show that it is possible to get rid of the $b_{\rho\sigma}^k$ by prolonging $\mathcal{P}$ to a manifold of larger dimension. Set

$$U_i := \frac{\partial}{\partial \omega^i} + \sum_{\rho=1}^{p} a_i^\rho \frac{\partial}{\partial \pi^\rho} \quad \text{for } i = 1, \dots, n,$$

where the a_i^ρ now are regarded as unknowns—rather than as functions of $x = (x^1, \dots, x^r)$. The U_i form an involution if $d\theta^k(U_i, U_j) = 0$ for

$k = 1, \ldots, s$ and $i, j = 1, \ldots, n$, that is,

$$c_{ij}^k + \sum_{\rho=1}^{p} (a_{i\rho}^k a_j^\rho - a_{j\rho}^k a_i^\rho) + \sum_{\rho=1}^{p} b_{\rho\sigma}^k a_i^\rho a_j^\sigma = 0.$$

Here c_{ij}^k, $a_{i\rho}^k$ and $b_{\rho\sigma}^k$ are known functions of x. We thus arrive at a system

$$F_{ij}^k(x; a_i^\rho) = 0 \quad \text{for } k = 1, \ldots, s \text{ and } 1 \le i < j \le n. \tag{*}$$

We restrict ourselves to the generic case where the implicit function theorem may be applied so as to solve the equations first with respect to as many a_i^ρ as possible, and then perhaps also with respect to some of the x^i. Three different cases are conceivable:

(i) *some* of the a_i^ρ (the principal ones) are solved as functions of the remaining parametric a_i^ρ and the x^i;
(ii) *all* a_i^ρ are solved as functions of the x^i;
(iii) *all* a_i^ρ and *some* of the x^i are solved as functions of the remaining parametric x^i.

In the third case there are certain relations between the x^i only, which determine a subset M of $\mathbb{R}^r$.

If $\omega^1 \wedge \cdots \wedge \omega^n{}_{|M} = 0$, then $\mathcal{P}$ admits *no* involution of the wanted form.

If $\omega^1 \wedge \cdots \wedge \omega^n{}_{|M} \ne 0$, $\mathbb{R}^r$ is replaced by M (so that the dimension r is diminished), and one starts all over again.

In the second case the a_i^ρ are determinate functions of x (i.e., there are *no parametric coefficients*), and hence give rise to a determinate vector field system $(U_1, \ldots, U_n)$. If this is complete, we are through. Otherwise the congruences

$$[U_i, U_j] \equiv 0 \pmod{U_1, \ldots, U_n} \quad \text{for } 1 \le i < j \le n$$

determine a subset M of $\mathbb{R}^r$, and as in the case above one then checks whether $\omega^1 \wedge \cdots \wedge \omega^n{}_{|M}$ vanishes or not.

In the first case *the parametric a_i^ρ are replaced by new variables* $y^1, \ldots, y^t$, so that all a_i^ρ are expressed as functions of the x^i and the y^j:

$$a_i^\rho = \alpha_i^\rho(x^1, \ldots, x^r, y^1, \ldots, y^t) \quad \text{for } i = 1, \ldots, n \text{ and } \rho = 1, \ldots, p.$$

Then the dual vector field system

$$\mathcal{V} = (\partial/\partial\omega^1, \ldots, \partial/\partial\omega^n, \partial/\partial\pi^1, \ldots, \partial/\partial\pi^p) \text{ on } \mathbb{R}^r$$

is prolonged to the vector field system

$$\mathcal{V}^{(1)} := (X_1^{(1)}, \ldots, X_n^{(1)}, \partial/\partial y^1, \ldots, \partial/\partial y^t) \text{ on } \mathbb{R}_x^r \times \mathbb{R}_y^t = \mathbb{R}^{r+t},$$

where

$$X_i^{(1)} := \partial/\partial \omega^i + \sum_{\rho=1}^p \alpha_i^\rho(x, y) \left(\partial/\partial \pi^\rho\right) \quad \text{for } i = 1, \ldots, n.$$

$\mathcal{P}$ is prolonged to the dual of $\mathcal{V}^{(1)}$, i.e.,

$$\mathcal{P}^{(1)} = (\theta^1, \ldots, \theta^s, \theta^{s+1}, \ldots, \theta^{s+p}),$$

with $\theta^1, \ldots, \theta^s$ now being considered as one-forms on $\mathbb{R}^{r+t}$, and

$$\theta^{s+j} := \pi^j - \sum_{k=1}^n \alpha_k^j(x, y) \, \omega^k \quad \text{for } j = 1, \ldots, p.$$

Suppose for a moment that $\omega^i = dx^i$ for $i = 1, \ldots, n$. Then any n-dimensional integral manifold of $\mathcal{P}^{(1)}$ on which $\omega^1 \wedge \cdots \wedge \omega^n \neq 0$ is of the form

$$\begin{cases} x^{n+i} = \zeta^{n+i}(x^1, \ldots, x^n) & \text{for } i = 1, \ldots, r-n, \\ y^j = \eta^j(x^1, \ldots, x^n) & \text{for } j = 1, \ldots, t. \end{cases}$$

Clearly the first $r - n$ equations define an integral manifold of $\mathcal{P}$ on which $\omega^1 \wedge \cdots \wedge \omega^n \neq 0$. In this way the search for n-dimensional integrals of $(\mathbb{R}^r, \mathcal{P})$ on which $\omega^1 \wedge \cdots \wedge \omega^n \neq 0$ can be replaced by the corresponding problem for $(\mathbb{R}^{r+t}, \mathcal{P}^{(1)})$.

By construction $\{\theta^1, \ldots, \theta^{s+p}, \omega^1, \ldots, \omega^n, dy^1, \ldots, dy^t\}$ is a local coframe for $\mathbb{R}^{r+t}$. Therefore it is possible to express $d\theta^1, \ldots, d\theta^{s+p}$ as linear combinations of $\omega^i \wedge \omega^j$, $\omega^i \wedge dy^k$ and $dy^k \wedge dy^l$ modulo $\mathcal{P}^{(1)}$.

Claim. The $dy^k \wedge dy^l$ do not enter into the $d\theta^i$.

Note first that the $X_i^{(1)}$ have been defined so that $d\theta^k(\mathcal{V}^{(1)}, \mathcal{V}^{(1)}) = 0$, i.e.,

$$d\theta^k \equiv 0 \pmod{\mathcal{P}^{(1)}} \quad \text{for } k = 1, \ldots, s.$$

Furthermore

$$d\theta^{s+j} = d\pi^j - \sum_{k=1}^n (d\alpha_k^j \wedge \omega^k + \alpha_k^j \, d\omega^k) \quad \text{for } j = 1, \ldots, p.$$

But since

$$d\pi^j \equiv 0 \text{ and } d\omega^k \equiv 0 \pmod{\theta^1, \ldots, \theta^s, \omega^1, \ldots, \omega^n, \pi^1, \ldots, \pi^p},$$

also

$$d\pi^j \equiv 0 \text{ and } d\omega^k \equiv 0 \quad (\text{mod } \theta^1, \ldots, \theta^s, \theta^{s+1}, \ldots, \theta^{s+p}, \omega^1, \ldots, \omega^n),$$

and therefore

$$d\theta^{s+j} \equiv 0 \quad (\text{mod } \theta^1, \ldots, \theta^{s+p}, \omega^1, \ldots, \omega^n) \text{ for } j = 1, \ldots, p.$$

Conclusion. *Using a prolongation if necessary it may be assumed that we are dealing with a pfaffian system* $\mathcal{P} = (\theta^1, \ldots, \theta^s)$ *on* $\mathbb{R}^r$ *with the structure equations*

$$d\theta^k \equiv \frac{1}{2} \sum_{i,j=1}^{n} c_{ij}^k \, \omega^i \wedge \omega^j + \sum_{i=1}^{n}\sum_{\rho=1}^{p} a_{i\rho}^k \, \omega^i \wedge \pi^\rho \quad (\text{mod } \mathcal{P}) \quad \text{for } k = 1, \ldots, s,$$

—*with no* $\pi^i \wedge \pi^j$ *being present.*

Replacing π^ρ by $\bar{\pi}^\rho := \pi^\rho - \sum_{j=1}^{n} b_j^\rho \, \omega^j$, where the b_j^ρ are arbitrary functions, $\{\theta^1, \ldots, \theta^s, \omega^1, \ldots, \omega^n, \bar{\pi}^1, \ldots, \bar{\pi}^p\}$ is also a coframe for $\mathbb{R}^r$, and the structure equations go over into

$$d\theta^k \equiv \sum_{1 \le i < j \le n} \left(c_{ij}^k + \sum_{\rho=1}^{p} (a_{i\rho}^k b_j^\rho - a_{j\rho}^k b_i^\rho) \right) \omega^i \wedge \omega^j + \sum_{i=1}^{n}\sum_{\rho=1}^{p} a_{i\rho}^k \, \omega^i \wedge \bar{\pi}^\rho.$$

Choosing the b_j^ρ suitably it may therefore be possible to kill some of the coefficients in front of $\omega^i \wedge \omega^j$.

This is in particular true if $\mathcal{P}$ admits *at least one* involution $(U_1, \ldots, U_n)$ with $U_i = \partial/\partial\omega^i + \sum_{\rho=1}^{p} a_i^\rho \, (\partial/\partial\pi^\rho)$ for $i = 1, \ldots, n$. For setting $b_j^\rho := a_j^\rho$, it then follows that $\bar{\pi}^\rho(U_i) = 0$, and therefore the equations

$$d\theta^k(U_i, U_j) = 0 \quad \text{for } k = 1, \ldots, s \text{ and } i, j = 1, \ldots, n$$

imply that

$$c_{ij}^k + \sum_{\rho=1}^{p} (a_{i\rho}^k a_j^\rho - a_{j\rho}^k a_i^\rho) = 0 \quad \text{for } k = 1, \ldots, s \text{ and } 1 \le i < j \le n.$$

Hence

$$d\theta^k \equiv \sum_{i=1}^{n}\sum_{\rho=1}^{p} a_{i\rho}^k \, \omega^i \wedge \bar{\pi}^\rho \quad (\text{mod } \mathcal{P}) \quad \text{for } k = 1, \ldots, s.$$

Final conclusion. *If* $\mathcal{P}$ *admits* **at least one** *involution of the wanted form, it is possible to choose the complementary one-forms* π^ρ *such that the structure* *equations for* $\mathcal{P} = (\theta^1, \ldots, \theta^s)$ *are reduced to*

$$d\theta^k \equiv \sum_{i=1}^n \sum_{\rho=1}^p a_{i\rho}^k \, \omega^i \wedge \pi^\rho \quad (\text{mod } \mathcal{P}) \quad \text{for } k = 1, \ldots, s.$$

In the light of this we now make the following *assumptions*:

(i) $\mathcal{P} = (\theta^1, \ldots, \theta^s)$ is a pfaffian system on $\mathbb{R}^r$ with the involutive genus n;

(ii) $\omega^1, \ldots, \omega^n$ and $\pi^1, \ldots, \pi^p$ are one-forms which together with $\theta^1, \ldots, \theta^s$ constitute a local coframe for $\mathbb{R}^r$;

(iii) the structure equations of $\mathcal{P}$ are

$$d\theta^k \equiv \sum_{i=1}^n \sum_{\rho=1}^p a_{i\rho}^k \, \omega^i \wedge \pi^\rho \quad (\text{mod } \mathcal{P}) \quad \text{for } k = 1, \ldots, s.$$

We then investigate the possibility of finding the general n^{th} order involution $\mathcal{I}_n = (U_1, \ldots, U_n)$ in the form

$$U_i = \frac{\partial}{\partial \omega^i} + \sum_{\rho=1}^n a_i^\rho \frac{\partial}{\partial \pi^\rho} \quad \text{for } i = 1, \ldots, n.$$

The general 1-dimensional involution is $\mathcal{I}_1 = (X)$, where X is an arbitrary vector field in $\mathcal{V} = \mathcal{P}^\perp$, i.e.,

$$X = \sum_{i=1}^n u^i \frac{\partial}{\partial \omega^i} + \sum_{\rho=1}^p v^\rho \frac{\partial}{\partial \pi^\rho}$$

with arbitrary coefficients u^i and v^ρ. The first character of $\mathcal{P}$ is

$$s_1 = \max_{X \in \mathcal{V}} \dim(X \rfloor d\theta^1, \ldots, X \rfloor d\theta^s),$$

where

$$X \rfloor d\theta^k = \sum_{i=1}^n \sum_{\rho=1}^p a_{i\rho}^k (u^i \pi^\rho - v^\rho \omega^i) \quad \text{for } k = 1, \ldots, s.$$

Those X for which this maximal number is attained are the regular ones. Let

$$X_1 = \sum_{i=1}^n u_1^i \frac{\partial}{\partial \omega^i} + \sum_{\rho=1}^p v_1^\rho \frac{\partial}{\partial \pi^\rho}$$

be regular. Then there are s_1 linearly independent conditions for

$$X_2 = \sum_{i=1}^{n} u_2^i \frac{\partial}{\partial \omega^i} + \sum_{\rho=1}^{p} v_2^\rho \frac{\partial}{\partial \pi^\rho}$$

to be in involution with X_1, namely s_1 of

$$X_1 \rfloor d\theta^k(X_2) = 0, \qquad k = 1, \ldots, s,$$

or, more explicitly, s_1 of the equations

$$\sum_{i=1}^{n} \sum_{\rho=1}^{p} a_{i\rho}^k (u_1^i v_2^\rho - v_1^\rho u_2^i) = 0, \qquad k = 1, \ldots, s. \tag{1}$$

A first condition for $\mathcal{P}$ to be involutive with respect to $\omega^1 \wedge \cdots \wedge \omega^n$ is that (1) leaves all the u_2^i free. That is, it must be possible to solve (1) with respect to s_1 of the v_2^ρ for any regular X_1, i.e., the rank of the $s \times p$ matrix

$$\begin{pmatrix} \sum_{i=1}^{n} a_{i1}^1 u_1^i & \cdots & \sum_{i=1}^{n} a_{ip}^1 u_1^i \\ \vdots & \ddots & \vdots \\ \sum_{i=1}^{n} a_{i1}^s u_1^i & \cdots & \sum_{i=1}^{p} a_{ip}^s u_1^i \end{pmatrix}$$

equals s_1. Suppose this to be true. Then the general 2-dimensional involution $\mathcal{I}_2$ equals (X_1, X_2), where all u_1^i, v_1^ρ, u_2^i and $p - s_1$ of the v_2^ρ are parametric.

The conditions for

$$X_3 = \sum_{i=1}^{n} u_3^i \frac{\partial}{\partial \omega^i} + \sum_{\rho=1}^{p} v_3^\rho \frac{\partial}{\partial \pi^\rho}$$

to be in involution with $\mathcal{I}_2$ are

$$X_1 \rfloor d\theta^k(X_3) = X_2 \rfloor d\theta^k(X_3) = 0 \quad \text{for } k = 1, \ldots, s. \tag{2}$$

The second character s_2 is defined by

$$s_1 + s_2 := \max_{\mathcal{I}_2} \dim(X_1 \rfloor d\theta^1, \ldots, X_1 \rfloor d\theta^s, X_2 \rfloor d\theta^1, \ldots, X_2 \rfloor d\theta^s).$$

$\mathcal{I}_2 = (X_1, X_2)$ is regular if this maximum number is attained. In that case (2) gives $s_1 + s_2$ linearly independent equations for the coefficients u_3^i, v_3^ρ of X_3. A second condition for $\mathcal{P}$ to be involutive with respect to $\omega^1 \wedge \cdots \wedge \omega^n$ is that these do not impose any relation between the u_3^i

only, or equivalently that $s_1 + s_2$ of the v_3^ρ can be solved from (2). This is true precisely if the rank of the $2s \times p$ matrix

$$\begin{pmatrix} \sum a_{i1}^1 u_1^i & \cdots & \sum a_{ip}^1 u_1^i \\ \vdots & \ddots & \vdots \\ \sum a_{i1}^s u_1^i & \cdots & \sum a_{ip}^s u_1^i \\ \sum a_{i1}^1 u_2^i & \cdots & \sum a_{ip}^1 u_2^i \\ \vdots & \ddots & \vdots \\ \sum a_{i1}^s u_2^i & \cdots & \sum a_{ip}^s u_2^i \end{pmatrix}$$

equals $s_1 + s_2$. If so, (2) determines the general 3-dimensional involution $\mathcal{I}_3 = (X_1, X_2, X_3)$ of $\mathcal{P}$ with respect to $\omega^1 \wedge \cdots \wedge \omega^n$, where all u_1^i, u_2^i, u_3^i, v_1^ρ, $p - s_1$ of the v_2^ρ and $p - s_1 - s_2$ of the v_3^ρ are parametric.

Continuing in this way—if possible—we end up with

$$X_1 \rfloor d\theta^k(X_n) = \cdots = X_{n-1} \rfloor d\theta^k(X_n) = 0 \quad \text{for } k = 1, \ldots, s,$$

which for a regular $\mathcal{I}_{n-1} = (X_1, \ldots, X_{n-1})$ gives $s_1 + \cdots + s_{n-1}$ linearly independent equations for the coefficients u_n^i and v_n^ρ of X_n. The last condition for $\mathcal{P}$ to be involutive with respect to $\omega^1 \wedge \cdots \wedge \omega^n$ is that these equations can be solved with respect to $s_1 + \cdots + s_{n-1}$ of the v_n^i, or equivalently that the rank of the $(n-1)s \times p$ matrix

$$\begin{pmatrix} \sum a_{i1}^1 u_1^i & \cdots & \sum a_{ip}^1 u_1^i \\ \vdots & \ddots & \vdots \\ \sum a_{i1}^s u_{n-1}^i & \cdots & \sum a_{ip}^s u_{n-1}^i \end{pmatrix}$$

is $s_1 + \cdots + s_{n-1}$. So in particular $s_1 + \cdots + s_{n-1} \le p$.

Since all u_k^i are parametric, we can specialize them to $u_k^i = \delta_k^i$, and thereby obtain the general n^{th} order involution $\mathcal{I}_n$ with

$$U_i = \frac{\partial}{\partial \omega^i} + \sum_{\rho=1}^p a_i^j \frac{\partial}{\partial \pi^j} \quad \text{for } i = 1, \ldots, n.$$

Out of the total number np of coefficients a_i^j appearing here,

$$p + (p - s_1) + (p - s_1 - s_2) + \cdots + (p - s_1 - \cdots - s_{n-1})$$
$$= np - \sum_{k=1}^{n-1} (n-k) s_k$$

are parametric.

The transition from involutions to complete vector field systems then proceeds as before, with repeated applications of the Cauchy–Kowalewski theorem.

Although this test for involutivity is straightforward in principle, it is often rather tedious. Therefore Cartan invented a shortcut. To prepare for this we first define the *reduced characters* of $\mathcal{P}$.

With

$$X_1 = \sum_{i=1}^{n} u_1^i \frac{\partial}{\partial \omega^i} + \sum_{\rho=1}^{p} v_1^\rho \frac{\partial}{\partial \pi^\rho}$$

and

$$d\theta^k \equiv \sum_{i=1}^{n} \sum_{\rho=1}^{p} a_{i\rho}^k \, \omega^i \wedge \pi^\rho \quad (\text{mod } \mathcal{P}),$$

it follows that

$$X_1 \rfloor d\theta^k = \sum_{i,\rho} a_{i\rho}^k \, (u_1^i \, \pi^\rho - v_1^\rho \, \omega^i)$$

$$\equiv \sum_{i,\rho} a_{i\rho}^k \, u_1^i \, \pi^\rho \quad (\text{mod } \omega^1, \ldots, \omega^n).$$

Let us now forget about the ω^i and set

$$[X_1 \rfloor d\theta^k] := \sum_{i,\rho} a_{i\rho}^k u_1^i \, \pi^\rho.$$

Then if $X_2 = \sum_{i=1}^{n} u_2^i \, (\partial/\partial \omega^i) + \sum_{\rho=1}^{p} v_2^\rho \, (\partial/\partial \pi^\rho)$,

$$[X_1 \rfloor d\theta^k](X_2) = \sum_{i,\rho} a_{i\rho}^k u_1^i v_2^\rho.$$

With X_1 given, $[X_1, d\theta^k](X_2) = 0$ yields a linear system for the v_2^ρ:

$$\sum_{i,\rho} a_{i\rho}^k u_1^i v_2^\rho = 0 \quad \text{for } k = 1, \ldots, s.$$

Define the first reduced character σ_1 as the maximal rank of such systems for different X_1:

$$\sigma_1 := \max_{u_1} \text{rank} \begin{pmatrix} \sum a_{i1}^1 u_1^i & \cdots & \sum a_{ip}^1 u_1^i \\ \vdots & \ddots & \vdots \\ \sum a_{i1}^s u_1^i & \cdots & \sum a_{ip}^s u_1^i \end{pmatrix},$$

where $u_1 := (u_1^1, \ldots, u_1^n)$.

Next

$$[X_2 \rfloor d\theta^k] = \sum_{i,\rho} a_{i\rho}^k u_2^i \, \pi^\rho.$$

If $X_3 = \sum_{i=1}^n u_3^i \left(\partial/\partial\omega^i\right) + \sum_{\rho=1}^p v_2^\rho \left(\partial/\partial\pi^\rho\right)$, the equations

$$[X_1 \rfloor d\theta^k](X_3) = [X_2 \rfloor d\theta^k](X_3) = 0 \quad \text{for } k = 1,\dots,s$$

are equivalent to

$$\begin{cases} \sum_{i,\rho} a_{i\rho}^k u_1^i v_3^\rho = 0, \\ \sum_{i,\rho} a_{i\rho}^k u_2^i v_3^\rho = 0 \end{cases} \quad \text{for } k = 1,\dots,s.$$

Define the second reduced character σ_2 by

$$\sigma_2 := \max_{(u_1,u_2)} \text{rank} \begin{pmatrix} \sum a_{i1}^1 u_1^i & \cdots & \sum a_{ip}^1 u_1^i \\ \vdots & \ddots & \vdots \\ \sum a_{i1}^s u_1^i & \cdots & \sum a_{ip}^s u_1^i \\ \sum a_{i1}^1 u_2^i & \cdots & \sum a_{ip}^1 u_2^i \\ \vdots & \ddots & \vdots \\ \sum a_{i1}^s u_2^i & \cdots & \sum a_{ip}^s u_2^i \end{pmatrix},$$

whereupon the remaining reduced characters $\sigma_3, \dots, \sigma_n$ are defined in an analogous manner.

Theorem 4.1.1 (Cartan's test for involutivity). *Let $\mathcal{P} = (\theta^1,\dots,\theta^s)$ be a pfaffian system on $\mathbb{R}^r$ with the involutive genus n, let Ω^n be a decomposable n-form, let the one-forms $\pi^1, \dots, \pi^p$ with $p = r - n - s$ satisfy $\theta^1 \wedge \cdots \wedge \theta^n \wedge \Omega^n \wedge \pi^1 \cdots \wedge \pi^p \neq 0$, and assume that the structure equations are*

$$d\theta^k \equiv \sum_{i=1}^n \sum_{\rho=1}^p a_{i\rho}^k \, \omega^i \wedge \pi^\rho \pmod{\mathcal{P}} \quad \text{for } k = 1,\dots,s,$$

where $\omega^1 \wedge \cdots \wedge \omega^n$ is a nonzero multiple of Ω^n. Let

$$U_i = \frac{\partial}{\partial\omega^i} + \sum_{\rho=1}^p b_i^\rho \frac{\partial}{\partial\pi^\rho} \quad \text{for } i = 1,\dots,n,$$

with arbitrary coefficients b_i^ρ. Then

$\mathcal{I}_n = (U_1,\dots,U_n)$ *is an involution*
$\Longleftrightarrow d\theta^k(U_i, U_j) = 0$ *for $k = 1,\dots,s$ and $1 \leq i < j \leq n$*
$\Longleftrightarrow \sum_{\rho=1}^p (a_{i\rho}^k b_j^\rho - a_{j\rho}^k b_i^\rho) = 0$ *for $k = 1,\dots,s$ and $1 \leq i < j \leq n$.*

From the linear system in the last line it is possible to solve a certain number of principal b_i^ρ as linear combinations of the remaining parametric b_i^ρ.

*Then **the number of parametric coefficients is at most equal to***

$$np - \sum_{k=1}^{n-1}(n-k)\sigma_k.$$

*Moreover, **if $\mathcal{P}$ does not admit any Cauchy characteristic vector field** (which may be assumed by considering the reduced system $\mathcal{P}_{\mathrm{red}}$ instead of $\mathcal{P}$), **and this maximal number is attained, then $\mathcal{P}$ is involutive with respect to Ω^n.** That is, the general n^{th} order involution depending on $np - \sum_{k=1}^{n-1}(n-k)s_k$ arbitrary parameters can be written as $(U_1, \ldots, U_n)$ with U_i as above. In particular $s_i = \sigma_i$ for $i = 1, \ldots, n-1$.*

Remark. It is clear from the theorem that the involutivity of $\mathcal{P}$ depends solely on the structure functions $a_{i\rho}^k$. The structure $\{a_{i\rho}^k\}$ is accordingly said to be involutive if and only if $\mathcal{P}$ is.

Proof of the theorem. Introduce the one-forms

$$\pi_i^k := \sum_{\rho=1}^{p} a_{i\rho}^k \pi^\rho \quad \text{for } k = 1, \ldots, s \text{ and } i = 1, \ldots, n,$$

so that the structure equations for $\mathcal{P}$ get simplified to

$$d\theta^k \equiv \sum_{i=1}^{n} \omega^i \wedge \pi_i^k \pmod{\mathcal{P}} \quad \text{for } k = 1, \ldots, s.$$

Then

$$[U_1 \rfloor d\theta^k] = \pi_1^k \quad \text{for } k = 1, \ldots, s.$$

By definition

$$\sigma_1 = \max_{U}\{\text{number of independent one-foms } [U, \rfloor d\theta^k]\}.$$

The vector fields U for which this maximal number is attained form a dense open subset of the set of all vector fields. If U_1 is not among these, this can be remedied by making a linear transformation $\omega^i \mapsto \bar{\omega}^i$ close to the identity, and replacing U_1 by

$$\bar{U}_1 = \frac{\partial}{\partial \bar{\omega}_1} + \sum_{\rho=1}^{p} b_i^\rho \frac{\partial}{\partial \pi^\rho}.$$

Thus we may assume that σ_1 of the π_1^k—say $\pi_1^1, \ldots, \pi_1^{\sigma_1}$—are linearly independent, while the others are linear combinations of these. Next

$$[U_2 \rfloor d\theta^k] = \pi_2^k \quad \text{for } k = 1, \ldots, s,$$

and we may similarly assume that there are precisely σ_2 of the π_2^k—say $\pi_2^1, \ldots, \pi_2^{\sigma_2}$—which together with $\pi_1^1, \ldots, \pi_1^{\sigma_1}$ are linearly independent.

Continuing in this way we find that $\sigma_1 + \sigma_2 + \cdots + \sigma_n$ of the π_i^k are linearly independent—namely the π_i^k with $k \leq \sigma_i$.

Because

$$\mathcal{I}_n^\perp = \left(\theta^1, \ldots, \theta^s, \pi^1 - \sum_{j=1}^n b_j^1 \omega^j, \ldots, \pi^p - \sum_{j=1}^n b_j^p \omega^j \right),$$

we have

$$\pi_i^k = \sum_{\rho=1}^p a_{i\rho}^k \pi^\rho \equiv \sum_{j=1}^n \left(\sum_{\rho=1}^p a_{i\rho}^k b_j^\rho \right) \omega^j \quad (\text{mod } \mathcal{I}_n^\perp).$$

Introducing

$$b_{ij}^k := \sum_{\rho=1}^p a_{i\rho}^k b_j^\rho \quad \text{for } k = 1, \ldots, s \text{ and } i, j = 1, \ldots, n,$$

it follows that

$$\pi_i^k \equiv \sum_{j=1}^n b_{ij}^k \omega^j \quad (\text{mod } \mathcal{I}_n^\perp).$$

Since there are $\sigma_1 + \cdots + \sigma_n$ independent $\pi_i^k = \sum_{\rho=1}^p a_{i\rho}^k \pi^\rho$,

$$c := p - \sigma_1 - \cdots - \sigma_n$$

of the p one-forms π^ρ are independent of the π_i^k—call these $\varpi^1, \ldots, \varpi^c$. Note that with

$$d\theta^k \equiv \sum_{i=1}^n \omega^i \wedge \pi_i^k \quad (\text{mod } \mathcal{P}) \quad \text{for } k = 1, \ldots, s,$$

the vector fields $\partial/\partial \varpi^1, \ldots, \partial/\partial \varpi^c$ in $\mathcal{P}^\perp$ are Cauchy characteristic. The ϖ^l are linear combinations of the ω^i modulo $\mathcal{I}_n^\perp$:

$$\varpi^l \equiv \sum_{i=1}^n c_i^l \omega^i \quad (\text{mod } \mathcal{I}_n^\perp) \quad \text{for } l = 1, \ldots, c.$$

Using the π_i^k with $k \leq \sigma_i$ and the ϖ^l instead of $\pi^1, \dots, \pi^p$, we have

$$\mathcal{I}_n^{\perp} = \left(\theta^m ; \pi_i^k - \sum_{j=1}^n b_{ij}^k \, \omega^j ; \varpi^l - \sum_{i=1}^n c_i^l \, \omega^i \right)$$

with $m = 1, \dots, s$, $i = 1, \dots, n$, $k = 1, \dots, \sigma_i$ and $l = 1, \dots, s$.

Then

$$\mathcal{I}_n = (U_1, \dots, U_n)$$

where

$$U_i = \frac{\partial}{\partial \omega^i} + \sum_{k,j} b_{ji}^k \frac{\partial}{\partial \pi_j^k} + \sum_l c_i^l \frac{\partial}{\partial \varpi^l} \qquad \text{for } i = 1, \dots, n,$$

and we are interested in the number of parametric coefficients entering here.

Because

$$\pi_i^k \equiv \sum_{j=1}^n b_{ij}^k \, \omega^j \quad (\bmod \; \mathcal{I}_n^{\perp})$$

and $\omega^1, \dots, \omega^n$ are linearly independent, any linear relation between the π_i^k gives rise for each j to a similar relation between the b_{ij}^k.

Furthermore

$$d\theta^k \equiv \sum_{i=1}^n \omega^i \wedge \pi_i^k \equiv \sum_{i,j=1}^n b_{ij}^k \, \omega^i \wedge \omega^j \quad (\bmod \; \mathcal{I}_n^{\perp})$$

$$= \sum_{1 \leq i < j \leq n} (b_{ij}^k - b_{ji}^k) \omega^i \wedge \omega^j,$$

and this applied to $(\mathcal{I}_n, \mathcal{I}_n)$ shows that

$$b_{ij}^k = b_{ji}^k.$$

Thus the b_{ij}^k satisfy *at least* the following conditions:

(1) $b_{ij}^k = b_{ji}^k$,
(2) any linear relation for the π_i^k gives rise for each j to a similar relation for the b_{ij}^k.

Since only the π_i^k with $k \leq \sigma_i$ are linearly independent, the parametric b_{ij}^k occur among those $b_{ij}^k = b_{ji}^k$ for which $k \leq \sigma_i$ and $k \leq \sigma_j$. Assuming $i \leq j$ and using the straightforward fact that $\sigma_1 \geq \sigma_2 \geq \cdots \geq \sigma_n$, there are as many such as there are triples (k, i, j) with $1 \leq i \leq j \leq n$ and $k \leq \sigma_j$, i.e.,

- for $j = n$ there are $n \cdot \sigma_n$,
- for $j = n - 1$ there are $(n - 1) \cdot \sigma_{n-1}$,
- ...
- for $j = 1$ there are $1 \cdot \sigma_1$,

so that altogether there are

$$n \sigma_n + (n - 1) \sigma_{n-1} + \cdots + \sigma_1$$

b_{ij}^k with $k \le \sigma_j$. Hence there are *at most* so many parametric b_{ij}^k.
    Besides these there are

$$n c = n (p - \sigma_1 - \cdots - \sigma_n)$$

coefficients c_i^l. Thus the total number of parametric coefficients in $\mathcal{I}_n$ is
at most

$$n (p - \sigma_1 - \cdots - \sigma_n) + n \sigma_n + (n - 1) \sigma_{n-1} + \cdots + \sigma_1 = np - \sum_{k=1}^{n-1} (n - k) \sigma_k,$$

which proves the first part of the theorem.

    Suppose next that $\mathcal{P}$ admits no Cauchy characteristic vector fields—
so that $c = 0$—and that the number of parametric coefficients in the
congruences

$$\pi_i^k \equiv \sum_{j=1}^{n} b_{ij}^k \, \omega^j \quad (\mathrm{mod} \ \mathcal{I}_n^{\perp})$$

equals the largest possible number $np - \sum_{k=1}^{n-1}(n - k) \sigma_k$—which means
that the b_{ij}^k satisfy no relations besides **(1)** and **(2)**. We will then show
that $\mathcal{P}$ is involutive with respect to $\omega^1 \wedge \cdots \wedge \omega^n$. To do this we have to
prove that the general n-dimensional involution $\mathcal{I}_n$ can be constructed in
such a way that the ω^i are left independent.
    Modulo a linear transformation of the ω^i we may assume the generic
vector field X_1 of $\mathcal{V} = \mathcal{P}^{\perp}$ to be determined by the pfaffian equations

$$\omega^1(X_1) = 1, \ \omega^2(X_1) = \cdots = \omega^n(X_1) = 0 \ \text{and} \ \pi_i^k(X_1) = b_{i1}^k,$$

where the b_{i1}^k are bound to satisfy the same linear relations as the π_i^k, but
are otherwise arbitrary. With

$$d\theta^k \equiv \sum_{i=1}^{n} \omega^i \wedge \pi_i^k \quad (\mathrm{mod} \ \mathcal{P}),$$

the next vector field X_2 is to satisfy the pfaffian equations

$$X_1 \rfloor d\theta^k \equiv \pi_1^k - \sum_{i=1}^{n} b_{i1}^k \omega^i = 0 \quad \text{for } k = 1, \dots, s.$$

Suppose that these *do* establish a relation between the ω^i. In that case there are functions a_k such that

$$\sum_{k=1}^{s} a_k \pi_1^k = 0, \quad \text{while} \quad \sum_{k=1}^{s} a_k b_{i1}^k \neq 0 \quad \text{for at least one } i.$$

But by (1) necessarily $b_{i1}^k = b_{1i}^k$, and then (2) gives a contradiction. Hence the equations $X_1 \rfloor d\theta^k = 0$ leave the ω^i free, and consequently $s_1 = \sigma_1$.

Then the general second order involution is (X_1, X_2), with X_2 being determined by the pfaffian equations

$$\omega^1(X_2) = 0, \ \omega^2(X_2) = 1, \ \omega^3(X_2) = \cdots = \omega^n(X_2) = 0 \text{ and } \pi_i^k(X_2) = b_{i2}^k,$$

where the b_{i2}^k satisfy the same linear relations as the π_i^k do, and also

$$0 = (X_1 \rfloor d\theta^k)(X_2) = b_{12}^k - b_{21}^k.$$

Now if the equations

$$X_1 \rfloor d\theta^k = X_2 \rfloor d\theta^k = 0 \quad \text{for } k = 1, \dots, s$$

imply a relation between the ω^i, there is at least one relation involving b_{i1}^k and b_{i2}^k which is not satisfied by π_1^k and π_2^k. But by (1), $b_{i1}^k = b_{1i}^k$ and $b_{i2}^k = b_{2i}^k$, and from this one gets a contradiction to (2). Thus the ω^i are not affected, and therefore $s_2 = \sigma_2$.

Continuing in this manner it is seen that all ω^i are left independent, and consequently also that $s_i = \sigma_i$ for $i = 1, \dots, n-1$.   □

4.2 Prolongations

Let $S \subset J^q(\mathbb{R}_x^n, \mathbb{R}_z^m)$ be a PDE system, and let $z(x) = (z^1(x), \dots, z^m(x))$ be a solution defined on some open subset of $\mathbb{R}_x^n$, which by our usual conventions is identified with $\mathbb{R}_x^n$ itself. Let α be the source projection:

$$\alpha : J^q(\mathbb{R}_x^n, \mathbb{R}_z^m) \to \mathbb{R}_x^n$$

$$(x, z, p) \mapsto x.$$

Then since $\text{graph}^{(q)} z \subset S$,

$$\alpha : S \longrightarrow \mathbb{R}_x^n$$

is *onto*. So

a necessary condition for S to be solvable is that S projects *onto* the space of independent variables under the source projection.

Let $F(x^1,\ldots,x^n;z^1,\ldots,z^m;\ldots,p^j_{i_1\ldots i_k},\ldots) = 0$ be one of the defining equations for S. Then

$$F\left(x^1,\ldots,x^n;z^1(x),\ldots,z^m(x);\ldots,\frac{\partial^k z^j}{\partial x^{i_1}\ldots\partial x^{i_k}}(x),\ldots\right)$$

vanishes identically, so that differentiation with respect to x^i gives

$$0 = \frac{\partial F}{\partial x^i}\left(x;z;\frac{\partial z}{\partial x}\right) + \sum_{j=1}^{m}\frac{\partial F}{\partial z^j}\left(x;z;\frac{\partial z}{\partial x}\right)\frac{\partial z^j}{\partial x^i}$$

$$+ \cdots + \sum_{j=1}^{m}\frac{\partial F}{\partial p^j_{i_1\ldots i_k}}\left(x;z;\frac{\partial z}{\partial x}\right)\frac{\partial^{k+1} z^j}{\partial x^i\,\partial x^{i_1}\ldots\partial x^{i_k}} + \cdots.$$

Thus $z(x)$ is automatically also a solution of

$$F^{(1)}_i := \frac{\partial F}{\partial x^i} + \sum_{j=1}^{m}\left(\frac{\partial F}{\partial z^j}p^j_i + \sum_{i_1=1}^{n}\frac{\partial F}{\partial p^j_{i_1}}p^j_{ii_1} + \cdots + \sum_{i_1,\ldots,i_q=1}^{n}\frac{\partial F}{\partial p^j_{i_1,\ldots,i_q}}p^j_{ii_1\ldots i_q}\right)$$

$$= 0.$$

Note that $F^{(1)}_i$ is defined on $J^{q+1}(\mathbb{R}^n_x,\mathbb{R}^m_z)$.

Definition. The *total derivative* with respect to x^i is the mapping

$$\frac{d}{dx^i} : C^\infty(J^q(\mathbb{R}^n,\mathbb{R}^m)) \longrightarrow C^\infty(J^{q+1}(\mathbb{R}^n,\mathbb{R}^m))$$

defined by

$$\frac{d}{dx^i} = \frac{\partial}{\partial x^i} + \sum_{j=1}^{m}\left\{p^j_i\frac{\partial}{\partial z^j} + \sum_{i_1=1}^{n}p^j_{ii_1}\frac{\partial}{\partial p^j_{i_1}} + \cdots\right.$$

$$\left. + \sum_{i_1,\ldots,i_m=1}^{n}p^j_{ii_1\ldots i_m}\frac{\partial}{\partial p^j_{i_1\ldots i_m}} + \cdots\right\}.$$

Thus d/dx^i in itself is of infinite order, but this does not create any problems since it only acts on jet bundles of finite order, i.e., $J^q(\mathbb{R}^n,\mathbb{R}^m)$ with $q < \infty$.

Remark. Here is another way of arriving at d/dx^i. Let F be a smooth function on the jet bundle $J^q(\mathbb{R}^n, \mathbb{R}^m)$; since

$$
\begin{cases}
dz^j \equiv \sum_{i_1}^n p_{i_1}^j \, dx^{i_1}, \\
dp_{i_1}^j \equiv \sum_{i_2=1}^n p_{i_1 i_2}^j \, dx^{i_2}, \\
\cdots \\
dp_{i_1 \ldots i_{q-1}}^j \equiv \sum_{i_q=1}^n p_{i_1 \ldots i_q}^j \, dx^{i_q}
\end{cases}
\qquad (\mathrm{mod} \ {}^q Ct^{n,m}),
$$

it follows that

$$
dF = \sum_{i=1}^n \frac{\partial F}{\partial x^i} \, dx^i + \sum_{j=1}^m \frac{\partial F}{\partial z^j} \, dz^j + \cdots + \sum_{j=1}^m \sum_{i_1,\ldots,i_k=1}^n \frac{\partial F}{\partial p_{i_1 \ldots i_k}^j} \, dp_{i_1 \ldots i_k}^j + \cdots
$$

$$
\equiv \sum_{i=1}^n \left(\frac{\partial F}{\partial x^i} + \sum_{j=1}^m \left(\frac{\partial F}{\partial z^j} p_i^j + \cdots + \sum_{i_1,\ldots,i_k=1}^n \frac{\partial F}{\partial p_{i_1 \ldots i_k}^j} p_{i i_1 \ldots i_k}^j + \cdots \right) \right) dx^i
$$

$$
(\mathrm{mod} \ {}^q Ct^{n,m}),
$$

whence

$$
dF \equiv \sum_{i=1}^n \frac{dF}{dx^i} \, dx^i \quad (\mathrm{mod} \ {}^q Ct^{n,m}).
$$

Suppose now that $S \subset J^q(\mathbb{R}^n, \mathbb{R}^m)$ is defined by

$$
F^a(x; z; p) = 0 \quad \text{for } a = 1, \ldots, A.
$$

Definition. The *first prolongation* $S^{(1)}$ of S is the following PDE system in $J^{q+1}(\mathbb{R}^n, \mathbb{R}^m)$:

$$
\begin{cases}
F^a = 0, \\
dF^a / dx^i = 0
\end{cases}
\quad \text{for } a = 1, \ldots, A \text{ and } i = 1, \ldots, n.
$$

The *second prolongation* $S^{(2)} \subset J^{q+2}(\mathbb{R}^n, \mathbb{R}^m)$ is defined by

$$
\begin{cases}
F^a = 0, \\
dF^a / dx^i = 0, \\
d^2 F^a / dx^i dx^j = 0
\end{cases}
\quad \text{for } a = 1, \ldots, A \text{ and } i, j = 1, \ldots, n,
$$

and so on.

Remark. From the way d/dx^i is introduced it follows that if $z(x)$ is a solution of S, it is also a solution of $S^{(1)}$, $S^{(2)}$, $\ldots$, and the converse is obvious. Thus *S and all of its prolongations have the same solution sets.*

The following example indicates the importance of this observation.

Example 1. Consider the innocent looking system $\mathcal{S} \subset J^1(\mathbb{R}_x^2, \mathbb{R}_z)$ defined by

$$\begin{cases} \partial z/\partial x^1 = 1, \\ \partial z/\partial x^2 = z. \end{cases}$$

Then $\mathcal{S}^{(1)}$ is given by

$$\begin{cases} \partial z/\partial x^1 = 1, \\ \partial z/\partial x^2 = z, \\ \partial^2 z/(\partial x^1)^2 = 0, \\ \partial^2 z/\partial x^2 \partial x^1 = 0, \\ \partial^2 z/\partial x^1 \partial x^2 = \partial z/\partial x^1, \\ \partial^2 z/(\partial x^2)^2 = \partial z/\partial x^2. \end{cases}$$

So in particular

$$0 = \frac{\partial^2 z}{\partial x^2 \partial x^1} = \frac{\partial^2 z}{\partial x^1 \partial x^2} = \frac{\partial z}{\partial x^1} = 1$$

—which is a contradiction. Therefore $\mathcal{S}^{(1)}$—and hence also $\mathcal{S}$—*admits no solution.*

More generally, let $\mathcal{S} \subset J^q(\mathbb{R}^n, \mathbb{R}^m)$ be defined by

$$F^a(x; z; p) = 0 \quad \text{for } a = 1, \dots, A,$$

where $F^a \in C^\infty(J^q(\mathbb{R}^n, \mathbb{R}^m))$. Then the prolongations $\mathcal{S}^{(1)}$, $\mathcal{S}^{(2)}$, ..., are given by further equations

$$G^b = 0,$$

where $G^b \in C^\infty(J^{q+k}(\mathbb{R}^n, \mathbb{R}^m))$ for $k = 1, 2, \dots$. By an elimination process it might be possible to extract an equation $F = 0$ from these with $F \in C^\infty(J^q(\mathbb{R}^n, \mathbb{R}^m))$, and $dF^1 \wedge \cdots \wedge dF^a \wedge dF \neq 0$ on $J^q(\mathbb{R}^n, \mathbb{R}^m)$. Then

any solution of $\mathcal{S}$ must satisfy the further q^{th} order PDE $F = 0$!

Such an extra condition is called an *integrability condition*. The complete disaster occurs if there is an integrability condition which only involves the independent variables—for then S admits no solution at all.

Example 2—taken from [Janet 1929], page 76–7 (by the way, this book still seems to be one of the best ever written on prolongation theory);

for more detailed discussions of this example, see Janet's book, as well as [Pommaret 1978], page 135–6, and [Pommaret 1994], page 108–12.

Consider $S \subset J^2(\mathbb{R}_x^3, \mathbb{R}_z)$, given by

$$\begin{cases} p_{33} = x^2 \, p_{11}, \\ p_{22} = 0 \end{cases} \quad \text{or} \quad \begin{cases} \partial^2 z / \partial (x^3)^2 = x^2 \, \partial^2 z / \partial (x^1)^2, \\ \partial^2 z / \partial (x^2)^2 = 0. \end{cases}$$

The first prolongation $S^{(1)}$ is obtained by adding the equations

$$\begin{cases} p_{333} = x^2 \, p_{311}, \\ p_{332} = p_{11} + x^2 \, p_{211}, \\ p_{331} = x^2 \, p_{111}, \\ p_{322} = 0, \\ p_{222} = 0, \\ p_{221} = 0 \end{cases}$$

to those of S. It is clearly not possible to extract any second order PDE from these third order PDEs, so there is no second order integrability condition. However, the second order prolongation $S^{(2)}$ contains the fourth order equations

$$\begin{cases} p_{3322} = 2p_{211} + x^2 \, p_{2211}, \\ p_{3322} = 0, \\ p_{2211} = 0, \end{cases}$$

which immediately yield the third order equation

$$p_{211} = 0.$$

Since this is not a consequence of the equations constituting $S^{(1)}$, $p_{211} = 0$ is a third order integrability condition which any solution of S must satisfy.

The moral of this example is the following: From the fact that there are no integrability conditions of a certain order, one cannot conclude that there are no integrability conditions of higher orders. So to find *all* integrability conditions it seems necessary to make an infinite number of prolongations and eliminations.

The natural projection mappings

$$\pi_q^{q+1} : J^{q+1}(\mathbb{R}^n, \mathbb{R}^m) \longrightarrow J^q(\mathbb{R}^n, \mathbb{R}^m)$$

induce mappings

$$\cdots \longrightarrow \mathcal{S}^{(2)} \longrightarrow \mathcal{S}^{(1)} \longrightarrow \mathcal{S} \xrightarrow[\text{projection}]{\text{source}} \mathbb{R}^n.$$

The statement 'it is impossible to extract any integrability condition for $\mathcal{S}^{(k)}$ from the defining equations of $\mathcal{S}^{(k+1)}$' then means that $\mathcal{S}^{(k+1)}$ projects *onto* $\mathcal{S}^{(k)}$:

$$\mathcal{S}^{(k+1)} \twoheadrightarrow \mathcal{S}^{(k)}.$$

Definition. The PDE system $\mathcal{S} \subset J^q(\mathbb{R}^n_x, \mathbb{R}^m_z)$ is said to be *involutive* with respect to the independent variables $x^1, \ldots, x^n$ if all the projections

$$\cdots \longrightarrow \mathcal{S}^{(3)} \longrightarrow \mathcal{S}^{(2)} \longrightarrow \mathcal{S}^{(1)} \longrightarrow \mathcal{S} \longrightarrow \mathbb{R}^n_x$$

are *surjective*.

So in this case prolongations and eliminations will never ever give rise to any integrability conditions.

Theorem 4.2.1. *Involutivity implies formal solvability.*

The idea is this: starting from some $x_0 \in \mathbb{R}^n$ one can go to the left in the above chain of surjections and find Taylor polynomials (using the natural identification of points in jet bundles with Taylor polynomials) $P_q \in \mathcal{S}$, $P_{q+1} \in \mathcal{S}^{(1)}, \ldots$, centred at x_0, such that for any $l > k$,

$$P_l = P_k + \text{ terms of order } > k.$$

Then

$$f = \lim_{q \to \infty} P_q$$

is a formal power series solution of $\mathcal{S}$. $\qquad\qquad\square$

A priori this concept of involutivity looks rather pointless, since in order to test it one has to consider infinitely many prolongations and try an infinite number of eliminations. But fortunately it is not that bad.

The prolongation theorem for PDE systems. *Under certain general conditions the following is true: after having made a **finite** number of prolongations one arrives at a prolongation $\bar{\mathcal{S}}$ such that*

- *either $\bar{\mathcal{S}}$ is involutive (and hence formally solvable),*
- *or $\bar{\mathcal{S}} \longrightarrow \mathbb{R}^n_x$ is **not surjective**, so that $\mathcal{S}$ admits no solution.* $\qquad\square$

A proof of the pfaffian version of this theorem is sketched in the next section.

Remark. The connection between involutivity for PDE systems and Cartan's test for involutivity of pfaffian systems is the following.

Suppose that $S^{(k)} \subset J^{q+k}(\mathbb{R}^n, \mathbb{R}^m)$, and that $S^{(k+1)} \longrightarrow S^{(k)}$ is *not* surjective. Then the solutions of S have to satisfy a larger number of $(q+k)^{\text{th}}$ order equations than expected—namely besides those of $S^{(k)}$ also all integrability conditions. Therefore the number of 'arbitrary parameters' is smaller than it should be.

We have already discussed how to prolong a pfaffian system when preparing for Cartan's test. Let us here just consider the pfaffian version of the PDE system

$$\begin{cases} \partial z/\partial x^1 = 1, \\ \partial z/\partial x^2 = z. \end{cases}$$

That is,

$$\theta := dz - dx^1 - z\, dx^2 = 0,$$

with the independence condition $dx^1 \wedge dx^2 \neq 0$ on the wanted integral manifolds. If $\mathcal{N}$ is one of these, not only θ but also $d\theta$ vanishes on $\mathcal{N}$. However,

$$d\theta = -dz \wedge dx^2 \equiv -(dx^1 + z\, dx^2) \wedge dx^2 = -dx^1 \wedge dx^2 \pmod{\theta},$$

and we cannot have both $d\theta_{|\mathcal{N}} = 0$ and $dx^1 \wedge dx^2_{|\mathcal{N}} \neq 0$.

Let us next consider prolongations from the vector field angle. If $\mathcal{P} = (\theta^1, \dots, \theta^s)$ is a pfaffian sytem on $\mathbb{R}^n$ with the independence condition $\omega^1 \wedge \cdots \wedge \omega^n \neq 0$, choose $p := r - s - n$ one-forms $\pi^1, \dots, \pi^p$ such that $\{\theta^1, \dots, \theta^s, \omega^1, \dots, \omega^n, \pi^1, \dots, \pi^p\}$ constitutes a local coframe for $\mathbb{R}^r$. Then the wanted complete subsystems of $\mathcal{P}^{\perp}$ are of the form $\mathcal{W} = (U_1, \dots, U_n)$ with

$$U_i = \frac{\partial}{\partial \omega^i} + \sum_{j=1}^{p} a_i^j \frac{\partial}{\partial \pi^j}, \qquad i = 1, \dots, n,$$

for suitable smooth functions a_i^j on $\mathbb{R}^r$. Setting

$$X_i := \frac{\partial}{\partial \omega^i} \qquad \text{for } i = 1, \dots, n$$

and

$$X_{n+j} := \frac{\partial}{\partial \pi^j} \quad \text{for } j = 1, \dots, p,$$

we have

$$U_i = X_i + \sum_{j=1}^{p} a_i^j X_{n+j}, \quad l = 1, \dots, n.$$

Furthermore the dual vector field system of $\mathcal{P}$ is

$$\mathcal{V} = (X_1, \dots, X_{n+p}).$$

If the derived system is

$$\mathcal{V}' = (X_1, \dots, X_{n+p}; Z_1, \dots, Z_d),$$

there are structure functions c_{ij}^k such that

$$[X_i, X_j] \equiv \sum_{k=1}^{d} c_{ij}^k Z_k \pmod{\mathcal{V}} \quad \text{for } i, j = 1, \dots, n+p.$$

Then

$$[U_i, U_j] \equiv \sum_{m=1}^{d} \left\{ c_{ij}^m + \sum_{k=1}^{p} (a_j^k c_{i,n+k}^m - a_i^k c_{j,n+k}^m) \right.$$
$$\left. + \sum_{k,l=1}^{p} a_i^k a_j^l c_{n+k,n+l}^m \right\} Z_m \pmod{\mathcal{V}},$$

so that $[U_i, U_j] \equiv 0 \pmod{\mathcal{V}}$ if and only if

$$c_{ij}^m + \sum_{k=1}^{p} (a_j^k c_{i,n+k}^m - a_i^k c_{j,n+k}^m) + \sum_{k,l=1}^{p} a_i^k a_j^l c_{n+k,n+l}^m = 0 \quad \text{for } m = 1, \dots, d.$$

Here the structure functions c_{ij}^k are known functions of $x = (x^1, \dots, x^r)$, while the a_i^j are unknowns. Let us assume (by shrinking neighbourhoods etcetera) that the implicit function theorem can be used unrestrictedly. Then this system is solved with respect to as many a_i^j as possible, and perhaps also with respect to some of the x^k. Three different cases are conceivable:

    (i) *Some* a_i^j—the principal ones—are solved as functions of the remaining parametric a_i^j and the x^k.

    (ii) *All* a_i^j are expressed as functions of the x^k, with the latter being independent.

(iii) *All a_i^j and some of the x^k are solved as functions of the remaining x^k.*

In the third case we choose indices such that

$$\begin{cases} a_i^j = \alpha_i^j(x^1,\ldots,x^m), \\ x^{m+k} = \xi^{m+k}(x^1,\ldots,x^m) & \text{for } k = 1,\ldots,r-m. \end{cases}$$

Then all integral manifolds of $\mathcal{V}$ are contained in the subset M of $\mathbb{R}^r$ defined by the second set of equations. If $\omega^1 \wedge \cdots \wedge \omega^n{}_{|M} = 0$, there are *no* integral manifolds satisfying the given independence condition. Otherwise, the original problem is replaced by that of integrating $\mathcal{P}_{|M}$ on M.

In the second case we obtain a determinate vector field system $\mathcal{W} = (U_1,\ldots,U_n)$. If this is complete, everything is just fine. Otherwise the congruences $[U_i, U_j] \equiv 0 \pmod{\mathcal{W}}$ define a subset M of $\mathbb{R}^r$, to which the problem is restricted.

In the first case there is a certain number—say t—of parametric a_i^j. If we replace these by new variables $y^1, \ldots, y^t$, all a_i^j can be expressed as functions of $x^1,\ldots,x^r,y^1,\ldots,y^t$:

$$a_i^j = \alpha_i^j(x^1,\ldots,x^r,y^1,\ldots,y^t).$$

Then the vector fields

$$U_i = X_i + \sum_{j=1}^p \alpha_i^j(x,y)\,X_{n+j}, \qquad i = 1,\ldots,n,$$

can be thought of either as vector fields on $\mathbb{R}^r$ depending on t parameters, or as determinate vector fields living on $\mathbb{R}^{r+t} = \mathbb{R}_x^r \times \mathbb{R}_y^t$. Adopting the latter point of view we set

$$X_i^{(1)} := X_i + \sum_{j=1}^p \alpha_i^j(x,y)\,X_{n+j} \qquad \text{for } i = 1,\ldots,n,$$

and define the first prolongation of $\mathcal{V}$ to be the vector field system

$$\mathcal{V}^{(1)} := \left(X_1^{(1)},\ldots,X_n^{(1)}, \frac{\partial}{\partial y^1},\ldots,\frac{\partial}{\partial y^t} \right)$$

on $\mathbb{R}^{r+t}$. Then instead of integrating $\mathcal{V}$ on $\mathbb{R}^r$ we look for complete subsystems of $\mathcal{V}^{(1)}$ on $\mathbb{R}^{r+t}$ of the form $\mathcal{W}^{(1)} = (U_1^{(1)},\ldots,U_n^{(1)})$, with

$$U_i^{(1)} = X_i^{(1)} + \sum_{k=1}^t b_i^k \frac{\partial}{\partial y^k}, \qquad i = 1,\ldots,n,$$

for suitable smooth functions b_i^k on $\mathbb{R}^{r+t}$. If the coefficients $\alpha_i^j(x, y)$ and $b_i^k(x, y)$ do depend on the y-variables, $\mathcal{W}^{(1)}$ cannot be projected so as to give a complete subsystem of $\mathcal{V}$ on $\mathbb{R}^r$—however, the integral manifolds of $\mathcal{W}^{(1)}$ admit projections to $\mathbb{R}^r$, which are integral manifolds of $\mathcal{V}$.

To see this, let $\mathcal{N}^{(1)}$ be an n-dimensional integral manifold of $\mathcal{W}^{(1)}$ on which $\omega^1 \wedge \cdots \wedge \omega^n \neq 0$, with the ω^i now being considered as one-forms on $\mathbb{R}^{r+t}$. Since the ω^i originally live on $\mathbb{R}^r$, we may suppose that the local coordinates $x^1, \ldots, x^r$ of $\mathbb{R}^r$ have been chosen such that $x^1, \ldots, x^n$ serve as local coordinates on $\mathcal{N}^{(1)}$. Then $\mathcal{N}^{(1)}$ is expressed as

$$\begin{cases} x^{n+j} = \xi^{n+j}(x^1, \ldots, x^n) & \text{for } j = 1, \ldots, r - n, \\ y^k = \eta^k(x^1, \ldots, x^n) & \text{for } k = 1, \ldots, t. \end{cases}$$

Let $x = (x^1, \ldots, x^n)$, $\xi = (\xi^{n+1}, \ldots, \xi^r)$, $\eta = (\eta^1, \ldots, \eta^t)$, and let $(x, \xi(x), \eta(x))$ be an arbitrary point of $\mathcal{N}^{(1)}$. The tangent space of $\mathcal{N}^{(1)}$ at this point is then spanned by the vectors

$$T_i := X_i(x, \xi(x)) + \sum_{j=1}^p \alpha_i^j(x, \xi(x), \eta(x)) \, X_{n+j}(x, \xi(x))$$

$$+ \sum_{k=1}^t b_i^k(x, \xi(x), \eta(x)) \frac{\partial}{\partial y^k} \quad \text{for } i = 1, \ldots, n.$$

With π denoting the natural projection $\mathbb{R}_x^r \times \mathbb{R}_y^t \to \mathbb{R}_x^r$, $\mathcal{N} := \pi(\mathcal{N}^{(1)})$ is given by

$$x^{n+j} = \xi^{n+j}(x^1, \ldots, x^n) \quad \text{for } j = 1, \ldots, r - n.$$

The tangent space of $\mathcal{N}$ at $(x, \xi(x))$ is spanned by the vectors

$$X_i(x, \xi(x)) + \sum_{j=1}^p \alpha_i^j(x, \xi(x), \eta(x)) \, X_{n+j}(x, \xi(x)) \quad \text{for } i = 1, \ldots, n.$$

Since these are linear combinations of $X_1, \ldots, X_{n+p}$, it follows that $T_n(\mathcal{N}) \subseteq \mathcal{V}_n$ for each $n \in \mathcal{N}$, and hence $\mathcal{N}$ is really an n-dimensional integral maifold of $\mathcal{V}$.

Note that while the integral manifolds of $\mathcal{W}^{(1)}$ form a nice foliation of $\mathbb{R}^{r+t}$, their projections to $\mathbb{R}^r$ in general do not.

Conclusion. *If $\mathcal{V}^{(1)}$ can be integrated, it is possible to obtain integral manifolds of $\mathcal{V}$ by projections.*

Example. Consider a system $S \subset J^2(\mathbb{R}^2, \mathbb{R})$ of two PDEs:

$$F^k(x^1, x^2, z, p_1, p_2, p_{11}, p_{12}, p_{22}) = 0 \quad \text{for } k = 1, 2.$$

If we assume these equations to be solvable with respect to p_{11} and p_{22}, S can be written in the form

$$\begin{cases} p_{11} + F(x^1, x^2, z, p_1, p_2, p_{12}) = 0, \\ p_{22} + G(x^1, x^2, z, p_1, p_2, p_{12}) = 0, \end{cases}$$

so that $x^1, x^2, z, p_1, p_2, p_{12}$ serve as local coordinates on S. The contact vector field system $\mathcal{V}$ of S is generated by the vector fields

$$X_1 := \frac{\partial}{\partial x^1} + p_1 \frac{\partial}{\partial z} - F \frac{\partial}{\partial p_1} + p_{12} \frac{\partial}{\partial p_2},$$

$$X_2 := \frac{\partial}{\partial x^2} + p_2 \frac{\partial}{\partial z} + p_{12} \frac{\partial}{\partial p_1} - G \frac{\partial}{\partial p_2},$$

$$X_3 := \frac{\partial}{\partial p_{12}}.$$

Complete 2-dimensional subsystems of $\mathcal{V}$ on which $dx^1 \wedge dx^2 \neq 0$ are then of the form $\mathcal{W} = (U_1, U_2)$ with

$$\begin{cases} U_1 = X_1 + a X_3, \\ U_2 = X_2 + b X_3, \end{cases}$$

where a and b are suitable smooth functions on S. The commutation relations for the X_i are

$$[X_1, X_2] = X_2 F \frac{\partial}{\partial p_1} - X_1 G \frac{\partial}{\partial p_2},$$

$$[X_1, X_3] = \frac{\partial F}{\partial p_{12}} \frac{\partial}{\partial p_1} - \frac{\partial}{\partial p_2},$$

$$[X_2, X_3] = \frac{\partial G}{\partial p_{12}} \frac{\partial}{\partial p_2} - \frac{\partial}{\partial p_1}.$$

Therefore

$$[U_1, U_2] = [X_1, X_2] + b [X_1, X_3] + a [X_3, X_2] + (U_1 b - U_2 a) X_3$$

$$= \left(X_2 F + b \frac{\partial F}{\partial p_{12}} + a \right) \frac{\partial}{\partial p_1} - \left(X_1 G + b + a \frac{\partial G}{\partial p_{12}} \right) \frac{\partial}{\partial p_2}$$

$$+ (U_1 b - U_2 a) X_3.$$

Hence (U_1, U_2) is an involution if and only if

$$\begin{pmatrix} 1 & \partial F/\partial p_{12} \\ \partial G/\partial p_{12} & 1 \end{pmatrix} \begin{pmatrix} a \\ b \end{pmatrix} = \begin{pmatrix} -X_2 F \\ -X_1 G \end{pmatrix}. \tag{*}$$

We want to solve a and b from this system.

1. If $(\partial F/\partial p_{12})(\partial G/\partial p_{12}) \neq 1$, a and b are solved from (*) so as to give a determinate vector field system $\mathcal{W} = (U_1, U_2)$ with

$$[U_1, U_2] = (U_1 b - U_2 a) X_3.$$

(i) $\mathcal{W}$ is complete if $U_1 b - U_2 a$ vanishes identically.

(ii) Otherwise the equation $U_1 b - U_2 a = 0$ defines a subset M of S. If $dx^1 \wedge dx^2{}_{|M} = 0$, there are no integral manifolds on which $dx^1 \wedge dx^2 \neq 0$. If $dx^1 \wedge dx^2{}_{|M} \neq 0$, S is replaced by M, and one starts all over again.

2. If $(\partial F/\partial p_{12})(\partial G/\partial p_{12}) = 1$, (*) is equivalent to

$$a + \frac{\partial F}{\partial p_{12}} b = -X_2 F,$$

$$a + \frac{\partial F}{\partial p_{12}} b = -\frac{\partial F}{\partial p_{12}} X_1 G.$$

(i) If $X_2 F - (\partial F/\partial p_{12}) \cdot X_1 G$ does not vanish identically, define $M \subset S$ by the equation

$$X_2 F - \frac{\partial F}{\partial p_{12}} X_1 G = 0,$$

and reason as in **1(ii)**.

(ii) If $X_2 F - (\partial F/\partial p_{12}) \cdot X_1 G \equiv 0$ on S, (*) is equivalent to the one equation

$$a + \frac{\partial F}{\partial p_{12}} b = -X_2 F.$$

Thus

$$\begin{cases} U_1 = X_1 - \left(X_2 F + (\partial F/\partial p_{12}) \cdot b\right) X_3, \\ U_2 = X_2 + b X_3. \end{cases}$$

Replacing the unknown b by a new variable y, the prolonged system $\mathcal{V}^{(1)}$ on $S \times \mathbb{R}_y$ is generated by the three vector fields

$$X_1^{(1)} := X_1 - \left(X_2 F + y \frac{\partial F}{\partial p_{12}}\right) X_3,$$

$$X_2^{(1)} := X_2 + y X_3,$$

$$X_3^{(1)} := \frac{\partial}{\partial y}.$$

Then we look for 2-dimensional complete subsystems of $\mathcal{V}^{(1)}$ with generators of the form

$$\begin{cases} U_1^{(1)} := X_1^{(1)} + a^{(1)} \, \partial/\partial y, \\ U_2^{(1)} := X_2^{(1)} + b^{(1)} \, \partial/\partial y, \end{cases}$$

where $a^{(1)}$, $b^{(1)}$ are smooth functions on $S \times \mathbb{R}_y$—i.e., we start from the beginning again, but with $(S, \mathcal{V})$ replaced by $(S \times \mathbb{R}_y, \mathcal{V}^{(1)})$.

Remark. The easiest way to solve the last case is by means of Cauchy characteristics.

In fact, let us investigate under what circumstances $\mathcal{V} = (X_1, X_2, X_3)$ admits Cauchy charateristic vector fields!

Letting $A = \sum_{i=1}^{3} a^i X_i$ and $B = \sum_{i=1}^{3} b^i X_i$, the commutation relations for the X_i show that

$$[A, B] \equiv 0 \quad (\text{mod } \mathcal{V}) \iff \mathsf{M}(\mathbf{a}) \cdot \mathbf{b} = \mathbf{0},$$

where $\mathbf{a} = (a^1, a^2, a^3)^T$, $\mathbf{b} = (b^1, b^2, b^3)^T$ and the 2×3 matrix $\mathsf{M}(\mathbf{a})$ is given by

$$\begin{pmatrix} -a^2 X_2 F - a^3 \, \partial F/\partial p_{12} & a^1 X_2 F + a^3 & a^1 \, \partial F/\partial p_{12} - a^2 \\ -a^2 X_1 G - a^3 & a^2 X_1 G + a^3 \, \partial F/\partial p_{12} & a^1 - a^2 \, \partial G/\partial p_{12} \end{pmatrix}.$$

A is Cauchy characteristic precisely if $\mathsf{M}(\mathbf{A})$ is the zero matrix, i.e.,

$$a^1 = \frac{\partial G}{\partial p_{12}} a^2, \quad a^2 = \frac{\partial F}{\partial p_{12}} a^1 \quad \text{and} \quad a^3 = -X_2 F \cdot a^1 = -X_1 G \cdot a^2,$$

or

$$\frac{\partial F}{\partial p_{12}} \frac{\partial G}{\partial p_{12}} = 1,$$

$$X_2 F = \frac{\partial F}{\partial p_{12}} X_1 G,$$

—which is case **2(ii)** above.

Setting $a^1 = 1$ we have $a^2 = \partial F/\partial p_{12}$ and $a^3 = -X_2 F$, so that the Cauchy characteristic subsystem is 1-dimensional, with the generator

$$C = X_1 + \frac{\partial F}{\partial p_{12}} X_2 - X_2 F \cdot X_3.$$

The dual pfaffian system of $\mathcal{V}$ is generated by the one-forms

$$\begin{cases} \theta^0 = dz - p_1 \, dx^1 - p_2 \, dx^2, \\ \theta^1 = dp_1 + F \, dx^1 - p_{12} \, dx^2, \\ \theta^2 = dp_2 - p_{12} \, dx^1 + G \, dx^2. \end{cases}$$

Therefore

$$\mathcal{N}_1: \quad x^i = \xi^i(t), \; z = \zeta(t), \; p_i = \pi_i(t), \; p_{12} = \pi_{12}(t) \quad \text{for } i = 1, 2$$

is an integral curve of $\mathcal{V}$ if

$$\frac{d\zeta}{dt} = \pi_1(t) \frac{d\xi^1}{dt} + \pi_2(t) \frac{d\xi^2}{dt},$$

$$\frac{d\pi_1}{dt} = -F(\xi^i(t), \pi_i(t), \pi_{12}(t)) \frac{d\xi^1}{dt} + \pi_{12}(t) \frac{d\xi^2}{dt},$$

$$\frac{d\pi_2}{dt} = \pi_{12}(t) \frac{d\xi^1}{dt} - G(\xi^i(t), \pi_i(t), \pi_{12}(t)) \frac{d\xi^2}{dt}.$$

Here $\xi^1(t)$, $\xi^2(t)$ and $\pi_{12}(t)$ may be chosen arbitrarily, whereupon $\zeta(t)$, $\pi_1(t)$ and $\pi_2(t)$ can be solved because of the existence theorem for local solutions of first order ODE systems.

To each integral curve $\mathcal{N}_1$ found in this way and not tangent to the Cauchy characteristic vector field C, we then obtain a 2-dimensional integral manifold $\mathcal{N}_2$ as the union of all Cauchy characteristics passing through $\mathcal{N}_1$. Accordingly the general solution found by means of this procedure depends on three arbitrary functions of one variable.

After this remark we return to our prolongations. The general idea is that the prolonged system should be easier to deal with than the original one.

Let us state the decisive result in the realm of pfaffian systems.

Theorem 4.2.2 (The prolongation theorem). *If the pfaffian system $\mathcal{P}$ is not involutive with respect to the independence condition $\omega^1 \wedge \cdots \wedge \omega^n \neq 0$, it is under certain general conditions possible by means of a **finite** number of prolongations*

- *either to realize that $\mathcal{P}$ admits no integral manifold on which $\omega^1 \wedge \cdots \wedge \omega^n \neq 0$,*
- *or to find a prolongation $\mathcal{P}^{(m)}$ which is involutive with respect to $\omega^1 \wedge \cdots \wedge \omega^n$.*

This is explained in the next section by combining ideas from [Cartan 1904] and [Janet 1929].

4.3 Explanation of the prolongation theorem

The prolongation theorem is essentially a *finiteness* theorem, and Janet's idea was to reduce it to Hilbert's basis theorem for ideals of homogeneous polynomials.

One ingredient in the transition from PDEs to polynomials is the association

$$\frac{\partial^k z}{\partial x^{i_1} \dots \partial x^{i_k}} \longleftrightarrow x^{i_1} \cdots x^{i_k},$$

which is based upon the fact that the order of the differentiations in the left hand side does not matter.

The corresponding transition for one-forms is not so obvious, but Cartan found a way to accomplish this too.

The following sketch of the prolongation theorem is a blending of Cartan's and Janet's ideas. The interested reader should consult [Janet 1929] for further details, and also for a very efficient algorithm; moreover, take a look at [Pommaret 1978] and [Pommaret 1994] for a modern and generalized version of Janet's theory.

According to the results in the first section we assume that we have a pfaffian system $\mathcal{P} = (\theta^1, \dots, \theta^s)$ on $\mathbb{R}^r$ and the independence condition $\omega^1 \wedge \cdots \wedge \omega^n \neq 0$, with $\mathcal{P}$ satisfying the structure equations

$$d\theta^\sigma \equiv \sum_{i=1}^{n} \sum_{\rho=1}^{r-s-n} a_{i\rho}^\sigma \, \omega^i \wedge \pi^\rho \quad (\text{mod } \mathcal{P}) \quad \text{for } \sigma = 1, \dots, s,$$

where the $a_{i\rho}^\sigma$ are smooth functions on $\mathbb{R}^r$, and $\theta^1 \wedge \dots \theta^s \wedge \omega^1 \wedge \cdots \wedge \omega^n \wedge \pi^1 \wedge \cdots \wedge \pi^{r-s-n} \neq 0$. Setting

$$\pi_i^\sigma := \sum_{\rho=1}^{r-s-n} a_{i\rho}^\sigma \, \pi^\rho \quad \text{for } \sigma = 1, \dots, s \text{ and } i = 1, \dots, n,$$

this is simplified to

$$d\theta^\sigma \equiv \sum_{i=1}^{n} \omega^i \wedge \pi_i^\sigma \quad (\text{mod } \mathcal{P}) \quad \text{for } \sigma = 1, \dots, s.$$

The $s n$ one-forms π_i^σ are in general not linearly independent, but satisfy a certain number of linear relations—say

$$\sum_{i,\sigma} A_{\sigma\alpha}^i \, \pi_i^\sigma = 0 \quad \text{for } \alpha = 1, \dots, a,$$

with smooth functions $A_{\sigma\alpha}^i$.

If

$$\text{span} \{\pi_i^\sigma \mid \sigma = 1, \dots s \text{ and } i = 1, \dots, n\} \subsetneqq \text{span} \{\pi^\rho \mid \rho = 1, \dots, r - s - n\},$$

$\mathcal{P}$ admits Cauchy characteristic vector fields. As usual, this can be avoided by considering the reduced pfaffian system $\mathcal{P}_{\text{red}}$ instead. The fact that $\mathcal{P}_{\text{red}}$ does not have any Cauchy characteristic vector fields also means that Cartan's test for involutivity applies.

The most general n-dimensional involution of $\mathcal{P}$ on which $\omega^1 \wedge \cdots \wedge \omega^n \neq 0$ is defined by a pfaffian system

$$\left(\theta^\sigma, \pi_i^\sigma - \sum_{j=1}^n t_{ij}^\sigma \omega^j \right), \qquad \sigma = 1, \ldots, s, \ i = 1, \ldots, n,$$

satisfying $d\theta^\sigma \equiv 0 \pmod{\theta^\sigma, \pi_i^\sigma - \sum_{j=1}^n t_{ij}^\sigma \omega^j}$, i.e.,

$$0 = \sum_{i=1}^n \omega^i \wedge \left(\sum_{j=1}^n t_{ij}^\sigma \omega^j \right).$$

With $\omega^i \wedge \omega^j \neq 0$ this implies that

$$t_{ij}^\sigma = t_{ji}^\sigma \quad \text{for } i, j = 1, \ldots, n.$$

Moreover,

$$0 = \sum_{i,\sigma} A_{\sigma\alpha}^i \pi_i^\sigma \equiv \sum_{j=1}^n \left(\sum_{i,\sigma} A_{\sigma\alpha}^i t_{ij}^\sigma \right) \omega^j \pmod{\theta^\sigma, \pi_i^\sigma - \sum_{j=1}^n t_{ij}^\sigma \omega^j},$$

showing that

$$\sum_{i,\sigma} A_{\sigma\alpha}^i(x) t_{ij}^\sigma = 0 \quad \text{for } \alpha = 1, \ldots, a \text{ and } j = 1, \ldots, n.$$

From these relations a maximal number of *principal* coefficients $\hat{t}_{ij}^\sigma$ are solved as functions of the local coordinates $x^1, \ldots, x^r$ of $\mathbb{R}^r$ and the remaining *parametric* coefficients $\bar{t}_{ij}^\sigma$.

If the number of parametric coefficients is $r_1 - r$, the *prolonged system* $\mathcal{P}^{(1)}$ is defined on $\mathbb{R}^{r_1} = \mathbb{R}_x^r \times \mathbb{R}_t^{r_1-r}$ by

$$\mathcal{P}^{(1)} = (\theta^\sigma, \varpi_i^\sigma) \quad \text{for } \sigma = 1, \ldots, s \text{ and } i = 1, \ldots, n,$$

where

$$\varpi_i^\sigma := \pi_i^\sigma - \sum_{j=1}^n t_{ij}^\sigma \omega^j.$$

Since

$$d\theta^\sigma \equiv \sum_i \omega^i \wedge \pi_i^\sigma = \sum_i \omega^i \wedge \varpi_i^\sigma \pmod{\mathcal{P}},$$

we have

$$d\theta^\sigma \equiv 0 \quad (\text{mod } \mathcal{P}^{(1)}).$$

In order to determine the structure equations for $\mathcal{P}^{(1)}$ it therefore suffices to investigate $d\varpi_i^\sigma$ $(\text{mod } \mathcal{P}^{(1)})$. Now $0 = d^2\theta^\sigma \equiv \sum_i \omega^i \wedge d\varpi_i^\sigma$ $(\text{mod } \mathcal{P}^{(1)})$ and $0 = \sum_{i,\sigma} A_{\sigma\alpha}^i \pi_i^\sigma = \sum_{i,\sigma} A_{\sigma\alpha}^i \varpi_i^\sigma$, so that the $d\varpi_i^\sigma$ satisfy

$$\begin{cases} \sum_{i=1}^n \omega^i \wedge d\varpi_i^\sigma \equiv 0, \\ \sum_{i=1}^n \sum_{\sigma=1}^s A_{\sigma\alpha}^i d\varpi_i^\sigma \equiv 0 \end{cases} \quad (\text{mod } \mathcal{P}^{(1)}).$$

Lemma 4.3.1. *The most general two-forms Π_i^σ that satisfy*

$$\begin{cases} \sum_{i=1}^n \omega^i \wedge \Pi_i^\sigma \equiv 0, \\ \sum_{i=1}^n \sum_{\sigma=1}^s A_{\sigma\alpha}^i \Pi_i^\sigma \equiv 0 \text{ for } \alpha = 1,\dots,a \end{cases} \quad (\text{mod } \mathcal{P}^{(1)}) \qquad (*)$$

can be written as

$$\Pi_i^\sigma \equiv \sum_{j=1}^n \omega^j \wedge \pi_{ij}^\sigma \quad (\text{mod } \mathcal{P}^{(1)}),$$

where the π_{ij}^σ are one-forms which are arbitrary except for satisfying

$$\pi_{ij}^\sigma = \pi_{ji}^\sigma \quad \text{and} \quad \sum_{i,\sigma} A_{\sigma\alpha}^i \pi_{ij}^\sigma = 0 \quad \text{for } \alpha = 1,\dots,a.$$

Proof. It is evident that such Π_i^σ are solutions of (*). The converse is also clear, except for $\pi_{ij}^\sigma = \pi_{ji}^\sigma$.

Assume inductively that $\pi_{ij}^\sigma = \pi_{ji}^\sigma$ for $i,j \le h$. Then modulo $\mathcal{P}^{(1)}$ and $\omega^{h+2}, \dots, \omega^n$,

$$0 \equiv \sum_{i=1}^{h+1} \omega^i \wedge \Pi_i^\sigma \equiv \sum_{i,j=1}^{h+1} \omega^i \wedge \omega^j \wedge \pi_{ij}^\sigma$$

$$\equiv 0 + \omega^{h+1} \wedge \sum_{j=1}^{h+1} \omega^j \wedge \pi_{h+1,j}^\sigma + \sum_{i=1}^{h+1} \omega^i \wedge \omega^{h+1} \wedge \pi_{i,h+1}^\sigma$$

$$= -\omega^{h+1} \wedge \left(\sum_{i=1}^h \omega^i \wedge (\pi_{i,h+1}^\sigma - \pi_{h+1,i}^\sigma) \right).$$

So modulo $\mathcal{P}^{(1)}$ and $\omega^{h+2}, \dots, \omega^n$,

$$\pi_{i,h+1}^\sigma \equiv \pi_{h+1,i}^\sigma + \sum_{j=1}^h l_{ij}^\sigma \omega^j$$

for certain smooth functions l_{ij}^σ with $l_{ij}^\sigma = l_{ji}^\sigma$. Consequently

$$\Pi_i^\sigma \equiv \sum_{j=1}^{h+1} \omega^j \wedge \pi_{ij}^\sigma \equiv \sum_{j=1}^{h} \omega^j \wedge \pi_{ij}^\sigma + \omega^{h+1} \wedge \pi_{h+1,i}^\sigma - \sum_{j=1}^{h} l_{ij}^\sigma \, \omega^j \wedge \omega^{h+1}$$

$$= \sum_{j=1}^{h} \omega^j \wedge (\pi_{ij}^\sigma - l_{ij}^\sigma \, \omega^{h+1}) + \omega^{h+1} \wedge \pi_{h+1,i}^\sigma \quad (\text{mod } \mathcal{P}^{(1)}, \omega^{h+2}, \dots, \omega^n).$$

From $\sum_{i,\sigma} A_{\sigma\alpha}^i \pi_{ij}^\sigma = 0$ a certain number of principal π_{ij}^σ are expressed as linear combinations of the remaining parametric ones. Replacing the latter by $\pi_{ij}^\sigma - l_{ij}^\sigma \, \omega^{h+1}$ we may assume that the corresponding l_{ij}^σ vanish. This replacement affects the principal π_{ij}^σ for $i, j = 1, \dots, h$—although they will still satisfy $\pi_{ij}^\sigma - \pi_{ji}^\sigma = 0$—and may add some multiples of ω^{h+1} to the $\pi_{h+1,i}^\sigma$. But because the latter only appear in wedge products with ω^{h+1}, the last fact is of no importance. The two-forms $\sum_{j=1}^{h} l_{ij}^\sigma \, \omega^j \wedge \omega^{h+1}$ will obviously satisfy the same linear relations as the Π_{ij}^σ. Therefore the vanishing of the parametric l_{ij}^σ forces *all* l_{ij}^σ to vanish, and so

$$\sum_{j=1}^{h+1} \omega^j \wedge \pi_{ij}^\sigma \equiv \Pi_i^\sigma \equiv \sum_{j=1}^{h} \omega^j \wedge \pi_{ij}^\sigma + \omega^{h+1} \wedge \pi_{h+1,i}^\sigma \quad (\text{mod } \mathcal{P}^{(1)}, \omega^{h+2}, \dots, \omega^n).$$

Thus it may indeed be supposed that $\pi_{i,h+1}^\sigma = \pi_{h+1,i}^\sigma$ for $i = 1, \dots, h+1$. □

As a consequence the structure equations of $\mathcal{P}^{(1)}$ are

$$\begin{cases} d\theta^\sigma \equiv 0, \\ d\varpi_i^\sigma \equiv \sum_{j=1}^{n} \omega^j \wedge \pi_{ij}^\sigma \end{cases} \quad (\text{mod } \mathcal{P}^{(1)}),$$

where the one-forms π_{ij}^σ are symmetric with respect to the lower indices, and satisfy the same linear relations as the π_i^σ and $t_{ij}^\sigma = t_{ji}^\sigma$ do.

The most general n-dimensional involution of $\mathcal{P}^{(1)}$ on which $\omega^1 \wedge \cdots \wedge \omega^n \neq 0$ is defined by adding pfaffian equations

$$\pi_{ij}^\sigma - \sum_{k=1}^{n} t_{ijk}^\sigma \, \omega^k = 0$$

to $\mathcal{P}^{(1)}$, with the t_{ijk}^σ being smooth functions on $\mathbb{R}^{r_1}$ such that the $d\varpi_i^\sigma$ vanish modulo $\mathcal{P}^{(1)}$ because of these equations, i.e.,

$$\sum_{j,k=1}^{n} \omega^j \wedge t_{ijk}^\sigma \, \omega^k = 0.$$

But then $t_{ijk}^\sigma = t_{ikj}^\sigma$. Furthermore, because $\pi_{ij}^\sigma = \pi_{ji}^\sigma$, also $t_{ijk}^\sigma = t_{jik}^\sigma$, so that the t_{ijk}^σ are symmetric with respect to *all* the lower indices.

Because the t^σ_{ijk} for each k satisfy the same linear relations as the t^σ_{ij} do, it is for each principal t^σ_{ij} and each k possible to solve the corresponding t^σ_{ijk} as function of x and the parametric t^ρ_{ab}, t^ρ_{abc}. Hence

$$t^\sigma_{ij} \text{ principal} \implies t^\sigma_{ijk} \text{ principal for } k = 1,\ldots,n.$$

If we make successive prolongations in this way there will for each $\sigma \in \{1,\ldots,s\}$ arise one-forms $\pi^\sigma_{i_1}$, $\pi^\sigma_{i_1 i_2}$, ... and coefficients $t^\sigma_{i_1 i_2}$, $t^\sigma_{i_1 i_2 i_3}$, ... with the following properties:

- they are symmetric with respect to the lower indices;
- for each i_k, the $t^\sigma_{i_1 \ldots i_k}$ satisfy the same linear relations as the $t^\sigma_{i_1 \ldots i_{k-1}}$ do, and analogously for the one-forms $\pi^\sigma_{i_1 \ldots i_k}$;
- if $t^\sigma_{i_1 \ldots i_k}$ is principal, so are $t^\sigma_{i_1 \ldots i_k i_{k+1} \ldots i_{k+l}}$ for all $i_{k+1}, \ldots, i_{k+l}$.

Let us now fix a σ. We start by solving the principal coefficients $\hat{t}^\sigma_{ij} = \hat{t}^\sigma_{ji}$ as functions of x and the parametric $\bar{t}^\rho_{kl}$ for $\rho = 1,\ldots,s$. To each principal $\hat{t}^\sigma_{ij}$ we associate the monomial $x^i x^j$. For $k_1, k_2, \ldots$, $\hat{t}^\sigma_{ijk_1 k_2 \ldots}$ is also principal, and to this coefficient corresponds the monomial $x^i x^j x^{k_1} x^{k_2} \ldots$. In this way *an ideal* $\mathfrak{M}^\sigma_1$ *of monomials* arises, generated by those monomials $x^i x^j$ which are associated to the principal coefficients $\hat{t}^\sigma_{ij}$.

Let $\hat{t}^\rho_{ij}$ and $\hat{t}^\rho_{ik}$ be two principal coefficients. Then $\hat{t}^\rho_{ijk}$ and $\hat{t}^\rho_{ikj}$ are also principal, and can thus be expressed by means of the x and the parametric $\bar{t}$:

$$\begin{cases} \hat{t}^\rho_{ijk} = F^\rho_{ijk}(x,\bar{t}), \\ \hat{t}^\rho_{ikj} = F^\rho_{ikj}(x,\bar{t}). \end{cases}$$

But $\hat{t}^\rho_{ijk} = \hat{t}^\rho_{ikj}$, and therefore

$$F^\rho_{ijk}(x,\bar{t}) = F^\rho_{ikj}(x,\bar{t}). \tag{†}$$

As in analogous cases treated before, we assume that the implicit function theorem can be applied to the equations (†), and find three possibilities:

(i) The relations (†) are identities.

(ii) It is possible to solve certain $\bar{t}$ from (†), thereby turning them into *principal coefficients* $\hat{t}$.

(iii) The equations (†) give rise to relations between the x-variables, which define a subset M of $\mathbb{R}^r$. Let us assume M to be a manifold locally. If $\omega^1 \wedge \cdots \wedge \omega^n|_M = 0$ there is no integral manifold of $\mathcal{P}$ on which $\omega^1 \wedge \cdots \wedge \omega^n \neq 0$. If $\omega^1 \wedge \cdots \wedge \omega^n|_M \neq 0$, $\mathbb{R}^r$ is replaced by M and $\mathcal{P}$ by $\mathcal{P}_{|M}$, and we start all over again.

In the second case we obtain integrability conditions, and it is these that cause inequality in the test for involutivity.

Note that similarly there may arise linear relations involving the π^σ_{ijk} which are not valid for the corresponding π^σ_{ij}.

Excluding the third case, let $\mathfrak{M}^\sigma_2$ be the monomial ideal generated by those monomials $x^i x^j$ and $x^i x^j x^k$ that are associated to *principal coefficients* $\hat{t}^\sigma_{ij}$ and $\hat{t}^\sigma_{ijk}$. If by the second case at least one parametric coefficient is changed into a principal one,

$$\mathfrak{M}^\sigma_1 \subsetneq \mathfrak{M}^\sigma_2,$$

but otherwise

$$\mathfrak{M}^\sigma_1 = \mathfrak{M}^\sigma_2.$$

Continuing in this way and supposing that no relation involving only the x-variables ever arises, we obtain an ascending chain of monomial ideals

$$\mathfrak{M}^\sigma_1 \subseteq \mathfrak{M}^\sigma_2 \subseteq \mathfrak{M}^\sigma_3 \subseteq \cdots.$$

According to Hilbert's basis theorem (which by the way is much easier to prove for monomial ideals than in the general case) there is an index $h(\sigma)$ such that

$$\mathfrak{M}^\sigma_h = \mathfrak{M}^\sigma_{h(\sigma)} \quad \text{for } h \geq h(\sigma).$$

Having determined $h(\sigma)$ for each $\sigma \in \{1, 2, \ldots, s\}$, we set

$$h := \max_\sigma h(\sigma).$$

Then

$$\mathfrak{M}^\sigma_k = \mathfrak{M}^\sigma_h \quad \text{for all } k \geq h \text{ and for } \sigma = 1, 2, \ldots, s.$$

Consequently, after h prolongations no more integrability conditions emerge, and therefore the number of parametric coefficients has attained its maximal value—so that $\mathcal{P}^{(h)}$ is *involutive*.

5

Drach's classification, second order PDEs in one dependent variable, and Monge characteristics

Suppose that $\mathcal{P}$ is a pfaffian system which is not involutive, but whose m^{th} prolongation $\mathcal{P}^{(m)}$ is. Then Cartan's local existence theorem gives all the integral manifolds of $\mathcal{P}$ which arise from *regular* involutions of $\mathcal{P}^{(m)}$—but there may be other, singular, integral manifolds that are not reached in this way. And how to obtain these?

In 1957 Kuranishi proved his prolongation theorem which roughly says that if $\mathcal{N}$ is an integral manifold (possibly singular) of a pfaffian system $\mathcal{P}$, then there is a prolongation $\mathcal{P}^{(m)}$ such that the corresponding prolongation $\mathcal{N}^{(m)}$ of $\mathcal{N}$ is got as a *regular* integral manifold of $\mathcal{P}^{(m)}$ by means of Cartan's existence theorem.

Ironically, in the very same year Hans Lewy found his celebrated counterexample to the Cauchy–Kowalewski theorem in the C^∞ category—which makes the question of local solvability for C^∞ PDE systems wide open.

But in the analytic category Kuranishi's prolongation theorem combined with Cartan's existence theorem in principle yields *all* local solutions—in spite of possible singular vector fields and singular involutions.

There is however another angle from which to regard singular vector fields. To wit, let the structure equations for a vector field system $\mathcal{V} = (X_1, \ldots, X_q)$ be

$$[X_i, X_j] \equiv \sum_{k=1}^{p} c_{ij}^k Z_k \quad (\text{mod } \mathcal{V}).$$

Then it is natural to look for a basis of $\mathcal{V}$ making as many as possible of the structure functions c_{ij}^k vanish—which means that we should look for vector fields in $\mathcal{V}$ commuting modulo $\mathcal{V}$ with a maximal number of vector fields in $\mathcal{V}$. But then these are as singular as possible!

So from this point of view the singular vector fields (and singular involutions) seem to be the good guys after all.

Anyway, it is clearly of interest to investigate *to what extent singular vector fields do occur* in vector field systems appearing in real life.

In order to do this we use a provisional classification of PDE systems due to Jules Drach (see [Drach 1897]), which says that any PDE system is equivalent to either a first or a second order PDE system in *one* dependent variable.

A first order PDE system in one dependent variable is equivalent to a submanifold S of $J^1(\mathbb{R}^n, \mathbb{R})$, equipped with the contact form

$$\theta = \left(dz - \sum_{i=1}^{n} p_i \, dx^i \right) \Big|_S.$$

The *class* of a pfaffian equation $\theta = 0$ is by definition the smallest number of variables needed to express it. It can be shown that the class of a pfaffian equation is always an odd integer, and that a pfaffian equation of class $2s + 1$ can be reduced to the canonical form

$$dw - \sum_{j=1}^{s} q_j \, dy^j = 0.$$

In this way the problem of integrating a first order PDE system in one dependent variable is reduced to that of finding the generalized 1-graphs of the jet bundle $J^1(\mathbb{R}^s, \mathbb{R})$—which is easy (see section **6.5**).

Thus the real problem commences with second order PDE systems in one dependent variable. Here we start modestly by looking at a *single* second order PDE, and especially investigate the occurrence of singular vector fields.

The results obtained motivate the introduction of the concept of *Monge characteristic subsystems*, which will be one of the main issues in the following.

5.1 The classification of Drach

Before examining PDE systems more closely, we want to put them in some sort of order.

As a first step, any PDE system can be written as a *first order* PDE system by considering the derivatives up to next highest order as new variables. A somewhat more sophisticated way of achieving this is to use the associated pfaffian system:

$$\sum_{k=1}^{r} a_k^i(x^1,\ldots,x^r)\,dx^k = 0, \qquad \text{for } i = 1,\ldots,s.$$

Supposing that the original independent variables for the PDE system are $x^1, \ldots, x^n$, we consider

$$z^j := x^{n+j}, \qquad j = 1,\ldots,r-n =: m,$$

as functions of these, whence

$$\sum_{k=1}^{n} a_k^i(x,z)\,dx^k + \sum_{l=1}^{m} a_{n+l}^i(x,z)\left(\sum_{k=1}^{n} \frac{\partial z^l}{\partial x^k}\,dx^k\right) = 0,$$

or

$$a_k^i(x,z) + \sum_{l=1}^{m} a_{n+l}^i(x,z)\,\frac{\partial z^l}{\partial x^k} = 0 \quad \text{for } k = 1,\ldots,n \text{ and } i = 1,\ldots,s.$$

In this way any PDE system is equivalent to a quasi-linear *first order* PDE system.

Having a first order PDE system

$$F^a(x^1,\ldots,x^n;z^1,\ldots,z^m;\ldots,\partial z^j/\partial x^k,\ldots) = 0, \qquad a = 1,\ldots,A, \qquad (S)$$

with $m > 1$ unknowns, we next want to reduce this number to 1.

To this end, consider the second order PDE system $\mathcal{D}' \subset J^2(\mathbb{R}_x^{n+m}, \mathbb{R}_z)$ given by

$$\frac{\partial^2 z}{\partial x^{n+s}\partial x^{n+t}} = 0 \quad \text{for } s,t = 1,\ldots,m. \tag{$\mathcal{D}'$}$$

The general solution of $\mathcal{D}'$ is

$$z = \sum_{j=1}^{m} z^j(x^1,\ldots,x^n)\,x^{n+j} + z^0(x^1,\ldots,x^n),$$

where $z^0(x^1,\ldots,x^n), \ldots, z^m(x^1,\ldots,x^n)$ are arbitrary functions of n variables; moreover

$$z^j(x^1,\ldots,x^n) = \frac{\partial z}{\partial x^{n+j}} \quad \text{for } j = 1,\ldots,m.$$

Next we require $z^j = \partial z/\partial x^{n+j}$ to be solutions of S, i.e.,

$$F^a\left(x^1,\ldots,x^n;\frac{\partial z}{\partial x^{n+1}},\ldots,\frac{\partial z}{\partial x^{n+m}};\ldots,\frac{\partial^2 z}{\partial x^k \partial x^{n+j}},\ldots\right) = 0. \qquad (\mathcal{D}'')$$

Definition. The *Drach system* $\mathcal{D}$ associated to the first order PDE system S is $\mathcal{D}' + \mathcal{D}'' \subset J^2(\mathbb{R}^{n+m}_x, \mathbb{R}_z)$, i.e.,

$$\begin{cases} \partial^2 z/\partial x^{n+s} x^{n+t} = 0 & \text{for } s,t = 1,\ldots,m, \\ F^a\left(x^1,\ldots,x^n;\partial z/\partial x^{n+1},\ldots,\partial z/\partial x^{n+m};\ldots,\partial^2 z/\partial x^k \partial x^{n+j},\ldots\right) & (\mathcal{D}) \\ \quad = 0 & \text{for } a = 1,\ldots,A, \ j = 1,\ldots m \text{ and } k = 1,\ldots,n. \end{cases}$$

By construction the general solution of $\mathcal{D}$ is

$$z = \sum_{j=1}^{m} z^j(x^1,\ldots,x^n)\, x^{n+j} + z^0(x^1,\ldots,x^n),$$

where $(z^1(x^1,\ldots,x^n),\ldots,z^m(x^1,\ldots,x^n))$ constitutes the general solution of S, and $z^0(x^1,\ldots,x^n)$ is arbitrary.

And with this we have demonstrated the **theorem of Drach**.

Theorem 5.1.1. *Solving S is equivalent to solving $\mathcal{D}$.* $\qquad\square$

In this way any PDE system is equivalent to either a first or a second order PDE system in a single unknown. Of course the reduction performed is utterly trivial, but it is nonetheless a convenient triviality.

5.2 Second order PDEs in one unknown and their singular vector fields

So according to Drach, second order PDE systems in one unknown are of a special interest. Let us here look at a *single* second order PDE

$$f\left(x^1,\ldots,x^n;z;\ldots,\frac{\partial z}{\partial x^i},\ldots;\ldots,\frac{\partial^2 z}{\partial x^i \partial x^j},\ldots\right) = 0, \qquad (S)$$

and especially investigate the occurrence of singular vector fields, following [Vessiot 1930].

As usual, S is considered as a submanifold of a jet bundle—which in this case is $J^2(\mathbb{R}^n,\mathbb{R})$:

$$f(x^i, z, p_i, p_{ij}) = 0. \qquad (S)$$

The contact pfaffian system $\mathcal{P}_0$ of $J^2(\mathbb{R}^2, \mathbb{R})$, is generated by

$$
\begin{cases}
dz - \sum_{i=1}^{n} p_i \, dx^i \\
\text{and} \\
dp_i - \sum_{j=1}^{n} p_{ij} \, dx^j & \text{for } i = 1, \ldots, n.
\end{cases}
$$

According to the general recipe this should be restricted to S to give $\mathcal{P}_S := \mathcal{P}_{0|S}$, whereupon we are to look for maximal complete subsystems of $\mathcal{P}_S^\perp$. However, $\mathcal{P}_S$ is in general much more complicated than $\mathcal{P}_0$, so we prefer to keep the latter by the following trick: replace S by the foliation

$$
f(x^i, z, p_i, p_{ij}) = c \qquad (c \in \mathbb{R}) \tag{fol}
$$

of $J^2(\mathbb{R}^n, \mathbb{R})$, with the 0-leaf being equal to S; note that this foliation is equivalently defined by the pfaffian equation

$$
df = 0.
$$

Consider now the pfaffian system $\mathcal{P} = (\mathcal{P}_0, df)$ on $J^2(\mathbb{R}^n, \mathbb{R})$, and its dual vector field system $\mathcal{V} = \mathcal{P}^\perp$. If $\mathcal{W}$ is a complete subsystem of $\mathcal{V}$, $\mathcal{W}$ defines a subfoliation of *fol*, and gives in particular a foliation of the 0-leaf S, consisting of integral manifolds of $\mathcal{P}_S$.

Therefore we might as well study the vector field system $\mathcal{V}$ on $J^2(\mathbb{R}^n, \mathbb{R})$. The dual $\mathcal{V}_0$ of $\mathcal{P}_0$ consists of those vector fields

$$
X = \sum_{i=1}^{n} a^i \frac{\partial}{\partial x^i} + b \frac{\partial}{\partial z} + \sum_{i=1}^{n} b^i \frac{\partial}{\partial p_i} + \sum_{i,j=1}^{n} c^{ij} \frac{\partial}{\partial p_{ij}}
$$

which satisfy

$$
\begin{cases}
0 = \langle dz - \sum_{k=1}^{n} p_k \, dx^k, X \rangle = b - \sum_{k=1}^{n} a^k \, p_k, \\
0 = \langle dp_k - \sum_{l=1}^{n} p_{kl} \, dx^l, X \rangle = b^k - \sum_{l=1}^{n} p_{kl} \, a^l,
\end{cases}
$$

i.e.,

$$
X = \sum_{i=1}^{n} a^i \left(\frac{\partial}{\partial x^i} + p_i \frac{\partial}{\partial z} + \sum_{j=1}^{n} p_{ij} \frac{\partial}{\partial p_j} \right) + \sum_{i,j=1}^{n} c^{ij} \frac{\partial}{\partial p_{ij}}.
$$

Hence $\mathcal{V}_0$ is generated by the $m := n + \frac{1}{2}n(n+1)$ vector fields

$$
X_i := \frac{\partial}{\partial x^i} + p_i \frac{\partial}{\partial z} + \sum_{j=1}^{n} p_{ij} \frac{\partial}{\partial p_j} \qquad \text{for } i = 1, \ldots, n,
$$

and

$$P_{ij} := \frac{\partial}{\partial p_{ij}} \quad \text{for } i,j = 1,\ldots,n.$$

Note that $P_{ij} = P_{ji}$, so that precisely $\frac{1}{2}n(n+1)$ of the P_{ij} are independent. The commutation relations are very simple:

$$[P_{ij}, X_i] = \frac{\partial}{\partial p_j}, \quad \text{while all other brackets vanish.}$$

The contact vector field system $\mathcal{V}$ associated to the foliation *fol* is then

$$\mathcal{V} = \{X \in \mathcal{V}_0 \mid \langle df, X \rangle = 0\},$$

so that dim $\mathcal{V} = m - 1$.

Let

$$A = \sum_{i=1}^{n} a^i X_i + \sum_{i,j=1}^{n} a^{ij} P_{ij} \quad \text{and} \quad B = \sum_{i=1}^{n} b^i X_i + \sum_{i,j=1}^{n} b^{ij} P_{ij}$$

be vector fields in $\mathcal{V}_0$ with smooth coefficients that satisfy $a^{ij} = a^{ji}$ and $b^{ij} = b^{ji}$, but are arbitrary otherwise. Then

$$[A, B] \equiv -\sum_{i,j=1}^{n} (a^i b^{ij} - b^i a^{ij}) \frac{\partial}{\partial p_j} \quad (\text{mod } \mathcal{V}_0),$$

so that

$$[A, B] \equiv 0 \quad (\text{mod } \mathcal{V}_0) \iff \sum_{i=1}^{n} (a^i b^{ij} - b^i a^{ij}) = 0 \quad \text{for } j = 1,\ldots,n.$$

Specialize now to the case $A, B \in \mathcal{V}$. Since $Af = Bf = 0$, we also have $[A, B]f = 0$. Therefore, if

$$[A, B] \equiv 0 \quad (\text{mod } \mathcal{V}_0),$$

$[A, B]$ will automatically belong to $\mathcal{V} \subsetneq \mathcal{V}_0$.
    So

$$[A, B] \equiv 0 \quad (\text{mod } \mathcal{V}_0) \iff [A, B] \equiv 0 \quad (\text{mod } \mathcal{V}).$$

Fix a nonzero $A \in \mathcal{V}$—assume for instance that $a^1 \neq 0$—and let

$$B = \sum_{i=1}^{n} b^i X_i + \sum_{i,j=1}^{n} b^{ij} P_{ij} \quad (\text{with } b^{ij} = b^{ji})$$

belong to $\mathcal{V}$, i.e.,

$$\sum_{i=1}^{n} b^i X_i f + \sum_{i,j=1}^{n} b^{ij} P_{ij} f = 0. \tag{1}$$

Then $[A, B] \equiv 0 \pmod{\mathcal{V}_0}$ if the coefficients of B satisfy the following n independent equations:

$$b^{1j} = \sum_{i=1}^{n} \frac{a^{ij}}{a^1} b^i - \sum_{i=2}^{n} \frac{a^i}{a^1} b^{ij} \quad \text{for } j = 1, \ldots, n. \tag{2}$$

So if (1) is independent of (2), the requirement $[A, B] \equiv 0 \pmod{\mathcal{V}}$ gives n independent relations for the coefficients of B—this is the regular case. The singular case occurs if (1) is a consequence (2), so that $[A, B] \equiv 0 \pmod{\mathcal{V}}$ only yields $n - 1$ independent conditions for the b^i and b^{ij}. Thus A is a singular vector field in $\mathcal{V}$ precisely when there are factors c^j such that

$$\sum_{i=1}^{n} b^i X_i f + \sum_{i,j=1}^{n} b^{ij} P_{ij} f = \sum_{j=1}^{n} c^j \left(\sum_{i=1}^{n} (a^i b^{ij} - b^i a^{ij}) \right)$$

for all b^i and b^{ij}, or

$$\sum_{i=1}^{n} \left(X_i f + \sum_{j=1}^{n} a^{ij} c^j \right) b^i + \sum_{i,j=1}^{n} P_{ij} f \cdot b^{ij} - \sum_{i,j=1}^{n} a^i c^j b^{ij} = 0.$$

Thus the coefficients a^{ij} are to satisfy

$$\sum_{j=1}^{n} a^{ij} c^j + X_i f = 0 \quad \text{for } i = 1, \ldots, n. \tag{3}$$

With $b^{ij} = b^{ji}$, the b^{ij} may be identified with second order monomials in n unknowns $u^1, \ldots, u^n$:

$$b^{ij} \longleftrightarrow u^i u^j.$$

Then the a^i must satisfy

$$\sum_{i,j=1}^{n} P_{ij} f \cdot u^i u^j = \left(\sum_{i=1}^{n} a^i u^i \right) \left(\sum_{j=1}^{n} c^j u^j \right). \tag{4}$$

As a consequence we have the following result.

Theorem 5.2.1. *The contact vector field system $\mathcal{V}$ associated to the foliation*

$$f(x^i, z, p_i, p_{ij}) = c, \qquad c \in \mathbb{R},$$

of $J^2(\mathbb{R}^n, \mathbb{R})$ admits singular vector fields only if the quadratic form

$$\mathcal{Q}_f := \sum_{i,j=1}^{n} \frac{\partial f}{\partial p_{ij}} u^i u^j$$

is a product of two linear forms—which rarely is the case when $n > 2$. □

Remark. To remain in the real domain we should also require the linear forms to be real—for instance, $(u^1)^2 + (u^2)^2$ has the complex factorization $(u^1 + iu^2)(u^1 - iu^2)$, but no real one. However, in the following it will turn out that the theory will be much simpler if we accept *complex variables*—which we also will do eventually.

If we assume $\mathcal{V}$ to admit singular vector fields, there are two different cases to consider.

The hyperbolic case: *the linear forms are different.* Then

$$\sum_{i,j=1}^{n} P_{ij}f \cdot u^i u^j = \left(\sum_{i=1}^{n} \alpha_i u^i \right) \left(\sum_{j=1}^{n} \beta_j u^j \right),$$

with different factors in the right hand side. Note that α_i and β_j are determined up to a factor only—since $c\alpha_i$, $c^{-1}\beta_j$ would work as well for an arbitrary nonzero c.

Disregarding this c, there are two choices for the a^i: either $a^i = \alpha_i$, or $a^i = \beta_i$. With $a^i = \alpha_i$ we have $c^j = \beta_j$, and then (3) shows that

$$\sum_{j=1}^{n} a^{ij} \beta_j = -X_i f \quad \text{for } i = 1, \ldots, n. \tag{5}$$

If $\beta_1 \neq 0$ for instance,

$$a^{i1} = -\frac{1}{\beta_1} \left(X_i f + \sum_{j=2}^{n} a^{ij} \beta_j \right) \quad \text{for } i = 1, \ldots, n,$$

while the a^{ij} for which $j \geq i > 1$ are arbitrary. The corresponding vector fields $A = \sum_{i=1}^{n} \alpha^i X_i + \sum_{i,j=1}^{n} a^{ij} P_{ij}$ belong not only to $\mathcal{V}_0$, but also to $\mathcal{V}$, since

$$Af = \sum_{i=1}^{n} \alpha^i \, X_i f + \sum_{i,j=1}^{n} a^{ij} \, P_{ij} f$$

$$= \sum_{i=1}^{n} \alpha_i \left(X_i f + \sum_{j=1}^{n} a^{ij} \beta_j \right) \quad \text{(using } P_{ij} f + P_{ji} f = \alpha_i \beta_j + \alpha_j \beta_i \text{)}$$

$$= 0 \quad \text{(because of (5))}.$$

In this way we obtain a *singular sub vector field system* $\mathcal{F}$ of $\mathcal{V}$ with the dimension $\frac{1}{2}n(n-1)+1$, where the last 1 is due to the arbitrary factor c.

The choice $a^i = \beta_i$ analogously gives another singular vector field system $\mathcal{G}$ of $\mathcal{V}$ of the dimension $\frac{1}{2}n(n-1)+1$.

The parabolic case: *the linear forms coincide.* Then

$$\sum_{i,j=1}^{n} P_{ij} f \cdot u^i u^j = \left(\sum_{i=1}^{n} \alpha_i u^i \right)^2,$$

so that $a^i = c\alpha_i$, $c^i = c^{-1}\alpha_i$ for an arbitrary $c \neq 0$ and $i = 1,\ldots,n$. Thereupon n of the a^{ij} are solved from (3).

So in this case there is one (but 'double') singular vector field system.

The independence condition $dx^1 \wedge \cdots \wedge dx^n \neq 0$ shows that the wanted complete subsystems of $\mathcal{V}$ are generated by the n vector fields

$$U_i = X_i + \sum_{k,l=1}^{n} \pi_i^{kl} \, P_{kl}, \qquad i = 1,\ldots,n,$$

where the smooth coefficients π_i^{kl} may be supposed to be symmetric with respect to the upper indices, and furthermore satisfy the conditions resulting from

- $[U_i, U_j] \equiv 0 \pmod{\mathcal{V}_0}$,
- $U_i f = 0$.

The first shows that

$$0 = \left[X_i + \sum_{k,l} \pi_i^{kl} \, P_{kl}, \; X_j + \sum_{s,t} \pi_j^{st} \, P_{st} \right]$$

$$= \sum_{s} \pi_j^{si} \, P_{si} - \sum_{k} \pi_i^{kj} \, P_{kj} = \sum_{k=1}^{n} (\pi_j^{ki} - \pi_i^{kj}) \, P_{kj},$$

so that

$$\pi_j^{ki} = \pi_i^{kj},$$

and hence the π are symmetric with respect to *all* indices. The second says that

$$X_i f + \sum_{k,l=1}^{n} \pi_i^{kl} P_{kl} f = 0 \quad \text{for } i = 1, \dots, n. \tag{6}$$

Theorem 5.2.2. *The general involution* $\mathcal{I}_n = (U_1, \dots, U_n)$ *contains one vector field from each singular subsystem.*

Proof. Let us consider the singular subsystem $\mathcal{F}$ of the hyperbolic case for instance. Using notations as above

$$A \in \mathcal{F} \iff \begin{cases} A = \sum_{i=1}^{n} \alpha_i X_i + \sum_{k,l=1}^{n} a^{kl} P_{kl}, \text{ where the } a^{kl} \\ \text{satisfy } X_k f + \sum_{l=1}^{n} a^{kl} \beta_l = 0 \text{ for } k = 1, \dots, n. \end{cases}$$

Therefore there is but one possible choice for $V \in \mathcal{I}_n \cap \mathcal{F}$:

$$V := \sum_{i=1}^{n} \alpha_i U_i = \sum_{i=1}^{n} \alpha_i X_i + \sum_{k,l=1}^{n} \left(\sum_{i=1}^{n} \alpha_i \pi_i^{kl} \right) P_{kl}.$$

This belongs to $\mathcal{F}$ if

$$X_k f + \sum_{l=1}^{n} \left(\sum_{i=1}^{n} \alpha_i \pi_i^{kl} \right) \beta^l = 0 \quad \text{for } k = 1, \dots, n.$$

With $\alpha_i \beta_l + \alpha_l \beta_i = P_{il} f + P_{li} f$, this is equivalent to

$$X_k f + \sum_{i,l=1}^{n} \pi_i^{kl} P_{il} f = 0 \quad \text{for } k = 1, \dots, n,$$

—which is true because of (6) and the symmetry of the indices. $\square$

So in the hyperbolic case the general n-dimensional involution of $\mathcal{V}$ can be expressed as

$$\mathcal{I}_n = (U_1, \dots, U_{n-2}, F, G),$$

where $F \in \mathcal{F}$ and $G \in \mathcal{G}$ are singular vector fields. This indicates that the singular vector fields should be particularly helpful when $n = 2$—a case which is considered in detail in chapter **12**.

5.3 Monge characteristic subsystems

Let $V = (X_1, \ldots, X_q)$ be a vector field system, and let $U = (Y_1, \ldots, Y_m)$ be an m-dimensional subsystem of V. Set

$$\dim([U, V] \bmod V) := \text{the number of vector fields } [Y_i, X_j]$$
$$\text{that are independent} \bmod V,$$

and let

$$d_m := \max_{\dim U = m} \dim([U, V] \bmod V).$$

Definition. An m-dimensional subsystem U of V is said to be *regular* if $\dim([U, V] \bmod V) = d_m$, and is called *singular* otherwise.

When $m = 1$ this reduces to the concepts of regular and singular vector fields.

The considerations of the preceding section motivate the following definition.

Definition. A subsystem M of a vector field system V is *Monge characteristic* if

(i) M is singular in the above sense,
(ii) $M \cap I \neq 0$ for any maximal involution I of V,
(iii) $M \cap W$ is *complete* for any maximal complete subsystem W of V.

The integral manifolds of the complete system $M \cap W$ are called *Monge characteristics*.

For the case treated in the previous section, $\dim(W \cap I_n) = 1$, and hence $\dim(M \cap W) = 1$, so that the third requirement is automatically satisfied. But this need not be true if $\dim(M \cap W) > 1$.

Let V be an integrable vector field system with the Cauchy characteristic subsystem C. Then

$$\dim([C, V] \bmod V) = 0,$$

so C is maximally singular. Moreover, for any maximal complete subsystem W of V,

$$C \cap W = C,$$

which is always complete. So C is a Monge system.

We have earlier seen that the possible Cauchy characteristic vector fields are very useful for the integration of $\mathcal{V}$. One of the main points of this monograph will be to show that Monge characteristic vector field systems also are of great interest for the integration problem.

6

Integration of vector field systems $\mathcal{V}$
satisfying $\dim \mathcal{V}' = \dim \mathcal{V} + 1$

Let $S \subset J^1(\mathbb{R}^n_x, \mathbb{R}_z)$ be a first order PDE system in one unknown. Suppose that S has the dimension r as a submanifold of $J^1(\mathbb{R}^n_x, \mathbb{R}_z)$, and let

$$\theta := \left(dz - \sum_{i=1}^{n} p_i \, dx^i \right)\Big|_S.$$

Then the generalized solutions of S are by definition the n-dimensional integral manifolds of θ on S.

If $\mathcal{V} = \theta^{\perp}$ is the dual vector field system, $\dim \mathcal{V} = r - 1$. Hence there are only two possibilities for $\dim \mathcal{V}'$ (which by our usual assumptions has the same dimension at each point of S):

- either $\dim \mathcal{V}' = \dim \mathcal{V}$, and $\mathcal{V}$ is immediately integrated by means of the Frobenius theorem,
- or $\dim \mathcal{V}' = \dim \mathcal{V} + 1 = r$, in which case the integration problem coincides with the classical problem of Pfaff.

Following [Vessiot 1928] we will in this chapter solve the problem of determining the maximal complete subsystems of a vector field system $\mathcal{V} = (X_1, \ldots, X_q)$ on $\mathbb{R}^r$ satisfying $\dim \mathcal{V}' = \dim \mathcal{V} + 1$, without the latter necessarily being equal to r.

A surprising consequence of this study will be that the theory of first order PDEs in one unknown essentially is equivalent to the theory of contact transformations!

6.1 Maximal involutions

So let $\mathcal{V}$ be a q-dimensional vector field system on $\mathbb{R}^r$ with $\dim \mathcal{V}' = q + 1$; say that

$$\mathcal{V} = (X_1, \ldots, X_q) \quad \text{and} \quad \mathcal{V}' = (X_1, \ldots, X_q; Z).$$

Then there are structure functions $c_{ij} = -c_{ji}$ such that

$$[X_i, X_j] \equiv c_{ij} Z \pmod{\mathcal{V}} \quad \text{for } i, j = 1, \ldots, q.$$

Let $U = \sum_{i=1}^q u^i X_i$ and $V = \sum_{i=1}^q v^i X_i$ be two vector fields in $\mathcal{V}$, and set $\mathbf{u} = (u^1, \ldots, u^q)^T$, $\mathbf{v} = (v^1, \ldots, v^q)^T$, where T denotes transpose. Then

$$[U, V] \equiv \Phi(\mathbf{u}; \mathbf{v}) Z \pmod{V},$$

with

$$\Phi(\mathbf{u}; \mathbf{v}) := \sum_{i,j=1}^q c_{ij} u^i v^j = \sum_{1 \le i < j \le q} c_{ij} (u^i v^j - u^j v^i)$$

being an alternating linear form. Or using matrix formulation,

$$\Phi(\mathbf{u}; \mathbf{v}) = \mathbf{u}^T C \mathbf{v},$$

where C is a skew-symmetric matrix, which consequently has an even rank—say

$$\text{rank } C = 2s.$$

As usual we assume this rank to be equal at all points of $\mathbb{R}^r$.

The normal form of such a matrix has s ones and $q - s$ zeros just above the main diagonal, $1, \ldots, 1, 0, \ldots, 0$, and correspondingly $-1, \ldots, -1, 0, \ldots, 0$ just under, with all the other entries being equal to 0.

For any vector field system it is desirable to choose a basis for its derived system such that the set of structure functions becomes as simple as possible—and for our $\mathcal{V}$ this means that the matrix C should attain this normal form.

A trivial but useful observation is that if

$$Z \equiv \rho \hat{Z} \pmod{\mathcal{V}} \qquad \text{with } \rho \ne 0,$$

then

$$[X_i, X_j] \equiv c_{ij} Z \equiv \rho c_{ij} \hat{Z} \pmod{V},$$

so that

$$Z \mapsto \hat{Z} \implies c_{ij} \mapsto \hat{c}_{ij} = \rho \, c_{ij}.$$

Let us next define a new basis $\{X'_1, \ldots, X'_q\}$ for $\mathcal{V}$ by means of a nonsingular $q \times q$ matrix

$$M = \begin{pmatrix} m_1^1 & \cdots & m_1^q \\ \vdots & \ddots & \vdots \\ m_q^1 & \cdots & m_q^q \end{pmatrix},$$

by setting

$$X'_k := \sum_{i=1}^{q} m^i_k X_i \quad \text{for } k = 1, \ldots, q.$$

Expressing U in the new basis,

$$U = \sum_{k=1}^{q} u'^k X'_k = \sum_{i,k=1}^{q} u'^k m^i_k X_i,$$

so that

$$u^i = \sum_{k=1}^{q} m^i_k u'^k, \quad \text{or} \quad \mathbf{u} = \mathsf{M}^T \mathbf{u}'.$$

Then

$$\Phi(\mathbf{u}; \mathbf{v}) = \mathbf{u}^T \mathsf{C} \mathbf{v} = \mathbf{u}'^T \mathsf{MCM}^T \mathbf{v}' = \mathbf{u}'^T \mathsf{C}' \mathbf{v}'$$

with

$$\mathsf{C}' = \mathsf{MCM}^T.$$

Theorem 6.1.1. *It is possible to choose the transformation matrix* M *such that*

$$c'_{1,2} = -c'_{2,1} = c'_{3,4} = -c'_{4,3} = \cdots = c'_{2s-1,2s} = -c'_{2s,2s-1} = 1,$$

with all other $c'_{i,j}$ *vanishing. Let*

$$A_1 := X'_1, \ A_2 := X'_3, \ldots, A_s := X'_{2s-1},$$
$$B_1 := X'_2, \ B_2 := X'_4, \ldots, B_s := X'_{2s},$$
$$C_1 := X'_{2s+1}, \ldots, C_{q-2s} := X'_q,$$

so that $\mathcal{V} = (A_1, \ldots, A_s; B_1, \ldots, B_s; C_1, \ldots C_{q-2s})$. *Then*

$$[A_i, B_i] = Z \pmod{V} \quad \text{for } i = 1, \ldots, s,$$

while all other brackets involving these generators vanish modulo $\mathcal{V}$. *A basis for* $\mathcal{V}$ *with these simple commutation relations is said to be* **canonical**.

Proof. If rank $\mathsf{C} = 0$, all c_{ij} vanish. Otherwise it may be assumed that $c_{12} \neq 0$ and even—exchanging the indices 1 and 2 if necessary—that $c_{12} > 0$. Because

$$\Phi(\mathbf{u}; \mathbf{v}) = \sum_{i,j=1}^{q} c_{ij} u^i v^j,$$

we have

$$\frac{\partial\Phi}{\partial u^i} = \sum_{j=1}^q c_{ij}v^j \quad \text{and} \quad \frac{\partial\Phi}{\partial v^i} = -\sum_{j=1}^q c_{ij}u^j.$$

Let

$$\Psi(\mathbf{u};\mathbf{v}) := \Phi(\mathbf{u},\mathbf{v}) - \frac{1}{c_{12}}\left(\frac{\partial\Phi}{\partial u^1}\frac{\partial\Phi}{\partial v^2} - \frac{\partial\Phi}{\partial u^2}\frac{\partial\Phi}{\partial v^1}\right).$$

Then Ψ is an alternating bilinear form, satisfying

$$\frac{\partial\Psi}{\partial u^i} = \frac{\partial\Phi}{\partial u^i} - \frac{1}{c_{12}}\left(\frac{\partial\Phi}{\partial u^1}c_{i2} - \frac{\partial\Phi}{\partial u^2}c_{i1}\right).$$

In particular

$$\frac{\partial\Psi}{\partial u^1} = \frac{\partial\Psi}{\partial u^2} = 0.$$

Analogously $\partial\Psi/\partial v^1 = \partial\Psi/\partial v^2 = 0$, and therefore Ψ does not contain u^1, u^2, v^1, v^2. With

$$u'^1 := -c_{12}^{-\frac{1}{2}}\frac{\partial\Phi}{\partial v^1} = c_{12}^{-\frac{1}{2}}\sum_{j=1}^q c_{1j}u^j,$$

$$u'^2 := -c_{12}^{-\frac{1}{2}}\frac{\partial\Phi}{\partial v^2} = c_{12}^{-\frac{1}{2}}\sum_{j=1}^q c_{2j}u^j,$$

$$v'^1 := c_{12}^{-\frac{1}{2}}\frac{\partial\Phi}{\partial u^1} = c_{12}^{-\frac{1}{2}}\sum_{j=1}^q c_{1j}v^j,$$

$$v'^2 := c_{12}^{-\frac{1}{2}}\frac{\partial\Phi}{\partial u^2} = c_{12}^{-\frac{1}{2}}\sum_{j=1}^q c_{2j}v^j$$

it follows that

$$\Phi(\mathbf{u};\mathbf{v}) = u'^1v'^2 - u'^2v'^1 + \Psi(u^3,\ldots,u^q;v^3,\ldots,v^q).$$

Repeating this procedure again and again, one ends up with

$$\mathbf{u}^T\mathbf{Cv} = \Phi(\mathbf{u};\mathbf{v}) = (u'^1v'^2 - u'^2v'^1) + \cdots + (u'^{2m-1}v'^{2m} - u'^{2m}v'^{2m-1})$$
$$= \mathbf{u}'^T\mathbf{C}'\mathbf{v}'^T$$

for some m. But

$$2s = \text{rank}\,\mathbf{C} = \text{rank}\,\mathbf{C}' = 2m,$$

so that $m = s$. This change of coefficients for U and V determines the transition matrix $\mathbf{M}$, which then provides the wanted canonical basis for $\mathcal{V}$. $\qquad\square$

Corollary 6.1.1. *The Cauchy characteristic vector field system of $\mathcal{V}$ is*

$$\mathcal{C}(\mathcal{V}) = (C_1, \dots, C_{q-2s}),$$

and

$$c := \dim \mathcal{C}(\mathcal{V}) = q - 2s.$$

$\square$

Corollary 6.1.2. *If $\dim \mathcal{V} = q$ is odd, $\mathcal{V}$ admits at least one Cauchy characteristic vector field.* $\square$

Example. Let $\mathcal{S} \subset J^1(\mathbb{R}^n, \mathbb{R})$ be a single PDE. Then $\dim \mathcal{S} = (2n + 1) - 1 = 2n$, and the dimension of the associated vector field system $\mathcal{V}$ is $\dim \mathcal{V} = \dim \mathcal{S} - 1 = 2n - 1$, which is odd.

It is not hard to see by direct calculations that actually $\dim \mathcal{C}(\mathcal{V}) = 1$. Giving Cauchy data for $\mathcal{S}$ amounts to giving an $(n - 1)$-dimensional integral manifold of $\mathcal{V}$. If this is transverse to the Cauchy charateristic vector field, the wanted n-dimensional integral manifold is obtained as the union of all Cauchy characteristics passing through the Cauchy data.

Let us next determine the maximal involutions of $\mathcal{V} = (A_1, \dots, A_s; B_1, \dots, B_s; C_1 \dots, C_c)$. Since all these contain the Cauchy characteristic vector fields $C_1, \dots, C_c$, we may for a moment assume that there are no such, so that $\mathcal{V} = (A_1, \dots, A_s; B_1, \dots, B_s)$. Let $\mathcal{I}_m = (U_1, \dots, U_m)$ be an m-dimensional involution, where

$$U_k = \sum_{i=1}^{s} (a_k^i A_i + b_k^i B_i) \quad \text{for } k = 1, \dots, m.$$

Being an involution, $\mathcal{I}_m$ is contained in

$$\bar{\mathcal{I}}_m := \{U \in \mathcal{V} \mid [U, U_k] \equiv 0 \pmod{\mathcal{V}} \quad \text{for } k = 1, \dots, m\}.$$

Clearly

$$U = \sum_{i=1}^{s}(a^i A_i + b^i B_i) \in \bar{\mathcal{I}}_m \iff \sum_{i=1}^{s}(a^i b_k^i - b^i a_k^i) = 0 \quad \text{for } k = 1, \dots, m.$$

With $U_1, \dots, U_m$ being linearly independent, the latter is a system of m linearly independent equations for the coefficients $a^1, \dots, a^s, b^1, \dots, b^s$. Therefore $\dim \bar{\mathcal{I}}_m = 2s - m$, so that

$$m = \dim \mathcal{I}_m \leq \dim \bar{\mathcal{I}}_m = 2s - m.$$

Consequently $m \leq s$. If $m < s$ it is possible to add a vector field

$U_{m+1} \in \bar{\mathcal{I}}_m \setminus \mathcal{I}_m$ to the basis of $\mathcal{I}_m$ so as to obtain an $(m+1)$-dimensional involution $\mathcal{I}_{m+1}$. If $m+1 < s$, one more vector field—from $\bar{\mathcal{I}}_{m+1} \setminus \mathcal{I}_{m+1}$—can be added, giving an $(m+2)$-dimensional involution. And so on, until we end up with an s-dimensional involution $\mathcal{I}_s$. To this we finally add the Cauchy characteristic vector fields $C_1, \ldots, C_c$.

Theorem 6.1.2. *The maximal involutions of $\mathcal{V}$ have the dimension*

$$s + c = s + (q - 2s) = q - s,$$

where $s = \text{rank } C$, $c = \dim \mathcal{C}(\mathcal{V})$ and $q = \dim \mathcal{V}$. $\qquad\square$

Hence the involutive genus of $\mathcal{V}$ is $s + c$. According to Cartan's local existence theorem this means that there are $(s + c)$-dimensional complete subsystems of $\mathcal{V}$ in the analytic category.

However we would like to avoid the Cauchy–Kowalewski theorem so as to be able to establish the existence of $(s + c)$-dimensional complete subsystems also in the C^∞ case—and this is done in the next section.

6.2 Complete subsystems

Given a vector field system $\mathcal{V}$ on $\mathbb{R}^r$ with $\dim \mathcal{V} = q$, $\dim \mathcal{V}' = q + 1$ and $\dim \mathcal{C}(\mathcal{V}) = c$, we now want to determine complete subsystems $\mathcal{W}$ of $\mathcal{V}$ of the largest possible dimension $s + c$, with $2s$ being equal to the rank of the skew-symmetric matrix of structure functions.

Such a $\mathcal{W}$ admits $r - s - c$ functionally independent first integrals ϕ^1, $\ldots, \phi^{r-s-c}$. These in turn determine $\mathcal{W}$ by

$\mathcal{W} = $ the system of all vector fields on $\mathbb{R}^r$ killing $\phi^1, \ldots, \phi^{r-s-c}$.

Since

$$\mathcal{C}(\mathcal{V}) \subseteq \mathcal{W} \subseteq \mathcal{V},$$

all first integrals of $\mathcal{W}$ are also first integrals of $\mathcal{C}(\mathcal{V})$, and among the first integrals of $\mathcal{W}$ are those of $\mathcal{V}$. Let us consider the latter first.

Lemma 6.2.1. *$\mathcal{V}$ and $\mathcal{V}'$ have the same first integrals.*

Proof. For let ϕ be a first integral of $\mathcal{V}$. Then for $X, Y \in \mathcal{V}$,

$$[X, Y]\phi = X(Y\phi) - Y(X\phi) = 0,$$

so that ϕ is a first integral of $\mathcal{V}'$ too. And the converse is evident. $\qquad\square$

Supposing that all vector field systems under consideration have constant pointwise dimension, the chain

$$\mathcal{V} \subseteq \mathcal{V}' \subseteq \mathcal{V}'' \subseteq \cdots$$

must for dimension reasons terminate with a complete vector field system $\bar{\mathcal{V}}$, called the *completion* of $\mathcal{V}$.

According to the lemma, the first integrals of $\mathcal{V}$ are the same as those of $\bar{\mathcal{V}}$—and the latter are obtained by means of the Frobenius theorem. So we may assume that we know

$$\tau := r - \dim \bar{\mathcal{V}}$$

functionally independent first integrals of $\mathcal{V}$; let us call them $t^1, \ldots, t^\tau$.

$\mathcal{C}(\mathcal{V})$ being complete, its first integrals are also found by using the Frobenius theorem. Choose one—say ϕ^1—such that $d\phi^1 \wedge dt^1 \wedge \cdots \wedge dt^\tau \neq 0$, and set

$$\mathcal{V}_1 := \{X \in \mathcal{V} \mid X\phi^1 = 0\}.$$

Then $\mathcal{C}(\mathcal{V}) \subseteq \mathcal{V}_1$ and $\dim \mathcal{V}_1 = \dim \mathcal{V} - 1$. For $C \in \mathcal{C}(\mathcal{V})$ and $X_1 \in \mathcal{V}_1$,

$$[C, X_1] \in \mathcal{V} \quad \text{and} \quad [C, X_1]\phi^1 = 0,$$

so $[C, X_1] \in \mathcal{V}_1$, whence $\mathcal{C}(\mathcal{V}) \subseteq \mathcal{C}(\mathcal{V}_1)$—i.e., $\dim \mathcal{C}(\mathcal{V}_1) \geq c$.

If $s = 1$, $\dim \mathcal{V}_1 = q - 1 = (q - 2s) + 1 = c + 1$, and so $\mathcal{V}_1 = (X_0, C_1, \ldots, C_c)$ for some $X_0 \in \mathcal{V} \setminus \mathcal{C}$ with $X_0(\phi^1) = 0$. As just shown, $[C_i, X_0] \in \mathcal{V}_1$ for $i = 1, \ldots, c$, and this fact together with the completeness of $\mathcal{C}(\mathcal{V})$ establishes that $\mathcal{V}_1$ is complete—and we are done.

Let us therefore suppose that $s > 1$ in the following!

Lemma 6.2.2. $\dim \mathcal{C}(\mathcal{V}_1) \leq c + 1$.

Proof. If $\dim \mathcal{C}(\mathcal{V}_1) = c + 2$ there are vector fields $X_1, X_2, \ldots$ such that $\mathcal{C}(\mathcal{V}_1) = (X_1, X_2, C_1, \ldots, C_c)$, $\mathcal{V}_1 = (X_1, \ldots, X_{2s-1}, C_1, \ldots, C_c)$ and $\mathcal{V} = (X_1, \ldots, X_{2s}, C_1, \ldots, C_c)$. Now there are smooth functions u^1, u^2 such that

$$[u^1 X_1 + u^2 X_2, X_{2s}] \equiv 0 \pmod{\mathcal{V}}$$

—since this simply means that $c_{1,2s} u^1 + c_{2,2s} u^2 = 0$. But then $u^1 X_1 + u^2 X_2 \in \mathcal{C}(\mathcal{V})$—which is a contradiction. $\square$

So $\dim \mathcal{C}(\mathcal{V}_1)$ is equal either to $c = \dim \mathcal{C}(\mathcal{V})$ or to $c + 1$.

Lemma 6.2.3. $\dim \mathcal{V}_1' = \dim \mathcal{V}_1 + 1 \quad (= q)$.

Proof. Recall that a vector field system and its derived system have the same first integrals. Because ϕ^1 is a first integral of V_1' but not of V', $V_1' \subsetneq V'$, and hence $\dim V_1' < \dim V' = \dim V + 1 = q + 1$. On the other hand, since $s > 1$ we have $\dim V_1 = q-1 > q-2s+1 = c+1 \geq \dim C(V_1)$. Therefore V_1 is not complete, and hence $\dim V_1' > \dim V_1 = q - 1$. $\quad\square$

Lemma 6.2.4. $\dim C(V_1) = c + 1$.

Proof. Since $\dim V_1' = \dim V_1 + 1$ it follows that

$$\dim V_1 - \dim C(V_1) = \text{the rank of the skew-symmetric matrix}$$
$$\text{of structure functions for } V_1$$

is an *even integer* $2s_1$. According to lemma **6.2.2** there are two possibilities:

- either $\dim C(V_1) = c$, and then $\dim V_1 - \dim C(V_1) = q - 1 - (q - 2s) = 2s - 1$ is odd—a contradiction,
- or $\dim C(V_1) = c+1$, and then $\dim V_1 - \dim C(V_1) = q - 1 - (q - 2s + 1) = 2s - 2$ is even—OK.

$$\square$$

By construction V_1 admits at least one more first integral than V does—namely ϕ^1. But this is all there is.

Lemma 6.2.5. *If V_1 admits m functionally independent first integrals, V admits $m - 1$.*

Proof. Introduce the m first integrals of V_1 as coordinates $x^1, \dots, x^m$ for $\mathbb{R}^r$, with $x^1 = \phi^1$. Then V_1 has a basis of the form

$$X_i = \frac{\partial}{\partial x^{m+i}} + \sum_{j=q+m}^{r} f_i^j \frac{\partial}{\partial x^j} \qquad \text{for } i = 1, \dots, q - 1.$$

And then $V = (X_1, \dots, X_{q-1}, X_q)$ for a suitable

$$X_q = \frac{\partial}{\partial x^1} + \sum_{j=2}^{m} g^j \frac{\partial}{\partial x^j} + \sum_{j=q+m}^{r} f_q^j \frac{\partial}{\partial x^j}.$$

Because $\dim V_1' = \dim V_1 + 1$ there is a vector field Y not involving $\partial/\partial x^{m+1}, \dots, \partial/\partial x^{m+q-1}$ such that $V_1' = (X_1, \dots, X_{q-1}, Y)$. Furthermore, with $x^1, \dots, x^m$ being first integrals of V_1', Y does not contain $\partial/\partial x^1, \dots, \partial/\partial x^m$ either, and hence must be of the form

$$Y = \sum_{j=q+m}^{r} h^j \frac{\partial}{\partial x^j}.$$

With Y belonging to $\mathcal{V}_1' \subset \mathcal{V}'$, but not to $\mathcal{V}$, it follows that $\mathcal{V}' = (X_1, \ldots , X_q, Y)$. For $i = 1, \ldots , q-1$, $[X_i, X_q] \in \mathcal{V}'$ does not contain $\partial/\partial x^1$, and so must be a linear combination of $X_1, \ldots , X_{q-1}$ and Y. But then $\partial/\partial x^2, \ldots , \partial/\partial x^m$ are also absent from $[X_i, X_q]$, i.e., $X_i(g^j) = 0$ for $i = 1, \ldots , q-1$ and $j = 2, \ldots , m$. This means that the g^j are first integrals of $\mathcal{V}_1$, and consequently are functions of $x^1, \ldots , x^m$ only.

The first integrals of $\mathcal{V}_1$ consists of all smooth functions of $x^1, \ldots , x^m$, and the first integrals of $\mathcal{V} \supset \mathcal{V}_1$ are those functions f of $x^1, \ldots , x^m$ that satisfy $X_q(f) = 0$—that is,

$$\frac{\partial f}{\partial x^1} + \sum_{j=2}^{m} g^j(x^1, \ldots , x^m) \frac{\partial f}{\partial x^j} = 0.$$

Thus there are indeed $m - 1$ functionally independent first integrals of $\mathcal{V}$. $\square$

Conclusion.

- $\dim \mathcal{V}_1' = \dim \mathcal{V}_1 + 1$—i.e., $\mathcal{V}_1$ inherits the defining property of $\mathcal{V}$;
- $\dim \mathcal{V}_1 = \dim \mathcal{V} - 1$;
- $\dim \mathcal{C}(\mathcal{V}_1) = \dim \mathcal{C}(\mathcal{V}) + 1$;
- $s_1 = \frac{1}{2}(\dim \mathcal{V}_1 - \dim \mathcal{C}(\mathcal{V}_1)) = s - 1$;
- $\mathcal{V}_1$ admits one more first integral than $\mathcal{V}$ does, namely ϕ^1;
- the dimension of the maximal involutions of $\mathcal{V}_1$ equals $s_1 + \dim \mathcal{C}(\mathcal{V}_1) = (s-1) + (q - 2s + 1) = q - s = $ the dimension of the maximal involutions of $\mathcal{V}$.

Because $\dim \mathcal{V}_1' = \dim \mathcal{V}_1 + 1$, the procedure which gives the subsystem $\mathcal{V}_1$ of $\mathcal{V}$ applies to $\mathcal{V}_1$ as well if $s_1 > 1$, and is the starting point of an obvious induction argument.

Setting $\mathcal{V}_0 := \mathcal{V}$, the kth step—for $k = 1, \ldots , s$—consists in:

choose ϕ^k which is a first integral of $\mathcal{C}(\mathcal{V}_{k-1})$ but not of $\mathcal{V}_{k-1}$, and set $\mathcal{V}_k := \{X \in \mathcal{V}_{k-1} \mid X\phi^k = 0\}$.

Then $\dim \mathcal{V}_k = \dim \mathcal{V}_{k-1} - 1 = \cdots = \dim \mathcal{V} - k = q - k$, and $\dim \mathcal{C}(\mathcal{V}_k) = \dim \mathcal{C}(\mathcal{V}_{k-1}) + 1 = \cdots = \dim \mathcal{C}(\mathcal{V}) + k = c + k$. For $k = s$ it follows that $\dim \mathcal{V}_s = q - s = (q - 2s) + s = c + s = \dim \mathcal{C}(\mathcal{V}_s)$. Hence $\mathcal{V}_s = \mathcal{C}(\mathcal{V}_s)$, and so $\mathcal{V}_s$ is complete.

Consequently,

$\mathcal{V}$ does admit $(s + c)$-dimensional complete subsystems!

And this has been achieved without using the Cauchy–Kowalewski theorem.

However, we want to know more about these complete subsystems than just that they exist.

6.3 The generalized contact bracket

Let us now refine the construction of the previous section. Let $\phi^1, \ldots, \phi^s$ be the first integrals of $\mathcal{C}(\mathcal{V})$ used in order to define the complete subsystem $\mathcal{V}_s$, and set

$$\mathcal{V}_{\phi^i} := \{X \in \mathcal{V} \mid X\phi^i = 0\}.$$

Fix an integer $k \in \{1, \ldots, s\}$. Then

$$\mathcal{C}(\mathcal{V}_k) \supseteq \mathcal{C}(\mathcal{V}_{\phi^k}) \text{ for } k = 1, \ldots, s, \text{ and } \mathcal{C}(\mathcal{V}_k) \supseteq \mathcal{C}(\mathcal{V}_{k-1}),$$

so that

$$\mathcal{C}(\mathcal{V}_k) \supseteq \mathcal{C}(\mathcal{V}_{k-1}) + \mathcal{C}(\mathcal{V}_{\phi^k}).$$

Lemma 6.3.1. $\mathcal{C}(\mathcal{V}_k) = \mathcal{C}(\mathcal{V}_{k-1}) + \mathcal{C}(\mathcal{V}_{\phi^k})$ *for* $k = 1, \ldots, s$.

Proof. $\mathcal{C}(\mathcal{V}_k)$ (respectively $\mathcal{C}(\mathcal{V}_{k-1})$) is obtained by adding k (respectively $k-1$) independent vector fields to $\mathcal{C}(\mathcal{V})$, while $\mathcal{C}(\mathcal{V}_{\phi^k})$ is got by adding a single vector field to $\mathcal{C}(\mathcal{V})$. So for dimension reasons it suffices to show that $\mathcal{C}(\mathcal{V}_{\phi^k}) \not\subseteq \mathcal{C}(\mathcal{V}_{k-1})$.

Since $\mathcal{V}_k = \mathcal{V}_{k-1} \cap \mathcal{V}_{\phi^k}$,

$$\begin{aligned} \dim(\mathcal{V}_{k-1} + \mathcal{V}_{\phi^k}) &= \dim \mathcal{V}_{k-1} + \dim \mathcal{V}_{\phi^k} - \dim \mathcal{V}_k \\ &= (q - k + 1) + (q - 1) - (q - k) = q = \dim \mathcal{V}, \end{aligned}$$

and hence $\mathcal{V} = \mathcal{V}_{k-1} + \mathcal{V}_{\phi^k}$.

Suppose that $\mathcal{C}(\mathcal{V}_{\phi^k}) \subseteq \mathcal{C}(\mathcal{V}_{k-1})$. Then $[\mathcal{C}(\mathcal{V}_{\phi^k}), \mathcal{V}_{k-1}] \subseteq \mathcal{V}_{k-1}$, and by definition $[\mathcal{C}(\mathcal{V}_{\phi^k}), \mathcal{V}_{\phi^k}] \subseteq \mathcal{V}_{\phi^k}$. Thus it would follow that $[\mathcal{C}(\mathcal{V}_{\phi^k}), \mathcal{V}] \subseteq \mathcal{V}$, i.e., $\mathcal{C}(\mathcal{V}_{\phi^k}) \subseteq \mathcal{C}(\mathcal{V})$—a contradiction. $\square$

The lemma shows that

$$\begin{aligned} \mathcal{C}(\mathcal{V}_k) &= \mathcal{C}(\mathcal{V}_{k-1}) + \mathcal{C}(\mathcal{V}_{\phi^k}) = \mathcal{C}(\mathcal{V}_{k-2}) + \mathcal{C}(\mathcal{V}_{\phi^{k-1}}) + \mathcal{C}(\mathcal{V}_{\phi^k}) \\ &= \cdots = \mathcal{C}(\mathcal{V}_{\phi^1}) + \mathcal{C}(\mathcal{V}_{\phi^2}) + \cdots + \mathcal{C}(\mathcal{V}_{\phi^k}) \quad \text{for } k = 1, \ldots, s. \end{aligned}$$

Now $\dim C(\mathcal{V}_{\phi^k}) = \dim C(\mathcal{V}) + 1$, so $C(\mathcal{V}_{\phi^i}) = (T_i, C_1, \ldots, C_c)$ for a vector field $T_i \in C(\mathcal{V}_{\phi^i}) \setminus C(\mathcal{V})$. Consequently

$$\mathcal{V}_s = C(\mathcal{V}_s) = \sum_{i=1}^{s} C(\mathcal{V}_{\phi^i}) = (T_1, \ldots, T_s, C_1, \ldots, C_c);$$

incidentally this formula also shows that $T_1, \ldots, T_s, C_1, \ldots, C_c$ are linearly independent.

Conclusion. *In order to determine the complete vector field system $\mathcal{V}_s$ it suffices to find vector fields $T_i \in C(\mathcal{V}_{\phi^i}) \setminus C(\mathcal{V})$ for $i = 1, \ldots, s$.*

So let ϕ be a first integral of $C(\mathcal{V})$ but not of $\mathcal{V}$, let $\mathcal{V}_\phi := \{X \in \mathcal{V} \mid X\phi = 0\}$, and let us determine a vector field $T \in C(\mathcal{V}_\phi) \setminus C(\mathcal{V})$.

For $T \in \mathcal{V}$,

$$T \in C(\mathcal{V}_\phi) \iff T \in \mathcal{V}_\phi \text{ and } [T, \mathcal{V}_\phi] \equiv 0 \pmod{\mathcal{V}_\phi}$$
$$\iff T\phi = 0 \text{ and } [T, U] \equiv \rho_U \cdot U\phi \cdot Z \pmod{\mathcal{V}} \text{ for all } U \in \mathcal{V},$$

where ρ_U is some smooth function. Note that if $C \in C(\mathcal{V})$, $(T + C)\phi = 0$ and $[T + C, U] \equiv [T, U] \pmod{\mathcal{V}}$, so that $T \in C(\mathcal{V}_\phi) \Longrightarrow T + C \in C(\mathcal{V}_\phi)$ as well. But we will see that T is going to be well determined modulo a multiplicative factor and modulo $C(\mathcal{V})$.

In fact, let $\mathcal{V} := (X_1, \ldots, X_{2s}, C_1, \ldots, C_c)$. $\mathcal{V}_\phi$ consists of $C(\mathcal{V})$ and all vector fields

$$U = \sum_{i=1}^{2s} u^i X_i \quad \text{which satisfy} \quad \sum_{i=1}^{2s} u^i X_i \phi = 0.$$

Let the wanted $T_\phi \in C(\mathcal{V}_\phi) \setminus C(\mathcal{V})$ be

$$T_\phi = \sum_{i=1}^{2s} v^i X_i.$$

Then $[T_\phi, \mathcal{V}_\phi] \equiv 0 \pmod{\mathcal{V}_\phi}$, and in particular

$$[T_\phi, U] \equiv 0 \pmod{\mathcal{V}}$$

for all $U = \sum_{i=1}^{2s} u^i X_i$ satisfying $\sum_{i=1}^{2s} u^i X_i \phi = 0$, i.e.,

$$\sum_{i,j=1}^{2s} c_{ij} u^i v^j = \sum_{i=1}^{2s} u^i \left(\sum_{j=1}^{2s} c_{ij} v^j \right) = 0$$

for all $2s$-tuples $(u^1,\ldots,u^{2s})$ which satisfy $\sum_{i=1}^{2s} u^i X_i \phi = 0$. So for some function ρ,

$$\sum_{j=1}^{2s} c_{ij} v^j = \rho \cdot X_i \phi \qquad \text{for } i = 1,\ldots, 2s.$$

Clearly T_ϕ can be multiplied by an arbitrary nonzero factor, and therefore we may choose $\rho = 1$ in the following.

With

$$\delta := \det \begin{pmatrix} c_{1,1} & \cdots & c_{1,2s} \\ \vdots & \ddots & \vdots \\ c_{2s,1} & \cdots & c_{2s,2s} \end{pmatrix},$$

Cramer's rule says that

$$v_k = \delta^{-1} \cdot \det \begin{pmatrix} c_{1,1} & \cdots & c_{1,k-1} & X_1\phi & c_{1,k+1} & \cdots & c_{1,2s} \\ \vdots & \ddots & \vdots & \vdots & \vdots & \ddots & \vdots \\ c_{2s,1} & \cdots & c_{2s,k-1} & X_{2s}\phi & c_{2s,k+1} & \cdots & c_{2s,2s} \end{pmatrix}$$

$$= (-1)^{k-1} \delta^{-1} \cdot \det \begin{pmatrix} c_{1,1} & \cdots & c_{1,k-1} & c_{1,k+1} & \cdots & c_{1,2s} & -X_1\phi \\ \vdots & \ddots & \vdots & \vdots & \ddots & \vdots & \vdots \\ c_{2s,1} & \cdots & c_{2s,k-1} & c_{2s,k+1} & \cdots & c_{2s,2s} & -X_{2s}\phi \end{pmatrix}.$$

But then

$$T_\phi = \sum_{k=1}^{2s} v^k X_k = \delta^{-1} \cdot \det \begin{pmatrix} c_{1,1} & \cdots & c_{1,2s} & -X_1\phi \\ \vdots & \ddots & \vdots & \vdots \\ c_{2s,1} & \cdots & c_{2s,2s} & -X_{2s}\phi \\ X_1 & \cdots & X_{2s} & 0 \end{pmatrix},$$

with the second equality being seen by expansion of the last row.

By construction $T_\phi \in \mathcal{V}$ and $[T_\phi, \mathcal{V}_\phi] \subseteq \mathcal{V}$. To ensure that $T_\phi \in \mathcal{C}(\mathcal{V}_\phi)$—as we want—it suffices to show that $T_\phi \phi = 0$. For then $T_\phi \in \mathcal{V}_\phi$ and $[T_\phi, \mathcal{V}_\phi]\phi = 0$, so that indeed $[T_\phi, \mathcal{V}_\phi] \in \mathcal{V}_\phi$. Consider

$$T_\phi f = \delta^{-1} \cdot \det \begin{pmatrix} c_{1,1} & \cdots & c_{1,2s} & -X_1\phi \\ \vdots & \ddots & \vdots & \vdots \\ c_{2s,1} & \cdots & c_{2s,2s} & -X_{2s}\phi \\ X_1 f & \cdots & X_{2s}f & 0 \end{pmatrix}$$

for an arbitrary smooth function f. Making the changes

$$\text{rows} \longleftrightarrow \text{columns}, \quad c_{ij} \longleftrightarrow c_{j,i} = -c_{i,j}, \quad \phi \longleftrightarrow -\phi, \quad f \longleftrightarrow -f$$

we obtain an expression which equals both $(-1)^{2s+1} T_\phi f$ and $T_f \phi$. Therefore $T_f \phi = -T_\phi f$, and so in particular $T_\phi \phi = 0$. Thus T_ϕ is the wanted vector field in $\mathcal{C}(\mathcal{V}_\phi) \setminus \mathcal{C}(\mathcal{V})$.

Definition. The vector field T_ϕ is denoted by $\{\phi, \cdot\}$, so that $T_\phi f = \{\phi, f\}$; $\{\cdot, \cdot\}$ is called the *generalized contact bracket*—for reasons that will appear in the next section.

The generalized contact bracket $\{\phi, \cdot\}$ becomes particularly simple if $\mathcal{V}$ is given a canonical basis, i.e., $\mathcal{V} = (A_1, \ldots, A_s, B_1, \ldots, B_s, C_1, \ldots, C_c)$ with $[A_i, B_i] \equiv Z \pmod{\mathcal{V}}$, and all other brackets vanishing modulo $\mathcal{V}$. Indeed, repeating the argument above we set

$$U := \sum_{i=1}^{s} (a^i A_i + b^i B_i), \quad T := \sum_{i=1}^{s} (u^i A_i + v^i B_i),$$

and require that $[T, U] \equiv \sum_{i=1}^{s} (u^i b^i - v^i a^i) \cdot Z \pmod{\mathcal{V}}$ shall vanish modulo $\mathcal{V}$ whenever $U\phi = 0$. That is, for some smooth function ρ,

$$[T, U] \equiv \rho \cdot U\phi \cdot Z \pmod{V}.$$

$\rho = 1$ gives the equations

$$\sum_{i=1}^{s} (u^i b^i - v^i a^i) = \sum_{i=1}^{s} (a^i A_i \phi + b^i B_i \phi),$$

which are to be satisfied for all a^i and b^i. Hence

$$u^i = B_i \phi \quad \text{and} \quad v^i = -A_i \phi,$$

so that

$$\{\phi, \cdot\} = T = \sum_{i=1}^{s} (B_i \phi \cdot A_i - A_i \phi \cdot B_i).$$

Using this expression it is easy to derive the following connection between generalized contact brackets and Lie brackets.

Theorem 6.3.1. $[\{\phi, \cdot\}, \{\psi, \cdot\}] \equiv \{\phi, \psi\} \cdot Z \pmod{\mathcal{V}}$.

Proof. In terms of the canonical basis for $\mathcal{V}$,

$$[\{\phi,\cdot\},\{\psi,\cdot\}] = \left[\sum_{i=1}^{s}(B_i\phi\cdot A_i - A_i\phi\cdot B_i), \sum_{j=1}^{s}(B_j\psi\cdot A_j - A_j\psi\cdot B_j)\right]$$

$$\equiv \sum_{i=1}^{s}(B_i\phi\cdot A_i\psi - A_i\phi\cdot B_i\psi)\cdot Z \quad (\mathrm{mod}\ \mathcal{V})$$

$$\equiv \{\phi,\psi\}\cdot Z \quad (\mathrm{mod}\ V).$$

Changing the basis but keeping Z fixed will only add a Cauchy characteristic vector field to $\{\phi,\cdot\}$ and $\{\psi,\cdot\}$ respectively by an earlier remark, and therefore the equivalence above still remains valid. $\qquad\square$

Let us now summarize the results of this section.

Recipe for obtaining complete $(s+c)$-dimensional subsystems of $\mathcal{V}$.

- Choose a function ϕ^1 which is a first integral of $\mathcal{C}(\mathcal{V})$ but not of $\mathcal{V}$, set $\mathcal{C}_1 := \mathcal{C}(\mathcal{V}) + \{\phi^1,\cdot\}$ and $\mathcal{V}_1 := \{X \in \mathcal{V} \mid X\phi^1 = 0\}$.
- Then choose a function ϕ^2 which is a first integral of $\mathcal{C}_1$ but not of $\mathcal{V}_1$, set $\mathcal{C}_2 := \mathcal{C}(\mathcal{V}) + \{\phi^1,\cdot\} + \{\phi^2,\cdot\}$ and $\mathcal{V}_2 := \{X \in \mathcal{V}_1 \mid X\phi^2 = 0\}$.
- Continue in this way until $\mathcal{C}_{s-1}$ and $\mathcal{V}_{s-1}$ are obtained.
- Finally choose a function ϕ^s which is a first integral of $\mathcal{C}_{s-1}$ but not of $\mathcal{V}_{s-1}$, and set $\mathcal{C}_s := \mathcal{C}(\mathcal{V}) + \{\phi^1,\cdot\} + \cdots + \{\phi^s,\cdot\}$. Then $\mathcal{C}_s$ is a complete $(s+c)$-dimensional subsystem of $\mathcal{V}$.

6.4 Reduction to a canonical form and systems of contact coordinates

By considering the reduced vector field system $\mathcal{V}_{\mathrm{red}}$ if necessary we assume in the following that $\mathcal{V}$ admits no Cauchy characteristic vector field, so that $\dim \mathcal{V} = 2s$.

Suppose that we have found *one* s-dimensional complete subsystem

$$\mathcal{W}_s = (\{\phi_1,\cdot\},\ldots,\{\phi_s,\cdot\})$$

using the procedure of the last section. By means of $\mathcal{W}_s$ we will then find a simple *canonical form for* $\mathcal{V}$, from which all other s-dimensional complete subsystems of $\mathcal{V}$ are determined in a very explicit way.

With $\mathcal{W}_s$ living on $\mathbb{R}^r$, $\mathcal{W}_s$ admits $r-s$ functionally independent first integrals. Among these are $\phi^1, \ldots, \phi^s$, and the possible first integrals of $\mathcal{V}$; since $\dim \mathcal{V}' = 2s+1$, there are at most $r-2s-1$ independent ones.

Let

$$\{\phi^1, \ldots, \phi^s, \phi^{s+1}, \ldots, \phi^{s+p}\}, \quad \text{where } p := r - 2s,$$

be a system of functionally independent first integrals of $\mathcal{W}_s$, with the first integrals of $\mathcal{V}$ being placed rearmost in this list, and thus occurring somewhere to the right of ϕ^{s+1}.

Complete these first integrals of $\mathcal{W}_s$ with functions $x^1, \ldots, x^s$ having the property $dx^1 \wedge \cdots \wedge dx^s \wedge d\phi^1 \wedge \cdots \wedge d\phi^{s+p} \neq 0$, and use these $2s + p = r$ functions as local coordinates for $\mathbb{R}^r$. Then

$$\mathcal{W}_s = (\partial/\partial x^1, \ldots, \partial/\partial x^s).$$

Now $\mathcal{W}_s \subset \mathcal{V} \subset \mathcal{V}'$, and we wish to complete our basis for $\mathcal{W}_s$ to bases for $\mathcal{V}$ and $\mathcal{V}'$.

Using Gaussian elimination, the missing generators of $\mathcal{V}$ can be written as

$$\Phi_j = \frac{\partial}{\partial \phi^j} + \sum_{l=1}^{p} \theta_j^l \frac{\partial}{\partial \phi^{s+l}} \quad \text{for } j = 1, \ldots, s;$$

note that $\theta_j^l = 0$ if ϕ^{s+l} happens to be a first integral of $\mathcal{V}$.

Finally, the vector field $Z \in \mathcal{V}' \setminus \mathcal{V}$ can be chosen as

$$Z = \frac{\partial}{\partial \phi^{s+1}} + \sum_{k=1}^{p} \theta^k \frac{\partial}{\partial \phi^{s+k}}.$$

Then

$$[\partial/\partial x^i, \Phi_j] = \sum_{l=1}^{p} \frac{\partial \theta_j^l}{\partial x^i} \frac{\partial}{\partial \phi^{s+l}} \in \mathcal{V}' = \left(\frac{\partial}{\partial x^1}, \ldots, \frac{\partial}{\partial x^s}, \Phi_1, \ldots, \Phi_s, Z \right),$$

and hence this bracket must be a multiple of Z:

$$[\partial/\partial x^i, \Phi_j] = \frac{\partial \theta_j^1}{\partial x^i} Z \quad \text{for } i, j = 1, \ldots, s.$$

Setting

$$X_i := \partial/\partial x^i \text{ and } X_{s+i} := \Phi_i \quad \text{for } i = 1, \ldots, s,$$

and defining the structure functions c_{ij} by

$$[X_i, X_j] \equiv c_{ij} Z \pmod{\mathcal{V}} \quad \text{for } i, j = 1, \ldots, 2s,$$

it follows that

$$\det \left((c_{ij})_{i,j=1}^{2s} \right) = \left(\frac{\partial(\theta_1^1, \ldots, \theta_s^1)}{\partial(x^1, \ldots, x^s)} \right)^2.$$

Since the determinant in the left hand side does not vanish, this shows that $\theta_1^1, \ldots, \theta_s^1$ can be introduced as new local coordinates instead of x^1, $\ldots, x^s$. Denoting them by $\psi^1, \ldots, \psi^s$ and putting

$$\Psi_i := \frac{\partial}{\partial \psi^i},$$

we then have

$$\mathcal{W}_s = (\Psi_1, \ldots, \Psi_s), \quad \mathcal{V} = (\Phi_1, \ldots, \Phi_s, \Psi_1, \ldots, \Psi_s)$$

and

$$\mathcal{V}' = (\Phi_1, \ldots, \Phi_s, \Psi_1, \ldots, \Psi_s, Z).$$

Furthermore the Φ_j and Z can be written as

$$\Phi_j = \frac{\partial}{\partial \phi^j} + \psi^j \frac{\partial}{\partial \phi^{s+1}} + R_j \quad \text{for } j = 1, \ldots, s, \quad \text{and} \quad Z = \frac{\partial}{\partial \phi^{s+1}} + R,$$

where R_j and R are linear combinations of the vector fields $\partial/\partial \phi^{s+2}, \ldots,$ $\partial/\partial \phi^{s+p}$ only.

Because $[\Psi_i, \Phi_j] = \delta_j^i \cdot \partial/\partial \phi^{s+1} + [\partial/\partial \psi^i, R_j]$ is a multiple of Z,

$$[\partial/\partial \psi^i, R_j] = 0 \text{ for } i \neq j \text{ and } [\partial/\partial \psi^i, R_i] = R.$$

The case $s = 1$. Here

$$\Phi_1 = \frac{\partial}{\partial \phi^1} + \psi^1 \frac{\partial}{\partial \phi^2} + R_1,$$

$$\Psi_1 = \frac{\partial}{\partial \psi^1},$$

$$Z = \frac{\partial}{\partial \phi^2} + \left[\frac{\partial}{\partial \psi^1}, R_1 \right],$$

or with a change of notation:

$$\mathcal{V} = \left(\frac{\partial}{\partial x^1} + x^1 \frac{\partial}{\partial x^0} + Y, \frac{\partial}{\partial x^1} \right), \quad \text{and} \quad Z = \frac{\partial}{\partial x^0} + \left[\frac{\partial}{\partial x^1}, Y \right],$$

where the vector field Y does not involve $\partial/\partial x^0$ and $\partial/\partial x^1$. In order to be more specific as regards Y we must make more assumptions on $\mathcal{V}$—which we will do in section **6.7**.

Assume in the rest of this section that $s > 1$! Then for $i = 1, \ldots, s$,

$$
\left[\frac{\partial}{\partial \psi^i}, R \right] = \left[\frac{\partial}{\partial \psi^i}, \left[\frac{\partial}{\partial \psi^j}, R_j \right] \right] \quad \text{for some } j \neq i
$$

$$
= \left[\frac{\partial}{\partial \psi^j}, \left[\frac{\partial}{\partial \psi^i}, R_j \right] \right] \quad \text{by Jacobi}
$$

$$
= 0,
$$

so that R does not depend on $\psi^1, \ldots, \psi^s$—that is,

$$
R = \sum_{i=2}^{p} a^i(\phi) \frac{\partial}{\partial \phi^{s+i}}.
$$

Moreover, if

$$
R_i = \sum_{j=2}^{p} b_i^j(\phi, \psi) \frac{\partial}{\partial \phi^{s+j}},
$$

it follows that

$$
R = \left[\frac{\partial}{\partial \psi^i}, R_i \right] = \sum_{j=2}^{p} \frac{\partial b_i^j}{\partial \psi^i} \frac{\partial}{\partial \phi^{s+j}},
$$

and thus

$$
\frac{\partial b_i^j}{\partial \psi^i} = a^i(\phi), \quad \text{or} \quad b_i^j = \psi^i \cdot a^j(\phi) + c_i^j(\phi).
$$

Hence

$$
R_i = \psi^i \cdot R + S_i,
$$

where both R and S_i are of the form $\sum_{j=2}^{p} f^j(\phi) \left(\partial / \partial \phi^{s+j} \right)$. But then

$$
\Phi_i = \frac{\partial}{\partial \phi^i} + \psi^i \frac{\partial}{\partial \phi^{s+1}} + \psi^i R + S_i
$$

$$
= \frac{\partial}{\partial \phi^i} + \psi^i Z + S_i \quad \text{for } i = 1, \ldots, s,
$$

where

$$
Z = \frac{\partial}{\partial \phi^{s+1}} + \sum_{k=2}^{p} a^k(\phi) \frac{\partial}{\partial \phi^{s+k}}.
$$

Consequently $[\Phi_i, \Phi_j]$—which belongs to $\mathcal{V}'$—does not contain $\partial / \partial \phi^1$,

$\dots$, $\partial/\partial\phi^{s+1}$, and so must vanish:

$$0 = [\Phi_i, \Phi_j]$$

$$= \left[\frac{\partial}{\partial\phi^i} + S_i, \frac{\partial}{\partial\phi^j} + S_j\right] + \psi^j\left[\frac{\partial}{\partial\phi^i} + S_i, Z\right] - \psi^i\left[\frac{\partial}{\partial\phi^j} + S_j, Z\right].$$

In particular

$$\left[\frac{\partial}{\partial\phi^i} + S_i, Z\right] = 0,$$

showing that

$$[\Phi_i, Z] = 0 \quad \text{for } i = 1, \dots, s.$$

Combining this with the obvious equalities

$$[\Phi_i, \Psi_j] = \delta_i^j \cdot Z \quad \text{and} \quad [\Psi_i, Z] = 0,$$

we have arrived at the following important result.

Theorem 6.4.1. $\mathcal{V}'$ *is complete if* $s > 1$, *i.e.,* $\mathcal{V}$ *admits* $r - 2s - 1$ *functionally independent first integrals.* □

With the notation used here it follows that ϕ^{s+2}, $\dots$, ϕ^{s+p} are first integrals of $\mathcal{V}$, so that Φ_i and Z do not contain $\partial/\partial\phi^{s+2}, \dots, \partial/\partial\phi^{s+p}$—i.e., R and the R_i vanish. Hence

$$\mathcal{V} = \left(\frac{\partial}{\partial\phi^1} + \psi^1\frac{\partial}{\partial\phi^{s+1}}, \dots, \frac{\partial}{\partial\phi^s} + \psi^s\frac{\partial}{\partial\phi^{s+1}}, \frac{\partial}{\partial\psi^1}, \dots, \frac{\partial}{\partial\psi^s}\right),$$

and $\mathcal{V}'$ is obtained from $\mathcal{V}$ by adding the vector field

$$Z = \frac{\partial}{\partial\phi^{s+1}}.$$

In order to recognize what has happened, set

$$\begin{cases} x^i := \phi^i & \text{for } i = 1, \dots, s, \\ z := \phi^{s+1}, \\ p_i := \psi^i & \text{for } i = 1, \dots, s. \end{cases}$$

Then

$$\mathcal{V} = \left(\frac{\partial}{\partial x^1} + p_1\frac{\partial}{\partial z}, \dots, \frac{\partial}{\partial x^s} + p_s\frac{\partial}{\partial z}, \frac{\partial}{\partial p_1}, \dots, \frac{\partial}{\partial p_s}\right),$$

—which is nothing but the contact vector field system of the jet bundle $J^1(\mathbb{R}^s_x, \mathbb{R}_z)$. Moreover, $Z = \partial/\partial z$, and the complete subsystem W_s which we started from equals $(\partial/\partial p_1, \dots, \partial/\partial p_s)$.

Theorem 6.4.2. *Let $\mathcal{V}$ be a vector field system on $\mathbb{R}^r$ having no Cauchy characteristic vector field, and satisfying $\dim \mathcal{V} > 2$, $\dim \mathcal{V}' = \dim \mathcal{V} + 1$. Then there is a structure preserving local diffeomorphism*

$$(\mathbb{R}^r, \mathcal{V}) \cong J^1(\mathbb{R}^s_x, \mathbb{R}_z) \times \mathbb{R}^{r-2s-1}_t,$$

where $J^1(\mathbb{R}^s_x, \mathbb{R}_z)$ is equipped with its contact vector field system $(\partial/\partial x^1 + p_1 \partial/\partial z, \ldots, \partial/\partial x^s + p_s \partial/\partial z, \partial/\partial p_1, \ldots, \partial/\partial p_s)$, dual to the contact form $\theta^0 = dz - \sum_{i=1}^{s} p_i \, dx^i$, and $\mathbb{R}^{r-2s-1}_t$ is a parameter space. $\qquad\square$

Corollary 6.4.1. *Let $\theta = 0$ be a pfaffian equation on $\mathbb{R}^r$ with $d\theta \neq 0$. Setting $\mathcal{V} := \theta^{\perp}$, it then follows that $\dim \mathcal{V} = r - 1$ and $\dim \mathcal{V}' = r$. Applying the theorem to $\mathcal{V}_{\mathrm{red}}$ we can find local coordinates $x^1, \ldots, x^s, z$, $p_1, \ldots, p_s, u^1, \ldots, u^c$ for $\mathbb{R}^r$ with $2s + 1 + c = r$, such that $\mathcal{V}_{\mathrm{red}} = (\partial/\partial x^1 + p_1 \partial/\partial z, \ldots, \partial/\partial x^s + p_s \partial/\partial z, \partial/\partial p_1, \ldots, \partial/\partial p_s)$ and $C(\mathcal{V}) = (\partial/\partial u^1, \ldots, \partial/\partial u^c)$. Then*

$$\theta = 0 \iff dz - \sum_{i=1}^{s} p_i \, dx^i = 0.$$

$\qquad\square$

Setting

$$\begin{cases} X_i := \partial/\partial x^i + p_i \, \partial/\partial z, \\ P_i := \partial/\partial p_i \end{cases} \qquad \text{for } i = 1, \ldots, s,$$

we have

$$\mathcal{V} = (X_1, \ldots, X_s, P_1, \ldots, P_s).$$

This basis for $\mathcal{V}$ is canonical, since $[P_i, X_i] = \partial/\partial z = Z$, while all other brackets vanish. Employing it for the generalized contact bracket, we get

$$\{f, g\} = \sum_{i=1}^{s} (P_i f \cdot X_i g - P_i g \cdot X_i f)$$

$$= \sum_{i=1}^{s} \left(\frac{\partial f}{\partial p_i} \left(\frac{\partial g}{\partial x^i} + p_i \frac{\partial g}{\partial z} \right) - \frac{\partial g}{\partial p_i} \left(\frac{\partial f}{\partial x^i} + p_i \frac{\partial f}{\partial z} \right) \right),$$

which is the usual contact bracket.

Note. Suppressing the z variable, $\{f, g\}$ is transformed into the *canonical* or *Poisson bracket* known from symplectic geometry and classical mechanics.

In particular,

$$
\begin{cases}
\{z,\cdot\} = -\sum_{i=1}^{s} p_i \left(\partial/\partial p_i\right), & \\
\{x^i,\cdot\} = -\partial/\partial p_i & \text{for } i = 1,\ldots,s, \\
\{p_i,\cdot\} = \partial/\partial x^i + p_i \,\partial/\partial z & \text{for } i = 1,\ldots,s,
\end{cases}
$$

so that

$$
\{z,\cdot\} = \sum_{i=1}^{s} p_i \{x^i,\cdot\}.
$$

This shows how the functions $x^1, \ldots, x^s, z, p_1, \ldots, p_s$ are connected to each other. The p_i are said to be *polar functions* with respect to $x^1, \ldots,$ x^s, and z.

Since $x^1, \ldots, x^s, z, p_1, \ldots, p_s$ form a system of coordinates for the jet bundle $J^1(\mathbb{R}^s_x, \mathbb{R}_z)$, they are said to constitute a *system of contact coordinates*.

The formulas above show that the contact coordinates satisfy the following first order PDE system:

$$
\begin{cases}
\{z,x^i\} = 0, \quad \{x^i,x^j\} = 0, \quad \{p_i,p_j\} = 0, \\
\{x^i,p_j\} = -\delta^i_j, \quad \{z,p_i\} = -p_i,
\end{cases}
$$

for $i,j = 1,\ldots,s$.

If we use a noncanonical basis of $\mathcal{V}$ the corresponding generalized contact brackets will be multiples of the contact brackets used here, so that the contact coordinates instead satisfy

$$
\begin{cases}
\{z,x^i\} = 0, \quad \{x^i,x^j\} = 0, \quad \{p_i,p_j\} = 0, \\
\{x^i,p_j\} = -\rho\,\delta^i_j, \quad \{z,p_i\} = -\rho\,p_i,
\end{cases}
\qquad (S(^1Cont^{s,1}))
$$

for $i,j = 1,\ldots,s$, where ρ is an unspecified nonzero function. (Recall that the family $^1Cont^{s,1}$ of contact transformations on $J^1(\mathbb{R}^s, \mathbb{R})$ was introduced in section **2.2**—we will encounter it again in section **6.6**.)

Conversely, let $\mathcal{V}$ be a $2s$-dimensional vector field system on $\mathbb{R}^{2s+p}$ with $s > 1$ and $p \geq 1$ such that

- $\dim \mathcal{V}' = \dim \mathcal{V} + 1$,
- $C(\mathcal{V}) = \{0\}$.

Then it is possible to define a generalized contact bracket $\{\cdot,\cdot\}$ associated to a basis of $\mathcal{V}$; moreover $\mathcal{V}$ admits a set $\{t^1,\ldots,t^{p-1}\}$ of functionally independent first integrals, which may be assumed to be known thanks to the Frobenius theorem.

Theorem 6.4.3. *Let* $\{x^1, \ldots, x^s, z, p_1, \ldots, p_s\}$ *be a* $(2s+1)$*-tuple of functions on* $\mathbb{R}^{2s+p}$ *satisfying the PDE system* $\mathcal{S}(^1\mathit{Cont}^{s,1})$ *and* $dx^1 \wedge \cdots \wedge x^s \wedge dz \wedge dp_1 \wedge \cdots \wedge dp_s \wedge dt^1 \wedge \cdots \wedge dt^{p-1} \neq 0$. *Then*

$$\mathcal{V} = (X_1, \ldots, X_s, P_1, \ldots, P_s), \quad \text{with } X_i = \partial/\partial x^i + p_i \partial/\partial z \text{ and } P_i = \partial/\partial p_i,$$

so that $\mathcal{V}$ *is brought to a canonical form by solving* $\mathcal{S}(^1\mathit{Cont}^{s,1})$.

Proof. Use x^i, z, p_i and t^j as local coordinates for $\mathbb{R}^r$. Since $C(\mathcal{V}) = \{0\}$, any function is a first integral of $C(\mathcal{V})$. Let $f(x,z,p)$ be an arbitrary nonconstant function of x^i, z, and p_i. Then f is a first integral of $C(\mathcal{V})$, but not of $\mathcal{V}$. It follows that $\{f, \cdot\} \in \mathcal{V}$, and by the chain rule,

$$\{f, \cdot\} = \sum_{i=1}^{s} \{f, x^i\} \frac{\partial}{\partial x^i} + \{f, z\} \frac{\partial}{\partial z} + \sum_{i=1}^{s} \{f, p_i\} \frac{\partial}{\partial p_i}.$$

Replacing f by $x^1, \ldots, x^s, z, p_1, \ldots, p_s$ respectively and using $\mathcal{S}(^1\mathit{Cont}^{s,1})$ it follows that

$$\{x^i, \cdot\} = -\rho \, P_i, \quad \{z, \cdot\} = -\rho \sum_{i=1}^{s} p_i \, P_i \text{ and } \{p_i, \cdot\} = \rho \, X_i \text{ for } i = 1, \ldots, s,$$

so that $X_1, \ldots, X_s, P_1, \ldots, P_s \in \mathcal{V}$. Being linearly independent they necessarily form a basis for $\mathcal{V}$. $\square$

6.5 How to find all maximal complete subsystems of $\mathcal{V}$

If $\mathcal{V}$ is a vector field system on $\mathbb{R}^r$ such that $\dim \mathcal{V} > 2$, $C(\mathcal{V}) = \{0\}$ and $\dim \mathcal{V}' = \dim \mathcal{V} + 1$, $(\mathbb{R}^r, \mathcal{V})$ can be identified with

$$J^1(\mathbb{R}^s_x, \mathbb{R}_z) \times \mathbb{R}^{r-2s-1}_t,$$

where the first factor is endowed with its canonical contact vector field system

$$\mathcal{K}_0 = \left(\frac{\partial}{\partial x^1} + p_1 \frac{\partial}{\partial z}, \ldots, \frac{\partial}{\partial x^s} + p_s \frac{\partial}{\partial z}, \frac{\partial}{\partial p_1}, \ldots, \frac{\partial}{\partial p_s} \right)$$

$$= (X_1, \ldots, X_s, P_1, \ldots, P_s), \text{ dual to the contact form } \theta^0 = dz - \sum_{i=1}^{s} p_i \, dx^i,$$

and the second factor is a parameter space provided with the vector field system (0), dual to the pfaffian system $(dt^1, \ldots, dt^{r-2s-1})$; the integer s appearing here is half the rank of the skew-symmetric matrix of structure functions for $\mathcal{V}$.

The maximal integral manifolds of $\mathcal{K}_0$ on $J^1(\mathbb{R}^s, \mathbb{R})$ are by definition the generalized 1-graphs, so that the maximal integral manifolds of $\mathcal{V}$ are given by

$$\{\text{generalized 1-graphs}\} \times \{t\},$$

where $\{t\}$ is a point in $\mathbb{R}_t^{r-2s-1}$. Thus we can forget about the latter, and our task is reduced to the following:

determine all complete s-dimensional subsystems of $\mathcal{K}_0$!

Such a subsystem is of the form $\mathcal{W} = (Y_1, \ldots, Y_s)$, with $Y_i = \sum_{j=1}^{s}(a_i^j X_j + b_i^j P_j)$ satisfying $[Y_i, Y_j] \equiv 0 \pmod{\mathcal{W}}$ for $i, j = 1, \ldots, s$.

First special case. All $a_i^j = 0$, so that

$$W = (P_1, \ldots, P_s) = \left(\frac{\partial}{\partial p_1}, \ldots, \frac{\partial}{\partial p_s} \right).$$

Then $\mathcal{W}$ is precisely the complete subsystem of $\mathcal{V}$ from which its canonical form was derived. Its first integrals are $x^1, \ldots, x^s, z$, so that the integral manifolds are given by

$$x^1 = \text{constant}, \ldots, x^s = \text{constant}, \ z = \text{constant}.$$

Second special case. If the matrix $\left((a_i^j)_{i,j=1}^{s} \right)$ is nonsingular, $\mathcal{W}$ admits generators of the form

$$Y_i = X_i + \sum_{j=1}^{s} b_i^j P_j = \frac{\partial}{\partial x^i} + p_i \frac{\partial}{\partial z} + \sum_{j=1}^{s} b_i^j \frac{\partial}{\partial p_j} \quad \text{for } i = 1, \ldots, s.$$

Then $x^1, \ldots, x^s$ can be used as local coordinates on the s-dimensional leaves of the foliation defined by $\mathcal{W}$, so that these leaves are of the form

$$\begin{cases} z = W^0(x^1, \ldots, x^s; a^0, \ldots, a^s), \\ p_i = W_i(x^1, \ldots, x^s; a^0, \ldots, a^s) \quad \text{for } i = 1, \ldots, s, \end{cases}$$

where $a^0, \ldots, a^s$ are arbitrary parameters which conversely can be solved from this system:

$$a^j = \alpha^j(x^1, \ldots, x^s, z, p_1, \ldots, p_s) \quad \text{for } j = 0, 1, \ldots, s.$$

Since the functions $z - W^0$ and $p_j - W_j$ vanish on each leaf we have

$$\begin{cases} 0 = Y_i(z - W^0) = p_i - \partial W^0 / \partial x^i, \\ 0 = Y_i(p_j - W_j) = b_i^j - \partial W_j / \partial x^i, \end{cases} \quad \text{for } i = 1, \ldots, s,$$

so that

$$p_i = W_i = \frac{\partial W^0}{\partial x^i} \quad \text{and} \quad b_i^j = \frac{\partial W_j}{\partial x^i} = \frac{\partial^2 W^0}{\partial x^i \partial x^j}.$$

Hence

$$Y_i = \frac{\partial}{\partial x^i} + \frac{\partial W^0}{\partial x^i} \cdot \frac{\partial}{\partial z} + \sum_{j=1}^{s} \frac{\partial^2 W^0}{\partial x^i \partial x^j} \cdot \frac{\partial}{\partial p_j} \quad \text{for } i = 1, \ldots, s,$$

where the smooth function W^0 is arbitrary except for the condition that it should be possible to solve $a^0, \ldots, a^s$ from the system

$$z = W^0(x^1, \ldots, x^s; a^0, \ldots, a^s),$$

$$p_i = \frac{\partial W^0}{\partial x^i}(x^1, \ldots, x^s; a^0, \ldots, a^s), \quad i = 1, \ldots, s.$$

Since it is readily seen from the expression for Y_i that $[Y_i, Y_j] = 0$ for $i, j = 1, \ldots, s$, $W = (Y_1, \ldots, Y_s)$ is indeed complete.

The general case. If rank $\left((a_i^j)_{i,j=1}^s \right) = t$ with $0 < t < s$, the generators of W can after a suitable change of indices be presented as

$$\begin{cases} A_i = X_i + \sum_{k=1}^{s-t} a_i^{t+k} X_{t+k} + \sum_{k=1}^{s} c_i^k P_k & \text{for } i = 1, \ldots, t, \\ B_{t+j} = \sum_{k=1}^{s} b_{t+j}^k P_k & \text{for } j = 1, \ldots, s - t. \end{cases}$$

Furthermore the B_{t+j} can be chosen as

$$B_{t+j} = P_{t+j} + \sum_{k=1}^{t} b_{t+j}^k P_k \quad \text{for } j = 1, \ldots, s - t,$$

—because otherwise there is a nontrivial linear combination $B = \sum_{j=1}^{t} c^j P_j$ belonging to W, and then

$$[A_i, B] \equiv c^i Z \pmod{\mathcal{K}_0} \quad \text{for } i = 1, \ldots, t,$$

which contradicts the completeness of W.

Thus $W = (A_1, \ldots, A_t, B_{t+1}, \ldots, B_s)$ with

$$A_i = \frac{\partial}{\partial x^i} + p_i \frac{\partial}{\partial z} + \sum_{k=1}^{s-t} a_i^{t+k} \left(\frac{\partial}{\partial x^{t+k}} + p_{t+k} \frac{\partial}{\partial z} \right) + \sum_{k=1}^{t} c_i^k \frac{\partial}{\partial p_k}$$

for $i = 1, \ldots, t$, and

$$B_{t+j} = \frac{\partial}{\partial p_{t+j}} + \sum_{k=1}^{t} b_{t+k}^k \frac{\partial}{\partial p_k} \quad \text{for } j = 1, \ldots, s - t.$$

It follows from this that $x^1, \ldots, x^t, p_{t+1}, \ldots, p_s$ can be used as local coordinates on the leaves of the foliation defined by $\mathcal{W}$, so that these leaves are given by

$$
\begin{cases}
z = W^0(x^1, \ldots, x^t, p_{t+1}, \ldots, p_s; a^0, \ldots, a^s), \\
x^{t+j} = W^{t+j}(x^1, \ldots, x^t, p_{t+1}, \ldots, p_s; a^0, \ldots, a^s), & j = 1, \ldots, s-t, \\
p_i = V_i(x^1, \ldots, x^t, p_{t+1}, \ldots, p_s; a^0, \ldots, a^s), & i = 1, \ldots, t,
\end{cases}
$$

where $a^0, \ldots, a^s$ are arbitrary parameters which conversely may be solved from this system, and the functions W^0, W^{t+j} and V_i are to satisfy

$$
\begin{cases}
0 = A_i(z - W^0) = p_i - \partial W^0/\partial x^i + \sum_{k=1}^{s-t} a_i^{t+k} p_{t+k}, \\
0 = A_i(x^{t+j} - W^{t+j}) = a_i^{t+j} - \partial W^{t+j}/\partial x^i, \\
0 = A_i(p_j - V_j) = c_i^j - \partial V_j/\partial x^i, \\
0 = B_{t+j}(z - W^0) = -\partial W^0/\partial p_{t+j}, \\
0 = B_{t+j}(x^{t+i} - W^{t+i}) = -\partial W^{t+i}/\partial p_{t+j}, \\
0 = B_{t+j}(p_i - V_i) = b_{t+j}^i - \partial V_i/\partial p_{t+j}.
\end{cases}
$$

Hence

$$
\frac{\partial W^0}{\partial p_{t+j}} = \frac{\partial W^{t+i}}{\partial p_{t+j}} = 0, \quad a_i^{t+j} = \frac{\partial W^{t+j}}{\partial x^i}, \quad c_i^j = \frac{\partial V_j}{\partial x^i}, \quad b_{t+j}^i = \frac{\partial V_i}{\partial p_{t+j}},
$$

and

$$
p_i = V_i = \frac{\partial W^0}{\partial x^i} - \sum_{k=1}^{s-t} \frac{\partial W^{t+k}}{\partial x^i} p_{t+k} = \frac{\partial}{\partial x^i}\left(W^0 - \sum_{k=1}^{s-t} p_{t+k} W^{t+k}\right),
$$

so that

$$
c_i^j = \frac{\partial^2}{\partial x^i \partial x^j}\left(W^0 - \sum_{k=1}^{s-t} p_{t+k} W^{t+k}\right), \quad \text{and} \quad b_{t+j}^i = -\frac{\partial W^{t+j}}{\partial x^i}.
$$

Then our foliation has the form

$$
\begin{cases}
z = W^0(x^1, \ldots, x^t; a^0, \ldots, a^s), \\
x^{t+j} = W^{t+j}(x^1, \ldots, x^t; a^0, \ldots, a^s), & j = 1, \ldots, s-t, \\
p_i = (\partial/\partial x^i)\left(W^0 - \sum_{k=1}^{s-t} p_{t+k} W^{t+k}\right), & i = 1, \ldots, t.
\end{cases}
$$

Introducing

$$
W(x^1, \ldots, x^t, p_{t+1}, \ldots, p_s; a^0, \ldots, a^s)
$$

$$
:= W^0(x^1, \ldots, x^t; a^0, \ldots, a^s) - \sum_{k=1}^{s-t} p_{t+k} W^{t+k}(x^1, \ldots, x^t; a^0, \ldots, a^s)
$$

—which accordingly is *linear* in $p_{t+1}, \ldots, p_s$—we have $W^{t+k} = -\partial W / \partial p_{t+k}$, and the formulas above simplify to

$$
\begin{cases}
z = W - \sum_{k=1}^{s-t} p_{t+k} \left(\partial W / \partial p_{t+k} \right), \\
x^{t+j} = -\partial W / \partial p_{t+j}, & j = 1, \ldots, s - t, \\
p_i = \partial W / \partial x^i, & i = 1, \ldots, t.
\end{cases}
\tag{1}
$$

Furthermore the coefficients in the vector fields A_i and B_{t+j} have now been reduced to

$$
\begin{cases}
a_i^{t+j} = -\partial^2 W / \partial x^i \partial p_{t+j}, & i = 1, \ldots, t, \ j = 1, \ldots, s - t, \\
b_{t+j}^k = \partial^2 W / \partial x^k \partial p_{t+j}, & j = 1, \ldots, s - t, \ k = 1, \ldots, t, \\
c_i^k = \partial^2 W / \partial x^i \partial x^k, & i, k = 1, \ldots, t.
\end{cases}
\tag{2}
$$

Thus to each complete subsystem of V of the form $(A_1, \ldots, A_t, B_{t+1}, \ldots, B_s)$ corresponds a certain function

$$
W(x^1, \ldots, x^t, p_{t+1}, \ldots, p_s; a^0, \ldots, a^s),
$$

from which the coefficients in the A_i and B_{t+j} are calculated as in (2), and with the corresponding foliation being given by (1).

Conversely: start with a smooth function

$$
W(x^1, \ldots, x^t, p_{t+1}, \ldots, p_s; a^0, \ldots, a^s),
$$

which is arbitrary except for the possibility of solving $a^0, \ldots, a^s$ as functions of $x^1, \ldots, x^s, z, p_1, \ldots, p_s$ from (1). Then (1) defines a foliation of $\mathbb{R}^{2s+1}$ with s-dimensional leaves. Let

$$
\mathcal{W} := \{\text{vector fields in } \mathcal{K}_0 \text{ which are everywhere tangent to this foliation}\}.
$$

Then $\mathcal{W}$ is an s-dimensional complete subsystem of $\mathcal{K}_0$, with the corresponding foliation given by (1).

Indeed, $U = \sum_{i=1}^s (\alpha^i X_i + \beta^i P_i)$ in $\mathcal{K}_0$ belongs to $\mathcal{W}$ if and only if

$$
\begin{cases}
U(z - W + \sum_{k=1}^{s-t} p_{t+k} \left(\partial W / \partial p_{t+k} \right)) = 0 \\
U(x^{t+j} + \partial W / \partial p_{t+j}) = 0 & \text{for} \quad j = 1, \ldots, s - t, \\
U(p_i - \partial W / \partial x^i) = 0 & \text{for} \quad i = 1, \ldots, t.
\end{cases}
$$

These equations are equivalent to

$$\sum_{i=1}^{s} \alpha^i p_i = \sum_{i=1}^{t} \alpha^i \left(\frac{\partial W}{\partial x^i} - \sum_{k=1}^{s-t} p_{t+k} \frac{\partial^2 W}{\partial x^i \partial p_{t+k}} \right) \tag{3}$$

$$- \sum_{i=t+1}^{s} \beta^i \sum_{k=1}^{s-t} p_{t+k} \frac{\partial^2 W}{\partial p_i \partial p_{t+k}},$$

$$\alpha^{t+j} = - \sum_{i=1}^{t} \alpha^i \frac{\partial^2 W}{\partial x^i \partial p_{t+j}} - \sum_{i=t+1}^{s} \beta^i \frac{\partial^2 W}{\partial p_i \partial p_{t+j}}, \qquad j = 1, \ldots, s-t, \tag{4}$$

$$\beta^i = \sum_{j=1}^{t} \alpha^j \frac{\partial^2 W}{\partial x^i \partial x^j} + \sum_{j=t+1}^{s} \beta^j \frac{\partial^2 W}{\partial x^i \partial p_j}, \qquad i = 1, \ldots, t, \tag{5}$$

respectively.

Inserting α^{t+j} from (4) together with $p_i = \partial W/\partial x^i$ into (3), (3) becomes identically satisfied. Hence $\alpha^1, \ldots, \alpha^t, \beta^{t+1}, \ldots, \beta^s$ are left arbitrary, while $\alpha^{t+1}, \ldots, \alpha^s, \beta^1, \ldots, \beta^t$ are linear combinations of these. But then $\dim \mathcal{W} = s$, and $\mathcal{W}$ has generators of the form

$$\begin{cases} X_i + \sum_{k=1}^{s-t} a_i^k X_{t+k} + \sum_{k=1}^{s} c_i^k P_k & \text{for } i = 1, \ldots, t, \\ P_{t+j} + \sum_{k=1}^{t} b_{t+j}^k P_k & \text{for } j = t+1, \ldots, s. \end{cases}$$

By definition $\mathcal{W}$—which lives on $\mathbb{R}_{x,z,p}^{2s+1}$—admits the $s+1$ functionally independent functions $a^0, \ldots, a^s$ solved from (1), and hence must be complete.

Thus we have found that the most general complete s-dimensional subsystem $\mathcal{W}$ of $\mathcal{K}_0$ is characterized by an integer $t \in \{0, 1, \ldots, s\}$ and a smooth function $W(x^1, \ldots, x^t, p_{t+1}, \ldots, p_s; a^0, \ldots, a^s)$, which satisfies the nondegeneracy condition that the parameters $a^0, \ldots, a^s$ can be solved as functions of $x^1, \ldots, x^s, z, p_1, \ldots, p_s$ from (1), but otherwise is arbitrary. $\mathcal{W}$ is generated by the vector fields

$$A_i = X_i - \sum_{k=1}^{s-t} \frac{\partial^2 W}{\partial x^i \partial p_{t+k}} X_k + \sum_{l=1}^{t} \frac{\partial^2 W}{\partial x^i \partial x^l} P_l \qquad \text{for } i = 1, \ldots, t,$$

$$B_{t+j} = P_{t+j} + \sum_{l=1}^{t} \frac{\partial^2 W}{\partial x^l \partial p_{t+j}} P_l \qquad \text{for } j = 1, \ldots, s-t,$$

and the leaves of the foliation corresponding to $\mathcal{W}$ are given by fixing $a^0, \ldots, a^s$ in (1).

Equivalently, $\mathcal{W}$ is the complete vector field system which has $a^0(x, z, p)$, $\ldots, a^s(x, z, p)$ as first integrals.

6.6 Contact transformations and Lie pseudogroups

Suppose that $s > 1$, and let $\mathcal{K}$ be a $2s$-dimensional vector field system on $\mathbb{R}^{2s+1}$ with $C(\mathcal{K}) = \{0\}$ and $\dim \mathcal{K}' = 2s + 1$. By the methods in section **6.3** it is possible to find a complete s-dimensional subsystem $\mathcal{W}_0$ of $\mathcal{K}$; starting from this it is explained in section **6.4** how to derive the canonical form

$$\mathcal{K} = (X_1, \ldots, X_s, P_1, \ldots, P_s)$$

for $\mathcal{K}$, where

$$\begin{cases} X_i = \partial/\partial x^i + p_i \, \partial/\partial z, \\ P_i = \partial/\partial p_i \end{cases} \qquad \text{for } i = 1, \ldots, s,$$

and $\mathcal{W}_0 = (P_1, \ldots, P_s)$. Finally, in section **6.5** it is shown how to find the *most general* complete s-dimensional subsystem of $\mathcal{K}$ by using this canonical form. As explained there it is characterized by a function

$$W(x^1, \ldots, x^t, p_{t+1}, \ldots, p_s; a^0, \ldots, a^s)$$

having the property that $a^0, \ldots, a^s$ can be solved from the system

$$\begin{cases} z = W + \sum_{k=1}^{s-t} p_{t+k} \, x^{t+k}, \\ x^{t+j} = -\partial W/\partial p_{t+j} & \text{for } j = 1, \ldots, s - t, \\ p_i = \partial W/\partial x^i & \text{for } i = 1, \ldots, t, \end{cases}$$

so that $x^1, \ldots, x^t, p_{t+1}, \ldots, p_s, a^0, \ldots, a^s$ can be used as coordinates for $\mathbb{R}^{2s+1}$ instead of $x^1, \ldots, x^s, z, p_1, \ldots, p_s$. The complete subsystem $\mathcal{W}_W$ associated to W is

$$\mathcal{W}_W = \{\text{all vector fields annihilating } a^0(x, z, p), \ldots, a^s(x, z, p)\}.$$

To complete the circle we can then bring $\mathcal{K}$ to a canonical form proceeding from $\mathcal{W}_W$—as explained in section **6.4**. Since we already have $s + 1$ functionally independent first integrals $a^0(x, z, p), \ldots, a^s(x, z, p)$ for $\mathcal{W}_W$, there only remains to determine the polar functions $b_1, \ldots, b_s$ associated to these a^i. This can be done by means of the equality

$$\{a^0, \cdot\} = \sum_{i=1}^{s} b_i \{a^i, \cdot\}. \tag{*}$$

On the one hand,

$$\{z, \cdot\} = \sum_{i=1}^{s} p_i \{x^i, \cdot\},$$

and on the other

$$\{z, \cdot\} = \left\{ W + \sum_{j=1}^{s-t} p_{t+j}\, x^{t+j}, \cdot \right\}$$

$$= \sum_{i=1}^{t} \frac{\partial W}{\partial x^i}\, \{x^i, \cdot\} + \sum_{k=1}^{s-t} \frac{\partial W}{\partial p_{t+k}}\, \{p_{t+k}, \cdot\} + \sum_{j=0}^{s} \frac{\partial W}{\partial a^j}\, \{a^j, \cdot\}$$

$$+ \sum_{j=1}^{s-t} x^{t+j}\, \{p_{t+j}, \cdot\} + \sum_{j=1}^{s-t} p_{t+j}\, \{x^{t+j}, \cdot\}$$

$$= \sum_{i=1}^{s} p_i\, \{x^i, \cdot\} + \sum_{j=0}^{s} \frac{\partial W}{\partial a^j}\, \{a^j, \cdot\}.$$

Hence $\sum_{j=0}^{s} \left(\partial W / \partial a^j \right) \{a^j, \cdot\} = 0$, or

$$\frac{\partial W}{\partial a^0}\, \{a^0, \cdot\} = \sum_{i=1}^{s} \left(-\frac{\partial W}{\partial a^i} \right) \{a^i, \cdot\}.$$

It thus follows from (*) that the polar functions b_i are given by

$$b_i = -\frac{\partial W / \partial a^i}{\partial W / \partial a^0} \qquad \text{for } i = 1, \ldots, s.$$

According to section **6.4**, the functions $a^0, \ldots, a^s, b_1, \ldots, b_s$ define a system of contact coordinates for $\mathbb{R}^{2s+1}$, with respect to which $\mathcal{K}$ acquires a canonical form. In fact, setting

$$\begin{cases} y^i := a^i & \text{for } i = 1, \ldots, s, \\ w := a^0, \\ q_i := b_i & \text{for } i = 1, \ldots, s, \end{cases}$$

and

$$Y_i := \frac{\partial}{\partial y^i} + q_i \frac{\partial}{\partial w}, \quad Q_i := \frac{\partial}{\partial q_i} \qquad \text{for } i = 1, \ldots, s,$$

we have

$$\mathcal{K} = (Y_1, \ldots, Y_s, Q_1, \ldots, Q_s) \quad \text{and} \quad \mathcal{W}_W = (Q_1, \ldots, Q_s).$$

Recall that the jet bundle $J^1(\mathbb{R}^s_x, \mathbb{R}_z)$ is a *structured manifold*, with the structure given either by the contact vector field system

$$\mathcal{K}_0 = \left(\frac{\partial}{\partial x^1} + p_1 \frac{\partial}{\partial z}, \ldots, \frac{\partial}{\partial x^s} + p_s \frac{\partial}{\partial z}, \frac{\partial}{\partial p_1}, \ldots, \frac{\partial}{\partial p_s} \right),$$

or by the dual pfaffian equation

$$\theta^0 := dz - \sum_{i=1}^{s} p_i \, dx^i = 0.$$

The considerations above show that the function $W(x^1, \ldots, x^t, p_{t+1}, \ldots, p_s, w, y^1, \ldots, y^s)$ defines a *structure preserving local diffeomorphism*

$$\phi_W : J^1(\mathbb{R}_x^s, \mathbb{R}_z) \overset{\cong}{\longrightarrow} J^1(\mathbb{R}_y^s, \mathbb{R}_w)$$

—i.e., a *contact transformation*—by solving the system

$$\begin{cases} z = W - \sum_{k=1}^{s-t} p_{t+k} \left(\partial W / \partial p_{t+k} \right), \\ x_{t+j} = -\partial W / \partial p_{t+j} & \text{for } j = 1, \ldots, s-t, \\ p_i = \partial W / \partial x^i & \text{for } i = 1, \ldots, t, \\ q_i \, \partial W / \partial w + \partial W / \partial y^i = 0 & \text{for } i = 1, \ldots, s. \end{cases}$$

with respect to $y^1, \ldots, y^s, w, q_1, \ldots, q_s$.

Conversely, let

$$\phi : J^1(\mathbb{R}_x^s, \mathbb{R}_z) \overset{\cong}{\longrightarrow} J^1(\mathbb{R}_y^s, \mathbb{R}_w)$$

be a contact transformation, and let $(\partial/\partial q_1, \ldots, \partial/\partial q_s)$ be the distinguished complete subsystem of the contact vector field system of the right hand side $J^1(\mathbb{R}_y^s, \mathbb{R}_w)$. Since local diffeomorphisms preserve Lie brackets,

$$(\phi^{-1})_* \left(\frac{\partial}{\partial q_1}, \ldots, \frac{\partial}{\partial q_s} \right)$$

is a complete subsystem of the contact vector field system of $J^1(\mathbb{R}_x^s, \mathbb{R}_z)$, and therefore equals $\mathcal{W}_W$ for some smooth function W on $J^1(\mathbb{R}_x^s, \mathbb{R}_z)$. But then $\phi = \phi_W$, so that we have proved the following result.

Theorem 6.6.1. *Every contact transformation is of the form ϕ_W for a suitable function W.* $\square$

Recall from section **2.2** that the family of contact transformations is denoted by $^1Cont^{s,1}$.

Because there to each complete s-dimensional subsystem $\mathcal{W}$ of the contact vector field system $\mathcal{K}_0$ corresponds a contact transformation converting $\mathcal{W}$ into the distinguished subsystem

$$(\partial/\partial q_1, \ldots, \partial/\partial q_s),$$

we deduce the following result.

Theorem 6.6.2. *The family $^1Cont^{s,1}$ of contact transformations acting on $J^1(\mathbb{R}^s, \mathbb{R})$ induces a **transitive 'group' of automorphisms** of the set of s-dimensional complete subsystems of the contact vector field system $\mathcal{K}$.* $\square$

More generally, let M be a manifold with some structure S, and let $\mathfrak{G}$ be the family of all local diffeomorphisms of M that preserve S. Then it is readily seen that

$$\phi_1, \phi_2 \in \mathfrak{G} \text{ and } \phi_2 \circ \phi_1 \text{ is defined} \implies \phi_2 \circ \phi_1 \in \mathfrak{G},$$
$$\phi \in \mathfrak{G} \implies \phi^{-1} \in \mathfrak{G} \text{ too,}$$

so that $\mathfrak{G}$ has a grouplike character; in general it is not an honest group since $\phi_2 \circ \phi_1$ need not be defined for all $\phi_1, \phi_2 \in \mathfrak{G}$.

Definition. A family $\mathfrak{G}$ of local diffeomorphisms on a manifold M is called a *pseudogroup* if the following hold:

(i) if U, V, W are open subsets of M and $g_1 : U \longrightarrow V$, $g_2 : V \longrightarrow W$ belong to $\mathfrak{G}$, so does $g_2 \circ g_1 : U \longrightarrow W$,

(ii) if $g : U \longrightarrow V$ belongs to $\mathfrak{G}$, so does $g^{-1} : V \longrightarrow U$,

(iii) id $: M \longrightarrow M$ belongs to $\mathfrak{G}$,

(iv) if $g \in \mathfrak{G}$ and U is an open subset of the domain of g, also $g_{|U} \in \mathfrak{G}$,

(v) if g is a local diffeomorphism with the property that each point in the domain of g has a neighbourhood such that the restriction of g to this neighbourhood belongs to $\mathfrak{G}$, then g itself belongs to $\mathfrak{G}$.

Clearly the family $^1Cont^{s,1}$ of contact transformations acting on $J^1(\mathbb{R}^s, \mathbb{R})$ is an example of a pseudogroup.

Unfortunately the concept of pseudogroup turns out to be too general to admit any interesting theory. However, recall from section **6.4** that our prototypical example $^1Cont^{s,1}$ has one more property: any contact transformation

$$\phi : J^1(\mathbb{R}^s_x, \mathbb{R}_z) \xrightarrow{\cong} J^1(\mathbb{R}^s_y, \mathbb{R}_w)$$
$$(x, y, p) \mapsto (y, w, q)$$

satisfies the PDE system

$$\begin{cases} \{y^i, y^j\} = 0, \ \{w, y^i\} = 0, \ \{q_i, q_j\} = 0, \\ \{q_i, y^j\} = \rho \, \delta_i^j, \ \{q_i, w\} = \rho \, q_i \end{cases} \quad i, j = 1, \ldots, s. \quad (\mathcal{S}(^1Cont^{s,1}))$$

This is called the *defining PDE system* for $^1Cont^{s,1}$. (Actually $\mathcal{S}(^1Cont^{s,1})$

is not involutive, but becomes so if we add the integrability condition

$$\frac{\partial(y^1,\ldots,y^s,w,q_1,\ldots,q_s)}{\partial(x^1,\ldots,x^s,z,p_1,\ldots,p_s)} = \rho^{s+1}.)$$

More generally, if the pseudogroup $\mathfrak{G}$ is defined by *leaving some local structure invariant*, then expressing this invariance in local coordinates will almost automatically lead to a PDE system.

Definition: A *Lie pseudogroup* $\mathfrak{G}$ on a manifold M is a pseudogroup having the property that the elements of $\mathfrak{G}$ constitute the general solution of a PDE system $S(\mathfrak{G}) \subseteq J^q(M,M)$ for some $q \geq 0$. $S(\mathfrak{G})$ is called the *defining PDE system* of $\mathfrak{G}$.

The solutions of $S(\mathfrak{G})$ are local diffeomorphisms which are to satisfy properties 1.– 5. in the definition of a pseudogroup, and this puts *heavy restrictions* on $S(\mathfrak{G})$—and it is natural to ask which PDE systems can occur as defining systems of Lie pseudogroups. The answer is given by the three fundamental theorems of Lie and Cartan, which we will encounter in sections **14.1–14.3**.

Example. Let $M = \mathbb{R}$, and let $\mathbb{R}_x$ and $\mathbb{R}_y$ be two copies of $\mathbb{R}$. If $\mathfrak{G}$ is a Lie pseudogroup on $\mathbb{R}$, its elements can be regarded as locally defined functions

$$g : \mathbb{R}_x \to \mathbb{R}_y$$
$$x \mapsto y(x)$$

which make up the general solution of an ordinary differential equation. The largest such group is that consisting of *all* local diffeomorphisms of $\mathbb{R}$, with the defining equation $0 = 0$; the smallest consists of the *identity transformation only*, with the defining equation $y = x$. In between these there are up to isomorphism only three different groups (see the examples in section **1.1**):

(i) the *group of translations*, $y = x + a$, defined by $y' = 1$,
(ii) the *affine group*, $y = ax + b$, defined by $y'' = 0$,
(iii) the *projective group*, $y = (ax + b)/(cx + d)$, defined by $y' y''' - \frac{3}{2}(y'')^2 = 0$.

These depend on one, two and three parameters respectively.

More generally, whenever the first character s_1 of the defining PDE system $S(\mathfrak{G})$ vanishes, the general solution will depend on *a finite number of parameters* only. In this case $\mathfrak{G}$ is said to be a *finite dimensional Lie pseudogroup*, or a *local Lie group*.

6.7 Explicitly integrable systems

The canonical form for $\mathcal{V}$ derived in section **6.4** presupposes that $s > 1$; here we consider the exceptional case $s = 1$ under certain further assumptions.

So let $\mathcal{V}$ be a 2-dimensional vector field system on $\mathbb{R}^r$ with $C(\mathcal{V}) = \{0\}$. In section **6.4** we found the following semi-canonical form for $\mathcal{V}$:

$$\mathcal{V} = (X, X_1) \quad \text{with} \quad \begin{cases} X = \partial/\partial x + x^1 \, \partial/\partial x^0 + Y, \\ X_1 = \partial/\partial x^1, \end{cases}$$

where the vector field Y does not contain $\partial/\partial x$, $\partial/\partial x^0$ and $\partial/\partial x^1$. As local coordinates for $\mathbb{R}^r$ we have x, x^0, x^1, the τ functionally independent first integrals $t^1, \ldots, t^\tau$ of $\mathcal{V}$, and $l := r - 3 - \tau$ further variables $y^1, \ldots, y^l$. Hence it is possible to choose Y as

$$Y = \sum_{i=1}^{l} b^i(x, y, t) \frac{\partial}{\partial y^i}.$$

The derived system $\mathcal{V}'$ is obtained by adding

$$Z := [X_1, X] = \frac{\partial}{\partial x^0} + \left[\frac{\partial}{\partial x^1}, Y\right] = \frac{\partial}{\partial x^0} + \sum_{i=1}^{l} \frac{\partial b^i}{\partial x^1} \frac{\partial}{\partial y^i}$$

to $\mathcal{V}$. Then $\mathcal{V}''$ is got by adding $[X, Z]$ and $[X_1, Z]$ to $\mathcal{V}'$, so that $\dim \mathcal{V}''$ equals 4 or 5.

Assuming that $\dim \mathcal{V}'' = \dim \mathcal{V}' + 1$, the results of section **6.1** can be applied, and in particular

$$0 < \dim \mathcal{V}' - \dim C(\mathcal{V}') = 3 - \dim C(\mathcal{V}') \quad \text{is even,}$$

so that necessarily

$$\dim C(\mathcal{V}') = 1.$$

By the rectification lemma the Cauchy characteristic vector field of $\mathcal{V}'$ can be assumed to have the form $\partial/\partial z$. $\mathcal{V}'_{\text{red}}$ lives on $\mathbb{R}^{r-1}$, and has the semi-canonical form (X', X_1') with

$$\begin{cases} X' = \partial/\partial x' + (x^1)' \, \partial/\partial (x^0)' + Y', \\ X_1' = \partial/\partial (x^1)', \end{cases} \quad \text{where } Y' = \sum_{i=1}^{l-1} b^i(x', y', t) \frac{\partial}{\partial (y^i)'};$$

note that $t^1, \ldots, t^\tau$ are functionally independent first integrals of $\mathcal{V}'$ too.

Since $\mathcal{V} \subset \mathcal{V}' = (X', X'_1, \partial/\partial z)$,

- either $\mathcal{V} = (X' + a\,\partial/\partial z, X'_1 + b\,\partial/\partial z)$ for suitable functions a, b,
- or $\mathcal{V} = (X' + c\,X'_1, \partial/\partial z)$ for a suitable function c.

In the first case $[X' + a\,\partial/\partial z, X'_1 + b\,\partial/\partial z]$ contains $\partial/\partial(x^0)'$ but not $\partial/\partial x'$, and hence cannot belong to $\mathcal{V}'$—so this is to be rejected. In the second case

$$\left[X' + c\,X'_1, \frac{\partial}{\partial z} \right] = -\frac{\partial c}{\partial z}\, X'_1,$$

which shows that $\partial c/\partial z \neq 0$, for otherwise $\mathcal{V}$ would be complete. Introducing $x^2 := c$ as a new variable instead of z, V has been reduced to the form

$$\mathcal{V} = \left(\frac{\partial}{\partial x} + x^1 \frac{\partial}{\partial x^0} + x^2 \frac{\partial}{\partial x^1} + Y, \frac{\partial}{\partial x^2} \right), \qquad \text{where}$$

$$Y = \sum_{i=1}^{l-1} b^i(x, y, t) \frac{\partial}{\partial y^i} \qquad \text{and} \qquad \frac{\partial b^i}{\partial x^2} = 0 \text{ for } i = 1, \dots, l-1.$$

Then

$$\mathcal{V}' = \left(\frac{\partial}{\partial x} + x^1 \frac{\partial}{\partial x^0} + Y, \frac{\partial}{\partial x^2}, \frac{\partial}{\partial x^1} \right) \qquad \text{with} \qquad \frac{\partial}{\partial x^2} \in \mathcal{C}(\mathcal{V}').$$

Lemma 6.7.1. *Suppose that $\dim \mathcal{V}^{(\alpha)} = \dim \mathcal{V}^{(\alpha-1)} + 1$ for $\alpha = 1, 2, \dots, h$. Then it is possible to find local coordinates $x, x^0, x^1, \dots, x^h, y^1, \dots, y^{l-h+1}$, $t^1, \dots, t^\tau$ (with the t's being independent first integrals of V) for $\mathbb{R}^r$ such that*

$$\mathcal{V} = \left(\frac{\partial}{\partial x} + x^1 \frac{\partial}{\partial x^0} + x^2 \frac{\partial}{\partial x^1} + \cdots + x^h \frac{\partial}{\partial x^{h-1}} + Y, \frac{\partial}{\partial x^h} \right),$$

where

$$Y = \sum_{i=1}^{l-h+1} b^i(x, x^0, x^1, y^1, \dots, y^{l-h+1}, t^1, \dots, t^\tau) \frac{\partial}{\partial y^i}.$$

Proof. Assume the lemma to be true for $h = k$, and let us prove it for $h = k + 1$. The reasoning above, combined with the result for $h = k$, shows that $\mathcal{V}'_{\text{red}}$ can be written as

$$\mathcal{V}'_{\text{red}} = \left(\frac{\partial}{\partial x'} + (x^1)' \frac{\partial}{\partial (x^0)'} + \cdots + (x^k)' \frac{\partial}{\partial (x^{k-1})'} + Y', \frac{\partial}{\partial (x^k)'} \right),$$

with

$$Y' = \sum_{i=1}^{l-k+1} b^i \frac{\partial}{\partial (y^i)'}.$$

Then $\mathcal{V}' = \mathcal{V}'_{\text{red}} + \partial/\partial z$, with $\partial/\partial z$ being Cauchy characteristic for $\mathcal{V}'$. So $\mathcal{V} \subset \mathcal{V}'$ has generators which are linear combinations of

$$\frac{\partial}{\partial x'} + \cdots + (x^k)' \frac{\partial}{\partial (x^{k-1})'} + Y', \quad \frac{\partial}{\partial (x^k)'} \quad \text{and} \quad \frac{\partial}{\partial z}.$$

The bracket of the first two contains $\partial/\partial (x^{k-1})'$, but not $\partial/\partial x'$, and so does not belong to $\mathcal{V}'$. Thus $\mathcal{V}$ is generated by the vector fields

$$\frac{\partial}{\partial x'} + \cdots + (x^k)' \frac{\partial}{\partial (x^{k-1})'} + Y' + c \frac{\partial}{\partial (x^k)'} \quad \text{and} \quad \frac{\partial}{\partial z},$$

where $\partial c/\partial z \neq 0$, since $\mathcal{V}$ would be complete otherwise. Introducing c as a new variable x^{k+1} instead of z,

$$V = \left(\frac{\partial}{\partial x} + x^1 \frac{\partial}{\partial x^0} + \cdots + x^k \frac{\partial}{\partial x^{k-1}} + Y + x^{k+1} \frac{\partial}{\partial x^k}, \frac{\partial}{\partial x^{k+1}} \right),$$

with $[\partial/\partial x^{k+1}, Y] = 0$, which proves the induction step. $\qquad \square$

The most favourable case occurs if $h = l + 1 = r - 2 - \tau$, for then there are no y-variables, and

$$V = \left(\frac{\partial}{\partial x} + x^1 \frac{\partial}{\partial x^0} + \cdots + x^{l+1} \frac{\partial}{\partial x^l}, \frac{\partial}{\partial x^{l+1}} \right).$$

In this case

$$V' = \left(\frac{\partial}{\partial x} + x^1 \frac{\partial}{\partial x^0} + \cdots + x^l \frac{\partial}{\partial x^{l-1}}, \frac{\partial}{\partial x^l}, \frac{\partial}{\partial x^{l+1}} \right)$$

$$\text{with} \quad \frac{\partial}{\partial x^{l+1}} \in C(V'),$$

$$V'' = \left(\frac{\partial}{\partial x} + x^1 \frac{\partial}{\partial x^0} + \cdots + x^{l-1} \frac{\partial}{\partial x^{l-2}}, \frac{\partial}{\partial x^{l-1}}, \frac{\partial}{\partial x^l}, \frac{\partial}{\partial x^{l+1}} \right)$$

$$\text{with} \quad \frac{\partial}{\partial x^l}, \frac{\partial}{\partial x^{l+1}} \in C(V''),$$

and so on, until after $l + 1$ steps

$$V^{(l+1)} = \left(\frac{\partial}{\partial x}, \frac{\partial}{\partial x^0}, \ldots, \frac{\partial}{\partial x^{l+1}} \right), \quad \text{which is complete.}$$

Let us now integrate $V = (\partial/\partial x + x^1 \partial/\partial x^0 + \cdots + x^{l+1} \partial/\partial x^l, \partial/\partial x^{l+1})$.

Not being complete, $\mathcal{V}$ admits only 1-dimensional integral manifolds. The most general such on which $dx \neq 0$ are integral curves of the vector field

$$X := \frac{\partial}{\partial x} + x^1 \frac{\partial}{\partial x^0} + \cdots + x^{l+1} \frac{\partial}{\partial x^l} + F(x, x^0, x^1, \ldots, x^{l+1}, t^1, \ldots, t^\tau) \frac{\partial}{\partial x^{l+1}},$$

where F is an arbitrary smooth function. The dual pfaffian system of (X) consists of all one-forms

$$\theta = a\,dx + \sum_{i=0}^{l+1} a_i\,dx^i + \sum_{j=1}^{\tau} b_j\,dt^j$$

satisfying $\theta(X) = 0$—that is,

$$0 = a + x^1 a_0 + \cdots + x^{l+1} a_l + + F a_{l+1} \iff a = -\sum_{j=0}^{l} x^{i+1} a_i - F a_{l+1}.$$

Thus

$$(X)^\perp = (dx^0 - x^1\,dx, dx^1 - x^2\,dx, \ldots, dx^l - x^{l+1}\,dx, dx^{l+1} - F\,dx,$$
$$dt^1, \ldots, dt^\tau),$$

and therefore

$$\begin{cases} t^i = \text{constant} & \text{for } i = 1, \ldots, \tau, \\ dx^j = x^{j+1}\,dx & \text{for } j = 0, \ldots, l, \\ dx^{l+1} = F\,dx, \end{cases}$$

on the wanted integral curves. Thus the t^i appear as parameters rather than as variables, and can therefore be forgotten in the following.

   Now $dx^0 = x^1\,dx$ means that x^0 is a function of x, $x^0 = f(x)$, and then

$$x^1 = \frac{dx^0}{dx} = f'(x).$$

After that,

$$dx^1 = x^2\,dx \iff f''(x)\,dx = x^2\,dx, \quad \text{and so } x^2 = f''(x).$$

Continuing in this way,

$$x^0 = f(x), \quad x^1 = f'(x), \ldots, x^{l+1} = f^{(l+1)}(x),$$

and finally

$$dx^{l+1} = F\,dx \iff f^{(l+2)}(x) = F(x, f(x), f'(x), \ldots, f^{(l+1)}(x));$$

since the choice of F was arbitrary, we can regard $f(x)$ as an arbitray function.

Conclusion. *Integrating* V *is equivalent to solving the ordinary differential equations*

$$y^{(l+2)} = F(x, y, y', \ldots, y^{(l+1)})$$

for arbitrary smooth functions F.

Let us recapitulate what we have done.

Theorem 6.7.1. *Let* V *be a 2-dimensional vector field system on* $\mathbb{R}^r$ *such that* $C(V) = \{0\}$ *and* $\dim V^{(\alpha)} = \dim V^{(\alpha-1)} + 1$ *for* $\alpha = 1, \ldots, l+1$, *while* V^{l+1} *is complete. Since* $\dim V^{(l+1)} = l+3$, V *admits* $\tau := r-l-3$ *functionally independent first integrals* $t^1, \ldots, t^\tau$.

Then it is possible to introduce $l + 3$ variables $x, x^0, x^1, \ldots, x^{l+1}$ *which together with* $t^1, \ldots, t^\tau$ *form a system of local coordinates for* $\mathbb{R}^r$ *such that* V *is brought to the canonical form*

$$V = \left(\frac{\partial}{\partial x} + x^1 \frac{\partial}{\partial x^0} + \cdots + x^{l+1} \frac{\partial}{\partial x^l}, \frac{\partial}{\partial x^{l+1}} \right).$$

Moreover, the integral curves of V on which $dx \neq 0$ are given by

$$\begin{cases} x^0 = f(x), \ x^1 = f'(x), \ldots, \ x^{l+1} = f^{(l+1)}(x), \\ t^1 = constant, \ldots, \ t^\tau = constant, \end{cases}$$

where $f(x)$ is an arbitrary smooth function. □

If the original coordinates for $\mathbb{R}^r$ were $y^1, \ldots, y^r$, the introduction of the new coordinates $x, x^0, x^1, \ldots, x^{l+1}, t^1, \ldots, t^\tau$ is made by means of a local diffeomorphism

$$\Phi : \mathbb{R}^r \xrightarrow{\cong} \mathbb{R}^r$$

$$(x, t) \mapsto y^i = \Phi^i(x, t) \quad \text{for } i = 1, \ldots, r.$$

In the old coordinates the integral curves of V are expressed by

$$\begin{cases} y^i = \Phi^i(x, x^0, x^1, \ldots, x^{l+1}, t^1, \ldots, t^\tau) \quad \text{for } i = 1, \ldots, r, \\ \quad \text{where } x^0 = f(x), \ x^1 = f'(x), \ldots, x^{l+1} = f^{(l+1)}(x), \quad (*) \\ \quad \text{and } t^1 = constant, \ldots, t^\tau = constant. \end{cases}$$

Definition. A system of character 1 which has a general solution of the form (*) is said to be *explicitly integrable*.

The theorem above has a straightforward converse.

Theorem 6.7.2. *If the 2-dimensional vector field system V is explicitly integrable there is an integer l such that* $\dim V^{(\alpha)} = \dim V^{(\alpha-1)+1}$ *for $\alpha = 1,\ldots,l+1$, and $V^{(l+1)}$ is complete.*

Proof. Introduce new local coordinates $x, x^0, x^1, \ldots, x^{l+1}, t^1, \ldots, t^\tau$ by inverting the equations $y^i = \Phi^i(x, x^0, \ldots, x^{l+1}, t^1, \ldots, t^\tau)$ $(i = 1, \ldots, r)$ in (*). Then the dual pfaffian system $V^\perp$ of V acquires the form

$$\begin{cases} dx^0 - x^1\, dx = 0, \ dx^1 - x^2\, dx = 0, \ldots, dx^l - x^{l+1}\, dx = 0, \\ dt^1 = 0, \ldots, dt^\tau = 0. \end{cases}$$

And so

$$V = \left(\frac{\partial}{\partial x} + x^1 \frac{\partial}{\partial x^0} + \cdots + x^{l+1} \frac{\partial}{\partial x^l}, \frac{\partial}{\partial x^{l+1}} \right),$$

from which the conclusion is evident. $\qquad\qquad\qquad\square$

The concept of explicitly integrable systems was introduced by Hilbert, who proved that the general solution of the ODE

$$\frac{dz}{dx} = \left(\frac{d^2 y}{dx^2} \right)^2$$

cannot be written in the form

$$\begin{cases} x = \phi(t, w(t), w'(t), \ldots, w^{(n)}(t)), \\ y = \psi(t, w(t), w'(t), \ldots, w^{(n)}(t)), \\ z = \chi(t, w(t), w'(t), \ldots, w^{(n)}(t)), \end{cases}$$

where ϕ, ψ and χ are determinate functions, and w is an arbitrary smooth function of t.

Then Cartan found—in [Cartan 1914]—a very simple characterization of explicitly integrable pfaffian systems P—namely that $\dim P' = \dim P - 1$, $\dim P'' = \dim P' - 1$, and so on until a complete system is reached.

So the results above merely constitute a dual version of Cartan's theory.

7

Higher order contact transformations

The results of the preceding chapter make it possible to sketch the proof of Lie's rectification theorem for first order PDE systems, mentioned at the end of section **2.2**. This can be interpreted as saying that the Lie pseudogroup $^1Cont^{n,1}$ is sufficiently large for allowing local rectifications of submanifolds of $J^1(\mathbb{R}^n, \mathbb{R})$ corresponding to involutive first order PDE systems.

Having achieved this, Drach's classification shows that in order to solve arbitrary PDE systems it then would be sufficient to demonstrate the corresponding theorem for the second order jet bundle $J^2(\mathbb{R}^n, \mathbb{R})$. In this context Lie posed the following questions.

(i) Is $^2Cont^{n,1}$ really larger than $^1Cont^{n,1}$?
(ii) Given $\mathcal{S} \subset J^2(\mathbb{R}^n, \mathbb{R})$, is it possible to find a local diffeomorphism $\Phi: J^2(\mathbb{R}^n, \mathbb{R}) \longrightarrow J^2(\mathbb{R}^n, \mathbb{R})$ which without being a contact transformation induces a bijection $\mathrm{sol}(\mathcal{S}) \leftrightarrow \mathrm{sol}(\Phi(\mathcal{S}))$?

The background of the first question is that Lie—correctly—thought that $^2Cont^{n,1}$ is far too small for flattening out submanifolds of $J^2(\mathbb{R}^n, \mathbb{R})$ corresponding to involutive second order PDE systems. On the other hand Lie thought that there might exist coordinate transformations of $J^2(\mathbb{R}^n, \mathbb{R})$ which are *not* contact transformations, but nevertheless can be used for simplification purposes—whence the second question.

In section **7.2** we explain Bäcklund's answer to the first question, which says that $^qCont^{n,1} \cong {}^1Cont^{n,1}$ for $q \geq 1$. Then section **7.3** considers related questions for $^qCont^{n,m}$.

Lie's second question was answered with *no* in [Cartan 1915]. However, Bäcklund had earlier shown that *certain* PDE systems admit *correspondences* that establish bijections of the wanted kind. Since Bäcklund's theory can be applied to a very restricted class of PDE systems only, we do not discuss it here, but merely mention that [Goursat 1918] still is a very good reference for the original version of Bäcklund transformations.

7.1 Lie's rectification theorem for first order PDE systems in one dependent variable

Consider $\mathcal{S} \subset J^1(\mathbb{R}^n_x, \mathbb{R}_z)$ given by

$$F_i(x, z, p) := p_i - f_i(x^1, \ldots, x^n, z, p_{m+1}, \ldots, p_n) = 0 \quad \text{for } i = 1, \ldots, m.$$

Its first prolongation is

$$\mathcal{S}^{(1)}: \quad F_i = 0 \quad \text{and} \quad dF_i/dx^j = 0 \quad \text{for } i = 1, \ldots, m, \ j = 1, \ldots, n,$$

where $d/dx^j = \partial/\partial x^j + p_j \, \partial/\partial z + \sum_{k=1}^n p_{jk} \, (\partial/\partial p_k)$. That is,

$$\frac{dF_i}{dx^j} = p_{ji} - \frac{\partial f_i}{\partial x^j} - p_j \frac{\partial f_i}{\partial z} - \sum_{k=m+1}^n p_{jk} \frac{\partial f_i}{\partial p_k}.$$

So from $dF_i/dx^j = 0$ we deduce that

$$p_{ji} = \frac{\partial f_i}{\partial x^j} + p_j \frac{\partial f_i}{\partial z} + \sum_{k=m+1}^n \left(\frac{\partial f_j}{\partial x^k} + p_k \frac{\partial f_j}{\partial z} + \sum_{l=m+1}^n p_{kl} \frac{\partial f_j}{\partial p_l} \right) \frac{\partial f_i}{\partial p_k}.$$

From these second order conditions we can extract equations only involving first order jet coordinates. To wit, using $p_{ji} = p_{ij}$ when $i, j = 1, \ldots, m$ we find that

$$0 = p_{ji} - p_{ij} = \left(\frac{\partial f_i}{\partial x^j} + p_j \frac{\partial f_i}{\partial z} \right) - \left(\frac{\partial f_j}{\partial x^i} + p_i \frac{\partial f_j}{\partial z} \right)$$

$$+ \sum_{k=m+1}^n \left(\left(\frac{\partial f_j}{\partial x^k} + p_k \frac{\partial f_j}{\partial z} \right) \frac{\partial f_i}{\partial p_k} - \left(\frac{\partial f_i}{\partial x^k} + p_k \frac{\partial f_i}{\partial z} \right) \frac{\partial f_j}{\partial p^k} \right).$$

The expression after the second equality sign looks a lot like a contact bracket—and in fact,

$$\{p_i - f_i, p_j - f_j\} = \left(\frac{\partial f_i}{\partial x^j} + p_j \frac{\partial f_i}{\partial z} \right) - \left(\frac{\partial f_j}{\partial x^i} + p_i \frac{\partial f_j}{\partial z} \right) +$$

$$\sum_{k=m+1}^n \left(\left(\frac{\partial f_j}{\partial x^k} + p_k \frac{\partial f_j}{\partial z} \right) \frac{\partial f_i}{\partial p_k} - \left(\frac{\partial f_i}{\partial x^k} + p_k \frac{\partial f_i}{\partial z} \right) \frac{\partial f_j}{\partial p_k} \right)$$

Conclusion. *The solutions of $\mathcal{S}$ have to satisfy the following first order* *equations:*

$$\begin{cases} p_i = f_i(x^1,\ldots,x^n,z,p_{m+1},\ldots,p_n), \\ \{p_i - f_i, p_j - f_j\} = 0 \end{cases} \quad for\ i,j = 1,\ldots,m.$$

So the first order integrability conditions for S are

$$\{p_i - f_i, p_j - f_j\}_{|S} = 0 \quad for\ i,j = 1,\ldots,m.$$

Now $\{p_i - f_i, p_j - f_j\}$ can for $i,j = 1,\ldots,m$ be rewritten as

$$\{p_i - f_i, p_j - f_j\} = (p_j - f_j)\frac{\partial f_i}{\partial z} - (p_i - f_i)\frac{\partial f_j}{\partial z} + \text{ something which}$$
$$\text{only depends on } x^1,\ldots,x^n,z,p_{m+1},\ldots,p_n,$$

and therefore

$$\{p_i - f_i, p_j - f_j\}_{|S} = 0 \iff$$
$$\{p_i - f_i, p_j - f_j\} = (p_j - f_j)\frac{\partial f_i}{\partial z} - (p_i - f_i)\frac{\partial f_j}{\partial z}.$$

Thus we have obtained the following result.

Lemma 7.1.1. *A necessary condition for*

$$S: \quad p_i = f_i(x^1,\ldots,x^n,z,p_{m+1},\ldots,p_n), \quad i = 1,\ldots,m,$$

to be involutive is that

$$\{p_i - f_i, p_j - f_j\} = (p_j - f_j)\frac{\partial f_i}{\partial z} - (p_i - f_i)\frac{\partial f_j}{\partial z} \quad for\ i,j = 1,\ldots,m.$$

$\square$

In order to apply the results of the preceding chapter we want even more, namely

$$\{p_i - f_i, p_j - f_j\} = 0 \quad for\ i,j = 1,\ldots,m,$$

which is true if the f_i do not depend on z. And there is indeed a classical trick for getting rid of the dependent variable.

Recall that an n-dimensional complete subsystem of the vector field system associated to S defines a foliation of S with leaves that are generalized solutions. Projecting this down to $\mathbb{R}^n_x \times \mathbb{R}_z$ will in general *not* give a nice foliation there. But nonetheless we may by brute force introduce the concept of 'foliation of $\mathbb{R}^n_x \times \mathbb{R}_z$ with leaves that are graphs of solutions of S'.

Definition. Let S be a first order PDE system in $J^1(\mathbb{R}^n, \mathbb{R})$ defined by

$$F_k(x^1, \ldots, x^n, z, p_1, \ldots, p_n) = 0 \quad \text{for } k = 1, \ldots, m.$$

A 'complete solution' of S is a 1-parameter foliation

$$u(x^1, \ldots, x^n, z) = t \quad (t \in \mathbb{R}) \tag{*}$$

of $\mathbb{R}^n_x \times \mathbb{R}_z$ with $\partial u / \partial z \neq 0$, such that the function $z = \zeta_t(x^1, \ldots, x^n)$, obtained from (*) by means of the implicit function theorem, is a solution of S for each value of the parameter t.

Let L_t be the leaf defined by $u(x, z) = t$. On L_t we have

$$z = \zeta_t(x^1, \ldots, x^n) \quad \text{and} \quad u(x^1, \ldots, x^n, \zeta_t) \equiv t,$$

so that

$$\frac{\partial u}{\partial x^i} + \frac{\partial u}{\partial z} \cdot \frac{\partial z}{\partial x^i} = 0 \quad \text{for } i = 1, \ldots, n.$$

Hence

$$F_k\left(x^1, \ldots, x^n, z, -\frac{\partial u / \partial x^1}{\partial u / \partial z}, \ldots, -\frac{\partial u / \partial x^n}{\partial u / \partial z}\right) = 0 \quad \text{for } k = 1, \ldots, n$$

on *each* L_t, i.e., on all of $\mathbb{R}^n_x \times \mathbb{R}_z$. Replacing z by an *independent variable* x^{n+1} and considering u as a new dependent variable, u will satisfy the PDE system

$$F_k\left(x^1, \ldots, x^{n+1}, -\frac{\partial u / \partial x^1}{\partial u / \partial x^{n+1}}, \ldots, -\frac{\partial u / \partial x^n}{\partial u / \partial x^{n+1}}\right) = 0 \quad \text{for } k = 1, \ldots, m.$$

By this artifice S can be solved in two steps: first determine all 'complete solutions' u of S by solving the PDE system

$$F_k\left(x^1, \ldots, x^{n+1}, -\frac{p_1}{p_{n+1}}, \ldots, -\frac{p_n}{p_{n+1}}\right) = 0 \quad \text{for } k = 1, \ldots, m,$$

which does not involve the dependent variable, and then obtain solutions $z = \zeta(x^1, \ldots, x^n)$ by means of the implicit function theorem.

Accepting this we are reduced to systems not involving the dependent variable:

$$p_i = f_i(x^1, \ldots, x^n, p_{m+1}, \ldots, p_n) \quad \text{for } i = 1, \ldots, m.$$

Then the integrability conditions are given by

$$\{p_i - f_i, p_j - f_j\} = 0 \quad \text{for } i, j = 1, \ldots, m.$$

Theorem 7.1.1 (Lie's rectification theorem). *Let $S \subset J^1(\mathbb{R}^n_x, \mathbb{R}_z)$ be given by*

$$p_i = f_i(x^1, \ldots, x^n, p_{m+1}, \ldots, p_n) \quad for\ i = 1, \ldots, m,$$

where

$$\{p_i - f_i, p_j - f_j\} = 0 \quad for\ i, j = 1, \ldots, m.$$

Then there is a contact transformation $\Phi : J^1(\mathbb{R}^n_x, \mathbb{R}_z) \longrightarrow J^1(\mathbb{R}^n_y, \mathbb{R}_w)$ such that

$$\Phi(S) = \{(y, w, q) \in J^1(\mathbb{R}^n_y, \mathbb{R}_w) \mid y^1 = \cdots = y^m = 0\}.$$

Proof. The contact vector field system

$$\left(\frac{\partial}{\partial x^1} + p_1 \frac{\partial}{\partial z}, \ldots, \frac{\partial}{\partial x^n} + p_n \frac{\partial}{\partial z}, \frac{\partial}{\partial p_1}, \ldots, \frac{\partial}{\partial p_n} \right)$$

of $J^1(\mathbb{R}^n_x, \mathbb{R}_z)$ admits no Cauchy characteristic vector field, and has no first integrals. Following the method of the last chapter we can therefore introduce a new contact coordinate system $y^1, \ldots, y^n, w, q_1, \ldots, q_n$ as follows:

- y^1 can be chosen arbitrarily; we take $y^1 := p_1 - f_1(x, p)$;
- y^2 is chosen such that $\{y^1, y^2\} = 0$ and y^2 is not a function of y^1; let us take $y^2 := p_2 - f_2(x, p)$;
- y^3 is chosen such that $\{y^1, y^3\} = \{y^2, y^3\} = 0$ and y^3 is not a function of y^1 and y^2; let us take $y^3 := p_3 - f_3(x, p)$;
- continue in this way until y^{m-1} is defined;
- since $\{p_i - f_i, p_j - f_j\} = 0$ for all $i, j = 1, \ldots, m$, we can finally choose $y^m := p_m - f_m(x, p)$.

After that, $y^{m+1}, \ldots, y^n$ are determined successively as functionally independent solutions of the Frobenius systems

$$\{y^1, y^{m+j}\} = \cdots = \{y^{m+j-1}, y^{m+j}\} = 0 \quad for\ j = 1, \ldots, n - m.$$

This gives the complete subsystem $(\partial/\partial y^1, \ldots, \partial/\partial y^n)$, whereupon w, q_1, $\ldots$, q_n are obtained by the reduction procedure in section **6.4**. The resulting

$$\Phi : J^1(\mathbb{R}^n_x, \mathbb{R}_z) \to J^1(\mathbb{R}^n_y, \mathbb{R}_w)$$

$$(x, z, p) \mapsto (y, w, q)$$

is then a contact transformation with the property

$$\Phi(S) = \{(y, w, q) \in J^1(\mathbb{R}^n_y, \mathbb{R}_w) \mid y^1 = \cdots = y^m = 0\}.$$

$\square$

7.2 Bäcklund's theorems

Definitions. The *contact pfaffian system* $^qCt^{n,1}$ on $J^q(\mathbb{R}^n_x, \mathbb{R}_z)$ is generated by the one-forms

$$
\begin{cases}
\theta^0 = dz - \sum_{i_1=1}^n p_{i_1}\, dx^{i_1}, \\
\theta^{i_1} = dp_{i_1} - \sum_{i_2=1}^n p_{i_1 i_2}\, dx^{i_2}, \\
\cdots \\
\theta^{i_1 \cdots i_{q-1}} = dp_{i_1 \cdots i_{q-1}} - \sum_{i_q=1}^n p_{i_1 \cdots i_q}\, dx^{i_q}
\end{cases}
\qquad \text{for } 1 \le i_1, \ldots, i_q \le n.
$$

The Lie pseudogroup $^qCont^{n,1}$ of qth *order contact transformations* is the family of all local diffeomorphisms of $J^q(\mathbb{R}^n_x, \mathbb{R}_z)$ which preserve $^qCt^{n,1}$. In particular a contact transformation of $J^1(\mathbb{R}^n_x, \mathbb{R}_z)$ is the same as a first order contact transformation.

The purpose of this section is to prove that $^qCont^{n,1}$ and $^1Cont^{n,1}$ are essentially isomorphic for $q \ge 2$, which unfortunately has the consequence that $^qCont^{n,1}$ is far too small for enabling rectifications of involutive q^{th} order PDE systems, considered as submanifolds of $J^q(\mathbb{R}^n, \mathbb{R})$.

Groups of contact transformations of different orders may be connected to each other by means of the natural projections

$$
\pi^q_r : J^q(\mathbb{R}^n, \mathbb{R}) \longrightarrow J^r(\mathbb{R}^n, \mathbb{R}) \quad \text{for } q > r,
$$

since these induce injections

$$
(\pi^q_r)^* : {}^rCt^{n,1} \hookrightarrow {}^qCt^{n,1},
$$

where $(\pi^q_r)^*({}^rCt^{n,1})$ is the module generated by the one-forms of $^rCt^{n,1}$, lifted to $J^q(\mathbb{R}^n, \mathbb{R})$, over the ring of smooth functions on $J^q(\mathbb{R}^n, \mathbb{R})$. Identifying $(\pi^q_r)^*({}^rCt^{n,1})$ with $^rCt^{n,1}$, there arises a chain of inclusions:

$$
{}^1Ct^{n,1} \hookrightarrow {}^2Ct^{n,1} \hookrightarrow \cdots \hookrightarrow {}^qCt^{n,1}.
$$

In this situation **Bäcklund's first theorem** says the following.

Theorem 7.2.1. *If* $\Phi \in {}^qCont^{n,1}$, Φ *preserves not only* $^qCt^{n,1}$, *but also each of* $^{q-1}Ct^{n,1}, \ldots, {}^1Ct^{n,1}$.

In order to prove this we first recall that the derivative $\mathcal{P}'$ of a pfaffian system $\mathcal{P}$ is defined by

$$
\mathcal{P}' := \{\theta \in \mathcal{P} \mid d\theta \equiv 0 \pmod{\mathcal{P}}\}.
$$

Let us for instance determine $(^rCt^{n,1})'$.

$$\theta^0 = dz - \sum_{i=1}^{n} p_i \, dx^i \implies$$

$$d\theta^0 = -\sum dp_i \wedge dx^i \equiv -\sum p_{ij} \, dx^j \wedge dx^i \pmod{^2Ct^{n,1}}.$$

But $p_{ij} = p_{ji}$ and $dx^i \wedge dx^j = -dx^j \wedge dx^i$, and so

$$d\theta^0 = 0 \pmod{^2Ct^{n,1}}.$$

Next

$$\theta^i = dp_i - \sum_{j=1}^{n} p_{ij} \, dx^j \implies d\theta^i = -\sum_{j=1}^{n} dp_{ij} \wedge dx^j$$

$$\equiv -\sum_{j,k=1}^{n} p_{ijk} \, dx^k \wedge dx^j \pmod{^3Ct^{n,1}},$$

and hence $d\theta^i \equiv 0 \pmod{^3Ct^{n,1}}$. In the same way one sees that

$$d\theta^{i_1 \ldots i_k} \equiv 0 \pmod{^{k+2}Ct^{n,1}} \quad \text{for } k \le r-2.$$

However,

$$d\theta^{i_1 \ldots i_{r-1}} = -\sum_{i_r=1}^{n} dp_{i_1 \ldots i_r} \wedge dx^{i_r},$$

which is *not* $\equiv 0 \pmod{^rCt^{n,1}}$. Thus $(^rCt^{n,1})'$ is generated by θ^0, θ^{i_1}, ... , $\theta^{i_1 \ldots i_{r-2}}$, i.e.,

$$(^rCt^{n,1})' = {}^{r-1}Ct^{n,1}.$$

Bäcklund's first theorem is a consequence of this observation and the following result.

Lemma 7.2.1. *If a local diffeomorphism ϕ preserves a pfaffian system $\mathcal{P}$— that is, $\phi^*(\mathcal{P}) = \mathcal{P}$—it also preserves $\mathcal{P}'$.*

Proof.

$$\theta \in \mathcal{P}' \iff \theta \in \mathcal{P} \text{ and } d\theta \equiv 0 \pmod{\mathcal{P}}$$

$$\iff \phi^*(\theta) \in \phi^*(\mathcal{P}) \text{ and } \phi^*(d\theta) \equiv 0 \pmod{\phi^*(\mathcal{P})}$$

$$\iff \phi^*(\theta) \in \mathcal{P} \text{ and } d(\phi^*(\theta)) \equiv 0 \pmod{\mathcal{P}}$$

$$\iff \phi^*(\theta) \in \mathcal{P}'.$$

$\square$

Thus $\Phi^q \in {}^qCont^{n,1}$ automatically preserves ${}^{q-1}Ct^{n,1}$, ${}^{q-2}Ct^{n,1}$, ... and ${}^1Ct^{n,1}$, and so is a contact transformation of order $q - 1$, ..., 1 as well—which is precisely the content of Bäcklund's first theorem.

But more is true: under rather general conditions Φ^q is a *prolongation* of a contact transformation Φ acting on $J^1(\mathbb{R}^n, \mathbb{R})$—where the notion of prolongation is introduced in the following manner.

Definition. Let M and M' be manifolds with $\dim M = m$ and $\dim M' = m + n$, and let $\pi: M' \longrightarrow M$ be a submersion mapping given by $(x; y) \mapsto x$, where $(x; y) = (x^1, \ldots, x^m; y^1, \ldots, y^n)$ and $x = (x^1, \ldots, x^m)$ are local coordinates for M' and M respectively. A local diffeomorphism g' acting on M' is *projectable* if it is of the form $(x; y) \mapsto (g(x); h(x; y))$, in which case its *projection* $\pi_*(g')$ on M is given by $x \mapsto g(x)$. Conversely g' is said to be a *prolongation* of g in this situation. A pseudogroup $\mathfrak{G}'$ on M' is an *isomorphic prolongation* of a pseudogroup $\mathfrak{G}$ acting on M if each $g' \in \mathfrak{G}'$ is projectable and π induces an isomorphism

$$\pi_*: \mathfrak{G}' \xrightarrow{\cong} \mathfrak{G}$$
$$g' \mapsto \pi_*(g').$$

In our case we will consider the projection

$$\pi_1^q: J^q(\mathbb{R}^n, \mathbb{R}) \longrightarrow J^1(\mathbb{R}^n, \mathbb{R}) \quad \text{for } q > 1.$$

With $\hat{J}^q(\mathbb{R}^n_{\hat{x}}, \mathbb{R}_{\hat{z}})$ being a copy of $J^q(\mathbb{R}^n_x, \mathbb{R}_z)$, a qth order contact transformation can be thought of as a mapping

$$\Phi^q: J^q(\mathbb{R}^n, \mathbb{R}) \longrightarrow \hat{J}^q(\mathbb{R}^n, \mathbb{R})$$

with $(\Phi^q)^*({}^q\hat{C}t^{n,1}) \cong {}^qCt^{n,1}$. Assume that Φ^q is given by

$$\begin{cases} \hat{x}^k = X^k(x^i; z; p_i; p_{ij}; \ldots), \\ \hat{z} = Z(x^i; z; p_i; p_{ij}; \ldots), \\ \hat{p}_k = P_k(x^i; z; p_i; p_{ij}; \ldots), \\ \ldots. \end{cases}$$

Notations. Let $\mathfrak{G}^q := {}^qCont^{n,1}$ and let $\mathfrak{G}_0^q := \{\Phi^q \in \mathfrak{G}^q \mid \hat{x}^k = X^k(x^i; z; p_i)$ and $\det(\partial X^k/\partial x^i + p_i\, \partial X^k/\partial z) \neq 0\}$.

It is now possible to formulate **Bäcklund's second theorem**.

Theorem 7.2.2. 1. *Any $\Phi^q \in \mathfrak{G}_0^q$ admits a projection $\Phi^1 \in \mathfrak{G}_0^1$ such that Φ^q is a prolongation of Φ^1.*

2. *Any contact transformation*

$$\Phi^1 : \begin{cases} \hat{x}^k = X^k(x^i; z; p_i), \\ \hat{z} = Z(x^i; z; p_i), \\ \hat{p}_k = P_k(x^i; z; p_i) \end{cases}$$

with $\det(dX^j/dx^i) \neq 0$—*which in particular is satisfied if* Φ^1 *is close to the identity— has a uniquely defined prolongation to a* q^{th} *order contact transformation* Φ^q.

Proof. Let $\Phi^q \in \mathfrak{G}_0^q$. According to the first theorem, $(\Phi^q)^*({}^rCt^{n,1}) \cong {}^rCt^{n,1}$ for $r = 1, \ldots, q$. For $r = 1$ this amounts to

$$dZ - \sum_{j=1}^{n} P_j \, dX^j = \rho^0 \cdot \left(dz - \sum_{i=1}^{n} p_i \, dx^i \right)$$

for some nonvanishing function ρ^0 on $J^q(\mathbb{R}^n, \mathbb{R})$—that is,

$$\sum_{i=1}^{n} \left(\frac{\partial Z}{\partial x^i} - \sum_{j=1}^{n} P_j \frac{\partial X^j}{\partial x^i} + \rho^0 p_i \right) dx^i$$

$$+ \left(\frac{\partial Z}{\partial z} - \sum_{j=1}^{n} P_j \frac{\partial X^j}{\partial z} - \rho^0 \right) dz$$

$$+ \sum_{i=1}^{n} \left(\frac{\partial Z}{\partial p_i} - \sum_{j=1}^{n} P_j \frac{\partial X^j}{\partial p_i} \right) dp_i$$

$$+ \sum_{i,j=1}^{n} \frac{\partial Z}{\partial p_{ij}} \, dp_{ij} + \sum_{i,j,k=1}^{n} \frac{\partial Z}{\partial p_{ijk}} \, dp_{ijk} + \cdots = 0.$$

The last line shows that

$$\frac{\partial Z}{\partial p_{ij}} = \frac{\partial Z}{\partial p_{ijk}} = \cdots = 0,$$

and hence

$$\hat{z} = Z(x^i; z; p_i).$$

According to the second line,

$$\rho^0 = \frac{\partial Z}{\partial z} - \sum_{j=1}^{n} P_j \frac{\partial X^j}{\partial z}.$$

Inserting this into the coefficient of dx^i, we get

$$\frac{\partial Z}{\partial x^i} + p_i \frac{\partial Z}{\partial z} = \sum_{j=1}^{n} P_j \left(\frac{\partial X^j}{\partial x^i} + p_i \frac{\partial X^j}{\partial z} \right) \qquad \text{for } i = 1, \ldots, n.$$

By assumption $\det(\partial X^j/\partial x^i + p_i \, \partial X^j/\partial z) \neq 0$, and therefore it is possible to solve the P_j from this system so as to express them by means of derivatives of Z and X^j. But then the P_j are functions of x^i, z, p_i only, so it is possible to define $\Phi^1 \in \mathfrak{G}_0^1$ by

$$\begin{cases} \hat{x}^k = X^k(x^i; z; p_i), \\ \hat{z} = Z(x^i; z; p_i), \\ \hat{p}_k = P_k(x^i; z; p_i). \end{cases}$$

Conversely, starting with an arbitrary $\Phi^1 \in \mathfrak{G}^1$ we want to find a prolongation $\Phi^2 \in \mathfrak{G}^2$ by adding suitable relations

$$\hat{p}_{kl} = P_{kl}(x^i; z; p_i; p_{ij})$$

to those defining Φ^1. Since Φ^2 is to preserve $^2Ct^{n,1}$, the P_{kl} must satisfy

$$dP_k - \sum_{j=1}^{n} P_{kj} \, dX^j = \rho_k^0 \left(dz - \sum_{i=1}^{n} p_i \, dx^i \right) + \sum_{l=1}^{n} \rho_k^l \left(dp_l - \sum_{i=1}^{n} p_{li} \, dx^i \right)$$

for certain functions $\rho_k^0, \rho_k^1, \ldots, \rho_k^n$ living on $J^2(\mathbb{R}^n, \mathbb{R})$. Thus

$$\sum_{i=1}^{n} \left(\frac{\partial P_k}{\partial x^i} - \sum_{j=1}^{n} P_{kj} \frac{\partial X^j}{\partial x^i} + \rho_k^0 p_i + \sum_{l=1}^{n} \rho_k^l \, p_{li} \right) dx^i$$

$$+ \left(\frac{\partial P_k}{\partial z} - \sum_{j=1}^{n} P_{kj} \frac{\partial X^j}{\partial z} - \rho_k^0 \right) dz$$

$$+ \sum_{i=1}^{n} \left(\frac{\partial P_k}{\partial p_i} - \sum_{j=1}^{n} P_{kj} \frac{\partial X^j}{\partial p_i} - \rho_k^i \right) dp_i = 0,$$

showing that

$$\rho_k^0 = \frac{\partial P_k}{\partial z} - \sum_{j=1}^{n} P_{kj} \frac{\partial X^j}{\partial z}$$

and

$$\rho_k^i = \frac{\partial P_k}{\partial p_i} - \sum_{j=1}^{n} P_{kj} \frac{\partial X^j}{\partial p_i} \qquad \text{for } i = 1, \ldots, n.$$

Inserting this into the coefficient of dx^i,

$$\frac{dP_k}{dx^i} = \sum_{j=1}^{n} P_{kj} \frac{dX^j}{dx^i} \quad \text{for } i,k = 1,\ldots,n.$$

Because $\det(dX^j/dx^i) \neq 0$, the P_{kj} can be solved from this system and are thus *uniquely determined* by the given X^j and P_k. Hence Φ^1 has a *unique contact prolongation* $\Phi^2 \in \mathfrak{G}_0^2$.

Once we have found Φ^2, the next prolongation $\Phi^3 \in \mathfrak{G}_0^3$ is analogously determined by the condition that Φ^3 preserves ${}^3Ct^{n,1}$, which implies that

$$\frac{dP_{ij}}{dx^k} = \sum_l P_{ijl} \frac{dX^l}{dx^k}, \quad \text{whence } P_{ijl} = \ldots,$$

and so on. $\qquad\qquad\qquad\qquad\qquad\qquad\qquad\qquad\qquad\qquad\qquad\square$

7.3 Contact prolongations of local diffeomorphisms

Definitions. The *contact pfaffian system* ${}^qCt^{n,m}$ on $J^q(\mathbb{R}^n_x, \mathbb{R}^m_y)$ is generated by the one-forms

$$\begin{cases} \theta_0^j = dy^j - \sum_{i_1=1}^n p_{i_1}^j\, dx^{i_1}, \\ \theta_{i_1}^j = dp_{i_1}^j - \sum_{i_2=1}^n p_{i_1 i_2}^j\, dx^{i_2}, \\ \ldots \\ \theta_{i_1\ldots i_{q-1}}^j = dp_{i_1\ldots i_q}^j - \sum_{i_q=1}^n p_{i_1\ldots i_q}^j\, dx^{i_q}, \end{cases}$$

where $j = 1,\ldots,m$ and $1 \le i_1,\ldots,i_q \le n$.

The Lie pseudogroup ${}^qCont^{n,m}$ of qth order contact transformations consists of all local diffeomorphisms of $J^q(\mathbb{R}^n_x, \mathbb{R}^m_y)$ that preserve ${}^qCt^{n,m}$.

We will need the following **prolongation theorem for local diffeomorphisms** when studying Lie pseudogroups later on.

Theorem 7.3.1. *Let the local diffeomorphism*

$$\Phi : \mathbb{R}^n_x \times \mathbb{R}^m_y \longrightarrow \mathbb{R}^n_{\hat{x}} \times \mathbb{R}^m_{\hat{y}}$$

be given by

$$\begin{cases} \hat{x}^i = X^i(x^1,\ldots,x^n; y^1,\ldots,y^m) & \text{for } i = 1,\ldots,n, \\ \hat{y}^j = Y^j(x^1,\ldots,x^n; y^1,\ldots,y^m) & \text{for } j = 1,\ldots,m. \end{cases}$$

If $\det(dX^i/dx^k) \neq 0$—which in particular is true if Φ is close to the *identity—there is for each $q \geq 1$ a unique q^{th} order contact transformation $\Phi^q \in {}^q Cont^{n,m}$ which prolongs Φ with respect to the projection*

$$\pi_0^q : J^q(\mathbb{R}^n, \mathbb{R}^m) \longrightarrow J^0(\mathbb{R}^n, \mathbb{R}^m) \cong \mathbb{R}^n \times \mathbb{R}^m.$$

Proof. Φ^1 is defined by adding suitable equations

$$\hat{p}_k^j = P_k^j(x^1, \ldots, x^n; y^1, \ldots, y^m; p_1^1, \ldots, p_n^1; \ldots; p_1^m, \ldots, p_n^m)$$

to those defining Φ in such a way that Φ^1 preserves the contact pfaffian system ${}^1 Ct^{n,m}$ of $J^1(\mathbb{R}^n, \mathbb{R}^m)$. That is, there are to be functions ρ_l^j on $J^1(\mathbb{R}^n, \mathbb{R}^m)$ such that

$$dY^j - \sum_{i=1}^n P_i^j \, dX^i = \sum_{l=1}^n \rho_l^j \left(dy^l - \sum_{k=1}^n p_k^l \, dx^k \right) \qquad \text{for } j = 1, \ldots, m.$$

Rewriting this as

$$\sum_{k=1}^n \left(\frac{\partial Y^j}{\partial x^k} - \sum_{i=1}^n P_i^j \frac{\partial X^i}{\partial x^k} + \sum_{l=1}^m \rho_l^j p_k^l \right) dx^k$$

$$+ \sum_{l=1}^m \left(\frac{\partial Y^j}{\partial y^l} - \sum_{i=1}^n P_i^j \frac{\partial X^i}{\partial y^l} - \rho_l^j \right) dy^l = 0,$$

we see that

$$\rho_l^j = \frac{\partial Y^j}{\partial y^l} - \sum_{i=1}^n P_i^j \frac{\partial X^i}{\partial y^l} \qquad \text{for } j, l = 1, \ldots, m,$$

and

$$\frac{\partial Y^j}{\partial x^k} - \sum_{i=1}^n P_i^j \frac{\partial X^i}{\partial x^k} + \sum_{l=1}^m p_k^l \left(\frac{\partial Y^j}{\partial y^l} - \sum_{i=1}^n P_i^j \frac{\partial X^i}{\partial y^l} \right) = 0$$

for $j = 1, \ldots, m$ and $k = 1, \ldots, n$. The last line can be regrouped as

$$\frac{\partial Y^j}{\partial x^k} + \sum_{l=1}^m p_k^l \frac{\partial Y^j}{\partial y^l} = \sum_{l=1}^n P_i^j \left(\frac{\partial X^i}{\partial x^k} + \sum_{l=1}^m p_k^l \frac{\partial X^i}{\partial y^l} \right),$$

i.e.,

$$\frac{dY^i}{dx^k} = \sum_{i=1}^n P_i^j \frac{dX^i}{dx^k} \qquad \text{for } j = 1, \ldots, m \text{ and } k = 1, \ldots, n.$$

So if $\det(dX^i/dx^k) \neq 0$, the wanted P_i^j are uniquely determined by the functions X^i and Y^j.

In the same way it can be seen that $\Phi^1 \in {}^1 Cont^{n,m}$ has a unique prolongation $\Phi^2 \in {}^2 Cont^{n,m}$, and so on.   □

8

Local Lie groups

When solving first order PDE systems in one unknown we encountered the Lie pseudogroup of contact transformations. Recall that a Lie pseudogroup is a pseudogroup the elements of which are solutions of a defining PDE system. The simplest case occurs when the first Cartan character of this system vanishes, so that the general solution depends on a number of parameters only (and not on arbitrary functions). The corresponding Lie pseudogroups are called *local Lie groups*, and the theory of these will play a decisive role in the study of second order PDE systems in one dependent and two independent variables.

Mainly following the classics [Lie–Scheffers 1893] and [Pontryagin 1966] we here present those aspects of local Lie groups that will be needed later. To illustrate what happens we will use the affine group in one variable as an example throughout the chapter.

The ground field is supposed to be $\mathbb{R}$, but $\mathbb{C}$ would work just as well. Later on we will switch to $\mathbb{C}$, since this makes everything much easier.

8.1 The parameter group and its structure constants

Let G be a Lie pseudogroup acting on $\mathbb{R}^n$ with $S(G)$ as its defining PDE system, so that $G =$ *the set of local solutions of* $S(G)$. If the Cartan character s_1 of $S(G)$ vanishes, the general element $g \in G$ is a local diffeomorphism of the form

$$g : \mathbb{R}^n \xrightarrow{\cong} \mathbb{R}^n$$
$$x \mapsto g(x; a) = g_a(x),$$

where $a = (a^1, \ldots, a^r)$ belongs to a parameter set. Or more precisely, g is given by

$$x = (x^1, \ldots, x^n) \mapsto (g^1(x^1, \ldots, x^n; a^1, \ldots, a^r), \ldots, g^n(x^1, \ldots, x^n; a^1, \ldots, a^r)).$$

The group operation is composition of local diffeomorphisms—*when defined*, that is—and so is automatically associative:

$$g_a \circ (g_b \circ g_c) = (g_a \circ g_b) \circ g_c, \quad \text{whenever the two sides are defined.}$$

The pseudogroup properties are the following:

- $g_a \circ g_b$ is defined $\implies \exists c = \gamma(a; b)$ such that $g_a \circ g_b = g_c$; the $\gamma^i(a; b)$ are called *composition functions*,
- $\exists e$ such that $g_e = \mathrm{id}$, i.e., $g_e(x) = x$ for $\forall x$; we choose coordinates in the parameter space such that $e = (0, \ldots, 0)$,
- for each $g_a \in G$ there is an $\bar{a}$ such that $g_a \circ g_{\bar{a}} = g_{\bar{a}} \circ g_a = \mathrm{id}$, that is $g_a^{-1} = g_{\bar{a}}$.

Clearly these demands put heavy restrictions on the possible defining PDE systems. So a basic question is

which PDE systems are defining systems for local Lie groups?

The answer is given by the **fundamental theorems of Lie**, which will be considered in sections **8.5** and **8.6**.

Note conversely that if G is a local r-parameter group acting on $\mathbb{R}^n$ by $x \mapsto g(x; a)$, G certainly satisfies a PDE system. For set $y := g(x; a)$, and form sufficiently many derivatives

$$\frac{\partial y^i}{\partial x^j} = \frac{\partial g^i}{\partial x^j}, \quad \frac{\partial^2 y^i}{\partial x^j \partial x^k} = \frac{\partial^2 g^i}{\partial x^j \partial x^k}, \quad \cdots$$

that all the parameters $a^1, \ldots, a^r$ can be eliminated among the equations thus obtained. When integrating the resulting PDE system, the a^i will reappear as integration constants.

Another basic question is this: suppose that G_1 acts on a manifold M_1, and G_2 on a manifold M_2 (with $\dim M_1 \neq \dim M_2$ in general); how should the concept of isomorphism be defined for such groups?

To resolve this we first of all want to get rid of the manifold on which the local Lie group G happens to act, so as to be able to study 'G itself'. But we have already noted that G induces a multiplication operation in the parameter space:

$$(a, b) \mapsto \gamma(a; b) =: a \cdot b.$$

And it is easy to see that γ in fact gives the parameter space $\mathbb{R}^r$ the structure of a local group, which is invariantly tied to G. We denote this parameter group by $\mathfrak{G}$, and consider the original G as a realization of $\mathfrak{G}$ in the form of a transformation group.

Definition. Two local Lie groups G_1 and G_2 are *locally isomorphic* if and only if the corresponding parameter groups $\mathfrak{G}_1$ and $\mathfrak{G}_2$ are. The latter means that there is a local diffeomorphism ϕ of the underlying manifolds of $\mathfrak{G}_1$ and $\mathfrak{G}_2$ such that

$$c = a \cdot b \text{ in } \mathfrak{G}_1 \iff \phi(c) = \phi(a) \cdot \phi(b) \text{ in } \mathfrak{G}_2.$$

Example: the affine group A acting on $\mathbb{R}$—(1). This group is defined by

$$x \mapsto a^1 x + a^2 =: g(x; a^1, a^2) = g(x; a) = g_a(x).$$

As a function of x, $g(x; a^1, a^2)$ satisfies the ODE

$$\frac{d^2 g}{dx^2} = 0.$$

Conversely *any* solution of this is clearly of the form $g(x; a^1, a^2)$, and thus belongs to the affine group A. Let us check the group property:

$$g_a \circ g_b(x) = g_a(b^1 x + b^2) = a^1 (b^1 x + b^2) + a^2$$
$$= a^1 b^1 x + (a^1 b^2 + a^2) = g_c(x),$$

with

$$c = (c^1, c^2) = (\gamma^1(a; b), \gamma^2(a; b)) := (a^1 b^1, a^1 b^2 + a^2).$$

The identity transformation is g_e with $e = (1, 0)$ (rather than $(0, 0)$—but this does not matter). The inverse $\bar{a} = (\bar{a}^1, \bar{a}^2)$ of $a = (a^1, a^2)$ is obtained from

$$\begin{cases} \gamma^1(a; \bar{a}) = a^1 \bar{a}^1 = 1, \\ \gamma^2(a; \bar{a}) = a^1 \bar{a}^2 + a^2 = 0, \end{cases}$$

i.e.,

$$\begin{cases} \bar{a}^1 = 1/a^1, \\ \bar{a}^2 = -a^2/a^1 \end{cases} \quad \text{when } a^1 \neq 0 \text{ (which is the case near } e = (1, 0)).$$

So the parameter group $\mathfrak{A}$ corresponding to the affine group is $(\mathbb{R}^2, \cdot)$, where $\mathbb{R}^2$ stands for a suitable neighbourhood of $e = (1,0)$, and the multiplication $\cdot$ is given by

$$x \cdot y = (x^1, x^2) \cdot (y^1, y^2) = (\gamma^1(x;y), \gamma^2(x;y)) \text{ with } \begin{cases} \gamma^1(x;y) = x^1 y^1, \\ \gamma^2(x;y) = x^1 y^2 + x^2. \end{cases}$$

Let us expand the γ^i in Taylor polynomials at $e = (1,0)$:

$$\begin{cases} \gamma^1(x;y) = 1 + (x^1 - 1) + (y^1 - 1) + (x^1 - 1)(y^1 - 1), \\ \gamma^2(x;y) = x^2 + y^2 + (x^1 - 1)y^2. \end{cases}$$

Return to the case of a general parameter group with $\mathbb{R}^r$ as parameter space, $e = (0,\dots,0)$ as identity, and the multiplication

$$x \cdot y = \gamma(x;y).$$

Then

$$x = x \cdot e = \gamma(x;e),$$

so that

$$\gamma^i(x^1,\dots,x^r;0,\dots,0) = x^i,$$

and analogously

$$\gamma^i(0,\dots,0;y^1,\dots,y^r) = y^i.$$

Therefore the Taylor expansion of $\gamma^i(x;y)$ at $(e;e)$ has the following form (recall that all functions are supposed to be smooth):

$$\gamma^i(x;y) = x^i + y^i + \sum_{j,k=1}^r \alpha^i_{jk} x^j y^k + \sum_{j_1,j_2,k=1}^r \alpha^i_{j_1 j_2 k} x^{j_1} x^{j_2} y^k$$

$$+ \sum_{j,k_1,k_2=1}^r \alpha^i_{j k_1 k_2} x^j y^{k_1} y^{k_2} + \cdots .$$

It turns out that the numbers

$$c^i_{jk} := \alpha^i_{jk} - \alpha^i_{kj}$$

are extremely important—in fact we will later see that $\mathfrak{G}$ is uniquely determined by these so called *structure constants*!

Theorem 8.1.1. *The structure constants c^i_{jk} of any parameter group $\mathfrak{G}$ satisfy the two relations*

$$c^i_{jk} = -c^i_{kj}, \tag{*}$$

$$\sum_{s=1}^{r} \{ c^l_{is}c^s_{jk} + c^l_{js}c^s_{ki} + c^l_{ks}c^s_{ij} \} = 0. \tag{**}$$

In fact, (*) is an immediate consequence of the definition, while (**) is seen by a careful comparison of the third order terms of both sides in

$$x \cdot (y \cdot z) = (x \cdot y) \cdot z \iff \gamma(x; \gamma(y; z)) = \gamma(\gamma(x; y); z).$$

$\square$

One of the main points of this chapter will be the proof of the converse statement—**the third fundamental theorem of Lie** (see section **8.6**).

Theorem 8.1.2. *Let $\{c^i_{jk}\}^r_{i,j,k=1}$ be numbers which satisfy (*) and (**). Then there is a unique local parameter group $\mathfrak{G}$ having the c^i_{jk} as its structure constants.*

Example: the affine group—(2). The Taylor expansions of the composition functions γ^i at $e = (1, 0)$ show that there is but one nonzero structure constant:

$$c^2_{12} = 1.$$

8.2 The left- and right-invariant parameter groups

We started by looking at a local Lie group G acting on $\mathbb{R}^n_x$:

$$G \ni g : \mathbb{R}^n_x \overset{\cong}{\to} \mathbb{R}^n_x$$

$$x \mapsto g(x; a) = g_a(x),$$

where $a = (a^1, \ldots, a^r)$ are parameters which have arisen as integration constants when solving the defining PDE system of G. Then we were led to consider the associated parameter group $\mathfrak{G}$, acting on the parameter space $\mathbb{R}^r$ by

$$\mathbb{R}^r \times \mathbb{R}^r \to \mathbb{R}^r$$

$$(a, b) \mapsto a \cdot b = \gamma(a; b).$$

$\mathfrak{G}$ can be considered as a local Lie group in two ways: either $a^1, \ldots, a^r$ are considered as variables $x^1, \ldots, x^r$, and the b^i as parameters—

$$x \mapsto x \cdot b = \gamma(x; b)—$$

or the $b^1, \ldots, b^r$ may be considered as variables $y^1, \ldots, y^r$, while the a^i are thought of as parameters—

$$y \mapsto a \cdot y = \gamma(a; y).$$

The first is the group of *right multiplications*, and the second the group of *left multiplications*. To confuse matters they are, however, usually characterized in a different manner.

Definition. A local diffeomorphism ϕ of $\mathbb{R}^r$ is *left-invariant* if ϕ commutes with left multiplications, i.e.,

$$\phi(a \cdot x) = a \cdot \phi(x) \iff \phi(\gamma(a; x)) = \gamma(a; \phi(x)).$$

Right-invariance is defined analogously.

Theorem 8.2.1. *A local diffeomorphism ϕ is left-invariant if and only if ϕ is a right multiplication.*

Proof. Suppose first that ϕ is a right multiplication by b, i.e., $\phi(x) = x \cdot b$. Then

$$\phi(a \cdot x) = (a \cdot x) \cdot b = a \cdot (x \cdot b) = a \cdot \phi(x)$$

by associativity. Conversely, suppose that

$$\phi(a \cdot x) = a \cdot \phi(x) \quad \text{for all } a, x \in \mathbb{R}^r.$$

Then $x := e$ yields

$$\phi(a) = a \cdot \phi(e),$$

that is,

$$\phi(x) = x \cdot \phi(e) = \text{right multiplication by } \phi(e).$$

$\square$

Therefore the group of right multiplications is called the *left-invariant parameter group*, and is denoted by $\mathfrak{G}_L$.

Analogously the group of left multiplications consists of right-invariant diffeomorphisms, and is therefore called the *right-invariant parameter group*, and is denoted by $\mathfrak{G}_R$.

Incidentally this shows that $\mathfrak{G}_L$ consists of all local diffeomorphisms which commute with all elements in $\mathfrak{G}_R$, and conversely. In view of this $\mathfrak{G}_L$ and $\mathfrak{G}_R$ are said to be *reciprocal* (or opposite) *Lie groups*.

Mostly one is interested in only one of these groups, but for our applications to hyperbolic second order PDEs it is essential to be acquainted with both, and to know how one passes from the one to the other.

Let us return to the original transformation group acting on $\mathbb{R}^n$ by $x \mapsto g_a(x)$. If a and b belong to the parameter space,

$$g_a \circ g_b = g_c \quad \text{with } c = \gamma(a;b),$$

and there is an $\bar{a}$ such that

$$g_a^{-1} = g_{\bar{a}}.$$

Inverting the local diffeomorphisms g,

$$g_c = g_a \circ g_b \implies g_c^{-1} = g_b^{-1} \circ g_a^{-1} \iff g_{\bar{c}} = g_{\bar{b}} \circ g_{\bar{a}},$$

we see that

$$\bar{c} = \gamma(\bar{b};\bar{a}).$$

But this means that the order of multiplication has been reversed, so that right multiplication by b goes over into left multiplication by $\bar{b}$, and left multiplication by a goes over into right multiplication by $\bar{a}$.

Lemma 8.2.1. *Inversion in the parameter space maps* $\mathfrak{G}_L$ *to* $\mathfrak{G}_R$ *and vice versa.* □

Example: the affine group—(3). Since

$$(a^1, a^2) \cdot (b^1, b^2) = (a^1 b^1, a^1 b^2 + a^2),$$

the left-invariant parameter group (= the group of right multiplications) is

$$\mathfrak{A}_L : \begin{cases} x^1 \mapsto x^1 b^1, \\ x^2 \mapsto x^1 b^2 + x^2, \end{cases}$$

and the right-invariant parameter group is

$$\mathfrak{A}_R : \begin{cases} y^1 \mapsto a^1 y^1, \\ y^2 \mapsto a^1 y^2 + a^2. \end{cases}$$

$\mathfrak{A}$ has arisen from A acting on $\mathbb{R}$ by $x \mapsto g_a(x) = a^1 x + a^2$; by earlier results

$$g_a \circ g_b = g_c \quad \text{with } c = (a^1 b^1, a^1 b^2 + a^2)$$

and

$$g_a^{-1} = g_{\bar{a}} \text{ with } \bar{a} = \left(\frac{1}{a^1}, -\frac{a^2}{a^1} \right).$$

Inversion yields

$$g_{\bar{c}}^{-1} = g_{\bar{b}}^{-1} \circ g_a^{-1} \iff g_{\bar{c}} = g_{\bar{b}} \circ g_{\bar{a}}.$$

But

$$g_{\bar{b}}(g_{\bar{a}}(x)) = g_{\bar{b}} \left(\frac{x}{a^1} - \frac{a^2}{a^1} \right) = \frac{x}{b^1 a^1} + \left(-\frac{a^2}{b^1 a^1} - \frac{b^2}{b^1} \right) = g_{\bar{c}}(x),$$

with

$$\bar{c} = \left(\frac{1}{b^1 a^1}, \frac{1}{b^1} \left(-\frac{a^2}{a^1} \right) - \frac{b^2}{b^1} \right),$$

which indeed equals

$$\bar{b} \cdot \bar{a} = \left(\frac{1}{b^1}, -\frac{b^2}{b^1} \right) \cdot \left(\frac{1}{a^1}, -\frac{a^2}{a^1} \right).$$

8.3 Left- and right-invariant vector fields and their dual Maurer–Cartan forms

The study of local Lie groups is greatly simplified by the intimate connection with their Lie algebras of vector fields. Unfortunately this happy state of affairs does not generalize to general Lie pseudogroups, which we will encounter in chapter **14**.

Let $\mathfrak{G}_L$ be a left-invariant parameter group with the multiplication

$$x \mapsto x \cdot b = \gamma(x; b),$$

and let $\beta = (\beta^1, \ldots, \beta^r) \in T_e \mathbb{R}^r$, where $\mathbb{R}^r$ is the parameter space. Then to β is associated a certain vector field X^β, defined as follows.

Let $t \mapsto b(t)$ be a curve passing through e when $t = 0$, and having the tangent vector β there, i.e.,

$$\frac{db^i}{dt}(0) = \beta^i \quad \text{for } i = 1, \ldots, r.$$

Then $t \mapsto \gamma(x; b(t))$ is a curve through x, which at this point has a tangent vector with the components

$$\frac{d\gamma^i(x; b(t))}{dt}(0) = \sum_{k=1}^{r} \frac{\partial \gamma^i}{\partial b^k}(x; e) \frac{db^k}{dt}(0) = \sum_{k=1}^{r} \beta^k {}_R \psi_k^i(x) \quad \text{for } i = 1, \ldots, r,$$

where the functions

$$_R\psi_k^i(x) := \frac{\partial \gamma^i}{\partial b^k}(x;e)$$

are called *right transformation functions* for $\mathfrak{G}$.

Letting x vary we obtain the vector field

$$X^\beta := \sum_{i=1}^r \left(\sum_{k=1}^r \beta^k \,_R\psi_k^i(x) \right) \frac{\partial}{\partial x^i} = \sum_{k=1}^r \beta^k X_k,$$

where

$$X_k = \sum_{i=1}^r \,_R\psi_k^i(x) \frac{\partial}{\partial x^i} \qquad \text{for } k = 1,\ldots,r.$$

These X^β are called *infinitesimal right multiplications*.

Remark. Taylor expansions of the composition functions $\gamma(a;b)$ reveal that

$$_R\psi_k^i(e) = \delta_k^i,$$

so that the matrix $(_R\psi_k^i(x))$ is nonsingular near e. Furthermore,

$$X_k(e) = \frac{\partial}{\partial x^k} \qquad \text{for } k = 1,\ldots,r.$$

Notation. $L_x = \gamma(x;\cdot)$ denotes *left multiplication by* x, i.e., $L_x(y) = \gamma(x;y) = x \cdot y$.

Using this, X^β and X_k are clearly constructed such that

$$X^\beta(x) = (L_x)_{*e}(\beta) \quad \text{and} \quad X_k(x) = (L_x)_{*e}\left(\frac{\partial}{\partial x^k}\right),$$

with $\partial/\partial x^k$ being considered as a vector in $T_e\mathbb{R}^r$.

Theorem 8.3.1. *The infinitesimal right multiplications X^β are **left-invariant** in the sense that*

$$X^\beta(a \cdot x) = (L_{a \cdot x})_{*e}(\beta) = (L_a)_{*x} \circ (L_x)_{*e}(\beta)$$
$$= (L_a)_{*x}(X^\beta(x)).$$

Conversely, if the vector field X is left-invariant,

$$X(x) = X(x \cdot e) = (L_x)_{*e}(X(e)) = X^\beta, \qquad \text{with } \beta = X(e),$$

so that X is an infinitesimal right multiplication. $\qquad\square$

Definition. $\mathfrak{g}_L :=$ the vector space of left-invariant vector fields

$$= \operatorname{span}_{\mathbb{R}}\{X_1, \ldots, X_r\}.$$

By construction we have a vector space isomorphism

$$T_e\mathbb{R}^r \xrightarrow{\cong} \mathfrak{g}_L$$

$$\beta \mapsto X^\beta = \sum_{k=1}^r \beta^k X_k.$$

Since the Lie bracket $[\,,\,]$ commutes with push-forward, also $[X_i, X_j] \in \mathfrak{g}_L$ for $i, j = 1, \ldots, r$—that is, there are numbers $\tilde{c}^k_{ij} = -\tilde{c}^k_{ji}$ such that

$$[X_i, X_j] = \sum_{k=1}^r \tilde{c}^k_{ij} X_k \quad \text{for } i, j = 1, \ldots, r.$$

This means that $\mathfrak{g}_L$ is a *Lie algebra* with the structure constants $\tilde{c}^k_{ij}$.

Remark. We will later see that actually $\tilde{c}^k_{ij} = c^k_{ij} =$ the structure constants of the parameter group $\mathfrak{G}$.

Definition. The *Maurer–Cartan forms* $\lambda^1, \ldots, \lambda^r$ are dual to the basis vector fields $X_1, \ldots, X_r$ of $\mathfrak{g}_L$, and are accordingly defined by

$$\langle \lambda^i, X_j \rangle = \delta^i_j \quad \text{for } i, j = 1, \ldots, r.$$

Note that by duality $d\lambda^i = \sum_{j,k=1}^r (-\tilde{c}^i_{jk}) \lambda^j \wedge \lambda^k$.

To any local diffeomorphism ϕ is associated a push-forward mapping, defined by

ϕ_* on vector fields,

and

$(\phi^{-1})^*$ on differential forms (including zero-forms — functions).

Applying this push-forward of $\phi = L_a$, $\phi^{-1} = L_{\bar{a}}$ to $\delta^i_j = \langle \lambda^i, X_j \rangle$ we get

$$\delta^i_j = \langle L_{\bar{a}}^* \lambda^i, (L_a)_* X_j \rangle = \langle L_{\bar{a}}^* \lambda^i, X_j \rangle,$$

so that

$$L_{\bar{a}}^* \lambda^i = \lambda^i \quad \text{for } i = 1, \ldots, r.$$

Letting a (and thereby also $\bar{a}$) run through $\mathbb{R}^r$ we see that the λ^i are left-invariant too.

Theorem 8.3.2. *The Maurer–Cartan forms associated to $\mathfrak{g}_L$ are invariant under left multiplications.* □

Later on we will see that the converse is also true, i.e.,

$$\phi \in \mathfrak{G}_L \iff \phi^*(\lambda^i) = \lambda^i \quad \text{for } i = 1, \ldots, r.$$

The Lie algebra $\mathfrak{g}_R$ of right-invariant vector fields and the corresponding right-invariant Maurer–Cartan forms are defined analogously from $\mathfrak{G}_R$. Thus $\mathfrak{g}_R = \operatorname{span}_\mathbb{R}\{Y_1, \ldots, Y_r\}$, where

$$Y_k = \sum_{i=1}^r {}_L\psi_k^i(y) \frac{\partial}{\partial y^i} \quad \text{for } k = 1, \ldots, r,$$

with the *left transformation functions* ${}_L\psi_k^i$ being defined by

$$_L\psi_k^i(y) = \frac{\partial \gamma^i}{\partial a^k}(e; y).$$

Example: the affine group—(4). $\mathfrak{A}_L$ has the composition functions

$$\begin{cases} \gamma^1 = x^1 b^1, \\ \gamma^2 = x^1 b^2 + x^2, \end{cases}$$

so that the right transformation functions ${}_R\psi_j^i = \frac{\partial \gamma^i}{\partial b^j}(x; e)$ are given by

$$\begin{cases} {}_R\psi_1^1 = x^1, & {}_R\psi_2^1 = 0, \\ {}_R\psi_1^2 = 0, & {}_R\psi_2^1 = x^1. \end{cases}$$

Therefore $\mathfrak{a}_L$ is generated by

$$X_1 = \sum_{i=1}^2 {}_R\psi_1^i(x) \frac{\partial}{\partial x^i} = x^1 \frac{\partial}{\partial x^1} \quad \text{and} \quad X_2 = \sum_{i=1}^2 {}_R\psi_2^i(x) \frac{\partial}{\partial x^i} = x^1 \frac{\partial}{\partial x^2}.$$

Note that

$$[X_1, X_2] = x^1 \frac{\partial}{\partial x^2} = X_2 = \sum_{i=1}^2 c_{12}^i X_i$$

with $c_{12}^1 = 0$ and $c_{12}^2 = 1$—i.e., the structure constants of $\mathfrak{A}$.

The dual Maurer–Cartan forms are defined by $\langle \lambda^i, X_j \rangle = \delta_j^i$, and thus

$$\lambda^1 = \frac{dx^1}{x^1}, \quad \lambda^2 = \frac{dx^2}{x^1};$$

observe that they are well-defined near $e = (1, 0)$. Moreover

$$d\lambda^1 = 0 \quad \text{and} \quad d\lambda^2 = -\frac{dx^1 \wedge dx^2}{(x^1)^2} = -\lambda^1 \wedge \lambda^2.$$

Let us check that these λ^i really are invariant with respect to the left multiplications $(x^1, x^2) \mapsto (a^1 x^1, a^1 x^2 + a^2)$:

$$\lambda^1 = \frac{dx^1}{x^1} \mapsto \frac{a^1 dx^1}{a^1 x^1} = \lambda^1,$$

$$\lambda^2 = \frac{dx^2}{x^1} \mapsto \frac{a^1 dx^2}{a^1 x^1} = \lambda^2.$$

Let us conversely determine the local diffeomorphisms

$$\mathbb{R}^2_x \overset{\cong}{\to} \mathbb{R}^2_\xi$$
$$(x^1, x^2) \mapsto (\xi^1(x^1, x^2), \xi^2(x^1, x^2))$$

which leave λ^1 and λ^2 invariant. Invariance of λ^1 amounts to

$$\frac{d\xi^1}{\xi^1} = \frac{dx^1}{x^1} \iff d\xi^1 = \frac{\xi^1}{x^1} dx^1,$$

so that ξ^1 is a function of x^1 satisfying $d\xi^1/dx^1 = \xi^1/x^1$; hence

$$\xi^1 = a^1 x^1,$$

where a^1 is an arbitrary constant. Then invariance of λ^2 means that

$$\frac{d\xi^2}{\xi^1} = \frac{dx^2}{x^1} \iff d\xi^2 = \frac{\xi^1}{x^1} dx^2 = a^1 dx^2,$$

so that

$$\xi^2 = a^1 x^2 + a^2$$

for another constant a^2.

Conclusion. *The local diffeomorphisms preserving the Maurer–Cartan forms associated to u_L are precisely the left multiplications!*

The composition functions of $\mathfrak{A}_R$ are

$$\begin{cases} y^1 = a^1 y^1, \\ y^2 = a^1 y^2 + a^2, \end{cases}$$

so that the left transformation functions $_L\psi^i_j(y) = \frac{\partial y^i}{\partial a^j}(e; y)$ become

$$\begin{cases} _L\psi^1_1 = y^1, & _L\psi^1_2 = 0, \\ _L\psi^2_1 = y^2, & _L\psi^2_2 = 1. \end{cases}$$

Thus the Lie algebra $\mathfrak{a}_R$ of right-invariant vector fields (or infinitesimal left translations) is generated by

$$Y_1 = \sum_{i=1}^{2} {}_L\psi_1^i(y) \frac{\partial}{\partial y^i} = y^1 \frac{\partial}{\partial y^1} + y^2 \frac{\partial}{\partial y^2},$$

$$Y_2 = \sum_{i=1}^{2} {}_L\psi_2^i(y) \frac{\partial}{\partial y^i} = \frac{\partial}{\partial y^2}.$$

Then

$$[Y_1, Y_2] = -\frac{\partial}{\partial y^2} = -Y_2 = \sum_{i=1}^{2} (-c_{12}^i) Y_i,$$

where c_{12}^i are the structure constants of $\mathfrak{A}$.

The dual Maurer–Cartan forms are in this case

$$\omega^1 = \frac{dy^1}{y^1} \quad \text{and} \quad \omega^2 = dy^2 - \frac{y^2}{y^1} dy^1.$$

They satisfy the structure equations

$$d\omega^1 = 0 \quad \text{and} \quad d\omega^2 = -\frac{1}{y^1} dy^2 \wedge dy^1 = \omega^1 \wedge \omega^2.$$

We have that ω^1 and ω^2 are invariant with respect to the right multiplications $(y^1, y^2) \mapsto (y^1 b^1, y^1 b^2 + y^2)$ since

$$\omega^1 \mapsto \frac{dy^1 b^1}{y^1 b^1} = \omega^1,$$

$$\omega^2 \mapsto dy^1 b^2 + dy^2 - \frac{y^1 b^2 + y^2}{y^1 b^1} dy^1 b^1 = dy^1 \left(b^2 - b^2 - \frac{y^2}{y^1} \right) + dy^2 = \omega^2.$$

Conversely it is easily seen that the local diffeomorphisms preserving ω^1 and ω^2 are precisely the right multiplications.

We saw earlier that $\mathfrak{G}_L$ is turned into $\mathfrak{G}_R$ and vice versa by inversion in the parameter space. Let us check that the inversion also switches the Lie algebras, although with an extra minus sign added.

To $\beta \in T_e \mathbb{R}^r$ is associated the left-invariant vector field

$$X^\beta = \sum_{i=1}^{r} \frac{dy^i(x; b(t))}{dt} \Big|_{t=0} \cdot \frac{\partial}{\partial x^i},$$

where $t \mapsto b(t)$ satisfies $b(0) = e$ and $db/dt(0) = \beta$.

The inversion $x \mapsto \bar{x}$ is defined by

$$\gamma(x;\bar{x}) = e \iff \gamma^i(x;\bar{x}) = 0 \quad \text{for } i = 1,\ldots,r.$$

Now d/dt applied to $\gamma^i(b(t);\bar{b}(t)) = 0$ gives

$$\sum_{k=1}^{r} \left(\frac{\partial \gamma^i}{\partial a^k}(b(t);\bar{b}(t))\frac{db^k}{dt} + \frac{\partial \gamma^i}{\partial b^k}(b(t);\bar{b}(t))\frac{d\bar{b}^k}{dt} \right) = 0,$$

which for $t = 0$ turns into

$$\sum_{k=1}^{r} \left(\delta_k^i \frac{db^k}{dt}(0) + \delta_k^i \frac{d\bar{b}^k}{dt} \right) = 0,$$

i.e.,

$$\frac{d\bar{b}^i}{dt}(0) = -\frac{db^i}{dt}(0) = -\beta^i \quad \text{for } i = 1,\ldots,r.$$

So under inversion the coefficient $\frac{d}{dt}\gamma^i(x;b(t))|_{t=0}$ of X^β goes over into

$$\frac{d}{dt}\gamma^i(\bar{b}(t);\bar{x})|_{t=0} = \sum_{k=1}^{r} \frac{\partial \gamma^i}{\partial a^k}(\bar{b}(t);\bar{x})|_{t=0} \cdot \frac{d\bar{b}^k}{dt}(0)$$

$$= -\sum_{k=1}^{r} \beta^k \frac{\partial \gamma^i}{\partial a^k}(e;\bar{x}) = -\sum_{k=1}^{r} \beta^k \, {}_L\psi_k^i(\bar{x}),$$

whence

$$X^\beta(x) \mapsto -Y^\beta(\bar{x}),$$

where Y^β is the right-invariant vector field associated to β. Thus inversion in the parameter space induces a mapping

$$\mathfrak{g}_L = \operatorname{span}_{\mathbb{R}}\{X_1,\ldots,X_r\} \mapsto \operatorname{span}_{\mathbb{R}}\{-Y_1,\ldots,-Y_r\} = -\mathfrak{g}_R,$$

with the last minus sign reminding us of the fact that the structure constants c_{ij}^k of $\mathfrak{g}_L$ are mapped to the structure constants $-c_{ij}^k$ of $\mathfrak{g}_R$:

$$[X_i, X_j] = \sum_{k=1}^{r} c_{ij}^k X_k \mapsto [-Y_i, -Y_j] = \sum_{k=1}^{r} c_{ij}^k(-Y_k),$$

so that

$$[Y_i, Y_j] = \sum_{k=1}^{r}(-c_{ij}^k) Y_k.$$

Example: the affine group—(5). Here

$$\mathfrak{a}_L = \operatorname{span}_{\mathbb{R}}\left\{x^1 \frac{\partial}{\partial x^1}, x^1 \frac{\partial}{\partial x^2}\right\}, \qquad \mathfrak{a}_R = \operatorname{span}_{\mathbb{R}}\left\{y^1 \frac{\partial}{\partial y^1} + y^2 \frac{\partial}{\partial y^2}, \frac{\partial}{\partial y^2}\right\},$$

and the inversion is given by

$$(x^1, x^2) \mapsto \left(\frac{1}{x^1}, -\frac{x^2}{x^1}\right) = (\bar{x}^1, \bar{x}^2) =: (y^1, y^2).$$

Then

$$x^1 \frac{\partial}{\partial x^1} \mapsto -\bar{x}^1 \frac{\partial}{\partial \bar{x}^1} - \bar{x}^2 \frac{\partial}{\partial \bar{x}^2} = -\left(y^1 \frac{\partial}{\partial y^1} + y^2 \frac{\partial}{\partial y^2}\right)$$

and

$$x^1 \frac{\partial}{\partial x^2} \mapsto -\frac{\partial}{\partial \bar{x}^2} = -\frac{\partial}{\partial y^2},$$

so that indeed $\mathfrak{a}_L \mapsto -\mathfrak{a}_R$.

8.4 One-parameter subgroups and the exponential mapping

Let $\mathfrak{G} = (\mathbb{R}^r, \cdot)$ be a local parameter group with multiplication given by $a \cdot b = \gamma(a; b)$. A 1-parameter subgroup is a mapping

$$\mathbb{R}_t \to \mathfrak{G}$$
$$t \mapsto g(t) = (g^1(t), \ldots, g^r(t))$$

which satisfies

- $g(0) = e = (0, \ldots, 0)$,
- $g(s + t) = g(s) \cdot g(t) = \gamma(g(s); g(t))$.

Note that since $s + t = t + s$, we also have

$$g(s + t) = g(t + s) = g(t) \cdot g(s) = \gamma(g(t); g(s)),$$

although $\gamma(a; b) \neq \gamma(b; a)$ in general.

Therefore dg^i/dt can be computed in two ways: either as

$$\frac{dg^i}{dt} = \lim_{\Delta t \to 0} \frac{g^i(t + \Delta t) - g^i(t)}{\Delta t} = \lim_{\Delta t \to 0} \frac{\gamma^i(g(t); g(\Delta t)) - \gamma^i(g(t); e)}{\Delta t}$$

$$= \sum_{j=1}^r \frac{\partial \gamma^i}{\partial b^j}(g(t); e) \frac{dg^j}{dt}(0) = \sum_{j=1}^r R\psi^i_j(g(t)) \alpha^j,$$

where $_R\psi^i_j(x)$ are the right transformation functions and $\alpha = (\alpha^1, \ldots, \alpha^r)$ is the tangent vector of $t \mapsto g(t)$ at e, or as

$$\frac{dg^i}{dt} = \lim_{\Delta t \to 0} \frac{\gamma^i(g(\Delta t); g(t)) - \gamma^i(e; g(t))}{\Delta t}$$

$$= \sum_{j=1}^r \frac{\partial \gamma^i}{\partial a^j}(g(t); e) \frac{dg^j}{dt}(0) = \sum_{j=1}^r {_L\psi^i_j}(g(t))\, \alpha^j,$$

where $_L\psi^i_j$ are the left transformation functions of $\mathfrak{G}$.

Recall that the Lie algebra $\mathfrak{g}_L$ of left-invariant vector fields is generated by

$$X_j = \sum_{i=1}^r {_R\psi^i_j}(x) \frac{\partial}{\partial x^i} \qquad \text{for } j = 1, \ldots, r.$$

Consequently $g(t)$ is the integral curve of $\sum_{j=1}^r \alpha^j X_j$ passing through e when $t = 0$.

With $\mathfrak{g}_R = \operatorname{span}_{\mathbb{R}}\{Y_1, \ldots, Y_r\}$, where

$$Y_j = \sum_{i=1}^r {_L\psi^i_j} \frac{\partial}{\partial y^i} \qquad \text{for } j = 1, \ldots, r,$$

it analogously follows that $g(t)$ is the integral curve of $\sum_{j=1}^r \alpha^j Y_j$ passing through e when $t = 0$.

Example: the affine group—(6). The 1-parameter groups $t \mapsto g_\alpha(t)$ of $\mathfrak{A}$ are integral curves through $e = (1, 0)$ of both $\mathfrak{a}_L$ and $\mathfrak{a}_R$, and they are uniquely determined by the tangent vector

$$\alpha = (\alpha^1, \alpha^2) = \left(\frac{dg^1}{dt}(0), \frac{dg^2}{dt}(0) \right) \qquad \text{at } e.$$

Consider $\mathfrak{a}_L = \operatorname{span}\{X_1, X_2\}$ at first. Then

$$(g_\alpha)_* \left(\frac{d}{dt} \right) = \alpha^1 X_1 + \alpha^2 X_2 = \alpha^1 x^1 \frac{\partial}{\partial x^1} + \alpha^2 x^1 \frac{\partial}{\partial x^2},$$

so that $g_\alpha = (g^1_\alpha, g^2_\alpha)$ satisfies

$$\begin{cases} dg^1_\alpha/dt = \alpha^1 g^1_\alpha, \\ dg^2_\alpha/dt = \alpha^2 g^1_\alpha \end{cases} \qquad \text{and} \qquad \begin{cases} g^1_\alpha(0) = 1, \\ g^2_\alpha(0) = 0. \end{cases}$$

Hence

$$g^1_\alpha(t) = e^{\alpha^1 t},$$

while

$$\begin{cases} dg_\alpha^2/dt = \alpha^2 \, e^{\alpha^1 t}, \\ g_\alpha^2(0) = 0 \end{cases} \implies g_\alpha^2(t) = \alpha^2 \, \frac{e^{\alpha^1 t} - 1}{\alpha^1} \quad \text{if } \alpha^1 \neq 0;$$

when $\alpha^1 = 0$, $(e^{\alpha^1 t} - 1)/\alpha^1$ is to be replaced by

$$\lim_{\alpha^1 \to 0} \frac{e^{\alpha^1 t} - 1}{\alpha^1} = t.$$

Using $\mathfrak{a}_R$ instead,

$$(g_\alpha)_* \left(\frac{d}{dt} \right) = \alpha^1 \, Y_1 + \alpha^2 \, Y_2 = \alpha^1 y^1 \frac{\partial}{\partial y^1} + (\alpha^1 y^2 + \alpha^2) \frac{\partial}{\partial y^2},$$

so that

$$\begin{cases} dg_\alpha^1/dt = \alpha^1 \, g_\alpha^1, \\ dg_\alpha^2/dt = \alpha^1 \, g_\alpha^2 + \alpha^2 \end{cases} \quad \text{and} \quad \begin{cases} g_\alpha^1(0) = 1, \\ g_\alpha^2(0) = 0. \end{cases}$$

Although the ODE system for g_α^2 appearing here is quite different from that above, this system has the same solutions anyway:

$$g_\alpha(t) = \left(e^{\alpha^1 t}, \alpha^2 \, \frac{e^{\alpha^1 t} - 1}{\alpha^1} \right).$$

With the left- and right-invariant cases being similar, we formulate the general theorem for the left-invariant situation.

Theorem 8.4.1. *Let $\mathfrak{G}$ be a local r-parameter group with the right transformation functions $_R\psi_j^i(x)$. There is a one-to-one mapping*

$$T_e\mathfrak{G} \to \text{local 1-parameter groups}$$
$$\alpha \mapsto g_\alpha(t) = (g_\alpha^1(t), \ldots, g_\alpha^r),$$

where $g_\alpha(t)$ is the unique local solution of the system

$$\begin{cases} dg_\alpha^i/dt = \sum_{j=1}^r {}_R\psi_j^i(g_\alpha(t)) \, \alpha^j, \\ g^i(0) = 0 \end{cases} \quad \text{for } i = 1, \ldots, r. \qquad (*)$$

Proof. There remains to prove that any solution of $(*)$ really is a 1-parameter group of $\mathfrak{G}$, i.e., that

$$g(s + t) = g(s) \cdot g(t) = \gamma(g(s); g(t))$$

whenever $g(s+t)$, $g(s)$ and $g(t)$ are all defined. Letting

$$h(s,t) := g(s) \cdot g(t),$$

we will do this by showing that $h(s,t) = g(s+t)$.

Clearly $h(s,0) = g(s) \cdot e = g(s)$. Next

$$h^i(t, \Delta t) = \gamma^i(g(t); g(\Delta t)) = g^i(t) + \sum_{j=1}^{r} R\psi_j^i(g(t))\, \alpha^j\, \Delta t + O(\Delta t^2),$$

and

$$g^i(t + \Delta t) = g^i(t) + \frac{dg^i}{dt}(t)\,\Delta t + O(\Delta t^2)$$

$$= g^i(t) + \sum_{j=1}^{r} R\psi_j^i(g(t))\, \alpha^j\, \Delta t + O(\Delta t^2),$$

so that $g^i(t + \Delta t) = h^i(t, \Delta t) + O(\Delta t^2)$. Therefore

$$
\begin{aligned}
h^i(s, t + \Delta t) &= \gamma^i(g(s); g(t + \Delta t)) \\
&= \gamma^i(g(s); h(t, \Delta t)) + O(\Delta t^2) \\
&= \gamma^i(g(s); \gamma(g(t); g(\Delta t))) + \cdots \\
&= \gamma^i(\gamma(g(s); g(t)); g(\Delta t)) + \cdots \quad \text{by associativity} \\
&= \gamma^i(h(s, t); g(\Delta t)) + \cdots \\
&= h^i(s, t) + \sum_{j=1}^{r} R\psi_j^i(h(s, t))\, \alpha^j\, \Delta t + \cdots,
\end{aligned}
$$

and so

$$\frac{dh^i}{dt} = \sum_{j=1}^{r} R\psi_j^i(h(s, t))\, \alpha^j.$$

But then $h(s,t)$ satisfies the same system

$$
\begin{cases}
dh^i/dt &= \sum_{j=1}^{r} R\psi_j^i(h(s, t))\, \alpha^j, \\
h^i(s, 0) &= g^i(s),
\end{cases}
\qquad \text{for } i = 1, \dots, r,
$$

as $g(s+t)$ does. The uniqueness theorem for solutions of ODE systems with a parameter then shows that $h(s,t) = g(s+t)$, i.e.,

$$g(s+t) = g(s) \cdot g(t).$$

$\square$

Lemma 8.4.1. *Let $t \mapsto g_\alpha(t)$ be the 1-parameter group satisfying $dg_\alpha/dt(0) = \alpha$, and let $a \in \mathbb{R}$. Then $t \mapsto g_\alpha(at)$ is the 1-parameter group associated to the tangent vector $a\alpha \in T_e\mathfrak{G}$, i.e., $g_\alpha(at) = g_{a\alpha}(t)$.*

Proof. We have $g_\alpha(as) \cdot g_\alpha(at) = g_\alpha(as + at) = g_\alpha(a(s + t))$, and $g_\alpha(a \cdot 0) = g_\alpha(0) = e$, so that $t \mapsto g_\alpha(at)$ indeed is a 1-parameter group. Its tangent vector at e has the components

$$\left.\frac{dg^i(at)}{dt}\right|_{t=0} = a \left.\frac{dg^i(at)}{d(at)}\right|_{t=0} = a\alpha^i.$$

But the unique 1-parameter group associated to $a\alpha \in T_e\mathfrak{G}$ is $t \mapsto g_{a\alpha}(t)$, and hence necessarily

$$g_\alpha(at) = g_{a\alpha}(t).$$

□

If we write $g_\alpha(t) = g(\alpha^1, \ldots, \alpha^r; t)$, the lemma says that

$$g(\alpha^1, \ldots, \alpha^r; at) = g(a\alpha^1, \ldots, a\alpha^r; t).$$

Let us for once be a bit specific about the domain of definition: assume that $g(\alpha^1, \ldots, \alpha^r; t)$ is defined for $|t| \leq \epsilon$ and $|\alpha^i| \leq \epsilon$. Because

$$g(\alpha^1, \ldots, \alpha^r; t) = g\left(\frac{\alpha^1 t}{\epsilon}, \ldots, \frac{\alpha^r t}{\epsilon}; \epsilon\right),$$

g is defined whenever $|\alpha^i t| \leq \epsilon^2$, and in particular for $t = 1$ and $|\alpha^i| \leq \epsilon^2$.

Notation. Let $\alpha = (\alpha^1, \ldots, \alpha^r) \in T_e\mathfrak{G}$ with $|\alpha^i|$ so small that $g_\alpha(t)$ is defined for $|t| \leq 1$ at least. Then let

$$\exp(\alpha) := g_\alpha(1),$$

so that exp maps a neighbourhood of $(0, \ldots, 0) \in T_e\mathfrak{G}$ into $\mathfrak{G}$.

Lemma 8.4.2. $\exp : T_e\mathfrak{G} \longrightarrow \mathfrak{G}$ *is a local diffeomorphism.*

Proof. Let us calculate the Jacobian of exp at $(0,\dots,0)$:

$$\frac{\partial \exp^i}{\partial \alpha^j}(0\dots,0) = \frac{\partial \exp^i}{\partial \alpha^j}(0,\dots,0,\alpha^j,0,\dots,0)_{|\alpha^j=0}$$

$$= \frac{1}{\alpha^j}\frac{d}{dt}\exp^i(0,\dots,0,\alpha^j t,0,\dots,0)_{|t=0}$$

$$= \frac{1}{\alpha^j}\frac{d}{dt}g^i(0,\dots,0,\alpha^j t,0,\dots,0;1)_{|t=0}$$

$$= \frac{1}{\alpha^j}\frac{d}{dt}g^i(0,\dots,0,\alpha^j,0,\dots,0;t)_{|t=0}$$

$$= \frac{1}{\alpha^j}\,\alpha^j\,\delta^i_j = \delta^i_j.$$

$\square$

Denoting the local inverse by log we thus have

$$T_e\mathfrak{G} \underset{\log}{\overset{\exp}{\rightleftarrows}} \mathfrak{G}.$$

As an r-dimensional vector space, $T_e\mathfrak{G}$ may be identified with $\mathbb{R}^r$—having $y^1,\dots,y^r$ as coordinates, say. Then

$$\mathfrak{G} \underset{\exp}{\overset{\log}{\rightleftarrows}} T_e\mathfrak{G} \overset{\cong}{\longleftrightarrow} \mathbb{R}^r_y,$$

and thus it is possible to use $y^1,\dots,y^r$ as local coordinates near $e \in \mathfrak{G}$. The connection between these new coordinates and the old $x^1,\dots,x^r$ is

$$x^i = \exp^i(y^1,\dots,y^r) \quad \text{for } i = 1,\dots,r.$$

Lemma 8.4.3. *Each 1-parameter subgroup has the very simple form*

$$y^i = \alpha^i t \quad \text{for } i = 1,\dots,r$$

in the new coordinates.

Proof.

$$y^i = \alpha^i t \iff x^i = \exp^i(\alpha^1 t,\dots,\alpha^r t)$$
$$= g^i(\alpha^1 t,\dots,\alpha^r t;1) = g^i(\alpha^1,\dots,\alpha^r;t)$$
$$= g^i_\alpha(t) \qquad \text{for } i = 1,\dots,r.$$

$\square$

Definition. A *canonical coordinate system of the first kind* for $\mathfrak{G}$ is a coordinate system near e in which the 1-parameter subgroups appear as straight lines through the origin.

So by the lemmas above it is always possible to introduce a canonical coordinate system of the first kind at $e \in \mathfrak{G}$.

We have earlier seen that any local diffeomorphism of the parameter group $\mathfrak{G}$ which commutes with all right multiplications must be a left multiplication and vice versa. This means that the left-invariant parameter group $\mathfrak{G}_L$ and the right-invariant parameter group $\mathfrak{G}_R$ are reciprocals of each other.

Let us now use the exponential mapping to prove the analogous fact for the left- and right-invariant Lie algebras $\mathfrak{g}_L$ and $\mathfrak{g}_R$.

Notation. Let X be a vector field on the parameter group. Then

$$t \mapsto e^{tX}(x)$$

denotes the (unique) local integral curve of X passing through $x \in \mathfrak{G}$ when $t = 0$.

Theorem 8.4.2. $e^{tX}(x) = x \cdot e^{tX}$ for all x and t if and only if X is left-invariant, and $e^{tX}(x) = e^{tX} \cdot x$ for all x and t if and only if X is right-invariant.

Proof. It clearly suffices to prove the first statement. $x \cdot e^{tX} = e^{tX}(x)$ means that $x \cdot e^{tX}$ is the integral curve of X passing through x when $t = 0$, i.e.,

$$\frac{d}{ds}\left(x \cdot e^{(t_0+s)X}\right)\Big|_{s=0} = X\left(x \cdot e^{t_0 X}\right) \qquad \text{for each } t_0.$$

But

$$\frac{d}{ds}\left(x \cdot e^{(t_0+s)X}\right)\Big|_{s=0} = \frac{d}{ds}\left(x \cdot e^{t_0 X} \cdot e^{sX}\right)\Big|_{s=0} = \frac{d}{ds}\gamma\left(x \cdot e^{t_0 X}; e^{sX}\right)\Big|_{s=0},$$

and $s \mapsto e^{sX}$ is a curve passing through e when $s = 0$ with $\frac{d}{ds}(e^{sX})\big|_{s=0} = X(e)$. Consequently

$$\frac{d}{ds}\left(x \cdot e^{(t_0+s)X}\right)\Big|_{s=0} = L_{*(x \cdot e^{t_0 X})}X(e),$$

which equals $X(x \cdot e^{t_0 X})$ if and only if X is left-invariant.   □

Expressed in a slightly different way,

e^{tX} is a right (left) multiplication if and only if X is an infinitesimal right (left) multiplication.

Let Y be a right-invariant and X a left-invariant vector field on $\mathfrak{G}$. With $a := e^Y$ and $b := e^X$, associativity shows that for any $x \in \mathfrak{G}$,

$$a \cdot x \cdot b = a \cdot (x \cdot b) = e^Y \cdot (x \cdot e^X) = e^Y(e^X(x))$$
$$= e^Y \circ e^X(x),$$

and

$$a \cdot x \cdot b = (a \cdot x) \cdot b = (e^Y \cdot x) \cdot e^X = e^X(e^Y(x))$$
$$= e^X \circ e^Y(x),$$

so that

$$e^Y \circ e^X = e^X \circ e^Y.$$

Next we need a result about local and infinitesimal commutativity.

Theorem 8.4.3. *Let X and Y be vector fields on $\mathbb{R}^r$. Then*

$$e^{sX} \circ e^{tY} = e^{tY} \circ e^{sX} \quad \text{for all } s \text{ and } t \iff [X, Y] = 0.$$

Proof. $\implies$: Consider the curve

$$t \mapsto e^{-\sqrt{t}Y} \circ e^{-\sqrt{t}X} \circ e^{\sqrt{t}Y} \circ e^{\sqrt{t}X}(x) =: c_x(t)$$

for $t \geq 0$, starting at x when $t = 0$. Straightforward but messy calculations show that

$$\lim_{t \searrow 0} \frac{d}{dt} c_x(t) = [X, Y](x).$$

Therefore

$$e^{tX} \circ e^{tY} = e^{tY} \circ e^{tX} \implies [X, Y] = 0.$$

$\impliedby$: By the rectification lemma for vector fields we may assume that $X = \partial/\partial x^1$. Then if $Y = \sum_{k=1}^r b^k(x) (\partial/\partial x^k)$,

$$0 = [X, Y] = \sum_{k=1}^r \frac{\partial b^k}{\partial x^1} \frac{\partial}{\partial x^k},$$

so that $b^k = b^k(x^2, \ldots, x^r)$ for $k = 1, \ldots, r$.
    Clearly

$$e^{sX}(x) = e^{s\partial/\partial x^1}(x) = (x^1 + s, x^2, \ldots, x^r) = x + (s, 0, \ldots, 0),$$

while $e^{tY}(x) =: g(t)$ is obtained as the unique solution of

$$\begin{cases} dg^k/dt = b^k(g^2(t), \dots, g^r(t)), \\ g^k(0) = x^k \end{cases} \qquad \text{for } k = 1, \dots, r.$$

We want to show that

$$e^{sX} \circ e^{tY}(x) = e^{tY} \circ e^{sX}(x),$$

i.e.,

$$e^{tY}(x) + (s, 0, \dots, 0) = e^{tY}(x^1 + s, x^2, \dots, x^r).$$

But since $\partial b^k / \partial x^1 \equiv 0$, both sides satisfy the ODE system

$$\frac{dz^k}{dt} = b^k(z^2, \dots, z^r) \quad \text{for } k = 1, \dots, r,$$

as well as the initial condition

$$z(0) = (x^1 + s, x^2, \dots, x^r),$$

—and therefore they indeed are equal.   □

Suppose next that X is a vector field on $\mathfrak{G}$ which *commutes with all right-invariant vector fields*—that is, $[X, \mathfrak{g}_R] = 0$.

Because the exponential mapping is a local diffeomorphism, any $a \approx e$ can be written as $a = e^Y$ for a right-invariant vector field Y. By the theorem,

$$e^X \circ e^Y(x) = e^Y \circ e^X(x),$$

or

$$e^X(a \cdot x) = a \cdot e^X(x),$$

whence e^X commutes with left multiplications—but then e^X must be a right multiplication and X is an infinitesimal right multiplication, that is a *left-invariant vector field*.

Conversely, if X is left-invariant, e^X commutes with left multiplications, and going backwards in the argument above we see that X commutes with all right-invariant vector fields, so that $[X, \mathfrak{g}_R] = 0$.

Definition. If $\mathfrak{g} = \text{span}_{\mathbb{R}}\{X_1, \dots, X_r\}$ is a Lie algebra of vector fields, its *reciprocal Lie algebra* is

$$\mathfrak{g}^{\text{rec}} = \{Y \mid [Y, X_i] = 0 \text{ for } i = 1, \dots, r\}.$$

Applying this to $\mathfrak{g}_L$ and $\mathfrak{g}_R$ we obtain the following important result.

Theorem 8.4.4. $\mathfrak{g}_L$ *and* $\mathfrak{g}_R$ *are reciprocals of each other.*   □

Remark. The theorem on the connection between local and infinitesimal commutativity can also be used to prove the Frobenius theorem. For instance, let $V = (X, Y)$ be a 2-dimensional vector field system on $\mathbb{R}^n$. In order to obtain a 2-dimensional integral manifold through a given point $x \in \mathbb{R}^n$, it is tempting to first form an integral curve of X through x,

$$t \mapsto e^{tX}(x),$$

and then through each point of this curve construct an integral curve of Y:

$$(s, t) \mapsto e^{sY}\left(e^{tX}(x)\right).$$

Then surely Y is everywhere tangent to the resulting 2-dimensional manifold, but X is in general not (except when $s = 0$).

However, if $[X, Y] = 0$,

$$(s, t) \mapsto e^{sY} \circ e^{tX}(x) = e^{tX} \circ e^{sY}(x),$$

and then both X and Y are tangent to this manifold, which therefore *is an integral manifold of* V.

Generalizing this to arbitrary dimensions we get the following result.

Theorem 8.4.5. *Let* V *be a* q-*dimensional complete vector field system on* $\mathbb{R}^n$. *Then by Gaussian elimination it is possible to find a basis* $\{X_1, \ldots, X_q\}$ *of* V *such that*

$$[X_i, X_j] = 0 \quad for \ i, j = 1, \ldots, q.$$

If we use this basis, the integral manifold through a given point $x \in \mathbb{R}^n$ *is given by*

$$\mathbb{R}_t^q \ni (t^1, \ldots, t^q) \mapsto e^{t^1 X_1} \circ e^{t^2 X_2} \circ \cdots \circ e^{t^q X_q}(x).$$

□

Example: the affine group—(7). The left- and right-invariant Lie algebras are

$$\mathfrak{a}_L = \left(x^1 \frac{\partial}{\partial x^1}, x^1 \frac{\partial}{\partial x^2}\right) \quad \text{and} \quad \mathfrak{a}_R = \left(x^1 \frac{\partial}{\partial x^1} + x^2 \frac{\partial}{\partial x^2}, \frac{\partial}{\partial x^2}\right).$$

Let us determine the reciprocal Lie algebra $\mathfrak{a}_L^{\text{rec}}$ of $\mathfrak{a}_L$ directly. $X = a\,\partial/\partial x^1 + b\,\partial/\partial x^2 \in \mathfrak{a}_L^{\text{rec}}$ if and only if

$$0 = \left[x^1 \frac{\partial}{\partial x^1}, a \frac{\partial}{\partial x^1} + b \frac{\partial}{\partial x^2}\right] = \left(x^1 \frac{\partial a}{\partial x^1} - a\right) \frac{\partial}{\partial x^1} + x^1 \frac{\partial b}{\partial x^1} \frac{\partial}{\partial x^2},$$

and

$$0 = \left[x^1 \frac{\partial}{\partial x^2}, a \frac{\partial}{\partial x^1} + b \frac{\partial}{\partial x^2} \right] = x^1 \frac{\partial a}{\partial x^2} \frac{\partial}{\partial x^1} + \left(x^1 \frac{\partial b}{\partial x^2} - a \right) \frac{\partial}{\partial x^2}.$$

That is,

$$\frac{\partial a}{\partial x^2} = 0 \Longrightarrow a = a(x^1), \qquad \frac{\partial b}{\partial x^1} = 0 \Longrightarrow b = b(x^2),$$

$$x^1 \frac{da}{dx^1} = a \Longrightarrow a = c_1 x^1, \qquad x^1 \frac{db}{dx^2} = a = c_1 x^1 \Longrightarrow b = c_1 x^2 + c_2,$$

so that

$$X = c_1 \left(x^1 \frac{\partial}{\partial x^1} + x^2 \frac{\partial}{\partial x^2} \right) + c_2 \frac{\partial}{\partial x^2},$$

i.e.,

$$\mathfrak{a}_L^{rec} = \mathrm{span}\,_{\mathbb{R}} \left\{ x^1 \frac{\partial}{\partial x^1} + x^2 \frac{\partial}{\partial x^2}, \frac{\partial}{\partial x^2} \right\} = \mathfrak{a}_R$$

—as expected.

Starting from the left-invariant parameter group $\mathfrak{G}_L$ one finds its left-invariant Lie algebra $\mathfrak{g}_L$ by differentiations. Conversely it is possible to pass from $\mathfrak{g}_L$ back to $\mathfrak{G}_L$ by an integration process. This will be considered in detail in connection with the third fundamental theorem, but let us first present a suggestive classical method for achieving this, with no claim that it always works.

If $\mathfrak{g}_L = \mathrm{span}\,\{X_1, \ldots, X_r\}$, the generic element in $\mathfrak{g}_L$ is

$$X = \sum_{i=1}^{r} c^i X_i,$$

where $c^1, \ldots, c^r$ are arbitrary parameters. Define a family of local diffeomorphisms acting on the parameter space $\mathbb{R}^r$ by

$$\mathbb{R}^r \ni x \mapsto e^{tX}(x) = e^{t \sum_{i=1}^{r} c^i X_i}(x).$$

Apparently there are $r + 1$ parameters present here—$c^1, \ldots, c^r$ and t. But if $c^m \neq 0$ say, we can replace c^i by c^i/c^m, and t by $t c^m$, and then only r parameters remain.

General claim. This r-parameter family of local diffeomorphisms is nothing but $\mathfrak{G}_L$.

Example: the affine group—(8). Consider the left-invariant Lie algebra
$\mathfrak{a}_L = \text{span}_{\mathbb{R}}\{x^1\partial/\partial x^1, x^1\partial/\partial x^2\}$ first. Let

$$X = x^1\frac{\partial}{\partial x^1} + sx^1\frac{\partial}{\partial x^2},$$

with s being an arbitrary parameter, and let

$$\xi(x^1, x^2; t) := e^{tX}(x^1, x^2) = e^{t(x^1\partial/\partial x^1 + sx^1\partial/\partial x^2)}(x^1, x^2).$$

Then

$$\begin{cases} d\xi^1/dt = \xi^1, & \xi^1(x^1, x^2; 0) = x^1, \\ d\xi^2/dt = s\,\xi^1, & \xi^2(x^1, x^2; 0) = x^2, \end{cases}$$

so that

$$\begin{cases} \xi^1 = x^1 \cdot e^t, \\ \xi^2 = x^2 + x^1 \cdot s(e^t - 1). \end{cases}$$

Replacing t, s by new parameters $b^1 := e^t$ and $b^2 := s(e^t - 1)$, we have

$$(x^1, x^2) \mapsto e^{tX}(x^1, x^2) = (x^1 b^1, x^1 b^2 + x^2),$$

which indeed is the defining equation of $\mathfrak{A}_L$.

Considering $\mathfrak{a}_R = \text{span}_{\mathbb{R}}\{y^1\partial/\partial y^1 + y^2\partial/\partial y^2, \partial/\partial y^2\}$ next, we let

$$Y = y^1\frac{\partial}{\partial y^1} + y^2\frac{\partial}{\partial y^2} + s\frac{\partial}{\partial y^2} = y^1\frac{\partial}{\partial y^1} + (y^2 + s)\frac{\partial}{\partial y^2},$$

where s is an arbitrary parameter. Then

$$\eta(y^1, y^2; t) := e^{tY}(y^1, y^2) = e^{t(y^1\partial/\partial y^1 + (y^2+s)\partial/\partial y^2)}(y^1, y^2)$$

is equivalent to

$$\begin{cases} d\eta^1/dt = \eta^1, & \eta^1(y^1, y^2; 0) = y^1, \\ d\eta^2/dt = \eta^2 + s, & \eta^2(y^1, y^2; 0) = y^2, \end{cases}$$

with the solutions

$$\begin{cases} \eta^1 = y^1 \cdot e^t, \\ \eta^2 = y^2 \cdot e^t + s(e^t - 1). \end{cases}$$

Setting $a^1 := e^t$ and $a^2 := s(e^t - 1)$, this means that

$$(y^1, y^2) \mapsto e^{tY}(y^1, y^2) = (a^1 y^1, a^1 y^2 + a^2),$$

which is the defining equation of $\mathfrak{A}_R$.

8.5 The first and second fundamental theorems

In this section we will characterize those PDE systems whose general solutions constitute local parameter groups. Since we know by now how to pass from right multiplications to left multiplications and vice versa, we concentrate on right multiplications here. That is, we assume $\mathfrak{G} = \mathfrak{G}_L$ to be a local parameter group acting on the parameter space $\mathbb{R}^r$ by *right multiplications*,

$$\mathbb{R}^r \ni x \mapsto x \cdot b = \gamma(x; b) = R_b(x) \quad \text{for } b \in \mathbb{R}^r,$$

and want to determine the PDE system which the γ^i satisfy. When solving this system, the parameters $b^1, \ldots, b^r$ will emerge as integration constants.

Setting

$$z^i := \gamma^i(x^1, \ldots, x^r; b^1, \ldots, b^r) \quad \text{for } i = 1, \ldots, r,$$

we are interested in the derivatives

$$\frac{\partial z^i}{\partial x^j} = \frac{\partial \gamma^i}{\partial x^j}(x; b), \quad \text{where the } b^k \text{ are kept constant.}$$

Notation. $\phi^i_j(x \cdot b, x) := \partial \gamma^i / \partial x^j(x; b)$.

Remarks.

   (i) $x = e \implies \phi^i_j(b; e) = \partial \gamma^i / \partial x^j(e; b) = {}_L \psi^i_j(b)$, where the latter are the *left transformation functions*;

   (ii) $b = e \implies \phi^i_j(x, x) = \partial \gamma^i / \partial x^j(x; e) = \delta^i_j$.

Lemma 8.5.1. *Let $x, y, z \in \mathbb{R}^r$. Then*

$$\phi^i_k(z, x) = \sum_{j=1}^r \phi^i_j(z, y) \cdot \phi^j_k(y, x).$$

Proof. Write $z = x \cdot x^{-1} y \cdot y^{-1} z = \gamma(\gamma(x; u); v)$ with $u := x^{-1} y$ and $v := y^{-1} z$ (we are tacitly assuming that x, y, z are so close to $e \in \mathbb{R}^r$ that this makes sense). With $\gamma(x; u) = x \cdot u = y$, the chain rule gives

$$\phi^i_k(z, x) = \frac{\partial}{\partial x^k} \gamma^i(\gamma(x; u); v) = \sum_{j=1}^r \frac{\partial \gamma^i}{\partial y^j}(y; v) \frac{\partial \gamma^j}{\partial x^k}(x; u)$$

$$= \sum_{j=1}^r \phi^i_j(z, y) \cdot \phi^j_k(y, x) \text{---since } y \cdot v = z \text{ and } x \cdot u = y.$$

$\square$

Special cases.   1. If $x = z = e$ and $y = x$, we have

$$\sum_{j=1}^{r} \phi^i_j(e, x)\, \phi^j_k(x, e) = \phi^i_k(e, e) = \delta^i_k \quad \text{or} \quad \sum_{j=1}^{r} \phi^i_j(e, x) \cdot {}_L\psi^j_k(x) = \delta^i_k,$$

so that the matrix $(\phi^i_j(e, x))$ is the inverse of the matrix $({}_L\psi^i_j(x))$. Therefore we set

$${}_L\overset{-1_i}{\psi}{}^i_j(x) := \phi^i_j(e, x).$$

2. $y = e \Longrightarrow \phi^i_k(z, x) = \sum_{j=1}^{r} \phi^i_j(z, e)\, \phi^j_k(e, x)$

$$= \sum_{j=1}^{r} {}_L\psi^i_j(z)\, {}_L\overset{-1_j}{\psi}{}^j_k(x).$$

This last result is the original version of **Lie's first fundamental theorem.**

Theorem 8.5.1. *Let $z^i = \gamma^i(x; b)$ be the composition functions of $\mathfrak{G}_L$. Then these functions satisfy the PDE system*

$$\frac{\partial z^i}{\partial x^k} = \sum_{j=1}^{r} {}_L\psi^i_j(z)\, {}_L\overset{-1_j}{\psi}{}^j_k(x) \quad \text{for } i, k = 1, \ldots, r. \tag{*}$$

$\square$

In connection with the second fundamental theorem we will see that the general solution of (*) depends on r arbitrary parameters, $z^i = \gamma^i(x^1, \ldots, x^r; b^1, \ldots, b^r)$, and indeed defines a parameter group.

Equation (*) can be rewritten as

$$\sum_{k=1}^{r} {}_L\overset{-1_i}{\psi}{}^i_k(z)\, \frac{\partial z^k}{\partial x^j} = {}_L\overset{-1_i}{\psi}{}^i_j(x);$$

taking wedge product with dx^j and summing over j, this gives

$$\sum_{k=1}^{r} {}_L\overset{-1_i}{\psi}{}^i_k(z)\, dz^k = \sum_{j=1}^{r} {}_L\overset{-1_i}{\psi}{}^i_j(x)\, dx^j. \tag{**}$$

Definition. The Maurer–Cartan forms for $\mathfrak{G}_L$ are

$$\omega^i := \sum_{j=1}^{r} {}_L\overset{-1_i}{\psi}{}^i_j(x)\, dx^j \quad \text{for } i = 1, \ldots, r.$$

Since $_L\bar{\psi}^i_j(e) = \delta^i_j$, $\{\omega^1,\ldots,\omega^r\}$ constitutes a coframe near e, called the *Cartan coframe*.

Recalling that $z = x \cdot b$, (**) shows that the ω^i do not change under right multiplication by b—so we have the following **coframe version of the first fundamental theorem.**

Theorem 8.5.2. *The Maurer–Cartan forms $\omega^1,\ldots,\omega^r$ are invariant with respect to the right multiplications $x \mapsto z = x \cdot b$.* □

Suppose conversely that ϕ is a local diffeomorphism of $\mathbb{R}^r$ which preserves the ω^i, i.e.,

$$\phi^*(\omega^i) = \omega^i \quad \text{for } i = 1,\ldots,r.$$

Letting $b := \phi(e)$ and setting $f := R_{b^{-1}} \circ \phi$,

$$\begin{cases} f^*(\omega^i) = \phi^* \circ R^*_{b^{-1}}(\omega^i) = \phi^*(\omega^i) = \omega^i & \text{for } i = 1,\ldots,r, \\ \text{and} \\ f(e) = R_{b^{-1}}(b) = b \cdot b^{-1} = e. \end{cases}$$

In connection with the second fundamental theorem we will see that these requirements force f to be the identity mapping. Hence

$$R_{b^{-1}} \circ \phi = \text{id}, \text{ so that } \phi = R_b$$

—which is the **converse of the first fundamental theorem.**

The vector fields X_j dual to the Maurer–Cartan forms are defined by

$$\langle \omega^i, X_j \rangle = \delta^i_j,$$

and thus

$$X_j = \sum_{k=1}^r {}_L\psi^k_j(x) \frac{\partial}{\partial x^k} \quad \text{for } j = 1,\ldots,r.$$

Because $_L\psi^i_j(e) = \delta^i_j$, $\{X_1,\ldots,X_r\}$ forms a frame for $\mathbb{R}^r$ near e. By duality we then get the following **frame version of the first fundamental theorem.**

Theorem 8.5.3. *The vector fields*

$$X_j = \sum_{k=1}^r {}_L\psi^k_j(x) \frac{\partial}{\partial x^k}, \qquad j = 1,\ldots,r,$$

are invariant with respect to right multiplications. □

We have earlier seen that these X_j generate $\mathfrak{g}_R$, so this theorem does not come as a big surprise.

Remark. In order to obtain the corresponding results for the group $\mathfrak{G}_R$ of left multiplications, it suffices to change the indices L and R throughout.

Let us stick to the coframe version in the following (and in particular later when we study general Lie pseudogroups), and introduce

$$\chi_j^i(x) := {}_L\bar\psi_j^{-1i}(x)$$

in order to facilitate the notation. Then the Maurer–Cartan forms are given by

$$\omega^i = \sum_{j=1}^r \chi_j^i(x)\,dx^j \quad \text{for } i = 1,\dots,r.$$

With $\omega^1,\dots,\omega^r$ forming a local coframe for $\mathbb{R}^r$ there are structure functions $C_{jk}^i(x) = -C_{kj}^i(x)$ such that

$$d\omega^i = \frac{1}{2}\sum_{j,k=1}^r C_{jk}^i(x)\,\omega^j \wedge \omega^k = \sum_{1 \le j < k \le r} C_{jk}^i(x)\,\omega^j \wedge \omega^k \quad \text{for } i = 1,\dots,r.$$

Applying right multiplication to this we get

$$d\omega^i = \sum_{j<k} R_b^*(C_{jk}^i(x))\,\omega^j \wedge \omega^k \quad \text{for } i = 1,\dots,r,$$

and hence the $C_{jk}^i(x)$ are invariant under right multiplications too. But with the group of right multiplications being transitive—for instance, $R_{x^{-1}}(x) = x \cdot x^{-1} = e$ for each x—this forces the C_{jk}^i to be *constants*.

Lemma 8.5.2. *The constants C_{jk}^i coincide with the structure constants c_{jk}^i of the parameter group $\mathfrak{G}$, which were defined by $c_{jk}^i := \alpha_{jk}^i - \alpha_{kj}^i$, with the α_{jk}^i being coefficients in the Taylor expansions of the composition functions:*

$$\gamma^i(x;y) = x^i + y^i + \sum_{j,k=1}^r \alpha_{jk}^i\, x^j y^k + \cdots.$$

Proof. Inserting the expressions $\omega^i = \sum_{j=1}^r \chi_j^i(x)\,dx^j$ into $d\omega^i = \sum_{j<k}\bigl(C_{jk}^i$ $\times\, \omega^j \wedge \omega^k\bigr)$, the left hand side goes over into

$$d\omega^i = \sum_{1 \le j < k \le r} \left(\frac{\partial \chi_k^i}{\partial x^j} - \frac{\partial \chi_j^i}{\partial x^k}\right) dx^j \wedge dx^k,$$

while the right hand side becomes

$$\frac{1}{2}\sum_{p,q=1}^{r} C^i_{pq}\,\omega^p \wedge \omega^q = \frac{1}{2}\sum_{p,q,j,k=1}^{r} C^i_{pq}\chi^p_j\chi^q_k\,dx^j \wedge dx^k$$

$$= \frac{1}{2}\sum_{j<k}\left(\sum_{p,q=1}^{r} C^i_{pq}(\chi^p_j\chi^q_k - \chi^p_k\chi^q_j)\right) dx^j \wedge dx^k$$

$$= \sum_{j<k}\left(\sum_{p,q=1}^{r} C^i_{pq}\chi^p_j\chi^q_k\right) dx^j \wedge dx^k,$$

so that

$$\frac{\partial \chi^i_k}{\partial x^j} - \frac{\partial \chi^i_j}{\partial x^k} = \sum_{p,q=1}^{r} C^i_{pq}\chi^p_j\chi^q_k. \tag{†}$$

The coframe version of the first fundamental theorem implies that

$$\sum_{k=1}^{r}\chi^i_k(z)\frac{\partial z^k}{\partial x^j} = \chi^i_j(x) \quad \text{for } z = x \cdot b.$$

If we set $x := e$ this reduces to

$$\sum_{k=1}^{r}\chi^i_k(b)\frac{\partial \gamma^k}{\partial x^j}(e;b) = \delta^i_j.$$

Applying $\partial/\partial b^l$ to this we get

$$\sum_{k=1}^{r}\left(\frac{\partial \chi^i_k}{\partial b^l}(b)\frac{\partial \gamma^k}{\partial x^j}(e;b) + \chi^i_k(b)\frac{\partial^2 \gamma^k}{\partial x^j \partial b^l}(e;b)\right) = 0,$$

which at $b = e$ simplifies to

$$\frac{\partial \chi^i_j}{\partial b^l}(e) + \frac{\partial^2 \gamma^i}{\partial x^j \partial b^l}(e;e) = 0,$$

so that

$$\frac{\partial \chi^i_j}{\partial b^l}(e) = -\alpha^i_{jl}.$$

Setting $x = e$ in (†) this shows that

$$-\alpha^i_{kj} + \alpha^i_{jk} = C^i_{jk},$$

and hence in fact

$$C^i_{jk} = c^i_{jk}.$$

$\square$

With this we have proved the **second fundamental theorem**.

Theorem 8.5.4. *The Maurer–Cartan forms $\omega^1, \dots, \omega^r$ satisfy the structure equations*

$$d\omega^i = \sum_{1 \leq j < k \leq r} c^i_{jk}\, \omega^j \wedge \omega^k \quad for\ i = 1, \dots, r,$$

where the c^i_{jk} are the structure constants of the parameter group. □

Let us conversely start from a local coframe $\{\omega^1, \dots, \omega^r\}$ satisfying

$$d\omega^i = \sum_{j<k} C^i_{jk}\, \omega^j \wedge \omega^k \quad \text{for } i = 1, \dots, r,$$

for certain constants C^i_{jk}, and determine the local diffeomorphisms of $\mathbb{R}^r$ that preserve each ω^i. If we introduce a copy $\mathbb{R}^r_{\hat{x}}$ of $\mathbb{R}^r_x$ and equip it with the coframe $\{\hat{\omega}^1, \dots, \hat{\omega}^r\}$, where $\hat{\omega}^i$ is obtained from ω^i by changing x to $\hat{x}$, such a diffeomorphism may be regarded as a mapping

$$\phi : \mathbb{R}^r_x \to \mathbb{R}^r_{\hat{x}}$$
$$x \mapsto \hat{x} = \phi(x)$$

for which $\phi^*(\hat{\omega}^i) = \omega^i$. The graph of ϕ is then an r-dimensional submanifold of $\mathbb{R}^r_x \times \mathbb{R}^r_{\hat{x}}$, on which the lifts of ω^i and $\hat{\omega}^i$ to $\mathbb{R}^r_x \times \mathbb{R}^r_{\hat{x}}$ are equal. Denoting these lifts by ω^i and $\hat{\omega}^i$ too, we find that the graphs of the local diffeomorphims preserving the Maurer–Cartan forms are r-dimensional integral manifolds of the pfaffian system generated by

$$\theta^i := \omega^i - \hat{\omega}^i \quad for\ i = 1, \dots, r$$

on $\mathbb{R}^r_x \times \mathbb{R}^r_{\hat{x}}$. Now

$$d\theta^i = d\omega^i - d\hat{\omega}^i = \sum_{j<k} C^i_{jk}\, (\omega^j \wedge \omega^k - \hat{\omega}^j \wedge \hat{\omega}^k) \equiv 0 \pmod{\theta^1, \dots, \theta^r},$$

which means that the Frobenius theorem applies. Consequently the integral manifolds of $(\theta^1, \dots, \theta^r)$ form an r-parameter family of r-dimensional submanifolds of $\mathbb{R}^r_x \times \mathbb{R}^r_{\hat{x}}$. On each one $\hat{\omega}^i = \omega^i$, so that both $\{\omega^1, \dots, \omega^r\}$ and $\{\hat{\omega}^1, \dots, \hat{\omega}^r\}$ can be used as local coframes on the submanifold in question. But then it is the graph of a local diffeomorphism, which by construction leaves the Maurer–Cartan forms invariant.—Note that the unique integral manifold passing through $(e, \hat{e})$ must be the diagonal of $\mathbb{R}^r_x \times \mathbb{R}^r_{\hat{x}}$, which is the graph of the identity mapping.

If ϕ and ψ are such local diffeomorphisms with $\phi \circ \psi$ being defined,

$$(\phi \circ \psi)^*(\omega^i) = \psi^* \circ \phi^*(\omega^i) = \omega^i,$$

whence $\phi \circ \psi$ preserves the ω^i too. And with $\phi \circ \phi^{-1} = \mathrm{id}$,

$$\omega^i = (\phi \circ \phi^{-1})^*(\omega^i) = (\phi^{-1})^* \circ \phi^*(\omega^i) = (\phi^{-1})^*(\omega^i),$$

so that also ϕ^{-1} preserves the ω^i. Thus we have proved the **converse of the second fundamental theorem**.

Theorem 8.5.5. *Let* $\{\omega^1, \ldots, \omega^r\}$ *be a local coframe on* $\mathbb{R}^r$ *satisfying the structure equations*

$$d\omega^i = \sum_{j<k} C^i_{jk}\, \omega^j \wedge \omega^k, \qquad i = 1, \ldots, r,$$

with **constant** C^i_{jk}. *Then the family of local diffeomorphisms leaving each of the* ω^i *invariant forms an r-parameter local Lie group.* □

The amazing fact is that by the fundamental theorems *all* local parameter groups arise in this extremely simple way.

8.6 The third fundamental theorem

The definition of the structure constants c^i_{jk} shows immediately that they are anti-symmetric with respect to the lower indices:

$$c^i_{jk} = -c^i_{kj} \qquad \text{for } i, j, k = 1, \ldots, r.$$

Moreover, by the second fundamental theorem the Maurer–Cartan forms ω^i satisfy the structure equations

$$d\omega^i = \frac{1}{2} \sum_{j,k=1}^r c^i_{jk}\, \omega^j \wedge \omega^k \qquad \text{for } i = 1, \ldots, r.$$

Differentiating these, we get

$$0 = d^2\omega^i = \frac{1}{2} \sum_{j,k=1}^r (c^i_{jk}\, d\omega^j \wedge \omega^k - c^i_{jk}\, \omega^j \wedge d\omega^k).$$

Inserting the above expressions for $d\omega^j$ and $d\omega^k$ and utilizing the fact that $\omega^i \wedge \omega^j \wedge \omega^k$ form a basis for the three-forms on $\mathbb{R}^r$ for $1 \le i < j < k \le r$, we find that

$$\sum_{s=1}^r (c^i_{js} c^s_{kl} + c^i_{ks} c^s_{lj} + c^i_{ls} c^s_{jk}) = 0 \qquad \text{for } i, j, k, l = 1, \ldots, r.$$

These two requirements on the structure constants c^i_{jk} make up **Lie's third fundamental theorem**.

Theorem 8.6.1. *The structure constants c_{jk}^i of a local parameter group satisfy*

- $c_{jk}^i = -c_{kj}^i$ *(anti-symmetry),*
- $\sum_{s=1}^{r}(c_{js}^i c_{kl}^s + c_{ks}^i c_{lj}^s + c_{ls}^i c_{jk}^s) = 0$ *(Jacobi's identity).*

$\square$

The main point of this section is to prove the converse: if $\{c_{jk}^i\}_{i,j,k=1,\ldots,r}$ is a collection of numbers satisfying these two conditions, it is possible to construct a local parameter group $\mathfrak{G}$ having the c_{jk}^i as its structure constants.

In order to prove this it suffices by the second fundamental theorem to create Maurer–Cartan forms

$$\omega^i = \sum_{j=1}^{r} \chi_j^i(x)\, dx^j \quad \text{for } i = 1,\ldots,r$$

satisfying

$$d\omega^i = \frac{1}{2} \sum_{j,k=1}^{r} c_{jk}^i\, \omega^j \wedge \omega^k,$$

or equivalently

$$\frac{\partial \chi_k^i}{\partial x^j} - \frac{\partial \chi_j^i}{\partial x^k} = \sum_{p,q=1}^{r} c_{pq}^i \chi_j^p(x) \chi_k^q(x).$$

It turns out to be convenient to do this in a canonical coordinate system of the first kind. The following lemma describes the functions $\chi_j^i = {}_L\bar{\psi}_j^{-1i}$ in such a system.

Lemma 8.6.1. *The following equivalence holds: $x^1,\ldots,x^r$ are canonical coordinates of the first kind for a local parameter group $\mathfrak{G} \iff$ the functions ${}_L\bar{\psi}_j^{-1i}(x)$ satisfy*

$$\sum_{j=1}^{r} {}_L\bar{\psi}_j^{-1i}(x) \cdot x^j = x^i \quad \text{for } i = 1,\ldots,r. \tag{*}$$

Proof. $\Longrightarrow$: Recall from section 8.4 that $t \mapsto x(t)$ is a 1-parameter group if and only if $x^i(0) = 0$ and

$$\frac{dx^i}{dt}(t) = \sum_{j=1}^{r} {}_L\psi_j^i(x(t)) \frac{dx^j}{dt}(0),$$

or equivalently

$$\frac{dx^i}{dt}(0) = \sum_{j=1}^{r} L \bar{\psi}^{1i}_{j}(x(t)) \frac{dx^j}{dt} \quad \text{for } i = 1,\ldots,r. \tag{†}$$

In a coordinate system of the first kind all 1-parameter groups are of the form $t \mapsto \alpha t = (\alpha^1 t,\ldots,\alpha^r t)$, whence (†) reduces to

$$\alpha^i = \sum_{j=1}^{r} L \bar{\psi}^{1i}_{j}(\alpha t) \, \alpha^j.$$

Equation (*) follows from this by setting $t = 1$.

$\Longleftarrow$: Let $t \mapsto x(t)$ be a 1-parameter group with $dx^i/dt(0) = \alpha^i$. Then $x(t)$ is determined by the ODE system

$$\begin{cases} \alpha^i = \sum_{j=1}^{r} L \bar{\psi}^{1i}_{j}(x(t)) \, (dx^j/dt), \\ x^i(0) = 0 \end{cases} \quad \text{for } i = 1,\ldots,r.$$

Now (*) shows that $x(t) := \alpha t$ satisfies

$$\sum_{j=1}^{r} L \bar{\psi}^{1i}_{j}(\alpha t) \, \alpha^j = \alpha^i,$$

so that $t \mapsto \alpha t$ really is a solution of the above system—and by uniqueness it is the only one. Hence $x(t) = \alpha t$. $\qquad\square$

The key idea for constructing the functions $\chi^i_j(t)$ from the set $\{c^i_{jk}\}$ is to prove the converse of the following result.

Lemma 8.6.2. *Let $(\chi^i_j(x))$ be the inverse of the matrix of left transformation functions for a local parameter group $\mathfrak{G}$ with the structure constants $\{c^i_{jk}\}$, let $x^1,\ldots,x^r$ be canonical coordinates of the first kind for $\mathfrak{G}$, and let $\alpha = (\alpha^1,\ldots,\alpha^r)$ be a fixed vector. Introduce the auxiliary functions*

$$f^i_j(t;\alpha) := t \chi^i_j(\alpha^1 t,\ldots,\alpha^r t) \equiv t \cdot \chi^i_j(\alpha t).$$

Then the $\chi^i_j(x)$ are obviously regained from the $f^i_j(t;\alpha)$ by

$$\chi^i_j(x) = f^i_j(1;x).$$

These auxiliary functions are uniquely determined as solutions of the ODE system

$$\begin{cases} (df^i_j/dt)(t;\alpha) = \delta^i_j + \sum_{k,l=1}^{r} c^i_{kl}\alpha^k f^l_j(t;\alpha), \\ f^i_j(0;\alpha) = 0. \end{cases}$$

Proof. Applying $\partial/\partial x^j$ to $\sum_{k=1}^{r} \chi_k^i(x) x^k = x^i$ gives

$$\sum_{k=1}^{r} \frac{\partial \chi_k^i}{\partial x^j}(x) x^k + \chi_j^i(x) = \delta_j^i.$$

If we multiply $\partial \chi_k^i/\partial x^j - \partial \chi_j^i/\partial x^k = \sum_{p,q=1}^{r} c_{pq}^i \chi_j^p \chi_k^q$ by x^k and sum over k, it next follows that

$$\sum_{k=1}^{r} \left(\frac{\partial \chi_k^i}{\partial x^j} x^k - \frac{\partial \chi_j^i}{\partial x^k} x^k \right) = \sum_{p,q,k=1}^{r} c_{pq}^i \chi_j^p \chi_k^q x^k$$

$$= \sum_{p,q=1}^{r} c_{pq}^i \chi_j^p x^q = - \sum_{p,q=1}^{r} c_{pq}^i x^p \chi_j^q.$$

Combining these two results, we get

$$\sum_{k=1}^{r} \frac{\partial \chi_j^i}{\partial x^k}(x) x^k - \sum_{p,q=1}^{r} c_{pq}^i x^p \chi_j^q(x) + \chi_j^i(x) = \delta_j^i.$$

Finally, setting $x = \alpha t$,

$$\sum_{k=1}^{r} \frac{\partial \chi_j^i}{\partial x^k}(\alpha t) \alpha^k t + \chi_j^i(\alpha t) = \delta_j^i + \sum_{p,q=1}^{r} c_{pq}^i \alpha^p t \chi_j^q(\alpha t),$$

or

$$\frac{df_j^i}{dt}(t;\alpha) = \delta_j^i + \sum_{p,q=1}^{r} c_{pq}^i \alpha^p f_j^q(t;\alpha).$$

□

This lemma suggests the following procedure in order to obtain Maurer–Cartan forms $\omega^1,\ldots,\omega^r$ with $d\omega^i = \frac{1}{2}\sum_{j,k=1}^{r} c_{jk}^i \omega^j \wedge \omega^k$ from a collection $\{c_{jk}^i\}$ of numbers satisfying the conditions of the third fundamental theorem:

$$\{c_{jk}^i\} \Longrightarrow \text{ the ODE system } \begin{cases} (df_j^i/dt)(t;\alpha) = \delta_j^i + \sum_{p,q=1}^{r} c_{pq}^i \alpha^p f_j^q(t;\alpha), \\ f_j^i(0;\alpha) = 0 \end{cases}$$

$\Longrightarrow$ solutions $f_j^i(t;\alpha) \Longrightarrow$ functions $\chi_j^i(x) = f_j^i(1;x)$

$\Longrightarrow$ Maurer–Cartan forms $\omega^i = \sum_{j=1}^{r} \chi_j^i(x) dx^j$ for $i = 1,\ldots,r$, satisfying

the structure equations $d\omega^i = \frac{1}{2} \sum_{j,k=1}^{r} c_{jk}^i \omega^j \wedge \omega^k$.

And this in fact gives us the **converse of the third fundamental theorem**.

Theorem 8.6.2. *Let* $\{c^i_{jk}\}^r_{i,j,k=1}$ *be a collection of numbers satisfying*

- $c^i_{jk} = -c^i_{kj}$,
- $\sum^r_{s=1}(c^i_{js}c^s_{kl} + c^i_{ks}c^s_{lj} + c^i_{ls}c^s_{jk}) = 0$,

let $\alpha = (\alpha^1,\ldots,\alpha^r)$ *be a constant vector, and consider the initial value problem*

$$\begin{cases} df^i_j/dt(t;\alpha) = \delta^i_j + \sum^r_{p,q=1} c^i_{pq}\alpha^p f^q_j(t;\alpha), \\ f^i_j(0;\alpha) = 0. \end{cases}$$

Because this is an ODE system with constant constant coefficients, its solutions $f^i_j(t;\alpha)$ *are analytic and defined for all t.*

The functions $\chi^i_j(x) := f^i_j(1;x)$ *have the following properties:*

$$\sum^r_{j=1}\chi^i_j(x)x^j = x^i, \tag{1}$$

$$\frac{\partial\chi^i_k}{\partial x^j}(x) - \frac{\partial\chi^i_j}{\partial x^k}(x) = \sum^r_{p,q=1}c^i_{pq}\chi^p_j(x)\chi^q_k(x), \tag{2}$$

$$\chi^i_j(e) = \delta^i_j. \tag{3}$$

Therefore the structure equations for the one-forms $\omega^i := \sum^r_{j=1}\chi^i_j(x)\,dx^j$ *are*

$$d\omega^i = \frac{1}{2}\sum^r_{j,k=1}c^i_{jk}\,\omega^j\wedge\omega^k \quad \text{for } i = 1,\ldots,r.$$

Finally the converse of the second fundamental theorem shows that these one-forms define a local parameter group with the c^i_{jk} *as structure constants.*

Proof. We must establish (1), (2) and (3).

For $\alpha = (0,\ldots,0) = e$ the initial value problem has the solution $f^i_j(t;0) = \delta^i_j \cdot t$, so that

$$\chi^i_j(e) = f^i_j(1;0) = \delta^i_j,$$

proving (3).

In order to demonstrate (1) we set

$$h^i(t) := \sum^r_{j=1}f^i_j(t;\alpha)\alpha^j - t\alpha^i$$

and prove that $h^i(t) \equiv 0$—which suffices, since (1) is equivalent to $h^i(1) = 0$.

Firstly $h^i(0) = 0$, and secondly

$$
\frac{dh^i}{dt} = \sum_{j=1}^{r} \left(\delta^i_j + \sum_{p,q=1}^{r} c^i_{pq} \alpha^p f^q_j(t;\alpha) \right) \alpha^j - \alpha^i
$$

$$
= \sum_{j,p,q=1}^{r} c^i_{pq} \alpha^p f^q_j(t;\alpha) \alpha^j
$$

$$
= \sum_{j,p,q=1}^{r} c^i_{pq} \alpha^p (f^q_j(t;\alpha)\alpha^j - t\alpha^q) \quad \text{using } c^i_{pq} = -c^i_{qp}
$$

$$
= \sum_{p,q=1}^{r} c^i_{pq} \alpha^p h^q(t).
$$

Thus the $h^i(t)$ satisfy the system

$$
\begin{cases} dh^i/dt = \sum_{p,q=1}^{r} c^i_{pq} \alpha^p h^q(t), \\ h^i(0) = 0 \end{cases} \qquad \text{for } i = 1,\dots,r,
$$

having $h^i(t) \equiv 0$ as its unique solution.

At last (2) is proved by a similar trick: this time we set

$$
h^i_{jk}(t) := \frac{\partial f^i_k}{\partial \alpha^j}(t;\alpha) - \frac{\partial f^i_j}{\partial \alpha^k}(t;\alpha) - \sum_{p,q=1}^{r} c^i_{pq} f^p_j(t;\alpha) f^q_k(t;\alpha),
$$

and show that $h^i_{jk}(t) \equiv 0$; note that (2) is equivalent to $h^i_{jk}(1) = 0$.

Firstly $f^i_j(0;\alpha) \equiv 0 \implies (\partial f^i_j/\partial \alpha^k)(0;\alpha) = 0$, so that the h^i_{jk} satisfy the initial condition $h^i_{jk}(0) = 0$.

Claim. The $h^i_{jk}(t)$ are solutions of the ODE system

$$
\frac{dh^i_{jk}}{dt} = \sum_{p,q=1}^{r} c^i_{pq} \alpha^p h^q_{jk}.
$$

This together with the initial condition above forces the $h^i_{jk}(t)$ to vanish identically.

In order to establish the claim, apply $\partial/\partial \alpha^j$ to

$$
\frac{df^i_k}{dt} = \delta^i_k + \sum_{p,q=1}^{r} c^i_{pq} \alpha^p f^q_k
$$

so as to get

$$
\frac{\partial^2 f^i_k}{\partial t \partial \alpha^j} = \sum_{q=1}^{r} c^i_{jq} f^q_k + \sum_{p,q=1}^{r} c^i_{pq} \alpha^p \frac{\partial f^q_k}{\partial \alpha^j}.
$$

But then

$$\frac{dh^i_{jk}}{dt} = \frac{\partial^2 f^i_k}{\partial t \partial \alpha^j} - \frac{\partial^2 f^i_j}{\partial t \partial \alpha^k} - \sum_{p,q=1}^{r} c^i_{pq} \left(\frac{\partial f^p_j}{\partial t} f^q_k + f^p_j \frac{\partial f^q_k}{\partial t} \right)$$

$$= \sum_{q=1}^{r} c^i_{jq} f^q_k + \sum_{p,q=1}^{r} c^i_{pq} \alpha^p \frac{\partial f^q_k}{\partial \alpha^j} - \sum_{q=1}^{r} c^i_{kq} f^q_j - \sum_{p,q=1}^{r} c^i_{pq} \alpha^p \frac{\partial f^q_j}{\partial \alpha^k}$$

$$- \sum_{p,q=1}^{r} c^i_{pq} \left(\delta^p_j + \sum_{m,n=1}^{r} c^p_{mn} \alpha^m f^n_j \right) f^q_k$$

$$- \sum_{p,q=1}^{r} c^i_{pq} f^p_j \left(\delta^q_k + \sum_{m,n=1}^{r} c^q_{mn} \alpha^m f^n_k \right).$$

If we make careful use of the conditions imposed on the c^i_{jk}, this complicated expression simplifies to

$$\frac{dh^i_{jk}}{dt} = \sum_{p,q=1}^{r} c^i_{pq} \alpha^p \left(\frac{\partial f^q_k}{\partial \alpha^j} - \frac{\partial f^q_j}{\partial \alpha^k} - \sum_{m,n=1}^{r} c^q_{mn} f^m_j f^n_k \right)$$

$$= \sum_{p,q=1}^{r} c^i_{pq} \alpha^p h^q_{jk},$$

and with this the claim is established. $\qquad\square$

Remark. As we have worked in the C^∞ category throughout it is a bit surprising that this theorem shows that the functions $\chi^i_j(x)$ in fact are *analytic* when expressed in canonical coordinates of the first kind. Hence without loss of generality all parameter groups may be assumed to be analytic, rather than just smooth.

8.7 Local transformation groups

We started this chapter by considering a local Lie group G acting on $\mathbb{R}^n_x$,

$$G \ni g : \mathbb{R}^n_x \times \mathbb{R}^r_a \to \mathbb{R}^n_x$$

$$(x, a) \mapsto g(x; a) = g_a(x),$$

and from G we derived its parameter group $\mathfrak{G}$, which acts on the parameter space $\mathbb{R}^r$,

$$\mathbb{R}^r \times \mathbb{R}^r \to \mathbb{R}^r$$

$$(a, b) \mapsto \gamma(a; b) = a \cdot b,$$

in such a way that $g_a(g_b(x)) = g_{a \cdot b}(x).$

In this section we turn this all around by starting with a parameter group $\mathfrak{G} \cong (\mathbb{R}^r, \cdot)$ and letting it act on $\mathbb{R}^n_x$:

$$\mathbb{R}^n_x \times \mathfrak{G} \rightarrow \mathbb{R}^n_x$$
$$(x, a) \mapsto \phi_a(x) = \phi(x; a).$$

Definition. $\mathfrak{G}$ defines a local Lie transformation group G acting on $\mathbb{R}^n_x$ by $x \mapsto \phi_a(x) = \phi(x; a)$ if ϕ is smooth and
1. $\phi_e = \mathrm{id}$,
2. $\phi_a(\phi_b(x)) = \phi_{a \cdot b}(x)$.
We also assume that the action is *effective* in the sense that
3. $\phi_a = \phi_e \iff a = e$;
this means that there are 'no unnecessary parameters' in the action.

Recall that the Lie algebra $\mathfrak{g}_L$ of left-invariant vector fields on $\mathfrak{G}$ is given by $\mathrm{span}_\mathbb{R}\{X_1, \ldots, X_r\}$, where

$$X_i(a) = (L_a)_{*e}\left(\frac{\partial}{\partial a^i}\right) = \sum_{j=1}^{r} {}_R\psi_i^j(a)\frac{\partial}{\partial a^j},$$

with $\partial/\partial a^i$ being considered as a tangent vector in $T_e\mathfrak{G}$.

We will now construct a family of vector fields on $\mathbb{R}^n_x$ in a similar manner. For any $\alpha \in T_e\mathfrak{G}$, let $t \mapsto a(t)$ be a curve in $\mathfrak{G}$ with $a(0) = (0, \ldots, 0) = e$, and $da^i/dt(0) = \alpha^i$ for $i = 1, \ldots, r$. Then for each $x \in \mathbb{R}^n$ consider the curve

$$t \mapsto \phi(x; a(t)) = (\phi^1(x; a(t)), \ldots, \phi^n(x; a(t)))$$

passing through x when $t = 0$. Its tangent vector at $t = 0$ is given by

$$\sum_{i=1}^{n}\sum_{j=1}^{r} \frac{\partial \phi^i}{\partial a^j}(x; e)\frac{da^j}{dt}(0)\frac{\partial}{\partial x^i} = \sum_{j=1}^{r}\alpha^j\left(\sum_{i=1}^{n}\frac{\partial \phi^i}{\partial a^j}(x; e)\frac{\partial}{\partial x^i}\right) = \sum_{j=1}^{r}\alpha^j K_j,$$

with

$$K_j := \sum_{i=1}^{n}\frac{\partial \phi^i}{\partial a^j}(x; e)\frac{\partial}{\partial x^i} \qquad \text{for } j = 1, \ldots, r.$$

Notations. Set $\lambda_j^i(x) := (\partial \phi^i/\partial a^j)(x; e)$; these functions correspond to the right transformation functions ${}_R\psi_j^i(a)$ of $\mathfrak{G}$. The vector field K_i is called the *Killing vector field* associated to $\partial/\partial a^i \in T_e\mathfrak{G}$.

Remark. Infinitesimally the effectiveness of the action ϕ means that

$$\phi(x; \cdot)_{*e} : T_e\mathfrak{G} \longrightarrow T_x\mathbb{R}^n$$

is *injective*, and this implies that the Killing vector fields $K_1, \ldots, K_r$ are linearly independent. Thus

$$\mathfrak{k} := \operatorname{span}_{\mathbb{R}}\{K_1, \ldots, K_r\}$$

is an r-dimensional vector space of vector fields on $\mathbb{R}_x^n$.

Definition. The *adjoint action* of the parameter group $\mathfrak{G}$ on itself is given by

$$\mathfrak{G} \ni a \mapsto \operatorname{Ad}_a : \mathfrak{G} \to \mathfrak{G}$$
$$x \mapsto a \cdot x \cdot a^{-1};$$

in particular $\operatorname{Ad}_a(e) = a \cdot e \cdot a^{-1} = e$.

Consider the right multiplication $x \mapsto x \cdot b =: z$. If x is multiplied by $\Delta x = (\Delta x^1, \ldots, \Delta x^r) \approx (0, \ldots, 0)$ on the right, z gets multiplied by a small increment Δz:

$$x \mapsto x \cdot \Delta x \Longrightarrow z = x \cdot b \mapsto x \cdot \Delta x \cdot b = z \cdot \Delta z.$$

But then

$$\operatorname{Ad}_z(\Delta z) = z \cdot \Delta z \cdot z^{-1} = x \cdot \Delta x \cdot b \cdot b^{-1} \cdot x^{-1} = \operatorname{Ad}_x(\Delta x).$$

Clearly there are functions $v_j^i(x)$ with $v_j^i(e) = \delta_j^i$ such that

$$\operatorname{Ad}_x^i(\Delta x) = \sum_{j=1}^r v_j^i(x) \Delta x^j + O((\Delta x)^2).$$

Applying this to $\operatorname{Ad}_z(\Delta z)$,

$$\operatorname{Ad}_z^i(\Delta z) = \sum_{j=1}^r v_j^i(z) \Delta z^j + \cdots$$

$$= \sum_{j=1}^r v_j^i(z) \left(\sum_{k=1}^r \frac{\partial z^j}{\partial x^k} \Delta x^k \right) + \cdots.$$

So equating the first order terms in $\operatorname{Ad}_z(\Delta z) = \operatorname{Ad}_x(\Delta x)$, we have

$$\sum_{j=1}^r v_j^i(z) \frac{\partial z^j}{\partial x^k} = v_k^i(x) \quad \text{for } i, j = 1, \ldots, r.$$

Comparing this with the first fundamental theorem for parameter groups

Local Lie groups

we deduce that actually

$$v^i_j(x) = \chi^i_j(x) = {}_L\overline{\psi}{}^{1i}_j(x).$$

Conclusion. $\mathrm{Ad}^i_x(\Delta x) = \sum^r_{j=1}\chi^i_j(x)\,\Delta x^j + O((\Delta x)^2).$

Let us now use this in order to determine the defining PDE system for the action functions $\phi^i(x;a)$, thereby obtaining the **first fundamental theorem for Lie group actions**.

Theorem 8.7.1. *The functions $\phi^i(x;a)$ are solutions of the system*

$$\begin{cases} \partial\phi^i/\partial a^j = \sum^r_{k=1}\lambda^i_k(\phi)\chi^k_j(a), & i=1,\dots,n \text{ and } j=1,\dots,r, \\ \phi^i(x;e) = x^i, & i=1,\dots,n. \end{cases}$$

Proof. Let $\Delta a \approx e = (0,\dots,0)$, and set $\delta a := \mathrm{Ad}_a(\Delta a) = a\cdot\Delta a\cdot a^{-1}$. Then

$$\phi(x;a\cdot\Delta a) = \phi(x;\delta a\cdot a) = \phi_{\delta a\cdot a}(x) = \phi_{\delta a}(\phi_a(x)) = \phi(\phi_a(x);\delta a).$$

With $z := \phi_a(x) = \phi(x;a)$,

$$\phi^i(x;a\cdot\Delta a) = \phi^i(z;\delta a) = \phi^i\left(z;\dots,\sum^r_{j=1}\chi^i_j(a)\Delta a^j + \cdots,\cdots\right)$$

$$= \phi^i(z;e) + \sum^r_{j,k=1}\frac{\partial\phi^i}{\partial a^k}(z;e)\,\chi^k_j(a)\,\Delta a^j + \cdots$$

$$= \phi^i(x;a) + \sum^r_{j,k=1}\lambda^i_k(z)\chi^k_j(a)\,\Delta a^j + \cdots,$$

whence

$$\frac{\partial\phi^i}{\partial a^j} = \lim_{\Delta a^j\to 0}\frac{\phi^i(x;a\cdot\Delta a^j) - \phi^i(x;a)}{\Delta a^j} = \sum^r_{k=1}\lambda^i_k(\phi)\chi^k_j(a).$$

$\square$

The second fundamental theorem specifies the integrability conditions for this defining system:

$$\frac{\partial^2 \phi^i}{\partial a^k \partial a^j} = \frac{\partial^2 \phi^i}{\partial a^j \partial a^k} \Longleftrightarrow$$

$$\sum_{l,p=1}^{r} \sum_{m=1}^{n} \frac{\partial \lambda_l^i}{\partial z^m}(z)\, \lambda_p^m(z) \chi_k^p(a) \chi_j^l(a) + \sum_{l=1}^{r} \lambda_l^i(z)\, \frac{\partial \chi_j^l}{\partial a^k}(a)$$

$$= \sum_{l,p=1}^{r} \sum_{m=1}^{n} \frac{\partial \lambda_l^i}{\partial z^m}(z)\, \lambda_p^m(z) \chi_j^p(a) \chi_k^l(a) + \sum_{l=1}^{r} \lambda_l^i(z)\, \frac{\partial \chi_k^l}{\partial a^j}(a) \Longleftrightarrow$$

$$\sum_{l,p=1}^{r} \sum_{m=1}^{n} \left(\frac{\partial \lambda_l^i}{\partial z^m}(z)\lambda_p^m(z) - \frac{\partial \lambda_p^i}{\partial z^m}\lambda_l^m(z) \right) \chi_k^p(a)\chi_j^l(a)$$

$$+ \sum_{q=1}^{r} \lambda_q^i \left(\frac{\partial \chi_j^q}{\partial a^k}(a) - \frac{\partial \chi_k^q}{\partial a^j}(a) \right) = 0.$$

According to the second fundamental theorem for $\mathfrak{G}$,

$$\frac{\partial \chi_j^q}{\partial a^k} - \frac{\partial \chi_k^q}{\partial a^j} = - \sum_{l,p=1}^{r} c_{lp}^q \chi_j^l \chi_k^p,$$

where the c_{jk}^i are the structure constants of $\mathfrak{G}$. Hence the above equality reduces to

$$\sum_{l,p=1}^{r} \sum_{m=1}^{n} \left(\frac{\partial \lambda_l^i}{\partial z^m}(z)\lambda_p^m(z) - \frac{\partial \lambda_p^i}{\partial z^m}(z)\lambda_l^m(z) - \sum_{q=1}^{r} \lambda_q^i(z)\, c_{lp}^q \right) \chi_k^p(a)\chi_j^l(a) = 0.$$

With the matrix $(\chi_j^i(a))$ being nonsingular for $a \approx e$, this shows that the integrability conditions take the form

$$\sum_{m=1}^{n} \left(\frac{\partial \lambda_l^i}{\partial z^m}\lambda_k^m - \frac{\partial \lambda_k^i}{\partial z^m}\lambda_l^m \right) = \sum_{q=1}^{r} c_{lk}^q\, \lambda_q^i.$$

Remark. Conversely it can be shown that when these integrability conditions are fulfilled, the defining system

$$\frac{\partial \phi^i}{\partial a^j} = \sum_{k=1}^{r} \lambda_k^i(\phi)\chi_j^k(a)$$

admits a unique solution satisfying $\phi^i(x; e) = x^i$.

If we recall the Killing vector fields

$$K_j = \sum_{l=1}^{r} \lambda_j^l(x)\, \frac{\partial}{\partial x^l} \quad \text{for } j = 1, \dots, r,$$

the integrability conditions above amount precisely to

$$[K_j, K_k] = \sum_{l=1}^{n} c_{jk}^l K_l \quad \text{for } j, k = 1, \ldots, r,$$

so that $\mathfrak{k} = \mathrm{span}_{\mathbb{R}}\{K_1, \ldots, K_r\}$ actually is a Lie algebra, called the *Killing algebra*, induced on $\mathbb{R}_x^n$ by the action of $\mathfrak{G}$. And this is the content of the **second fundamental theorem for Lie group actions**.

Theorem 8.7.2. *We have that $\mathfrak{k}$ is a Lie algebra of vector fields on $\mathbb{R}_x^n$ with the same structure constants as the parameter group $\mathfrak{G}$.* □

The isomorphisms

$$T_e\mathfrak{G} \xrightarrow{\cong} \mathfrak{g}_L$$

$$\partial/\partial a^i \mapsto X_i = \sum_{k=1}^{r} \chi_i^k(a)\,(\partial/\partial a^k)$$

and

$$T_e\mathfrak{G} \xrightarrow{\cong} \mathfrak{k}$$

$$\partial/\partial a^i \mapsto K_i = \sum_{l=1}^{n} \lambda_i^l(x)\,(\partial/\partial x^l)$$

induce an isomorphism of vector spaces:

$$\mathcal{K}: \mathfrak{g}_L \xrightarrow{\cong} \mathfrak{k}$$

$$\sum_{i=1}^{r} \alpha^i X_i \mapsto \sum_{i=1}^{r} \alpha^i K_i.$$

Theorem 8.7.3. $\mathcal{K}$ *is a Lie algebra isomorphism, i.e.,*

$$\mathcal{K}([X, Y]) = [\mathcal{K}(X), \mathcal{K}(Y)] \quad \text{for } X, Y \in \mathfrak{g}_L,$$

called the Killing isomorphism.

Proof. Because of the linearity it suffices to check this for the basis vector fields—which is immediate:

$$\mathcal{K}([X_j, X_k]) = \mathcal{K}\left(\sum_{i=1}^{r} c_{jk}^i X_i\right) = \sum_{i=1}^{r} c_{jk}^i K_i = [K_j, K_k].$$

□

Conversely *any* Lie algebra of vector fields on $\mathbb{R}_x^n$ arises in this way.

Theorem 8.7.4. *Let* $\mathfrak{h} := \operatorname{span}_{\mathbb{R}}\{Y_1, \ldots, Y_r\}$ *be an r-dimensional Lie algebra of vector fields on* $\mathbb{R}^n_x$ *with*

$$Y_i = \sum_{l=1}^{n} \lambda_i^l(x) \left(\partial/\partial x^l\right) \quad \text{and} \quad [Y_j, Y_k] = \sum_{i=1}^{r} c_{jk}^i Y_i \quad \text{for } i, j, k = 1, \ldots, r.$$

Then there is a unique local r-parameter group $\mathfrak{G}$ *acting on* $\mathbb{R}^n_x$ *such that* $\mathfrak{h}$ *equals the Killing algebra of vector fields associated to* $\mathfrak{G}$.

Proof. From the structure equations $[Y_j, Y_k] = \sum_{i=1}^{r} c_{jk}^i Y_i$ and from the Jacobi identity for vector fields it follows that the structure constants c_{jk}^i of $\mathfrak{h}$ satisfy

- $c_{jk}^i = -c_{kj}^i$,
- $\sum_{s=1}^{r}(c_{js}^i c_{kl}^s + c_{ks}^i c_{lj}^s + c_{ls}^i c_{jk}^s) = 0$.

Therefore the converse of the third fundamental theorem shows that there is a unique local parameter group $\mathfrak{G} = (\mathbb{R}^r, \cdot)$ with these c_{jk}^i as its structure constants.

Let $(\chi_j^i(a))$ be the inverse of the matrix of left transformation functions $(_L\psi_j^i(a))$ for $\mathfrak{G}$. Note that *if* $\mathfrak{G}$ acts on $\mathbb{R}^n_x$ by

$$\mathbb{R}^n_x \times \mathbb{R}^r_a \to \mathbb{R}^n_x$$
$$(x, a) \mapsto \phi(x; a),$$

and induces the Killing vector fields $Y_i = \sum_{l=1}^{r} \lambda_i^l(x) \left(\partial/\partial x^l\right)$ for $i = 1, \ldots, r$, then the ϕ^i are solutions of the system

$$\begin{cases} \partial\phi^i/\partial a^j = \sum_{k=1}^{r} \lambda_k^i(\phi)\chi_j^k(a), \\ \phi^i(x; e) = x^i \end{cases} \quad \text{for } i = 1, \ldots, n \text{ and } j = 1, \ldots, r. \quad (*)$$

Let us now turn this around—i.e., we start with the system (*) and show that its solutions do define the wanted action of $\mathfrak{G}$ on $\mathbb{R}^n_x$.

Firstly the integrability conditions for (*) correspond precisely to the fact that $\mathfrak{h}$ is a Lie algebra, and therefore (*) is really locally solvable.

Secondly we have to show that its solutions $\phi^i(x; a)$ define an effective action on $\mathbb{R}^n_x$, i.e.,

$$\phi_a(\phi_b(x)) = \phi_{a \cdot b}(x)$$

and

$$\phi_a = \operatorname{id} \iff a = e.$$

Let $a, b \in \mathfrak{G}$, and let $c := a \cdot b$, $z := \phi(x; b)$, $\zeta^* := \phi(z; a)$ and $\zeta := \phi(x; c)$; we then must prove that $\zeta^* = \zeta$.

Now by definition ζ^* satisfies

$$\frac{\partial \zeta^{*i}}{\partial a^j} = \sum_{k=1}^{r} \lambda_k^i(\zeta^*) \chi_j^k(a).$$

Next, according to the first fundamental theorem for parameter groups,

$$\sum_{k=1}^{r} \chi_k^i(c) \frac{\partial c^k}{\partial a^j} = \chi_j^i(a), \quad \text{with } \chi_j^i = {}_L \bar{\psi}_j^i.$$

Consequently

$$\frac{\partial c^k}{\partial a^j} = \sum_{l=1}^{r} \chi_l^k(c) {}_L \psi_j^l(a),$$

by means of which we can calculate

$$\frac{\partial \zeta^i}{\partial a^j} = \sum_{k=1}^{r} \frac{\partial \zeta^i}{\partial c^k} \frac{\partial c^k}{\partial a^j} = \sum_{k,l,m=1}^{r} \lambda_m^i(\zeta) \chi_k^m(c) {}_L \bar{\psi}_l^k(a) {}_L \psi_j^l(a)$$

$$= \sum_{m=1}^{r} \lambda_m^i(\zeta) \chi_j^m(a).$$

Therefore ζ and ζ^* both satisfy (*), and so must be equal.

Finally let

$$\mathfrak{N} := \{a \in \mathfrak{G} \mid \phi_a = \mathrm{id}\}.$$

Then $\mathfrak{N}$ is a normal subgroup of $\mathfrak{G}$, corresponding to an ideal $\mathfrak{n}$ of $\mathfrak{g}_L$ which is mapped to 0 by $\mathcal{K}$. Thus

$$r = \dim \mathfrak{h} = \dim \mathfrak{g}_L - \dim \mathfrak{n} = r - \dim \mathfrak{n},$$

showing that $\mathfrak{n} = \{0\}$ and $\mathfrak{N} = \{e\}$. Hence

$$\mathbb{R}_x^n \times \mathfrak{G} \to \mathbb{R}_x^n$$

$$(x, a) \mapsto \phi(x; a)$$

is an effective action of $\mathfrak{G}$ on $\mathbb{R}_x^n$, which by construction induces an isomorphism

$$\mathcal{K} : \mathfrak{g}_L \xrightarrow{\cong} \mathfrak{h}.$$

$\square$

9

Structural classification of 3-dimensional Lie algebras over the complex numbers

In chapters **12** and **13** we will consider Vessiot's classification of those hyperbolic second order PDEs in one dependent and two independent variables whose associated Monge systems admit at least two independent first integrals. This turns out to be based upon the classification of 3-dimensional Lie algebras, which we study in this chapter—thereby also obtaining concrete examples of the abstract theory presented in the preceding chapter.

Let

$$\mathfrak{g} = \operatorname{span}\{X_1,\ldots,X_r\}$$

be an r-dimensional Lie algebra with the structure equations

$$[X_i, X_j] = \sum_{k=1}^{r} c_{ij}^k X_k \quad \text{for } i, j = 1,\ldots,r.$$

Introducing a new basis $\{\bar{X}_1,\ldots,\bar{X}_r\}$ by means of a nonsingular matrix (γ_j^i),

$$\begin{pmatrix} \bar{X}_1 \\ \vdots \\ \bar{X}_r \end{pmatrix} = \begin{pmatrix} \gamma_1^1 & \cdots & \gamma_1^r \\ \vdots & \ddots & \vdots \\ \gamma_r^1 & \cdots & \gamma_r^r \end{pmatrix} \begin{pmatrix} X_1 \\ \vdots \\ X_r \end{pmatrix}$$

or

$$\bar{X} = \gamma X \Longleftrightarrow X = \gamma^{-1} \bar{X},$$

221

we get new structure equations:

$$[\bar{X}_i, \bar{X}_j] = \left[\sum_{a=1}^{r} \gamma_i^a X_a, \sum_{b=1}^{r} \gamma_j^b X_b \right] = \sum_{a,b=1}^{r} \gamma_i^a \gamma_j^b \left(\sum_{l=1}^{r} c_{ab}^l X_l \right)$$

$$= \sum_{k=1}^{r} \left(\sum_{a,b,l=1}^{r} \gamma_i^a \gamma_j^b c_{ab}^l (\gamma^{-1})_l^k \right) \bar{X}_k = \sum_{k=1}^{r} \bar{c}_{ij}^k \bar{X}_k,$$

with

$$\bar{c}_{ij}^k = \sum_{a,b,l=1}^{r} \gamma_i^a \gamma_j^b c_{ab}^l (\gamma^{-1})_l^k.$$

Definitions. Two sets $\{c_{ij}^k\}$ and $\{\bar{c}_{ij}^k\}$ of structure constants are said to be *equivalent* if they are connected by means of a nonsingular matrix (γ_j^i) as above.

In this case the corresponding Lie algebras are said to be *structurally equivalent*.

A natural task is to determine the equivalence classes of r-dimensional Lie algebras for a given r, and to find as simple representatives as possible for each.

In this chapter we will explain what happens when $r = 3$, following [Lie–Scheffers 1893].

Restriction. Having to solve *eigenvalue problems*, we consider only *Lie algebras over* $\mathbb{C}$ here.

Let us recall two convenient concepts.

Definitions A vector subspace $\mathfrak{h}$ of $\mathfrak{g} = \operatorname{span}_{\mathbb{C}}\{X_1, \ldots, X_r\}$ is a *subalgebra* if

$$[\mathfrak{h}, \mathfrak{h}] \subseteq \mathfrak{h}.$$

A subalgebra $\mathfrak{k}$ is an *ideal* if

$$[\mathfrak{g}, \mathfrak{k}] \subseteq \mathfrak{k}.$$

An example of an ideal is the *derived algebra* $\mathfrak{g}'$, generated by the brackets of the elements in $\mathfrak{g}$:

$$\mathfrak{g}' = \operatorname{span}_{\mathbb{C}}\{[X_i, X_j] \mid 1 \leq i < j \leq r\}.$$

This is really an ideal since for $a, i, j = 1, \ldots, r$,

$$[X_a, [X_i, X_j]] = \left[X_a, \sum_{k=1}^{r} c_{ij}^k X_k \right] = \sum_{k=1}^{r} c_{ij}^k [X_a, X_k] \in \mathfrak{g}'.$$

Lemma 9.0.1. *Let* $\mathfrak{g}$ *be an r-dimensional Lie algebra over* $\mathbb{C}$*, and let* X *be a nonzero element of* $\mathfrak{g}$*. Then there is a 2-dimensional subalgebra* $\mathfrak{h}$ *of* $\mathfrak{g}$ *such that* $X \in \mathfrak{h}$*.*

Proof. Setting $X_1 := X$ we may assume that

$$\mathfrak{g} = \operatorname{span}_{\mathbb{C}}\{X_1, \ldots, X_r\},$$

with the structure equations

$$[X_i, X_j] = \sum_{k=1}^{r} c_{ij}^k X_k \qquad (\text{where } c_{ij}^k \in \mathbb{C}).$$

Our task is to find

$$Y = \sum_{j=2}^{r} d^j X_j, \qquad \text{with } d^j \in \mathbb{C},$$

such that $[X_1, Y]$ is a linear combination of X_1 and Y:

$$[X_1, Y] = a X_1 + b Y \qquad \text{with } a, b \in \mathbb{C},$$

i.e.,

$$\sum_{j=2}^{r} \sum_{k=1}^{r} d^j c_{1j}^k X_k = a X_1 + b \sum_{k=2}^{r} d^k X_k.$$

With the X_k being linearly independent, this is equivalent to the system

$$\sum_{j=2}^{r} d^j c_{1j}^1 = a, \tag{1}$$

$$\sum_{j=2}^{r} d^j c_{1j}^k = b d^k \qquad \text{for } k = 2, \ldots, r. \tag{2_k}$$

The equations (2_k) amount to

$$\begin{pmatrix} c_{12}^2 & \cdots & c_{1r}^2 \\ \vdots & \ddots & \vdots \\ c_{12}^r & \cdots & c_{1r}^r \end{pmatrix} \begin{pmatrix} d^2 \\ \vdots \\ d^r \end{pmatrix} = b \begin{pmatrix} d^2 \\ \vdots \\ d^r \end{pmatrix},$$

so that $(d^2, \ldots, d^r)^T$ is an eigenvector of the matrix in the left hand side, and b is the corresponding eigenvalue. Because the ground field is $\mathbb{C}$, this eigenvalue problem has at least one solution.

Having found $d^2, \ldots, d^r$, equation (1) then determines a. $\qquad \square$

Let us now classify the 2-dimensional Lie algebras!

Lemma 9.0.2. *There are two structural equivalence classes of 2-dimensional Lie algebras* $\mathfrak{g} = \operatorname{span}_{\mathbb{C}}\{X_1, X_2\}$:

- *either* $[X_1, X_2] = 0$,
- *or* $[X_1, X_2] = X_1$.

Proof. We have $[X_1, X_2] = c^1 X_1 + c^2 X_2$ for certain $c^1, c^2 \in \mathbb{C}$.

- *First case*: $c^1 = c^2 = 0$.
- *Second case*: at least one of the c^k is nonzero—for instance $c^1 \neq 0$. Then we introduce new basis vectors for $\mathfrak{g}$ by

$$\bar{X}_1 = X_1 + \frac{c^2}{c^1} X_2,$$

$$\bar{X}_2 = \frac{1}{c^1} X_2,$$

so that

$$[\bar{X}_1, \bar{X}_2] = \frac{1}{c^1} [X_1, X_2] = \frac{1}{c^1} (c^1 X_1 + c^2 X_2) = \bar{X}_1.$$

$\square$

9.1 The classification

Let us now consider 3-dimensional Lie algebras $\mathfrak{g} = \operatorname{span}_{\mathbb{C}}\{X_1, X_2, X_3\}$. The structure equations

$$[X_i, X_j] = \sum_{k=1}^{3} c_{ij}^k X_k \quad \text{for } i, j = 1, 2, 3$$

can be written in matrix form as

$$\begin{pmatrix} [X_1, X_2] \\ [X_1, X_3] \\ [X_2, X_3] \end{pmatrix} = \begin{pmatrix} c_{12}^1 & c_{12}^2 & c_{12}^3 \\ c_{13}^1 & c_{13}^2 & c_{13}^3 \\ c_{23}^1 & c_{23}^2 & c_{23}^3 \end{pmatrix} \begin{pmatrix} X_1 \\ X_2 \\ X_3 \end{pmatrix},$$

or with evident notations, $\mathfrak{g}' = c\,\mathfrak{g}$, where the matrix c is called the *structure matrix*.

Thus

$$\dim \mathfrak{g}' = \operatorname{rank} c.$$

Recall that the matrix elements c_{ij}^k are anti-symmetric with respect to the lower indices, and satisfy the Jacobi identity.

First case: $\dim \mathfrak{g}' = \operatorname{rank} c = 3$. Because of the lemmas earlier it may be assumed that

$$\mathfrak{g} = \operatorname{span}_{\mathbb{C}}\{X_1, X_2, X_3\}$$

with $\mathfrak{h} := \operatorname{span}_{\mathbb{C}}\{X_1, X_2\}$ being a 2-dimensional subalgebra such that either $[X_1, X_2] = 0$ or $[X_1, X_2] = X_1$. The first alternative is ruled out by the fact that $\dim \mathfrak{g}' = 3$, and therefore the structure matrix reduces to

$$c = \begin{pmatrix} 1 & 0 & 0 \\ c_{13}^1 & c_{13}^2 & c_{13}^3 \\ c_{23}^1 & c_{23}^2 & c_{23}^3 \end{pmatrix}.$$

Inserting this into the Jacobi identity

$$[[X_1, X_2], X_3] + [[X_2, X_3], X_1] + [[X_3, X_1], X_2] = 0$$

it follows that

$$c_{13}^1 X_1 + c_{13}^2 X_2 + c_{13}^3 X_3 - c_{23}^1 X_1 - c_{23}^3(c_{13}^1 X_1 + c_{13}^2 X_2 + c_{13}^3 X_3)$$
$$-c_{13}^1 X_1 + c_{13}^3(c_{23}^1 X_1 + c_{23}^2 X_2 + c_{23}^3 X_3) = 0,$$

or

$$(-c_{23}^2 - c_{23}^3 c_{13}^1 + c_{13}^3 c_{23}^1) X_1 + (c_{13}^2 - c_{23}^3 c_{13}^2 + c_{13}^3 c_{23}^2) X_2 + c_{13}^3 X_3 = 0.$$

Hence

$$\begin{cases} c_{13}^3 = 0, \\ c_{13}^2(1 - c_{23}^3) = 0, \\ c_{23}^2 + c_{23}^3 c_{13}^1 = 0. \end{cases}$$

Now c_{13}^2 cannot vanish, because otherwise $[X_1, X_3] = c_{13}^1 X_1 = c_{13}^1 [X_1, X_2]$, so that $\dim \mathfrak{g}' < 3$. Thus $c_{23}^3 = 1$ and $c_{23}^2 = -c_{13}^1$, and so

$$c = \begin{pmatrix} 1 & 0 & 0 \\ c_{13}^1 & c_{13}^2 & 0 \\ c_{23}^1 & -c_{13}^1 & 1 \end{pmatrix}.$$

Replacing X_3 by $\bar{X}_3 := X_3 + \lambda^1 X_1 + \lambda^2 X_2$, we have

$$[X_1, \bar{X}_3] = c_{13}^1 X_1 + c_{13}^2 X_2 + \lambda^2 X_1 = (c_{13}^1 + \lambda^2) X_1 + c_{13}^2 X_2$$

and

$$[X_2, \bar{X}_3] = c_{23}^1 X_1 - c_{13}^1 X_2 + X_3 - \lambda^1 X_1$$
$$= (c_{23}^1 - 2\lambda^1) X_1 - (c_{13}^1 + \lambda^2) X_2 + \bar{X}_3.$$

Choosing $\lambda^2 := -c_{13}^1$ and $2\lambda^1 := c_{23}^1$, we simplify these brackets to

$$[X_1, \bar{X}_3] = c_{13}^2 X_2, \quad [X_2, \bar{X}_3] = \bar{X}_3.$$

We can get rid of c_{13}^2 too by setting $\tilde{X}_1 := aX_1$, $\tilde{X}_2 := X_2$ and $\tilde{X}_3 = a\bar{X}_3$ for some $a \in \mathbb{C}$, because then

$$[\tilde{X}_1, \tilde{X}_2] = \tilde{X}_1, \quad [\tilde{X}_1, \tilde{X}_3] = a^2 c_{13}^2 \tilde{X}_2, \quad [\tilde{X}_2, \tilde{X}_3] = \tilde{X}_3,$$

and choosing a such that $a^2 c_{13}^2 = 2$ we finally have (forgetting about the tildes)

$$\boxed{[X_1, X_2] = X_1, \quad [X_1, X_3] = 2X_2, \quad [X_2, X_3] = X_3}.$$

(The custom of using boxed formulas for Lie algebras goes back to Lie himself: 'Wir machen für die Gruppe ein Haus', he used to say.)

Second case: $\dim \mathfrak{g}' = \operatorname{rank} c = 2$. Here we choose generators X_1, X_2, X_3 for $\mathfrak{g}$ such that $\mathfrak{g}' = \operatorname{span}_{\mathbb{C}}\{X_1, X_2\}$ and $[X_1, X_2] = c_{12}^1 X_1$ with c_{12}^1 equal to 0 or 1. Then the structure equations become

$$\begin{cases} [X_1, X_2] = c_{12}^1 X_1, \\ [X_1, X_3] = c_{13}^1 X_1 + c_{13}^2 X_2, \\ [X_2, X_3] = c_{23}^1 X_1 + c_{23}^2 X_2. \end{cases}$$

Inserting this into the Jacobi identity

$$[[X_1, X_2], X_3] + [[X_2, X_3], X_1] + [[X_3, X_1], X_2] = 0$$

we get

$$c_{12}^1(c_{13}^1 X_1 + c_{13}^2 X_2) - c_{23}^2 c_{12}^1 X_1 - c_{13}^1 c_{12}^1 X_1 = 0, \quad \text{or} \quad c_{12}^1 c_{23}^2 = c_{12}^1 c_{13}^1 = 0.$$

If $c_{12}^1 \neq 0$, this forces $c_{13}^1 = c_{23}^2 = 0$, so that

$$c = \begin{pmatrix} c_{12}^1 & 0 & 0 \\ c_{13}^1 & 0 & 0 \\ c_{23}^1 & 0 & 0 \end{pmatrix}$$

and $\operatorname{rank} c < 2$—a contradiction. Thus $c_{12}^1 = 0$ and

$$c = \begin{pmatrix} 0 & 0 & 0 \\ c_{13}^1 & c_{13}^2 & 0 \\ c_{23}^1 & c_{23}^2 & 0 \end{pmatrix} \quad \text{with } \det \begin{pmatrix} c_{13}^1 & c_{13}^2 \\ c_{23}^1 & c_{23}^2 \end{pmatrix} \neq 0.$$

Let us now replace X_1 by a new generator $\bar{X}_1 := \lambda^1 X_1 + \lambda^2 X_2$ (with $\lambda^1 \neq 0$) so that

$$[\bar{X}_1, X_3] = \alpha \bar{X}_1 \quad \text{for some } \alpha \in \mathbb{C},$$

i.e.,

$$\lambda^1[X_1, X_3] + \lambda^2[X_2, X_3] \equiv (\lambda^1 c_{13}^1 + \lambda^2 c_{23}^1)X_1 + (\lambda^1 c_{13}^2 + \lambda^2 c_{23}^2)X_2$$
$$= \alpha(\lambda^1 X_1 + \lambda^2 X_2),$$

which is equivalent to

$$\begin{pmatrix} c_{13}^1 & c_{23}^1 \\ c_{13}^2 & c_{23}^2 \end{pmatrix} \begin{pmatrix} \lambda^1 \\ \lambda^2 \end{pmatrix} = \alpha \begin{pmatrix} \lambda^1 \\ \lambda^2 \end{pmatrix}.$$

This eigenvalue problem has a nonzero solution (λ^1, λ^2), and by interchanging the indices if necessary we may assume that $\lambda^1 \neq 0$.

The structure equations have now been simplified to

$$\begin{cases} [X_1, X_2] = 0, \\ [X_1, X_3] = \alpha X_1, \\ [X_2, X_3] = \beta^1 X_1 + \beta^2 X_2, \end{cases}$$

where $\alpha\beta^2 \neq 0$. Replacing X_2 by $\bar{X}_2 := X_2 + \lambda X_1$ we get $[X_1, \bar{X}_2] = 0$ and

$$[\bar{X}_2, X_3] = \beta^1 X_1 + \beta^2 X_2 + \lambda\alpha X_1$$
$$= (\beta^1 + \lambda(\alpha - \beta^2))X_1 + \beta^2 \bar{X}_2.$$

Subcase 1: $\alpha \neq \beta^2$, or $\alpha = \beta^2$ and $\beta^1 = 0$. Then we can choose λ such that $[\bar{X}_2, X_3] = \beta^2 \bar{X}_2$, whence the structure equations are simplified to

$$\begin{cases} [X_1, X_2] = 0, \\ [X_1, X_3] = \alpha X_1, \\ [X_2, X_3] = \beta^2 X_2, \end{cases}$$

with $\alpha\beta^2 \neq 0$. Replacing X_3 by $\alpha^{-1}X_3$ we get rid of α and arrive at the final form:

$$\boxed{[X_1, X_2] = 0, \quad [X_1, X_3] = X_1, \quad [X_2, X_3] = c X_2, \quad \text{where } c \neq 0}.$$

Subcase 2: $\alpha = \beta^2$ and $\beta^1 \neq 0$. Here

$$\begin{cases} [X_1, X_2] = 0, \\ [X_1, X_3] = \alpha X_1, \\ [X_2, X_3] = \beta^1 X_1 + \alpha X_2, \end{cases}$$

with $\alpha \neq 0$. Replacing X_3 by $\alpha^{-1} X_3$ again we reduce this to

$$\begin{cases} [X_1, X_2] = 0, \\ [X_1, X_3] = X_1, \\ [X_2, X_3] = cX_1 + X_2, \end{cases}$$

with $c \neq 0$. Finally, replacing X_1 by cX_1 we have

$$\boxed{[X_1, X_2] = 0, \quad [X_1, X_3] = X_1, \quad [X_2, X_3] = X_1 + X_2}.$$

Third case: $\dim \mathfrak{g}' = \operatorname{rank} c = 1$. If we choose generators X_1, X_2, X_3 for $\mathfrak{g}$ such that $\mathfrak{g}' = \operatorname{span}_{\mathbb{C}}\{X_1\}$, the structure equations for $\mathfrak{g}$ become

$$\begin{cases} [X_1, X_2] = \alpha X_1, \\ [X_1, X_3] = \beta X_1, \\ [X_2, X_3] = \gamma X_1, \end{cases}$$

with at least one of α, β, γ being nonzero.

Subcase 1: $\beta \neq 0$. If also $\alpha \neq 0$ we replace X_2 by $\bar{X}_2 := -(\beta/\alpha)X_2 + X_3$, so that

$$[X_1, \bar{X}_2] = -\beta X_1 + \beta X_1 = 0.$$

That is, we may assume that $\alpha = 0$. But then

$$\left[X_1, X_2 - \frac{\gamma}{\beta}X_1\right] = \left[X_2 - \frac{\gamma}{\beta}X_1, X_3\right] = 0,$$

so replacing X_2 by $X_2 - (\gamma/\beta)X_1$ we may assume that $\gamma = 0$ as well. Finally, replacing X_3 by $\beta^{-1}X_3$ we arrive at

$$\boxed{[X_1, X_2] = 0, \quad [X_1, X_3] = X_1, \quad [X_2, X_3] = 0}$$

—which happens to be the Lie algebra one gets by setting $c = 0$ in subcase **1** of the case $\dim \mathfrak{g}' = 2$.

Subcase 2: $\beta = 0$. If $\gamma = 0$ too, necessarily $\alpha \neq 0$. Replacing X_2 by αX_2 we have in this case

$$\begin{cases} [X_1, X_2] = X_1, \\ [X_1, X_3] = 0, \\ [X_2, X_3] = 0, \end{cases}$$

which is the same as the preceding subcase if the indices 2 and 3 are interchanged.

Let us now suppose that $\gamma \neq 0$, and consider what happens if $\alpha \neq 0$.

Then $[X_1, -(\gamma/\alpha)X_2 + X_3] = 0$, so replacing X_2 by $-(\gamma/\alpha)X_2 + X_3$, we actually are allowed to assume that $\alpha = 0$. Finally, replacing X_1 by γX_1 we get

$$\boxed{[X_1, X_2] = 0, \quad [X_1, X_3] = 0, \quad [X_2, X_3] = X_1}.$$

Fourth case: $\dim \mathfrak{g}' = \operatorname{rank} c = 0$. In this case the equivalence class reduces to the abelian algebra:

$$\boxed{[X_1, X_2] = 0, \quad [X_1, X_3] = 0, \quad [X_2, X_3] = 0}.$$

With this our *classification of 3-dimensional Lie algebras* is complete.

Theorem 9.1.1. *Choosing generators appropriately, the structure equations of a 3-dimensional Lie algebra over* $\mathbb{C}$ *can be reduced to one of the following six forms:*

I. $\boxed{[X_1, X_2] = X_1, \quad [X_1, X_3] = 2X_2, \quad [X_2, X_3] = X_3}$;

II. $\boxed{[X_1, X_2] = 0, \quad [X_1, X_3] = X_1, \quad [X_2, X_3] = c\, X_2 \quad \text{with } c \neq 0}$;

III. $\boxed{[X_1, X_2] = 0, \quad [X_1, X_3] = X_1, \quad [X_2, X_3] = X_1 + X_2}$;

IV. $\boxed{[X_1, X_2] = 0, \quad [X_1, X_3] = X_1, \quad [X_2, X_3] = 0}$ *(i.e.,* **II.** *with* $c = 0$*)*

;

V. $\boxed{[X_1, X_2] = 0, \quad [X_1, X_3] = 0, \quad [X_2, X_3] = X_1}$;

VI. $\boxed{[X_1, X_2] = 0, \quad [X_1, X_3] = 0, \quad [X_2, X_3] = 0}$. $\square$

9.2 Realizations as transformation groups

We started the study of local Lie groups by considering transformation groups, and saw that they give rise to local parameter groups, to which are associated left- and right-invariant Lie algebras of vector fields. It could be interesting to see examples of transformation groups which yield the Lie algebras found in the preceding section. Out of the many possible choices we pick those used by Vessiot in his classification of hyberbolic second order PDEs in one dependent and two independent variables, for which each of the Monge systems admits at least two independent first integrals.

I. $\boxed{[X_1, X_2] = X_1, \quad [X_1, X_3] = 2X_2, \quad [X_2, X_3] = X_3}$.

Example. The projective group in one variable, i.e.,

$$x \mapsto \frac{a_1 x + a_2}{a_3 x + 1} = \phi_a(x),$$

with the identity element $e = (1, 0, 0)$. The parameter group is defined by

$$\phi_c(x) = \phi_a \circ \phi_b(x) = \frac{\frac{a_1 b_1 + a_2 b_3}{a_3 b_2 + 1} x + \frac{a_1 b_2 + a_2}{a_3 b_2 + 1}}{\frac{a_3 b_1 + b_3}{a_3 b_2 + 1} x + 1},$$

that is,

$$\begin{cases} c_1 = (a_1 b_1 + a_2 b_3)/(a_3 b_2 + 1) =: \gamma_1(a; b), \\ c_2 = (a_1 b_2 + a_2)/(a_3 b_2 + 1) =: \gamma_2(a; b), \\ c_3 = (a_3 b_1 + b_3)/(a_3 b_2 + 1) =: \gamma_3(a; b). \end{cases}$$

The right-invariant group (or group of left multiplications) is

$$\mathfrak{G}_R : z^1 = \frac{a_1 y^1 + a_2 y^3}{a_3 y^2 + 1}, \quad z^2 = \frac{a_1 y^2 + a_2}{a_3 y^2 + 1}, \quad z^3 = \frac{a_3 y^1 + y^3}{a_3 y^2 + 1}.$$

The corresponding right-invariant Lie algebra is generated by

$$\begin{cases} Y_1 = y^1\, \partial/\partial y^1 + y^2\, \partial/\partial y^2, \\ Y_2 = y^3\, \partial/\partial y^1 + \partial/\partial y^2, \\ Y_3 = -y^1 y^2\, \partial/\partial y^1 - (y^2)^2\, \partial/\partial y^2 + (y^1 - y^2 y^3)\, \partial/\partial y^3. \end{cases}$$

The structure equations for these are

$$\boxed{[Y_1, Y_2] = -Y_2, \quad [Y_1, Y_3] = Y_3, \quad [Y_2, Y_3] = -2Y_1}.$$

So setting $X_1 := -Y_2$, $X_2 := Y_1$ and $X_3 := Y_3$ we do obtain the wanted type.

The left-invariant group (i.e., the group of right multiplications) is

$$\mathfrak{G}_L : z^1 = \frac{b_1 x^1 + b_3 x^2}{b_2 x^3 + 1}, \quad z^2 = \frac{b_2 x^1 + x^2}{b_2 x^3 + 1}, \quad z^3 = \frac{b_1 x^3 + b_3}{b_2 x^3 + 1},$$

with its left-invariant Lie algebra generated by

$$\begin{cases} X_1 = x^1\, \partial/\partial x^1 + x^3\, \partial/\partial x^3, \\ X_2 = -x^1 x^3 \partial/\partial x^1 + (x^1 - x^2 x^3)\partial/\partial x^2 - (x^3)^2\, \partial/\partial x^3, \\ X_3 = x^2\, \partial/\partial x^1 + \partial/\partial x^3. \end{cases}$$

The corresponding structure equations are

$$\boxed{[X_1, X_2] = X_2, \quad [X_1, X_3] = -X_3, \quad [X_2, X_3] = 2X_1}$$

—that is, the structure constants are the same as above for the Y_i except for a minus sign.

II. $\boxed{[X_1, X_2] = 0, \quad [X_1, X_3] = X_1, \quad [X_2, X_3] = cX_2 \quad \text{with } c \neq 0}$.

Example. Consider the following transformation group acting on $\mathbb{C}^2_{xy}$:

$$(x, y) \mapsto (a_1 x + a_2 y + a_3, (a_1)^m y) = \phi_a(x, y),$$

where m is a fixed number. The identity element is $e = (1, 0, 0)$. Composing ϕ_b and ϕ_a one gets

$$\phi_c(x, y) = \phi_a \circ \phi_b(x, y) = (a_1 b_1 x + (a_1 b_2 + a_2 (b_1)^m)y + a_1 b_3 + a_3, (a_1 b_1)^m y),$$

so that the parameter group $(a, b) \mapsto c$ is given by

$$\begin{cases} c_1 = a_1 b_1, \\ c_2 = a_1 b_2 + a_2 (b_1)^m, \\ c_3 = a_1 b_3 + a_3. \end{cases}$$

Thus the right-invariant parameter group takes the form

$$\mathfrak{G}_R : z^1 = a_1 y^1, \quad z^2 = a_1 y^2 + a_2 (y^1)^m, \quad z^3 = a_1 y^3 + a_3,$$

with the right-invariant Lie algebra generated by

$$\begin{cases} Y_1 = y^1 \, \partial/\partial y^1 + y^2 \, \partial/\partial y^2 + y^3 \, \partial/\partial y^3, \\ Y_2 = (y^1)^m \, \partial/\partial y^2, \\ Y_3 = \partial/\partial y^3. \end{cases}$$

Therefore the structure equations are

$$\boxed{[Y_1, Y_2] = (m - 1)Y_2, \quad [Y_1, Y_3] = -Y_3, \quad [Y_2, Y_3] = 0},$$

and the wanted form **II** is obtained by setting $X_1 := -Y_3$, $X_2 := Y_2$, $X_3 := Y_1$ and $c := 1 - m$.

The left-invariant group is

$$\mathfrak{G}_L : z^1 = b_1 x^1, \quad z^2 = b_2 x^1 + (b_1)^m x^2, \quad z^3 = b_3 x^1 + x^3,$$

and the corresponding Lie algebra $\mathfrak{g}_L$ is generated by

$$\begin{cases} X_1 = x^1 \, \partial/\partial x^1 + m x^2 \, \partial/\partial x^2, \\ X_2 = x^1 \, \partial/\partial x^2, \\ X_3 = x^1 \, \partial/\partial x^3. \end{cases}$$

Its structure equations are

$$\boxed{[X_1, X_2] = (1 - m)X_2, \quad [X_1, X_3] = X_3, \quad [X_2, X_3] = 0}.$$

III. $\boxed{[X_1, X_2] = 0, \quad [X_1, X_3] = X_1, \quad [X_2, X_3] = X_1 + X_2}.$

Example. This time we look at the following linear transformation group acting on $\mathbb{C}^2_{xy}$:

$$(x, y) \mapsto (a_1x + a_2y + a_3, y + \log a_1) = \phi_a(x, y).$$

Also here the identity element is $e = (1, 0, 0)$. The parameter group is obtained from

$$\phi_c(x, y) = \phi_a \circ \phi_b(x, y)$$
$$= (a_1b_1x + (a_1b_2 + a_2)y + a_1b_3 + a_2 \log b_1 + a_3, y + \log(a_1b_1)),$$

so that

$$\begin{cases} c_1 = a_1b_1, \\ c_2 = a_1b_2 + a_2, \\ c_3 = a_1b_3 + a_2 \log b_1 + a_3. \end{cases}$$

Hence the right-invariant parameter group is

$$\mathfrak{G}_R : z^1 = a_1y^1, \quad z^2 = a_1y^2 + a_2, \quad z^3 = a_1y^3 + a_2 \log y^1 + a_3,$$

with the right-invariant Lie algebra $\mathfrak{g}_R$ generated by

$$\begin{cases} Y_1 = y^1\,\partial/\partial y^1 + y^2\,\partial/\partial y^2 + y^3\,\partial/\partial y^3, \\ Y_2 = \partial/\partial y^2 + \log y^1\,\partial/\partial y^3, \\ Y_3 = \partial/\partial y^3. \end{cases}$$

Consequently the structure equations are

$$\boxed{[Y_1, Y_2] = Y_3 - Y_2, \quad [Y_1, Y_3] = -Y_3, \quad [Y_2, Y_3] = 0},$$

and the replacements $X_1 := Y_3$, $X_2 := -Y_2$ and $X_3 := Y_1$ give the wanted structure.

Analogously the left-invariant parameter group is

$$\mathfrak{G}_L : z^1 = b_1x^1, \quad z^2 = b_2x^1 + x^2, \quad z^3 = b_3x^1 + \log b_1\,x^2 + x^3.$$

Its Lie algebra $\mathfrak{g}_L$ has the generators

$$\begin{cases} X_1 = x^1\,\partial/\partial x^1 + x^2\,\partial/\partial x^3, \\ X_2 = x^1\,\partial/\partial x^2, \\ X_3 = x^1\,\partial/\partial x^3, \end{cases}$$

with the structure

$$\boxed{[X_1, X_2] = X_2 - X_3, \quad [X_1, X_3] = X_3, \quad [X_2, X_3] = 0}.$$

IV. $\boxed{[X_1, X_2] = 0, \quad [X_1, X_3] = X_1, \quad [X_2, X_3] = 0}$.

This is simply case **II** with $m = 1$ (or $c = 0$).

V. $\boxed{[X_1, X_2] = 0, \quad [X_1, X_3] = 0, \quad [X_2, X_3] = X_1}$.

Example. Here we start with a linear transformation group acting on $\mathbb{C}^2_{xy}$:

$$(x, y) \mapsto (x + a_3 y + a_1, y + a_2) = \phi_a(x),$$

having the identity element $e = (0, 0, 0)$. With

$$\phi_c(x, y) = \phi_a \circ \phi_b(x, y) = (x + (a_3 + b_3)y + a_1 + b_1 + a_3 b_2, y + a_2 + b_2),$$

the corresponding parameter group is given by

$$\begin{cases} c_1 = a_1 + b_1 + a_3 b_2, \\ c_2 = a_2 + b_2, \\ c_3 = a_3 + b_3. \end{cases}$$

Thus the right-invariant group $\mathfrak{G}_R$ is

$$z^1 = a_1 + y^1 + a_3 y^2, \quad z^2 = a_2 + y^2, \quad z^3 = a_3 + y^3.$$

The right-invariant Lie algebra $\mathfrak{g}_R$ is generated by the vector fields

$$\begin{cases} Y_1 = \partial/\partial y^1, \\ Y_2 = \partial/\partial y^2, \\ Y_3 = y^2 \, \partial/\partial y^1 + \partial/\partial y^3, \end{cases}$$

which satisfy the structure equations

$$\boxed{[Y_1, Y_2] = 0, \quad [Y_1, Y_3] = 0, \quad [Y_2, Y_3] = Y_1} \, ,$$

—i.e., type **V**.

The left-invariant Lie group $\mathfrak{G}_L$ is defined by

$$z^1 = x^1 + b_1 + b_2 x^3, \quad z^2 = x^2 + b_2, \quad z^3 = x^3 + b_3,$$

with the corresponding left-invariant Lie algebra generated by

$$\begin{cases} X_1 = \partial/\partial x^1, \\ X_2 = x^3 \, \partial/\partial x^1 + \partial/\partial x^2, \\ X_3 = \partial/\partial x^3. \end{cases}$$

The structure is

$$\boxed{[X_1, X_2] = 0, \quad [X_1, X_3] = 0, \quad [X_2, X_3] = -X_1} \, .$$

VI. $\boxed{[X_1, X_2] = 0, \quad [X_1, X_3] = 0, \quad [X_2, X_3] = 0}$.

Example. The translation group G acting on $\mathbb{C}^3_{xyz}$ by

$$(x, y, z) \mapsto (x + a_1, y + a_2, z + a_3) = \phi_a(x, y, z),$$

with the identity element $e = (0, 0, 0)$. Since

$$\phi_c(x, y, z) = \phi_a \circ \phi_b(x, y, z) = (x + b_1 + a_1, y + b_2 + a_2, z + b_3 + a_3),$$

the parameter group $\mathfrak{G}$ is given by

$$\begin{cases} c_1 = a_1 + b_1, \\ c_2 = a_2 + b_2, \\ c_3 = a_3 + b_3. \end{cases}$$

Thus $\mathfrak{G}_R = \mathfrak{G}_L = G =$ the translation group in three variables, and

$$\mathfrak{g}_R = \mathfrak{g}_L = \operatorname{span}_\mathbb{C}\{\partial/\partial x^1, \partial/\partial x^2, \partial/\partial x^3\}$$

is the 3-dimensional abelian Lie algebra.

10

Lie equations and Lie vector field systems

This chapter is essentially a preparation for the study of hyperbolic second order PDEs in chapter **12**. However, the subject of Lie equations originally arose out of a quite different context, which we want to sketch first.

Let $V = (X_1, \ldots, X_q)$ be a complete vector field system on $\mathbb{R}^n$ with

$$X_i = \sum_{j=1}^{n} \xi_i^j(x) \frac{\partial}{\partial x^j} \quad \text{for } i = 1, \ldots, q.$$

The theorem of Frobenius shows that V defines a local foliation of $\mathbb{R}^n$ with $(n-q)$-dimensional leaves. Furthermore, let $\mathfrak{G}$ be an r-dimensional Lie group acting on $\mathbb{R}^n$ by

$$x^i \mapsto f^i(x^1, \ldots, x^n; a^1, \ldots, a^r) \quad \text{for } i = 1, \ldots, n.$$

V is said to *admit the Lie group* $\mathfrak{G}$ if $\mathfrak{G}$ *leaves the associated foliation invariant*—that is, transforms leaves into leaves, or equivalently induces a mapping of first integrals into first integrals.

In his famous paper [Lie 1885], Lie studies to what extent the knowledge that a complete vector field system V admits a Lie group may be used in order to integrate V—without using the general existence theorem for local solutions of ODE systems, as Frobenius does.

So suppose that V admits the Lie parameter group $\mathfrak{G}$, and let

$$\mathfrak{g} = \operatorname{span}_{\mathbb{R}} \{A_1, \ldots, A_r\} \quad \text{with} \quad A_i = \sum_{j=1}^{n} \alpha_i^j(x) \frac{\partial}{\partial x^j} \quad \text{for } i = 1, \ldots, r$$

be the associated Lie algebra, having the structure constants c_{ij}^s:

$$[A_i, A_j] = \sum_{s=1}^{r} c_{ij}^s A_s \quad \text{for } i, j = 1, \ldots, r.$$

The fact that V admits $\mathfrak{G}$ means from an infinitesimal point of view that $\mathfrak{g}$ preserves V in the sense that $[\mathfrak{g}, V] \subseteq V$, or more precisely

$$[A_i, X_j] = \sum_{k=1}^{q} \lambda_{ij}^k(x) X_k \quad \text{for } i = 1, \ldots, r \text{ and } j = 1, \ldots, q.$$

If f is a first integral of V,

$$0 = [A_i, X_j]f = A_i(X_jf) - X_j(A_if) = -X_j(A_if) \quad \text{for } j = 1, \ldots, q,$$

so that A_if is a first integral of V too—and hence $\mathfrak{g}$ induces an action on the space of first integrals of V.

$X_1, \ldots, X_q$ are linearly independent over the ring of smooth functions on $\mathbb{R}^n$, but $A_1, \ldots, A_r$ are linearly independent over the constants only. If we suppose that there are precisely h relations involving the A_i and X_j over the smooth functions, these can therefore be resolved with respect to h of the A_i; say that

$$A_i = \sum_{k=h+1}^{r} \lambda_i^k(x) A_k + \sum_{j=1}^{q} \mu_i^j(x) X_j \quad \text{for } i = 1, \ldots, h,$$

while $A_{h+1}, \ldots, A_r, X_1, \ldots, X_q$ are linearly independent over the functions. Commuting these relations with X_l we see that

$$0 \equiv \sum_{k=h+1}^{r} X_l \lambda_i^k(x) A_k \pmod{V} \quad \text{for } l = 1, \ldots, q,$$

whence the coefficients $\lambda_i^k(x)$ must be first integrals of V. Further first integrals may then be found by letting $\mathfrak{g}$ act on these $\lambda_i^k(x)$.

If we assume that v functionally independent first integrals of V are found in this way, these are introduced as new local coordinates for $\mathbb{R}^n$, replacing $x^{n-v+1}, \ldots, x^n$ say. Then the λ_i^k only depend on $x^{n-v+1}, \ldots, x^n$, and

$$X_j = \sum_{k=1}^{n-v} \zeta_j^k(x^1, \ldots, x^{n-v}; x^{n-v+1}, \ldots, x^n) \frac{\partial}{\partial x^k} \quad \text{for } j = 1, \ldots, q.$$

With

$$A_i x^{n-v+l} = \text{function of } x^{n-v+1}, \ldots, x^n$$
$$= \alpha_i^{n-v+l}(x^{n-v+1}, \ldots, x^n) \quad \text{for } l = 1, \ldots, v,$$

the A_i are expressed as

$$A_i = \sum_{k=1}^{n-v} \alpha_i^k(x^1, \ldots, x^n) \frac{\partial}{\partial x^k} + \sum_{l=1}^{v} \alpha_i^{n-v+l}(x^{n-v+1}, \ldots, x^n) \frac{\partial}{\partial x^{n-v+l}}$$

for $i = 1, \ldots, r$. Eliminating $\partial/\partial x^{n-v+1}, \ldots, \partial/\partial x^n$ from these equations by forming suitable linear combinations with coefficients that are functions of $x^{n-v+1}, \ldots, x^n$ we find a certain number of vector fields

$$A_i' = \sum_{k=1}^{r} \gamma_i^k (x^{n-v+1}, \ldots, x^n) A_k, \quad i = 1, \ldots, r',$$

which *do not contain* $\partial/\partial x^{n-v+1}, \ldots, \partial/\partial x^n$, and which form a sort of Lie algebra over the ring of smooth functions in $x^{n-v+1}, \ldots, x^n$—rather than over a field—leaving V invariant. Since there no longer occur any differentiations with respect to $x^{n-v+1}, \ldots, x^n$, these variables appear as parameters. Fixing them in some arbitrary way we are reduced to a problem in $n - v$ variables only, with the only relations between the A_i' and X_j being of the form

$$A_i' = \sum_{k=h'+1}^{r'} \lambda_i'^k A_k' + \sum_{j=1}^{q} \mu_i'^j(x) X_j \quad \text{for } i = 1, \ldots, h',$$

where the $\lambda_i'^k$ now are *constants*. Replacing A_i' by $A_i' - \sum_{k=h'+1}^{r'} \lambda_i'^k A_k'$ when $i = 1, \ldots, h'$, these relations are simplified to

$$A_i' = \sum_{j=1}^{q} \mu_i'^j(x) X_j \quad \text{for } i = 1, \ldots, h'.$$

With a change of notation we have therefore arrived at the following situation:

> $V = (X_1, \ldots, X_q)$ *is a complete vector field system admitting the Lie algebra* $\mathfrak{g} = \mathrm{span}_\mathbb{R}\{A_1, \ldots, A_r\}$, *with* A_i *and* X_j *being related only by* $A_i = \sum_{j=1}^{q} \mu_i^j(x) X_j$ *for* $i = 1, \ldots, h$.

Now let $\mathfrak{h} := \mathrm{span}_\mathbb{R}\{A_1, \ldots, A_h\}$. Since for $k = 1, \ldots, r$ and $i = 1, \ldots, h$,

$$[A_k, A_i] = \sum_{j=1}^{q} (A_k \mu_i^j(x) \cdot X_j + \mu_i^j(x) [A_k, X_j]) \in V,$$

the $[A_k, A_i]$ are linear combinations of $X_1, \ldots, X_q$—but this is possible only if $[A_k, A_i] \in \mathfrak{h}$, and therefore

$$[\mathfrak{g}, \mathfrak{h}] \subseteq \mathfrak{h}, \quad \text{that is,} \quad \mathfrak{h} \text{ is an } \textit{ideal} \text{ of } \mathfrak{g}.$$

Moreover

$$\mathfrak{h} \subseteq V \quad \text{and} \quad [\mathfrak{h}, V] \subseteq V,$$

so that the vector field system $(A_1, \ldots, A_h)$ associated to $\mathfrak{h}$ is a *subsystem of the Cauchy characteristic vector field system* $C(\mathcal{V})$.

$(A_1, \ldots, A_h)$ is surely complete, and its first integrals can be found from the defining equations of the Lie group $\mathfrak{H}$ corresponding to $\mathfrak{h}$. If we suppose that $(A_1, \ldots, A_h)$ admits n' functionally independent first integrals, these are used as local coordinates for $\mathbb{R}^n$—instead of $x^1, \ldots, x^{n'}$ for instance.

Just as it is possible to reduce a vector field system to act on the space of first integrals of its Cauchy characteristic system, we can here reduce $\mathcal{V}$ to a vector field system acting on the coordinate functions $x^1, \ldots, x^{n'}$. Furthermore, for $i = 1, \ldots, n'$ the inclusion $[\mathfrak{g}, \mathfrak{h}] \subseteq \mathfrak{h}$ implies that

$$0 = [\mathfrak{g}, \mathfrak{h}](x^i) = \mathfrak{g} \circ \mathfrak{h}(x^i) - \mathfrak{h} \circ \mathfrak{g}(x^i) = -\mathfrak{h}(\mathfrak{g}(x^i)),$$

so that $\mathfrak{g}(x^i)$ is also a first integral of $\mathfrak{h}$—i.e., a function of $x^1, \ldots, x^{n'}$. Thus $\mathfrak{g}$ projects to a Lie algebra acting on $\mathbb{R}^{n'}_{x^1, \ldots, x^{n'}}$, in such a way that its ideal $\mathfrak{h}$ projects to 0.

With new notations we are therefore reduced to the following situation:

> $\mathcal{V} = (X_1, \ldots, X_q)$ is a complete vector field system on $\mathbb{R}^n$, which admits the Lie algebra $\mathfrak{g} = \operatorname{span}_{\mathbb{R}}\{A_1, \ldots, A_r\}$. Moreover the vector fields $X_1, \ldots, X_q, A_1, \ldots, A_r$ are linearly independent over the ring of smooth functions, and generate a $(q+r)$-dimensional complete vector field system $(X_1, \ldots, X_q, A_1, \ldots, A_r)$. So in particular $q + r \leq n$.

If $q + r < n$, Lie supposes that the latter complete system can be integrated in some way (recall that the whole point is that we want to avoid the Frobenius theorem), and if so its first integrals appear as parameters, which we are allowed to fix. In that way it may be assumed that $q + r = n$.

$(A_1, \ldots, A_r)$ is a complete system, and its first integrals—let us call them $t^1, \ldots, t^q$—can be found by means of the defining equations of the local Lie group corresponding to $\mathfrak{g}$. Introducing r more functions $x^1, \ldots, x^r$ satisfying $dt^1 \wedge \cdots \wedge dt^q \wedge dx^1 \wedge \cdots \wedge dx^r \neq 0$ we can utilize $t^1, \ldots, t^q, x^1, \ldots, x^r$ as local coordinates for $\mathbb{R}^n$. Using Gauss elimination it is possible to find a basis for $\mathcal{V}$ of the form

$$X_j = \frac{\partial}{\partial t^j} + \sum_{k=1}^{r} \xi_j^k(t, x) \frac{\partial}{\partial x^k} \qquad \text{for } j = 1, \ldots, q,$$

while the generators of $\mathfrak{g}$ may be written as

$$A_i = \sum_{k=1}^{r} \alpha_i^k(t, x) \frac{\partial}{\partial x^k} \qquad \text{for } i = 1, \ldots, r.$$

Then $[\mathfrak{g}, \mathcal{V}] \subseteq \mathcal{V}$ is equivalent to $[A_i, X_j] = 0$ for $i = 1, \ldots, r$ and $j = 1, \ldots, q$.

Finally, using the blowing-up method of Mayer described in section **2.5**, we can replace $\mathcal{V}$ by a *single* vector field

$$X = \frac{\partial}{\partial t} + \sum_{k=1}^{r} \xi^k(t, x) \frac{\partial}{\partial x^k},$$

—so that $q = 1$.

Hence the original problem has at last been reduced to the following simple form.

Lie's integration problem: *Suppose that the vector field* $X = \partial/\partial t + \sum_{k=1}^{r} \xi^k(t, x) \left(\partial/\partial x^k\right)$ *admits the Lie algebra*

$$\mathfrak{g} = \mathrm{span}_\mathbb{R}\{A_1, \ldots, A_r\}$$

in the sense that $[A_i, X] = 0$ *for* $i = 1, \ldots, r$. *Then use this information in order to find r functionally independent first integrals of X in the simplest possible way—and in particular without using the existence theorem for local solutions of ODE systems.*

The key idea for performing this is to look for *maximal ideals* of $\mathfrak{g}$.

1. Suppose first that $\mathfrak{g}$ admits a chain

$$\mathfrak{g} \equiv \mathfrak{g}_0 \supset \mathfrak{g}_1 \supset \cdots \supset \mathfrak{g}_r = \{0\},$$

where $\mathfrak{g}_i$ is an ideal of $\mathfrak{g}_{i-1}$ of codimension 1 for $i = 1, \ldots, r$. Then X *can be integrated by means of r quadratures.* For this reason Lie called such Lie algebras *solvable.*

2. If $\mathfrak{g}$ is not solvable it is possible to find a Jordan–Hölder decomposition

$$\mathfrak{g} \equiv \mathfrak{g}_0 \supset \mathfrak{g}_1 \supset \cdots \supset \mathfrak{g}_p \supset \mathfrak{g}_{p+1} = \{0\},$$

where $\mathfrak{g}_i$ is a *maximal ideal* of $\mathfrak{g}_{i-1}$ for $i = 1, \ldots, p$, and $\mathfrak{g}_p$ is *simple*—that is, admits no nontrivial ideal. Then $\mathfrak{g}_i/\mathfrak{g}_{i+1}$ are simple Lie algebras for $i = 0, \ldots, p$.

In this case Lie showed that the integration problem for X *can be reduced to analogous integration problems for the simple Lie algebras* $\mathfrak{g}_i/\mathfrak{g}_{i+1}$, and hence it suffices to solve the integration problem for *simple* Lie algebras only.

This then is the historical background for the problem of classifying *simple Lie algebras*, which later was accomplished by Killing and Cartan over the field of complex numbers. And by means of this classification Cartan could give the definitive solution to Lie's integration problem in [Cartan 1896]—to which the reader is referred for further information.

However, these considerations also give rise to another problem—solved by Vessiot, Guldberg and Lie himself—which is of greater interest for us, namely, the following. Let

$$\mathfrak{g} = \operatorname{span}_{\mathbb{R}}\{A_1, \dots, A_r\} \quad \text{with} \quad A_i = \sum_{j=1}^{r} \alpha_i^j(x) \frac{\partial}{\partial x^j} \quad \text{for } i = 1, \dots, r$$

be a Lie algebra of vector fields on $\mathbb{R}^r$ which leaves the vector field

$$X = \frac{\partial}{\partial t} + \sum_{k=1}^{r} \xi^k(t, x) \frac{\partial}{\partial x^k} \quad \text{on } \mathbb{R}_t \times \mathbb{R}_x^r$$

invariant, and let $\operatorname{span}_{\mathbb{R}}\{B_1, \dots, B_r\}$ be the reciprocal Lie algebra of $\mathfrak{g}$, so that $[A_i, B_j] = 0$ for $i, j = 1, \dots, r$. Then for each fixed t

$$\left[A_i, \sum_{k=1}^{r} \xi^k(t, x) \frac{\partial}{\partial x^k} \right] = 0 \quad \text{for } i = 1, \dots, r,$$

whence

$$\sum_{k=1}^{r} \xi^k(t, x) \frac{\partial}{\partial x^k} \in \operatorname{span}_{\mathbb{R}}\{B_1, \dots, B_r\}.$$

Letting t vary we deduce that

$$X = \frac{\partial}{\partial t} + \sum_{k=1}^{r} b^k(t) B_k \quad \text{for certain smooth functions } b^k(t).$$

Definition. Let $\mathfrak{g} = \operatorname{span}_{\mathbb{R}}\{A_1, \dots, A_r\}$ be a Lie algebra of vector fields on $\mathbb{R}_x^r$ with

$$A_i = \sum_{j=1}^{r} \alpha_i^j(x) \frac{\partial}{\partial x^j} \quad \text{for } i = 1, \dots, r.$$

Then a *Lie vector field associated to* $\mathfrak{g}$ is a vector field on $\mathbb{R}_t \times \mathbb{R}_x^n$ of the form

$$X = \frac{\partial}{\partial t} + \sum_{i=1}^{r} a^i(t) A_i \quad \left(= \frac{\partial}{\partial t} + \sum_{i,j=1}^{r} a^i(t)\alpha_i^j(x) \frac{\partial}{\partial x^j} \right),$$

where $a^1(t), \ldots, a^r(t)$ are smooth functions of t. The first order PDE $Xf = 0$ as well as the equivalent ODE system

$$\frac{dx^j}{dt} = \sum_{i=1}^{r} a^i(t) \alpha_i^j(x) \quad \text{for } j = 1, \ldots, r$$

is called a *Lie equation associated to* $\mathfrak{g}$.

Example. Let $\mathfrak{G}$ be the projective group in one variable:

$$x \mapsto \frac{ax + b}{cx + d}, \quad \text{where } ad - bc \neq 0.$$

Its Lie algebra is $\mathfrak{g} = \operatorname{span}_{\mathbb{R}}\{\partial/\partial x, x\, \partial/\partial x, x^2\, \partial/\partial x\}$, and hence an associated Lie vector field is of the form

$$
\begin{aligned}
X &= \frac{\partial}{\partial t} + a(t) \frac{\partial}{\partial x} + b(t) x \frac{\partial}{\partial x} + c(t) x^2 \frac{\partial}{\partial x} \\
&= \frac{\partial}{\partial t} + \left(a(t) + b(t) x + c(t) x^2 \right) \frac{\partial}{\partial x}.
\end{aligned}
$$

The corresponding ODE is

$$\frac{dx}{dt} = a(t) + b(t) x + c(t) x^2,$$

—that is, a *Riccati equation*.

By means of [Lie–Scheffers 1883], Kap. **24**, we will in the next section see that Lie equations satisfy a *superposition principle*.

Let us consider a sysem of *linear* ODEs in order to exemplify this property:

$$\frac{d\bar{x}}{dt} = A(t)\bar{x} + \bar{f}(t),$$

where $\bar{x}(t) = (x^1(t), \ldots, x^n(t))^T$, $A(t)$ is an $n \times n$ matrix, and $\bar{f}(t) = (f^1(t), \ldots, f^n(t))^T$. Its general solution can be written as

$$
\begin{aligned}
\bar{x}(t) &= \bar{x}_{\text{hom}}(t) + \bar{x}_{\text{part}}(t) \\
&= \sum_{i=1}^{n} C_i \bar{x}_{\text{hom}}^i(t) + \bar{x}_{\text{part}}(t),
\end{aligned}
$$

where the C_i are arbitrary constants, the $\bar{x}_{\text{hom}}^i$ are n linearly independent solutions of the associated homogeneous ODE system $d\bar{x}/dt = A(t)\bar{x}$, and $\bar{x}_{\text{part}}$ is some particular solution of the original ODE. If we rewrite $\bar{x}(t)$ as

$$\bar{x}(t) = \sum_{i=1}^{n} C_i \left(\bar{x}_{\text{hom}}^i(t) + \bar{x}_{\text{part}}(t) \right) + \left(1 - \sum_{i=1}^{n} C_i \right) \bar{x}_{\text{part}},$$

this general solution is expressed as a *function of n+1 particular solutions and n arbitrary constants.*

A **natural problem** arising out of this is

> which ODE systems have the property that the general solution can be given as a function of a number of particular solutions and a number of arbitrary constants?

And the answer is *Lie equations*!—as will be shown in section **10.1**.

Following [Vessiot 1937] we then study Vessiot's generalization of this theory to 2-dimensional complete Lie vector field systems in section **10.2**.

10.1 Characterization of ODE systems with fundamental solutions

Definition. Let $f(x,t) = (f^1(x,t),\ldots,f^n(x,t))^T$ be given, and let $x(t) = (x^1(t),\ldots,x^n(t))^T$ be unknowns. The ODE system

$$\frac{dx}{dt} = f(x,t) \tag{*}$$

is called a *system with fundamental solutions* or a *system admitting a superposition principle* if the general solution can be expressed as a determinate function of a certain number of *particular solutions*

$$^{(k)}x(t) = \left(^{(k)}x^1(t),\ldots,{}^{(k)}x^n(t)\right), \qquad k = 1,\ldots,m,$$

and n arbitrary parameters $a = (a^1,\ldots,a^n)^T$,

$$x(t) = F(^{(1)}x(t),\ldots,{}^{(m)}x(t); a), \tag{†}$$

with the parameters $a^1,\ldots,a^n$ being essential in the sense that they conversely can be solved from the solution formula (†):

$$a = J(^{(1)}x(t),\ldots,{}^{(m)}x(t),x(t)), \qquad \text{where } J = (J^1,\ldots,J^n)^T. \tag{‡}$$

Main problem: *Characterize the set of ODE systems with fundamental solutions!*

In order to do this we first observe that (‡) implies that

$$J^i(^{(1)}x^1(t),\ldots,{}^{(1)}x^n(t),\ldots,{}^{(m)}x^1(t),\ldots,{}^{(m)}x^n(t),x^1(t),\ldots,x^n(t))$$

are *constants* whenever $^{(1)}x(t), \ldots,{}^{(m)}x(t)$ and $x(t)$ satisfy (*). Therefore the derivatives

$$\frac{dJ^i}{dt} = \sum_{k=1}^{m}\left(\sum_{l=1}^{n}\frac{\partial J^i}{\partial^{(k)}x^l}f^l(^{(k)}x,t)\right) + \sum_{l=1}^{n}\frac{\partial J^i}{\partial x^l}f^l(x,t)$$

vanish *identically* in all the $(m+1)n+1$ variables $^{(1)}x^1, \ldots, {}^{(1)}x^n, \ldots,$ $^{(m)}x^1, \ldots, {}^{(m)}x^n, x^1, \ldots, x^n, t$. Introducing the $m+1$ vector fields

$$^{(k)}Y = \sum_{l=1}^{n} f^l({}^{(k)}x, t) \frac{\partial}{\partial^{(k)}x^l}, \qquad \text{living on } \mathbb{R}^n_{(k)x} \text{ for } k = 1, \ldots, m,$$

and

$$Y = \sum_{l=1}^{n} f^l(x, t) \frac{\partial}{\partial x^l}, \qquad \text{living on } \mathbb{R}^n_x,$$

it follows that

$$\left(\sum_{k=1}^{m} {}^{(k)}Y + Y \right) J^i = 0 \quad \text{for } i = 1, \ldots, n.$$

Conclusion. *The vector field*

$$U({}^{(1)}x, \ldots, {}^{(m)}x, x, t) := \sum_{k=1}^{m} {}^{(k)}Y + Y$$

admits n *first integrals* $J^1({}^{(1)}x, \ldots, {}^{(m)}x, x), \ldots, J^n({}^{(1)}x, \ldots, {}^{(m)}x, x)$ *which do not depend on* t. *Moreover, from the fact that* ($\ddagger$) *can be solved with respect to* $x^1, \ldots, x^n$ *so as to give* ($\dagger$), *we have*

$$\frac{\partial(J^1, \ldots, J^n)}{\partial(x^1, \ldots, x^n)} \neq 0.$$

U does not involve $\partial/\partial t$, whence t merely appears as a parameter—which we express by writing

$$U({}^{(1)}x, \ldots, {}^{(m)}x, x, t) \equiv U_t({}^{(1)}x, \ldots, {}^{(m)}x, x).$$

Freezing t to a fixed value t_0, U_{t_0} is decomposed as

$$U_{t_0} = \sum_{k=1}^{m} \left(\sum_{l=1}^{n} f^l({}^{(k)}x, t_0) \frac{\partial}{\partial^{(k)}x^l} \right) + \sum_{l=1}^{n} f^l(x, t_0) \frac{\partial}{\partial x^l}$$

$$= \text{a sum of vector fields living on } \mathbb{R}^n_{(1)x}, \ldots, \mathbb{R}^n_{(m)x} \text{ and } \mathbb{R}^n_x \text{ respectively.}$$

Notation. Let us set $^{(m+1)}x := x$.

Definition. A vector field X on $\mathbb{R}^{(m+1)n}$ is said to be *projectable* if

$$X = \sum_{k=1}^{m+1} {}^{(k)}X,$$

where each $^{(k)}X$ is a vector field on $\mathbb{R}^n_{(k)x}$, called the *projection of X on* $\mathbb{R}^n_{(k)x}$.

Hence our U_t is projectable for each fixed t. Different values $t_1, t_2, \ldots$ of t give rise to projectable vector fields $U_{t_1}, U_{t_2}, \ldots$ on $\mathbb{R}^{(m+1)n}$. Since they all admit $J^1, \ldots, J^n$ as first integrals, there can be at most $(m+1)n - n = mn$ among them which are linearly independent (locally—as always). Let

$$\{U_1, \ldots, U_s\} \quad \text{for some } s \le mn$$

be a maximal set of linearly independent vector fields of this form.

With the U_i being projectable it immediately follows that all brackets $[U_i, U_j]$ are projectable too. Letting

$\mathcal{V} :=$ the completion of $(U_1, \ldots, U_s)$ with respect to Lie brackets,

it follows that $r := \dim \mathcal{V} \le mn$, and that $\mathcal{V}$ can be expressed as

$$\mathcal{V} = (X_1, \ldots, X_r),$$

where all the X_i are projectable vector fields on $\mathbb{R}^{(m+1)n}$: $X_i = \sum_{k=1}^{m+1} {}^{(k)}X_i$ for $i = 1, \ldots, r$. The completeness of $\mathcal{V}$ means that there are structure functions $c_{ij}^l(^{(1)}x, \ldots, ^{(m+1)}x)$ such that

$$[X_i, X_j] = \sum_{l=1}^r c_{ij}^l X_l \quad \text{for } i, j = 1, \ldots, r.$$

If we consider the different projections of the left hand sides it follows from this that the c_{ij}^l actually must be *constants*.

Conclusion. $X_1, \ldots, X_r$ span a *Lie algebra* $\mathfrak{g}$ on $\mathbb{R}^{(m+1)n}$ with the structure constants c_{ij}^l: $[X_i, X_j] = \sum_{l=1}^r c_{ij}^l X_l$. Furthermore, for each $k \in \{1, \ldots, m+1\}$,

$$^{(k)}\mathfrak{g} := \operatorname{span}_{\mathbb{R}}\{^{(k)}X_1, \ldots, ^{(k)}X_r\}$$

is a *Lie algebra on* $\mathbb{R}^n_{(k)x}$ with the same structure constants c_{ij}^l.

Since $U_1, \ldots, U_s \in \mathcal{V} = (X_1, \ldots, X_r)$, the U_i can be written as $U_i = \sum_{j=1}^r \mu_i^j X_j$. But with the U_i and X_j being projectable, this is possible only if all the μ_i^j are *constants*. Hence

$$U_i = \sum_{j=1}^r \mu_i^j X_j \quad \text{for } i = 1, \ldots, s \text{ and certain constants } \mu_i^j.$$

Let us now return to the vector field $U = U_t(^{(1)}x, \ldots, ^{(m+1)}x)$. For each

t this is a linear combination of $U_1, \ldots, U_s$ with constant coefficients; letting t vary we have

$$U = \sum_{i=1}^{s} \alpha^i(t) U_i = \sum_{i=1}^{s} \alpha^i(t) \left(\sum_{j=1}^{r} \mu_i^j X_j \right) = \sum_{j=1}^{r} a^j(t) X_j$$

with $a^j(t) := \sum_{i=1}^{s} \alpha^i(t) \mu_i^j$. Setting $A_j := {}^{(m+1)}X_j$ for $j = 1, \ldots, r$, projection onto $\mathbb{R}_t \times {}^{(m+1)}\mathbb{R}^n \equiv \mathbb{R}_t \times \mathbb{R}_x^n$ shows that

$$Y \equiv \sum_{l=1}^{n} f^l(x, t) \frac{\partial}{\partial x^l} = \sum_{j=1}^{r} a^j(t) A_j,$$

where $A_1, \ldots, A_r$ are vector fields on $\mathbb{R}_x^n$ which span a Lie algebra $\mathfrak{a}$ there with the structure constants c_{ij}^l.

Final conclusion. *Let*

$$\frac{dx^l}{dt} = f^l(x, t) \quad \text{for } l = 1, \ldots, n$$

be an ODE system with fundamental solutions. Then the associated vector field

$$Y = \sum_{l=1}^{n} f^l(x, t) \frac{\partial}{\partial x^l}$$

can be written as

$$Y = \sum_{i=1}^{r} a^i(t) A_i,$$

with $\operatorname{span}_{\mathbb{R}}\{A_1, \ldots, A_r\} = \mathfrak{a}$ *being a **Lie algebra** on* $\mathbb{R}_x^n$.

Note. In Lie's integration problem there was a Lie algebra involved from the beginning. However, ODE systems with fundamental solutions are a priori not related to Lie algebras at all—but the astonishing fact is that they turn out to be so anyway.

In order to prove the converse of the above conclusion we need a result that is also of some interest in itself.

Lemma 10.1.1. *Let* $\mathfrak{g} = \operatorname{span}_{\mathbb{R}}\{X_1, \ldots, X_r\}$ *be an r-dimensional Lie algebra of vector fields on* $\mathbb{R}_x^n$ *for some* $r \geq 1$, *and let* $\mathcal{V}(\mathfrak{g}) := (X_1, \ldots, X_r)$ *be the associated vector field system, the dimension of which necessarily satisfies*

$$1 \leq \dim \mathcal{V}(\mathfrak{g}) \leq \dim \mathfrak{g} = r \quad \text{and} \quad \dim \mathcal{V}(\mathfrak{g}) \leq n.$$

Starting from the X_i it is then possible to construct vector fields W_i living on $\mathbb{R}^r$ such that $\mathfrak{h} := \operatorname{span}_\mathbb{R}\{W_1,\ldots,W_r\}$ is an r-dimensional Lie algebra on $\mathbb{R}^r$ with the same structure constants as $\mathfrak{g}$, and with $\dim V(\mathfrak{h}) = r$.

Definition. A Lie algebra $\mathfrak{h}$ on $\mathbb{R}^r$ satisfying $\dim \mathfrak{h} = \dim V(\mathfrak{h}) = r$ is said to be *simply transitive*.

Proof of the lemma. Let $X_i = \sum_{j=1}^n \xi_i^j(x)\,(\partial/\partial x^j)$ for $i = 1,\ldots,r$, and let $^{(k)}x = (^{(k)}x^1,\ldots,\,^{(k)}x^n)$ for $k = 1,\ldots,r$ be coordinates for r copies of $\mathbb{R}^n$. Set

$$^{(k)}X_i := \sum_{j=1}^n \xi_i^j(^{(k)}x)\,\frac{\partial}{\partial^{(k)}x^j} \qquad \text{for } i,k = 1,\ldots,r,$$

and define the projectable vector fields

$$Z_i := \sum_{k=1}^r {}^{(k)}X_i \qquad \text{for } i = 1,\ldots,r, \text{ living on } \mathbb{R}^n \times \cdots \times \mathbb{R}^n = \mathbb{R}^{rn}.$$

Claim. $Z_1,\ldots,Z_r$ are linearly independent over the ring of smooth functions on $\mathbb{R}^{rn}$—i.e., $\dim(Z_1,\ldots,Z_r) = r$.

For assume that there is some relation

$$\sum_{j=1}^r \phi^j(^{(1)}x^1,\ldots,\,^{(1)}x^n,\ldots,\,^{(r)}x^1,\ldots,\,^{(r)}x^n)\,Z_j = 0.$$

Then

$$\sum_{j=1}^r \phi^j(^{(1)}x,\ldots,\,^{(r)}x)\,\xi_j^k(^{(l)}x) = 0 \qquad \text{for } k = 1,\ldots,n \text{ and } l = 1,\ldots,r,$$

that is

$$\begin{pmatrix} \xi_1^1(^{(l)}x) & \cdots & \xi_r^1(^{(l)}x) \\ \vdots & \ddots & \vdots \\ \xi_1^n(^{(l)}x) & \cdots & \xi_r^n(^{(l)}x) \end{pmatrix} \begin{pmatrix} \phi^1 \\ \vdots \\ \phi^r \end{pmatrix} = \begin{pmatrix} 0 \\ \vdots \\ 0 \end{pmatrix} \qquad \text{for } l = 1,\ldots,r.$$

Or shorter:

$$^{(l)}\Xi\,\phi = 0 \qquad \text{for } l = 1,\ldots,r.$$

Since at least one of the columns in the $n \times r$ matrix $^{(l)}\Xi$ is nonvanishing, $\operatorname{rank}{}^{(l)}\Xi \geq 1$ for each l.

Let $v_1 := \operatorname{rank}{}^{(1)}\Xi$. If $v_1 = r$, all the ϕ^j vanish, and the claim is proved. So assume that $1 \leq v_1 < r$. Then v_1 of the rows in $^{(1)}\Xi$ are linearly independent over the ring of smooth functions on $\mathbb{R}^{rn}$, while the other are linear combinations of these. Moreover $^{(1)}\Xi\,\phi = 0$ has a nontrivial solution depending on $^{(1)}x^1,\ldots,{}^{(1)}x^n$ only: $\phi = \phi(^{(1)}x)$.

Next let $v_1 + v_2$ be the rank of the $2n \times r$ matrix formed by the $2n$ rows in $^{(1)}\Xi$ and $^{(2)}\Xi$. If $v_2 = 0$, all the rows in $^{(2)}\Xi$ are linear combinations of the rows in $^{(1)}\Xi$ with coefficients that are functions on $\mathbb{R}^{rn}$, and then

$$\sum_{j=1}^{n} \phi^j(^{(1)}x)\,{}^{(2)}X_j = 0.$$

Fixing a generic $^{(1)}x$ it follows that $^{(2)}X_1, \ldots, {}^{(2)}X_r$ are *linearly dependent over the constants*, so that dim $\mathfrak{g} < r$—a contradiction. Hence $v_2 \geq 1$.

Continuing in this manner we see that the number of rows in $^{(1)}\Xi, \ldots,$ $^{(r)}\Xi$ that are linearly independent over the functions on $\mathbb{R}^{rn}$ is $v_1 + v_2 + \cdots + v_r$ with each $v_i \geq 1$. But then all the ϕ^j must vanish, and this proves the claim.

Because $\mathfrak{g}$ is a Lie algebra there are structure constants c_{ij}^s such that

$$[X_i, X_j] = \sum_{s=1}^{r} c_{ij}^s X_s \quad \text{for } i, j = 1, \ldots, r.$$

As a consequence

$$[Z_i, Z_j] = \left[\sum_{k=1}^{r} {}^{(k)}X_i, \sum_{l=1}^{r} {}^{(l)}X_j \right] = \sum_{k=1}^{r} [{}^{(k)}X_i, {}^{(k)}X_j]$$

$$= \sum_{k=1}^{r} \left(\sum_{s=1}^{r} c_{ij}^s\,{}^{(k)}X_s \right) = \sum_{s=1}^{r} c_{ij}^s Z_s,$$

so that span $_\mathbb{R}\{Z_1,\ldots,Z_r\}$ is an r-dimensional Lie algebra on $\mathbb{R}^{rn}$. Thus $(Z_1,\ldots,Z_r)$ is an r-dimensional complete vector field system admitting $rn - r$ functionally independent first integrals $t^1,\ldots,t^{rn-r}$.

Add r functions $y^1,\ldots,y^r$ to the t^i so that $y^1,\ldots,y^r,t^1,\ldots,t^{rn-r}$ form a system of local coordinates for $\mathbb{R}^{rn}$. Expressed in these the Z_i acquire the form

$$Z_i = \sum_{k=1}^{r} \zeta_i^k(y^1,\ldots,y^r;t^1,\ldots,t^{rn-r})\,\frac{\partial}{\partial y^k}, \quad i = 1,\ldots,r.$$

In forming the Lie brackets $[Z_i, Z_j] = \sum_{s=1}^{r} c_{ij}^s Z_s$ there are no differentiations with respect to the t^i, and therefore $t = (t^1,\ldots,t^{rn-r})$ can be fixed to some generic t_0. Setting

$$W_i := \sum_{k=1}^{r} \zeta_i^k(y^1,\ldots,y^r;t_0^1,\ldots,t_0^{rn-r})\,\frac{\partial}{\partial y^k} \quad \text{for } i = 1,\ldots,r$$

it follows that $W_1, \ldots, W_r$ are vector fields on $\mathbb{R}^r_y$ which are linearly independent over the ring of smooth functions on $\mathbb{R}^r_y$, and which satisfy

$$[W_i, W_j] = \sum_{s=1}^r c^s_{ij} W_s \quad \text{for } i, j = 1, \ldots, r.$$

Then $\mathfrak{h} := \operatorname{span}_{\mathbb{R}}\{W_1, \ldots, W_r\}$ is the wanted simply transitive Lie algebra with the same dimension and structure as the given $\mathfrak{g}$. $\qquad\square$

Let us now check that a Lie equation associated to a Lie algebra $\mathfrak{a} = \operatorname{span}_{\mathbb{R}}\{A_1, \ldots, A_r\}$ is an ODE system with fundamental solutions. Assuming that

$$A_i = \sum_{j=1}^n \alpha^j_i(x) \frac{\partial}{\partial x^j} \quad \text{for } i = 1, \ldots, r,$$

we let

$$X = \frac{\partial}{\partial t} + \sum_{i=1}^r a^i(t) A_i = \frac{\partial}{\partial t} + \sum_{j=1}^n \left(\sum_{i=1}^r a^i(t) \alpha^j_i(x) \right) \frac{\partial}{\partial x^j}$$

be a Lie vector field associated to $\mathfrak{a}$. The corresponding ODE system is then given by

$$\frac{dx^j}{dt} = \sum_{i=1}^r a^i(t) \alpha^j_i(x) \quad \text{for } j = 1, \ldots, n.$$

The first part of the proof above shows that choosing the integer m large enough and introducing m copies $^{(1)}\mathbb{R}^n, \ldots, ^{(m)}\mathbb{R}^n$ of $\mathbb{R}^n$, the prolonged vector fields

$$\tilde{A}_i := \sum_{k=1}^m \left(\sum_{j=1}^n \alpha^i_j(^{(k)}x) \frac{\partial}{\partial^{(k)} x^j} \right) + A_i$$

$$= {}^{(1)}A_i + \cdots + {}^{(m)}A_i + A_i, \quad \text{for } i = 1, \ldots, r,$$

will generate a complete r-dimensional vector field system $\tilde{\mathcal{V}}$ on $\mathbb{R}^n_{(1)_x} \times \cdots \times \mathbb{R}^n_{(m)_x} \times \mathbb{R}^n_x$. Thus $\tilde{\mathcal{V}}$ admits precisely $(m+1)n - r$ functionally independent first integrals.

Claim. It is possible to choose m such that $\tilde{\mathcal{V}} = (\tilde{A}_1, \ldots, \tilde{A}_r)$ admits n first integrals $J^1, \ldots, J^n$ satisfying

$$\frac{\partial(J^1, \ldots, J^n)}{\partial(x^1, \ldots, x^n)} \neq 0.$$

To see this, consider the vector fields

$$\bar{A}_i := \sum_{l=1}^{m} {}^{(l)}A_i \quad \text{on } \mathbb{R}^n_{(1)x} \times \cdots \times \mathbb{R}^n_{(m)x} \text{ for } i = 1, \ldots, r.$$

With m large enough $\bar{\mathcal{V}} := (\bar{A}_1, \ldots, \bar{A}_r)$ is a complete r-dimensional vector field system on $\mathbb{R}^{mn}$ admitting $mn - r > 0$ functionally independent first integrals $f^1, \ldots, f^{mn-r}$. If we lift these to functions on $\mathbb{R}^{mn} \times \mathbb{R}^n_x$ which are independent of $x^1, \ldots, x^n$, they are also functionally independent first integrals of $\tilde{\mathcal{V}}$. Now $\tilde{\mathcal{V}}$, having the same dimension as $\bar{\mathcal{V}}$ but being defined on $\mathbb{R}^{mn+n}$, admits n more functionally independent first integrals than $\bar{\mathcal{V}}$ does, and these necessarily have the property in the claim—for otherwise $\tilde{\mathcal{V}}$ would admit a first integral f which is not a function of $f^1, \ldots, f^{mn-r}$ and does not depend on $x^1, \ldots, x^n$, so that f would be a further functionally independent first integral of $\bar{\mathcal{V}}$. $\qquad\square$

Suppose now that ${}^{(1)}x(t), \ldots, {}^{(m)}x(t), x(t)$ are $m + 1$ solutions of the Lie equation

$$\frac{dx^j}{dt} = \sum_{i=1}^{r} a^i(t)\alpha_i^j(x) \quad \text{for } j = 1, \ldots, n. \tag{*}$$

Then

$$\frac{d}{dt} J^i \left({}^{(1)}x(t), \ldots, {}^{(m)}x(t), x(t) \right) = \sum_{j=1}^{r} a^j(t)\, \tilde{A}_j J^i = 0,$$

so that

$$J^i \left({}^{(1)}x(t), \ldots, {}^{(m)}x(t), x(t) \right) = \text{constant} =: c^i \quad \text{for } i = 1, \ldots, n.$$

Solving these equations with respect to $x^1(t), \ldots, x^n(t)$ we get

$$x(t) = F({}^{(1)}x(t), \ldots, {}^{(m)}x(t); c),$$

and hence (*) is indeed an ODE system with fundamental slutions.

Let us summarize the results obtained.

Theorem 10.1.1. *A necessary and sufficient condition for*

$$\frac{dx^i}{dt} = f^i(x, t), \qquad i = 1, \ldots, n,$$

to be an ODE system with fundamental solutions is that there is a Lie algebra $\mathfrak{a} = \operatorname{span}_{\mathbb{R}}\{A_1,\ldots,A_r\}$ *with* $A_i = \sum_{j=1}^{n} \alpha_i^j(x) \, (\partial/\partial x^j)$ *for* $i = 1,\ldots,r$ *such that*

$$f^i(x,t) = \sum_{j=1}^{r} a^j(t)\alpha_j^i(x)$$

for certain smooth functions $a^1(t),\ldots,a^r(t)$. $\qquad\qquad\qquad\square$

Above we encountered the concept of *simply transitive* Lie algebras. In the next section we will need a result about the equivalence of such Lie algebras, and therefore we might as well look at this right now.

Definition. Let $\mathfrak{g}$ and $\mathfrak{h}$ be r-dimensional Lie algebras of vector fields on $\mathbb{R}_x^n$ and $\mathbb{R}_y^n$ respectively. Then $\mathfrak{g}$ and $\mathfrak{h}$ are *equivalent* if there is a local diffeomorphism ϕ mapping a neighbourhood of $0 \in \mathbb{R}_x^n$ onto a neighbourhood of $0 \in \mathbb{R}_y^n$ such that

$$\phi_*(\mathfrak{g}) = \mathfrak{h} \quad \text{and} \quad (\phi^{-1})_*(\mathfrak{h}) = \mathfrak{g}.$$

Or more down-to-earth: $\mathfrak{g}$ and $\mathfrak{h}$ are really identical, but happen to be expressed with respect to different local coordinates.

The structure constants of $\mathfrak{g} = \operatorname{span}_{\mathbb{R}}\{X_1,\ldots,X_r\}$ are defined by $[X_i, X_j] = \sum_{k=1}^{r} c_{ij}^k X_k$. But these depend not only on $\mathfrak{g}$ but also on the the basis chosen.

Recall from the preceding chapter that two r-dimensional Lie algebras $\mathfrak{g}$ and $\mathfrak{h}$ are said to *have the same structure* if it is possible to find bases $\{X_1,\ldots,X_r\}$ for $\mathfrak{g}$ and $\{Y_1,\ldots,Y_r\}$ for $\mathfrak{h}$ such that

$$[X_i, X_j] = \sum_{k=1}^{r} c_{ij}^k X_k \quad \text{and} \quad [Y_i, Y_j] = \sum_{k=1}^{r} c_{ij}^k Y_k$$

with the *same structure constants* c_{ij}^k.

Example. If $\mathfrak{g}$ and $\mathfrak{h}$ are equivalent, they certainly have the same structure. For if $\mathfrak{g} = \operatorname{span}_{\mathbb{R}}\{X_1,\ldots,X_r\}$ with $[X_i, X_j] = \sum_{k=1}^{r} c_{ij}^k X_k$ and ϕ is the local diffeomorphism that provides the equivalence,

$$[\phi_*(X_i), \phi_*(X_j)] = \phi_*([X_i, X_j]) = \sum_{k=1}^{r} c_{ij}^k \, \phi_*(X_k),$$

so that the basis $\{\phi_*(X_1),\ldots,\phi_*(X_r)\}$ for $\mathfrak{h}$ gives the same structure constants c_{ij}^k.

Theorem 10.1.2. *Let* $\mathfrak{g}$ *and* $\mathfrak{h}$ *be simply transitive Lie algebras on* $\mathbb{R}_x^n$ *and* $\mathbb{R}_y^n$ *with the same structure. Then* $\mathfrak{g}$ *and* $\mathfrak{h}$ *are equivalent.*

Proof. With $\mathfrak{g}$ and $\mathfrak{h}$ defined near $0 \in \mathbb{R}^n_x$ and $0 \in \mathbb{R}^n_y$ respectively we have to construct a local diffeomorphism $\phi : \mathbb{R}^n_x \xrightarrow{\cong} \mathbb{R}^n_y$ such that $\phi(0) = 0$, $\phi_*(\mathfrak{g}) = \mathfrak{h}$ and $(\phi^{-1})_*(\mathfrak{h}) = \mathfrak{g}$.

If $\mathfrak{g} = \operatorname{span}_{\mathbb{R}}\{X_1, \ldots, X_n\}$ and $[X_i, X_j] = \sum_{k=1}^n c^k_{ij} X_k$, it is by assumption possible to find a basis $\{Y_1, \ldots, Y_n\}$ for $\mathfrak{h}$ such that $[Y_i, Y_j] = \sum_{k=1}^n c^k_{ij} Y_k$. The vector fields

$$X_i = \sum_{j=1}^n \xi^j_i(x) \frac{\partial}{\partial x^j} \quad \text{and} \quad Y_i = \sum_{j=1}^n \eta^j_i(y) \frac{\partial}{\partial y^j}$$

occurring here can be considered as vector fields on $\mathbb{R}^n_x \times \mathbb{R}^n_y$, with the X_i not depending on the y-coordinates and the Y_i not depending on the x-coordinates.

If we can find such a ϕ, the natural projections $\mathbb{R}^n_x \times \mathbb{R}^n_y \longrightarrow \mathbb{R}^n_x$ and $\mathbb{R}^n_x \times \mathbb{R}^n_y \longrightarrow \mathbb{R}^n_y$ induce local diffeomorphisms $\pi_x : \operatorname{graph}(\phi) \xrightarrow{\cong} \mathbb{R}^n_x$ and $\pi_y : \operatorname{graph}(\phi) \xrightarrow{\cong} \mathbb{R}^n_y$,

$$\mathbb{R}^n_x \times \mathbb{R}^n_y$$
$$\cup$$
$$\operatorname{graph}(\phi)$$
$$\pi_x \swarrow \qquad\qquad \searrow \pi_y$$
$$\mathbb{R}^n_x \qquad\qquad \mathbb{R}^n_y,$$

such that $\phi = \pi_y \circ (\pi_x^{-1})$, $\phi^{-1} = \pi_x \circ (\pi_y^{-1})$, and

$$(\pi_y)_* \circ (\pi_x^{-1})_* X_i = Y_i, \quad (\pi_x)_* \circ (\pi_y^{-1})_* Y_i = X_i \quad \text{for } i = 1, \ldots, n.$$

This works if

$$(\pi_x^{-1})_* X_i = (\pi_y^{-1})_* Y_i = X_i + Y_i \quad \text{for } i = 1, \ldots, n,$$

which requires that the r linearly independent vector fields $X_1 + Y_1, \ldots, X_n + Y_n$ living on $\mathbb{R}^n_x \times \mathbb{R}^n_y$ may be restricted to the n-dimensional submanifold $\operatorname{graph}(\phi)$ of $\mathbb{R}^n_x \times \mathbb{R}^n_y$—that is, $\operatorname{graph}(\phi)$ should be an integral manifold of the vector field system $\mathcal{V} = (X_1 + Y_1, \ldots, X_n + Y_n)$. Now

$$[X_i + Y_i, X_j + Y_j] = \sum_{k=1}^n c^k_{ij} X_k + \sum_{k=1}^n c^k_{ij} Y_k = \sum_{k=1}^n c^k_{ij} (X_k + Y_k),$$

so that $\mathcal{V}$ is *complete*, and accordingly provides a foliation of $\mathbb{R}^n_x \times \mathbb{R}^n_y$ with n-dimensional leaves. And then we have but one choice for ϕ:

$\operatorname{graph}(\phi)$ is the *leaf of $\mathcal{V}$ passing through* $(0,0) \in \mathbb{R}^n_x \times \mathbb{R}^n_y$. $\qquad\qquad\square$

Remark. Lie's idea of finding a certain local diffeomorphism by considering its graph as an integral manifold of a complete vector field system was later frequently used by Cartan (although employing one-forms rather than vector fields of course), and therefore this method is nowadays called 'Cartan's technique of the graph'.

10.2 Lie vector field systems associated to Lie groups

Let $\mathfrak{G}$ be an r-dimensional local parameter group with the composition functions γ^i:

$$\mathbb{R}^r \times \mathbb{R}^r \to \mathbb{R}^r$$
$$(a, b) \mapsto \gamma(a; b).$$

To $\mathfrak{G}$ is associated the left-invariant group $\mathfrak{G}_L$ of right multiplications,

$$x \mapsto \gamma(x; b),$$

and the right-invariant group $\mathfrak{G}_R$ of left multiplications,

$$x \mapsto \gamma(a; x).$$

The corresponding infinitesimal transformations make up the left-invariant Lie algebra $\mathfrak{g}_L = \operatorname{span}_{\mathbb{R}}\{L_1, \ldots, L_r\}$ and the right-invariant Lie algebra $\mathfrak{g}_R = \operatorname{span}_{\mathbb{R}}\{R_1, \ldots, R_r\}$ respectively.

Recall that a Lie equation associated to $\mathfrak{g}_L$ is the ODE system corresponding to a vector field of the form

$$\frac{\partial}{\partial t} + \sum_{i=1}^{r} a^i(t) L_i \quad \text{on } \mathbb{R}_t \times \mathbb{R}_x^r,$$

where the $a^i(t)$ are arbitrary smooth functions.

Definition. A *Lie vector field system associated to* $\mathfrak{G}$ is a 2-dimensional vector field system (U, V) on $\mathbb{R}_u \times \mathbb{R}_v \times \mathbb{R}_x^r$ of the form

$$\begin{cases} U = \partial/\partial u + \sum_{i=1}^{r} \phi^i(u) R_i, \\ V = \partial/\partial v + \sum_{i=1}^{r} \psi^i(v) L_i, \end{cases}$$

where the $\phi^i(u)$ and $\psi^i(v)$ are arbitrary functions of one variable.

From $[\mathfrak{g}_L, \mathfrak{g}_R] = 0$ it immediately follows that $[U, V] = 0$, so that a Lie vector field system is automatically *complete*.

Let us more generally consider a complete 2-dimensional vector field system $\mathcal{V}$ on $\mathbb{R}_u \times \mathbb{R}_v \times \mathbb{R}_x^n$ in Jacobian form, i.e., $\mathcal{V} = (U, V)$ with

$$\begin{cases} U = \partial/\partial u + \sum_{i=1}^n a^i(x^1, \dots, x^n; u, v) \, (\partial/\partial x^i) \equiv \partial/\partial u + A, \\ V = \partial/\partial v + \sum_{i=1}^n b^i(x^1, \dots, x^n; u, v) \, (\partial/\partial x^i) \equiv \partial/\partial v + B. \end{cases}$$

The completeness of $\mathcal{V}$ means that

$$0 = [U, V] = [\partial/\partial u, B] - [\partial/\partial v, A] + [A, B].$$

$\mathcal{V}$ is said to be of *Vessiot type* if *each of these three terms vanishes*, i.e.,

$$a^i = a^i(x^1, \dots, x^n; u), \ b^i = b^i(x^1, \dots, x^n; v) \text{ for } i = 1, \dots, n, \text{ and } [A, B] = 0.$$

In this section we will demonstrate two facts:

- *To any system $\mathcal{V}$ of Vessiot type belongs a local parameter group $\mathfrak{G}$, having the property that the system can be reduced to a Lie vector field system associated to $\mathfrak{G}$.*
- *Using the composition functions of $\mathfrak{G}$, we may find the integral manifolds of $\mathcal{V}$ on which $du \wedge dv \neq 0$ by solving Lie equations associated to $\mathfrak{g}_L$ and $\mathfrak{g}_R$ respectively.*

Remark. Of course the integral manifolds of any complete vector field system can be found by means of the Frobenius theorem. But the game played here is that we want to avoid this (and the general local existence theorem for solutions of ODE systems on which it is based) as far as possible, and use more elementary methods instead.

So let us consider a complete 2-dimensional vector field system $\mathcal{V} = (U, V)$ of Vessiot type:

$$\begin{cases} U = \partial/\partial u + \sum_{i=1}^n a^i(x; u) \, (\partial/\partial x^i) \equiv \partial/\partial u + A, \\ V = \partial/\partial v + \sum_{i=1}^n b^i(x; v) \, (\partial/\partial x^i) \equiv \partial/\partial v + B, \end{cases} \qquad \text{where } [A, B] = 0.$$

First we will show that

A may be expressed as $A = \sum_{i=1}^l \phi^i(x; u) R_i$, where $R_1, \dots, R_l$ are vector fields on $\mathbb{R}_x^n$ satisfying $[B, R_i] = 0$ for $i = 1, \dots, l$, and the coefficients $\phi^1(x; u), \dots, \phi^l(x; u)$ are first integrals of V.

To this end note that since B commutes with $\partial/\partial u$ and A, the Jacobi identity shows that B (and V) also commutes with the vector fields

$$\left[\frac{\partial}{\partial u}, A\right] = \sum_{i=1}^{n} \frac{\partial a^i}{\partial u}(x;u) \frac{\partial}{\partial x^i} =: A^{(1)},$$

$$\left[\frac{\partial}{\partial u}, A^{(1)}\right] = \sum_{i=1}^{n} \frac{\partial^2 a^i}{\partial u^2}(x;u) \frac{\partial}{\partial x^i} =: A^{(2)},$$

and so on. Considering these on a suitable neighborhood of the origin in $\mathbb{R}_u \times \mathbb{R}_x^n$ there is—for dimension reasons—an integer l such that $A^{(0)} := A, A^{(1)}, \ldots, A^{(l-1)}$ are linearly independent over the ring of smooth functions on $\mathbb{R}_u \times \mathbb{R}_x^n$, while the next one is a linear combination of these:

$$A^{(l)} \equiv \sum_{i=1}^{n} \frac{\partial^l a^i}{\partial u^l}(x;u) \frac{\partial}{\partial x^i} = \sum_{j=0}^{l-1} c_{(j)}(x;u) A^{(j)}.$$

Bracketing both sides with V, we get

$$0 = \sum_{j=0}^{l-1} V(c_{(j)}) A^{(j)} \implies V(c_{(j)}) = 0 \text{ for } j = 0, \ldots, l-1,$$

so that all the $c_{(j)}$ are first integrals of V. Moreover the coefficients $a^i(x;u)$ appearing in A are solutions of the lth order homogeneous ODE

$$\frac{\partial^l y}{\partial u^l} = \sum_{j=0}^{l-1} c_{(j)}(x;u) \frac{\partial^j y}{\partial u^j}, \qquad (*)$$

where $x^1, \ldots, x^n$ are to be considered as parameters.

Fixing $x = (x^1, \ldots, x^n)$, the general solution of (*) can be written as

$$y(x;u) = \sum_{i=1}^{l} C_i y^i(x;u),$$

where $y^1, \ldots, y^l$ are linearly independent solutions of (*) and $C_1, \ldots, C_l$ are arbitrary constants. Varying x, we have

$$y(x;u) = \sum_{i=1}^{l} C_i(x) y^i(x;u),$$

and therefore

$$\begin{pmatrix} y \\ \partial y/\partial u \\ \vdots \\ \partial^{l-1} y/\partial u^{l-1} \end{pmatrix} = \begin{pmatrix} y^1 & \cdots & y^l \\ \partial y^1/\partial u & \cdots & \partial y^l/\partial u \\ \vdots & \ddots & \vdots \\ \partial^{l-1} y^1/\partial u^{l-1} & \cdots & \partial^{l-1} y^l/\partial u^{l-1} \end{pmatrix} \begin{pmatrix} C_1 \\ \vdots \\ C_l \end{pmatrix},$$

or $\bar{y}(x;u) = W(x;u)\,\bar{C}(x)$, where $W(x;u)$ is the Wronskian matrix of the linearly independent solutions $y^1(x;u),\ldots,y^l(x;u)$.

For $u = 0$ we can choose $\bar{C}(x)$ so that $\bar{y}(x;0) =$ a given constant vector $\bar{k} \in \mathbb{R}^l$:

$$\bar{k} = \bar{y}(x;0) = W(x;0)\bar{C}(x) \Longleftrightarrow \bar{C}(x) = W^{-1}(x;0)\bar{k}.$$

In this way it is possible to find l linearly independent solutions $\phi^1(x;u),\ldots$ $\phi^l(x;u)$ of (*) such that

$$\begin{pmatrix} \phi^i(x;0) \\ \frac{\partial \phi^i}{\partial u}(x;0) \\ \vdots \\ \frac{\partial^{l-1}\phi^i}{\partial u^{l-1}}(x;0) \end{pmatrix} \qquad \text{are linearly independent } constant \text{ vectors for } i = 1,\ldots,l.$$

Letting V act on both sides of the identities

$$\frac{\partial^l \phi^i}{\partial u^l} = \sum_{j=0}^{l-1} c_{(j)}(x;u)\frac{\partial^j \phi^i}{\partial u^j},$$

and recalling that the $c_{(j)}(x;u)$ are first integrals of V, we get

$$\frac{\partial^l(V\phi^i)}{\partial u^l} = \sum_{j=0}^{l-1} c_{(j)}(x;u)\frac{\partial^j(V\phi^i)}{\partial u^j} \qquad \text{for } i = 1,\ldots,l.$$

But when $u = 0$ we know that $V\phi^i$ and its derivatives up to and including the order $l - 1$ all vanish, whence

$$V\phi^i = 0 \quad \text{for } i = 1,\ldots,l,$$

and so $\phi^1(x;u),\ldots,\phi^l(x;u)$ indeed are *first integrals of V*.

Since the coefficients $a^i(x;u)$ of A are solutions of (*), they can be expressed as

$$a^i(x;u) = \sum_{j=1}^{l} \rho_j^i(x)\phi^j(x;u),$$

and therefore

$$A = \sum_{i=1}^{n} a^i(x;u)\frac{\partial}{\partial x^i} = \sum_{i=1}^{n}\sum_{j=1}^{l} \rho_j^i(x)\phi^j(x;u)\frac{\partial}{\partial x^i}$$

$$= \sum_{j=1}^{l} \phi^j(x;u)\left(\sum_{i=1}^{n} \rho_j^i(x)\frac{\partial}{\partial x^i}\right) \equiv \sum_{j=1}^{l} \phi^j(x;u)\,R_j,$$

where the $R_j = \sum_{i=1}^{n} \rho_j^i(x)\,(\partial/\partial x^i)$ are vector fields on $\mathbb{R}_x^n$.

Claim. $R_1, \ldots, R_l$ are linearly independent, and they all commute with B.

For the $\partial^i \phi^j / \partial u^i$ are first integrals of B and the matrix $(\partial^i \phi^j / \partial u^i)$ is nonsingular, so that the R_i can be solved from the system

$$A^{(i)} = \sum_{j=1}^{l} \frac{\partial^i \phi^j}{\partial u^i} R_j \qquad i = 0, \ldots, l-1,$$

so as to be expressed as linear combinations of the $A^{(i)}$ with coefficients that are first integrals of B.

Conclusion. $A = \sum_{j=1}^{l} \phi^j(x; u) R_j$, where $B\phi^j = 0$ and $[B, R_j] = 0$ for $j = 1, \ldots, l$.

Analogously $B = \sum_{k=1}^{m} \psi^k(x; v) L_k$, where $A\psi^k = 0$ and $[A, L_k] = 0$ for $k = 1, \ldots, m$.

Let $\phi(x; u)$ be a first integral of B—as for instance each of the $\phi^j(x; u)$ is. Then

$$0 = B\phi = \sum_{k=1}^{m} \psi^k(x; v) L_k \phi,$$

whence the linear independence of the ψ^k as functions of v implies that $L_k \phi = 0$ for $k = 1, \ldots, m$.

In the same way, if $\psi(x; v)$ is a first integral of A (as each $\psi^k(x; v)$ is), then $R_j \psi = 0$ for $j = 1, \ldots, l$.

Conclusion. $L_k \phi^j(x; u) = 0$ for $k = 1, \ldots, m$, and $R_j \psi^k(x; v) = 0$ for $j = 1, \ldots, l$.

As a consequence

$$0 = [R_j, B] = \sum_{k=1}^{m} \psi^k(x; v) [R_j, L_k],$$

whereupon the linear independence of the $\psi^k(x; v)$ as functions of v shows that

$$[R_j, L_k] = 0 \qquad \text{for } j = 1, \ldots, l \text{ and } k = 1, \ldots, m.$$

By Jacobi's identity also

$$[[R_i, R_j], L_k] = 0 \qquad \text{for } i, j = 1, \ldots, l \text{ and } k = 1, \ldots, m.$$

If we let

$$\mathcal{R} := \text{ the completion of } (R_1, \ldots, R_l) \text{ with respect to Lie brackets}$$
$$= (R_1, \ldots, R_p) \text{ say, where } R_1, \ldots, R_p \text{ are linearly independent,}$$

it follows that

$$[R_i, L_k] = 0 \quad \text{for } i = 1, \ldots, p \text{ and } k = 1, \ldots, m.$$

Analogously, if

$\mathcal{L} := $ the completion of $(L_1, \ldots, L_m)$ with respect to Lie brackets
$= (L_1, \ldots, L_q)$ with $L_1, \ldots, L_q$ linearly independent,

we have

$$[R_i, L_j] = 0 \quad \text{for } i = 1, \ldots, p \text{ and } j = 1, \ldots, q.$$

This last fact means that $\mathcal{R}$ and $\mathcal{L}$ are *reciprocal vector field systems*.

With R_i and L_j living on $\mathbb{R}^n_x$, the structure functions of $\mathcal{R}$ and $\mathcal{L}$ are functions of $x^1, \ldots, x^n$—for instance,

$$[R_i, R_j] = \sum_{k=1}^{p} c_{ij}^k(x) R_k \quad \text{for } i, j = 1, \ldots, p.$$

Bracketing both sides with L_l, we find

$$0 = \sum_{k=1}^{p} L_l c_{ij}^k R_k,$$

whence the linear independence of the R_k implies that *the structure functions of $\mathcal{R}$ are first integrals of $\mathcal{L}$.* In the same way *the structure functions of $\mathcal{L}$ are first integrals of $\mathcal{R}$.*

For the further analysis we have to consider three different cases:

(i) $p = q = n$.
(ii) $p < n$, $q = n$.
(iii) $p < n$, $q < n$.

First case: $p = q = n$. Here $\mathcal{R} = \mathcal{L} = (\partial/\partial x^1, \ldots, \partial/\partial x^n)$, and hence the only functions of $x^1, \ldots, x^n$ that are first integrals of $\mathcal{R}$ or $\mathcal{L}$ are the *constants*. In particular the structure functions of $\mathcal{R}$ and $\mathcal{L}$ are constants, and therefore $\text{span}_{\mathbb{R}}\{R_1, \ldots, R_n\}$ and $\text{span}_{\mathbb{R}}\{L_1, \ldots, L_n\}$ are *simply transitive Lie algebras*, which moreover are reciprocals of each other. But then the discussion in chapter **8** shows that there is a Lie parameter group $\mathfrak{G}$ such that

$$\text{span}_{\mathbb{R}}\{R_1, \ldots, R_n\} = \mathfrak{g}_R \quad \text{and} \quad \text{span}_{\mathbb{R}}\{L_1, \ldots, L_n\} = \mathfrak{g}_L.$$

Let us now return to the complete vector field system $\mathcal{V} = (U, V)$ of Vessiot type. We know that

$$U = \frac{\partial}{\partial u} + \sum_{j=1}^{l} \phi^j(x; u)\, R_j \quad \text{and} \quad V = \frac{\partial}{\partial v} + \sum_{k=1}^{m} \psi^k(x; v)\, L_k,$$

where the ϕ^j are first integrals of $\mathcal{L}$ and the ψ^k are first integrals of $\mathcal{R}$. Thus

$$\phi^j = \phi^j(u) \quad \text{and} \quad \psi^k = \psi^k(v),$$

and setting $\phi^{l+1} = \cdots = \phi^n = \psi^{m+1} = \cdots = \psi^n = 0$ we can therefore write U and V as

$$U = \frac{\partial}{\partial u} + \sum_{j=1}^{n} \phi^j(u)\, R_j \quad \text{and} \quad V = \frac{\partial}{\partial v} + \sum_{k=1}^{n} \psi^k(v)\, L_k.$$

Consequently $\mathcal{V}$ is a *Lie vector field system* associated to the Lie parameter group $\mathfrak{G}$.

Second case: $p < n$ and $q = n$. Then $\mathcal{R} = (R_1, \ldots, R_p)$ admits $s := n - p$ functionally independent first integrals $v^1, \ldots, v^s$, which preferably should be found without using the full strength of the Frobenius theorem. Let us use them as local coordinates for $\mathbb{R}^n$ together with $x^1, \ldots, x^p$, say.

Since $\mathcal{L} = (\partial/\partial x^1, \ldots, \partial/\partial x^p, \partial/\partial v^1, \ldots, \partial/\partial v^s)$, its only first integrals living on $\mathbb{R}^n$ are the constants. Therefore

$$[R_i, R_j] = \sum_{k=1}^{p} c_{ij}^k R_k \quad \text{for } i, j = 1, \ldots, p$$

with *constant* structure functions. Here

$$R_i = \sum_{j=1}^{p} r_i^j(x^1, \ldots, x^p; v^1, \ldots, v^s)\, \frac{\partial}{\partial x^j} \quad \text{for } i = 1, \ldots, p,$$

so that $\operatorname{span}_{\mathbb{R}}\{R_1, \ldots, R_p\}$ can be regarded as a 'simply transitive Lie algebra on $\mathbb{R}^p_x$, depending on the s parameters $v^1, \ldots, v^{s}$', called $\mathfrak{g}(v)$. Fixing $v = (v^1, \ldots, v^s)$ to $v_0 = (v_0^1, \ldots, v_0^s)$ we get an honest simply transitive Lie algebra $\mathfrak{g}(v_0)$. Now for any v, $\mathfrak{g}(v)$ and $\mathfrak{g}(v_0)$ have the same structure $\{c_{ij}^k\}$, and by the last theorem of the preceding section they are therefore *equivalent*, in the sense that there is a coordinate transformation

$$x^k \mapsto \xi^k(x^1, \ldots, x^p; v^1, \ldots, v^s) \quad \text{for } k = 1, \ldots, p$$

which converts $\mathfrak{g}(v)$ into $\mathfrak{g}(v_0)$.

Consequently we can get rid of the v^i by such a change of coordinates, and may assume that the R_i depend on $x^1, \ldots, x^p$ only:

$$R_i = \sum_{j=1}^{p} r_i^j(x^1, \ldots, x^p) \frac{\partial}{\partial x^j},$$

so that $\mathrm{span}_{\mathbb{R}}\{R_1, \ldots, R_p\}$ is a *simply transitive Lie algebra of vector fields on* $\mathbb{R}^p_x$. This can be written as

$\mathfrak{g}_R =$ the Lie algebra of right-invariant vector fields asociated to a Lie parameter group $\mathfrak{G}$ on $\mathbb{R}^p_x$.

Let us next look at $\mathcal{L} = (L_1, \ldots, L_n)$, where

$$L_i = \sum_{j=1}^{p} l_i^j(x^1, \ldots, x^p; v^1, \ldots, v^s) \frac{\partial}{\partial x^j} + \sum_{k=1}^{s} v_i^k(x^1, \ldots, x^p; v^1, \ldots, v^s) \frac{\partial}{\partial v^k}$$

for $i = 1, \ldots, n$. Now

$$0 = [R_i, L_j] \equiv \sum_{k=1}^{s} R_i v_j^k \frac{\partial}{\partial v^k} \quad (\mathrm{mod}\ \partial/\partial x^1, \ldots, \partial/\partial x^p),$$

showing that the v_j^k are first integrals of $\mathcal{R}$, i.e. functions of $v^1, \ldots, v^s$ only: $v_j^k = v_j^k(v^1, \ldots, v^s)$ for $k = 1, \ldots, s$ and $j = 1, \ldots, n$.

Using Gaussian elimination over the ring of smooth functions in the variables $v^1, \ldots, v^s$ it is therefore possible to find a basis of $\mathcal{L}$ of the form

$$L_i = \sum_{j=1}^{p} \lambda_i^j(x^1, \ldots, x^p; v^1, \ldots, v^s) \frac{\partial}{\partial x^j} \qquad \text{for } i = 1, \ldots, p$$

and

$$L_{p+j} = \frac{\partial}{\partial v^j} + \sum_{k=1}^{p} \lambda_{p+j}^k(x^1, \ldots, x^p; v^1, \ldots, v^s) \frac{\partial}{\partial x^k} \qquad \text{for } j = 1, \ldots, s;$$

since the R_i only involve differentiations with respect to the x-variables the formation of linear combinations with coefficients that are functions of $v^1, \ldots, v^s$ will preserve the commutation relations $[R_i, L_j] = 0$ for $i = 1, \ldots, p$ and $j = 1, \ldots, n$.

The structure functions of $\mathcal{L}$ are first integrals of $\mathcal{R}$, and therefore depend on $v^1, \ldots, v^s$ only. In particular

$$[L_i, L_j] = \sum_{k=1}^{p} b_{ij}^k(v^1, \ldots, v^s) L_k \qquad \text{for } i, j = 1, \ldots, p,$$

so that $\operatorname{span}_{\mathbb{R}}\{L_1,\ldots,L_p\}$ is a 'simply transitive Lie algebra on $\mathbb{R}_x^p$ depending on s parameters'. For each $v = (v^1,\ldots,v^s)$ this Lie algebra is reciprocal to $\mathfrak{g}_R = \operatorname{span}_{\mathbb{R}}\{R_1,\ldots,R_p\}$, whence $L_1,\ldots,L_p$ are linear combinations with constant coefficients of the generators of $\mathfrak{g}_R^{\mathrm{rec}} = \mathfrak{g}_L$; letting v vary we instead get coefficients depending on v. Hence we may replace $L_1,\ldots,L_p$ by the generators of $\mathfrak{g}_L$, and in this way get rid of the parameters v^i:

$$L_i = \sum_{j=1}^{p} l_i^j(x^1,\ldots,x^p)\frac{\partial}{\partial x^j} \quad \text{for } i = 1,\ldots,p, \text{ with } \operatorname{span}_{\mathbb{R}}\{L_1,\ldots,L_p\} = \mathfrak{g}_L.$$

There then remains to consider

$$L_{p+j} = \frac{\partial}{\partial v^j} + \sum_{k=1}^{p} \lambda_{p+j}^k(x^1,\ldots,x^p;v^1,\ldots,v^s)\frac{\partial}{\partial x^k} \quad \text{for } j = 1,\ldots,s.$$

From $[R_i, L_{p+j}] = 0$ and $\mathfrak{g}_R^{\mathrm{rec}} = \mathfrak{g}_L$ we infer that $\sum_{k=1}^{p} \lambda_{p+j}^k(x;v)\,(\partial/\partial x^k)$ is a linear combination of the generators of $\mathfrak{g}_L$ with coefficients that may depend on the parameters $v^1,\ldots,v^s$, i.e.,

$$L_{p+j} = \frac{\partial}{\partial v^j} + \sum_{k=1}^{p} l_{p+j}^k(v^1,\ldots,v^s)\,L_k \quad \text{for } j = 1,\ldots,s.$$

Going back to $\mathcal{V} = (U,V) = (\partial/\partial u + \sum_{j=1}^{p}\phi^j R_j, \partial/\partial v + \sum_{j=1}^{n}\psi^j L_j)$, where the ϕ^j are first integrals of $(L_1,\ldots,L_n)$ and the ψ^j are first integrals of $(R_1,\ldots,R_p)$, we see that U and V can be written as

$$U = \frac{\partial}{\partial u} + \sum_{j=1}^{p} \phi^j(u)\,R_j,$$

$$V = \frac{\partial}{\partial v} + \sum_{j=1}^{p} \psi^j(v^1,\ldots,v^s;v)\,L_j + \sum_{k=1}^{s} \chi^k(v^1,\ldots,v^s;v)\frac{\partial}{\partial v^k}.$$

The vector field

$$V_0 := \frac{\partial}{\partial v} + \sum_{k=1}^{s} \chi^k(v^1,\ldots,v^s;v)\frac{\partial}{\partial v^k} \quad \text{on } \mathbb{R}_v \times \mathbb{R}_{v^1,\ldots,v^s}^s$$

admits s functionally independent first integrals

$$w^k = w^k(v^1,\ldots,v^s;v) \quad \text{for } k = 1,\ldots,s;$$

note that regarded as functions on $\mathbb{R}_u \times \mathbb{R}_v \times \mathbb{R}_x^p \times \mathbb{R}_{v^1,\ldots,v^s}^s$ not depending on u and $x^1,\ldots,x^p$, they are automatically first integrals of $\mathcal{V}$ too.

Having found the w^k in some way, we use them as local coordinates instead of $v^1, \ldots, v^s$, and end up with

$$\begin{cases} U = \partial/\partial u + \sum_{j=1}^{p} \phi^j(u) \, R_j, \\ V = \partial/\partial v + \sum_{j=1}^{p} \theta^j(w^1, \ldots, w^s; v) \, L_j, \end{cases}$$

where $\operatorname{span}_{\mathbb{R}}\{R_1, \ldots, R_p\} = \mathfrak{g}_R$ and $\operatorname{span}_{\mathbb{R}}\{L_1, \ldots, L_p\} = \mathfrak{g}_L$ are reciprocal Lie algebras of vector fields on $\mathbb{R}^p_x$, and the w^k play the role of parameters only.

So also in this case it is possible to reduce V to a *Lie vector field system* associated to a *Lie parameter group* $\mathfrak{G}$.

Third case: $p < n$ and $q < n$. If $\dim(R_1, \ldots, R_p, L_1, \ldots, L_q) < n$, there are $n - p - q$ functionally independent first integrals of the R_i and L_j, only depending on $x^1, \ldots, x^n$. Introducing these as new local coordinates instead of $x^{p+q+1}, \ldots, x^n$, for instance, they will merely appear as parameters for V, and can therefore be left out of account.

So we suppose that $\dim(R_1, \ldots, R_p, L_1, \ldots, L_q) = n$, which means that the only common first integrals of $(R_1, \ldots, R_p)$ and $(L_1, \ldots, L_q)$ defined on $\mathbb{R}^n_x$ are the constants.

Assume furthermore that we can find $t := n - p$ functionally independent first integrals $v^1, \ldots, v^t$ of $(R_1, \ldots, R_p)$ and $s := n - q$ functionally independent first integrals $u^1, \ldots, u^s$ of $(L_1, \ldots, L_q)$ without using the full force of the Frobenius theorem. By the remark above the u^i and the v^j are functionally independent, so that we can introduce them as local coordinates for $\mathbb{R}^n$, together with $r := p + q - n$ further functions $x^1, \ldots, x^r$, and obtain

$$R_i = \sum_{k=1}^{r} r_i^k(x^1, \ldots, x^r; u^1, \ldots, u^s; v^1, \ldots, v^t) \frac{\partial}{\partial x^k}$$

$$+ \sum_{l=1}^{s} \mu_i^l(x^1, \ldots, x^r; u^1, \ldots, u^s; v^1, \ldots, v^t) \frac{\partial}{\partial u^l} \quad \text{for } i = 1, \ldots, p,$$

and

$$L_j = \sum_{k=1}^{r} l_j^k(x^1, \ldots, x^r; u^1, \ldots, u^s; v^1, \ldots, v^t) \frac{\partial}{\partial x^k}$$

$$+ \sum_{m=1}^{t} v_j^m(x^1, \ldots, x^r; u^1, \ldots, u^s; v^1, \ldots, v^t) \frac{\partial}{\partial v^m} \quad \text{for } j = 1, \ldots, q.$$

From $[L_j, R_i] = 0$ we conclude that the μ_i^l are first integrals of $(L_1, \ldots, L_q)$, and therefore functions of $u^1, \ldots, u^s$ only: $\mu_i^l = \mu_i^l(u^1, \ldots, u^s)$.

Analogously $v_j^m = v_j^m(v^1, \ldots, v^t)$.

Making a Gaussian elimination over the ring of smooth functions in $u^1, \ldots, u^s$ we can replace $R_1, \ldots R_p$ by new basis vector fields

$$\bar{R}_i = \sum_{k=1}^r \rho_i^k(x^1, \ldots, x^r; u^1, \ldots, u^s; v^1, \ldots, v^t) \, (\partial/\partial x^k) \quad \text{for } i = 1, \ldots, r$$

and

$$\bar{R}_{r+k} = \partial/\partial u^k + \sum_{l=1}^r \rho_{r+k}^l(x^1, \ldots, x^r; u^1, \ldots, u^s; v^1, \ldots, v^t) \, (\partial/\partial x^l)$$

for $k = 1, \ldots, s$; in the same way $L_1, \ldots, L_q$ may be replaced by

$$\bar{L}_j = \sum_{k=1}^r \lambda_j^k(x^1, \ldots, x^r; u^1, \ldots, u^s; v^1, \ldots, v^t) \, (\partial/\partial x^k) \quad \text{for } j = 1, \ldots, r$$

and

$$\bar{L}_{r+k} = \partial/\partial v^k + \sum_{l=1}^r \lambda_{r+k}^l(x^1, \ldots, x^r; u^1, \ldots, u^s; v^1, \ldots, v^t) \, (\partial/\partial x^l)$$

for $k = 1, \ldots, t$. Then $(\bar{R}_1, \ldots, \bar{R}_r)$ and $(\bar{L}_1, \ldots, \bar{L}_r)$ are complete vector field systems. Moreover, since e.g.

$$\left[L_j, \sum_{i=1}^p f^i(u^1, \ldots, u^s) R_i \right] = \sum_{i=1}^p f^i(u^1, \ldots, u^s) \, [L_j, R_i] = 0,$$

the $\bar{R}_i$ and $\bar{L}_j$ will also commute with each other:

$$[\bar{R}_i, \bar{L}_j] = 0 \quad \text{for } i = 1, \ldots, p \text{ and } j = 1, \ldots, q.$$

Let us write R_i, L_j instead of $\bar{R}_i, \bar{L}_j$ in the following.

Bracketing the structure equations $[R_i, R_j] = \sum_{k=1}^r c_{ij}^k R_k$ of the complete system $(R_1, \ldots, R_r)$ with each of $L_1, \ldots, L_q$ we deduce that the structure functions c_{ij}^k are first integrals of $(L_1, \ldots, L_q)$, i.e., functions of $u^1, \ldots, u^s$ only:

$$[R_i, R_j] = \sum_{k=1}^r c_{ij}^k(u^1, \ldots, u^s) R_k \quad \text{for } i, j = 1, \ldots, r.$$

Therefore $\operatorname{span}_{\mathbb{R}}\{R_1, \ldots, R_r\}$ may be regarded as a 'simply transitive Lie algebra on $\mathbb{R}_x^r$, depending on the s parameters $u^1, \ldots, u^{s}$. Similarly $\operatorname{span}_{\mathbb{R}}\{L_1, \ldots, L_r\}$ appears as a 'simply transitive Lie algebra on $\mathbb{R}_x^r$

depending on the t parameters $v^1,\dots,v^{t'}$; moreover these two Lie algebras are reciprocals of each other. But reciprocal Lie algebras are equivalent and have the same structure—so that the structure functions of $\operatorname{span}_{\mathbb{R}}\{R_1,\dots,R_r\}$ cannot depend on $u^1,\dots,u^s$, and those of $\operatorname{span}_{\mathbb{R}}\{L_1,\dots,L_r\}$ cannot depend on $v^1,\dots,v^t$, and hence they must all be constants,

Having realized this we can get rid of the parameters altogether by a suitable coordinate transformation, just as we did in a similar situation in the preceding case. Thus it may be assumed that

$$
\begin{cases}
R_i = \sum_{k=1}^r \rho_i^k(x^1,\dots,x^r)\,(\partial/\partial x^k), \\
L_i = \sum_{k=1}^r \lambda_i^k(x^1,\dots,x^r)\,(\partial/\partial x^k)
\end{cases}
\qquad \text{for } i = 1,\dots,r,
$$

and that

$$
\operatorname{span}_{\mathbb{R}}\{R_1,\dots,R_r\} = \mathfrak{g}_R, \qquad \operatorname{span}_{\mathbb{R}}\{L_1,\dots,L_r\} = \mathfrak{g}_L,
$$

where $\mathfrak{g}_R$ and $\mathfrak{g}_L$ are the right- and left-invariant Lie algebras of vector fields associated to a Lie parameter group $\mathfrak{G}$ on $\mathbb{R}^r_x$.

Since $(R_1,\dots,R_r) = (\partial/\partial x^1,\dots,\partial/\partial x^r)$, we can write R_{r+k} as

$$
R_{r+k} = \frac{\partial}{\partial u^k} + \sum_{l=1}^r a_{r+k}^l(x^1,\dots,x^r;u^1,\dots,u^s;v^1,\dots,v^t)\,R_l
$$

for $k = 1,\dots,s$. Bracketing both sides with each of $L_1,\dots,L_r$ we infer that the coefficients a_{r+k}^l are first integrals of $(L_1,\dots,L_r) = (\partial/\partial x^1,\dots,\partial/\partial x^r)$ and therefore do not depend on $x^1,\dots,x^r$. Hence

$$
R_{r+k} = \frac{\partial}{\partial u^k} + \sum_{l=1}^r a_{r+k}^l(u^1,\dots,u^s;v^1,\dots,v^t)\,R_l \quad \text{for } k = 1,\dots,s.
$$

Likewise

$$
L_{r+k} = \frac{\partial}{\partial v^k} + \sum_{l=1}^s b_{r+k}^l(u^1,\dots,u^s;v^1,\dots,v^t)\,L_l \quad \text{for } k = 1,\dots,t.
$$

Putting all this together we can express the generators U and V of the complete system $\mathcal{V}$ that we started from as

$$
U = \frac{\partial}{\partial u} + \sum_{i=1}^s \mu^i(u^1,\dots,u^s;u)\,\frac{\partial}{\partial u^i} + \sum_{k=1}^r f^k(u^1,\dots,u^s,v^1,\dots,v^t;u)\,R_k
$$

and

$$V = \frac{\partial}{\partial v} + \sum_{j=1}^{t} v^j(v^1,\ldots,v^t;v)\frac{\partial}{\partial v^j} + \sum_{k=1}^{r} g^k(u^1,\ldots,u^s,v^1,\ldots,v^t;v)\,L_k.$$

To simplify this we replace—if possible—$u^1,\ldots,u^s$ by functionally independent first integrals $\bar{u}^1(u^1,\ldots,u^s;u),\ldots,\bar{u}^s(u^1,\ldots,u^s;u)$ of the vector field

$$\frac{\partial}{\partial u} + \sum_{i=1}^{s} \mu^i(u^1,\ldots,u^s;u)\frac{\partial}{\partial u^i} \qquad \text{on } \mathbb{R}_u \times \mathbb{R}^s_{u^1,\ldots,u^s},$$

and $v^1,\ldots,v^t$ by functionally independent first integrals $\bar{v}^1(v^1,\ldots,v^t;v)$, $\ldots,\bar{v}^t(v^1,\ldots,v^t;v)$ of the vector field

$$\frac{\partial}{\partial v} + \sum_{i=1}^{t} v^i(v^1,\ldots,v^t;v)\frac{\partial}{\partial v^i} \qquad \text{on } \mathbb{R}_v \times \mathbb{R}^t_{v^1,\ldots,v^t}.$$

Note that the $\bar{u}^i$ and $\bar{v}^j$ are first integrals of V too, so they merely appear as parameters, and can therefore be disregarded. Then we are left with

$$U = \frac{\partial}{\partial u} + \sum_{k=1}^{r} \phi^k(u,v)\,R_k \quad \text{and} \quad V = \frac{\partial}{\partial v} + \sum_{k=1}^{r} \psi^k(u,v)\,L_k.$$

Because the ϕ^k are first integrals of V, the ψ^k are first integrals of U, and $[R_i, L_j] = 0$, the commutativity of U and V reduces to

$$\sum_{k=1}^{r}\left(\frac{\partial \phi^k}{\partial v}\,R_k - \frac{\partial \psi^k}{\partial u}\,L_k\right) = 0.$$

Subcase 1: $\mathfrak{g}_R \cap \mathfrak{g}_L = \{0\}$. Then $R_1,\ldots,R_r$ and $L_1,\ldots,L_r$ are vector fields on $\mathbb{R}^r_x$, which are linearly independent over $\mathbb{R}$. The equation above forces

$$\frac{\partial \phi^k}{\partial v} = \frac{\partial \psi^k}{\partial u} = 0,$$

whence $\phi^k = \phi^k(u)$, $\psi^k = \psi^k(v)$, and

$$\begin{cases} U = \partial/\partial u + \sum_{k=1}^{r} \phi^k(u)\,R_k, \\ V = \partial/\partial v + \sum_{k=1}^{r} \psi^k(v)\,L_k. \end{cases}$$

So $\mathcal{V} = (U,V)$ is indeed a *Lie vector field system* associated to the *Lie parameter group* $\mathfrak{G}$.

Subcase 2: dim $\mathfrak{g}_R \cap \mathfrak{g}_L = m > 0$. With $\mathfrak{g}_R \cap \mathfrak{g}_L = \mathrm{span}_{\mathbb{R}}\{N_1, \ldots, N_m\}$ we write $\mathfrak{g}_R$ and $\mathfrak{g}_L$ as

$$\begin{cases} \mathfrak{g}_R = \mathrm{span}_{\mathbb{R}}\{R_1, \ldots, R_{r-m}; N_1, \ldots, N_m\}, \\ \mathfrak{g}_L = \mathrm{span}_{\mathbb{R}}\{L_1, \ldots, L_{r-m}; N_1, \ldots, N_m\}, \end{cases}$$

and express U and V as

$$\begin{cases} U = \partial/\partial u + \sum_{k=1}^{r-m} \phi^k(u,v) R_k + \sum_{l=1}^{m} f^l(u,v) N_l, \\ V = \partial/\partial v + \sum_{k=1}^{r-m} \psi^k(u,v) L_k + \sum_{l=1}^{m} g^l(u,v) N_l. \end{cases}$$

But then $[U, V] = 0$ is equivalent to

$$\frac{\partial \phi^k}{\partial v} = \frac{\partial \psi^k}{\partial u} = 0 \text{ for } k = 1, \ldots, r - m \quad \text{and} \quad \frac{\partial f^l}{\partial v} = \frac{\partial g^l}{\partial u} \text{ for } l = 1, \ldots, m.$$

The last condition means that there are functions $h^l(u,v)$ (locally) such that

$$f^l = \frac{\partial h^l}{\partial u} \quad \text{and} \quad g^l = \frac{\partial h^l}{\partial v}.$$

Thus

$$\begin{cases} U = \partial/\partial u + \sum_{k=1}^{r-m} \phi^k(u) R_k + \sum_{l=1}^{m} \left(\partial h^l / \partial u\right)(u,v) \cdot N_l, \\ V = \partial/\partial v + \sum_{k=1}^{r-m} \psi^k(v) L_k + \sum_{l=1}^{m} \left(\partial h^l / \partial v\right)(u,v) \cdot N_l. \end{cases}$$

Since $[\mathfrak{g}_R, \mathfrak{g}_L] = \{0\}$, each N_l commutes with all R_k and L_k. From this it follows that the vector field system $(N_1, \ldots, N_m)$ associated to $\mathfrak{g}_R \cap \mathfrak{g}_L$ is the Cauchy characteristic subsystem of the vector field system $(R_1, \ldots, R_{r-m};$ $L_1, \ldots, L_{r-m}; N_1, \ldots, N_m)$ associated to $\mathfrak{g}_R + \mathfrak{g}_L$. Therefore it is possible (but one should try to do it without using the full strength of the Frobenius theorem) to introduce new local coordinates $x^1, \ldots, x^{r-m}, z^1, \ldots,$ z^m for $\mathbb{R}^r$ such that $(N_1, \ldots, N_m) = (\partial/\partial z^1, \ldots, \partial/\partial z^m)$. Because all the N_l commute, $\mathrm{span}_{\mathbb{R}}\{N_1, \ldots, N_m\}$ is isomorphic to the Lie algebra of infinitesimal translations, and hence we may suppose that $N_l = \partial/\partial z^l$ for $l = 1, \ldots, m$. If we express R_i, L_j and N_l in these new coordinates, the coefficients in R_i and L_j cannot depend on the z^l since $[\partial/\partial z^l, R_i] =$

$[\partial/\partial z^l, L_j] = 0$, so that we may set

$$R_i = \sum_{k=1}^{r-m} \rho_i^k(x^1,\ldots,x^{r-m}) \frac{\partial}{\partial x^k} + \sum_{l=1}^{m} \hat{\rho}_i^l(x^1,\ldots,x^{r-m}) \frac{\partial}{\partial z^l},$$

$$L_j = \sum_{k=1}^{r-m} \lambda_j^k(x^1,\ldots,x^{r-m}) \frac{\partial}{\partial x^k} + \sum_{l=1}^{m} \hat{\lambda}_j^l(x^1,\ldots,x^{r-m}) \frac{\partial}{\partial z^l},$$

$$N_l = \frac{\partial}{\partial z^l}$$

for $i, j = 1,\ldots,r-m$ and $l = 1,\ldots,m$.

With $N_l = \partial/\partial z^l$ it follows that $\hat{z}^k := z^k - h^k(u,v)$ are first integrals of U and V for $k = 1,\ldots,m$. If we replace the z^k by these first integrals, U and V are simplified to

$$\begin{cases} U = \partial/\partial u + \sum_{k=1}^{r-m} \phi^k(u) R_k, \\ V = \partial/\partial v + \sum_{k=1}^{r-m} \psi^k(v) L_k. \end{cases}$$

These are not yet in Lie form since the R_i and L_i involve linear combinations of $\partial/\partial\hat{z}^j$. However, R_i and L_i can obviously be projected to $\mathbb{R}_x^{r-m}$ as

$$\begin{cases} \bar{R}_i = \sum_{k=1}^{r-m} \rho_i^k(x^1,\ldots,x^{r-m})\,(\partial/\partial x^k), \\ \bar{L}_i = \sum_{k=1}^{r-m} \lambda_i^k(x^1,\ldots,x^{r-m})\,(\partial/\partial x^k) \end{cases} \qquad \text{for } i = 1,\ldots,m.$$

Correspondingly U and V can be projected to $\mathbb{R}_u \times \mathbb{R}_v \times \mathbb{R}_x^{r-m}$ as

$$\begin{cases} \bar{U} = \partial/\partial u + \sum_{k=1}^{r-m} \phi^k(u) \bar{R}_k, \\ \bar{V} = \partial/\partial v + \sum_{k=1}^{r-m} \psi^k(v) \bar{L}_k, \end{cases}$$

and the *projected vector field system* $\bar{\mathcal{V}} = (\bar{U}, \bar{V})$ is in *Lie form* with

$$\mathrm{span}\,_{\mathbb{R}}\{R_1,\ldots,R_{r-m}\} = \bar{\mathfrak{g}}_R \quad \text{and} \quad \mathrm{span}\,_{\mathbb{R}}\{L_1,\ldots,L_{r-m}\} = \bar{\mathfrak{g}}_L.$$

Lemma 10.2.1. *The integration of $\mathcal{V} = (U, V)$ reduces to the integration of the Lie vector field system $\bar{\mathcal{V}} = (\bar{U}, \bar{V})$, and to m quadratures.*

Proof. Indeed, suppose that we have found $r-m$ functionally independent first integrals of $(\bar{U}, \bar{V})$ on $\mathbb{R}_u \times \mathbb{R}_v \times \mathbb{R}_x^{r-m}$. Introducing these as new local coordinates instead of $x^1,\ldots,x^{r-m}$, U and V are changed into the form

$$\begin{cases} U = \partial/\partial u + \sum_{l=1}^{m} a^l(u,v)\,(\partial/\partial z^l), \\ V = \partial/\partial v + \sum_{l=1}^{m} b^l(u,v)\,(\partial/\partial z^l), \end{cases}$$

for certain functions a^l and b^l, which because of the commutativity of U and V must satisfy

$$\frac{\partial a^l}{\partial v} = \frac{\partial b^l}{\partial u} \quad \text{for } l = 1, \ldots, m.$$

The dual pfaffian system of (U, V) is then generated by the one-forms

$$dz^l - a^l(u, v)\, du - b^l(u, v)\, dv \quad \text{for } l = 1, \ldots, m.$$

Therefore the remaining m first integrals are given by

$$z^l = \int \left(a^l(u, v)\, du + b^l(u, v)\, dv \right) \quad \text{for } l = 1, \ldots, m.$$

$\square$

Summary. *In each case we have been reduced to integrating a Lie vector field system* (U, V) *associated to an r-parameter Lie group* $\mathfrak{G}$ *with the Lie algebras* $\mathfrak{g}_R = \operatorname{span}_{\mathbb{R}}\{R_1, \ldots, R_r\}$ *and* $\mathfrak{g}_L = \operatorname{span}_{\mathbb{R}}\{L_1, \ldots, L_r\}$:

$$\begin{cases} U = \partial/\partial u + \sum_{i=1}^{r} \phi^i(u)\, R_i, \\ V = \partial/\partial v + \sum_{i=1}^{r} \psi^i(v)\, L_i. \end{cases}$$

Let the composition functions for $\mathfrak{G}$ be $\gamma^i(a; b)$, i.e.,

$$\mathfrak{G} : \mathbb{R}^r \times \mathbb{R}^r \to \mathbb{R}^r$$

$$(a; b) \mapsto (\gamma^1(a; b), \ldots, \gamma^r(a; b)) = \gamma(a; b) = a \cdot b.$$

Theorem 10.2.1. *By means of these composition functions it is possible to reduce the integration of* (U, V) *to solving Lie equations in one variable associated to* $\mathfrak{g}_R$ *and* $\mathfrak{g}_L$ *respectively.*

Proof. Let us be a bit more specific: U and V are given by

$$\begin{cases} U = \partial/\partial u + \sum_{i,j=1}^{r} \phi^i(u)\rho_i^j(x^1, \ldots, x^r)\, (\partial/\partial x^j), \\ V = \partial/\partial v + \sum_{i,j=1}^{r} \psi^i(v)\lambda_i^j(x^1, \ldots, x^r)\, (\partial/\partial x^j), \end{cases}$$

and we are looking for complete integrals of (U, V) of the form

$$x^k = \xi^k(u, v; c^1, \ldots, c^r) \quad \text{for } k = 1, \ldots, r,$$

where $c^1, \ldots, c^r$ are arbitrary parameters, which are essential in the sense that they conversely can be solved from this system. Now if the vector field U is to be tangent to this 5-parameter family of 2-dimensional manifolds, U is bound to have the form $\partial/\partial u + \sum_{k=1}^{r} (\partial \xi^k/\partial u) (\partial/\partial x^k)$, so that $\partial \xi^k/\partial u = \sum_{i=1}^{r} \phi^i(u)\rho_i^k(\xi)$ for $k = 1, \ldots, r$.

Reasoning in the same way with V we infer that the ξ^k are solutions of the PDE system

$$\frac{\partial \xi^k}{\partial u} = \sum_{i=1}^{r} \phi^i(u)\rho_i^k(\xi), \qquad (\dagger)$$

$$\frac{\partial \xi^k}{\partial v} = \sum_{i=1}^{r} \psi^i(v)\lambda_i^k(\xi) \qquad (\ddagger)$$

for $k = 1, \ldots, r$.

Here $(\dagger)$ is associated to U, and just like U it is *invariant with respect to right multiplications* in the sense that if $\xi = (\xi^1, \ldots, \xi^r)$ is one solution of $(\dagger)$, $\xi \cdot a = \gamma(\xi; a)$ is also for any $a = (a^1, \ldots, a^r) \in \mathbb{R}^r$.

And in the same way $(\ddagger)$—associated to V—is *invariant with respect to left multiplications*.

Let us now tentatively set

$$\xi^k = \gamma^k(a^1(u), \ldots, a^r(u); b^1(v), \ldots, b^r(v)),$$

where the $a^i(u)$ and $b^i(v)$ are arbitrary smooth functions of *one* variable.

Since $(\dagger)$ does not contain $\partial/\partial v$, the $b^i(v)$ appear as parameters only in this PDE, and ξ results from multiplying $(a^1(u), \ldots, a^r(u))$ on the right by these parameters. So inserting ξ into $(\dagger)$ and using the invariance with respect to right multiplications, the resulting PDE is equivalent to

$$\frac{da^k}{du} = \sum_{i=1}^{r} \phi^i(u)\rho_i^k(a^1(u), \ldots, a^r(u))$$

—that is, a *Lie equation associated to* $\mathfrak{g}_R$.

Similarly, in inserting $\xi = \gamma(a(u); b(v))$ into $(\ddagger)$, $a(u)$ acts as a parameter by which $b(v)$ is multiplied on the left. Because $(\ddagger)$ is invariant under left multiplications this parameter can be deleted, and we end up with

$$\frac{db^k}{dv} = \sum_{i=1}^{r} \psi^i(v)\lambda_i^k(b^1(v), \ldots, b^r(v)) \quad \text{for } k = 1, \ldots, r$$

—that is, a *Lie equation associated to* $\mathfrak{g}_L$.

Note also that if $a(u)$ and $b(v)$ are *particular solutions* of these Lie equations,

$$\xi(u, v; c^1, \ldots, c^r) := a(u) \cdot c \cdot b(v)$$

$$= \gamma(\gamma(a(u); c); b(v)) = \gamma(a(u); \gamma(c; b(v)))$$

are also solutions of $(\dagger)$ and $(\ddagger)$ for arbitrary $c \in \mathbb{R}^r$. And knowing that the general solution of (†) + (‡) depends on r parameters we deduce that the general complete integral of (U, V) is given by

$$x = a(u) \cdot c \cdot b(v) \quad \text{with arbitrary } c \in \mathbb{R}^r.$$

The first integrals of (U, V) are found by extracting c from this equality:

$$c = \bar{a}(u) \cdot x \cdot \bar{b}(v) \Longleftrightarrow$$
$$c^i = \gamma^i(\gamma(\bar{a}(u); x); b(v)) = \gamma^i(\bar{a}(u); \gamma(x; \bar{b}(v))) \quad \text{for } i = 1, \ldots, r,$$

where $\bar{a}$ and $\bar{b}$ denote the inverses of a and b respectively: $a \cdot \bar{a} = e$ and $\bar{b} \cdot b = e.$ □

11

Second order PDEs in one dependent and two independent variables

By Drach's classification an arbitrary PDE system can be reduced to a first or second order system in one unknown, and first order systems in one unknown are completely understood thanks to the theory in chapter **6**. So the real difficulties start with second order systems in one unknown.

In this chapter we begin the study of *a single PDE in one dependent and two independent variables* with some generalities from [Vessiot 1924] and [Vessiot 1936]. Chapters **12–13** are then devoted to *hyperbolic PDEs*, and chapters **16–17** to *parabolic PDEs*. Finally, we study *PDE systems in one dependent and three independent variables* in chapter **18**—thus leaving the investigation of PDE systems in one dependent and more than three independent variables in suspense.

11.1 Second order PDEs and associated vector field systems

A second order PDE in one dependent and two independent variables is equivalent to a hypersurface in the jet bundle $J^2(\mathbb{R}^2, \mathbb{R})$. The latter has the dimension 8, and is classically endowed with the following coordinates:

> independent variables x and y,
> dependent variable z,
> first order jet coordinates $p \leftrightarrow \partial z/\partial x$ and $q \leftrightarrow \partial z/\partial y$,
> second order jet coordinates $r \leftrightarrow \partial^2 z/\partial x^2$, $s \leftrightarrow \partial^2 z/\partial x \partial y$ and $t \leftrightarrow \partial^2 z/\partial y^2$.

So our PDE S can be presented as

$$F\left(x, y, z, \frac{\partial z}{\partial x}, \frac{\partial z}{\partial y}, \frac{\partial^2 z}{\partial x^2}, \frac{\partial^2 z}{\partial x \partial y}, \frac{\partial^2 z}{\partial y^2}\right) = 0 \iff F(x, y, z, p, q, r, s, t) = 0.$$

The natural projection

$$\pi_1^2 : J^2(\mathbb{R}^2, \mathbb{R}) \to J^1(\mathbb{R}^2, \mathbb{R})$$

$$(x, y, z, p, q, r, s, t) \mapsto (x, y, z, p, q)$$

induces a projection $\pi : S \longrightarrow J^1(\mathbb{R}^2, \mathbb{R})$. If we assume the latter to be locally surjective, S can be expressed in the parametric form

$$\begin{cases} r = \rho(x, y, z, p, q, u, v), \\ s = \sigma(x, y, z, p, q, u, v), \\ t = \tau(x, y, z, p, q, u, v), \end{cases}$$

so that x, y, z, p, q, u, v serve as local coordiates on S.

The contact pfaffian system of $J^2(\mathbb{R}^2, \mathbb{R})$ is generated by the one-forms

$$\begin{cases} dz - p\,dx - q\,dy, \\ dp - r\,dx - s\,dy, \\ dq - s\,dx - t\,dy, \end{cases}$$

and its restriction to S by

$$\begin{cases} \theta^0 := dz - p\,dx - q\,dy, \\ \theta^1 := dp - \rho\,dx - \sigma\,dy, \\ \theta^2 := dq - \sigma\,dx - \tau\,dy. \end{cases}$$

The vector field

$$V = a\,\partial/\partial x + b\,\partial/\partial y + c\,\partial/\partial z + d\,\partial/\partial p + e\,\partial/\partial q + f\,\partial/\partial u + g\,\partial/\partial v,$$

where a, b, c, d, e, f, g are smooth functions on S, is dual to $(\theta^0, \theta^1, \theta^2)$ if and only if

$$\begin{cases} c - ap - bq = 0, \\ d - a\rho - b\sigma = 0, \\ e - a\sigma - b\tau = 0, \end{cases}$$

so that

$$V = a\left(\frac{\partial}{\partial x} + p\frac{\partial}{\partial z} + \rho\frac{\partial}{\partial p} + \sigma\frac{\partial}{\partial q}\right) + b\left(\frac{\partial}{\partial y} + q\frac{\partial}{\partial z} + \sigma\frac{\partial}{\partial p} + \tau\frac{\partial}{\partial q}\right)$$
$$+ f\frac{\partial}{\partial u} + g\frac{\partial}{\partial v},$$

where a, b, f and g are arbitrary smooth functions.

Conclusion. *The contact vector field system of S is*

$$V = \left(X, Y, \frac{\partial}{\partial u}, \frac{\partial}{\partial v} \right),$$

with

$$\begin{cases} X = \partial/\partial x + p\,\partial/\partial z + \rho\,\partial/\partial p + \sigma\,\partial/\partial q, \\ Y = \partial/\partial y + q\,\partial/\partial z + \sigma\,\partial/\partial p + \tau\,\partial/\partial q. \end{cases}$$

Definition. Let V_1 be a vector field system on a manifold M_1, and V_2 a vector field system on a manifold M_2, where $\dim M_1 = \dim M_2$. V_1 and V_2 are said to be (*locally*) *equivalent* if there is a local diffeomorpism $\phi : M_1 \xrightarrow{\cong} M_2$ such that $\phi_*(V_1) = V_2$ and $(\phi^{-1})_*(V_2) = V_1$.

Two PDE systems S_1 and S_2 are *equivalent* if their associated vector field systems are.

In the following we will only be interested in equivalence classes of vector field systems associated to second order PDEs in one dependent and two independent variables. Therefore it is possible to simplify the parametric representation of our $S \subset J^2(\mathbb{R}^2, \mathbb{R})$ somewhat. With S given by $F(x, y, z, p, q, r, s, t) = 0$, at least one of the derivatives $\partial F/\partial r$, $\partial F/\partial s$ and $\partial F/\partial t$ is to be nonzero—and we would like to have $\partial F/\partial r \neq 0$. If not, but instead $\partial F/\partial t \neq 0$, we just switch x and y in order to get $\partial F/\partial r \neq 0$. If $\partial F/\partial r = \partial F/\partial t = 0$, necessarily $\partial F/\partial s \neq 0$, in which case we may solve the defining equation of S with respect to s so as to obtain $s = \sigma(x, y, z, p, q)$. Then the change of variables

$$x' := x + y \quad \text{and} \quad y' := x - y$$

renders this last equation solvable with respect to $r' = \partial^2 z / \partial (x')^2$.

Therefore it is permissible to assume that S is given by

$$r = \rho(x, y, z, p, q, s, t),$$

so that the parameters u and v used above are replaced by s and t. The vector field system associated to this is $V = (X, Y, \partial/\partial s, \partial/\partial t)$, with

$$X = \frac{\partial}{\partial x} + p\frac{\partial}{\partial z} + \rho(x, y, z, p, q, s, t)\frac{\partial}{\partial p} + s\frac{\partial}{\partial q}$$

and

$$Y = \frac{\partial}{\partial y} + q\frac{\partial}{\partial z} + s\frac{\partial}{\partial p} + t\frac{\partial}{\partial q}.$$

The commutation relations for the generators of $\mathcal{V}$ are

$$[X, Y] = -Y\rho\frac{\partial}{\partial p}, \quad \left[X, \frac{\partial}{\partial s}\right] = -\frac{\partial\rho}{\partial s}\frac{\partial}{\partial p} - \frac{\partial}{\partial q}, \quad \left[X, \frac{\partial}{\partial t}\right] = -\frac{\partial\rho}{\partial t}\frac{\partial}{\partial p},$$

$$\left[Y, \frac{\partial}{\partial s}\right] = -\frac{\partial}{\partial p}, \quad \left[Y, \frac{\partial}{\partial t}\right] = -\frac{\partial}{\partial q}, \quad \left[\frac{\partial}{\partial s}, \frac{\partial}{\partial t}\right] = 0.$$

Therefore the derived system of $\mathcal{V}$ is

$$\mathcal{V}' = \left(\frac{\partial}{\partial x} + p\frac{\partial}{\partial z}, \frac{\partial}{\partial y} + q\frac{\partial}{\partial z}, \frac{\partial}{\partial p}, \frac{\partial}{\partial q}, \frac{\partial}{\partial s}, \frac{\partial}{\partial t}\right),$$

and so $\dim \mathcal{V}' = 6 = \dim \mathcal{V} + 2$. Note also that the Cauchy characteristic subsystem of $\mathcal{V}'$ is

$$C(\mathcal{V}') = \left(\frac{\partial}{\partial s}, \frac{\partial}{\partial t}\right),$$

which thus happens to be a subsystem of $\mathcal{V}$.

The second derivative is

$$\mathcal{V}'' = \left(\frac{\partial}{\partial x}, \frac{\partial}{\partial y}, \frac{\partial}{\partial z}, \frac{\partial}{\partial p}, \frac{\partial}{\partial q}, \frac{\partial}{\partial s}, \frac{\partial}{\partial t}\right),$$

i.e., the system of *all* vector fields on S. Consequently $\mathcal{V}$ admits no *nonconstant first integrals*.

Let us next determine the Cauchy characteristic subsystem $C(\mathcal{V})$ of $\mathcal{V}$:

$$C = c^1 X + c^2 Y + c^3\frac{\partial}{\partial s} + c^4\frac{\partial}{\partial t} \in C(\mathcal{V}) \Longleftrightarrow [C, \mathcal{V}] \equiv 0 \pmod{\mathcal{V}}.$$

In particular,

$$0 \equiv \left[\frac{\partial}{\partial s}, C\right] \equiv c^1\left(\frac{\partial\rho}{\partial s}\frac{\partial}{\partial p} + \frac{\partial}{\partial q}\right) + c^2\frac{\partial}{\partial p},$$

implying that $c^1 = c^2 = 0$; then

$$0 \equiv [Y, C] = \left[Y, c^3\frac{\partial}{\partial s} + c^4\frac{\partial}{\partial t}\right] \equiv -c^3\frac{\partial}{\partial p} - c^4\frac{\partial}{\partial q},$$

whence $c^3 = c^4 = 0$. Thus $C(\mathcal{V}) = \{0\}$.

Conclusion. $\dim \mathcal{V} = 4$, $\dim \mathcal{V}' = 6$, $\dim \mathcal{V}'' = 7$, $C(\mathcal{V}) = \{0\}$ *and* $C(\mathcal{V}')$ *is a 2-dimensional subsystem of* $\mathcal{V}$.

The transition from PDEs to vector field systems is immediate—but in general it is more difficult to go in the other direction, as the proof of the converse of this conclusion shows.

Theorem 11.1.1. *Let V be a vector field system on $\mathbb{R}^7$ with the following properties:*

(i) $\dim V = 4$, $\dim V' = 6$ *and* $\dim V'' = 7$;

(ii) $C(V) = \{0\}$;

(iii) $C(V')$ *is a 2-dimensional subsystem of* V.

Then V is associated to a second order PDE in one dependent and two independent variables.

Proof. Consider V' first. Since $\dim C(V') = 2$, we may introduce local coordinates $u, v, s^0, s^1, s^2, s^3, s^4$ for $\mathbb{R}^7$ such that $C(V') = (\partial/\partial u, \partial/\partial v)$, and V' is generated by $\partial/\partial u$, $\partial/\partial v$ and four vector fields of the form

$$A_i = \frac{\partial}{\partial s^i} + a_i(s) \frac{\partial}{\partial s^0} \quad \text{for } i = 1, 2, 3, 4,$$

where $s = (s^0, s^1, s^2, s^3, s^4)$. Then $V'_{\text{red}} = (A_1, A_2, A_3, A_4)$ is a 4-dimensional vector field system on $\mathbb{R}^5_s$.

If V'_{red} were complete, V' would also be—but this contradicts the fact that $\dim V'' = 7$. Hence

$$\dim(V'_{\text{red}})' = 5 = \dim V'_{\text{red}} + 1.$$

Therefore it is possible to apply the theory in section **6.4** and introduce local coordinates x, y, z, p, q for $\mathbb{R}^5$ such that V'_{red} is generated by the vector fields

$$\begin{cases} D_x := \partial/\partial x + p\,\partial/\partial z, \\ D_y := \partial/\partial y + q\,\partial/\partial z, \\ \partial/\partial p \quad \text{and} \quad \partial/\partial q, \end{cases}$$

that is,

$$(\mathbb{R}^5, V'_{\text{red}}) \cong (J^1(\mathbb{R}^2, \mathbb{R}), ({}^1Ct^{2,1})^\perp).$$

Important remark. *The introduction of the new local coordinates x, y, z, p, q is determined only up to a contact transformation of $J^1(\mathbb{R}^2, \mathbb{R})$!*

Because $\partial/\partial u, \partial/\partial v \in V \subset V' = (D_x, D_y, \partial/\partial p, \partial/\partial q, \partial/\partial u, \partial/\partial v)$, the 4-dimensional V can be expressed as

$$V = (A, B, \partial/\partial u, \partial/\partial v),$$

where

$$\begin{cases} A = a^1 D_x + a^2 D_y + a^3 \partial/\partial p + a^4 \partial/\partial q, \\ B = b^1 D_x + b^2 D_y + b^3 \partial/\partial p + b^4 \partial/\partial q \end{cases}$$

for suitable functions a^i and b^i satisfying

$$\operatorname{rank} \begin{pmatrix} a^1 & a^2 & a^3 & a^4 \\ b^1 & b^2 & b^3 & b^4 \end{pmatrix} = 2,$$

—that is, at least one of the 2×2 minors is to be nonvanishing. We *would like* to have

$$\det \begin{pmatrix} a^1 & a^2 \\ b^1 & b^2 \end{pmatrix} \neq 0.$$

Suppose this is not so, but for instance that instead

$$\det \begin{pmatrix} a^2 & a^3 \\ b^2 & b^3 \end{pmatrix} \neq 0.$$

Using the remark above we then introduce new coordinates for $J^1(\mathbb{R}^2, \mathbb{R})$ by means of *Ampère's contact transformation*:

$$\hat{x} := p, \quad \hat{p} := -x, \quad \hat{z} := z - px, \quad \hat{y} := y, \quad \hat{q} := q.$$

Because this coordinate transformation implies that

$$D_x \mapsto -\frac{\partial}{\partial \hat{p}}, \quad D_y \mapsto \frac{\partial}{\partial \hat{y}} + \hat{q}\frac{\partial}{\partial \hat{z}} = D_{\hat{y}},$$

$$\frac{\partial}{\partial p} \mapsto \frac{\partial}{\partial \hat{x}} + \hat{p}\frac{\partial}{\partial \hat{z}} = D_{\hat{x}} \quad \text{and} \quad \frac{\partial}{\partial q} \mapsto \frac{\partial}{\partial \hat{q}},$$

it really is a contact transformation. Moreover

$$\begin{cases} A \mapsto a^3 D_{\hat{x}} + a^2 D_{\hat{y}} - a^1 \partial/\partial\hat{p} + a^4 \partial/\partial\hat{q}, \\ B \mapsto b^3 D_{\hat{x}} + b^2 D_{\hat{y}} - b^1 \partial/\partial\hat{p} + b^4 \partial/\partial\hat{q}, \end{cases}$$

so that

$$a^1 b^2 - a^2 b^1 \mapsto a^3 b^2 - a^2 b^3, \quad \text{which is} \neq 0 \text{ by assumption.}$$

Therefore we may indeed assume that $a^1 b^2 - a^2 b^1 \neq 0$, and this means that A and B can be replaced by

$$X := D_x + \rho\frac{\partial}{\partial p} + \sigma'\frac{\partial}{\partial q} \quad \text{and} \quad Y := D_y + \sigma''\frac{\partial}{\partial p} + \tau\frac{\partial}{\partial q}$$

respectively. Now with $X, Y \in \mathcal{V}$ it follows that $[X, Y] \in \mathcal{V}'$—but

$$[X, Y] \equiv (\sigma' - \sigma'')\frac{\partial}{\partial z} \quad (\text{mod } \mathcal{V}'),$$

whence necessarily $\sigma' = \sigma''$. Consequently

$$\mathcal{V} = \left(D_x + \rho\frac{\partial}{\partial p} + \sigma\frac{\partial}{\partial q}, D_y + \sigma\frac{\partial}{\partial p} + \tau\frac{\partial}{\partial q}, \frac{\partial}{\partial u}, \frac{\partial}{\partial v} \right),$$

which is nothing but the vector field system associated to the second order PDE

$$\begin{cases} r = \rho(x, y, z, p, q, u, v), \\ s = \sigma(x, y, z, p, q, u, v), \\ t = \tau(x, y, z, p, q, u, v). \end{cases}$$

Let us finally show that u and v may be replaced by two of the second order jet coordinates r, s, t. In fact, otherwise

$$\frac{\partial(\rho, \sigma)}{\partial(u, v)}, \quad \frac{\partial(\rho, \tau)}{\partial(u, v)} \quad \text{and} \quad \frac{\partial(\sigma, \tau)}{\partial(u, v)} \quad \text{all vanish,}$$

or equivalently

$$\text{rank} \begin{pmatrix} \partial\rho/\partial u & \partial\sigma/\partial u & \partial\tau/\partial u \\ \partial\rho/\partial v & \partial\sigma/\partial v & \partial\tau/\partial v \end{pmatrix} = 1.$$

Thus the rows in the latter matrix are linearly dependent, which means that there are smooth nonzero functions f and g such that

$$f \frac{\partial\rho}{\partial u} + g \frac{\partial\rho}{\partial v} = f \frac{\partial\sigma}{\partial u} + g \frac{\partial\sigma}{\partial v} = f \frac{\partial\tau}{\partial u} + g \frac{\partial\tau}{\partial v} = 0.$$

But then the vector field $C := f \partial/\partial u + g \partial/\partial v$ kills each of ρ, σ and τ, which in turn implies that

$$[C, X] \equiv [C, Y] \equiv [C, \partial/\partial u] \equiv [C, \partial/\partial v] \equiv 0 \quad (\text{mod } \mathcal{V}).$$

Therefore C is a Cauchy characteristic vector field in $\mathcal{V}$—which contradicts the second assumption.

If for instance $\partial(\rho, \sigma)/\partial(u, v) \neq 0$, u and v may be replaced by $r := \rho$ and $s := \sigma$, in which case our PDE is reduced to

$$t = \tau(x, y, z, p, q, r, s).$$

$\square$

Definition. A *self-equivalence* or *symmetry* of a PDE system $\mathcal{S}$ is an equivalence of the associated vector field system $\mathcal{V}$ with itself. Clearly the set of self-equivalences forms a pseudogroup, which usually is called the *symmetry group* of $\mathcal{S}$.

Let $\mathcal{S}$ be a hypersurface in $J^2(\mathbb{R}^2, \mathbb{R})$, and let π denote the projection $\mathcal{S} \longrightarrow J^1(\mathbb{R}^2, \mathbb{R})$. The proof above shows that any self-equivalence ϕ of $\mathcal{S}$

is determined by a contact transformation κ making the diagram

$$
\begin{array}{ccc}
S & \xrightarrow{\phi} & S \\
{\scriptstyle \pi}\downarrow & & \downarrow{\scriptstyle \pi} \\
J^1(\mathbb{R}^2, \mathbb{R}) & \xrightarrow{\kappa} & J^1(\mathbb{R}^2, \mathbb{R})
\end{array}
$$

commutative. Or in other words: ϕ is a prolongation of a contact transformation.

Corollary 11.1.1. *The symmetry group of a second order PDE in one de-pendent and two independent variables is a subpseudogroup (which might be proper or improper) of the Lie pseudogroup* $^1\mathrm{Cont}^{2,1}$ *of contact trans-formations of* $J^1(\mathbb{R}^2, \mathbb{R})$. $\qquad\qquad\square$

11.2 Monge systems

To the PDE $r = \rho(x, y, z, p, q, s, t)$ is associated the 4-dimensional vector field system $\mathcal{V}$ generated by

$$
\begin{cases}
X = \partial/\partial x + p\,\partial/\partial z + \rho\,\partial/\partial p + s\,\partial/\partial q, \\
Y = \partial/\partial y + q\,\partial/\partial z + s\,\partial/\partial p + t\,\partial/\partial q, \\
\partial/\partial s \quad \text{and} \quad \partial/\partial t.
\end{cases}
$$

The derived system is

$$
\mathcal{V}' = \left(\frac{\partial}{\partial x} + p\frac{\partial}{\partial z}, \frac{\partial}{\partial y} + q\frac{\partial}{\partial z}, \frac{\partial}{\partial p}, \frac{\partial}{\partial q}, \frac{\partial}{\partial s}, \frac{\partial}{\partial t} \right).
$$

In this section we will determine the singular vector fields in $\mathcal{V}$ by the method of section **3.2**, thereby confirming the results of section **5.3** for this special case.

Temporarily setting

$$
X_1 := X, \; X_2 := Y, \; X_3 := \partial/\partial s, \; X_4 := \partial/\partial t \; \text{and} \; Z_1 := \partial/\partial p, \; Z_2 := \partial/\partial q,
$$

we have

$$
\mathcal{V} = (X_1, X_2, X_3, X_4) \quad \text{and} \quad \mathcal{V}' = (X_1, X_2, X_3, X_4; Z_1, Z_2).
$$

Let $A = \sum_{i=1}^4 a^i X_i$ and $B = \sum_{j=1}^4 b^j X_j$ be arbitrary vector fields in $\mathcal{V}$, with the a^i and b^j being smooth functions of the seven variables x, y, z, p, q, s, t. The commutation relations derived in the preceding section imply that modulo $\mathcal{V}$,

$$[A, B] \equiv \sum_{i,j=1}^{4} a^i b^j \, [X_i, X_j] = \sum_{1 \le i < j \le 4} (a^i b^j - a^j b^i) \, [X_i, X_j]$$

$$= (a^1 b^2 - a^2 b^1)(-Y\rho) Z_1 + (a^1 b^3 - a^3 b^1) \left(-\frac{\partial \rho}{\partial s} Z_1 - Z_2 \right)$$

$$+ (a^1 b^4 - a^4 b^1) \left(-\frac{\partial \rho}{\partial t} \right) Z_1 + (a^2 b^3 - a^3 b^2)(-Z_1)$$

$$+ (a^2 b^4 - a^4 b^2)(-Z_2)$$

$$= \left((a^2 b^1 - a^1 b^2) Y \rho + (a^3 b^1 - a^1 b^3) \frac{\partial \rho}{\partial s} \right.$$

$$\left. + (a^4 b^1 - a^1 b^4) \frac{\partial \rho}{\partial t} + a^3 b^2 - a^2 b^3 \right) Z_1$$

$$+ (a^3 b^1 - a^1 b^3 + a^4 b^2 - a^2 b^4) Z_2 \quad (\text{mod } \mathcal{V}).$$

Remark. The appearance of the 2×2 determinants $a^i b^j - a^j b^i$ suggests a connection with Plücker's line geometry—which will be contemplated in the next section.

Hence $[A, B] \equiv 0 \pmod{\mathcal{V}}$ if and only if $M(\mathbf{a})\mathbf{b} = \mathbf{0}$, where $\mathbf{a} = (a^1, a^2, a^3, a^4)^T$, $\mathbf{b} = (b^1, b^2, b^3, b^4)^T$ and

$$M(\mathbf{a}) = \begin{pmatrix} a^2 \, Y\rho + a^3 \frac{\partial \rho}{\partial s} + a^4 \frac{\partial \rho}{\partial t} & -a^1 \, Y\rho + a^3 & -a^1 \frac{\partial \rho}{\partial s} - a^2 & -a^1 \frac{\partial \rho}{\partial t} \\ a^3 & a^4 & -a^1 & -a^2 \end{pmatrix}.$$

Suppose that $A \ne 0$—for instance that $a^1 \ne 0$. Then $1 \le \operatorname{rank} M(\mathbf{a}) \le 2$, which in particular means that $\mathcal{V}$ admits no Cauchy characteristic vector field. But there may be singular ones:

$$A \text{ is singular} \iff \operatorname{rank} M(\mathbf{a}) = 1 \iff \text{all } 2 \times 2 \text{ minors vanish}.$$

The vanishing of the 2×2 determinant formed from the last two columns is equivalent to

$$(a^2)^2 + a^1 a^2 \frac{\partial \rho}{\partial s} - (a^1)^2 \frac{\partial \rho}{\partial t} = 0.$$

If we set $\lambda := a^2/a^1$ this gives the *characteristic equation* for the singular vector fields:

$$\lambda^2 + \lambda \frac{\partial \rho}{\partial s} - \frac{\partial \rho}{\partial t} = 0.$$

Important message. *We surely want to be able to solve this characteristic* equation, just as we wanted to be able to solve eigenvalue problems when classifying 3-dimensional Lie algebras. Therefore from now on we introduce the convention that **the ground field is** $\mathbb{C}$—rather than $\mathbb{R}$.

In particular this means that our smooth functions automatically become analytic. But **we do not want to use power series arguments!** So we wish to avoid the Cauchy–Kowalewski theorem, and thereby also Cartan's local existence theorem.

Solving the characteristic equation we obtain

$$\lambda = \begin{cases} \lambda^1, \\ \lambda^2, \end{cases} \quad \text{with } \lambda^1 \text{ and } \lambda^2 \text{ being functions of } x, y, z, p, q, s, t.$$

Note moreover that $\lambda^1 + \lambda^2 = -\partial\rho/\partial s$ and $\lambda^1 \cdot \lambda^2 = -\partial\rho/\partial t$.

The determinant given by the second and third columns is zero if and only if

$$a^3 = Y\rho \cdot a^1 + \left(\lambda + \frac{\partial\rho}{\partial s}\right) a^4.$$

With $a^2 = \lambda a^1$ and this a^3 it is then easily checked that *all* 2×2 minors of $M(\mathbf{a})$ vanish.

Choosing $\lambda := \lambda^1$ we have $a^2 = \lambda^1 a^1$ and $a^3 = Y\rho \cdot a^1 - \lambda^2 a^4$, giving the singular vector field

$$\begin{aligned}
A &= a^1 (X_1 + \lambda^1 X_2 + Y\rho \cdot X_3) + a^4 (X_4 - \lambda^2 X_3) \\
&= a^1 (X + \lambda^1 Y + Y\rho \cdot \partial/\partial s) + a^4 (\partial/\partial t - \lambda^2 \partial/\partial s),
\end{aligned}$$

with a^1 and a^4 being arbitrary functions. Note that since the set of singular vector fields is closed, we may let a^1 turn into zero and still remain in this set.

With $\lambda := \lambda^2$ we get $a^2 = \lambda^2 a^1$ and $a^3 = Y\rho \cdot a^1 - \lambda^1 a^4$, which gives the singular vector field

$$\begin{aligned}
A &= a^1 (X_1 + \lambda^2 X_2 + Y\rho \cdot X_3) + a^4 (X_4 - \lambda^1 X_3) \\
&= a^1 (X + \lambda^2 Y + Y\rho \cdot \partial/\partial s) + a^4 (\partial/\partial t - \lambda^1 \partial/\partial s).
\end{aligned}$$

Correspondingly we obtain *two singular subsystems* $\mathcal{F} = (F_1, F_2)$ and $\mathcal{G} = (G_1, G_2)$ of $\mathcal{V}$, where

$$F_1 = X + \lambda^1 Y + Y\rho \frac{\partial}{\partial s}, \qquad F_2 = \frac{\partial}{\partial t} - \lambda^2 \frac{\partial}{\partial s},$$

$$G_1 = X + \lambda^2 Y + Y\rho \frac{\partial}{\partial s}, \qquad G_2 = \frac{\partial}{\partial t} - \lambda^1 \frac{\partial}{\partial s}.$$

As in section **5.2** we say that our PDE is *hyperbolic* if $\mathcal{F} \neq \mathcal{G}$ (i.e., $\lambda^1 \neq \lambda^2$), and *parabolic* if $\mathcal{F} = \mathcal{G}$ (i.e., $\lambda^1 = \lambda^2$).

The singularity of the vector fields F_1, F_2, G_1, G_2 means that their commutation relations are maximally simple. Straightforward computations show that

$$
\begin{cases}
[F_i, G_j] \equiv 0 \text{ for } i, j = 1, 2 \iff [\mathcal{F}, \mathcal{G}] \equiv 0, \\
[F_1, F_2] \equiv (\lambda^2 - \lambda^1)(\partial/\partial q - \lambda^2 \, \partial/\partial p) =: F_3, \qquad (\mathrm{mod}\ \mathcal{V}). \\
[G_1, G_2] \equiv (\lambda^1 - \lambda^2)(\partial/\partial q - \lambda^1 \, \partial/\partial p) =: G_3
\end{cases}
$$

The parabolic case: $\lambda^1 = \lambda^2 = \lambda$. As basis for $\mathcal{V}$ we choose the four vector fields

$$
F_1 = \frac{\partial}{\partial x} + \lambda \frac{\partial}{\partial y} + (p + \lambda q) \frac{\partial}{\partial z} + (\rho + \lambda s) \frac{\partial}{\partial p} + (s + \lambda t) \frac{\partial}{\partial q} + Y \rho \frac{\partial}{\partial s},
$$

$$
Y = \frac{\partial}{\partial y} + q \frac{\partial}{\partial z} + s \frac{\partial}{\partial p} + t \frac{\partial}{\partial q},
$$

$$
F_2 = \frac{\partial}{\partial t} - \lambda \frac{\partial}{\partial s} \quad \text{and} \quad \frac{\partial}{\partial s}.
$$

The Lie structure of $\mathcal{V}$ is given by the commutation relations of these modulo $\mathcal{V}$ (note that $\partial \rho / \partial s + \lambda = -\lambda$),

$$
[F_1, Y] \equiv 0, \quad [F_1, F_2] \equiv 0, \quad \left[F_1, \frac{\partial}{\partial s} \right] \equiv \lambda \frac{\partial}{\partial p} - \frac{\partial}{\partial q},
$$

$$
[Y, F_2] \equiv \lambda \frac{\partial}{\partial p} - \frac{\partial}{\partial q}, \quad \left[Y, \frac{\partial}{\partial s} \right] \equiv -\frac{\partial}{\partial p}, \quad \left[F_2, \frac{\partial}{\partial s} \right] = 0,
$$

confirming that $\dim \mathcal{V}' = \dim \mathcal{V} + 2$.

Following the pattern of Cartan's local existence theorem we first look for 2-dimensional *involutions* of $\mathcal{V}$. Requiring that $dx \wedge dy \neq 0$ on these, they are of the form (A, B) with

$$
\begin{cases}
A = F_1 + a^1 F_2 + a^2 \, \partial/\partial s, \\
B = Y + b^1 F_2 + b^2 \, \partial/\partial s,
\end{cases}
$$

where the functions a^1, a^2, b^1, b^2 are to be chosen such that $[A, B] \equiv 0$ (mod $\mathcal{V}$), i.e.,

$$
0 \equiv [A, B] \equiv b^2 \left(\lambda \frac{\partial}{\partial p} - \frac{\partial}{\partial q} \right) - a^1 \left(\lambda \frac{\partial}{\partial p} - \frac{\partial}{\partial q} \right) + a^2 \frac{\partial}{\partial p}.
$$

Thus $b^2 = a^1$ and $a^2 = 0$, so that

$$\begin{cases} A = F_1 + a\,F_2, \\ B = Y + b\,F_2 + a\,\partial/\partial s, \end{cases}$$

where a and b are arbitrary functions. In particular this shows that $A \in (F_1, F_2) = \mathcal{F}$, and hence $\mathcal{F}$ is a *Monge characteristic subsystem of* $\mathcal{V}$—as defined in section **5.3**.

There then remains to determine a and b such that (A, B) becomes complete. One possible approach is the method of Darboux, which is described in section **11.4** for hyperbolic PDEs. We will however postpone the further study of the parabolic case to chapters **16** and **17**, where we examine Cartan's 'five variable paper'.

The hyperbolic case: $\lambda^1 \ne \lambda^2$. Here the four singular vector fields F_1, F_2, G_1, G_2 are linearly independent, and therefore define a basis for $\mathcal{V}$. Setting

$$F_3 := (\lambda^2 - \lambda^1)\left(\frac{\partial}{\partial q} - \lambda^2 \frac{\partial}{\partial p}\right) \quad \text{and} \quad G_3 := (\lambda^1 - \lambda^2)\left(\frac{\partial}{\partial q} - \lambda^1 \frac{\partial}{\partial p}\right)$$

we have $\mathcal{V} = (F_1, F_2, G_1, G_2)$, $\mathcal{V}' = (F_1, F_2, G_1, G_2; F_3, G_3)$ and the following Lie structure:

$$\begin{cases} [F_i, G_j] \equiv 0 \quad \text{for } i, j = 1, 2, \\ [F_1, F_2] \equiv F_3, \quad [G_1, G_2] \equiv G_3 \end{cases} \quad (\text{mod } \mathcal{V}).$$

The 2-dimensional involutions on which $dx \wedge dy \ne 0$ are generated by vector fields of the form

$$\begin{cases} F = F_1 + a\,F_2 + \alpha\,G_2, \\ G = G_1 + \beta\,F_2 + b\,G_2, \end{cases}$$

where the functions a, α, β and b are to satisfy $[F, G] \equiv 0 \pmod{\mathcal{V}}$, i.e.,

$$0 = \beta\,F_3 - \alpha\,G_3 \iff \alpha = \beta = 0.$$

Consequently the general 2-dimensional involution of $\mathcal{V}$ on which $dx \wedge dy \ne 0$ is generated by

$$\begin{cases} F = F_1 + a\,F_2, \\ G = G_1 + b\,G_2, \end{cases}$$

where a and b are arbitrary functions of x, y, z, p, q, s, t. Because $F \in \mathcal{F}$ and $G \in \mathcal{G}$, $\mathcal{F}$ and $\mathcal{G}$ are *Monge characteristic subsystems of* $\mathcal{V}$ (recalling what we did in section **5.3**).

In section **11.4** we will discuss Darboux's method for determining a and b such that (F, G) becomes a complete subsystem of $\mathcal{V}$.

11.3 A connection with line geometry

We have seen that second order PDEs in one dependent and two independent variables give rise to 4-dimensional vector field systems on 7-dimensional manifolds, having derivatives of the dimension 6.

Let us here be slightly more general and consider a 4-dimensional vector field system $\mathcal{V}$ on a manifold M of arbitrary dimension, and satisfying $\dim \mathcal{V}' = 6$.

Remark. The general case with $\dim \mathcal{V}' = \dim \mathcal{V} + 2$ is treated in the very interesting paper [Cartan 1901b].

So let $\mathcal{V} = (X_1, X_2, X_3, X_4)$, $\mathcal{V}' = (X_1, X_2, X_3, X_4; Z_1, Z_2)$, and suppose that $\mathcal{V}$ has the structure

$$[X_i, X_j] \equiv c_{ij}^1 Z_1 + c_{ij}^2 Z_2 \quad (\text{mod } \mathcal{V}) \quad \text{for } i, j = 1, \dots, 4,$$

where $c_{ij}^k + c_{ji}^k = 0$. Further let $A = \sum_{i=1}^4 a^i X_i$ and $B = \sum_{j=1}^4 b^j X_j$ be arbitrary vector fields in $\mathcal{V}$, with the a^i and b^j being functions on M. Then

$$A \text{ and } B \text{ are in involution} \iff [A, B] \equiv 0 \quad (\text{mod } \mathcal{V})$$

$$\iff \sum_{i,j=1}^4 a^i b^j (c_{ij}^1 Z_1 + c_{ij}^2 Z_2) = 0$$

$$\iff \sum_{1 \le i < j \le 4} c_{ij}^1 (a^i b^j - a^j b^i) Z_1 + \sum_{1 \le i < j \le 4} c_{ij}^2 (a^i b^j - a^j b^i) Z_2 = 0$$

$$\iff \sum_{1 \le i < j \le 4} c_{ij}^1 p^{ij} = 0 \text{ and } \sum_{1 \le i < j \le 4} c_{ij}^2 p^{ij} = 0, \text{ where } p^{ij} := a^i b^j - a^j b^i.$$

If we restrict the attention to some generic point $m \in M$, the coefficients a^i and b^j go over into the complex numbers $a^i(m)$ and $b^j(m)$, and in this way we obtain a one-to-one correspondence

$$\mathcal{V}_m \longleftrightarrow \mathbb{C}^4$$

$$A = \sum_{i=1}^4 a^i X_i \longleftrightarrow (a^1, a^2, a^3, a^4).$$

However we really are more interested in the 1-dimensional vector field system (A) generated by A than by A itself, and therefore the a^i should rather be regarded as *homogeneous coordinates of the 3-dimensional projective space* $\mathbb{P}^3(\mathbb{C})$:

$$(A) = \left(\sum_{i=1}^{4} a^i X_i \right) \longleftrightarrow a = (a^1, a^2, a^3, a^4) \in \mathbb{P}^3(\mathbb{C}).$$

In this way the 1-dimensional vector field systems (A) and (B) are identified with points a and b in $\mathbb{P}^3(\mathbb{C})$.

The key idea in the following is to study the *manifold of all lines in* $\mathbb{P}^3(\mathbb{C})$—which was a popular subject in Lie's days. Our exposition is based upon Kap. 7 in the very imaginative book [Lie–Scheffers 1896].

Let $a, b \in \mathbb{P}^3(\mathbb{C})$, and let $\ell(a, b)$ be the line through these points. Then

$$x \in \ell(a,b) \Longleftrightarrow x \text{ is a linear combination of } a \text{ and } b$$

$$\Longleftrightarrow \text{rank} \begin{pmatrix} x^1 & x^2 & x^3 & x^4 \\ a^1 & a^2 & a^3 & a^4 \\ b^1 & b^2 & b^3 & b^4 \end{pmatrix} = 2$$

$$\Longleftrightarrow \text{ each } 3 \times 3 \text{ minor vanishes.}$$

Laplace expansions of the first rows in these minors yield the following homogeneous linear system for the x^i, where $p^{ij} = a^i b^j - a^j b^i = -p^{ji}$:

$$\begin{pmatrix} 0 & p^{34} & -p^{24} & p^{23} \\ -p^{34} & 0 & p^{14} & -p^{13} \\ p^{24} & -p^{14} & 0 & p^{12} \\ -p^{23} & p^{13} & -p^{12} & 0 \end{pmatrix} \begin{pmatrix} x^1 \\ x^2 \\ x^3 \\ x^4 \end{pmatrix} = \begin{pmatrix} 0 \\ 0 \\ 0 \\ 0 \end{pmatrix}. \qquad (*)$$

Since there are infinitely many points x on the line $\ell(a, b)$, the determinant of the matrix $M(p)$ in the left hand side must vanish. But with $M(p)$ being an even-ordered skew-symmetric matrix, its determinant is the square of a polynomial in the variables p^{ij} (called the Pfaffian of $M(p)$)—and in this particular case

$$\det M(p) = (p^{12}p^{34} - p^{13}p^{24} + p^{14}p^{23})^2.$$

Because $x \in \ell(a, b)$ if and only if $(*)$ is satisfied, the quantities p^{ij} somehow characterize the line $\ell(a, b)$. Changing (a, b) to another couple (a', b') of points on $\ell(a, b)$ we easily see that there is a common factor ρ such that

$$(p')^{ij} := (a')^i (b')^j - (a')^j (b')^i = \rho \cdot p^{ij} \quad \text{for } i, j = 1, \dots, 4.$$

Thus what really makes sense is $(p^{12}, p^{13}, p^{14}, p^{23}, p^{24}, p^{34})$ *regarded as a point in* $\mathbb{P}^5(\mathbb{C})$. And by the remark above, this point belongs to the submanifold

$$\mathbb{L}^4(\mathbb{C}) := \{(p^{12}, p^{13}, p^{14}, p^{23}, p^{24}, p^{34}) \in \mathbb{P}^5(\mathbb{C}) \mid p^{12}p^{34} - p^{13}p^{24} + p^{14}p^{23} = 0\}.$$

And conversely any point in the 4-dimensional manifold $\mathbb{L}^4(\mathbb{C})$ gives rise to a line in $\mathbb{P}^5(\mathbb{C})$ by means of the equation (*).

Theorem 11.3.1. *There is a one-to-one correspondence between lines in $\mathbb{P}^3(\mathbb{C})$ and points in $\mathbb{L}^4(\mathbb{C})$.*   □

With $p^{ij} = a^i b^j - a^j b^i$, the line $\ell(a, b)$ is also denoted by ℓ_p. Let $\ell(c, d) = \ell_q$ be another line, with $q^{ij} = c^i d^j - c^j d^i$. The lines ℓ_p and ℓ_q intersect if and only if the four points a, b, c, d lie in the same plane, that is if

$$\begin{vmatrix} a^1 & a^2 & a^3 & a^4 \\ b^1 & b^2 & b^3 & b^4 \\ c^1 & c^2 & c^3 & c^4 \\ d^1 & d^2 & d^3 & d^4 \end{vmatrix} = 0,$$

which boils down to

$$p^{12}q^{34} + q^{12}p^{34} - p^{13}q^{24} - q^{13}p^{24} + p^{14}q^{23} + q^{14}p^{23} = 0.$$

Note incidentally that setting $(c, d) = (a, b)$, this equation gets specialized to the defining equation of $\mathbb{L}^4(\mathbb{C})$.

We have thus proved the following result.

Lemma 11.3.1. *Two lines ℓ_p and ℓ_q intersect if and only if*

$$q^{34}p^{12} - q^{24}p^{13} + q^{23}p^{14} + q^{14}p^{23} - q^{13}p^{24} + q^{12}p^{34} = 0.$$

□

If ℓ_q is given, this lemma explains how to find all lines ℓ_p which intersect ℓ_q.

Generalizing this we let let $c_{12}, c_{13}, c_{14}, c_{23}, c_{24}, c_{34}$ be given complex numbers. Then the lines ℓ_p for which the p^{ij} satisfy the linear homogeneous equation

$$c_{12}p^{12} + c_{13}p^{13} + c_{14}p^{14} + c_{23}p^{23} + c_{24}p^{24} + c_{34}p^{34} = 0$$

are said to form a *linear complex*.

A linear complex is called *special* if the coefficients c_{ij} satisfy

$$c_{12}c_{34} - c_{13}c_{24} + c_{14}c_{23} = 0 \text{—that is, } (c_{12}, \ldots, c_{34}) \in \mathbb{L}^4(\mathbb{C}).$$

If we set

$$d^{12} := c_{34}, \ d^{13} := -c_{24}, \ d^{14} := c_{23}, \ d^{23} := c_{14}, \ d^{24} := -c_{13}, \ d^{34} := c_{12},$$

also $(d^{12}, \ldots, d^{34}) \in \mathbb{L}^4(\mathbb{C})$, and thus defines a line ℓ_d which is called the *directrix* of the special linear complex. Since the equation for this complex equivalently can be written as

$$d^{34}p^{12} - d^{24}p^{13} + d^{23}p^{14} + d^{14}p^{23} - d^{13}p^{24} + d^{12}p^{34} = 0,$$

a *special linear complex* consists of all lines ℓ_p intersecting the directrix ℓ_d.

Let us now return to our vector fields $A = \sum_{i=1}^{4} a^i X_i$ and $B = \sum_{j=1}^{4} b^j X_j$ in $\mathcal{V}$. We saw earlier that the 1-dimensional vector field systems (A) and (B) are in involution if and only if the quantities $p^{ij} = a^i b^j - a^j b^i$ satisfy

$$\sum_{1 \le i < j \le 4} c_{ij}^k p^{ij} = 0 \quad \text{for } k = 1, 2,$$

where c_{ij}^k are the structure functions of $\mathcal{V}$.

Definition. The set of lines ℓ_p satisfying two independent linear homogeneous equations

$$\sum_{1 \le i < j \le 4} c_{ij}^1 p^{ij} = 0 \quad \text{and} \quad \sum_{1 \le i < j \le 4} c_{ij}^2 p^{ij} = 0,$$

with c_{ij}^1 and c_{ij}^2 being given complex numbers, is said to constitute a *linear congruence*.

Identifying (A) and (B) with points a and $b \in \mathbb{P}^3(\mathbb{C})$ we thus see that

(A) and (B) are in involution $\Longleftrightarrow$ the line $\ell(a, b)$ through a and b belongs to the linear congruence defined by the structure functions of $\mathcal{V}$.

Our task has therefore been reduced to that of understanding linear congruences.

Note first that the line ℓ_p belongs to the linear congruence

$$\sum_{1 \le i < j \le 4} c_{ij}^k p^{ij} = 0 \quad \text{for } k = 1, 2$$

if and only if ℓ_p for each $(e_1, e_2) \in \mathbb{C}^2$ is a member of the linear complex

$$e_1 \sum_{1 \le i < j \le 4} c_{ij}^1 p^{ij} + e_2 \sum_{1 \le i < j \le 4} c_{ij}^2 p^{ij} = 0.$$

And conversely, if ℓ_p belongs to two linearly independent linear complexes in this 2-parameter family, it belongs to all of them, and in particular to the linear congruence.

The linear complex $\sum_{1 \leq i < j \leq 4}(e_1 c_{ij}^1 + e_2 c_{ij}^2)p^{ij} = 0$ is special if and only if

$$
\begin{aligned}
E := &(e_1 c_{12}^1 + e_2 c_{12}^2)(e_1 c_{34}^1 + e_2 c_{34}^2) - (e_1 c_{13}^1 + e_2 c_{13}^2)(e_1 c_{24}^1 + e_2 c_{24}^2) \\
&+ (e_1 c_{14}^1 + e_2 c_{14}^2)(e_1 c_{23}^1 + e_2 c_{23}^2) = 0.
\end{aligned}
$$

If we assume that $e_1 \neq 0$ and set $\lambda := e_2/e_1$, this is a quadratic equation for λ, which we for obvious reasons call the *characteristic equation* of the linear congruence.

There are three different cases to consider.

1. *The generic (or hyperbolic) case.* The characteristic equation has *two different roots* λ^1 and λ^2, giving rise to two different special linear complexes. Thus the linear congruence consists of all lines which belong to both these special linear complexes, i.e., those that intersect both the directrices ℓ_{d^1} and ℓ_{d^2} of the latter.

The directrices ℓ_{d^1} and ℓ_{d^2} cannot intersect, because otherwise they would lie in a plane and *all* lines in this plane passing through the point of intersection would correspond to special linear complexes in our 2-parameter family—see case **3.** below—while we know that just two of them are special.

Take a point a outside of ℓ_{d^1} and ℓ_{d^2}. Through a there passes *exactly one line* intersecting both ℓ_{d^1} and ℓ_{d^2}. Thus the 1-dimensional vector field system (A) corresponding to a is in involution with all 1-dimensional vector field systems corresponding to points of this line. Let (B) be one such. Then the 2-dimensional vector field system (A, B) is the largest involution containing A. Consequently the involutive genus of the vector field system $\mathcal{V}$ equals 2.

But if $a \in \ell_{d^1}$, the set of lines through a which intersect ℓ_{d^1} and ℓ_{d^2} forms the plane defined by a and ℓ_{d^2}. So the vector field A corresponding to a is in involution with a 2-dimensional subsystem of $\mathcal{V}$—and hence A is singular. Accordingly the vector fields corresponding to the points of a directrix define a *2-dimensional singular subsystem of* $\mathcal{V}$.

2. *The singular (or parabolic) case.* Here the characteristic equation has a *double root*. It can be thought of as a limiting case where the two different roots—and the associated directrices—of the former case coalesce.

3. *The degenerate case.* In this case the characteristic equation is of the form $0 \cdot \lambda^2 + 0 \cdot \lambda + 0 = 0$, i.e., is identically satisfied. The vanishing of the first and third coefficients means that the two original linear complexes are special, and the vanishing of the second coefficient implies that the corresponding directrices ℓ_{d^1} and ℓ_{d^2} intersect in a point c—and thus span a plane Π.

Therefore the linear congruence consists of all lines situated in the plane Π, and of all lines passing through the point c of intersection.

Through a point a outside of Π there passes exactly one line in the linear congruence—namely $\ell(a, c)$. So again the involutive genus of $\mathcal{V}$ equals 2.

Let a_1, a_2 and a_3 be three noncollinear points in Π. Then the three lines $\ell(a_1, a_2)$, $\ell(a_1, a_3)$ and $\ell(a_2, a_3)$ all belong to the linear congruence, and so the corresponding vector fields A_1, A_2 and A_3 span a *3-dimensional involution* (A_1, A_2, A_3). Consequently each point in Π corresponds to a singular vector field in $\mathcal{V}$.

Finally any line through c belongs to the linear congruence. Therefore the vector field C corresponding to c is *Cauchy characteristic*.

Comparing these three cases with the results of the preceding section we see that the first two are analogous to hyperbolic and parabolic PDEs respectively. However, since the vector field system associated to a second order PDE does not admit any Cauchy characteristic vector field, the third case cannot possibly arise from such a PDE.

In the following two chapters we will concentrate on *hyperbolic PDEs* in one dependent and two independent variables. Let us therefore take a closer look at the vector field systems associated to such PDEs.

We know that such a vector field system $\mathcal{V}$ contains two Monge characteristic subsystems $\mathcal{F} = (F_1, F_2)$ and $\mathcal{G} = (G_1, G_2)$. Since F_1, F_2, G_1, G_2 are linearly independent, they constitute a basis: $\mathcal{V} = (F_1, F_2, G_1, G_2)$. With suitable F_3, G_3 the derivative can be written as $\mathcal{V}' = (F_1, F_2, G_1, G_2; F_3, G_3)$, so as to give the following structure:

$$
\begin{cases}
[F_i, G_j] \equiv 0 \text{ for } i, j = 1, 2, \\
[F_1, F_2] \equiv F_3, \qquad\qquad (\text{mod } \mathcal{V}). \qquad (\dagger) \\
[G_1, G_2] \equiv G_3
\end{cases}
$$

Furthermore $C(\mathcal{V}) = \{0\}$ and $\dim C(\mathcal{V}') = 2$.

Let us next abstract from the last properties concerning $C(\mathcal{V})$ and $C(\mathcal{V}')$, and consider a general vector field system $\mathcal{V} = (F_1, F_2, G_1, G_2)$ on a

7-dimensional manifold with a noncomplete $\mathcal{V}' = (F_1, F_2, G_1, G_2; F_3, G_3)$ and the structure (†), and discuss the different possibilities for $\dim C(\mathcal{V})$.

Note first that by Jacobi's identity

$$[G_i, F_3] \equiv [G_i, [F_1, F_2]] = [F_2, [F_1, G_i]] - [F_1, [F_2, G_i]] \quad \equiv 0 \pmod{\mathcal{V}'}$$

for $i = 1, 2$, so that $[G_i, F_3] \in \mathcal{V}'$. Analogously $[F_i, G_3] \in \mathcal{V}'$ for $i = 1, 2$. Writing the second deriative as $\mathcal{V}'' = (F_1, F_2, G_1, G_2; F_3, G_3; H)$, we get that $\mathcal{V}'$ has the structure

$$\begin{cases} [F_i, F_3] \equiv a_i\, H, \quad [G_i, G_3] \equiv b_i\, H, \quad [F_3, G_3] \equiv c\, H \quad \text{for} \quad i = 1, 2, \\ \text{and all other brackets vanishing} \quad (\text{mod } \mathcal{V}') \end{cases}$$

for certain functions a_1, a_2, b_1, b_2, c, which cannot all be zero, since $\mathcal{V}'$ would be complete otherwise. Then $C = \sum_{i=1}^{3} u^i\, F_i + \sum_{i=1}^{3} v^i\, G_i$ belongs to $C(\mathcal{V}')$ if and only if $[C, F_j] \equiv 0$ and $[C, G_j] \equiv 0 \pmod{\mathcal{V}'}$ for $j = 1, 2, 3$, i.e.,

$$\begin{cases} u^3 a_1 = 0, \quad u^3 a_2 = 0, \quad u^1 a_1 + u^2 a_2 - v^3 c = 0, \\ v^3 b_1 = 0, \quad v^3 b_2 = 0, \quad u^3 c + v^1 b_1 + v^2 b_2 = 0. \end{cases}$$

1. The generic case. At least one of a_1, a_2 and at least one of b_1, b_2 is nonzero. Then $u^3 = v^3 = 0$, and

$$a_1 u^1 + a_2 u^2 = 0, \quad b_1 v^1 + b_2 v^2 = 0,$$

so that $C(\mathcal{V}')$ is generated by the vector fields

$$a_2 F_1 - a_1 F_2 \in \mathcal{F} \quad \text{and} \quad b_2 G_1 - b_1 G_2 \in \mathcal{G}.$$

Thus $\dim C(\mathcal{V}') = 2$, and moreover $C(\mathcal{V}') \subset \mathcal{V}$.

Definition. A nonzero vector field $X \in \mathcal{V} \cap C(\mathcal{V}')$ is said to be *principal*. The principal vector fields (if there are any) span the *principal subsystem* of $\mathcal{V}$.

So in this case $\mathcal{V}$ has a 2-dimensional principal subsystem.

2. The singular case. Here $a_1 = a_2 = 0$, but at least one of b_1, b_2, c is nonzero. Then $v^3 = 0$ and $cu^3 + b_1 v^1 + b_2 v^2 = 0$, so that $\dim C(\mathcal{V}') = 4$.

And the same conclusion if $b_1 = b_2 = 0$, but at least one of a_1, a_2, c is nonzero.

So only the first case can be associated to a hyperbolic PDE. Combining this result with the theorem in section **11.1**, we arrive at the following conclusion.

Theorem 11.3.2. *Let $\mathcal{V} = (F_1, F_2, G_1, G_2)$ be a 4-dimensional vector field system on a 7-dimensional manifold with the derivative $\mathcal{V}' = (F_1, F_2, G_1, G_2; F_3, G_3)$, and the structure*

$$\begin{cases} [F_i, G_j] \equiv 0 & for\ i, j = 1, 2, \\ [F_1, F_2] \equiv F_3, & [G_1, G_2] \equiv G_3 \end{cases} \quad (\text{mod } \mathcal{V}).$$

Then $\mathcal{V}$ is the vector field system associated to a hyperbolic second order PDE in one dependent and two independent variables if and only if $\dim C(\mathcal{V}') = 2$—and then necessarily $C(\mathcal{V}') \subset \mathcal{V}$. $\quad\square$

In the following we only consider vector field systems $\mathcal{V}$ associated to hyperbolic PDEs.

Such a $\mathcal{V}$ does not admit any nonconstant first integral, because otherwise this would be a first integral of $\mathcal{V}'$ too, and then $\mathcal{V}'$—being a 6-dimensional vector field system on a 7-dimensional manifold—would be complete, and hence $\dim C(\mathcal{V}') = \dim \mathcal{V}' = 6$, which is impossible.

However, the Monge characteristic subsystems $\mathcal{F}$ and $\mathcal{G}$ might admit first integrals—but how many?

Consider $\mathcal{F} = (F_1, F_2)$ for instance. The number of functionally independent first integrals is $7 - \dim \bar{\mathcal{F}}$, where $\bar{\mathcal{F}}$ is the completion of $\mathcal{F}$. Because $[F_1, F_2] \equiv F_3 \notin \mathcal{F}$, $\dim \bar{\mathcal{F}} \geq 3$. And

$$\dim \bar{\mathcal{F}} = 3 \iff (F_1, F_2, F_3)' = (F_1, F_2, F_3) \subset \mathcal{V}' \iff a_1 = a_2 = 0,$$

with the notations above. But then $\dim C(\mathcal{V}') = 4$, so this cannot happen.

Thus $\dim \bar{\mathcal{F}} \geq 4$, and analogously $\dim \bar{\mathcal{G}} \geq 4$.

Theorem 11.3.3. *The Monge characteristic subsystems $\mathcal{F}$ and $\mathcal{G}$ of $\mathcal{V}$ admit at most three functionally independent first integrals.* $\quad\square$

11.4 Darboux's method for hyperbolic PDEs

Let $\mathcal{V}$ be the vector field system associated to a hyperbolic second order PDE in one dependent and two independent variables. Choosing the *singular* vector fields F_1, F_2, G_1, G_2 as basis for $\mathcal{V}$, the structure will become maximally simple, and the general 2-dimensional involution of $\mathcal{V}$ on which $dx \wedge dy \neq 0$ will contain *just two unknown functions a and b*:

$$\mathcal{I}_2 = (F_1 + a F_2, G_1 + b G_2).$$

What remains then is 'only' to determine a and b such that $\mathcal{I}_2$ becomes complete, i.e.,

$$[F_1 + a F_2, G_1 + b G_2] \equiv 0 \quad (\text{mod } (F_1 + a F_2, G_1 + b G_2)).$$

However, from the point of view of Cartan's local existence theorem the fact that there are only two unknown functions is not good at all—for the passage from involutions to complete subsystems is performed by solving a number of *underdetermined* Cauchy–Kowalewski systems, and in order for these to really be underdetermined it is essential to have *as many unknowns as possible*, which one gets by using *regular* vector fields and involutions. But this feature of Cartan's theory makes it look rather artificial (or perhaps even stupid), and therefore it was not altogether tragic when Hans Lewy discovered his famous counterexample to the Cauchy–Kowalewski theorem in the C^∞ category—for after that there is no longer any need to restrict oneself to regular vector fields.

But how to profit from singular vector fields and involutions? Unfortunately there seems to be no simple and concise answer to this question, although the following chapters surely will demonstrate how beneficial they are.

It appears that Darboux was the first to realize the usefulness of Monge systems and their first integrals for solving second order PDEs in one dependent and two independent variables in a report to l'Académie des Sciences from 1870. His method was further elaborated in chapter VII of the second part of [Goursat 1896–98]. Following up an idea of Goursat, Parsons then extended this method to the case of three independent variables in [Parsons 1960].

In principle Darboux's method may be applied to any vector field system admitting Monge characteristic subsystems. There are two main ideas:

- exploit the possible first integrals of each Monge system;
- if there are not enough first integrals from the beginning, prolongations might give Monge systems having a sufficient number of first integrals.

As a preparation for the heavier stuff in later chapters we will here take a look at how Darboux's method works for our hyperbolic PDEs, following [Vessiot 1924]. If we set

$$F := F_1 + a F_2 \quad \text{and} \quad G := G_1 + b G_2,$$

where a and b are unknown functions, the task is to determine the latter such that (F, G) becomes complete.

Using the notations of section **11.2**, straightforward calculations reveal that

$$[F, G] = \frac{F\lambda^2 - G\lambda^1}{\lambda^2 - \lambda^1}(G - F) + (\alpha - Ga)\,F_2 + (Fb - \beta)\,G_2,$$

where

$$\alpha = a(\lambda^2 - \lambda^1)^{-1}(F\lambda^2 - G\lambda^1) - \gamma, \quad \beta = b(\lambda^2 - \lambda^1)^{-1}(F\lambda^2 - G\lambda^1) - \gamma$$

and

$$\gamma = (\lambda^2 - \lambda^1)^{-1}\{(G - F) \circ Y\rho - bF\lambda^1 + aG\lambda^2\}.$$

Hence (F, G) is complete if and only if a and b are solutions of the PDE system

$$\begin{cases} Ga = \alpha, \\ Fb = \beta. \end{cases}$$

Recall from the preceding section that although V does not admit any first integrals, the Monge systems $\mathcal{F}$ and $\mathcal{G}$ might well do so; or to be more precise, each may admit at most three functionally independent ones.

1. Suppose first that $\mathcal{F}$ admits one first integral f with $G_2 f \neq 0$. Then we look for complete systems (F, G) for which $Gf = 0$, i.e., $G_1 f + b G_2 f = 0$—so that $b = -G_1 f / G_2 f$. Because $F_1 f = F_2 f = (G_1 + b G_2) f = 0$, also

$$[F, G]f = [F_1 + a F_2, G_1 + b G_2]f = 0,$$

whereupon the previous formula for $[F, G]$ implies that

$$(Fb - \beta)G_2 f = 0.$$

Thus b indeed satisfies the desired equality $Fb = \beta$, and there only remains to find an a satisfying $Ga = \alpha$, i.e.,

$$(G_1 + b G_2)a = (\lambda^2 - \lambda^1)^{-1}\{a(F_1 + a F_2)\lambda^2 - a(G_1 + b G_2)\lambda^1$$
$$+ (F_1 + a F_2 - G_1 - b G_2) \circ Y\rho + b(F_1 + a F_2)\lambda^1 - a(G_1 + b G_2)\lambda^2\}.$$

By the rectification lemma for vector fields it is possible to introduce new local coordinates $t^1, \ldots, t^7$ for the underlying manifold such that the known vector field $G_1 + b G_2$ goes over into $\partial/\partial t^1$. With $t = (t^1, \ldots, t^7)$, a is then a solution of an ODE depending on the six parameters $t^2, \ldots, t^7$ and involving known fuctions g_2, g_1 and g_0:

$$\frac{\partial a}{\partial t^1} = a^2 g_2(t) + a g_1(t) + g_0(t).$$

So in this way it is possible to find those complete vector field systems (F, G) for which $Gf = 0$, without invoking the Cauchy–Kowalewski theorem.

2. Suppose next that $\mathcal{F}$ admits two functionally independent first integrals f^1 and f^2. Let $\mathcal{N}_2$ be a 2-dimensional integral manifold belonging to the foliation given by a complete subsystem $(F_1 + a\,F_2, G_1 + b\,G_2)$ of $\mathcal{V}$. The vector field $F_1 + a\,F_2$ has $7 - 1 = 6$ functionally independent first integrals, of which f^1 and f^2 are two; call the others $f^3,\ldots,f^6$. The complete system $(F_1 + a\,F_2, G_1 + b\,G_2)$ admits $7 - 2 = 5$, which we denote by $g^1,\ldots,g^5$. Because the latter in particular are first integrals of $F_1 + a\,F_2$, they are functions of $f^1,\ldots,f^6$:

$$g^k = \gamma^k(f^1,\ldots,f^6) \quad \text{for } k = 1,\ldots,5.$$

The defining equations of $\mathcal{N}_2$ are

$$\gamma^k(f^1,\ldots,f^6) = c^k, \quad k = 1,\ldots,5,$$

for certain constants c^k. Elimination of $f^3,\ldots,f^6$ leaves one equation

$$\gamma(f^1,f^2) = 0$$

defining a hypersurface which contains $\mathcal{N}_2$. With f^1 and f^2 being first integrals of $\mathcal{F}$, $f := \gamma(f^1,f^2)$ also is. Using this first integral f in point **1.** above we obtain a family of 2-dimensional integral manifolds, including the $\mathcal{N}_2$ we started with. Thus:

> If $\mathcal{F}$ admits two functionally independent first integrals f^1 and f^2, each integral manifold of $\mathcal{V}$ on which $dx \wedge dy \neq 0$ is found by using a first integral of the form $\gamma(f^1, f^2)$ in **1.**, where γ is an arbitrary function. Thus the general integral manifold depends on 'one arbitrary function of two variables'.

3. If $\mathcal{F}$ admits a first integral f with $G_2 f \neq 0$, and $\mathcal{G}$ admits a first integral g with $F_2 g \neq 0$, a and b can be determined in such a way that $(F_1 + a\,F_2, G_1 + b\,G_2)$ becomes complete by requiring $F_1 + a\,F_2$ to kill g, and $G_1 + b\,G_2$ to kill f:

$$(F_1 + a\,F_2)g = 0 \Longrightarrow a = -\frac{F_1 g}{F_2 g} \quad \text{and} \quad (G_1 + b\,G_2)f = 0 \Longrightarrow b = -\frac{G_1 f}{G_2 f}.$$

With $F := F_1 + a\,F_2$ and $G := G_1 + b\,G_2$ we then have

$$[F, G]f = 0 \quad \text{and} \quad [F, G]g = 0,$$

whence the general formula for $[F, G]$ reveals that

$$(\alpha - Ga)F_2 g = 0 \quad \text{and} \quad (Fb - \beta)G_2 f = 0.$$

Therefore $Ga = \alpha$ and $Fb = \beta$, so that (F, G) really is complete.

The general conclusion is

> *the more functionally independent first integrals that the Monge systems admit, the greater is the chance of integrating $\mathcal{V}$!*

But if neither $\mathcal{F}$ nor $\mathcal{G}$ admits any nonconstant first integrals? Then the idea is to *prolong* $\mathcal{V}$, and hope that the prolonged Monge systems might admit enough first integrals.

The general 2-dimensional involution $\mathcal{I}_2$ of $\mathcal{V}$ is generated by

$$F = F_1 + a\,F_2 \quad \text{and} \quad G = G_1 + b\,G_2,$$

where a and b are arbitrary functions on a 7-dimensional manifold which we locally identify with $\mathbb{C}^7$. To prolong $\mathcal{V}$ we consider a and b as *new variables* instead, so that F and G become *determined vector fields* on $\mathbb{C}^7 \times \mathbb{C}^2_{ab} \cong \mathbb{C}^9$. The *first prolongation* $\mathcal{V}^{(1)}$ of $\mathcal{V}$ is defined as the vector field system on $\mathbb{C}^9$ generated by

$$F = F_1 + a\,F_2, \quad G = G_1 + b\,G_2, \quad \partial/\partial a \quad \text{and} \quad \partial/\partial b.$$

The structure of $\mathcal{V}^{(1)}$ is then, with the same α and β as before,

$$[F, G] = (\lambda^2 - \lambda^1)^{-1}(F\lambda^2 - F\lambda^1)(G - F) + \alpha F_2 - \beta G_2,$$
$$[\partial/\partial a, F] = F_2, \quad [\partial/\partial b, G] = G_2,$$

and the other brackets being equal to 0.

The general 2-dimensional involution $\mathcal{I}_2^{(1)}$ of $\mathcal{V}^{(1)}$ on which $dx \wedge dy \neq 0$ is then generated by

$$F + a^1\,\partial/\partial a + b^1\,\partial/\partial b \quad \text{and} \quad G + a^2\,\partial/\partial a + b^2\,\partial/\partial b,$$

where the coefficients a^1, a^2, b^1, b^2 are subject to the condition that the bracket of these vector fields has to vanish modulo $\mathcal{V}^{(1)}$:

$$0 = \left[F + a^1\frac{\partial}{\partial a} + b^1\frac{\partial}{\partial b}, G + a^2\frac{\partial}{\partial a} + b^2\frac{\partial}{\partial b}\right] \equiv \alpha F_2 - \beta G_2 - a^2 F_2 + b^1 G_2.$$

Hence $a^2 = \alpha$, $b^1 = \beta$, while the functions a^1 and b^2 are left arbitrary. Setting

$$F^{(1)} := F + \beta\,\partial/\partial b \quad \text{and} \quad G^{(1)} := G + \alpha\,\partial/\partial a$$

we can write $\mathcal{V}^{(1)} = (F^{(1)}, G^{(1)}, \partial/\partial a, \partial/\partial b)$, and it is easily seen that $\mathcal{V}^{(1)}$ has the Monge characteristic subsystems $\mathcal{F}^{(1)} = (F^{(1)}, \partial/\partial a)$ and $\mathcal{G}^{(1)} = (G^{(1)}, \partial/\partial b)$. Furthermore the general 2-dimensional involution $\mathcal{I}_2^{(1)}$ of $\mathcal{V}^{(1)}$ on which $dx \wedge dy \neq 0$ is generated by

$$F^{(1)} + c\,\partial/\partial a \quad \text{and} \quad G^{(1)} + d\,\partial/\partial b,$$

where c and d are arbitrary functions on $\mathbb{C}^9$. The complete 2-dimensional subsystems of $\mathcal{V}$ are then obtained by determining c and d such that

$$\left[F^{(1)} + c\frac{\partial}{\partial a}, G^{(1)} + d\frac{\partial}{\partial b}\right] \equiv 0 \quad (\text{mod } (F^{(1)} + c\,\partial/\partial a, G^{(1)} + d\,\partial/\partial b)).$$

But this situation is completely analogous to that treated before, with F_1, F_2, G_1, G_2 replaced by $F^{(1)}$, $\partial/\partial a$, $G^{(1)}$ and $\partial/\partial b$ respectively. So if $\mathcal{F}^{(1)}$ (or $\mathcal{G}^{(1)}$) admits first integrals, these can be used in order to determine 2-dimensional subsystems of $\mathcal{V}^{(1)}$. The corresponding leaves are prolongations to $\mathbb{C}^9$ of the wanted integral manifolds of $\mathcal{V}$, and the latter are obtained by means of the projection $\mathbb{C}^7 \times \mathbb{C}^2_{ab} \longrightarrow \mathbb{C}^7$—in analogy with what we did in section **4.2**.

If f is a first integral of $\mathcal{F}$ (or $\mathcal{G}$)—and hence lives on $\mathbb{C}^7$—it might as well be regarded as a function on $\mathbb{C}^7 \times \mathbb{C}^2_{ab}$ satisfying $\partial f/\partial a = \partial f/\partial b = 0$, and as such it is a first integral of $\mathcal{F}^{(1)}$ (or $\mathcal{G}^{(1)}$) too. So all first integrals of $\mathcal{F}$ (or $\mathcal{G}$) are also first integrals of $\mathcal{F}^{(1)}$ (or $\mathcal{G}^{(1)}$)—but the converse need of course not be true. Because of this it can be easier to solve the prolonged problem than the original one.

If $\mathcal{F}^{(1)}$ and $\mathcal{G}^{(1)}$ do not admit enough first integrals we continue the prolongation process. Next time c and d are considered as new variables (instead of being regarded as functions on $\mathbb{C}^9$), and we define the second prolongation

$$\mathcal{V}^{(2)} := (F^{(1)} + c\,\partial/\partial a, G^{(1)} + d\,\partial/\partial b, \partial/\partial c, \partial/\partial d)$$

as a vector field system on $\mathbb{C}^9 \times \mathbb{C}^2_{cd} \cong \mathbb{C}^{11}$. The structure of $\mathcal{V}^{(2)}$ is

$$\left[F^{(1)} + c\frac{\partial}{\partial a}, G^{(1)} + d\frac{\partial}{\partial b}\right] = \frac{F\lambda^2 - F\lambda^1}{\lambda^2 - \lambda^1}\left(G^{(1)} + d\frac{\partial}{\partial b} - F^{(1)} - c\frac{\partial}{\partial a}\right)$$

$$+ \alpha^1\frac{\partial}{\partial u} - \beta^1\frac{\partial}{\partial b} \quad \text{with } \alpha^1 = (\lambda^2 - \lambda^1)^{-1}(F\lambda^2 - F\lambda^1)(c - \alpha) + F^{(1)}\alpha$$

$$\text{and } \beta^1 = (\lambda^2 - \lambda^1)^{-1}(F\lambda^2 - F\lambda^1)(d - \beta) + F^{(1)}\beta,$$

$$\left[\frac{\partial}{\partial c}, F^{(1)} + c\frac{\partial}{\partial a}\right] = \frac{\partial}{\partial a}, \quad \left[\frac{\partial}{\partial d}, G^{(1)} + d\frac{\partial}{\partial b}\right] = \frac{\partial}{\partial b},$$

and all other brackets being equal to 0.

Using this it easily follows that the general 2-dimensional involution of $\mathcal{V}^{(2)}$ on which $dx \wedge dy \neq 0$ is generated by

$$F^{(1)} + \beta^1\,\partial/\partial d + e\,\partial/\partial c \quad \text{and} \quad G^{(1)} + \alpha^1\,\partial/\partial c + f\,\partial/\partial d,$$

where e and f are arbitrary functions on $\mathbb{C}^{11}$. Setting

$$F^{(2)} := F^{(1)} + \beta^1 \, \partial/\partial d \quad \text{and} \quad G^{(2)} := G^{(1)} + \alpha^1 \, \partial/\partial c$$

we obtain the Monge characteristic subsystems $\mathcal{F}^{(2)} = (F^{(2)}, \, \partial/\partial c)$ and $\mathcal{G}^{(2)} = (G^{(2)}, \partial/\partial d)$ of $\mathcal{V}^{(2)} = (F^{(2)}, G^{(2)}, \partial/\partial c, \partial/\partial d)$—and the situation is of the same type as encountered before.

If $\mathcal{F}^{(2)}$ or $\mathcal{G}^{(2)}$ does not admit sufficiently many first integrals, the same kind of prolongation procedure is repeated once again.

Although it is explained only in a special case here it should be clear that this prolongation method in principle can be applied whenever there are any Monge systems. But—is it sure that it ultimately leads to prolonged Monge systems having enough first integrals?

Unfortunately this is *not* the case; there is for instance a counterexample in section **51** of [Cartan 1911].

12

Hyperbolic PDEs with Monge systems admitting two or three first integrals

We have seen that each of the Monge systems associated to a hyperbolic PDE S in one dependent and two independent variables admits at most three functionally independent first integrals, and also that it is advantageous to have many first integrals. Following [Vessiot 1939] rather closely we will in this chapter investigate the most favourable cases—namely when each Monge system has two or three functionally independent first integrals.

Quite unexpectedly the theory of the corresponding hyberbolic PDEs turns out to be totally dominated by Lie groups—for which reason Cartan described Vessiot's discoveries as 'tout à fait remarquable' in [Cartan 1946].

Since the journey towards the main results is rather long, we present the highlights at once.

1. Suppose first that the Monge system $\mathcal{F} = (F_1, F_2)$ admits the first integrals v, v^0, and that the Monge system $\mathcal{G} = (G_1, G_2)$ admits the first integrals u, u^0. If we introduce these as local coordinates for S together with appropriate functions w^1, w^2, w^3, the generators of the vector field system $\mathcal{V}$ can be simplified to

$$\begin{cases} F_1 = \partial/\partial u + \sum_{i=1}^3 \phi^i(u, u^0) R_i, & F_2 = \partial/\partial u^0, \\ G_1 = \partial/\partial v + \sum_{i=1}^3 \psi^i(v, v^0) L_i, & G_2 = \partial/\partial v^0, \end{cases}$$

where $\operatorname{span}_{\mathbb{C}}\{R_1, R_2, R_3\} = \mathfrak{g}_R$ and $\operatorname{span}_{\mathbb{C}}\{L_1, L_2, L_3\} = \mathfrak{g}_L$ are the right- and left-invariant Lie algebras of a local Lie group $\mathfrak{G}$ acting on $\mathbb{C}^3_w$, and ϕ^i, ψ^i are arbitrary functions.

And just as in the case of 2-dimensional Lie vector field systems, the integration of $\mathcal{V}$ is reduced to solving Lie equations associated to $\mathfrak{g}_R$ and $\mathfrak{g}_L$.

2. Suppose next that $\mathcal{F}$ admits the first integrals v, v^0, and that $\mathcal{G}$ admits the first integrals u, u^0, u^1. Using these as local coordinates together with suitably chosen functions w^1 and w^2, the generators of $\mathcal{V}$ can be reduced to

$$\begin{cases} F_1 = \partial/\partial u + u^1\,\partial/\partial u^0 + \sum_{i=1}^2 \phi^i(u, u^0, u^1)\,R_i, & F_2 = \partial/\partial u^1, \\ G_1 = \partial/\partial v + v^0 \sum_{i=1}^2 \psi^i(v)\,L_i, & G_2 = \partial/\partial v^0, \end{cases}$$

where $\operatorname{span}_{\mathbb{C}}\{R_1, R_2\} = \mathfrak{g}_R$ and $\operatorname{span}_{\mathbb{C}}\{L_1, L_2\} = \mathfrak{g}_L$ are the right- and left-invariant Lie algebras of a 2-dimensional Lie group $\mathfrak{G}$ acting on $\mathbb{C}^2_w$, and ϕ^i, ψ^i are arbitrary functions.

3. Suppose finally that each Monge system admits three functionally independent first integrals. Then it is possible to introduce local coordinates x, y, z, p, q, r, t for $\mathcal{S}$ such that $\mathcal{V}$ gets generated by

$$\begin{cases} F_1 = \partial/\partial x + p\,\partial/\partial z + r\,\partial/\partial p, & F_2 = \partial/\partial r, \\ G_1 = \partial/\partial y + q\,\partial/\partial z + t\,\partial/\partial q, & G_2 = \partial/\partial t, \end{cases}$$

which is nothing but the vector field system associated to the wave equation $s = 0$.

There is still a lot of arbitrariness in the formulas above because of the functions ϕ^i and ψ^i. But these latter can be made more specific if there are first integrals of the *first order*—a concept which is explained in the next section.

12.1 First integrals of the first order

Let $\mathcal{S}$ be a hypersurface in $J^2(\mathbb{C}^2_{xy}, \mathbb{C}_z)$ which projects onto $J^1(\mathbb{C}^2_{xy}, \mathbb{C}_z)$, and hence can be described in the following parametric form:

$$\mathcal{S}: \quad \begin{cases} r = \rho(x, y, z, p, q, u, v), \\ s = \sigma(x, y, z, p, q, u, v), \\ t = \tau(x, y, z, p, q, u, v). \end{cases}$$

Here x, y, z, p, q, u, v are used as local coordinates for $\mathcal{S}$, and x, y, z, p, q as local coordinates for $J^1(\mathbb{C}^2_{xy}, \mathbb{C}_z)$. If we set

$$X := \partial/\partial x + p\,\partial/\partial z + \rho\,\partial/\partial p + \sigma\,\partial/\partial q$$

and

$$Y := \partial/\partial y + q\,\partial/\partial z + \sigma\,\partial/\partial p + \tau\,\partial/\partial q,$$

the vector field system $\mathcal{V}$ associated to S is given by

$$\mathcal{V} = (X, Y, \partial/\partial u, \partial/\partial v).$$

Its derivative

$$\mathcal{V}' = (\partial/\partial x + p\,\partial/\partial z, \partial/\partial y + q\,\partial/\partial z, \partial/\partial p, \partial/\partial q, \partial/\partial u, \partial/\partial v)$$

has the 2-dimensional Cauchy characteristic subsystem

$$C(\mathcal{V}') = (\partial/\partial u, \partial/\partial v).$$

Note that $C(\mathcal{V}')$ is a subsystem of $\mathcal{V}$, and as such it is called the *principal subsystem* of $\mathcal{V}$. Moreover

$$\mathcal{V}'_{red} = (\partial/\partial x + p\,\partial/\partial z, \partial/\partial y + q\,\partial/\partial z, \partial/\partial p, \partial/\partial q)$$
$$= \text{ the contact vector field system of } J^1(\mathbb{C}^2_{xy}, \mathbb{C}^1_z).$$

In order to make life as easy as possible we prefer to use the *singular vector fields* F_1, F_2, G_1, G_2 as basis for $\mathcal{V}$ instead of $X, Y, \partial/\partial u, \partial/\partial v$. Recall from section **11.2** that when S is given by $r = \rho(x, y, z, p, q, s, t,)$— i.e., $u = s$ and $v = t$—these are given by

$$\begin{cases} F_1 = X + \lambda^1 Y + Y\rho\,\partial/\partial s, & F_2 = \partial/\partial t - \lambda^2\,\partial/\partial s, \\ G_1 = X + \lambda^2 Y + Y\rho\,\partial/\partial s, & G_2 = \partial/\partial t - \lambda^1\partial/\partial s, \end{cases}$$

where $\lambda^1 \neq \lambda^2$ are the characteristic roots. The Lie structure is

$$\begin{cases} [F_i, G_j] \equiv 0 & \text{for } i, j \neq 0, \\ [F_1, F_2] \equiv (\lambda^2 - \lambda^1)(\partial/\partial q - \lambda^2\,\partial/\partial p) = F_3, & (\text{mod } \mathcal{V}), \\ [G_1, G_2] \equiv (\lambda^1 - \lambda^2)(\partial/\partial q - \lambda^1\partial/\partial p) = G_3 \end{cases}$$

so that $\mathcal{V}' = (F_1, F_2, G_1, G_2; F_3, G_3)$ with $C(\mathcal{V}') = (F_2, G_2) \subset \mathcal{V}$.

This generalizes immediately to the general case where S is given by means of the parameters u and v—we can find singular vector fields F_1, F_2, G_1, G_2, such that $\mathcal{V} = (F_1, F_2, G_1, G_2)$, with $\mathcal{V}$ having the Monge systems $\mathcal{F} = (F_1, F_2)$, $\mathcal{G} = (G_1, G_2)$, and the structure

$$\begin{cases} [F_i, G_i] \equiv 0 & \text{for } i, j = 1, 2 \quad (\text{mod } \mathcal{V}), \\ [F_1, F_2] = F_3, & [G_1, G_2] = G_3, \end{cases}$$

where F_3 and G_3 are linearly independent modulo $\mathcal{V}$; moreover $\mathcal{V}' = (F_1, F_2, G_1, G_2; F_3, G_3)$ with $C(\mathcal{V}') = (F_2, G_2) = (\partial/\partial u, \partial/\partial v) \subset \mathcal{V}$.

The 7-dimensional $S \subset J^2(\mathbb{C}^2_{xy}, \mathbb{C}_z)$ gets mapped onto $J^1(\mathbb{C}^2_{xy}, \mathbb{C}_z)$ under the natural projection $J^2(\mathbb{C}^2_{xy}, \mathbb{C}_z) \longrightarrow J^1(\mathbb{C}^2_{xy}, \mathbb{C}_z)$. The coordinates

x, y, z, p, q for $J^1(\mathbb{C}^2_{xy}, \mathbb{C}_z)$ are called *first order* jet coordinates, while r, s and t are of the *second order*.

Let f be a first integral of one of the Monge systems. Then f is said to be of the *first order* if $\partial f / \partial u = \partial f / \partial v = 0$, so that f can be thought of as living on $J^1(\mathbb{C}^2_{xy}, \mathbb{C}_z)$. More intrinsically this may be expressed as follows.

Definition. Let f be a first integral of $\mathcal{F}$ or $\mathcal{G}$. Then f is of the *first order* if f is also a first integral of the principal subsystem $C(\mathcal{V}')$ of $\mathcal{V}$.

In the preceding chapter it was shown that if we conversely start from a vector field system $\mathcal{V} = (F_1, F_2, G_1, G_2)$ defined on a 7-dimensional manifold and having the properties above, then it is possible to introduce local coordinates x, y, z, p, q, u, v so that $\mathcal{V}$ becomes associated to a hyperbolic PDE $\mathcal{S} \subset J^2(\mathbb{C}^2_{xy}, \mathbb{C}_z)$ in parameter form.

We now want to show that *if one of the Monge systems admits a first integral of the first order, then $\mathcal{S}$ can be expressed by means of an equation involving only two of the second order jet coordinates r, s and t.*

Furthermore, *if each of the Monge systems admits a first integral of the first order, then it is possible to find a defining equation of $\mathcal{S}$ which only contains one of the second order jet coordinates.*

To do this we must first find local coordinates x, y, z, p, q, u, v such that $\mathcal{V}' = (F_1, F_2, G_1, G_2; F_3, G_3)$ goes over into $\mathcal{V}' = (\partial/\partial x + p\,\partial/\partial z, \partial/\partial y + q\,\partial/\partial z, \partial/\partial p, \partial/\partial q, \partial/\partial u, \partial/\partial v)$.

To begin with we note that since $C(\mathcal{V}') = (F_2, G_2)$ is the Cauchy characteristic subsystem of $\mathcal{V}'$, it is surely complete—that is, there are functions f and g such that

$$[F_2, G_2] = f\, F_2 + g\, G_2.$$

Modifying F_2 and G_2 to $\tilde{F}_2 := a F_2$ and $\tilde{G}_2 := b G_2$, we want to determine the functions a and b such that $\tilde{F}_2$ and $\tilde{G}_2$ commute:

$$[\tilde{F}_2, \tilde{G}_2] = [a F_2, b G_2] = b\,(af - G_2 a)\, F_2 + a\,(bg + F_2 b)\, G_2 = 0.$$

Using the rectification lemma for vector fields the equations $G_2 a = fa$ and $F_2 b = -gb$ go over into ODEs, and can thus be solved. Therefore we may assume that F_2 and G_2 span an abelian 2-dimensional Lie algebra, and accordingly can be written as $F_2 = \partial/\partial u$ and $G_2 = \partial/\partial v$.

The remaining coordinates x, y, z, p, q can be found by means of the theory in chapter **6** since $\dim \mathcal{V}'' = \dim \mathcal{V}' + 1$, and $\mathcal{V}''$ is complete.

First we choose x as a nonconstant first integral of $C(\mathcal{V}')$. With $\mathcal{V}'_x :=$

$\{V \in \mathcal{V} \mid Vx = 0\}$ we then choose y as a first integral of $C(\mathcal{V}'_x)$ satisfying $dx \wedge dy \neq 0$. Setting $\mathcal{V}'_{xy} := \{V \in \mathcal{V} \mid Vx = Vy = 0\}$, we choose z as a first integral of $C(\mathcal{V}'_{xy})$ which satisfies $dx \wedge dy \wedge dz \neq 0$. After that, p and q are determined as the polar functions of x, y and z.

Expressing the F_i and G_i in these new coordinates they must be linear combinations of X, Y, $\partial/\partial u$ and $\partial/\partial v$. In particular

$$\begin{cases} F_1 = f^1 X + f^2 Y + f^3 \partial/\partial u + f^4 \partial/\partial v, \\ G_1 = g^1 X + g^2 Y + g^3 \partial/\partial u + g^4 \partial/\partial v, \end{cases}$$

where the functions f^i and g^i are to satisfy $f^1 g^2 - f^2 g^1 \neq 0$.

Suppose now that $\mathcal{G} = (G_1, G_2)$ admits a first integral of the first order. The last property means precisely that it is a first integral of $C(\mathcal{V}') = (\partial/\partial u, \partial/\partial v)$, and hence this function may be taken as our x. Because $0 = G_1 x = g^1$, G_1 is reduced to $g^2 Y + g^3 \partial/\partial u + g^4 \partial/\partial v$. The requirement $f^1 g^2 - f^2 g^1 \neq 0$ forces g^2 to be nonzero, and therefore G_1 may be simplified to

$$G_1 = Y + a\,\partial/\partial u + b\,\partial/\partial v.$$

Next the condition $[G_1, F_2] \equiv 0 \pmod{\mathcal{V}}$ implies that

$$0 \equiv [F_2, G_1] = \left[\frac{\partial}{\partial u}, \frac{\partial}{\partial y} + q\frac{\partial}{\partial z} + \sigma\frac{\partial}{\partial p} + \tau\frac{\partial}{\partial q} + a\frac{\partial}{\partial u} + b\frac{\partial}{\partial v}\right]$$

$$= \frac{\partial\sigma}{\partial u}\frac{\partial}{\partial p} + \frac{\partial\tau}{\partial u}\frac{\partial}{\partial q} + \text{terms not involving } \frac{\partial}{\partial p} \text{ or } \frac{\partial}{\partial q},$$

so that $\partial\sigma/\partial u = \partial\tau/\partial u = 0$. Consequently

$$\begin{cases} s = \sigma(x, y, z, p, q, v), \\ t = \tau(x, y, z, p, q, v). \end{cases}$$

Elimination of v shows that S is defined by an equation $F(x, y, z, p, q, s, t) = 0$, where r is absent.

And an analogous conclusion if $\mathcal{F}$ admits a first integral of the first order.

Theorem 12.1.1. *If one of the Monge systems admits a first integral of the first order, the defining equation of the PDE S can be expressed with only two of the second order jet coordinates r, s and t.* □

Still assuming that x is a first integral of the first order for $\mathcal{G}$, and that

$$G_1 = Y + a\,\partial/\partial u + b\,\partial/\partial v,$$

we may assume that

$$F_1 = X + cY + d\,\partial/\partial u + e\,\partial/\partial v$$

for suitable functions c, d and e. Then

$$F_3 = [F_1, F_2] = [\partial/\partial x + \cdots, \partial/\partial u]$$

does not contain $\partial/\partial x$, and neither does G_3. Hence

$$V'_x = (F_2, G_1, G_2, F_3, G_3).$$

By the general theory $\dim C(V'_x) = \dim C(V') + 1 = 3$, so we have to find three generators for $C(V'_x)$. It is immediately seen that $F_2 = \partial/\partial u$ commutes with G_1, G_2, F_3 and G_3 (mod V'_x), and that $G_2 = \partial/\partial v$ commutes with F_2, G_1, F_3, G_3 (mod V'_x), whence $F_2, G_2 \in C(V'_x)$.

Claim. F_3 belongs to $C(V'_x)$ too, so that $C(V'_x) = (F_2, G_2, F_3)$.

Proof. By the above $[F_2, F_3] \equiv [G_2, F_3] \equiv 0$ (mod V'_x), and at the end of section **11.3** it was seen that $[G_1, F_3] \equiv 0$ (mod V'); since neither G_1 nor F_3 contains $\partial/\partial x$, this also shows that $[G_1, F_3] \equiv 0$ (mod V'_x).

So there remains to show that $[G_3, F_3] \equiv 0$ (mod V'_x). By Jacobi's identity,

$$[G_3, F_3] = [G_3, [F_1, F_2]] = [F_2, [F_1, G_3]] - [F_1, [F_2, G_3]].$$

Here $[F_1, G_3] \in V'$ by section **11.3**, and since $F_2 \in C(V')$, also $[F_2, [F_1, G_3]] \in V'$. In the last term above

$$-G_3 = [G_2, G_1] = \left[\frac{\partial}{\partial v}, Y + a\frac{\partial}{\partial u} + b\frac{\partial}{\partial v}\right]$$

$$= \frac{\partial\sigma}{\partial v}\frac{\partial}{\partial p} + \frac{\partial\tau}{\partial v}\frac{\partial}{\partial q} + \frac{\partial a}{\partial v}\frac{\partial}{\partial u} + \frac{\partial b}{\partial v}\frac{\partial}{\partial v}.$$

Because $\partial\sigma/\partial u = \partial\tau/\partial u = 0$, this gives

$$-[F_2, G_3] = \left[\frac{\partial}{\partial u}, -G_3\right] = \frac{\partial^2 a}{\partial u\partial v}\frac{\partial}{\partial u} + \frac{\partial^2 b}{\partial u\partial v}\frac{\partial}{\partial v} \in V,$$

whence finally

$$[F_1, [F_2, G_3]] \in V'.$$

Thus $[G_3, F_3] \equiv 0$ (mod V'), and since none of G_3, F_3 contains $\partial/\partial x$, $[G_3, F_3]$ leaves x invariant—so that $[G_3, F_3] \equiv 0$ (mod V'_x). ☐

Suppose now also that $\mathcal{F}$ has a first integral of the first order. Being a first integral of $\mathcal{F}$, it is killed by F_1, F_2 and F_3, and being of first order it is killed by $G_2 = \partial/\partial v$. Thus it is a first integral of $C(\mathcal{V}_x)$, and may therefore be chosen as the wanted y.

Because $0 = F_1 y = c$,

$$F_1 = X + d\,\partial/\partial u + e\,\partial/\partial v.$$

The condition $[G_2, F_1] = [\partial/\partial v, F_1] \equiv 0 \pmod{\mathcal{V}}$ then implies that

$$\partial\rho/\partial v = \partial\sigma/\partial v = 0.$$

Combining this with the earlier property $\partial\sigma/\partial u = \partial\tau/\partial u = 0$ we have

$$\partial\sigma/\partial u = \partial\sigma/\partial v = 0,$$

so that the PDE $\mathcal{S}$ is given by

$$s = \sigma(x, y, z, p, q).$$

Theorem 12.1.2. *If each Monge system admits a first integral of the first order, the defining equation of the PDE $\mathcal{S}$ can be written as*

$$s = \sigma(x, y, z, p, q).$$

□

Let us conversely start with a PDE $s = \sigma(x, y, z, p, q)$. The associated vector field system $\mathcal{V}$ is generated by

$$X_1 := X = \partial/\partial x + p\,\partial/\partial z + r\,\partial/\partial p + \sigma\,\partial/\partial q,$$
$$X_2 := Y = \partial/\partial y + q\,\partial/\partial z + \sigma\,\partial/\partial p + t\,\partial/\partial q,$$
$$X_3 := \partial/\partial r \quad \text{and} \quad X_4 := \partial/\partial t.$$

The structure of this $\mathcal{V}$ is

$$[X_1, X_2] = X\sigma\,\frac{\partial}{\partial p} - Y\sigma\,\frac{\partial}{\partial q}, \quad [X_1, X_3] = -\frac{\partial}{\partial p}, \quad [X_1, X_4] = 0,$$

$$[X_2, X_3] = 0, \quad [X_2, X_4] = -\frac{\partial}{\partial q} \quad \text{and} \quad [X_3, X_4] = 0.$$

Let $A = \sum_{i=1}^{4} a^i X_i$ and $B = \sum_{i=1}^{4} b^i X_i$. Then modulo $\mathcal{V}$,

$$[A, B] \equiv \sum_{1 \le i < j \le 4} (a^i b^j - a^j b^i)[X_i, X_j]$$

$$= ((a^3 - a^2 X \sigma)b^1 + a^1 b^2 X \sigma - a^1 b^3) \frac{\partial}{\partial p}$$

$$+ (-a^2 b^1 Y \sigma + (a^1 Y \sigma - a^4)b^2 + a^2 b^4) \frac{\partial}{\partial q}.$$

If we set $\mathbf{a} := (a^1, a^2, a^3, a^4)^T$ and $\mathbf{b} := (b^1, b^2, b^3, b^4)^T$ this shows that $[A, B] \equiv 0 \pmod{\mathcal{V}} \Longleftrightarrow M(\mathbf{a})\mathbf{b} = \mathbf{0}$, where

$$M(\mathbf{a}) = \begin{pmatrix} a^3 - a^2 X \sigma & a^1 X \sigma & -a^1 & 0 \\ -a^2 Y \sigma & a^1 Y \sigma - a^4 & 0 & a^2 \end{pmatrix}.$$

A is *singular* if and only if rank $M(\mathbf{a}) = 1$—which can happen in two ways:

1. Either $a^1 \ne 0$, in which case the second row in $M(\mathbf{a})$ vanishes, and then $a^2 = 0$ and $a^4 = Y \sigma \cdot a^1$. Therefore

$$A = a^1 \left(X + Y \sigma \frac{\partial}{\partial t} \right) + a^3 \frac{\partial}{\partial r},$$

and thus there are two singular vector fields,

$$F_1 = X + Y \sigma \frac{\partial}{\partial t} \quad \text{and} \quad F_2 = \frac{\partial}{\partial r},$$

which together span the Monge system $\mathcal{F}$.

2. Or $a^2 \ne 0$, so that the first row in $M(\mathbf{a})$ vanishes. Then $a^1 = 0$ and $a^3 = X \sigma \cdot a^2$, whence we obtain the singular vector fields

$$G_1 = Y + X \sigma \frac{\partial}{\partial r} \quad \text{and} \quad G_2 = \frac{\partial}{\partial t},$$

spanning the Monge system $\mathcal{G}$.

From this very explicit description of $\mathcal{F}$ and $\mathcal{G}$ it is immediately clear that

> y is a first integral of the first order for $\mathcal{F}$, and x is a first integral of the first order for $\mathcal{G}$!

So the Monge systems do admit first integrals of the first order in this case.

Notation. Hyperbolic PDEs $s = \sigma(x, y, z, p, q)$ having the property that each Monge system admits two or three functionally independent first integrals are called *hyperbolic Goursat equations*.

This name comes from Goursat's remarkable classification of such equations under coordinate transformations of the form

$$x \mapsto \xi(x), \quad y \mapsto \eta(y), \quad z \mapsto \zeta(x,y,z)$$

in [Goursat 1899]. Unfortunately the ad hoc methods used do not give any satisfactory explanation of *why* there is such a classification at all.

The riddle was solved in [Vessiot 1942], where the classification of hyperbolic Goursat equations under the equivalence described in the preceding chapter essentially is reduced to that of classifying 2- and 3-dimensional Lie algebras (!)—as will be explained in the next chapter.

12.2 Two first integrals for each Monge system

In this section it is supposed that each of the Monge systems $\mathcal{F} = (F_1, F_2)$ and $\mathcal{G} = (G_1, G_2)$ admits two functionally independent first integrals, and the goal is to find simple canonical forms for the singular vector fields F_1, F_2, G_1 and G_2.

Assume firstly that u, u^0 are first integrals of $\mathcal{G}$, and that $x^1, \ldots, x^5$ together with u, u^0 form a system of local coordinates for the underlying 7-dimensional manifold. If $\partial/\partial u$ (or $\partial/\partial u^0$) did not occur in F_1 and F_2, u (or u^0) would be a nonconstant first integral of $\mathcal{V} = (F_1, F_2, G_1, G_2)$, which is not possible. More precisely our vector fields must be of the following form:

$$\begin{cases} F_1 = \partial/\partial u + \sum_{i=1}^{5} f_1^i \left(\partial/\partial x^i\right), & F_2 = \partial/\partial u^0 + \sum_{i=1}^{5} f_2^i \left(\partial/\partial x^i\right), \\ G_1 = \sum_{i=1}^{5} g_1^i \left(\partial/\partial x^i\right), & G_2 = \sum_{i=1}^{5} g_2^i \left(\partial/\partial x^i\right). \end{cases}$$

For otherwise we have for instance

$$F_1 = \frac{\partial}{\partial u} + a \frac{\partial}{\partial u^0} + \sum_{i=1}^{5} f_1^i \frac{\partial}{\partial x^i} \quad \text{and} \quad F_2 = \sum_{i=1}^{5} f_2^i \frac{\partial}{\partial x^i},$$

where a must be nonzero—for if not, u^0 would be a first integral of $\mathcal{V}$. But then

$$[G_i, F_1] = G_i a \frac{\partial}{\partial u^0} + \cdots \equiv 0 \pmod{\mathcal{V}} \implies G_i a = 0 \text{ for } i = 1, 2,$$

so that a is a first integral of $\mathcal{G}$, i.e., a function of u and u^0: $a = \alpha(u, u^0)$.

However, in that case V does admit first integrals, namely all solutions $f(u, u^0)$ of the PDE

$$\left(\frac{\partial}{\partial u} + \alpha(u, u^0) \frac{\partial}{\partial u^0}\right) f = 0.$$

Next assume that $\mathcal{F}$ admits the first integrals v, v^0, and use u, u^0, v, v^0 as local coordinates together with three more functions x^1, x^2, x^3. Applying the reasoning above one more time it follows that V has generators of the form

$$\begin{cases} F_1 = \partial/\partial u + \sum_{i=1}^3 f_1^i \left(\partial/\partial x^i\right), & F_2 = \partial/\partial u^0 + \sum_{i=1}^3 f_2^i \left(\partial/\partial x^i\right), \\ G_1 = \partial/\partial v + \sum_{i=1}^3 g_1^i \left(\partial/\partial x^i\right), & G_2 = \partial/\partial v^0 + \sum_{i=1}^3 g_2^i \left(\partial/\partial x^i\right). \end{cases}$$

Because the coefficients in front of $\partial/\partial u$, $\partial/\partial u^0$, $\partial/\partial v$ and $\partial/\partial v^0$ all equal 1, the equivalences $[F_i, G_j] \equiv 0 \pmod{V}$ for $i, j = 1, 2$ go over into the equalities $[F_i, G_j] = 0$.

In particular $[F_2, G_2] = 0$—so that (F_2, G_2) is complete, and therefore admits $7 - 2 = 5$ functionally independent first integrals, including u and v; call the others w^1, w^2, w^3. Using $u, u^0, v, v^0, w^1, w^2, w^3$ as local coordinates we have

$$\begin{cases} F_1 = \partial/\partial u + \sum_{i=1}^3 \rho^i(u, u^0, v, v^0, w^1, w^2, w^3) \left(\partial/\partial w^i\right), & F_2 = \partial/\partial u^0, \\ G_1 = \partial/\partial v + \sum_{i=1}^3 \lambda^i(u, u^0, v, v^0, w^1, w^2, w^3) \left(\partial/\partial w^i\right), & G_2 = \partial/\partial v^0. \end{cases}$$

Then $[F_1, G_2] = 0 \implies \partial \rho^i/\partial v^0 = 0$, $[F_2, G_1] = 0 \implies \partial \lambda^i/\partial u^0 = 0$, and so

$$F_1 = \frac{\partial}{\partial u} + R \quad \text{and} \quad G_1 = \frac{\partial}{\partial v} + L$$

with

$$\begin{cases} R = \sum_{i=1}^3 \rho^i(u, u^0, v, w^1, w^2, w^3) \left(\partial/\partial w^i\right), \\ L = \sum_{i=1}^3 \lambda^i(u, v, v^0, w^1, w^2, w^3) \left(\partial/\partial w^i\right). \end{cases}$$

Let us now make the dependence of R on u^0 more explicit by employing the same device as used when analysing Lie vector field systems in section **10.2**.

From $[F_1, G_1] = 0$ it trivially follows that $[F_2, [F_1, G_1]] = 0$, and so

$$0 = [F_2, [F_1, G_1]] = -[G_1, [F_2, F_1]] - [F_1, [G_1, F_2]] = -[G_1, [\partial/\partial u^0, F_1]],$$

where

$$\left[\frac{\partial}{\partial u^0}, F_1\right] = \sum_{i=1}^3 \frac{\partial \rho^i}{\partial u^0} \frac{\partial}{\partial w^i} =: R^{(1)}.$$

Consequently $[G_1, R^{(1)}] = 0$, and by repeating this process $[G_1, R^{(k)}] = 0$ for $k = 1, 2, 3, \ldots$, with

$$R^{(k)} := \sum_{i=1}^{3} \frac{\partial^k \rho^i}{\partial (u^0)^k} \frac{\partial}{\partial w^i}.$$

And since the ρ^i are independent of v^0, also $[G_2, R^{(k)}] = 0$ for $k = 1, 2, 3, \ldots$.

Recall that we are only interested in local and generic properties. Considering the $R^{(k)}$ as vector fields on $\mathbb{C}^3_w$ depending on the parameters u, u^0 and v, we therefore conclude that at most three of the $R^{(k)}$ can be linearly independent; note moreover that $R^{(1)} = [F_2, F_1] = -F_3 \neq 0$.

Let us assume that $R^{(1)}, \ldots, R^{(m)}$ (with $m \le 3$) are linearly independent, while $R^{(m+1)}$ is a linear combination of these:

$$R^{(m+1)} = \sum_{k=1}^{m} a_k(u, u^0, v, w^1, w^2, w^3) \, R^{(k)}.$$

Then for $i = 1$ and 2,

$$0 = [G_i, R^{(m+1)}] = \sum_{k=1}^{m} G_i a_k \, R^{(k)} + \sum_{k=1}^{m} a_k \cdot 0,$$

so that all the coefficients a_k are first integrals of $\mathcal{G}$, i.e., $a_k = a_k(u, u^0)$ for $k = 1, \ldots, m$.

Without loss of generality we may assume that $m = 3$, for applying $\partial^2/\partial(u^0)^2$ and $\partial/\partial u^0$ to

$$R^{(2)} = a_1(u, u^0) \, R^{(1)} \quad \text{and} \quad R^{(3)} = \sum_{k=1}^{2} a_k(u, u^0) \, R^{(k)} \quad \text{respectively}$$

we end up with $R^{(4)} = \sum_{k=1}^{3} a_k(u, u^0) \, R^{(k)}$ anyway. Since this last equality is equivalent to

$$\frac{\partial^4 \rho^i}{\partial (u^0)^4} = \sum_{k=1}^{3} a_k(u, u^0) \frac{\partial^k \rho^i}{\partial (u^0)^k} \quad \text{for } i = 1, 2, 3,$$

the ρ^i are *solutions of a fourth order linear homogeneous ODE*. Denoting four linearly independent solutions of this by

$$\phi^0 \equiv 1, \quad \phi^1(u, u^0), \quad \phi^2(u, u^0) \quad \text{and} \quad \phi^3(u, u^0),$$

we can write the ρ^i as

$$\rho^i = \sum_{j=0}^{3} r^i_j(u, v, w^1, w^2, w^3) \phi^j(u, u^0), \qquad i = 1, 2, 3,$$

and therefore

$$R = \sum_{i=1}^{3} \rho^i \frac{\partial}{\partial w^i} = \sum_{j=0}^{3} \phi^j(u, u^0) \left(\sum_{i=1}^{3} r^i_j \frac{\partial}{\partial w^i} \right).$$

So with $R_j := \sum_{i=1}^{3} r^i_j(u, v, w^1, w^2, w^3) \left(\partial/\partial w^i \right)$ for $j = 0, 1, 2, 3$, we have

$$R = R_0 + \sum_{j=1}^{3} \phi^j(u, u^0) R_j.$$

That is, the dependence on u^0 has been moved to the coefficients $\phi^j(u, u^0)$. Moreover

$$F_1 = \frac{\partial}{\partial u} + R = \frac{\partial}{\partial u} + R_0 + \sum_{j=1}^{3} \phi^j(u, u^0) R_j.$$

This can be further simplified by considering

$$\frac{\partial}{\partial u} + R_0 = \frac{\partial}{\partial u} + \sum_{j=1}^{3} r^i_0(u, v, w^1, w^2, w^3) \frac{\partial}{\partial w^i},$$

which for each value of the parameter v may be thought of as a vector field on $\mathbb{C}^4_{u,w^1,w^2,w^3}$ and as such admits three functionally independent first integrals

$$\tilde{w}^i = \omega^i(u, v, w^1, w^2, w^3), \qquad i = 1, 2, 3.$$

Replacing w^1, w^2, w^3 by $\tilde{w}^1, \tilde{w}^2, \tilde{w}^3$, $\partial/\partial u + R_0$ goes over into $\partial/\partial u$, while the vector fields R_1, R_2, R_3, F_2, G_1 and G_2 preserve their form. So in these new coordinates

$$F_1 = \partial/\partial u + \sum_{j=1}^{3} \phi^j(u, u^0) R_j$$

with the R_j being independent of u^0. Since G_2 is independent of u^0 too, and the ϕ^j are linearly independent as functions of u^0 for $j = 0, 1, 2, 3$,

$$[F_1, G_1] = 0 \iff 1 \cdot [\partial/\partial u, G_1] + \sum_{i=1}^{3} \phi^j(u, u^0) [R_j, G_1] = 0$$

$$\iff [\partial/\partial u, G_1] = 0 \quad \text{and} \quad [R_j, G_1] = 0 \quad \text{for } j = 1, 2, 3.$$

Because $G_1 = \partial/\partial v + L$, the equality $[\partial/\partial u, G_1] = 0$ means that L is independent of u, and therefore

$$L = \sum_{i=1}^{3} \lambda^i(v, v^0, \tilde{w}^1, \tilde{w}^2, \tilde{w}^3) \frac{\partial}{\partial \tilde{w}^i}.$$

The next item is to make the dependence on v^0 here more explicit by using the same trick as before: from $[F_1, G_1] = 0$ we infer that $[F_1, L] = 0$; bracketing this with $\partial/\partial v^0$ repeatedly, we have

$$[L^{(k)}, F_1] = 0, \quad \text{where } L^{(k)} := \sum_{i=1}^{3} \frac{\partial^k \lambda^i}{\partial(v^0)^k} \frac{\partial}{\partial \tilde{w}^i}.$$

Reasoning as with the $R^{(k)}$ we can express L as

$$L = L_0 + \sum_{j=1}^{3} \psi^j(v, v^0) L_j$$

where the ψ^j are independent as functions of v^0, and where

$$L_j = \sum_{i=1}^{3} l_j^i(v, \tilde{w}^1, \tilde{w}^2, \tilde{w}^3) \left(\partial/\partial \tilde{w}^i\right) \text{ for } j = 0, 1, 2, 3.$$

Hence $G_1 = \partial/\partial v + L_0 + \sum_{j=1}^{3} \psi^j(v, v^0) L_j$, where the L_j do not depend on v^0.

$\partial/\partial v + L_0$ is a vector field on $\mathbb{C}^4_{v, \tilde{w}^1, \tilde{w}^2, \tilde{w}^3}$, which hence admits three functionally independent first integrals $\hat{w}^1, \hat{w}^2, \hat{w}^3$ locally. Introducing these as new local coordinates instead of the $\tilde{w}^i$, we reduce $\partial/\partial v + L_0$ to $\partial/\partial v$. Since the $\hat{w}^i$ do not depend on u, u^0 and v^0, F_2 and G_2 are left invariant, and F_1 and the R_j preserve their form. Forgetting the hats we then have

$$G_1 = \frac{\partial}{\partial v} + \sum_{j=1}^{3} \psi^j(v, v^0) L_j.$$

Because the ψ^j are independent as functions of v^0 for $j = 0, 1, 2, 3$, the commutativity of F_1 and G_1 implies that

$$[\partial/\partial v, F_1] = 0 \quad \text{and} \quad [L_j, F_1] = 0 \quad \text{for } j = 1, 2, 3.$$

The first equality shows that $\partial r_j^i/\partial v = 0$ for all i and j.

Let us summarize what we have achieved thus far: *The generators of*

$\mathcal{V}$ can be written as

$$\begin{cases} F_1 = \partial/\partial u + \sum_{j=1}^{3} \phi^j(u, u^0) \, R_j, & F_2 = \partial/\partial u^0, \\ G_1 = \partial/\partial v + \sum_{j=1}^{3} \psi^j(v, v^0) \, L_j, & G_2 = \partial/\partial v^0, \end{cases}$$

where

$$R_j = \sum_{i=1}^{3} r_j^i(u, w^1, w^2, w^3) \, (\partial/\partial w^i)$$

and

$$L_j = \sum_{i=1}^{3} l_j^i(v, w^1, w^2, w^3) \, (\partial/\partial w^i) \quad \text{for } j = 1, 2, 3.$$

Moreover $[R_j, G_i] = 0$ *and* $[L_j, F_i] = 0$ *for* $j = 1, 2, 3$ *and* $i = 1, 2$.

In the final step we want to express the R_j and L_j as linear combinations of vector fields on $\mathbb{C}_w^3$ with coefficients which depend on u and v respectively. Since these cases are analogous it suffices to consider the R_j—and of course we use the same trick as before. Setting

$$R_j^{(k)} := \sum_{i=1}^{3} \frac{\partial^k r_j^i}{\partial u^k} \frac{\partial}{\partial w^i}$$

we infer from $[R_j, G_i] = 0$ that $[R_j^{(k)}, G_i] = 0$ for $j = 1, 2, 3$, $i = 1, 2$ and $k = 0, 1, 2, \dots$. Considering $R_j^{(0)}, R_j^{(1)}, \dots$ as vector fields on $\mathbb{C}_w^3$ depending on the parameter u,

$$R_j^{(3)} = \sum_{k=0}^{2} a_{jk}(u, w^1, w^2, w^3) \, R_j^{(k)},$$

where the a_{jk} turn out to be first integrals of $\mathcal{G}$, and hence only depend on u. But then the r_j^i satisfy a third order linear homogeneous ODE with respect to u for $i = 1, 2, 3$. If $\phi_j^1(u)$, $\phi_j^2(u)$ and $\phi_j^3(u)$ are fundamental solutions of this, the r_j^i can be written as

$$r_j^i = \sum_{k=1}^{3} \rho_{jk}^i(w^1, w^2, w^3) \phi_j^k(u).$$

Hence

$$R_j = \sum_{i=1}^{3} r_j^i \left(\partial/\partial w^i\right) = \sum_{k=1}^{3} \phi_j^k(u) \left(\sum_{i=1}^{3} \rho_{jk}^i(w^1, w^2, w^3) \left(\partial/\partial w^i\right)\right)$$

$$= \sum_{k=1}^{3} \phi_j^k(u) R_{jk},$$

where the R_{jk} now are genuine vector fields on $\mathbb{C}_w^3$. Because G_1 does not depend on u, and $\phi_j^1(u)$, $\phi_j^2(u)$ and $\phi_j^3(u)$ are linearly independent, the equality $[R_j, G_1] = 0$ implies that

$$[R_{jk}, G_1] = 0 \quad \text{for } j, k = 1, 2, 3.$$

And clearly $[R_{jk}, G_2] = 0$ too. Now at most three of the R_{jk} can be linearly independent locally. Letting $\{\tilde{R}_l\}$ be a basis for these vector fields, we can write

$$R_{jk} = \sum_l b_{jk}^l(w^1, w^2, w^3)\, \tilde{R}_l \quad \text{for } j, k = 1, 2, 3.$$

Commuting both sides with G_1 and G_2 it follows that the b_{jk}^l are first integrals of $\mathcal{G}$, i.e., functions of u and u^0—which is possible only if the b_{jk}^l are *constants*. Therefore

$$F_1 = \frac{\partial}{\partial u} + \sum_{j=1}^{3} \phi^j(u, u^0)\, R_j = \frac{\partial}{\partial u} + \sum_{j=1}^{3} \phi^j(u, u^0) \left(\sum_{k=1}^{3} \phi_j^k(u) \left(\sum_l b_{jk}^l \tilde{R}_l\right)\right)$$

$$= \frac{\partial}{\partial u} + \sum_j \tilde{\phi}^j(u, u^0)\, \tilde{R}_j,$$

where the $\tilde{R}_j$ are vector fields on $\mathbb{C}_w^3$.

If there were just one $\tilde{R}_j$, $\mathcal{F}$ would admit the two functionally independent first integrals of this on $\mathbb{C}_w^3$, which is impossible. And if there are two $\tilde{R}_j$ the vector field system $(\tilde{R}_1, \tilde{R}_2)$ cannot be complete, since it would admit a first integral on $\mathbb{C}_w^3$ otherwise, which also would be a first integral of $\mathcal{F}$.

Hence either there are three $\tilde{R}_j$, or there are only two, and in that case $[\tilde{R}_1, \tilde{R}_2] \notin (\tilde{R}_1, \tilde{R}_2)$.

The functions $\tilde{\phi}^j$ above may be assumed to be linearly independent, because otherwise the number of $\tilde{R}_j$ can be reduced. But then the equality $[F_1, G_1] = 0$ implies that $[\tilde{R}_j, G_1] = 0$ for each j. And obviously also $[\tilde{R}_j, G_2] = 0$.

In the same way G_1 can be simplified to

$$G_1 = \frac{\partial}{\partial v} + \sum_k \tilde{\psi}^k(v, v^0) \tilde{L}_k,$$

where the $\tilde{\psi}^k$ are linearly independent. From $[\tilde{R}_j, G_1] = 0$ it then follows that $[\tilde{R}_j, \tilde{L}_k] = 0$.

1. Suppose first that there are three $\tilde{R}_j$, which are linearly independent over the ring of functions. Being a 3-dimensional vector field system on $\mathbb{C}_w^3$, $(\tilde{R}_1, \tilde{R}_2, \tilde{R}_3)$ is locally complete and hence there are functions $c_{ij}^k(w^1, w^2, w^3)$ with the property that

$$[\tilde{R}_i, \tilde{R}_j] = \sum_{k=1}^{3} c_{ij}^k \tilde{R}_k \quad \text{for } i, j = 1, 2, 3.$$

Taking brackets of both sides with G_1 and G_2 it follows that the c_{ij}^k are first integrals of $\mathcal{G}$, and thus must be *constants*. Consequently $\operatorname{span}_\mathbb{C}\{\tilde{R}_1, \tilde{R}_2, \tilde{R}_3\}$ is a *Lie algebra* of vector fields on $\mathbb{C}_w^3$, which we denote by $\mathfrak{g}_R$.

2. Suppose next that there are only two $\tilde{R}_j$ present: $\tilde{R}_1$ and $\tilde{R}_2$. As remarked above, $\tilde{R}_3 := [\tilde{R}_1, \tilde{R}_2]$ is not a linear combination of $\tilde{R}_1$ and $\tilde{R}_2$, so also here we get a 3-dimensional vector field system on $\mathbb{C}_w^3$. And as in the first case, it follows that $\operatorname{span}_\mathbb{C}\{\tilde{R}_1, \tilde{R}_2, \tilde{R}_3\}$ is a *Lie algebra*, again called $\mathfrak{g}_R$.

The same conclusion is also valid for the $\tilde{L}_k$—that is, they give rise to a Lie algebra $\mathfrak{g}_L = \operatorname{span}_\mathbb{C}\{\tilde{L}_1, \tilde{L}_2, \tilde{L}_3\}$ of vector fields on $\mathbb{C}_w^3$.

If F_1 contains three $\tilde{R}_j$ and G_1 contains three $\tilde{L}_k$, the commutation relations $[\tilde{R}_j, \tilde{L}_k] = 0$ show that $\mathfrak{g}_R$ *and* $\mathfrak{g}_L$ *are reciprocal Lie algebras.*

If there are only two $\tilde{R}_j$ in F_1, $\tilde{R}_1$ and $\tilde{R}_2$ commute with each $\tilde{L}_k$—and by Jacobi's identity $\tilde{R}_3 = [\tilde{R}_1, \tilde{R}_2]$ also does. And conversely, if there are only two $\tilde{L}_k$ in G_1, these and their bracket commute with each $\tilde{R}_j$.

So in all cases $\mathfrak{g}_R$ *and* $\mathfrak{g}_L$ *are reciprocal Lie algebras.*

Theorem 12.2.1. *The generators of the vector field system $\mathcal{V}$ can be reduced to*

$$\begin{cases} F_1 = \partial/\partial u + \sum_{i=1}^{3} \phi^i(u, u^0) R_i, & F_2 = \partial/\partial u^0, \\ G_1 = \partial/\partial v + \sum_{i=1}^{3} \psi^i(v, v^0) L_i, & G_2 = \partial/\partial v^0, \end{cases}$$

where $\operatorname{span}_\mathbb{C}\{R_1, R_2, R_3\} = \mathfrak{g}_R$ *and* $\operatorname{span}_\mathbb{C}\{L_1, L_2, L_3\} = \mathfrak{g}_L$ *are reciprocal Lie algebras of vector fields on* $\mathbb{C}_w^3$. $\qquad\square$

Recalling from chapter **8** that because of the reciprocality there is a Lie parameter group $\mathfrak{G}$ such that $\mathfrak{g}_R$ = the right-invariant Lie algebra of $\mathfrak{G}$, and $\mathfrak{g}_L$ = the left-invariant Lie algebra of $\mathfrak{G}$, we see that the vector field system $\mathcal{V}$ associated to one of our hyperbolic PDEs is characterized firstly by a *3-dimensional Lie group* $\mathfrak{G}$, and secondly by *certain functions* $\phi^i(u, u^0)$ *and* $\psi^i(v, v^0)$.

Later on we will see that the dependence on the functions ϕ^i and ψ^i can be substantially reduced if $\mathcal{F}$ and $\mathcal{G}$ admit first integrals of the first order.

12.3 How to find integral manifolds

Since each Monge system is assumed to admit two functionally independent first integrals, the method of Darboux described in section **11.4** can be used. But here we will show that the integration of $\mathcal{V}$ actually can be reduced to solving Lie equations associated to the Lie algebras $\mathfrak{g}_R$ and $\mathfrak{g}_L$ belonging to the Monge systems.

$\mathcal{V}$ is generated by

$$\begin{cases} F_1 = \partial/\partial u + \sum_{i=1}^{3} \phi^i(u, u^0)\, R_i, & F_2 = \partial/\partial u^0, \\ G_1 = \partial/\partial v + \sum_{i=1}^{3} \psi^i(v, v^0)\, L_i, & G_2 = \partial/\partial v^0, \end{cases}$$

where $\operatorname{span}_{\mathbb{C}}\{R_1, R_2, R_3\} = \mathfrak{g}_R$ and $\operatorname{span}_{\mathbb{C}}\{L_1, L_2, L_3\} = \mathfrak{g}_L$ are reciprocal Lie algebras of vector fields on $\mathbb{C}_w^3$.

We are looking for complete 2-dimensional subsystems of $\mathcal{V}$. One immediate such is (F_2, G_2), the integral manifolds of which are obtained by setting u, v, w^1, w^2 and w^3 equal to constants.

However, we are more interested in integral manifolds on which $du \wedge dv \neq 0$. In section **11.2** it was seen that the most general 2-dimensional involution of $\mathcal{V}$ on which $du \wedge dv \neq 0$ is generated by

$$F = F_1 + a F_2 \quad \text{and} \quad G = G_1 + b G_2,$$

where a and b are arbitrary functions of $u, u^0, v, v^0, w^1, w^2, w^3$, and there remains to determine these functions such that (F, G) becomes complete.

Let us for a moment suppose that a and b have been determined in that way, and that the corresponding local foliation of $\mathbb{C}^7$ is given by

$$\begin{cases} u^0 = \mu(u, v; c^1, \dots, c^5), \\ v^0 = \nu(u, v; c^1, \dots, c^5), \\ w^i = \varpi^i(u, v; c^1, \dots, c^5) & \text{for } i = 1, 2, 3, \end{cases}$$

where the c^j are parameters that are essential in the sense that they conversely may be solved from the above system,

$$c^j = \gamma^j(u, v; u^0, v^0, w^1, w^2, w^3) \quad \text{for } j = 1, \ldots, 5,$$

so as to give a fundamental system of first integrals for (F, G).

Applying $F = \partial/\partial u + R + a\,\partial/\partial u^0$ to the identities $u^0 \equiv \mu(u, v; \gamma^1, \ldots, \gamma^5)$ and $v^0 \equiv v(u, v; \gamma^1, \ldots, \gamma^5)$ we get

$$a = Fu^0 = \frac{\partial\mu}{\partial u}Fu + \frac{\partial\mu}{\partial v}Fv + \sum_{i=1}^{5}\frac{\partial\mu}{\partial\gamma^i}F\gamma^i = \frac{\partial\mu}{\partial u} \quad \text{and} \quad 0 = Fv^0 = \frac{\partial v}{\partial u};$$

applying G instead we similarly have

$$0 = \frac{\partial\mu}{\partial v} = 0 \quad \text{and} \quad b = \frac{\partial v}{\partial v}.$$

Now $\partial\mu/\partial u = a$ and $\partial\mu/\partial v = 0$ together imply that $d\mu = a\,du$, so that

$$a = d\mu/du = \text{ a function of } u,$$

and analogously

$$b = dv/dv = \text{ a function of } v.$$

Thus our complete system is generated by

$$F = \frac{\partial}{\partial u} + R + \mu'(u)\frac{\partial}{\partial u^0} \quad \text{and} \quad G = \frac{\partial}{\partial v} + L + v'(v)\frac{\partial}{\partial v^0}.$$

Conversely: Letting $\phi(u)$ and $\psi(v)$ be arbitrary functions we immediately see that the vector field system $(\partial/\partial u + R + \phi(u)\,\partial/\partial u^0, \partial/\partial v + L + \psi(v)\,\partial/\partial v^0)$ is complete. So we have arrived at the following result.

Lemma 12.3.1. *The most general complete 2-dimensional subsystem of V on which $du \wedge dv \neq 0$ is generated by*

$$F = \partial/\partial u + \sum_{i=1}^{3}\phi^i(u, u^0)\,R_i + \phi(u)\,\partial/\partial u^0$$

and

$$G = \partial/\partial v + \sum_{i=1}^{3}\psi^i(v, v^0)\,L_i + \psi(v)\,\partial/\partial v^0,$$

where ϕ and ψ are arbitrary functions of one variable.

Moreover, on each integral manifold

$$u^0 = \int \phi(u) \, du \quad and \quad v^0 = \int \psi(v) \, dv.$$

$\square$

The differential forms $\omega^1 := du^0 - \phi(u) \, du$ and $\omega^2 := dv^0 - \psi(v) \, dv$ are exact and therefore generate a complete 2-dimensional pfaffian system (ω^1, ω^2), which in turn defines a local 2-parameter foliation of $\mathbb{C}^7$ with 5-dimensional leaves. The foliation associated to (F, G) that we are looking for is a subfoliation of this, with 2-dimensional leaves.

With $u^0 = \int \phi(u) \, du$ and $v^0 = \int \psi(v) \, dv$, the functions $\phi^i(u, u^0)$ and $\psi^i(v, v^0)$ go over into functions of u and v only, which we call $\Phi^i(u)$ and $\Psi^i(v)$ respectively. Then $F = \partial/\partial u + \sum_{i=1}^3 \Phi^i(u) \, R_i + \phi(u) \, \partial/\partial u^0$ and $G = \partial/\partial v + \sum_{i=1}^3 \Psi^i(v) \, L_i + \psi(v) \, \partial/\partial v^0$ can be projected to $\mathbb{C}^5_{u,v,w^1,w^2,w^3}$ so as to give

$$U = \frac{\partial}{\partial u} + \sum_{i=1}^3 \Phi^i(u) \, R_i \quad and \quad V = \frac{\partial}{\partial v} + \sum_{i=1}^3 \Psi^i(v) \, L_i.$$

Lemma 12.3.2. *U and V have **unique** prolongations to vector fields on $\mathbb{C}^7$ leaving the foliation defined by (ω^1, ω^2) invariant—and these prolongations are precisely F and G respectively:*

$$\{U, V\} \overset{projection}{\underset{prolongation}{\leftrightarrows}} \{F, G\}.$$

Proof. The most general prolongation of U to $\mathbb{C}^7$ is given by

$$\tilde{U} = U + f^0 \frac{\partial}{\partial u^0} + g^0 \frac{\partial}{\partial v^0},$$

where f^0 and g^0 are arbitrary functions on $\mathbb{C}^7$. $\tilde{U}$ is tangent to the foliation defined by (ω^1, ω^2) if

$$0 = \omega^1(\tilde{U}) = f^0 - \phi(u) \quad and \quad 0 = \omega^2(\tilde{U}) = g^0,$$

that is, $f^0 = \phi(u)$ and $g^0 = 0$. And analogously for V. $\square$

Clearly $[U, V] = [\partial/\partial u + \sum_{i=1}^3 \Phi^i(u) \, R_i, \partial/\partial v + \sum_{i=1}^3 \Psi^i(v) \, L_i] = 0$, so that (U, V) is complete and hence defines a local foliation of $\mathbb{C}^5_{u,v,w^1,w^2,w^3}$ with 2-dimensional leaves:

$$w^i = \varpi^i(u, v; c^1, c^2, c^3) \quad \text{for } i = 1, 2, 3, \text{ and arbitrary parameters } c^1, c^2, c^3.$$

And by the lemma the foliation of $\mathbb{C}^7$ associated to (F, G) that is our ultimate goal is obtained by prolonging this foliation of $\mathbb{C}^5$ to

$$
\begin{cases}
u^0 = \int \phi(u)\, du, \\
v^0 = \int \psi(v)\, dv, \\
\omega^i = \varpi^i(u, v; c^1, c^2, c^3) & \text{for } i = 1, 2, 3;
\end{cases}
$$

note that this does depend on five parameters.

Conclusion. *It suffices to integrate* (U, V).

Because $\operatorname{span}_{\mathbb{C}}\{R_1, R_2, R_3\}$ and $\operatorname{span}_{\mathbb{C}}\{L_1, L_2, L_3\}$ are the right- and left-invariant Lie algebras belonging to a 3-dimensional Lie parameter group $\mathfrak{G}$ acting on $\mathbb{C}^3_w$, $(U, V) = (\partial/\partial u + \sum_{i=1}^3 \Phi^i(u) R_i, \partial/\partial v + \sum_{i=1}^3 \Psi^i(v) L_i)$ is a *2-dimensional Lie vector field system associated to* $\mathfrak{G}$, and can thus be integrated by means of the methods in section **10.2**.

Denote the composition functions of $\mathfrak{G}$ by $\gamma^i(a, b)$:

$$
\mathfrak{G}: \mathbb{C}^3 \times \mathbb{C}^3 \to \mathbb{C}^3
$$

$$
(a, b) \mapsto (\gamma^1(a; b), \gamma^2(a; b), \gamma^3(a; b)) = \gamma(a; b) = a \cdot b.
$$

In order to obtain integral manifolds of (U, V) we tentatively set

$$
w^k = \gamma^k(a^1(u), a^2(u), a^3(u); b^1(v), b^2(v), b^3(v)) \iff w = a(u) \cdot b(v),
$$

where $a^i(u)$ and $b^i(v)$ are arbitrary functions of one variable. If $\bar{b}$ is the inverse of b,

$$
w = a(u) \cdot b(v) \iff a(u) = w \cdot \bar{b}(v) \iff a^k(u) = \gamma^k(w; \bar{b}(v)).
$$

The requirement $U(a(u) - w \cdot \bar{b}(v)) = 0$ is equivalent to

$$
\frac{da^k}{du} = \sum_{i=1}^3 \Phi^i(u)\, R_i \gamma^k(w; \bar{b}(v)).
$$

Since the vector fields $R_i = \sum_{j=1}^3 \rho_i^j(w^1, w^2, w^3)\, (\partial/\partial w^j)$ are *right-invariant*,

$$
R_i \gamma^k(w; \bar{b}(v)) = \sum_{j=1}^3 \rho_i^j(w)\, \frac{\partial}{\partial w^j}\, \gamma^k(w; \bar{b}(v))
$$

$$
= \sum_{j=1}^3 \rho_i^j(\gamma(w; \bar{b}(v)))\, \frac{\partial}{\partial w^j}\, w^k = \rho_i^k(a^1(u), a^2(u), a^3(u)),
$$

and therefore

$$\frac{da^k}{du} = \sum_{i=1}^{3} \Phi^i(u)\rho_i^k(a^1(u), a^2(u), a^3(u)) \quad \text{for } k = 1, 2, 3$$

—which is a *Lie equation associated to* $\mathfrak{g}_R$.

Similarly, because $L_i = \sum_{j=1}^{3} \lambda_i^j(w^1, w^2, w^3) \, (\partial/\partial w^j)$ is a left-invariant vector field, $V(w - a(u) \cdot b(v)) = 0$ if and only if

$$\frac{db^k}{dv} = \sum_{i=1}^{3} \Psi^i(v)\lambda_i^k(b^1(v), b^2(v), b^3(v)) \quad \text{for } k = 1, 2, 3.$$

Theorem 12.3.1. *The integral manifolds of* (U, V) *on which* $du \wedge dv \neq 0$ *are given by*

$$w^i = \gamma^i(a^1(u), a^2(u), a^3(u); b^1(v), b^2(v), b^3(v)) \quad \text{for } i = 1, 2, 3,$$

where the $a^k(u)$ *and* $b^k(v)$ *are the general solutions of the Lie equations*

$$\begin{cases} da^k/du = \sum_{i=1}^{3} \phi^i(u, \int \phi(u)\, du)\, \rho_i^k(a^1(u), a^2(u), a^3(u)), \\ db^k/du = \sum_{i=1}^{3} \psi^i(v, \int \psi(v)\, dv)\, \lambda_i^k(b^1(v), b^2(v), b^3(v)) \end{cases} \quad \text{for } k = 1, 2, 3,$$

involving the arbitrary functions $\phi(u)$ *and* $\psi(v)$. *The integral manifolds of* (F, G) *are then obtained by adding the equations*

$$\begin{cases} u^0 = \int \phi(u)\, du, \\ v^0 = \int \psi(v)\, dv. \end{cases}$$

$\square$

Letting $b := e = $ the identity element of $\mathfrak{G}$, we have $w^i = a^i(u)$ for $i = 1, 2, 3$. From the constructions made it is then clear that

$$\begin{cases} u^0 = \int \phi(u)\, du \quad \text{with an arbitrary } \phi(u), \\ w^i = a^i(u) \quad \text{for } i = 1, 2, 3 \end{cases}$$

constitutes the general integral curve of the Monge system $\mathcal{F} = (F_1, F_2)$ on $\mathbb{C}^5_{u,u^0,w^1,w^2,w^3}$.

And similarly the general integral curve of the Monge system $\mathcal{G} = (G_1, G_2)$ on $\mathbb{C}^5_{v,v^0,w^1,w^2,w^3}$ is given by

$$\begin{cases} v^0 = \int \psi(v)\, dv \quad \text{with } \psi(v) \text{ arbitrary}, \\ w^i = b^i(v) \quad \text{for } i = 1, 2, 3. \end{cases}$$

Theorem 12.3.2. *In order to find those integral manifolds of our vector field system $V = (F_1, F_2, G_1, G_2)$ on which $du \wedge dv \neq 0$, it suffices to integrate $\mathcal{F}$ and $\mathcal{G}$ separately, and then use the composition functions of the Lie group $\mathfrak{G}$.* □

12.4 Integrable systems

Recall from section **6.7** that $\mathcal{F} = (F_1, F_2)$ is said to be *explicitly integrable* if there is an integer p such that

$$\dim \mathcal{F}^{(k)} = \dim \mathcal{F}^{(k-1)} + 1 \text{ for } k = 1, \ldots, p, \text{ and } \mathcal{F}^{(p)} \text{ is complete.}$$

In that case the general integral curve of $\mathcal{F}$ can be written as

$$\begin{cases} u = \mu(x, \xi(x), \xi'(x), \ldots, \xi^{(p)}(x)), \\ u^0 = \mu^0(x, \xi(x), \xi'(x), \ldots, \xi^{(p)}(x)), \\ w^i = \alpha^i(x, \xi(x), \xi'(x), \ldots, \xi^{(p)}(x)) \quad \text{for } i = 1, 2, 3, \end{cases}$$

where $\xi(x)$ is an *arbitrary* function.

Notation. $x^k := \xi^{(k)}(x)$ for $k = 0, 1, \ldots, p$, and $\mathbf{x} := (x, x^0, x^1, \ldots, x^p) = (x, \xi(x), \xi'(x), \ldots, \xi^{(p)}(x))$.

If $\mathcal{G} = (G_1, G_2)$ is also explicitly integrable, its general integral curve may be written as

$$\begin{cases} v = v(\mathbf{y}), \\ v^0 = v^0(\mathbf{y}), \\ w^i = \beta^i(\mathbf{y}) \quad \text{for } i = 1, 2, 3, \end{cases}$$

where $\mathbf{y} = (y, y^0, y^1, \ldots, y^q) = (y, \eta(y), \eta'(y), \ldots, \eta^{(q)}(y))$ for an *arbitrary* function $\eta(y)$.

And then the general integral manifold of V is given by

$$\begin{cases} u = \mu(\mathbf{x}), \quad u^0 = \mu^0(\mathbf{x}), \\ v = v(\mathbf{y}), \quad v^0 = v^0(\mathbf{y}), \\ w^i = \gamma^i(\alpha^1(\mathbf{x}), \alpha^2(\mathbf{x}), \alpha^3(\mathbf{x}); \beta^1(\mathbf{y}), \beta^2(\mathbf{y}), \beta^3(\mathbf{y})) \\ \quad = \varpi^i(\mathbf{x}, \mathbf{y}) \quad \text{for } i = 1, 2, 3. \end{cases}$$

Example. Recall from section **12.1** that the vector field system V associated to the PDE $s = \sigma(x, y, z, p, q)$ is generated by

$$\begin{cases} F_1 = X + Y\sigma \, \partial/\partial t, \quad F_2 = \partial/\partial r, \\ G_1 = Y + X\sigma \, \partial/\partial r, \quad G_2 = \partial/\partial t, \end{cases}$$

where $X = \partial/\partial x + p\,\partial/\partial z + r\,\partial/\partial p + \sigma\,\partial/\partial q$, and $Y = \partial/\partial y + q\,\partial/\partial z + \sigma\,\partial/\partial p + t\,\partial/\partial q$. Then $[F_1, F_2] = -\partial/\partial p$, so that

$$\mathcal{F}' = \left(\frac{\partial}{\partial x} + p\frac{\partial}{\partial z} + \sigma\frac{\partial}{\partial q} + Y\sigma\,\frac{\partial}{\partial t}, \frac{\partial}{\partial p}, \frac{\partial}{\partial r} \right).$$

Next $[\partial/\partial x + p\,\partial/\partial z + \sigma\,\partial/\partial q + Y\sigma\partial/\partial t, \partial/\partial p] = -\partial/\partial z - (\partial\sigma/\partial p)\,\partial/\partial q - (\partial(Y\sigma)/\partial p)\,\partial/\partial t$, and hence

$$\mathcal{F}'' = \left(\frac{\partial}{\partial x} + \left(\sigma - p\frac{\partial\sigma}{\partial p} \right)\frac{\partial}{\partial q} + \left(Y\sigma - p\frac{\partial(Y\sigma)}{\partial p} \right)\frac{\partial}{\partial t}, \right.$$
$$\left. \frac{\partial}{\partial z} + \frac{\partial\sigma}{\partial p}\frac{\partial}{\partial q} + \frac{\partial(Y\sigma)}{\partial p}\frac{\partial}{\partial t}, \frac{\partial}{\partial p}, \frac{\partial}{\partial r} \right).$$

Assuming that $\mathcal{F}$ admits two functionally independent first integrals it follows that the completion $\bar{\mathcal{F}}$ of $\mathcal{F}$ has the dimension 5—and by the above this forces $\bar{\mathcal{F}}$ to be equal to $\mathcal{F}'''$. Hence

$$\dim \mathcal{F} = 2, \quad \dim \mathcal{F}' = 3, \quad \dim \mathcal{F}'' = 4 \quad \text{and} \quad \dim \mathcal{F}''' = 5,$$

so $\mathcal{F}$ is explicitly integrable. For analogous reasons $\mathcal{G}$ is also explicitly integrable.

Definition. The vector field system $\mathcal{V} = (F_1, F_2, G_1, G_2)$ is said to be *integrable* if the general 2-dimensional complete integral of $\mathcal{V}$ on which $du \wedge dv \neq 0$ can be written as

$$\begin{cases} u = \mu(\mathbf{x}, \mathbf{y}; \mathbf{c}), & u = \mu^0(\mathbf{x}, \mathbf{y}; \mathbf{c}), \\ v = \nu(\mathbf{x}, \mathbf{y}; \mathbf{c}), & v^0 = \nu^0(\mathbf{x}, \mathbf{y}; \mathbf{c}), \\ w^i = \varpi^i(\mathbf{x}, \mathbf{y}; \mathbf{c}) & \text{for } i = 1, 2, 3, \end{cases}$$

where $\mathbf{x} = (x, x^0, \ldots, x^p) = (x, \xi(x), \ldots, \xi^{(p)}(x))$ and $\mathbf{y} = (y, y^0, \ldots, y^q) = (y, \eta(y), \ldots, \eta^{(q)}(y))$ with $\xi(x)$ and $\eta(y)$ being arbitrary functions, and $\mathbf{c} = (c^1, \ldots, c^5)$ is a 5-tuple of arbitrary parameters.—Observe that this definition of *integrable* is quite different from that in section **3.5**!

Theorem 12.4.1. *If $\mathcal{V}$ is integrable, the general 2-dimensional complete integral of $\mathcal{V}$ on which $du \wedge dv \neq 0$ has the form*

$$\begin{cases} u = \mu(\mathbf{x}; \mathbf{c}), & u^0 = \mu^0(\mathbf{x}; \mathbf{c}), \\ v = \nu(\mathbf{y}; \mathbf{c}), & v^0 = \nu^0(\mathbf{y}; \mathbf{c}), \\ w^i = \varpi^i(\mathbf{x}, \mathbf{y}; \mathbf{c}) & \text{for } i = 1, 2, 3. \end{cases}$$

Moreover $\mathcal{V}$ is integrable if and only if each of the Monge systems $\mathcal{F}$ and $\mathcal{G}$ is explicitly integrable.

Proof. We have $du \wedge dv \neq 0$ and $dx \wedge dy \neq 0$ on each integral manifold. But $du \wedge du^0 = 0$, since u^0 is a function of u: $u^0 = \int \phi(u) \, du$, and therefore

$$0 = \frac{\partial(u, u^0)}{\partial(x, y)} = \frac{\partial u}{\partial x} \frac{\partial u^0}{\partial y} - \frac{\partial u}{\partial y} \frac{\partial u^0}{\partial x}.$$

Now

$$\frac{\partial u}{\partial x} = \frac{\partial \mu}{\partial x} + \sum_{i=0}^{p-1} \frac{\partial \mu}{\partial x^i} x^{i+1} + \frac{\partial \mu}{\partial x^p} \xi^{(p+1)}(x) = \left(A + \xi^{(p+1)} \frac{\partial}{\partial x^p} \right) \mu,$$

where $A = \partial/\partial x + \sum_{i=0}^{p-1} x^{i+1} (\partial/\partial x^i)$. Similarly

$$\frac{\partial u}{\partial y} = \left(B + \eta^{(q+1)} \frac{\partial}{\partial y^q} \right) \mu \quad \text{with} \quad B = \frac{\partial}{\partial y} + \sum_{i=0}^{q-1} y^{i+1} \frac{\partial}{\partial y^i},$$

and

$$\frac{\partial u^0}{\partial x} = \left(A + \xi^{(p+1)} \frac{\partial}{\partial x^p} \right) \mu^0, \qquad \frac{\partial u^0}{\partial y} = \left(B + \eta^{(q+1)} \frac{\partial}{\partial y^q} \right) \mu^0.$$

Thus

$$0 = \frac{\partial(u, u^0)}{\partial(x, y)} = \left(A\mu + \xi^{(p+1)} \frac{\partial \mu}{\partial x^p} \right) \left(B\mu^0 + \eta^{(q+1)} \frac{\partial \mu^0}{\partial y^q} \right)$$
$$- \left(B\mu + \eta^{(q+1)} \frac{\partial \mu}{\partial y^q} \right) \left(A\mu^0 + \xi^{(p+1)} \frac{\partial \mu^0}{\partial x^p} \right).$$

Since this is to be an identity in x and y for *arbitrary* functions $\xi(x)$ and $\eta(y)$, it splits into four equalities:

$$A\mu \, B\mu^0 - B\mu \, A\mu^0 = 0, \qquad A\mu \frac{\partial \mu^0}{\partial y^q} - \frac{\partial \mu}{\partial y^q} A\mu^0 = 0$$

$$\frac{\partial \mu}{\partial x^p} B\mu^0 - B\mu \frac{\partial \mu^0}{\partial x^p} = 0, \qquad \frac{\partial \mu}{\partial x^p} \frac{\partial \mu^0}{\partial y^q} - \frac{\partial \mu}{\partial y^q} \frac{\partial \mu^0}{\partial x^p} = 0.$$

1. Suppose first that $B\mu = \partial\mu/\partial y^q = 0$. Then μ is a first integral of the vector field system $\mathcal{B} = (B, \partial/\partial y^q)$, and automatically also of all its derivatives:

$$\mathcal{B}' = \left(\frac{\partial}{\partial y} + \sum_{i=0}^{q-2} y^{i+1} \frac{\partial}{\partial y^i}, \frac{\partial}{\partial y^{q-1}}, \frac{\partial}{\partial y^q} \right), \dots, \quad \mathcal{B}^{(q)} = \left(\frac{\partial}{\partial y}, \frac{\partial}{\partial y^0}, \dots, \frac{\partial}{\partial y^q} \right).$$

Consequently μ does not depend on the y^i.

In this case the four equalities above reduce to

$$A\mu \, B\mu^0 = 0, \quad A\mu \frac{\partial \mu^0}{\partial y^q} = 0, \quad \frac{\partial \mu}{\partial x^p} B\mu^0 = 0, \quad \frac{\partial \mu}{\partial x^q} \frac{\partial \mu^0}{\partial y^p} = 0.$$

If also $A\mu = \partial\mu/\partial x^p = 0$, μ does not depend on the x^i either, and so must be a constant—which contradicts the fact that $d\mu \neq 0$ on our integral manifolds. Therefore necessarily

$$B\mu^0 = \partial\mu^0/\partial y^q = 0,$$

whence also μ^0 is independent of the y^i.

Starting from the assumption $A\mu = \partial\mu/\partial x^p = 0$ instead, we will reach the same conclusion, but with x^i and y^i interchanged.

2. Then there remains to see what happens if $B\mu \neq 0$ or $\partial\mu/\partial y^q \neq 0$. These cases being analogous, we only consider the first one. If we set

$$\alpha := -\frac{A\mu}{B\mu} \quad \text{and} \quad \beta := -\frac{\partial\mu/\partial x^p}{B\mu}$$

the first and third of the four equalities above become

$$A\mu^0 + \alpha B\mu^0 = 0, \qquad \frac{\partial\mu^0}{\partial x^p} + \beta\, B\mu^0 = 0,$$

and then the second can be written as

$$\frac{\partial\mu}{\partial y^q}\, B\mu^0 - B\mu\, \frac{\partial\mu^0}{\partial y^q} = 0.$$

If we introduce the function γ by

$$\frac{\partial\mu}{\partial y^q} + \gamma\, B\mu = 0,$$

this last equality goes over into

$$\frac{\partial\mu^0}{\partial y^q} + \gamma\, B\mu^0 = 0.$$

But then μ and μ^0 are first integrals of the vector field system $W = (A + \alpha B, \partial/\partial x^p + \beta B, \partial/\partial y^q + \gamma B)$.

Here at least one of α and β must be nonzero, for otherwise $A\mu = \partial\mu/\partial x^p = 0$, and then we are back in the first case.

Now for instance

$$\left[A, \frac{\partial}{\partial x^p}\right] = \left[\frac{\partial}{\partial x} + \sum_{i=0}^{p-1} x^{i+1}\frac{\partial}{\partial x^i}, \frac{\partial}{\partial x^p}\right] = -\frac{\partial}{\partial x^{p-1}},$$

and similarly for $[B, \partial/\partial y^q]$. Therefore W' will contain vector fields of the form $\partial/\partial x^{p-1} + \cdots$, $\partial/\partial y^{q-1} + \cdots$. And analogously W'' will contain vector fields like $\partial/\partial x^{p-2} + \cdots$ and $\partial/\partial y^{q-2} + \cdots$.

Continuing in this way it follows that the first derivative of $\mathcal{W}$ which can be complete is of the form

$$\bar{\mathcal{W}} = \Big(\frac{\partial}{\partial x} + a \frac{\partial}{\partial y}, \frac{\partial}{\partial x^1} + a_1 \frac{\partial}{\partial y}, \dots, \frac{\partial}{\partial x^p} + a_p \frac{\partial}{\partial y},$$
$$\frac{\partial}{\partial y^1} + b_1 \frac{\partial}{\partial y}, \dots, \frac{\partial}{\partial y^q} + b_q \frac{\partial}{\partial y} \Big).$$

But then $\dim \bar{\mathcal{W}} = \dim (\text{underlying manifold}) - 1$, so that $\mathcal{W}$ admits just one functionally independent first integral. Because both μ and μ^0 are first integrals of $\mathcal{W}$, μ^0 is therefore a *determinate function* of μ. However, this cannot be true, since we know that u^0 may be chosen as an arbitrary function of u when forming the integral manifolds. Thus this second case is not possible.

The first case shows that we have either

$$u = \mu(\mathbf{x};\mathbf{c}), \quad u^0 = \mu^0(\mathbf{x};\mathbf{c}), \quad \text{or} \quad u = \mu(\mathbf{y};\mathbf{c}), \quad u^0 = \mu^0(\mathbf{y};\mathbf{c}),$$

and analogously for v and v^0. But since

$$\frac{\partial(u,v)}{\partial(x,y)} \neq 0 \quad \text{on each integral manifold,}$$

μ and v cannot depend on the same set of variables. Therefore our complete integral must be of the form

$$\begin{cases} u = \mu(\mathbf{x};\mathbf{c}), & u^0 = \mu^0(\mathbf{x};\mathbf{c}), \\ v = v(\mathbf{y};\mathbf{c}), & v^0 = v^0(\mathbf{y};\mathbf{c}), \\ w^i = \varpi^i(\mathbf{x},\mathbf{y};\mathbf{c}) & \text{for } i = 1,2,3. \end{cases}$$

These equations together define a local diffeomorphism

$$f : \mathbb{C}^2_{xy} \times \mathbb{C}^5_c \xrightarrow{\cong} \mathbb{C}^7_{u,u^0,v,v^0,w^1,w^2,w^3},$$

so that $f(\mathbb{C}^2_{xy} \times \{\mathbf{c}\})$ is an integral manifold of $\mathcal{V}$ for each $\mathbf{c} = (c^1,\dots,c^5)$. In particular we see that

$$f_* \Big(\frac{\partial}{\partial x} \Big) = \text{linear combination of } \frac{\partial}{\partial u}, \frac{\partial}{\partial u^0} \text{ and the } \frac{\partial}{\partial w^i}$$
$$= \text{linear combination of } F_1, F_2, G_1 \text{ and } G_2,$$

—which is possible only if $f_*(\partial/\partial x)$ is a linear combination of F_1 and F_2, i.e., $f_*(\partial/\partial x) \in \mathcal{F}$.

Running through all complete integrals of $\mathcal{V}$ we will obtain all vector fields belonging to $\mathcal{F}$ in this way, which means that the general integral curve of $\mathcal{F}$ can be written as

$$u = \mu(\mathbf{x}), \quad u^0 = \mu^0(\mathbf{x}), \quad w^i = \varpi^i(\mathbf{x}) \quad \text{for } i = 1, 2, 3.$$

But then $\mathcal{F}$ is *explicitly integrable*.

Considering $f_*(\partial/\partial y)$ instead we realize that $\mathcal{G}$ also *is explicitly integrable*. $\qquad\qquad\qquad\qquad\qquad\qquad\qquad\qquad\qquad\qquad\qquad\square$

Theorem 12.4.2. *If one of the Monge systems is explicitly integrable, the other admits a first integral of the first order.*

So if $\mathcal{V}$ is integrable, each Monge system admits a first integral of the first order.

Proof. Assume that $\mathcal{F}$, generated by

$$F_1 = \frac{\partial}{\partial u} + \sum_{i=1}^{3} \phi^i(u, u^0)\, R_i \quad \text{and} \quad F_2 = \frac{\partial}{\partial u^0},$$

is explicitly integrable. With $\mathcal{F}$ admitting the first integrals v and v^0, $\dim \bar{\mathcal{F}} = 7 - 2 = 5$, so that

$$\dim \mathcal{F}' = \dim \mathcal{F} + 1 = 3, \quad \dim \mathcal{F}'' = \dim \mathcal{F}' + 1 = 4,$$

and

$$\dim \mathcal{F}''' = \dim \mathcal{F}'' + 1 = 5 \text{—and therefore } \mathcal{F}''' = \bar{\mathcal{F}}.$$

$\mathcal{F}'$ is generated by F_1, F_2 and

$$[F_2, F_1] = \left[\frac{\partial}{\partial u^0}, F_1 \right] = \sum_{i=1}^{3} \frac{\partial \phi^i}{\partial u^0}\, R_i.$$

Since $\dim \mathcal{F}'' = 4$, only two of the vector fields

$$\left[F_1, \sum_{i=1}^{3} \frac{\partial \phi^0}{\partial u^0}\, R_i \right], \quad \sum_{i=1}^{3} \frac{\partial^2 \phi^i}{\partial (u^0)^2}\, R_i \quad \text{and} \quad \sum_{i=1}^{3} \frac{\partial \phi^i}{\partial u^0}\, R_i$$

can be linearly independent.

1. Suppose first that the second is a multiple of the third, i.e.,

$$\left[\frac{\partial}{\partial u^0}, \sum_{i=1}^{3} \frac{\partial \phi^i}{\partial u^0}\, R_i \right] = c \sum_{i=1}^{3} \frac{\partial \phi^i}{\partial u^0}\, R_i$$

for some function c. Then $F_2 = \partial/\partial u^0 \in C(\mathcal{F}')$, so that F_2 is the principal vector field of $\mathcal{V}$ belonging to $\mathcal{F}$. With u being a first integral of both $\mathcal{G}$

and F_2, it is also a first integral of $C(\mathcal{V}')$, and hence is a *first integral of* $\mathcal{G}$ *of the first order* (see the discussion in section **12.1**).

2. If the above does not occur, $[F_1, \sum_{i=1}^{3} (\partial \phi^i / \partial u^0) R_i]$ must be a linear combination of the other two vector fields modulo $\mathcal{F}'$, i.e.,

$$\left[F_1, \sum_{i=1}^{3} \frac{\partial \phi^i}{\partial u^0} R_i \right] \equiv \sum_{i=1}^{3} \left(a \frac{\partial \phi^i}{\partial u^0} + b \frac{\partial^2 \phi^i}{\partial (u^0)^2} \right) R_i \quad (\text{mod } \mathcal{F}')$$

for certain functions a and b. In this case we try to find a principal vector field belonging to $\mathcal{F}$ of the form $W = F_1 + g \, \partial / \partial u^0$ for a suitable function g. Now

$$\left[W, \sum_{i=1}^{3} \frac{\partial \phi^0}{\partial u^0} R_i \right] \equiv \sum_{i=1}^{3} \left(a \frac{\partial \phi^i}{\partial u^0} + (b+g) \frac{\partial^2 \phi^i}{\partial (u^0)^2} \right) R_i \quad (\text{mod } \mathcal{F}'),$$

so that choosing $g := -b$ we obtain

$$\left[W, \sum_{i=1}^{3} \frac{\partial \phi^i}{\partial u^0} R_i \right] \equiv 0 \quad (\text{mod } \mathcal{F}').$$

Since obviously $[W, F_1] \equiv [W, F_2] \equiv 0 \pmod{\mathcal{F}'}$, this shows that W does belong to $C(\mathcal{F}')$. And because W trivially commutes with $\mathcal{G}'$ modulo $\mathcal{V}'$, $W \in C(\mathcal{V}') \cap \mathcal{F}$, and thus really is a principal vector field.

Note next that

$$\left[F_1, \sum_{j=1}^{3} \frac{\partial \phi^j}{\partial u^0} R_j \right] = \left[\frac{\partial}{\partial u} + \sum_{i=1}^{3} \phi^i(u, u^0) R_i, \sum_{j=1}^{3} \frac{\partial \phi^j}{\partial u^0}(u, u^0) R_j \right]$$

must be of the form $\sum_{i=1}^{3} \chi^i(u, u^0) R_i$, and therefore a and b satisfy

$$\chi^i(u, u^0) = a \frac{\partial \phi^i}{\partial u^0} + b \frac{\partial^2 \phi^i}{\partial (u^0)^2} \quad \text{for } i = 1, 2, 3.$$

By assumption a and b can be solved from this system, and hence are functions of u and u^0 only. But then

$$W = \frac{\partial}{\partial u} - b(u, u^0) \frac{\partial}{\partial u^0} + \sum_{i=1}^{3} \phi^i(u, u^0) R_i$$

admits first integrals depending merely on u and u^0—namely any solution of $f(u, u^0)$ of

$$\left(\frac{\partial}{\partial u} - b(u, u^0) \frac{\partial}{\partial u^0} \right) f = 0.$$

Such an $f(u, u^0)$ is a first integral of $\mathcal{G}$, and also of the principal vector fields in $\mathcal{F}$ and $\mathcal{G}$—i.e., it is a *first integral of the first order for* $\mathcal{G}$.   □

Theorem 12.4.3. *If* $\mathcal{V}$ *is integrable,* $\mathcal{V}$ *is associated to a PDE of the form* $s = \sigma(x, y, z, p, q)$.

Proof. $\mathcal{V}$ is integrable $\iff$ $\mathcal{F}$ and $\mathcal{G}$ are explicitly integrable $\implies$ $\mathcal{F}$ and $\mathcal{G}$ both admit first integrals of the first order $\implies$ $\mathcal{V}$ is associated to a PDE $s = \sigma(x, y, z, p, q)$, according to section **12.1**.   □

Theorem 12.4.4. *If* $\mathcal{F}$ *is explicitly integrable, its generators can be simplified to*

$$F_1 = \frac{\partial}{\partial u} + u^0 \sum_{i=1}^{3} \phi^i(u) R_i \quad and \quad F_2 = \frac{\partial}{\partial u^0},$$

without changing G_1 *and* G_2.

Proof. Since $\mathcal{F}$ is explicitly integrable, $\mathcal{G}$ admits a first integral of the first order, and because of the discussion in section **12.1** we are allowed to choose this as our u. But then $\partial/\partial u^0$ is principal, so we have case **1** in the proof above—that is, there is a function $c(u, u^0)$ such that

$$\frac{\partial^2 \phi^i}{\partial(u^0)^2} = c \frac{\partial \phi^i}{\partial u^0} \quad \text{for } i = 1, 2, 3.$$

Let us regard this as a second order linear ODE with respect to u^0, having the fundamental solutions 1 and $\phi(u, u^0)$. Then the solutions ϕ^i can be written as

$$\phi^i(u, u^0) = a^i(u)\, \phi(u, u^0) + b^i(u) \cdot 1 \quad \text{for } i = 1, 2, 3.$$

Here $\phi(u, u^0)$ cannot be independent of u^0, for otherwise $F_1 = \partial/\partial u + \sum_{i=1}^{3} \phi^i(u) R_i$ commutes with $F_2 = \partial/\partial u^0$, so that $\mathcal{F}$ would be complete.

We may therefore choose $\tilde{u}^0 := \phi(u, u^0)$ as a new variable instead of u^0. Then

$$\phi^i = \tilde{u}^0\, a^i(u) + b^i(u) \quad \text{for } i = 1, 2, 3,$$

and F_2 goes over into a multiple of $\partial/\partial \tilde{u}^0$—so that F_2 may be redefined as $\partial/\partial \tilde{u}^0$. Therefore $\partial/\partial \tilde{u}^0$ need not occur in F_1, which hence can be chosen to be of the form

$$F_1 = \frac{\partial}{\partial u} + \sum_{i=1}^{3} b^i(u) R_i + \tilde{u}^0 \sum_{i=1}^{3} a^i(u) R_i.$$

Here $R_0 := \partial/\partial u + \sum_{i=1}^{3} b^i(u) R_i$ is a *Lie vector field* associated to $\mathfrak{g}_R$. Replacing w^1, w^2, w^3 by three functionally independent first integrals

$$\tilde{w}^k(w^1, w^2, w^3; u), \qquad k = 1, 2, 3,$$

of R_0, R_0 is reduced to $\partial/\partial u$. However, we want to do this *without affecting the L_i*, and for this we need the following result.

Lemma 12.4.1. *Let $\mathfrak{G}$ be an n-dimensional Lie parameter group acting on $\mathbb{C}_w^n$, and let $\mathfrak{g} = \mathrm{span}_{\mathbb{C}}\{R_1, \ldots, R_n\}$ be the associated Lie algebra with*

$$R_i = \sum_{j=1}^{n} \rho_i^j(w^1, \ldots, w^n) \, (\partial/\partial w^j) \qquad \text{for } i = 1, \ldots, n.$$

Also let

$$X = \frac{\partial}{\partial u} + \sum_{i=1}^{3} a^i(u) R_i$$

be a Lie vector field associated to $\mathfrak{g}$. By the local rectification lemma in section 2.1 it is possible to find first integrals $f^1(w^1, \ldots, w^n; u), \ldots, f^n(w^1, \ldots, w^n; u)$ of X such that $f^i(w^1, \ldots, w^n; 0) = w^i$ for $i = 1, \ldots, n$. Setting $\tilde{w}^i := f^i(w^1, \ldots, w^n; u)$ we get a local diffeomorphism

$$\mathbb{C}_w^n \times \mathbb{C}_u \xrightarrow{\cong} \mathbb{C}_{\tilde{w}}^n \times \mathbb{C}_u$$
$$(w^1, \ldots, w^n; u) \mapsto (f^1(w^1, \ldots, w^n; u), \ldots, f^n(w^1, \ldots, w^n; u); u),$$

rectifying X to $\partial/\partial u$. This rectified vector field induces the action of the translation group in one variable: $(\tilde{w}; u) \mapsto (\tilde{w}; u + t)$. If we go back to the original coordinates and consider u as a parameter, the mapping

$$f_u : (w^1, \ldots, w^n) \mapsto (f^1(w^1, \ldots, w^n; u), \ldots f^n(w^1, \ldots, w^n; u))$$

*defines the action of a **1-parameter subgroup** of $\mathfrak{G}$ on $\mathbb{C}_w^n$.*

Proof. Since $f_0(w^1, \ldots, w^n) = (w^1, \ldots, w^n)$, f_u defines for each $(w^1, \ldots, w^n) \in \mathbb{C}_w^n$ the tangent vector

$$T_w = \sum_{i=1}^{n} \frac{\partial f^i}{\partial u}(w^1, \ldots, w^n; 0) \frac{\partial}{\partial w^i}.$$

If we vary w, this defines a vector field on $\mathbb{C}_w^n$, and we must show that this belongs to $\mathrm{span}_{\mathbb{C}}\{R_1, \ldots, R_n\}$. By assumption

$$0 = Xf^i = \frac{\partial f^i}{\partial u} + \sum_{j=1}^{n} a^j(u) R_j f = \frac{\partial f^i}{\partial u} + \sum_{j=1}^{n} a^j(u) \sum_{k=1}^{n} \frac{\partial f^i}{\partial w^k} R_j w^k.$$

With

$$\frac{\partial f^i}{\partial w^k}(w^1,\ldots,w^n;0) = \delta^i_k$$

this shows that

$$\frac{\partial f^i}{\partial u}(w^1,\ldots,w^n;0) = -\sum_{j=1}^n a^j(0)\,R_j w^i = -\sum_{j=1}^n a^j(0)\rho^i_j(w^1,\ldots,w^n),$$

and therefore

$$T_w = -\sum_{i,j=1}^n a^j(0)\rho^i_j(w^1,\ldots,w^n)\,\frac{\partial}{\partial w^i} = -\sum_{j=1}^n a^j(0)\,R_j.$$

$\square$

Let $\tilde{w}^1$, $\tilde{w}^2$, $\tilde{w}^3$ be first integrals of R_0 as in the lemma, and make the change of coordinates

$$(w^1, w^2, w^3 ; u) \longmapsto (\tilde{w}^1, \tilde{w}^2, \tilde{w}^3 ; u).$$

According to the lemma this coordinate transformation induces the action of a 1-parameter subgroup

$$(w^1, w^2, w^3) \longmapsto (f^1(w^1, w^2, w^3 ; u), f^2(w^1, w^2, w^3 ; u), f^3(w^1, w^2, w^3 ; u))$$

belonging to $\mathfrak{G}_R$. But because of the reciprocality of $\mathfrak{g}_R$ and $\mathfrak{g}_L$ this change of variables does not affect $\mathfrak{g}_L = \operatorname{span}_{\mathbb{C}}\{L_1, L_2, L_3\}$, and hence leaves

$$G_1 = \partial/\partial v + \sum_{i=1}^3 \psi^i(v, v^0)\,L_i \quad\text{and}\quad G_2 = \partial/\partial v^0$$

invariant.

$\square$

If $\mathcal{G}$ is also explicitly integrable the same kind of argument can be used to simplify $\mathcal{G}$ too.

Corollary 12.4.1. *If we suppose both $\mathcal{F}$ and $\mathcal{G}$ to be explicitly integrable, the generators of $\mathcal{V}$ can be simplified to*

$$\begin{cases} F_1 = \partial/\partial u + u^0 \sum_{i=1}^3 \phi^i(u)\,R_i, & F_2 = \partial/\partial u^0, \\ G_1 = \partial/\partial v + v^0 \sum_{i=1}^3 \psi^i(v)\,L_i, & G_2 = \partial/\partial v^0, \end{cases}$$

so that the six functions ϕ^i and ψ^i now depend on one variable only. $\square$

Summary. The following statements are equivalent for a vector field system V having the property that each Monge subsystem admits two functionally independent first integrals:

(i) V is integrable.
(ii) V is associated to a PDE $s = \sigma(x, y, z, p, q)$.
(iii) Both Monge systems are explicitly integrable.
(iv) Each Monge system admits a first integral of the first order.

12.5 Two first integrals for one Monge system and three for the other

In this section it is assumed that $\mathcal{F} = (F_1, F_2)$ admits two functionally independent first integrals v, v^0, and that $\mathcal{G} = (G_1, G_2)$ admits three functionally independent first integrals u, u^0 and u^1.

Main result. F_1, F_2, G_1, G_2 *may be reduced to*

$$\begin{cases} F_1 = \partial/\partial u + u^1 \, \partial/\partial u^0 + \sum_{i=1}^2 \phi^i(u, u^0, u^1) \, R_i, & F_2 = \partial/\partial u^1, \\ G_1 = \partial/\partial v + v^0 \sum_{i=1}^2 \psi^i(v) \, L_i, & G_2 = \partial/\partial v^0, \end{cases}$$

where $\mathrm{span}_{\mathbb{C}}\{R_1, R_2\}$ *and* $\mathrm{span}_{\mathbb{C}}\{L_1, L_2\}$ *are reciprocal Lie algebras of vector fields on* $\mathbb{C}_w^2$.

Here $\mathcal{G}$ *is explicitly integrable. If $\mathcal{F}$ is also explicitly integrable, the above can be simplified to*

$$\begin{cases} F_1 = \partial/\partial u + u^1 \, \partial/\partial u^0 + u^0 \, (R_1 + u \, R_2), & F_2 = \partial/\partial u^1, \\ G_1 = \partial/\partial v + v^0 \, (L_1 + v \, L_2), & G_2 = \partial/\partial v^0. \end{cases}$$

Just as in the case of two first integrals for each Monge system, this is proved by a simple but rather tedious step-by-step process.

Lemma 12.5.1. *It is possible to find functions* w^1, w^2 *which together with* u, u^0, u^1, v *and* v^0 *form a system of local coordinates for the underlying manifold such that*

$$\begin{cases} F_1 = \partial/\partial u + u^1 \, \partial/\partial u^0 + R, & F_2 = \partial/\partial u^1, \\ G_1 = \partial/\partial v + L, & G_2 = \partial/\partial v^0, \end{cases}$$

where

$$R = \sum_{i=1}^2 \rho^i(u, u^0, u^1, v, w^1, w^2) \frac{\partial}{\partial w^i} \quad \text{and} \quad L = \sum_{i=1}^2 \lambda^i(u, u^0, v, v^0, w^1, w^2) \frac{\partial}{\partial w^i}.$$

Proof. Using u, u^0, u^1, v, v^0 and two more functions x^1, x^2 as local coordinates we know from before that G_1 and G_2 can be written as

$$G_1 = \frac{\partial}{\partial v} + \sum_{i=1}^{2} g_1^i \frac{\partial}{\partial x^i}, \quad G_2 = \frac{\partial}{\partial v^0} + \sum_{i=1}^{2} g_2^i \frac{\partial}{\partial x^i},$$

and we claim that F_1 and F_2 may be expressed as

$$F_1 = \frac{\partial}{\partial u} + \cdots, \quad F_2 = \frac{\partial}{\partial u^0} + \cdots \quad \text{or} \quad \frac{\partial}{\partial u^1} + \cdots .$$

If not we would have

$$F_1 = \frac{\partial}{\partial u} + a^0 \frac{\partial}{\partial u^0} + a^1 \frac{\partial}{\partial u^1} + \sum_{i=1}^{2} f_1^i \frac{\partial}{\partial x^i}, \quad F_2 = \sum_{i=1}^{2} f_2^i \frac{\partial}{\partial x^i},$$

and then $[G_i, F_1] = 0$ for $i = 1, 2$ implies that

$$G_1 a^0 = G_1 a^1 = 0 \quad \text{and} \quad G_2 a^0 = G_2 a^1 = 0,$$

so that a^0 and a^1 are first integrals of $\mathcal{G}$, i.e., functions of u, u^0 and u^1 only. But in that case $\mathcal{V}$ admits nonconstant first integrals—namely first integrals $f(u, u^0, u^1)$ of the vector field

$$\frac{\partial}{\partial u} + a^0(u, u^0, u^1) \frac{\partial}{\partial u^0} + a^1(u, u^0, u^1) \frac{\partial}{\partial u^1}.$$

And this is a contradiction.

Thus we may assume that

$$F_1 = \frac{\partial}{\partial u} + a \frac{\partial}{\partial u^0} + \sum_{i=1}^{2} f_1^i \frac{\partial}{\partial x^i} \quad \text{and} \quad F_2 = \frac{\partial}{\partial u^1} + b \frac{\partial}{\partial u^0} + \sum_{i=1}^{2} f_2^i \frac{\partial}{\partial x^i},$$

where by the same argument as above a and b are certain functions of u, u^0 and u^1. To simplify F_2 we replace u^0 by a first integral of the vector field

$$\partial/\partial u^1 + b(u, u^0, u^1) \partial/\partial u^0 \quad \text{on} \quad \mathbb{C}^3_{u,u^0,u^1}$$

so as to get

$$F_1 = \frac{\partial}{\partial u} + c(u, u^0, u^1) \frac{\partial}{\partial u^0} + \sum_{i=1}^{2} f_1^i \frac{\partial}{\partial x^i} \quad \text{and} \quad F_2 = \frac{\partial}{\partial u^1} + \sum_{i=1}^{2} f_2^i \frac{\partial}{\partial x^i}.$$

Here $\partial c / \partial u^1 \neq 0$, for otherwise $(\partial/\partial u + c \, \partial/\partial u^0, \partial/\partial u^1)$ would be a complete 2-dimensional vector field system on $\mathbb{C}^3_{u,u^0,u^1}$, admitting a nonconstant first integral, which also would be a first integral for $\mathcal{V}$. Thus it is possible to introduce $c(u, u^0, u^1)$ as a new variable which replaces u^1.

Because $[F_2, G_2] = 0$, (F_2, G_2) is a complete 2-dimensional vector field system on $\mathbb{C}^4_{u^1,v^0,x^1,x^2}$ with u, u^0, v as parameters. Replacing x^1 and x^2 by two functionally independent first integrals w^1 and w^2 of (F_2, G_2), we obtain the following preliminary simplification of the basis for $\mathcal{V}$:

$$F_1 = \frac{\partial}{\partial u} + u^1 \frac{\partial}{\partial u^0} + R, \qquad\qquad F_2 = \frac{\partial}{\partial u^1},$$

$$G_1 = \frac{\partial}{\partial v} + L, \qquad\qquad G_2 = \frac{\partial}{\partial v^0},$$

where R and L are certain linear combinations of $\partial/\partial w^1$ and $\partial/\partial w^2$. Since $0 = [G_2, F_1] = [\partial/\partial v^0, R]$ and $0 = [F_2, G_1] = [\partial/\partial u^1, L]$, R does not depend on v^0 and L does not depend on u^1. $\qquad\qquad\square$

Utilizing the equality $[F_1, G_1] = 0$ we next want to write R and L as linear combinations of vector fields on $\mathbb{C}^2_w$ without any parameters.

Lemma 12.5.2. *It is possible to introduce local coordinates with respect to which*

$$R = \sum_{i=1}^{2} r^i(u, u^0, u^1) R_i \quad \text{with} \quad R_i = \sum_{j=1}^{2} r_i^j(w^1, w^2) \frac{\partial}{\partial w^j},$$

and

$$L = \sum_{i=1}^{2} l^i(v, v^0) L_i \quad \text{with} \quad L_i = \sum_{j=1}^{2} l_i^j(w^1, w^2) \frac{\partial}{\partial w^j},$$

where $\operatorname{span}_{\mathbb{C}}\{R_1, R_2\} = \mathfrak{g}_R$ *and* $\operatorname{span}_{\mathbb{C}}\{L_1, L_2\} = \mathfrak{g}_L$ *are reciprocal Lie algebras, and such that moreover*

$$F_1 = \frac{\partial}{\partial u} + u^1 \frac{\partial}{\partial u^0} + \sum_{i=1}^{2} \phi^i(u, u^0, u^1) R_i, \qquad F_2 = \frac{\partial}{\partial u^1},$$

$$G_1 = \frac{\partial}{\partial v} + \sum_{i=1}^{2} \psi^i(v, v^0) L_i, \qquad\qquad G_2 = \frac{\partial}{\partial v^0}.$$

Proof. Because

$$[F_2, [F_2, F_1]] = \sum_{i=1}^{2} \frac{\partial^2 \rho^i}{\partial (u^1)^2} \frac{\partial}{\partial w^i} =: R^{(2)},$$

bracketing $[F_1, G_1] = 0$ repeatedly with F_2 gives

$$0 = [R^{(2)}, G_1] = [R^{(3)}, G_1] = [R^{(4)}, G_1] = \cdots,$$

where

$$R^{(k)} := \sum_{i=1}^{2} \frac{\partial^k \rho^i}{\partial(u^1)^k} \frac{\partial}{\partial w^i} \quad \text{for } k = 2, 3, 4, \cdots.$$

And clearly also $0 = [R^{(2)}, G_2] = [R^{(3)}, G_2] = \cdots$. Regarding the $R^{(k)}$ as vector fields on $\mathbb{C}_w^2$ depending on the parameters u, u^0, u^1, v, there are locally only two of them which can be linearly independent over the ring of functions. If $R^{(2)}$ and $R^{(3)}$ are linearly independent, $R^{(4)}$ is a linear combination of these:

$$R^{(4)} = a_2 R^{(2)} + a_3 R^{(3)}.$$

Bracketing both sides with G_1 and G_2 gives that a_2, a_3 are first integrals of $\mathcal{G}$, and hence are functions of u, u^0, u^1. If instead $R^{(3)} = a R^{(2)}$, differentiation with respect to u^1 will give the same kind of result.

Thus the coefficients ρ^i in R are solutions of the ODE

$$\frac{\partial^4 z}{\partial(u^1)^4} = a_2(u, u^0, u^1) \frac{\partial^2 z}{\partial(u^1)^2} + a_3(u, u^0, u^1) \frac{\partial^3 z}{\partial(u^1)^3},$$

with u and u^0 considered as parameters. This fourth order ODE has four solutions which are linearly independent with respect to u^1—1, $\phi^1(u, u^0, u^1)$, $\phi^2(u, u^0, u^1)$ and u^1, say—so that the ρ^i can be written as

$$\rho^i = \rho_0^i \cdot 1 + \rho_1^i \cdot \phi^1(u, u^0, u^1) + \rho_2^i \cdot \phi^2(u, u^0, u^1) + \rho_3^i \cdot u^1$$

with the ρ_j^i being functions of u, u^0, v, w^1, w^2 for $j = 0, 1, 2, 3$. Then

$$R = R_0 + \sum_{i=1}^{2} \phi^i(u, u^0, u^1) R_i + u^1 R_3,$$

where

$$R_i = \sum_{j=1}^{2} \rho_j^i(u, u^0, v, w^1, w^2) \frac{\partial}{\partial w^j} \quad \text{for } i = 0, 1, 2, 3,$$

so that

$$F_1 = \frac{\partial}{\partial u} + R_0 + u^1 \left(\frac{\partial}{\partial u^0} + R_3 \right) + \sum_{i=1}^{2} \phi^i(u, u^0, u^1) R_i.$$

If we replace w^1 and w^2 by two functionally independent first integrals of $\partial/\partial u^0 + R_3$, this latter vector field goes over into $\partial/\partial u^0$, whence F_1 can be written as follows with modified R_0, R_1, R_2:

$$F_1 = \frac{\partial}{\partial u} + R_0 + u^1 \frac{\partial}{\partial u^0} + \sum_{i=1}^{2} \phi^i(u, u^0, u^1) R_i.$$

But then $[F_1, G_1] = 0$ shows that $[\partial/\partial u^0, L] = 0$, so that the coefficients $\lambda^i(u, u^0, v, v^0, w^1, w^2)$ in L are independent of u^0. It also follows that

$$[G_1, \partial/\partial u + R_0] = 0 \quad \text{and} \quad [G_1, R_i] = 0 \quad \text{for } i = 1, 2.$$

Bracketing these equalities repeatedly with $\partial/\partial u^0$ and reasoning as above we infer that R_0, R_1, R_2 can be written as

$$R_0 = \tilde{R}_0 + \sum_{j=1}^{2} \phi_0^j(u, u^0)\, \tilde{R}_j \quad \text{and} \quad R_i = \sum_{j=1}^{2} \phi_i^j(u, u^0)\, \tilde{R}_j \quad \text{for } i = 1, 2,$$

where the vector fields $\tilde{R}_0, \tilde{R}_1$ and $\tilde{R}_2$ are independent of u^0. Forgetting about the tildes and modifying the ϕ^i,

$$R = R_0 + \sum_{i=1}^{2} \phi^i(u, u^0, u^1) R_i, \quad \text{where} \quad R_i = \sum_{j=1}^{2} \rho_i^j(u, v, w^1, w^2) \frac{\partial}{\partial w^j}.$$

If we replace w^1, w^2 by two functionally independent first integrals of $\partial/\partial u + R_0$, this vector field goes over into $\partial/\partial u$, and F_1 is simplified to

$$F_1 = \frac{\partial}{\partial u} + u^1 \frac{\partial}{\partial u^0} + \sum_{i=1}^{2} \phi^i(u, u^0, u^1) R_i.$$

Because u^1, ϕ^1 and ϕ^2 are linearly independent as functions of u^1, $[F_1, G_1] = 0$ implies that $[G_1, \partial/\partial u] = 0$ and $[G_1, R_i] = 0$ for $i = 1, 2$. Thus $G_1 = \partial/\partial v + L$ is independent of u:

$$L = \sum_{i=1}^{2} \lambda^i(v, v^0, w^1, w^2) \frac{\partial}{\partial w^i}.$$

Bracketing both sides of $[G_1, R_i] = 0$ with $\partial/\partial u$ and playing the usual game we can then write R as

$$R = \sum_{i=1}^{2} r^i(u, u^0, u^1) R_i \quad \text{with} \quad R_i = \sum_{j=1}^{2} r_i^j(v, w^1, w^2) \frac{\partial}{\partial w^j}.$$

By bracketing $[F_1, G_1] = 0$ with $\partial/\partial v^0$ instead, L can similarly be brought to the form

$$L = L_0 + \sum_{i=1}^{2} \psi^i(v,v^0)\, L_i \quad \text{with} \quad L_i = \sum_{j=1}^{2} \lambda_i^j(v,w^1,w^2)\, \frac{\partial}{\partial w^j},$$

where 1, $\psi^1(v,v^0)$ and $\psi^2(v,v^0)$ are linearly independent as functions of v^0, whence $G_1 = \partial/\partial v + L_0 + \sum_{i=1}^{2} \psi^i\, L_i$. Here L_0 can be made to vanish by replacing w^1, w^2 by two functionally independent first integrals of $\partial/\partial v + L_0$, and then

$$G_1 = \frac{\partial}{\partial v} + \sum_{i=1}^{2} \psi^i(v,v^0)\, L_i.$$

Because of the linear independence of 1, ψ^1 and ψ^2, the equality $[F_1, G_1] = 0$ is equivalent to the three equalities

$$[F_1, \partial/\partial v] = 0 \quad \text{and} \quad [F_1, L_i] = 0 \quad \text{for } i = 1, 2.$$

The first shows that the R_i are independent of v, and hence are genuine vector fields on $\mathbb{C}_w^2$. Then by bracketing $[F_1, L_i] = 0$ with $\partial/\partial v$ and arguing as usual the L_i can be expressed as linear combinations of vector fields which do not depend on v.

Conclusion.
$R = \sum_{i=1}^{2} r^i(u,u^0,u^1)\, R_i$ with $R_i = \sum_{j=1}^{2} r_i^j(w^1,w^2)\, (\partial/\partial w^j)$,
and $L = \sum_{i=1}^{2} l^i(v,v^0)\, L_i$ with $L_i = \sum_{j=1}^{2} l_i^j(w^1,w^2)\, (\partial/\partial w^j)$.

In analogy with the case of two first integrals for each Monge system it is finally realized from $[F_1, G_1] = 0$ that the R_i and L_i give rise to reciprocal Lie algebras of vector fields on $\mathbb{C}_w^2$. Thus

$$\begin{cases} F_1 = \partial/\partial u + u^1\, \partial/\partial u^0 + \sum_{i=1}^{2} \phi^i(u,u^0,u^1)\, R_i, & F_2 = \partial/\partial u^1, \\ G_1 = \partial/\partial v + \sum_{i=1}^{2} \psi^i(v,v^0)\, L_i, & G_2 = \partial/\partial v^0, \end{cases}$$

with $\text{span}_{\mathbb{C}}\{R_1, R_2\}$ and $\text{span}_{\mathbb{C}}\{L_1, L_2\}$ being reciprocal Lie algebras. $\square$

In order to simplify G_1 further we need the following result.

Lemma 12.5.3. $\mathcal{F}$ admits a first integral of the first order.

Proof. Let us first try to find a principal vector field in $\mathcal{G}$—say

$$W = a\, G_1 + b\, G_2 = a \left(\frac{\partial}{\partial v} + \sum_{i=1}^{2} \psi^i\, L_i \right) + b\, \frac{\partial}{\partial v^0}.$$

W is principal if $[W, \mathcal{V}'] \equiv 0 \pmod{\mathcal{V}'}$, where $\mathcal{V}'$ is generated by the six vector fields F_1, F_2, G_1, G_2, $F_3 := [F_2, F_1] = \partial/\partial u^0 + \sum_{i=1}^{2} (\partial \phi^i / \partial u^1) R_i$, and $G_3 := [G_2, G_1] = \sum_{i=1}^{2} (\partial \psi^i / \partial v^0) L_i$.

W clearly commutes with F_1, F_2, G_1, G_2 and F_3 modulo $\mathcal{V}'$, so we only need to check $[W, G_3]$:

$$[W, G_3] \equiv a\, [G_1, G_3] + b \sum_{i=1}^{2} \frac{\partial^2 \psi^i}{\partial(v^0)^2} L_i \quad (\mathrm{mod}\ \mathcal{V}').$$

Here $[G_1, G_3]$ is a linear combination of L_1 and L_2—say that $[G_1, G_3] = \sum_{i=1}^{2} \chi^i(v, v^0) L_i$. But then the whole right hand side is a linear combination of L_1 and L_2, and so

$$[W, G_3] \equiv 0 \quad (\mathrm{mod}\ \mathcal{V}') \iff a\, [G_1, G_3] + b \sum_{i=1}^{2} \frac{\partial^2 \psi^i}{\partial(v^0)^2} L_i = c\, G_3$$

for a suitable function c. Hence W is principal if and only if

$$\sum_{i=1}^{2} \left(a\,\chi^i + b \frac{\partial^2 \psi^i}{\partial(v^0)^2} \right) L_i = c \sum_{i=1}^{2} \frac{\partial \psi^i}{\partial v^0} L_i.$$

According to Cramer's rule this is satisfied with

$$a = \frac{\partial \psi^1}{\partial v^0} \frac{\partial^2 \psi^2}{\partial(v^0)^2} - \frac{\partial \psi^2}{\partial v^0} \frac{\partial^2 \psi^1}{\partial(v^0)^2}, \qquad b = \chi^1 \frac{\partial \psi^2}{\partial v^0} - \chi^2 \frac{\partial \psi^1}{\partial v^0}$$

and

$$c = \chi^1 \frac{\partial^2 \psi^2}{\partial(v^0)^2} - \chi^2 \frac{\partial^2 \psi^1}{\partial(v^0)^2}.$$

In particular we see that a, b and c are functions of v and v^0 only. But then

$$W = a(v, v^0) \left(\frac{\partial}{\partial v} + \sum_{i=1}^{2} \psi^i(v, v^0) L_i \right) + b(v, v^0) \frac{\partial}{\partial v^0}$$

will admit any first integral $f(v, v^0)$ of the vector field $a\, \partial/\partial v + b\, \partial/\partial v^0$ as a first integral. Since such an $f(v, v^0)$ is a first integral of both $\mathcal{F}$ and W, it is a *first integral of the first order* for $\mathcal{F}$. □

With this we are ready for the final simplification.

Theorem 12.5.1. $\mathcal{V}$ *has a canonical basis*

$$\begin{cases} F_1 = \partial/\partial u^0 + u^1\, \partial/\partial u^0 + \sum_{i=1}^{2} \phi^i(u, u^0, u^1)\, R_i, & F_2 = \partial/\partial u^1, \\ G_1 = \partial/\partial v + v^0 \sum_{i=1}^{2} \psi^i(v)\, L_i, & G_2 = \partial/\partial v^0. \end{cases}$$

Moreover $\mathcal{G} = (G_1, G_2)$ *is explicitly integrable.*

Proof. By the discussion in section **12.1** we may choose v as a first integral of the first order for $\mathcal{F}$. In that case the principal vector field $W \in C(\mathcal{V}') \cap \mathcal{G}$ does not contain $\partial/\partial v$—so $a = 0$ in the expression for W used above. But then the ψ^i are solutions of the ODE

$$\frac{\partial^2 z}{\partial (v^0)^2} = \frac{c(v, v^0)}{b(v, v^0)} \frac{\partial z}{\partial v^0},$$

with v being a parameter. By utilizing this in the usual way the ψ^i can be written as

$$\psi^i(v, v^0) = \alpha^i(v) \, \psi(v, v^0) + \beta^i(v) \quad \text{for } i = 1, 2,$$

where $\partial \psi / \partial v^0 \neq 0$. Choosing $\psi(v, v^0)$ as a new variable instead of v^0, G_1 takes the form

$$G_1 = \frac{\partial}{\partial v} + \sum_{i=1}^{2} \beta^i(v) L_i + v^0 \sum_{i=1}^{2} \alpha^i(v) L_i.$$

To get rid of the β^i we determine first integrals $f^i(w^1, w^2; v)$ of $\partial/\partial v + \sum_{i=1}^{2} \beta^i(v) L_i$ satisfying

$$f^i(w^1, w^2; 0) = w^i \quad \text{for } i = 1, 2,$$

and introduce these as new variables instead of the old w^1 and w^2. Then

$$(w^1, w^2) \mapsto (f^1(w^1, w^2; v), f^2(w^1, w^2; v))$$

defines a 1-parameter group of $\mathfrak{G}_L$. Because $\mathfrak{g}_L = \operatorname{span}_\mathbb{C}\{L_1, L_2\}$ and $\mathfrak{g}_R = \operatorname{span}_\mathbb{C}\{R_1, R_2\}$ are reciprocal Lie algebras, this change of coordinates does not alter the R_i—or F_1. So we finally have

$$F_1 = \frac{\partial}{\partial u} + u^1 \frac{\partial}{\partial u^0} + \sum_{i=1}^{2} \phi^i(u, u^0, u^1) \, R_i, \qquad F_2 = \frac{\partial}{\partial u^1},$$

$$G_1 = \frac{\partial}{\partial v} + v^0 \sum_{i=1}^{2} \psi^i(v) \, L_i, \qquad G_2 = \frac{\partial}{\partial v^0}.$$

From this we get $[G_2, G_1] = \sum_{i=1}^{2} \psi^i(v) \, L_i$, so that $\dim \mathcal{G}' = 3$. Because $\mathcal{G}$ admits precisely three functionally independent first integrals and $\mathcal{G}'$ lives on $\mathbb{C}^4_{v, v^0, w^1, w^2}$, $\mathcal{G}'' = (\partial/\partial v, \partial/\partial v^0, \partial/\partial w^1, \partial/\partial w^2)$ is 4-dimensional and complete—that is, $\mathcal{G}$ is *explicitly integrable*.   □

The next task is to determine complete 2-dimensional subsystems of $\mathcal{V}$

on which $du \wedge dv \neq 0$. From section **11.2** we know that the most general 2-dimensional involution of V satisfying this condition is generated by

$$F = F_1 + a F_2 \quad \text{and} \quad G = G_1 + b G_2,$$

with a and b being arbitrary functions. These are then to be determined in such a way that (F, G) becomes complete.

If

$$\begin{cases} u^0 = \mu^0(u, v; c^1, \dots, c^5), & u^1 = \mu^1(u, v; c^1, \dots, c^5), \\ v^0 = v^0(u, v; c^1, \dots, c^5), & \text{and} \quad w^i = \varpi^i(u, v; c^1, \dots, c^5) \quad \text{for } i = 1, 2 \end{cases}$$

is the foliation associated to such a complete subsystem, F and G will kill $u^0 - \mu^0$, $u^1 - \mu^1$ and $v^0 - v^0$, so that

$$\begin{aligned}
u^1 - \partial \mu^0 / \partial u &= 0, & \partial \mu^0 / \partial v &= 0, \\
a - \partial \mu^1 / \partial u &= 0, & \partial \mu^1 / \partial v &= 0, \\
\partial v^0 / \partial u &= 0, & b - \partial v^0 / \partial v &= 0,
\end{aligned}$$

Thus a is a function of u only, $a = \alpha(u)$, and b is a function of v only, $b = \beta(v)$. Morover

$$u^1 = \int \alpha(u) \, du, \quad u^0 = \int \left(\int \alpha(u) \, du \right) du \quad \text{and} \quad v^0 = \int \beta(v) \, dv.$$

Conversely, let $\alpha(u)$ and $\beta(v)$ be arbitrary functions, Then clearly $[F_1 + \alpha(u) F_2, G_1 + \beta(v) G_2] = 0$, so that $(F_1 + \alpha(u) F_2, G_1 + \beta(v) G_2)$ indeed is complete. Thereby we have obtained the following result.

Lemma 12.5.4. *The most general complete 2-dimensional subsystem of V on which $du \wedge dv \neq 0$ is generated by*

$$F = \frac{\partial}{\partial u} + u^1 \frac{\partial}{\partial u^0} + \alpha(u) \frac{\partial}{\partial u^1} + \sum_{i=1}^{2} \phi^i(u, u^0, u^1) R_i$$

and

$$G = \frac{\partial}{\partial v} + \beta(v) \frac{\partial}{\partial v^0} + v^0 \sum_{i=1}^{2} \psi^i(v) L_i,$$

where $\alpha(u)$ and $\beta(v)$ are arbitrary functions of one variable. Moreover, on *the integral manifolds*

$$u^1 = \int \alpha(u)\, du, \quad u^0 = \int \left(\int \alpha(u)\, du \right) du \quad and \quad v^0 = \int \beta(v)\, dv.$$

$\square$

What remains now is to determine the w^i as functions of u and v on the integral manifolds. Proceeding as we did in a similar situation in section **12.3**, we set

$$\Phi^i(u) := \phi^i \left(u, \int \left(\int \alpha(u)\, du \right) du, \int \alpha(u)\, du \right),$$

$$\Psi^i(v) := \int \beta(v)\, dv \cdot \psi^i(v) \quad for\ i = 1, 2,$$

and consider the vector fields

$$U = \frac{\partial}{\partial u} + \sum_{i=1}^{2} \Phi^i(u)\, R_i, \quad V = \frac{\partial}{\partial v} + \sum_{i=1}^{2} \Psi^i(v)\, L_i.$$

Then (U, V) is a *2-dimensional Lie vector field system* on $\mathbb{C}^4_{u,v,w^1,w^2}$, and if the integral manifolds of this on which $du \wedge dv \neq 0$ are given by

$$w^i = \varpi^i(u, v; c^1, c^2) \quad for\ i = 1, 2,$$

the integral manifolds of V that we are looking for are given by adding the equalities

$$u^0 = \int \left(\int \alpha(u)\, du \right) du, \quad u^1 = \int \alpha(u)\, du, \quad and \quad v^0 = \int \beta(v)\, dv.$$

If the composition functions of the underlying Lie parameter group $\mathfrak{G}$ are denoted by $\gamma^i(a; b)$ for $i = 1, 2$, the same type of arguments as in section **12.3** shows that

$$w^i = \gamma^i(a^1(u), a^2(u); b^1(v), b^2(v)),$$

where $a^i(u)$ and $b^i(v)$ are solutions of the Lie equations

$$\frac{da^i}{du} = \sum_{j=1}^{2} \Phi^j(u) r^i_j(a^1(u), a^2(u)) \quad and \quad \frac{db^i}{dv} = \sum_{j=1}^{2} \Psi^j(v) l^i_j(b^1(v), b^2(v))$$

respectively.

Note that

$$u^0 = \int \left(\int \alpha(u)\, du \right) du, \quad u^1 = \int \alpha(u)\, du, \quad w^i = a^i(u) \quad for\ i = 1, 2$$

is the most general integral curve of $\mathcal{F} = (F_1, F_2)$ on which $du \neq 0$, and likewise

$$v^0 = \int \beta(v) \, dv, \quad w^i = b^i(v) \quad \text{for } i = 1, 2$$

is the most general integral curve of $\mathcal{G} = (G_1, G_2)$ on which $dv \neq 0$.

Theorem 12.5.2. *In order to integrate* $\mathcal{V}$ *it suffices to integrate the Monge systems* $\mathcal{F}$ *and* $\mathcal{G}$ *separately, and this can be done by solving Lie equations associated to* $\mathfrak{g}_R$ *and* $\mathfrak{g}_L$ *respectively.* □

We saw earlier that $\mathcal{G}$ is explicitly integrable, and that $\mathcal{F}$ admits a first integral of the first order.

Lemma 12.5.5. *If* $\mathcal{F}$ *is also explicitly integrable,* $\mathcal{G}$ *admits two functionally independent first integrals of the first order.*

Proof. $\mathcal{F}'$ is generated by F_1, F_2 and

$$F_3 := [F_2, F_1] = \frac{\partial}{\partial u^0} + \sum_{i=1}^{2} \frac{\partial \phi^i}{\partial u^1} R_i,$$

and $\mathcal{F}''$ is generated by these and the two vector fields

$$[F_2, F_3] = \sum_{i=1}^{2} \frac{\partial^2 \phi^i}{\partial (u^1)^2} R_i, \quad [F_1, F_3] = \sum_{i=1}^{2} \theta^i R_i,$$

where

$$\theta^i = \frac{\partial^2 \phi^i}{\partial u \partial u^1} + u^1 \frac{\partial^2 \phi^i}{\partial u^0 \partial u^1} - \frac{\partial \phi^i}{\partial u^0} + \left(\phi^1 \frac{\partial \phi^2}{\partial u^2} - \phi^2 \frac{\partial \phi^1}{\partial u^1} \right) c_{12}^i \quad \text{for } i = 1, 2,$$

with c_{12}^1, c_{12}^2 being the structure constants of $\text{span}_{\mathbb{C}}\{R_1, R_2\}$. Since $\mathcal{F}$ is explicitly integrable, $\dim \mathcal{F}'' = 4$, and this forces the latter vector fields to be linearly dependent—i.e., there are functions $a(u, u^0, u^1)$ and $b(u, u^0, u^1)$ such that

$$a \theta^i + b \frac{\partial^2 \phi^i}{\partial (u^1)^2} = 0 \quad \text{for } i = 1, 2.$$

Setting $W := a F_1 + b F_2$ we then have

$$[W, F_3] \equiv a \sum_{i=1}^{2} \theta^i R_i + b \sum_{i=1}^{2} \frac{\partial^2 \phi^i}{\partial (u^1)^2} R_i = 0 \pmod{\mathcal{F}},$$

whence W is a principal vector field belonging to $\mathcal{F}$. But

$$W = a(u, u^0, u^1) \left(\frac{\partial}{\partial u} + u^1 \frac{\partial}{\partial u^0} + \sum_{i=1}^{2} \phi^i(u, u^0, u^1) R_i \right) + b(u, u^0, u^1) \frac{\partial}{\partial u^1}$$

admits two functionally independent first integrals which are functions of u, u^0, u^1 only—namely the first integrals of the vector field $a (\partial/\partial u + u^1 \, \partial/\partial u^0) + b \, \partial/\partial u^1$ on $\mathbb{C}^3_{u,u^0,u^1}$. And these are *first integrals of $\mathcal{G}$ of the first order*. $\qquad\Box$

Corollary 12.5.1. *$\mathcal{F}$ is explicitly integrable if and only if $\mathcal{V}$ is associated to a PDE of the form $s = \sigma(x, y, z, p, q)$.* $\qquad\Box$

In the rest of this section we suppose that $\mathcal{F}$ is explicitly integrable, and exploit this fact in order to simplify F_1 and G_1 further.

Lemma 12.5.6. *$F_1 = \partial/\partial u + u^1 \, \partial/\partial u^0 + \sum_{i=1}^{2} \phi^i(u, u^0, u^1) R_i$ can be reduced to*

$$F_1 = \frac{\partial}{\partial u} + u^1 \frac{\partial}{\partial u^0} + \sum_{i=1}^{2} \phi^i(u, u^0) R_i$$

without changing F_2 and the G_i.

Proof. If we choose u as one of the first integrals of the first order for $\mathcal{G}$, the principal vector field W does not contain $F_1 = \partial/\partial u + \cdots$, which means that the a used in the argument above will vanish. Then the ϕ^i are solutions of

$$\frac{\partial^2 \phi^i}{\partial (u^1)^2} = 0 \qquad \text{for } i = 1, 2,$$

and can accordingly be written as

$$\phi^i = g^i(u, u^0) \cdot 1 + h^i(u, u^0) \cdot u^1,$$

so that

$$F_1 = \frac{\partial}{\partial u} + \sum_{i=1}^{2} g^i(u, u^0) R_i + u^1 \left(\frac{\partial}{\partial u^0} + \sum_{i=1}^{2} h^i(u, u^0) R_i \right).$$

In order to get rid of the h^i we replace w^1, w^2 by two first integrals $\tilde{w}^i = f^i(w^1, w^2; u, u^0)$ of the vector field $\partial/\partial u^0 + \sum_{i=1}^{2} h^i(u, u^0) R_i$, satisfying $f^i(w^1, w^2; u, 0) = w^i$ for $i = 1, 2$. Then for fixed u,

$$f_{u,u^0} : \quad \mathbb{C}^2_w \to \mathbb{C}^2_w$$
$$(w^1, w^2) \mapsto (f^1(w^1, w^2; u, u^0), f^2(w^1, w^2; u, u^0))$$

defines a 1-parameter subgroup of $\mathfrak{G}_R$. So in particular $f_{u,u^0} \in \mathfrak{G}_R$ for each (u, u^0). If γ^i are the composition functions of $\mathfrak{G}_R$, the general element of $\mathfrak{G}_R$ is given by

$$(w^1, w^2) \mapsto (\gamma^1(a^1, a^2; w^1, w^2), \gamma^2(a^1, a^2; w^1, w^2)),$$

and therefore our $\tilde{w}^i$ can be written as

$$\tilde{w}^i = \gamma^i(\alpha^1(u, u^0), \alpha^2(u, u^0); w^1, w^2)$$

for suitable functions α^1 and α^2. With $R_i = \sum_{j=1}^{2} r_i^j(w^1, w^2) \left(\partial/\partial w^j\right)$, Lie's own version of the first fundamental theorem in section **8.5** shows that

$$\frac{\partial \gamma^i}{\partial a^j} = \sum_{k=1}^{2} \chi_j^k(a^1, a^2) r_k^i(\gamma^1, \gamma^2),$$

where the χ_j^k are the coefficients of the dual Maurer–Cartan forms. Thus

$$\frac{\partial \tilde{w}^i}{\partial u} = \sum_{j=1}^{2} \frac{\partial \gamma^i}{\partial a^j} \frac{\partial \alpha^j}{\partial u} = \sum_{k=1}^{2} \left(\sum_{j=1}^{2} \chi_j^k(\alpha^1(u, u^0), \alpha^2(u, u^0)) \frac{\partial \alpha^j}{\partial u} \right) r_k^i(\tilde{w}^1, \tilde{w}^2)$$

$$= \sum_{k=1}^{2} m^k(u, u^0) r_k^i(\tilde{w}^1, \tilde{w}^2).$$

So under the coordinate change $(w^1, w^2) \mapsto (\tilde{w}^1, \tilde{w}^2)$,

$$\frac{\partial}{\partial u} \mapsto \frac{\partial}{\partial u} + \sum_{k=1}^{2} m^k(u, u^0) R_k, \qquad \frac{\partial}{\partial u^0} + \sum_{i=1}^{2} h^i(u, u^0) R_i \mapsto \frac{\partial}{\partial u^0}$$

and the R_i go over into linear combinations of R_1 and R_2 with coefficients depending on u and u^0.

Hence F_1 gets reduced to

$$F_1 = \frac{\partial}{\partial u} + u^1 \frac{\partial}{\partial u^0} + \sum_{i=1}^{2} \phi^i(u, u^0) R_i,$$

and because of the reciprocality of $\mathfrak{g}_R$ and $\mathfrak{g}_L$ this coordinate transformation leaves the G_i invariant. $\square$

Next the dependence of the ϕ^i on u^0 can be simplified.

Lemma 12.5.7. *Changing u and u^0 suitably, F_1 can be brought to the form*

$$F_1 = \frac{\partial}{\partial u} + u^1 \frac{\partial}{\partial u^0} + u^0 \sum_{i=1}^{2} \phi^i(u) R_i.$$

Proof. $\mathcal{F}'$ is generated by F_1, $F_2 = \partial/\partial u^1$, and $F_3 = [F_2, F_1] = \partial/\partial u^0$. Consider the vector field system $\mathcal{H}$ generated by F_1 and F_3:

$$\mathcal{H} = (X_1, X_2) \quad \text{with} \quad X_1 = \partial/\partial u + \sum_{i=1}^{2} \phi^i(u, u^0)\, R_i \text{ and } X_2 = \partial/\partial u^0.$$

$\mathcal{H}'$ is generated by X_1, X_2 and

$$X_3 = [X_2, X_1] = \sum_{i=1}^{2} \frac{\partial \phi^i}{\partial u^0}(u, u^0)\, R_i.$$

Then $C = a^1 X_1 + a^2 X_2 \in C(\mathcal{H}')$ if and only if $[C, X_3] \equiv 0 \pmod{\mathcal{H}'}$. This condition is equivalent to

$$a^1 \frac{\partial^2 \phi^i}{\partial u \partial u^0} + a^2 \frac{\partial^2 \phi^i}{\partial(u^0)^2} + a^1 \left(\phi^1 \frac{\partial \phi^2}{\partial u^0} - \phi^2 \frac{\partial \phi^1}{\partial u^0} \right) c_{12}^i = b \frac{\partial \phi^i}{\partial u^0} \quad \text{for } i = 1, 2,$$

where c_{12}^i are the structure functions of $\mathfrak{g}_R$, and b is a suitable function. As in analogous cases treated earlier this system can be solved with a^1, a^2 and b only depending on u and u^0. But then C admits a nonconstant first integral of the form $f(u, u^0)$—namely a first integral of the vector field $a^1(u, u^0)\, \partial/\partial u + a^2(u, u^0)\, \partial/\partial u^0$ on $\mathbb{C}^2_{u, u^0}$. Such an $f(u, u^0)$ is clearly a first integral both of $\mathcal{G}$ and of $C(\mathcal{V}')$, and hence a first integral of the first order for $\mathcal{G}$. Let us choose u to be of this kind. Then $a^1 = 0$, and thus the ϕ^i satisfy an ODE system

$$\frac{\partial^2 \phi^i}{\partial(u^0)^2} = c(u, u^0) \frac{\partial \phi^i}{\partial u^0} \quad \text{for } i = 1, 2.$$

Consequently they can be written as

$$\phi^i(u, u^0) = \alpha^i(u)\, \psi(u, u^0) + \beta^i,$$

where $\partial \psi / \partial u^0 \neq 0$. By choosing this $\psi(u, u^0)$ as a new variable instead of u^0, and getting rid of the β^i as on similar occasions before, F_1 is reduced to the wanted form. $\square$

So by now the generators of $\mathcal{V}$ look like

$$\begin{cases} F_1 = \partial/\partial u + u^1\, \partial/\partial u^0 + u^0 \sum_{i=1}^{2} \phi^i(u)\, R_i, & F_2 = \partial/\partial u^1, \\ G_1 = \partial/\partial v + v^0 \sum_{i=1}^{2} \psi^i(v)\, L_i, & G_2 = \partial/\partial v^0. \end{cases}$$

But in fact it is possible to get rid of the arbitrary functions ϕ^i and ψ^i altogether.

Theorem 12.5.3. *If $\mathcal{F}$ is explicitly integrable, $\mathcal{V}$ has the canonical basis*

$$\begin{cases} F_1 = \partial/\partial u + u^1 \, \partial/\partial u^0 + u^0 \, (R_1 + u \, R_2), & F_2 = \partial/\partial u^1, \\ G_1 = \partial/\partial v + v^0 \, (L_1 + v \, L_2), & G_2 = \partial/\partial v^0. \end{cases}$$

Here u and u^0 are first integrals of the first order of $\mathcal{G}$, and v is a first integral of the first order of $\mathcal{F}$.

Proof. If $\phi^2(u) = c \, \phi^1(u)$ for a constant c, $\mathcal{F}$ would admit a nonconstant first integral only depending on w^1 and w^2—namely a first integral of the vector field $R_1 + c \, R_2$ on $\mathbb{C}^2_w$. But this is impossible.

Therefore u may be replaced by the quotient $\tilde{u} := \phi^2(u)/\phi^1(u)$, whereby F_1 is turned into a multiple of a vector field of the form

$$\frac{\partial}{\partial \tilde{u}} + u^1 a(\tilde{u}) \frac{\partial}{\partial u^0} + u^0 b(\tilde{u}) \, (R_1 + \tilde{u} \, R_2).$$

Setting $\tilde{u}^0 := u^0 b(\tilde{u})$, $\tilde{u}^1 := u^0 b'(\tilde{u}) + u^1 a(\tilde{u}) b(\tilde{u})$ and forgetting the tildes afterwards, we get

$$F_1 = \frac{\partial}{\partial u} + u^1 \frac{\partial}{\partial u^0} + u^0 \, (R_1 + u \, R_2),$$

while F_2 becomes a multiple of $\partial/\partial u^1$, and G_1, G_2 are not affected. Reasoning in the same way with the $\psi^i(v)$, we end up with the wanted result. $\qquad\square$

Corollary 12.5.2. *So in this case $\mathcal{V}$ is completely determined by a 2-dimensional Lie group—and by the discussion in chapter 9 there are only two such up to equivalence.* $\qquad\square$

Remark. The notion of *integrability* for $\mathcal{V}$ can be introduced in the same way as in the preceding section. With $\mathcal{G}$ automatically being explicitly integrable one then proves that

$$\mathcal{V} \text{ is integrable} \iff \mathcal{F} \text{ is explicitly integrable.}$$

12.6 Three first integrals for each Monge system

The last case to be studied is the one where each of the Monge systems $\mathcal{F} = (F_1, F_2)$ and $\mathcal{G} = (G_1, G_2)$ admits three functionally independent first integrals. Applying the same kind of arguments as before we readily see that the F_i and G_i can be written as

$$\begin{cases} F_1 = \partial/\partial u + u^1 \, \partial/\partial u^0 + \rho(u, u^0, u^1, v, v^0, w) \, \partial/\partial w, & F_2 = \partial/\partial u^1, \\ G_1 = \partial/\partial v + v^1 \, \partial/\partial v^0 + \lambda(u, u^0, v, v^0, v^1, w) \, \partial/\partial w, & G_2 = \partial/\partial v^1, \end{cases}$$

where u, u^0, u^1 are first integrals of $\mathcal{G}$, and v, v^0, v^1 are first integrals of $\mathcal{F}$. There then remains to simplify ρ and λ.

$\mathcal{F}'$ is generated by F_1, F_2 and

$$F_3 := [F_2, F_1] = \frac{\partial}{\partial u^0} + \frac{\partial \rho}{\partial u^1} \frac{\partial}{\partial w},$$

and therefore $W := b^1 F_1 + b^2 F_2$ is principal if and only if $[W, F_3] \equiv 0$ (mod $\mathcal{V}'$). Because

$$[F_1, F_3] \equiv \left(F_1 \frac{\partial \rho}{\partial u^1} - F_3 \rho \right) \frac{\partial}{\partial w} =: a_1 \frac{\partial}{\partial w} \quad (\text{mod } \mathcal{F})$$

and

$$[F_2, F_3] = \frac{\partial^2 \rho}{\partial (u^1)^2} \frac{\partial}{\partial w} =: a_2 \frac{\partial}{\partial w},$$

W is principal if and only if $a_1 b^1 + a_2 b^2 = 0$.

Claim. We may choose b^1 and b^2 to depend on u, u^0 and u^1 only.

For

$$0 = [G_1, [F_1, F_3]] = \left[G_1, a_1 \frac{\partial}{\partial w} \right] = \left(G_1 a_1 - a_1 \frac{\partial \lambda}{\partial w} \right) \frac{\partial}{\partial w}$$

$$\Longrightarrow G_1 a_1 = a_1 \frac{\partial \lambda}{\partial w},$$

and analogously

$$0 = [G_1, [F_2, F_3]] \Longrightarrow G_1 a_2 = a_2 \frac{\partial \lambda}{\partial w}.$$

But then

$$G_1 \left(\frac{a_1}{a_2} \right) = \frac{a_2 a_1 \partial \lambda / \partial w - a_1 a_2 \partial \lambda / \partial w}{(a_2)^2} = 0.$$

In the same way also $G_2(a_1/a_2) = 0$, which means that the quotient a_1/a_2 is a first integral of $\mathcal{G}$, and thus a function of u, u^0 and u^1 only—whence $a_1 b^1 + a_2 b^2 = 0$ may be satisfied with functions b^1 and b^2 merely depending on u, u^0 and u^1. $\qquad \square$

This being so, the principal vector field

$$W = b^1(u, u^0, u^1) \left(\frac{\partial}{\partial u} + u^1 \frac{\partial}{\partial u^0} + \rho \frac{\partial}{\partial w} \right) + b^2(u, u^0, u^1) \frac{\partial}{\partial u^1}$$

admits two functionally independent first integrals which only depend on u, u^0, u^1, and therefore are *first integrals of the first order for* $\mathcal{G}$.

Let us choose u as such a first integral of the first order for $\mathcal{G}$. Then

b^1 vanishes, so that $\partial/\partial u^1$ is a principal vector field. But then also u^0 is a first integral of the first order for $\mathcal{G}$.

With $b^1 = 0$ and $b^2 = 1$ we have $a_2 = 0$, i.e.,

$$\frac{\partial^2 \rho}{\partial (u^1)^2} = 0,$$

whence $\rho = \rho_0(u, u^0, v, v^0) + \rho_1(u, u^0, v, v^0)\, u^1$. Consequently F_1 is simplified to

$$F_1 = \frac{\partial}{\partial u} + \rho_0 \frac{\partial}{\partial w} + u^1 \left(\frac{\partial}{\partial u^0} + \rho_1 \frac{\partial}{\partial w} \right).$$

Analogously G_1 may be simplified to

$$G_1 = \frac{\partial}{\partial v} + \lambda_0 \frac{\partial}{\partial w} + v^1 \left(\frac{\partial}{\partial v^0} + \lambda_1 \frac{\partial}{\partial w} \right),$$

where λ_0, λ_1 are functions of u, u^0, v, v^0 only, and v, v^0 are first integrals of the first order for $\mathcal{F}$.

Note that since $\mathcal{F}$ and $\mathcal{G}$ admit first integrals of the first order, $\mathcal{V}$ is associated to a PDE $s = \sigma(x, y, z, p, q)$, i.e.,

$$\mathcal{V} = \left(\frac{\partial}{\partial x} + p \frac{\partial}{\partial z} + r \frac{\partial}{\partial p} + \sigma \frac{\partial}{\partial q}, \frac{\partial}{\partial y} + q \frac{\partial}{\partial z} + \sigma \frac{\partial}{\partial p} + t \frac{\partial}{\partial q}, \frac{\partial}{\partial r}, \frac{\partial}{\partial t} \right).$$

Our goal is to find suitable local coordinates making the F_i and G_i appear as these basis vector fields for $\mathcal{V}$.

Inserting the above expressions for F_1 and G_1 in $[F_1, G_1] = 0$ it follows that $(\partial/\partial u^0 + \rho_1 \partial/\partial w, \partial/\partial v^0 + \lambda_1 \partial/\partial w)$ is a complete 2-dimensional vector field system on $\mathbb{C}^5_{u, u^0, v, v^0, w}$. Replacing w by a first integral of this system we get rid of ρ_1 and λ_1, and thus

$$\begin{cases} F_1 = \partial/\partial u + \rho(u, u^0, v, v^0, w)\, \partial/\partial w + u^1 \partial/\partial u^0, & F_2 = \partial/\partial u^1, \\ G_1 = \partial/\partial v + \lambda(u, u^0, v, v^0, w)\, \partial/\partial w + v^1 \partial/\partial v^0, & G_2 = \partial/\partial v^1. \end{cases}$$

Then $F_3 = [F_2, F_1] = \partial/\partial u^0$, so that $\mathcal{F}' = (\partial/\partial u + \rho\, \partial/\partial w, \partial/\partial u^0, \partial/\partial u^1)$, and similarly $\mathcal{G}' = (\partial/\partial v + \lambda\, \partial/\partial w, \partial/\partial v^0, \partial/\partial v^1)$.

Let us now consider the following vector fields on $\mathbb{C}^5_{u, u^0, v, v^0, w}$:

$$\begin{cases} U_1 := \partial/\partial u + \rho\, \partial/\partial w, & U_2 := \partial/\partial u^0, \\ V_1 := \partial/\partial v + \lambda\, \partial/\partial w, & V_2 := \partial/\partial v^0. \end{cases}$$

Since $0 = [F_1, G_1] = [U_1 + u^1 \partial/\partial u^0, V_1 + v^1 \partial/\partial v^0]$,

$$[U_i, V_j] = 0 \quad \text{for } i, j = 1, 2,$$

from which it is in particular seen that $\partial \rho / \partial v^0 = 0$ and $\partial \lambda / \partial u^0 = 0$. The vector fields

$$[U_2, U_1] = \frac{\partial \rho}{\partial u^0} \frac{\partial}{\partial w} \quad \text{and} \quad [U_2, [U_2, U_1]] = \frac{\partial^2 \rho}{\partial (u^0)^2} \frac{\partial}{\partial w}$$

commute with V_1, V_2, and are clearly linearly dependent:

$$\frac{\partial^2 \rho}{\partial (u^0)^2} \frac{\partial}{\partial w} = m \frac{\partial \rho}{\partial u^0} \frac{\partial}{\partial w}, \quad \text{with } m = \frac{\partial^2 \rho}{\partial (u^0)^2} \bigg/ \frac{\partial \rho}{\partial u^0}.$$

Bracketing both sides with the V_i we deduce that m is a first integral of (V_1, V_2), that is, a function of u and u^0 only:

$$\frac{\partial^2 \rho}{\partial (u^0)^2} = m(u, u^0) \frac{\partial \rho}{\partial u^0}.$$

Therefore ρ can be written as

$$\rho = \rho_0(u, v, w) \cdot a(u, u^0) + \rho_1(u, v, w) \cdot 1$$

with $a(u, u^0)$ and 1 linearly independent as functions of u^0. Similarly

$$\lambda = \lambda_0(u, v, w) \cdot b(v, v_0) + \lambda_1(u, v, w) \cdot 1.$$

Let us now introduce the following new variables:

$a(u, u^0)$ instead of u^0, $b(v, v^0)$ instead of v^0,

$\partial a / \partial u + u^1 \, \partial a / \partial u^0$ instead of u^1 and $\partial b / \partial v + v^1 \, \partial b / \partial v^0$ instead of v^1.

Then

$$\begin{cases} F_1 = (\partial / \partial u + \rho_1 \, \partial / \partial w) + u^1 \, \partial / \partial u^0 + u^0 \rho_0 \, \partial / \partial w, \\ G_1 = (\partial / \partial v + \lambda_1 \, \partial / \partial w) + v^1 \, \partial / \partial v^0 + v^0 \lambda_0 \, \partial / \partial w. \end{cases}$$

From $[U_1, V_1] = 0$ it also follows that $[\partial / \partial u \mid \rho_1 \, \partial / \partial w, \partial / \partial v + \lambda_1 \, \partial / \partial w] = 0$. If we consider $(\partial / \partial u + \rho_1(u, v, w) \, \partial / \partial w, \partial / \partial v + \lambda_1(u, v, w) \, \partial / \partial w)$ as a complete 2-dimensional vector field system on $\mathbb{C}^3_{u,v,w}$, it admits a nonconstant first integral. Choosing w to be of this kind, we are left with

$$\begin{cases} F_1 = \partial / \partial u + u^1 \, \partial / \partial u^0 + u^0 \rho(u, v, w) \, \partial / \partial w, & F_2 = \partial / \partial u^1, \\ G_1 = \partial / \partial v + v^1 \, \partial / \partial v^0 + v^0 \lambda(u, v, w) \, \partial / \partial w, & G_2 = \partial / \partial v^1. \end{cases}$$

In this case $[F_1, G_1] = 0$ says that

$$v^0 \frac{\partial \lambda}{\partial u} - u^0 \frac{\partial \rho}{\partial v} + u^0 v^0 \left(\rho \frac{\partial \lambda}{\partial w} - \lambda \frac{\partial \rho}{\partial w} \right) = 0,$$

from which

$$\frac{\partial \lambda}{\partial u} = \frac{\partial \rho}{\partial v} = 0 \quad \text{and} \quad \rho(u, w) \frac{\partial \lambda}{\partial w}(v, w) - \lambda(v, w) \frac{\partial \rho}{\partial w}(u, w) = 0.$$

By the last equality

$$\frac{\partial \lambda / \partial w}{\lambda}(v, w) = \frac{\partial \rho / \partial w}{\rho}(u, w) = \text{a function of } w \text{ only} = \frac{f'(w)}{f(w)}, \text{ say.}$$

Thus ρ and λ can be written as $\rho(u, w) = f(w) \cdot \phi(u)$ and $\lambda(v, w) = f(w) \cdot \psi(v)$. If we replace w by $\int (1/f(w)) \, dw$, $\partial / \partial w$ goes over into $(1/f(w)) \partial / \partial w$, and our vector fields are accordingly simplified to

$$\begin{cases} F_1 = \partial / \partial u + u^1 \, \partial / \partial u^0 + u^0 \phi(u) \, \partial / \partial w, & F_2 = \partial / \partial u^1, \\ G_1 = \partial / \partial v + v^1 \, \partial / \partial v^0 + v^0 \psi(v) \, \partial / \partial w, & G_2 = \partial / \partial v^1. \end{cases}$$

From before, we know that $\mathcal{V}$ has a basis of the form

$$\begin{cases} \partial / \partial x + p \, \partial / \partial z + r \, \partial / \partial p + \sigma \, \partial / \partial q, & \partial / \partial r, \\ \partial / \partial y + q \, \partial / \partial z + \sigma \, \partial / \partial p + t \, \partial / \partial q, & \partial / \partial t, \end{cases}$$

for a suitable function $\sigma(x, y, z, p, q)$. In order to arrive at this form we first set $x := u$, $y := v$, $z := w$, $p := u^0 \phi(u)$ and $q := v^0 \psi(v)$, so that

$$\begin{cases} F_1 = \partial / \partial x + p \, \partial / \partial z + (u^0 \phi'(x) + u^1 \phi(x)) \, \partial / \partial p, & F_2 = \partial / \partial u^1, \\ G_1 = \partial / \partial y + q \, \partial / \partial z + (v^0 \psi'(y) + v^1 \psi(y)) \, \partial / \partial q, & G_2 = \partial / \partial v^1. \end{cases}$$

Next $r := u^0 \phi'(x) + u^1 \phi(x)$ and $t := v^0 \psi'(y) + v^1 \psi(y)$ will simplify this to

$$\begin{cases} F_1 = \partial / \partial x + p \, \partial / \partial z + r \, \partial / \partial p, & F_2 = \phi(x) \, \partial / \partial r, \\ G_1 = \partial / \partial y + q \, \partial / \partial z + t \, \partial / \partial q, & G_2 = \psi(y) \, \partial / \partial r. \end{cases}$$

However, the vector field system generated by these vector fields clearly coincides with the system associated to the wave equation $s = 0$.

Theorem 12.6.1. *If both Monge systems admit three functionally independent first integrals, $\mathcal{V}$ is associated to the wave equation $s = 0$.* $\square$

13

Classification of hyperbolic Goursat equations

Recall from section **12.1** that a PDE $s = \sigma(x, y, z, p, q)$ with an associated vector field system $\mathcal{V}$ for which each of the Monge systems $\mathcal{F}$ and $\mathcal{G}$ admits at least two functionally independent first integrals is called a *hyperbolic Goursat equation*. In this chapter we will discuss the classification of such Goursat equations given in [Vessiot 1942].

In the preceding chapter we saw that among the vector field systems considered, those which arise from Goursat equations are characterized as follows:

> $\mathcal{V}$ is associated to a hyperbolic Goursat equation $\Longleftrightarrow$ each Monge system is explicitly integrable $\Longleftrightarrow$ each Monge system admits at least one first integral of the first order.

Moreover these systems have simple canonical forms.

A. If each Monge system admits three functionally independent first integrals, $\mathcal{V}$ is the vector field system associated to the wave equation $s = 0$.

B. If $\mathcal{F} = (F_1, F_2)$ admits two functionally independent first integrals and $\mathcal{G} = (G_1, G_2)$ admits three, $\mathcal{V}$ has the canonical basis

$$\begin{cases} F_1 = \partial/\partial u + u^1\, \partial/\partial u^0 + u^0(R_1 + u\, R_2), & F_2 = \partial/\partial u^1, \\ G_1 = \partial/\partial v + v^0(L_1 + v\, L_2), & G_2 = \partial/\partial v^0. \end{cases}$$

where $\mathrm{span}_{\mathbb{C}}\{R_1, R_2\}$ and $\mathrm{span}_{\mathbb{C}}\{L_1, L_2\}$ are reciprocal Lie algebras of vector fields on $\mathbb{C}_w^2$. From chapter **9** we know that up to equivalence there are just two 2-dimensional Lie algebras—with one being abelian, and the other nonabelian.

In section **13.1** we will see that the abelian algebra gives rise to the

346

Goursat equation

$$s = \frac{p}{y - x}, \tag{B_I}$$

and the nonabelian to

$$s = \frac{pq}{z - x}. \tag{B_{II}}$$

C. If $\mathcal{F}$ and $\mathcal{G}$ each admit two functionally independent first integrals, the canonical basis for $\mathcal{V}$ is

$$\begin{cases} F_1 = \partial/\partial u + u^0 \sum_{i=1}^3 \phi^i(u) R_i, & F_2 = \partial/\partial u^0, \\ G_1 = \partial/\partial v + v^0 \sum_{i=1}^3 \psi^i(v) L_i, & G_2 = \partial/\partial v^0, \end{cases}$$

where $\mathrm{span}_{\mathbb{C}}\{R_1, R_2, R_3\}$ and $\mathrm{span}_{\mathbb{C}}\{L_1, L_2, L_3\}$ are reciprocal Lie algebras of vector fields on $\mathbb{C}_w^3$, and $\phi^i(u)$ and $\psi^i(v)$ are arbitrary functions.

In section **9.2** the 3-dimensional Lie algebras were classified into six distinct types, with one of them depending on an arbitrary parameter:

I.	$[X_1, X_2] = X_1, \quad [X_1, X_3] = 2X_2, \quad [X_2, X_3] = X_3$
II.	$[X_1, X_2] = 0, \quad [X_1, X_3] = X_1, \quad [X_2, X_3] = cX_2$ with $c \neq 0$
III.	$[X_1, X_2] = 0, \quad [X_1, X_3] = X_1, \quad [X_2, X_3] = X_1 + X_2$
IV.	$[X_1, X_2] = 0, \quad [X_1, X_3] = X_1, \quad [X_2, X_3] = 0$
V.	$[X_1, X_2] = 0, \quad [X_1, X_3] = 0, \quad [X_2, X_3] = X_1$
VI.	$[X_1, X_2] = 0, \quad [X_1, X_3] = 0, \quad [X_2, X_3] = 0$

The presence of the arbitrary functions ϕ^i and ψ^i makes it conceivable that to each type of Lie algebra there should correspond infinitely many nonequivalent Goursat equations—but surprisingly this occurs only for the abelian case (i.e., type **VI**). To type **I** belong three different Goursat equations, and to each of the types **II–V** only one Goursat equation. The classification is as follows.

I: The derived algebra is 3-dimensional. Then there are three different Goursat equations:

$$s + 2e^z = 0, \tag{C_I^1}$$

$$s = e^z \sqrt{p^2 - 1}, \tag{C_I^2}$$

$$s = \frac{\sqrt{p^2 - 1}\,\sqrt{q^2 - 1}}{\sinh z}. \tag{C_I^3}$$

II: The general case with a 2-dimensional derived algebra. The corresponding Goursat equation is

$$s + \frac{c}{2}\,\theta(p)\theta(q) = 0, \tag{C_{II}}$$

where the function $\theta(w)$ is implicitly defined by

$$(w - \theta)^c(w + c\theta) = 1.$$

III: The special case with a 2-dimensional derived algebra. The corresponding Goursat equation is

$$s = \frac{1}{z}\,\theta(p)\theta(q) \tag{C_{III}}$$

with the function $\theta(w)$ being defined by

$$\theta = (w + \theta)\log(w + \theta).$$

IV: One of the two cases where the derived Lie algebra is 1-dimensional. The corresponding Goursat equation is

$$s = \frac{\theta(p)\theta(q)}{x + y}, \tag{C_{IV}}$$

where the function $\theta(w)$ is defined by

$$\log(1 + \theta) = w + \theta.$$

V: The other case where the derived Lie algebra is 1-dimensional. The corresponding Goursat equation is

$$s + \frac{2}{x + y} \cdot \sqrt{pq} = 0. \tag{C_V}$$

VI: The abelian case. Here there are infinitely many Goursat equations which depend on the arbitrary functions ϕ^i and ψ^i—but they are all linear:

$$s - \frac{\partial \log A}{\partial y}\,p - \frac{\partial \log B}{\partial x}\,q + \frac{\partial \log A}{\partial y}\frac{\partial \log B}{\partial x}\,z = 0, \tag{C_{VI}}$$

with

$$A = \det \begin{pmatrix} \psi^1 & \phi^1 & (\phi^1)' \\ \psi^2 & \phi^2 & (\phi^2)' \\ \psi^3 & \phi^3 & (\phi^3)' \end{pmatrix} \quad \text{and} \quad B = \det \begin{pmatrix} \phi^1 & \psi^1 & (\psi^1)' \\ \phi^2 & \psi^2 & (\psi^2)' \\ \phi^3 & \psi^3 & (\psi^3)' \end{pmatrix},$$

where the ϕ^i are functions of x, and the ψ^i are functions of y.

Because the verifications of these results are performed along similar lines, we restrict ourselves to the cases B and C_I, referring to [Vessiot 1942] for the rest.

We saw in section **11.1** how to pass from a given vector field system to the corresponding PDE. In the present case this procedure can be somewhat simplified.

In fact, if we let

$$X := \frac{\partial}{\partial x} + p \frac{\partial}{\partial z} + r \frac{\partial}{\partial p} + \sigma \frac{\partial}{\partial q} \quad \text{and} \quad Y := \frac{\partial}{\partial y} + q \frac{\partial}{\partial z} + \sigma \frac{\partial}{\partial p} + t \frac{\partial}{\partial q},$$

the vector field system associated to $s = \sigma(x, y, z, p, q)$ takes the form $\mathcal{V} = (X, Y, \partial/\partial r, \partial/\partial t)$. Its Monge systems $\mathcal{F} = (F_1, F_2)$ and $\mathcal{G} = (G_1, G_2)$ are given by

$$\begin{cases} F_1 = X + Y\sigma \cdot \partial/\partial t, & F_2 = \partial/\partial r, \\ G_1 = Y + X\sigma \cdot \partial/\partial r, & G_2 = \partial/\partial t, \end{cases}$$

from which it is evident that x is *a first integral of the first order for $\mathcal{G}$*, and y is *a first integral of the first order for $\mathcal{F}$*. Moreover

$$F_3 := [F_2, F_1] = \partial/\partial p, \quad G_3 := [G_2, G_1] = \partial/\partial q,$$

and therefore $\mathcal{V}' = (\partial/\partial x + p\,\partial/\partial z, \partial/\partial y + q\,\partial/\partial z, \partial/\partial p, \partial/\partial q, \partial/\partial r, \partial/\partial t)$. From this it is clear that $(F_3, G_3, F_2, G_2) = (\partial/\partial p, \partial/\partial q, \partial/\partial r, \partial/\partial t)$ is a complete 4-dimensional subsystem of $\mathcal{V}'$, admitting x, y and z as first integrals.

We also have the relations

$$p = F_1 z, \quad q = G_1 z \quad \text{and} \quad \sigma = F_1 q = F_1 \circ G_1 z = G_1 p = G_1 \circ F_1 z.$$

In the cases **B** and **C** above, u is a first integral of the first order for $\mathcal{G}$, and v is a first integral of the first order for $\mathcal{F}$. Therefore we set

$$x := u \quad \text{and} \quad y := v.$$

Then x and y are first integrals of (F_3, G_3, F_2, G_2), and we determine z as a third functionally independent first integral of this complete system. After that the desired PDE is obtained from

$$p := F_1 z, \quad q := G_1 z \quad \text{and} \quad \sigma := F_1 \circ G_1 z = G_1 \circ F_1 z.$$

When the Goursat equations have been found, their solutions are obtained from the integrals of the explicitly integrable Monge systems $\mathcal{F}$ and $\mathcal{G}$ by means of the composition functions of the corresponding Lie group.

Recall from section **6.7** that the canonical form of an explicitly inte-grable 2-dimensional vector field system is

$$\mathcal{E}_n = \left(\frac{\partial}{\partial x} + \sum_{k=1}^{n} x^k \frac{\partial}{\partial x^{k-1}}, \frac{\partial}{\partial x^n} \right).$$

$\mathcal{E}_n$ has the derivatives

$$\mathcal{E}_n' = \left(\frac{\partial}{\partial x} + \sum_{k=1}^{n-1} x^k \frac{\partial}{\partial x^{k-1}}, \frac{\partial}{\partial x^{n-1}}, \frac{\partial}{\partial x^n} \right),$$

$$\dots$$

$$\mathcal{E}_n^{(n-1)} = \left(\frac{\partial}{\partial x} + x^1 \frac{\partial}{\partial x^0}, \frac{\partial}{\partial x^1}, \dots, \frac{\partial}{\partial x^n} \right),$$

$$\mathcal{E}_n^{(n)} = \left(\frac{\partial}{\partial x}, \frac{\partial}{\partial x^0}, \dots, \frac{\partial}{\partial x^n} \right),$$

so that

$$\mathcal{E}_n \subset \mathcal{E}_n' \subset \dots \subset \mathcal{E}_n^{(n-1)} \subset \mathcal{E}_n^{(n)},$$

where $\mathcal{E}_n^{(n)}$ is complete. The general integral curve of $\mathcal{E}_n$ on which $dx \neq 0$ is given by

$$x^0 = \phi(x), x^1 = \phi'(x), \dots, x^n = \phi^{(n)}(x),$$

with $\phi(x)$ being an arbitrary function.

So we are required to put the explicitly integrable Monge systems in this canonical form, read off their integral curves, and then use the composition functions in order to obtain the general solutions of the associated Goursat equations.

13.1 Goursat equations which are associated to 2-dimensional Lie groups

We are here looking for PDEs $s = \sigma(x, y, z, p, q)$ associated to vector field systems $\mathcal{V}$ generated by

$$\begin{cases} F_1 = \partial/\partial u + u^1 \, \partial/\partial u^0 + u^0 \, (R_1 + u \, R_2), & F_2 = \partial/\partial u^1, \\ G_1 = \partial/\partial v + v^0 \, (L_1 + v \, L_2), & G_2 = \partial/\partial v^0, \end{cases}$$

where $\text{span}_{\mathbb{C}}\{R_1, R_2\}$ and $\text{span}_{\mathbb{C}}\{L_1, L_2\}$ are the right- and left-invariant Lie algebras of one of the two types of 2-dimensional Lie groups.

As representative for the abelian Lie groups we choose

$$\mathfrak{G}_I: \qquad \mathbb{C}^2 \to \mathbb{C}^2$$
$$(z^1, z^2) \mapsto (z^1 + a^1, z^2 + a^2) =: \phi_a(z),$$

where a^1 and a^2 are arbitrary parameters. Then

$$\phi_a \circ \phi_b(z) = (z^1 + (a^1 + b^1), z^2 + (a^2 + b^2)) = \phi_c(z),$$

which means that the composition functions are given by

$$c^1 = \gamma^1(a; b) = a^1 + b^1, \quad c^2 = \gamma^2(a; b) = a^2 + b^2.$$

In this case the right- and left-invariant Lie algebras coincide:

$$R_1 = L_1 = \partial/\partial w^1, \quad R_2 = L_2 = \partial/\partial w^2.$$

As representative for the nonabelian 2-dimensional Lie groups we take the affine group in one variable—that is, the group which was used as a main example throughout chapter **8**:

$$\mathfrak{G}_{II}: \mathbb{C} \to \mathbb{C}$$
$$z \mapsto a^1 z + a^2 =: \phi_a(z).$$

Because

$$\phi_a \circ \phi_b(z) = \phi_a(b^1 z + b^2) = a^1 b^1 z + (a^1 b^1 + a^2) = \phi_c(z),$$

the composition functions are in this case given by

$$c^1 = \gamma^1(a; b) = a^1 b^1, \quad c^2 = \gamma^2(a; b) = a^1 b^2 + a^2,$$

and the identity element is $e = (1, 0)$. Recalling from section **8.3** that the right- and left-invariant Lie algebras are generated by

$$R_k = \sum_j \frac{\partial \gamma^j}{\partial a^k}(e; w) \frac{\partial}{\partial w^j} \quad \text{and} \quad L_k = \sum_j \frac{\partial \gamma^j}{\partial b^k}(w; e) \frac{\partial}{\partial w^j}$$

respectively, it follows that

$$\begin{cases} R_1 = w^1 \, \partial/\partial w^1 + w^2 \, \partial/\partial w^2, & R_2 = \partial/\partial w^2, \\ L_1 = w^1 \, \partial/\partial w^1, & L_2 = w^1 \, \partial/\partial w^2. \end{cases}$$

With the derived system $\mathcal{V}'$ being equal to

$$\mathcal{V}' = \left(\frac{\partial}{\partial u} + u^0 (R_1 + u R_2), \frac{\partial}{\partial u^0}, \frac{\partial}{\partial u^1}, \frac{\partial}{\partial v}, L_1 + v L_2, \frac{\partial}{\partial v^0} \right),$$

it is clear that $\mathcal{W} := (\partial/\partial u^0, \partial/\partial u^1, L_1 + v L_2, \partial/\partial v^0)$ is a complete 4-dimensional subsystem of $\mathcal{V}'$, admitting u and v as first integrals.

As explained earlier we set $x := u$ and $y := v$, and choose z as a third functionally independent first integral of $\mathcal{W}$. Thus z is independent of u^0, u^1 and v^0, and is a first integral of $L_1 + v L_2$. In case I,

$$L_1 + v L_2 = \frac{\partial}{\partial w^1} + v \frac{\partial}{\partial w^2} \qquad \text{has the first integral } z = vw^1 - w^2,$$

and in case II,

$$L_1 + v L_2 = w^1 \left(\frac{\partial}{\partial w^1} + v \frac{\partial}{\partial w^2} \right) \qquad \text{has the same first integral } z.$$

With $F_2 = \partial/\partial u^1$ and $G_2 = \partial/\partial v^0$ being the principal vector fields of $\mathcal{V}$, p and q are then obtained as

$$p = F_1 z, \quad q = G_1 z,$$

and σ as $\sigma = F_1 q = G_1 p$.

So in case I,

$$p = u^0 \left(\frac{\partial}{\partial w^1} + u \frac{\partial}{\partial w^2} \right) (vw^1 - w^2) = u^0 (v - u) = u^0 (y - x),$$

$$q = \left(\frac{\partial}{\partial v} + v^0 \frac{\partial}{\partial w^1} + v^0 v \frac{\partial}{\partial w^2} \right) (vw^1 - w^2) = w^1,$$

and

$$\sigma = F_1 q = F_1 w^1 = u^0.$$

Thus $p = \sigma (y - x)$—so the Goursat equation associated to the abelian 2-dimensional Lie group is given by

$$s = \frac{p}{y - x}. \tag{B_I}$$

In case II,

$$p = u^0 \left(w^1 \frac{\partial}{\partial w^1} + (w^2 + u) \frac{\partial}{\partial w^2} \right) (vw^1 - w^2) = u^0 (z - x),$$

$$q = \left(\frac{\partial}{\partial v} + v^0 \frac{\partial}{\partial w^1} + v^0 v \frac{\partial}{\partial w^2} \right) (vw^1 - w^2) = w^1,$$

and

$$\sigma = F_1 q = F_1 w^1 = u^0 w^1,$$

which gives the Goursat equation

$$s = \frac{pq}{z - x} \tag{B_{II}}$$

associated to the affine group.

In order to solve B_I we first have to put the explicitly integrable vector field systems

$$\mathcal{F} = \left(\frac{\partial}{\partial u} + u^1 \frac{\partial}{\partial u^0} + u^0 \frac{\partial}{\partial w^1} + u^0 u \frac{\partial}{\partial w^2}, \frac{\partial}{\partial u^1} \right)$$

and

$$\mathcal{G} = \left(\frac{\partial}{\partial v} + v^0 \frac{\partial}{\partial w^1} + v^0 v \frac{\partial}{\partial w^2}, \frac{\partial}{\partial v^0} \right)$$

into canonical form.

The last derivative of $\mathcal{F}$ which is *not* complete is

$$\mathcal{F}'' = (\partial/\partial u, \partial/\partial w^1 + u\, \partial/\partial w^2, \partial/\partial u^0, \partial/\partial u^1),$$

while the second derivative of the corresponding canonical explicitly integrable system $\mathcal{E}_3$ equals

$$\mathcal{E}_3'' = (\partial/\partial x + x^1\, \partial/\partial x^0, \partial/\partial x^1, \partial/\partial x^2, \partial/\partial x^3).$$

Therefore we replace w^2 by the first integral $a = uw^1 - w^2$ of $\partial/\partial w^1 + u\, \partial/\partial w^2$—for then the latter goes over into $\partial/\partial w^1$, and $\mathcal{F}$ receives its canonical form

$$\mathcal{F} = \left(\frac{\partial}{\partial u} + w^1 \frac{\partial}{\partial a} + u^0 \frac{\partial}{\partial w^1} + u^1 \frac{\partial}{\partial u^0}, \frac{\partial}{\partial u^1} \right).$$

The general integral curve of $\mathcal{F}$ on which $du \neq 0$ is given by

$$a = \xi(u), \quad w^1 = \xi'(u), \quad u^0 = \xi''(u) \quad \text{and} \quad u^1 = \xi'''(u),$$

where $\xi(u)$ is an arbitrary function. Therefore

$$w^1 = \xi'(u) =: w^1_{\mathcal{F}} \quad \text{and} \quad w^2 = uw^1 - a = u\xi'(u) - \xi(u) =: w^2_{\mathcal{F}}.$$

Similarly the last derivative of $\mathcal{G}$ which is not complete is

$$\mathcal{G}' = (\partial/\partial v, \partial/\partial w^1 + v\, \partial/\partial w^2, \partial/\partial v^0).$$

To put this into canonical form we replace w^2 by the first integral $b = vw^1 - w^2$ of $\partial/\partial w^1 + v\, \partial/\partial w^2$—which makes $\mathcal{G}$ go over into

$$\mathcal{G} = \left(\frac{\partial}{\partial v} + w^1 \frac{\partial}{\partial b} + v^0 \frac{\partial}{\partial w^1}, \frac{\partial}{\partial v^0} \right).$$

With $\eta(v)$ being an arbitrary function, the general integral curve of $\mathcal{G}$ on which $dv \neq 0$ is given by

$$b = \eta(v), \quad w^1 = \eta'(v), \quad \text{and} \quad v^0 = \eta''(v),$$

whence

$$w^1 = \eta'(v) =: w_{\mathcal{G}}^1, \quad \text{and} \quad w^2 = vw^1 - b = v\eta'(v) - \eta(v) =: w_{\mathcal{G}}^2.$$

If we utilize the composition functions $c^1 = a^1 + b^1$ and $c^2 = a^2 + b^2$ of $\mathfrak{G}_I$, the general integral surface of $\mathcal{V}$ on which $du \wedge dv \neq 0$ is given by

$$x = u, \quad y = v, \quad z = vw^1 - w^2, \quad w^1 = w_{\mathcal{F}}^1 + w_{\mathcal{G}}^1 = \xi'(u) + \eta'(v),$$

and

$$w^2 = w_{\mathcal{F}}^2 + w_{\mathcal{G}}^2 = u\xi'(u) + v\eta'(v) - \xi(u) - \eta(v).$$

Thus

$$z = (y - x)\xi'(x) + \xi(x) + \eta(y).$$

Theorem 13.1.1. *The general solution of the Goursat equation*

$$\frac{\partial^2 z}{\partial x \partial y} = \frac{\partial z/\partial x}{y - x} \tag{B_I}$$

associated to the abelian 2-dimensional Lie group is given by

$$z = (y - x)\xi'(x) + \xi(x) + \eta(y),$$

where $\xi(x)$ and $\eta(y)$ are arbitrary functions. □

Let us now solve $B_{II}: \ s = pq/(z - x)$ in the same way. Here

$$\mathcal{F} = (\partial/\partial u + u^1 \, \partial/\partial u^0 + u^0 \, (w^1 \, \partial/\partial w^1 + (w^2 + u) \, \partial/\partial w^2), \partial/\partial u^1),$$

with the second derivative

$$\mathcal{F}'' = (\partial/\partial u, w^1 \, \partial/\partial w^1 + (w^2 + u) \, \partial/\partial w^2, \partial/\partial u^0, \partial/\partial u^1).$$

If we replace w^2 by the first integral $a = (w^2 + u)/w^1$ of the vector field $w^1 \, \partial/\partial w^1 + (w^2 + u) \, \partial/\partial w^2$, $\mathcal{F}$ is changed into

$$\mathcal{F} = \left(\frac{\partial}{\partial u} + \frac{1}{w^1} \frac{\partial}{\partial a} + u^0 w^1 \frac{\partial}{\partial w^1} + u^1 \frac{\partial}{\partial u^0}, \frac{\partial}{\partial u^1} \right) =: (F_1, F_2).$$

Setting $\tilde{w}^1 := (w^1)^{-1}$ turns F_1 into

$$\frac{\partial}{\partial u} + \tilde{w}^1 \frac{\partial}{\partial a} - u^0 \tilde{w}^1 \frac{\partial}{\partial \tilde{w}^1} + u^1 \frac{\partial}{\partial u^0};$$

next replacing u^0 by $\tilde{u}^0 := -u^0 \tilde{w}^1$ we instead get

$$\frac{\partial}{\partial u} + \tilde{w}^1 \frac{\partial}{\partial a} + \tilde{u}^0 \frac{\partial}{\partial \tilde{w}^1} + ((u^0)^2 - u^1)\tilde{w}^1 \frac{\partial}{\partial \tilde{u}^0}.$$

Finally, with $\tilde{u}^1 := ((u^0)^2 - u^1)\tilde{w}^1$ replacing u^1, we put $\mathcal{F}$ into the canonical form

$$\mathcal{F} = \left(\frac{\partial}{\partial u} + \tilde{w}^1 \frac{\partial}{\partial a} + \tilde{u}^0 \frac{\partial}{\partial \tilde{w}^1} + \tilde{u}^1 \frac{\partial}{\partial \tilde{u}^0}, \frac{\partial}{\partial \tilde{u}^1} \right).$$

The general integral curve of $\mathcal{F}$ on which $du \neq 0$ is then given by

$$a = \xi(u), \quad \tilde{w}^1 = \frac{1}{w^1} = \xi'(u), \quad \tilde{u}^0 = \xi''(u) \quad \text{and} \quad \tilde{u}^1 = \xi'''(u),$$

where $\xi(u)$ is arbitrary. Therefore

$$w^1 = \frac{1}{\xi'(u)} =: w_{\mathcal{F}}^1 \quad \text{and} \quad w^2 = aw^1 - u = \frac{\xi(u)}{\xi'(u)} - u =: w_{\mathcal{F}}^2.$$

In the same way

$$\mathcal{G} = (\partial/\partial v + v^0 (w^1 \,\partial/\partial w^1 + vw^1 \,\partial/\partial w^2), \partial/\partial v^0)$$

is put into canonical form by first replacing w^2 by the first integral $b = vw^1 - w^2$ of $\partial/\partial w^1 + v \,\partial/\partial w^2$, and then letting $\tilde{v}^0 := v^0 w^1$ replace v^0:

$$\mathcal{G} = \left(\frac{\partial}{\partial v} + w^1 \frac{\partial}{\partial b} + \tilde{v}^0 \frac{\partial}{\partial w^1}, \frac{\partial}{\partial \tilde{v}^0} \right).$$

The general integral curve of $\mathcal{G}$ on which $dv \neq 0$ is therefore expressed as

$$b = \eta(v), \quad w^1 = \eta'(v) \quad \text{and} \quad \tilde{v}^0 = \eta''(v),$$

where $\eta(v)$ is an arbitrary function. Consequently

$$w^1 = \eta'(v) =: w_{\mathcal{G}}^1 \quad \text{and} \quad w^2 = vw^1 - b = v\eta'(v) - \eta(v) =: w_{\mathcal{G}}^2.$$

With the composition functions $c^1 = a^1 b^1$ and $c^2 = a^1 b^2 + a^2$ this means that the general integral surface of $\mathcal{V}$ on which $du \wedge dv \neq 0$ is given by

$$x = u, \quad y = v, \quad z = vw^1 - w^2$$

and

$$w^1 = w_{\mathcal{F}}^1 w_{\mathcal{G}}^1 = \frac{\eta'(v)}{\xi'(u)}, \quad w^2 = w_{\mathcal{F}}^1 w_{\mathcal{G}}^2 + w_{\mathcal{F}}^2 = \frac{v\eta'(v) - \eta(v)}{\xi'(u)} + \frac{\xi(u)}{\xi'(u)} - u.$$

Eliminating u, v, w^1 and w^2 we finally obtain the solution

$$z = y \frac{\eta'(y)}{\xi'(x)} - \frac{y\eta'(y) - \eta(y)}{\xi'(x)} - \frac{\xi(x)}{\xi'(x)} + x = \frac{\eta(y) - \xi(x)}{\xi'(x)} + x.$$

Theorem 13.1.2. *The general solution of the Goursat equation*

$$\frac{\partial^2 z}{\partial x \partial y} = \frac{\partial z/\partial x \cdot \partial z/\partial y}{z - x} \qquad (B_{II})$$

associated to the affine group in one variable is given by

$$z = \frac{\eta(y) - \xi(x)}{\xi'(x)} + x,$$

where $\xi(x)$ and $\eta(y)$ are arbitrary functions. $\square$

13.2 Goursat equations associated to the projective group in one variable

In this section we determine the Goursat equations associated to the vector field systems $\mathcal{V}$ generated by

$$\begin{cases} F_1 = \partial/\partial u + u^0 \sum_{i=1}^{3} \phi^i(u)\, R_i, & F_2 = \partial/\partial u^0, \\ G_1 = \partial/\partial v + v^0 \sum_{i=1}^{3} \psi^i(v)\, L_i, & G_2 = \partial/\partial v^0, \end{cases}$$

with $\mathfrak{g}_R = \operatorname{span}_{\mathbb{C}}\{R_1, R_2, R_3\}$ and $\mathfrak{g}_L = \operatorname{span}_{\mathbb{C}}\{L_1, L_2, L_3\}$ being reciprocal Lie algebras of *type* **I**—i.e., those for which the derived Lie algebras are 3-dimensional.

We saw in section **9.2** that this type is obtained from the projective group G in one variable:

$$G : \mathbb{C} \to \mathbb{C}$$

$$z \mapsto \frac{a_1 z + a_2}{a_3 z + 1},$$

where a_1, a_2, a_3 are parameters satisfying $a_1 - a_2 a_3 \neq 0$. The composition functions of the corresponding parameter group $\mathfrak{G}$ are

$$c_1 = \frac{a_1 b_1 + a_2 b_3}{a_3 b_2 + 1}, \quad c_2 = \frac{a_1 b_2 + a_2}{a_3 b_2 + 1} \quad \text{and} \quad c_3 = \frac{a_3 b_1 + b_3}{a_3 b_2 + 1},$$

so that the right-invariant Lie parameter group $\mathfrak{G}_R$ is defined by

$$w^1 \mapsto \frac{a_1 w^1 + a_2 w^3}{a_3 w^2 + 1}, \quad w^2 \mapsto \frac{a_1 w^2 + a_2}{a_3 w^2 + 1}, \quad w^3 \mapsto \frac{a_3 w^1 + w^3}{a_3 w^2 + 1},$$

and the left-invariant Lie parameter group $\mathfrak{G}_L$ by

$$w^1 \mapsto \frac{b_1 w^1 + b_3 w^2}{b_2 w^3 + 1}, \quad w^2 \mapsto \frac{b_2 w^1 + w^2}{b_2 w^3 + 1}, \quad w^3 \mapsto \frac{b_1 w^3 + b_3}{b_2 w^3 + 1}.$$

The corresponding right Lie algebra $\mathfrak{g}_R$ is generated by

$$R_1 = w^1 \frac{\partial}{\partial w^1} + w^2 \frac{\partial}{\partial w^2}, \quad R_2 = w^3 \frac{\partial}{\partial w^1} + \frac{\partial}{\partial w^2}$$

and

$$R_3 = w^1 \frac{\partial}{\partial w^3} - w^2 \sum_{i=1}^{3} w^i \frac{\partial}{\partial w^i},$$

and the left-invariant Lie algebra $\mathfrak{g}_L$ by

$$L_1 = w^1 \frac{\partial}{\partial w^1} + w^3 \frac{\partial}{\partial w^3}, \quad L_2 = w^1 \frac{\partial}{\partial w^2} - w^3 \sum_{i=1}^{3} w^i \frac{\partial}{\partial w^i}$$

and

$$L_3 = w^2 \frac{\partial}{\partial w^1} + \frac{\partial}{\partial w^3},$$

—see section **9.2**.

By the reciprocality each R_i is invariant under the action of $\mathfrak{G}_L$. On the other hand $\mathfrak{G}_R$ preserves the whole Lie algebra $\mathfrak{g}_R$, but not the individual R_i. In fact, under the action of $\mathfrak{G}_R$,

$$R_1 \mapsto \frac{a_1 + a_2 a_3}{a_1 - a_2 a_3} R_1 - \frac{a_1 a_2}{a_1 - a_2 a_3} R_2 + \frac{a_3}{a_1 - a_2 a_3} R_3,$$

$$R_2 \mapsto \frac{-2a_2 a_3}{a_1 - a_2 a_3} R_1 + \frac{a_1^2}{a_1 - a_2 a_3} R_2 - \frac{a_3^2}{a_1 - a_2 a_3} R_3,$$

$$R_3 \mapsto \frac{2a_2}{a_1 - a_2 a_3} R_1 - \frac{a_2^2}{a_1 - a_2 a_3} R_2 + R_3.$$

The idea now is to choose suitable functions $a_1(u)$, $a_2(u)$ and $a_3(u)$ with $a_1(u) - a_2(u)a_3(u) \neq 0$ such that the sum $\sum_{i=1}^{3} \phi^i(u) R_i$ in F_1 goes over into a multiple of *one* of the R_i under the corresponding action of $\mathfrak{G}_R$—say a multiple of R_1. For this to happen it is necessary that

$$\begin{cases} a_1 a_2 \phi^1 - a_1^2 \phi^2 + a_2^2 \phi^3 = 0, \\ a_3 \phi^1 - a_3^2 \phi^2 + \phi^3 = 0. \end{cases}$$

By the first equation

$$a_2 = a_1 \cdot \frac{-\phi^1 \pm \sqrt{(\phi^1)^2 + 4\phi^2 \phi^3}}{2\phi^3},$$

and by the second

$$a_3 = \frac{\phi^1 \pm \sqrt{(\phi^1)^2 + 4\phi^2\phi^3}}{2\phi^2}.$$

Choosing the same branches of the square root we will get $a_1 - a_2 a_3 = 0$—so this is ruled out. But with *different* branches—which presupposes that $(\phi^1)^2 + 4\phi^2\phi^3 \neq 0$—we do get an adequate solution.

In the singular case $(\phi^1)^2 + 4\phi^2\phi^3 = 0$ a similar argument shows that $\sum_{i=1}^{3} \phi^i(u) R_i$ can be replaced by a multiple of R_2.

With

$$\tilde{w}^1 = \frac{a_1(u)w^1 + a_2(u)w^2}{a_3(u)w^2 + 1}, \quad \tilde{w}^2 = \frac{a_1(u)w^2 + a_2(u)}{a_3(u)w^2 + 1}, \quad \tilde{w}^3 = \frac{a_3(u)w^1 + w^3}{a_3(u)w^2 + 1},$$

$\partial/\partial u$ is turned into

$$\frac{\partial}{\partial u} + \sum_{i=1}^{3} \frac{da_i}{du} \left(\sum_{j=1}^{3} \frac{\partial \tilde{w}^j}{\partial a_i} \frac{\partial}{\partial \tilde{w}^j} \right).$$

But according to the first fundamental theorem for Lie groups the vector fields

$$\sum_{j=1}^{3} \frac{\partial \tilde{w}^j}{\partial a_i} \frac{\partial}{\partial \tilde{w}^j} \quad \text{for } i = 1, 2, 3$$

are linear combinations of the vector fields R_1, R_2, R_3 expressed in the $\tilde{w}^j$, with coefficients that are functions of the $a_i(u)$. Therefore

$$\frac{\partial}{\partial u} \longmapsto \frac{\partial}{\partial u} + \sum_{i=1}^{3} m^i(u) R_i$$

for suitable functions $m^i(u)$. Redefining u^0 in an appropriate way, $\mathcal{F} = (F_1, F_2)$ is thereby brought to the form

$$F_1 = \partial/\partial u + m^2(u) R_2 + m^3(u) R_3 + u^0 R_1, \quad F_2 = \partial/\partial u^0$$

if $(\phi^1)^2 + 4\phi^2\phi^3 \neq 0$, and to

$$F_1 = \partial/\partial u + m^1(u) R_1 + m^3(u) R_3 + u^0 R_2, \quad F_2 = \partial/\partial u^0$$

if $(\phi^1)^2 + 4\phi^2\phi^3 = 0$.

Since $\mathfrak{G}_R$ and $\mathfrak{G}_L$ commute, the above change of coordinates does not affect $\mathcal{G} = (G_1, G_2)$. Using the same kind of argument the latter can similarly be brought either to the form

$$G_1 = \partial/\partial v + n^2(v) L_2 + n^3(v) L_3 + v^0 L_1, \quad G_2 = \partial/\partial v^0,$$

or to the form

$$G_1 = \partial/\partial v + n^1(v) L_1 + n^2(v) L_2 + v^0 L_3, \quad G_2 = \partial/\partial v^0.$$

This leads to three different cases as regards $\mathcal{V}$.

1. The simplest case occurs when $\mathcal{V}$ is generated by

$$\begin{cases} F_1 = \partial/\partial u + m^1(u) R_1 + m^3(u) R_3 + u^0 R_2, & F_2 = \partial/\partial u^0, \\ G_1 = \partial/\partial v + n^1(v) L_1 + n^2(v)L_2 + v^0 L_3, & G_2 = \partial/\partial v^0. \end{cases}$$

Then

$$\mathcal{V}' = (\partial/\partial u + m^1 R_1 + m^3 R_3, \partial/\partial v + n^1 L_1 + n^2 L_2, R_2, L_3, \partial/\partial u^0, \partial/\partial v^0)$$

with $\partial/\partial u^0, \partial/\partial v^0 \in C(\mathcal{V}')$. In order to reduce

$$\mathcal{V}'_{\text{red}} = (\partial/\partial u + m^1 R_1 + m^3 R_3, \partial/\partial v + n^1 L_1 + n^2 L_2, R_2, L_3)$$

to the canonical form

$$(\partial/\partial x + p\,\partial/\partial z, \partial/\partial y + q\,\partial/\partial z, \partial/\partial p, \partial/\partial q)$$

we utilize the fact that the complete 2-dimensional subsystem (R_2, L_3) of $\mathcal{V}'_{\text{red}}$ admits the three functionally independent first integrals

$$u, \quad v \quad \text{and} \quad a := w^1 - w^2 w^3.$$

The relations

$$R_1 a = a, R_2 a = 0, R_3 a = -2w^2 a; L_1 a = a, L_2 a = -2w^3 a, L_3 a = 0$$

are somewhat simplified if a is replaced by $\log a$:

$$R_1 \log a = 1, R_2 \log a = 0, R_3 \log a = -2w^2$$

and

$$L_1 \log a = 1, L_2 \log a = -2w^3, L_3 \log a = 0.$$

Let us therefore set

$$x := u, \quad y := v, \quad z := \log a = \log(w^1 - w^2 w^3),$$

and replace w^1 by z. Then

$$\mathcal{V}'_{\text{red}} \mapsto \left(\frac{\partial}{\partial x} + (m^1(x) - 2w^2 m^3(x)) \frac{\partial}{\partial z} + \cdots, \right.$$

$$\left. \frac{\partial}{\partial y} + (n^1(y) - 2w^3 n^2(y)) \frac{\partial}{\partial z} + \cdots, \frac{\partial}{\partial w^2}, \frac{\partial}{\partial w^3} \right),$$

so that

$$p = m^1(x) - 2w^2 m^3(x) \quad \text{and} \quad q = n^1(y) - 2w^3 n^2(y);$$

furthermore

$$\sigma = F_1 q = G_1 p = -2m^3(x)(w^1 - w^2 w^3) n^2(y) = -2m^3(x) n^2(y) e^z.$$

Finally, replacing x by $\int m^3(x) \, dx$ and y by $\int n^2(y) \, dy$ we obtain the Goursat equation

$$s + 2e^z = 0. \tag{C_I^1}$$

The final change of variables is equivalent to assuming that $m^3 = n^2 = 1$. Moreover, since m^1 and n^1 do not appear in C_I^1, we would have obtained this PDE by setting $m^1 = n^1 = 0$ from the beginning. So the canonical form for $\mathcal{V}$ in this case is

$$\begin{cases} F_1 = \partial/\partial u + R_3 + u^0 R_2, & F_2 = \partial/\partial u^0, \\ G_1 = \partial/\partial v + L_2 + v^0 L_3, & G_2 = \partial/\partial v^0, \end{cases}$$

with *no arbitrary functions left*.

Let us now determine the general solution of C_I^1. The first step consists in reducing the explicitly integrable system

$$\mathcal{F} = (\partial/\partial u + R_3 + u^0 R_2, \partial/\partial u^0) \quad \text{on} \quad \mathbb{C}^5_{u,u^0,w^1,w^2,w^3}$$

to its canonical form

$$\mathcal{E} = (\partial/\partial x + x^1 \, \partial/\partial x^0 + x^2 \, \partial/\partial x^1 + x^3 \, \partial/\partial x^2, \partial/\partial x^3).$$

Now

$$\mathcal{E}'' = (\partial/\partial x + x^1 \, \partial/\partial x^0, \partial/\partial x^1, \partial/\partial x^2, \partial/\partial x^3)$$

contains the complete 3-dimensional subsystem $(\partial/\partial x^1, \partial/\partial x^2, \partial/\partial x^3)$, which admits the first integrals x and x^0. Let us therefore consider the corresponding derivative of $\mathcal{F}$—noting that $[R_2, R_3] = -2R_1$:

$$\mathcal{F}'' = (\partial/\partial u + R_3, R_2, R_1, \partial/\partial u^0).$$

Here $(R_1, R_2, \partial/\partial u^0) = (w^1 \, \partial/\partial w^1 + w^2 \, \partial/\partial w^2, w^3 \partial/\partial w^1 + \partial/\partial w^2, \partial/\partial u^0)$ is the analogous complete 3-dimensional subsystem with u and w^3 as first integrals. We therefore set

$$x := u \quad \text{and} \quad x^0 := w^3,$$

whereupon

$$x^1 = F_1 x^0 = w^1 - w^2 w^3 = a, \quad x^2 = F_1 x^1 = -2w^2 a, \quad x^3 = F_1 x^2 = \cdots.$$

The general integral curve of $\mathcal{F}$ on which $dx \neq 0$ is given by

$$x^0 = w^3 = \phi(x), \quad x^1 = a = \phi'(x), \quad x^2 = -2w^2 a = \phi''(x), \quad x^3 = \phi'''(x),$$

where $\phi(x)$ is an arbitrary function. Thus

$$w_{\mathcal{F}}^3 = \phi(x), \quad w_{\mathcal{F}}^2 = -\frac{\phi''(x)}{2\phi'(x)}, \quad w_{\mathcal{F}}^1 = a + w_{\mathcal{F}}^2 w_{\mathcal{F}}^3 = \phi'(x) - \frac{\phi(x)\phi''(x)}{2\phi'(x)}.$$

In the same way we find the following relations on the general integral curve of $\mathcal{G}$:

$$w_{\mathcal{G}}^1 = \psi'(y) - \frac{\psi(y)\psi''(y)}{2\psi'(y)}, \quad w_{\mathcal{G}}^2 = \psi(y) \quad \text{and} \quad w_{\mathcal{G}}^3 = -\frac{\psi''(y)}{2\psi'(y)},$$

where $\psi(y)$ is an arbitrary function.

The composition functions of the projective group show that on the general integral surface of $\mathcal{V}$

$$w^1 = \frac{w_{\mathcal{F}}^1 w_{\mathcal{G}}^1 + w_{\mathcal{F}}^2 w_{\mathcal{G}}^3}{w_{\mathcal{F}}^3 w_{\mathcal{G}}^2 + 1}, \quad w^2 = \frac{w_{\mathcal{F}}^1 w_{\mathcal{G}}^2 + w_{\mathcal{F}}^2}{w_{\mathcal{F}}^3 w_{\mathcal{G}}^2 + 1}, \quad w^3 = \frac{w_{\mathcal{F}}^3 w_{\mathcal{G}}^1 + w_{\mathcal{G}}^3}{w_{\mathcal{F}}^3 w_{\mathcal{G}}^2 + 1}.$$

In particular

$$e^z = w^1 - w^2 w^3 = \frac{(w_{\mathcal{F}}^1 - w_{\mathcal{F}}^2 w_{\mathcal{F}}^3)(w_{\mathcal{G}}^1 - w_{\mathcal{G}}^2 w_{\mathcal{G}}^3)}{(w_{\mathcal{F}}^3 w_{\mathcal{G}}^2 + 1)^2} = \frac{\phi'(x)\psi'(y)}{(\phi(x)\psi(y) + 1)^2},$$

so that the general solution of the PDE $C_I^1 s + 2e^z = 0$, is given by

$$z = \log \phi'(x) + \log \psi'(y) - 2\log(\phi(x)\psi(y) + 1),$$

where $\phi(x)$ and $\psi(y)$ are arbitrary functions.

2. The second case to consider is when $\mathcal{V}$ is generated by

$$\begin{cases} F_1 = \partial/\partial u + m^2(u) R_2 + m^3(u) R_3 + u^0 R_1, & F_2 = \partial/\partial u^0, \\ G_1 = \partial/\partial v + n^1(v) L_1 + n^2(v) L_2 + v^0 L_3, & G_2 = \partial/\partial v^0. \end{cases}$$

Then

$$\mathcal{V}' = (\partial/\partial u + m^2 R_2 + m^3 R_3, \partial/\partial v + n^1 L_1 + n^2 L_2, R_1, L_3, \partial/\partial u^0, \partial/\partial v^0)$$

with $\partial/\partial u^0, \partial/\partial v^0 \in C(\mathcal{V}')$, so that

$$\mathcal{V}'_{\text{red}} = (\partial/\partial u + m^2 R_2 + m^3 R_3, \partial/\partial v + n^1 L_1 + n^2 L_2, R_1, L_3).$$

Here the 2-dimensional subsystem (R_1, L_3) is complete, and admits the first integrals

$$u, \quad v, \quad \text{and} \quad \frac{w^1}{w^2} - w^3.$$

Let us therefore in analogy with the preceding case set

$$x := u, \quad y := v \quad \text{and } z := \log \frac{w^1 - w^2 w^3}{w^3},$$

whereupon

$$p = F_1 z = -m^3(u)w^2 - \frac{m^2(u)}{w^2},$$

and

$$\sigma = G_1 p = \frac{m^2(u) - m^3(u)(w^2)^2}{w^2} \, n^2(v) \, \frac{w^1 - w^2 w^3}{w^2}.$$

Solving w^2 from the expression for p and using $e^z = (w^1 - w^2 w^3)/w^3$ we get

$$s = \sigma(x, y, z, p, q) = \sqrt{p^2 - 4m^2(x)m^3(x)} \, n^2(y) \, e^z.$$

By making the replacements

$$x \mapsto 2 \int \sqrt{m^2(x)m^3(x)} \, dx \quad \text{and} \quad y \mapsto \int n^2(y) \, dy$$

this is simplified to

$$s = e^z \sqrt{p^2 - 1}. \tag{C_I^2}$$

Note that this final form had been obtained directly with $4m^2(x)m^3(x) = 1$ and $n^2(y) = 1$. We may thus assume that

$$m^2(x) = m^3(x) = \frac{1}{2}, \quad n^1(y) = 0 \quad \text{and} \quad n^2(y) = 1$$

—which means that the vector field system $\mathcal{V}$ associated to the Goursat equation C_I^2 is generated by

$$F_1 = \frac{\partial}{\partial u} + \frac{1}{2}(R_2 + R_3) + u^0 R_1, \qquad F_2 = \frac{\partial}{\partial u^0},$$

$$G_1 = \frac{\partial}{\partial v} + L_2 + v^0 L_3, \qquad G_2 = \frac{\partial}{\partial v^0}.$$

In order to solve C_I^2 we put the explicitly integrable Monge systems $\mathcal{F}$ and $\mathcal{G}$ in canonical form, read off their solutions, and finally insert these into the composition functions for the projective group.

Let us consider $\mathcal{F}$ first. Its second derivative is

$$\mathcal{F}'' = \left(\frac{\partial}{\partial u} + \frac{1}{2}(R_2 + R_3), R_1, R_2 - R_3, \frac{\partial}{\partial u^0} \right).$$

The first vector field appearing here,

$$X := \frac{\partial}{\partial u} + \frac{1}{2}(R_2 + R_3)$$

$$= \frac{\partial}{\partial u} + \frac{1}{2}\left((w^3 - w^1 w^2)\frac{\partial}{\partial w^1} + (1 - (w^2)^2)\frac{\partial}{\partial w^2} + (w^1 - w^2 w^3)\frac{\partial}{\partial w^3} \right),$$

admits as first integral any first integral $f(u, w^2)$ of the vector field $\partial/\partial u + \frac{1}{2}(1 - (w^2)^2)\, \partial/\partial w^2$—for instance

$$f(u, w^2) = e^u \frac{w^2 - 1}{w^2 + 1}.$$

Recalling that $\mathfrak{G}_R$ acts on $\mathbb{C}_w^3$ by $w^i \mapsto \tilde{w}^i$, where

$$\tilde{w}^1 = \frac{a_1 w^1 + a_2 w^3}{a_3 w^2 + 1}, \quad \tilde{w}^2 = \frac{a_1 w^2 + a_2}{a_3 w^2 + 1} \quad \text{and} \quad \tilde{w}^3 = \frac{a_3 w^1 + w^3}{a_3 w^2 + 1},$$

we see that $f(u, w^2)$ will be equal to $\tilde{w}^2$ if we choose $a_1 := e^u$, $a_2 := -e^u$ and $a_3 := 1$. On replacing the w^i by the corresponding $\tilde{w}^i$ it follows that X will no longer contain $\partial/\partial \tilde{w}^2$. And according to the first fundamental theorem for Lie groups this change of coordinates will turn X into the form

$$\partial/\partial u + r^1(u)\, \tilde{R}_1 + r^2(u)\, \tilde{R}_2 + r^3(u)\, \tilde{R}_3.$$

Here the coefficient in front of $\partial/\partial \tilde{w}^2$ equals $r^1(u)\tilde{w}^2 + r^2(u) - r^3(u)(\tilde{w}^2)^2$—but we already know that this vanishes. Consequently $r^1(u) = r^2(u) = r^3(u) = 0$, and X has thereby been simplified to $\partial/\partial u$. Besides,

$$R_1 \mapsto \frac{1}{2}(e^u \tilde{R}_2 + e^{-u} \tilde{R}_3), \quad R_2 - R_3 \mapsto e^u \tilde{R}_2 - e^{-u} \tilde{R}_3,$$

so that $\mathcal{F}''$ has been transformed into

$$\mathcal{F}'' = \left(\frac{\partial}{\partial u}, \tilde{R}_2, \tilde{R}_3, \frac{\partial}{\partial u^0} \right),$$

and $F_1 = X + u^0 R_1$ into $\partial/\partial u + \frac{u^0}{2}(e^u \tilde{R}_2 + e^{-u} \tilde{R}_3)$.

Now $\tilde{R}_2 = \tilde{w}^3\, \partial/\partial \tilde{w}^1 + \partial/\partial \tilde{w}^2$ has the first integrals $\alpha := \tilde{w}^3$ and $\alpha^0 := \frac{1}{2}(\tilde{w}^2 \tilde{w}^3 - \tilde{w}^1)$, which we introduce as new variables instead of $\tilde{w}^3$ and $\tilde{w}^1$. Then

$$\tilde{R}_2 \mapsto \frac{\partial}{\partial \tilde{w}^2} \quad \text{and} \quad \tilde{R}_3 \mapsto -(\tilde{w}^2)^2 \frac{\partial}{\partial \tilde{w}^2} - 2\alpha^0 \left(\frac{\partial}{\partial \alpha} + \tilde{w}^2 \frac{\partial}{\partial \alpha^0} \right),$$

whence

$$F_1 \mapsto -u^0 \alpha^0 e^{-u} \left(\frac{\partial}{\partial \alpha} + \tilde{w}^2 \frac{\partial}{\partial \alpha^0} + \frac{(\tilde{w}^2)^2 - e^{2u}}{2\alpha^0} \frac{\partial}{\partial \tilde{w}^2} - \frac{e^u}{u^0 \alpha^0} \frac{\partial}{\partial u} \right).$$

Finally, by setting $\alpha^1 := \tilde{w}^2$, replacing u by

$$\alpha^2 := \frac{(\tilde{w}^2)^2 - e^{2u}}{2\alpha^0},$$

and u^0 by a suitable α^3, $\mathcal{F}$ is brought to the canonical form

$$\mathcal{F} = \left(\frac{\partial}{\partial \alpha} + \alpha^1 \frac{\partial}{\partial \alpha^0} + \alpha^2 \frac{\partial}{\partial \alpha^1} + \alpha^3 \frac{\partial}{\partial \alpha^2}, \frac{\partial}{\partial \alpha^3} \right).$$

The general integral curve of this on which $d\alpha \neq 0$ is given by

$$\alpha^0 = \phi(\alpha), \quad \alpha^1 = \phi'(\alpha), \quad \alpha^2 = \phi''(\alpha) \quad \text{and} \quad \alpha^3 = \phi'''(\alpha),$$

where $\phi(\alpha)$ is an arbitrary function. In particular

$$\tilde{w}^3 = \alpha, \quad \tilde{w}^2 = \phi'(\alpha), \quad \tilde{w}^1 = \tilde{w}^2 \tilde{w}^3 - 2\alpha^0 = \alpha\phi'(\alpha) - 2\phi(\alpha),$$

and u is related to α by

$$e^{2u} = (\tilde{w}^2)^2 - 2\alpha^0 \alpha^2 = (\phi'(\alpha))^2 - 2\phi(\alpha)\phi''(\alpha).$$

Inverting the transformation $w^i \mapsto \tilde{w}^i$ we get

$$w^1 = \frac{\tilde{a}_1 \tilde{w}^1 + \tilde{a}_2 \tilde{w}^3}{\tilde{a}_3 \tilde{w}^2 + 1}, \quad w^2 = \frac{\tilde{a}_1 \tilde{w}^2 + \tilde{a}_2}{\tilde{a}_3 \tilde{w}^2 + 1}, \quad w^3 = \frac{\tilde{a}_3 \tilde{w}^1 + \tilde{w}^3}{\tilde{a}_3 \tilde{w}^2 + 1}$$

with $\tilde{a}_1 = e^{-u}$, $\tilde{a}_2 = 1$ and $\tilde{a}_3 = -e^{-u}$. Setting $\phi_1(\alpha) := \alpha\phi'(\alpha) - 2\phi(\alpha)$ we therefore have

$$w_{\mathcal{F}}^1 = \frac{e^{-u}\phi_1(\alpha) + \alpha}{-e^{-u}\phi'(\alpha) + 1}, \quad w_{\mathcal{F}}^2 = \frac{e^{-u}\phi'(\alpha) + 1}{-e^{-u}\phi'(\alpha) + 1}, \quad w_{\mathcal{F}}^3 = \frac{-e^{-u}\phi_1(\alpha) + \alpha}{-e^{-u}\phi'(\alpha) + 1},$$

whence

$$w_{\mathcal{F}}^1 - w_{\mathcal{F}}^2 w_{\mathcal{F}}^3 = -\frac{4\phi(\alpha)e^{-u}}{(-e^{-u}\phi'(\alpha) + 1)^2}.$$

Since the Monge system $\mathcal{G}$ is the same as in the preceding case,

$$w_{\mathcal{G}}^1 = \psi'(v) - \frac{\psi(v)\psi''(v)}{2\psi'(v)}, \quad w_{\mathcal{G}}^2 = \psi(v), \quad w_{\mathcal{G}}^3 = -\frac{\psi''(v)}{2\psi'(v)}$$

and

$$w_{\mathcal{G}}^1 - w_{\mathcal{G}}^2 w_{\mathcal{G}}^3 = \psi'(v).$$

Using the composition functions we then obtain the following expressions for the w^i on the integral surfaces of $\mathcal{V}$:

$$w^1 = \frac{w^1_{\mathcal{F}} w^1_{\mathcal{G}} + w^2_{\mathcal{F}} w^3_{\mathcal{G}}}{w^3_{\mathcal{F}} w^2_{\mathcal{G}} + 1}, \qquad w^2 = \frac{w^1_{\mathcal{F}} w^2_{\mathcal{G}} + w^2_{\mathcal{F}}}{w^3_{\mathcal{F}} w^2_{\mathcal{G}} + 1}, \qquad w^3 = \frac{w^3_{\mathcal{F}} w^1_{\mathcal{G}} + w^3_{\mathcal{G}}}{w^3_{\mathcal{F}} w^2_{\mathcal{G}} + 1}.$$

Thus

$$e^z = \frac{w^1 - w^2 w^3}{w^2} = \frac{(w^1_{\mathcal{F}} - w^2_{\mathcal{F}} w^3_{\mathcal{F}})(w^1_{\mathcal{G}} - w^2_{\mathcal{G}} w^3_{\mathcal{G}})}{(w^1_{\mathcal{F}} w^2_{\mathcal{G}} + w^2_{\mathcal{F}})(w^3_{\mathcal{F}} w^2_{\mathcal{G}} + 1)}.$$

Setting $u = x$ and $v = y$ this takes the explicit form

$$2\psi'(y)e^{x-z} = \phi''(\alpha)(\alpha\psi(y) + 1)^2 - 2\phi'(\alpha)\psi(y)(\alpha\psi(y) + 1) + 2\phi(\alpha)\psi^2(y),$$

where α is a function of x, implicitly defined by

$$e^{2x} = (\phi'(\alpha))^2 - 2\phi(\alpha)\phi''(\alpha).$$

3. In the final case $\mathcal{V}$ is generated by

$$\begin{cases} F_1 = \partial/\partial u + m^2(u)\,R_2 + m^3(u)\,R_3 + u^0\,R_1, & F_2 = \partial/\partial u^0, \\ G_1 = \partial/\partial v + n^2(v)\,L_2 + n^3(v)\,L_3 + v^0\,L_1, & G_2 = \partial/\partial v^0, \end{cases}$$

from which it is seen that

$$\mathcal{V}_{\text{red}} = (\partial/\partial u + m^2 R_2 + m^3 R_3, \partial/\partial v + n^2 L_2 + n^3 L_3, R_1, L_1).$$

The complete subsystem

$$(R_1, L_1) = (w^1\,\partial/\partial w^1 + w^2\,\partial/\partial w^2,\, w^1\,\partial/\partial w^1 + w^3\,\partial/\partial w^3)$$

admits the first integrals u, v and $(w^1 - w^2 w^3)/w^1$. As before, we set $x := u$, $y := v$, and after that it turns out to be convenient to define z by

$$\cosh\frac{z}{2} = \sqrt{\frac{w^1}{w^1 - w^2 w^3}}.$$

Then

$$p = F_1 z = \frac{m^2(x)}{w^2}\tanh\frac{z}{2} + m^3(x)w^2\coth\frac{z}{2},$$

$$q = G_1 z = \frac{n^3(y)}{w^3}\tanh\frac{z}{2} + n^2(y)w^3\coth\frac{z}{2},$$

$$\sigma = F_1 q = G_1 p$$

$$= \left(\frac{m^2(x)}{w^2}\tanh\frac{z}{2} - m^3(x)w^2\coth\frac{z}{2}\right)\left(\frac{q}{\sinh z} - \frac{n^2(y)(w^1 - w^2 w^3)}{w^2}\right),$$

and these relations define the PDE

$$s \sinh z = \sqrt{p^2 - 4m^2(x)m^3(x)}\sqrt{q^2 - 4n^2(y)n^3(y)}.$$

If x and y are replaced by $2\int \sqrt{m^2(x)m^3(x)}\,dx$ and $2\int \sqrt{n^2(y)n^3(y)}\,dy$ respectively, this PDE gets simplified to the Goursat equation

$$s \sinh z = \sqrt{p^2 - 1}\sqrt{q^2 - 1}. \qquad (C_I^3)$$

We would have obtained the latter expression directly by setting $m^2 = m^3 = n^2 = n^3 = \frac{1}{2}$. Therefore the vector field system associated to C_I^3 is generated by

$$F_1 = \frac{\partial}{\partial u} + \frac{1}{2}(R_1 + R_3) + u^0 R_1, \qquad F_2 = \frac{\partial}{\partial u^0},$$

$$G_1 = \frac{\partial}{\partial v} + \frac{1}{2}(L_1 + L_3) + v^0 L_1, \qquad G_2 = \frac{\partial}{\partial v^0},$$

so that both $\mathcal{F}$ and $\mathcal{G}$ now look like the $\mathcal{F}$ of the preceding case. Therefore, letting $\phi(\alpha)$ and $\psi(\beta)$ be arbitrary functions and setting

$$\phi_1(\alpha) := \alpha\phi'(\alpha) - 2\phi(\alpha), \quad \psi_1(\beta) := \beta\psi'(\beta) - 2\psi(\beta),$$

we get

$$w_{\mathcal{F}}^1 = \frac{e^{-x}\phi_1(\alpha) + \alpha}{-e^{-x}\phi'(\alpha) + 1}, \quad w_{\mathcal{F}}^2 = \frac{e^{-x}\phi'(\alpha) + 1}{-e^{-x}\phi'(\alpha) + 1}, \quad w_{\mathcal{F}}^3 = \frac{-e^{-x}\phi_1(\alpha) + \alpha}{-e^{-x}\phi'(\alpha) + 1},$$

$$w_{\mathcal{G}}^1 = \frac{e^{-y}\psi_1(\beta) + \beta}{-e^{-y}\psi'(\beta) + 1}, \quad w_{\mathcal{G}}^2 = \frac{-e^{-y}\psi_1(\beta) + \beta}{-e^{-y}\psi'(\beta) + 1}, \quad w_{\mathcal{G}}^3 = \frac{e^{-y}\psi'(\beta) + 1}{-e^{-y}\psi'(\beta) + 1},$$

with $\alpha(x)$ and $\beta(y)$ implicitly defined by

$$e^{2x} = (\phi'(\alpha))^2 - 2\phi(\alpha)\phi''(\alpha) \quad \text{and} \quad e^{2y} = (\psi'(\beta))^2 - 2\psi(\beta)\psi''(\beta).$$

From the composition functions of the projective group it then follows that

$$\coth^2 \frac{z}{2} = \frac{(w_{\mathcal{F}}^1 w_{\mathcal{G}}^1 + w_{\mathcal{F}}^2 w_{\mathcal{G}}^3)(w_{\mathcal{F}}^3 w_{\mathcal{G}}^2 + 1)}{(w_{\mathcal{F}}^1 - w_{\mathcal{F}}^2 w_{\mathcal{F}}^3)(w_{\mathcal{G}}^1 - w_{\mathcal{G}}^2 w_{\mathcal{G}}^3)},$$

or more explicitly,

$$4\phi\psi e^{x+y}\left(\coth^2\frac{z}{2} - 2\right) = (\alpha\beta + 1)^2(4\phi\psi\phi''\psi'' - (\phi')^2(\psi')^2)$$

$$+ 4(\alpha\beta + 1)(2\phi\psi(\phi'\psi')' - \phi'\psi'(\phi\psi))$$

$$+ 8\phi\psi(\phi\psi'' + \psi\phi'') - 4(\phi^2(\psi')^2 + \psi^2(\phi')^2).$$

Thereby the Goursat equations associated to the first type in the classification of 3-dimensional Lie algebras have all been determined and solved.

And as demonstrated in [Vessiot 1942], the remaining types II–VI can be treated in a similar fashion—that is, the calculations are tedious, but there are no real difficulties.

14

Cartan's theory of Lie pseudogroups

The equivalence used for the classification of hyperbolic PDEs is based on the following principle.

Let S_1 and S_2 be two PDE systems. Identifying them with vector field systems (M_1, V_1) and (M_2, V_2) respectively, we say that S_1 and S_2 are *locally equivalent* if there is a local diffeomorphism $\phi : M_1 \longrightarrow M_2$ such that $\phi_*(V_1) = V_2$ and $(\phi^{-1})_*(V_2) = V_1$

Alternatively S_1 and S_2 can be written as pfaffian systems (M_1, P_1) and (M_2, P_2), and then they are locally equivalent if there is a local diffeomorphism $\phi : M_1 \longrightarrow M_2$ such that $\phi^*(P_2) = P_1$ and $(\phi^{-1})^*(P_1) = P_2$.

More generally, let M_1 and M_2 be two manifolds of the same dimension, and suppose that they are provided with *local structures* S_1 and S_2. Then (M_1, S_1) and (M_2, S_2) are *locally equivalent* if there is a local diffeomorphism $\phi : M_1 \longrightarrow M_2$ such that ϕ and ϕ^{-1} induce mappings which identify S_1 and S_2.

> The *equivalence problem* is: how is it possible to decide whether two local structures are equivalent or not?

In order to understand Cartan's solution of this problem, suppose that ϕ and ψ are two local equivalences such that $\psi^{-1} \circ \phi$ and $\psi \circ \phi^{-1}$ are defined. Then $\psi^{-1} \circ \phi$ and $\psi \circ \phi^{-1}$ are *self-equivalences*, or *symmetries*, of (M_1, S_1) and (M_2, S_2) respectively. Clearly the family of self-equivalences of a given local structure forms a *pseudogroup*. And mostly it even turns out to be a *Lie pseudogroup*—i.e., it is defined by a PDE system.

Notation. If (M, S) is a manifold with a local structure, sym(S) denotes the *pseudogroup of self-equivalences*, or the *symmetry pseudogroup* of (M, S).

Now if $\phi : M_1 \longrightarrow M_2$ is *one* local equivalence of (M_1, S_1) and (M_2, S_2), the most general can be written as

$$\phi \circ \phi_1 \qquad \text{where } \phi_1 \in \text{sym}(S_1) \text{ is arbitrary,}$$

or

$$\phi_2 \circ \phi \qquad \text{where } \phi_2 \in \text{sym}(S_2) \text{ is arbitrary}$$

—at least if we do not pay too much attention to the domains where these local diffeomorphisms are defined. In this way the symmetry pseudogroups of the two structures get connected with the equivalence problem. Let us in the following assume that all symmetry pseudogroups which are encountered are in fact Lie pseudogroups.

Cartan has shown that Lie pseudogroups can be given *a fairly simple canonical form*—which makes it possible to decide whether two Lie pseudogroups are locally equivalent in a straightforward manner. And clearly

(M_1, S_1) and (M_2, S_2) are locally equivalent $\Longrightarrow$ $\text{sym}(S_1)$ and $\text{sym}(S_2)$ are locally equivalent too.

Utilizing this, Cartan's idea for solving the equivalence problem is:

(i) First determine all local equivalences of the symmetry pseudogroups $\text{sym}(S_1)$ and $\text{sym}(S_2)$.
(ii) Then investigate which of these are also local equivalences of (M_1, S_1) and (M_2, S_2).

The purpose of the present chapter is to understand the necessary background from Cartan's theory of Lie pseudogroups, whereupon Cartan's solution of the equivalence problem is treated in the next chapter. After that we are prepared to tackle [Cartan 1910], where he exploits his solution of the equivalence problem in order to classify

• those systems of two PDEs in one dependent and two independent variables which admit a Cauchy charateristic vector field, and
• parabolic PDEs in one dependent and two independent variables for which the (double) Monge system is complete.

Vessiot's classification of hyberbolic PDEs in one dependent and two independent variables involves a lot of calculations, and unfortunately Cartan's method—although very different—also does. Therefore it seems hopeless to extend these results to PDE systems in one dependent and more than two independent variables.

However, the concept of equivalence can be weakened: let us say that two PDE systems are *structurally equivalent* if their associated vector field (or pfaffian) systems have the *same Lie structure*.

Using this type of equivalence Cartan classified all *PDE systems in one dependent and three independent variables*, and investigated in which cases it is possible to find solutions without using the general local existence theorem—and this is going to be the subject of the last chapter in this monograph.

Recall from section **6.6** that

> a Lie pseudogroup G acting on a manifold M is a pseudogroup of local diffeomorphisms which constitutes the general solution of its defining PDE system $S^q(G) \subset J^q(M, M)$.

In chapter **8** we treated the simplest case imaginable: that occurring when the general solution of the defining PDE system depends on a *finite number of parameters* only. Then G induces a *parameter group* $\mathfrak{G}$ acting on the parameter space. The first fundamental theorem states that if the dimension of the parameter space is n, then there are n one-forms ω^1, ..., ω^n on this space such that $\omega^1 \wedge \cdots \wedge \omega^n \neq 0$, and

> a local diffeomorphism g on the parameter space belongs to $\mathfrak{G}$ if and only if $g^*(\omega^i) = \omega^i$ for $i = 1, \ldots, n$.

Moreover, two finite dimensional Lie groups G and $\hat{G}$ (possibly acting on manifolds of different dimensions) are *locally isomorphic* if their parameter groups are, i.e., if the parameter spaces are of the same dimension and there is a local diffeomorphism ϕ between them such that

$$\phi^*(\hat{\omega}^i) = \omega^i \quad \text{for all } i.$$

But a general Lie pseudogroup G is 'infinite dimensional' in the sense that G—as the family of local solutions of the defining PDE system—depends on a *certain number of arbitrary functions depending on a certain number of variables*, in a sense made precise in section **3.3**.

Observing that the parameter space in the finite dimensional case can be identified with the set of all local solutions of the defining PDE system, it is tempting to replace it by the infinite dimensional space of local solutions in the general case. And this is the route taken in [Kuranishi 1959], [Kuranishi 1961].

However, since Lie pseudogroups are supposed to act on *finite dimensional manifolds* only, it is a bit awkward if their canonical forms do not. Now Cartan showed that it is indeed possible to define canonical forms on finite dimensional manifolds: the idea is that if the defining system

$S^q(G)$ is involutive, G induces a Lie pseudogroup $\mathfrak{G}^{(q)}$ acting on the set of qth order Taylor polynomials of local solutions of $S^q(G)$—that is, on $S^q(G)$ itself! Moreover it is clear from the construction that each local diffeomorphism of G has a *unique* prolongation to an element of $\mathfrak{G}^{(q)}$, and this makes it possible to identify G and $\mathfrak{G}^{(q)}$.

If $S^q(G)$ is involutive, the systems S^{q+k} also are for $k = 1, 2, 3, \ldots$, and G might be identified with $\mathfrak{G}^{(q+1)}$, $\mathfrak{G}^{(q+2)}$, $\ldots$ as well—and it turns out to be essential for the ensuing thery to consider *all* these Lie pseudogroups. Hence the *canonical form of G* is no longer a fixed parameter group, but instead

> a chain of Lie pseudogroups $\mathfrak{G}^{(q)}$, $\mathfrak{G}^{(q+1)}$, $\ldots$, acting on the prolongations of the defining PDE system of G.

This might seem a bit odd at first, but really it has not so much to do with Lie pseudogroups specifically, but is rather a common property of all PDE systems: in order to solve a general system it is often necessary to make a number of prolongations. Note also that $\lim_{q \to \infty} \mathfrak{G}^{(q)}$ can be thought of as the set of all *formal power series solutions* of $S^q(G)$—but this point of view is avoided in Cartan's theory.

With

$$n_{q+k} := \dim S^{q+k}(G) \quad \text{and} \quad p_{q+k+1} := n_{q+k+1} - n_{q+k},$$

the Lie pseudogroup $\mathfrak{G}^{(q+k)}$ can be characterized as follows:

> there are n_{q+k} one-forms $\omega^1, \ldots, \omega^{n_{q+k}}$ living on $S^{q+k}(G)$ such that $\omega^1 \wedge \cdots \wedge \omega^{n_{q+k}} \neq 0$, there are h_{q+k} essential invariants $I^1, \ldots, I^{h_{q+k}}$ on $S^{q+k}(G)$, and there is a matrix group $\mathfrak{St}^{(q+k)}$—called the stability group—which depends on these invariants and on the p_{q+k+1} fibre coordinates of the trivial fibre bundle
> $$S^{q+k+1}(G)$$
> $$\downarrow \pi_{q+k}^{q+k+1}$$
> $$S^{q+k}(G),$$

such that the local diffeomorphism g on $S^{q+k}(G)$ belongs to $\mathfrak{G}^{(q+k)}$ if and only if

$$g^*(I^i) = I^i \quad \text{for } i = 1, \ldots, h_{q+k},$$

and

$$g^* \begin{pmatrix} \omega^1 \\ \vdots \\ \omega^{n_{q+k}} \end{pmatrix} = M_g \begin{pmatrix} \omega^1 \\ \vdots \\ \omega^{n_{q+k}} \end{pmatrix} \quad \text{for some } M_g \in \mathfrak{St}^{(q+k)}.$$

Remarks. 1. $\pi_{q+k}^{q+k+1} : S^{q+k+1}(G) \longrightarrow S^{q+k}(G)$ is surjective because of the supposed involutivity of S^q.

2. In the finite dimensional case $S^q(G) = S^{q+1}(G) = \cdots$, so that all the p_{q+k+1} vanish; and besides, there are no essential invariants. Therefore the stability group $\mathfrak{St}$ reduces to the identity, and the above requirements boil down to the familiar

$$g^*(\omega^i) = \omega^i \quad \text{for all } i.$$

Suppose now that $\mathfrak{G}$ and $\hat{\mathfrak{G}}$ are two Lie pseudogroups presented in the above form, and that they have the same number h of invariants, the same number n of invariant one-forms, and have matrix groups depending on the same number p of parameters. Then $\mathfrak{G}$ and $\hat{\mathfrak{G}}$ are *locally equivalent* if there is a local diffeomorphism ϕ of the underlying manifolds such that

$$\phi^*(\hat{I}^i) = I^i \quad \text{for } i = 1, \dots, h,$$

and

$$\phi^* \begin{pmatrix} \hat{\omega}^1 \\ \vdots \\ \hat{\omega}^n \end{pmatrix} = M_\phi \begin{pmatrix} \omega^1 \\ \vdots \\ \omega^n \end{pmatrix} \quad \text{for some } M_\phi \in \mathfrak{St}.$$

And this is precisely Cartan's version of the equivalence problem—which a priori looks rather strange.

In the present chapter we discuss those parts of the theory of Lie pseudogroups which will be needed in the sequel, following [Cartan 1904], [Cartan 1905] and the unusually well-written review paper [Cartan 1937b].When the notations differ (as they sometimes do), we stick to the latter.

Remark. The theory presented here presupposes Cartan's local existence theorem, so we have to work in the analytic category. But we will be careful not to do so when analysing PDE systems in one dependent and two or three dependent variables in the sequel.

As regards the ground field, both $\mathbb{R}$ and $\mathbb{C}$ work. Let us continue to use the latter.

14.1 The first fundamental theorem

Let G be a Lie pseudogroup acting on $\mathbb{C}_x^r$. In order to be able to introduce the defining PDE system of G we let $\mathbb{C}_{\hat{x}}^r$ be a copy of $\mathbb{C}_x^r$, called the *target manifold*, while the original $\mathbb{C}_x^r$ is called the *source manifold*. Then an element $g \in G$ can be regarded either as a local diffeomorphism from source to target,

$$g : \mathbb{C}^r_x \to \mathbb{C}^r_{\hat{x}}$$
$$x \mapsto \hat{x} = g(x),$$

or as a *source transformation*,

$$g : \mathbb{C}^r_x \to \mathbb{C}^r_x$$
$$x \mapsto g(x).$$

Using the first point of view we can introduce the jet bundles $J^k(\mathbb{C}^r_x, \mathbb{C}^r_{\hat{x}})$ for $k = 0, 1, 2, \ldots$, and then the defining PDE system $S^q(G)$ of G is considered as a submanifold of $J^q(\mathbb{C}^r_x, \mathbb{C}^r_{\hat{x}})$.

Following Cartan we let each $h \in G$ define a 'pseudo-action' on G by

$$\Phi_h : G \to G$$
$$g \mapsto g \circ h$$

for all $g \in G$ such that $g \circ h$ is defined. Here h is regarded as a source transformation, while g and $g \circ h$ are thought of as mappings from the source to the target.

Alternatively the pseudo-action could have been defined by $g \to h \circ g$ instead, with h now being considered as a target transformation—and this is in fact what is done in [Vessiot 1903].

The action of G on $\mathbb{C}^r_x$ is trivially extended to an action on $J^0(\mathbb{C}^r_x, \mathbb{C}^r_{\hat{x}}) \cong \mathbb{C}^r_x \times \mathbb{C}^r_{\hat{x}}$ by

$$G \ni g \mapsto \tilde{g}^{(0)} := (g, \mathrm{id}) : \mathbb{C}^r_x \times \mathbb{C}^r_{\hat{x}} \to \mathbb{C}^r_x \times \mathbb{C}^r_{\hat{x}}$$
$$(x, \hat{x}) \mapsto (g(x), \hat{x}).$$

According to section **7.3** this $\tilde{g}^{(0)}$ has a *unique prolongation to a contact transformation* $\tilde{g}^{(q)}$ acting on $J^q(\mathbb{C}^r_x, \mathbb{C}^r_{\hat{x}})$. For instance,

$$\tilde{g}^{(1)} : J^1(\mathbb{C}^r_x, \mathbb{C}^r_{\hat{x}}) \longrightarrow J^1(\mathbb{C}^r_y, \mathbb{C}^r_{\hat{y}})$$

is defined as follows: $(x, \hat{x}, p) \mapsto (y, \hat{y}, q) = (g(x), \hat{x}, Q(x, \hat{x}, p))$, where— using a shorthand notation—the function Q is to satisfy $(\tilde{g}^{(1)})^*(d\hat{y} - q\,dy) = \rho\,(d\hat{x} - p\,dx)$ for some nonzero factor ρ—i.e.,

$$d\hat{x} - Q\frac{\partial g}{\partial x}dx = \rho\,(d\hat{x} - p\,dx).$$

Hence

$$\rho = 1 \quad \text{and} \quad Q(x, \hat{x}, p) = \frac{p}{\partial g/\partial x}.$$

So for $f: \mathbb{C}^r_x \to \mathbb{C}^r_{\hat{x}}$ we obtain

$$\tilde{g}^{(1)}(j^1_x(f)) = \tilde{g}^{(1)}\left(x, f(x), \frac{\partial f}{\partial x}\right) = \left(g(x), f(x), \frac{\partial f/\partial x}{\partial g/\partial x}\right).$$

Lemma 14.1.1. *We have $\tilde{g}^{(q)}(j^q_x(f)) = j^q_{g(x)}(f \circ g^{-1})$ for $q = 0, 1, 2, \ldots$.*

Proof. For $q = 0$ we are to find a function $\phi: \mathbb{C}^r_y \to \mathbb{C}^r_{\hat{y}}$ such that

$$\tilde{g}^{(0)}(j^0_x(f)) = j^0_y(\phi).$$

Or equivalently $(g(x), f(x)) = (y, \phi(y))$—i.e.,

$$x = g^{-1}(y) \quad \text{and} \quad \phi(y) = f(g^{-1}(y)), \quad \text{whence} \quad \phi = f \circ g^{-1}.$$

When $q = 1$,

$$j^1_y(\phi) = \left(y, \phi(y), \frac{\partial \phi}{\partial y}\right) = \left(g(x), f(x), \frac{\partial f}{\partial x}\frac{\partial x}{\partial y}\right)$$

$$= \left(g(x), f(x), \frac{\partial f/\partial x}{\partial g/\partial x}\right) = \tilde{g}^{(1)}(j^1_x(f)) \quad \text{by the above,}$$

or

$$\tilde{g}^{(1)}(j^1_x(f)) = j^1_{g(x)}(f \circ g^{-1}).$$

The corresponding results for $q > 1$ follow analogously. □

Lemma 14.1.2. *If the defining PDE system $S^q(G) \subset J^q(\mathbb{C}^r_x, \mathbb{C}^r_{\hat{x}})$ of G is involutive, the contact transformation $\tilde{g}^{(q)}$ on $J^q(\mathbb{C}^r_x, \mathbb{C}^r_{\hat{x}})$ restricts to a local diffeomorphism $g^{(q)}$ acting on $S^q(G)$.*

Since $\tilde{g}^{(q)}$ is a contact transformation it leaves the contact pfaffian system $^qCtr^r$ of $J^q(\mathbb{C}^r_x, \mathbb{C}^r_{\hat{x}})$ invariant, and therefore $g^{(q)}$ preserves its restriction $\mathcal{P}^{(q)}$ to $S^q(G)$—that is, the pfaffian system associated to $S^q(G)$.

Proof. Each point $s^q \in S^q(G)$ can be regarded as a qth order Taylor polynomial. Because of the involutivity we can apply Cartan's local existence theorem in order to extend s^q to a local solution of $S^q(G)$ Then $s^q = j^q_x(f)$ for some $f \in G$, where $x = \alpha(s^q)$ is the projection of s^q to the source manifold $\mathbb{C}^r_x$. If $j^q_x(f)$ belongs to the domain of definition of $\tilde{g}^{(q)}$, $f \circ g^{-1}$ is defined and belongs to G by the pseudogroup property. Thus

$$\tilde{g}^{(q)}(s^q) = \tilde{g}^{(q)}(j^q_x(f)) = j^q_{g(x)}(f \circ g^{-1}) \in S^q(G),$$

so that $\tilde{g}^{(q)}$ indeed restricts to a local diffeomorphism

$$g^{(q)}: S^q(G) \longrightarrow S^q(G).$$

□

Let us now consider the pfaffian system $\mathcal{P}^{(q)}$ more closely. $\mathcal{P}^{(q)}$ is the restriction to $S^q(G)$ of the contact pfaffian system ${}^qCt^{r,r}$ of $J^q(\mathbb{C}^r_x, \mathbb{C}^r_{\hat{x}})$:

$$
\begin{cases}
d\hat{x}^i = \sum_{j=1}^r p^i_j \, dx^j, \\
dp^i_j = \sum_{k=1}^r p^i_{jk} \, dx^k, \\
\cdots .
\end{cases}
$$

Firstly $S^q(G)$ may contain a number of zeroth order equations. We assume that these can be solved with respect to certain of the $\hat{x}^i$—say

$$\hat{x}^i = F^i(x^1, \ldots, x^r; \hat{x}^{h+1}, \ldots, \hat{x}^r) \quad \text{for } i = 1, \ldots, h.$$

Let us set $x = (x^1, \ldots, x^r)$ and $\hat{x} = (\hat{x}^{h+1}, \ldots, \hat{x}^r)$ in the following.

Because the prolongation of any $g \in G$ to $J^0(\mathbb{C}^r_x, \mathbb{C}^r_{\hat{x}})$ leaves both $\hat{x}^i$ and the equations $\hat{x}^i - F^i(x, \hat{x}) = 0$ invariant, it follows that the F^i are also preserved:

$$F^i(g(x), \hat{x}) = F^i(x, \hat{x}).$$

Replacing $\hat{x}^{h+1}, \ldots, \hat{x}^r$ by suitable constants $\hat{x}_0^{h+1}, \ldots, \hat{x}_0^r$ we find in this way h *invariants* $I^1, \ldots, I^h$ of G:

$$I^i(x^1, \ldots, x^r) := F^i(x^1, \ldots, x^r; \hat{x}_0^{h+1}, \ldots, \hat{x}_0^r) \quad \text{for } i = 1, \ldots, h.$$

Next look at the first order PDEs in $S^q(G)$. We assume that they can be solved with respect to certain of the p^i_j—which we call *principal*—in terms of x, $\hat{x}$ and the remaining *parametric* p^i_j, which we denote by ${}^{(1)}t^1$, $\ldots$, ${}^{(1)}t^{p_1}$. Then all the first order jet coordinates p^i_j can be written as

$$p^i_j = a^i_j(x, \hat{x}, {}^{(1)}t)$$

with ${}^{(1)}t := ({}^{(1)}t^1, \ldots, {}^{(1)}t^{p_1})$. Consequently

$$d\hat{x}^i = \sum_{j=1}^r a^i_j(x, \hat{x}, {}^{(1)}t) \, dx^j \quad \text{for } i = 1, \ldots, r,$$

so that $\mathcal{P}^{(q)}$ will contain the pfaffian equations $d\hat{x}^i - \sum_{j=1}^r a^i_j(x, \hat{x}, {}^{(1)}t) \, dx^j = 0$.

Since each $g^{(q)}$ leaves the $d\hat{x}^i$ and $\mathcal{P}^{(q)}$ invariant, we conclude that the one-forms

$${}^{(0)}\omega^i := \sum_{j=1}^r a^i_j(x, \hat{x}, {}^{(1)}t) \, dx^j \quad \text{for } i = 1, \ldots, r$$

are also *invariant*.

After that the second order PDEs in $S^q(G)$ are taken into account.

We assume that they can be solved with respect to a certain number of principal p^i_{jk}, so that each p^i_{jk} can be expressed as a function of $x, \hat{x}, {}^{(1)}t$ and the remaining parametric p^i_{jk}, which we denote by ${}^{(2)}t^1, \ldots, {}^{(2)}t^{p_2}$:

$$p^i_{jk} = a^i_{jk}(x, \hat{x}, {}^{(1)}t, {}^{(2)}t).$$

Then $dp^i_j = \sum_{k=1}^r a^i_{jk}(x, \hat{x}, {}^{(1)}t, {}^{(2)}t)\, dx^k$, and this is in particular true for the parametric p^i_j—i.e., the ${}^{(1)}t$:

$$d^{(1)}t^i - \sum_{j=1}^r {}^{(1)}a^i_j(x, \hat{x}, {}^{(1)}t, {}^{(2)}t)\, dx^j = 0 \quad \text{for } i = 1, \ldots, p.$$

For $k = 0, 1, \ldots, q-1$ we let $S^k(G)$ denote the submanifold of $J^k(\mathbb{C}^r_x, \mathbb{C}^r_{\hat{x}})$ defined by equations of order $\leq k$. Then

$x, \hat{x}$ constitute a system of local coordinates for $S^0(G)$,
$x, \hat{x}, {}^{(1)}t$ constitute a system of local coordinates for $S^1(G)$,
$x, \hat{x}, {}^{(1)}t, {}^{(2)}t$ constitute a system of local coordinates for $S^2(G)$.

Continuing in this manner we finally arrive at the equations of order q. As usual we assume that these can be solved with respect to a certain number of principal $p^i_{j_1,\ldots,j_q}$, so that all the qth order jet coordinates can be expressed as functions of $x, \hat{x}, {}^{(1)}t, {}^{(2)}t, \ldots, {}^{(q-1)}t$ and the parametric $p^i_{j_1,\ldots,j_q}$. Denoting the latter by ${}^{(q)}t^1, \ldots, {}^{(q)}t^{p_q}$ we thus have

$$p^i_{j_1,\ldots,j_q} = a^i_{j_1,\ldots,j_q}(x, \hat{x}, {}^{(1)}t, \ldots, {}^{(q)}t),$$

which gives the last pfaffian equations in $\mathcal{P}^{(q)}$:

$$d^{(q-1)}t^i - \sum_{j=1}^r {}^{(q-1)}a^i_j(x, \hat{x}, {}^{(1)}t, \ldots, {}^{(q)}t)\, dx^j = 0 \quad \text{for } i = 1, \ldots, p_{q-1}.$$

Conclusion. $\mathcal{P}^{(q)}$ *consists of the pfaffian equations*

$$\begin{cases} d\hat{x}^i - \sum_{j=1}^r a^i_j(x, \hat{x}, {}^{(1)}t)\, dx^j = 0 & \text{for } i = h+1, \ldots, r, \\ d^{(1)}t^i - \sum_{j=1}^r {}^{(1)}a^i_j(x, \hat{x}, {}^{(1)}t, {}^{(2)}t)\, dx^j = 0 & \text{for } i = 1, \ldots, p_1, \\ \ldots \\ d^{(q-1)}t^i - \sum_{j=1}^r {}^{(q-1)}a^i_j(x, \hat{x}, {}^{(1)}t, \ldots, {}^{(q)}t)\, dx^j = 0 & \text{for } i = 1, \ldots, p_{q-1}, \end{cases}$$

where $x, \hat{x}, {}^{(1)}t, \ldots, {}^{(q)}t$ *constitute a system of local coordinates for* $S^q(G)$.

From the discussion above we have got h invariant functions $I^1, \ldots, I^h$, and $r + (r - h)$ invariant one-forms ${}^{(0)}\omega^1, \ldots, {}^{(0)}\omega^r, d\hat{x}^{h+1}, \ldots, d\hat{x}^r$. Now Cartan's version of the first fundamental theorem for finite dimensional parameter groups states that such a group consists of all local diffeomorphisms of the parameter space that leave dim(parameter space) linearly independent one-forms invariant. In analogy with this we want to construct $\dim S^q(G) = r + (r - h) + p_1 + \cdots + p_q$ linearly independent one-forms on $S^q(G)$ such that the induced Lie pseudogroup on $S^q(G)$ is characterized by leaving $I^1, \ldots, I^h$ and these one-forms invariant. But in order to achieve this we also must have pfaffian equations involving $d^{(q)} t^1, \ldots, d^{(q)} t^{p_q}$ at our disposal—and these are obtained from $\mathcal{P}^{(q+1)} = $ the restriction of $^{q+1}Ct^{r,r}$ to the next prolongation $S^{q+1}(G)$. Denoting the parametric $p^i_{j_1, \ldots, j_{q+1}}$ by $u^1, \ldots, u^{p_{q+1}}$, the last set of equations in $\mathcal{P}^{(q+1)}$ has the form

$$d^{(q)} t^i - \sum_{j=1}^{r} {}^{(q)} a^i_j(x, \hat{x}, {}^{(1)}t, \ldots {}^{(q)}t; u) \, dx^j = 0 \quad \text{for } i = 1, \ldots, p_q.$$

Because of this last fact the analogue of the first fundamental theorem necessarily will become a bit complicated:

> G has a one-to-one prolongation to a Lie pseudogroup on $S^q(G)$ consisting of all local diffeomorphisms leaving $I^1, \ldots, I^h$ invariant, and having prolongations to $S^{q+1}(G)$ which leave $\dim S^q(G)$ one-forms—involving the differentials of the local coordinates for $S^q(G)$—invariant.

In order to demonstrate this we now construct invariant one-forms containing $d^{(1)} t^1, \ldots, d^{(1)} t^{p_q}$, starting from the invariant one-forms

$$^{(0)}\omega^i = \sum_{j=1}^{r} a^i_j(x, \hat{x}, {}^{(1)}t) \, dx^j \quad \text{for } i = 1, \ldots, r,$$

and the invariant pfaffian equations

$$d^{(1)} t^i - \sum_{j=1}^{r} {}^{(1)} a^i_j(x, \hat{x}, {}^{(1)}t, {}^{(2)}t) \, dx^j = 0 \quad \text{for } i = 1, \ldots, p_1.$$

The $^{(0)}\omega^i$ have arisen from the equations $d\hat{x}^i = \sum_{j=1}^{r} a^i_j \, dx^j$ that represent the first order PDEs for the local diffeomorphisms $\mathbb{C}^r_x \to \mathbb{C}^r_{\hat{x}}$ in G. Therefore the dx^i might conversely be solved so as to be expressed as linear combinations of the $d\hat{x}^i$—and hence $\det(a^i_j) \neq 0$. But then it is also possible to write $dx^1, \ldots, dx^n$ as linear combinations of $^{(0)}\omega^1, \ldots, ^{(0)}\omega^r$.

Because of this the differentials of the $^{(0)}\omega^i$ can be expressed as

$$d^{(0)}\omega^i = \sum_{j=1}^{r} da^i_j(x, \hat{x}, {}^{(1)}t) \wedge dx^j = \sum_{j=1}^{r} {}^{(0)}\omega^j \wedge \pi^i_j,$$

where the π_j^i are certain linear combinations of $d^{(0)}t^1, \ldots, d^{(0)}t^{p_1}, d\hat{x}^{k+1}$, $\ldots, d\hat{x}^r$ and the $^{(0)}\omega^i$. For the $(q+1)^{\text{th}}$ prolongation $g^{(q+1)}$ of $g \in G$,

$$g^{(q+1)*}(^{(0)}\omega^i) = {}^{(0)}\omega^i, \quad g^{(q+1)*}(d^{(0)}\omega^i) = d^{(0)}\omega^i,$$

and therefore

$$\sum_{j=1}^{r} {}^{(0)}\omega^j \wedge (g^{(q+1)*}(\pi_j^i) - \pi_j^i) = 0,$$

which in turn implies that

$$g^{(q+1)*}(\pi_j^i) \equiv \pi_j^i \pmod{{}^{(0)}\omega^1, \ldots, {}^{(0)}\omega^r}.$$

Now p_1 of the π_j^i are linearly independent modulo $^{(0)}\omega^1, \ldots, {}^{(0)}\omega^r$, $d\hat{x}^{h+1}, \ldots, d\hat{x}^r$—say

$$\bar{\pi}^i \equiv \sum_{j=1}^{p_1} c_j^i d^{(1)}t^j + \sum_{k=h+1}^{r} d_k^i d\hat{x}^k \pmod{{}^{(0)}\omega^1, \ldots, {}^{(0)}\omega^r} \quad \text{for } i = 1, \ldots, p_1,$$

with $\det(c_j^i) \neq 0$. By adding suitable multiples of the $^{(0)}\omega^i$ these can also be written in the following form, which involves the left hand sides of the pfaffian equations $d^{(1)}t^i - \sum {}^{(1)}a_j^i dx^j = 0$ and $d\hat{x}^i - \sum a_j^i dx^j = 0$:

$$\bar{\pi}^i \equiv \sum_{j=1}^{p_1} c_j^i \left(d^{(1)}t^j - \sum_{l=1}^{r} {}^{(1)}a_l^j dx^l \right) + \sum_{k=h+1}^{r} d_k^i (d\hat{x}^k - {}^{(0)}\omega^k)$$

$$\pmod{{}^{(0)}\omega^1, \ldots, {}^{(0)}\omega^k}.$$

Let us now define the one-forms

$$^{(1)}\omega^i := \sum_{j=1}^{p_1} c_j^i \left(d^{(1)}t^j - \sum_{l=1}^{r} {}^{(1)}a_l^j dx^l \right) + \sum_{k=h+1}^{r} d_k^i (d\hat{x}^k - {}^{(0)}\omega^k)$$

for $i = 1, \ldots, p_1$. For any $g^{(q+1)}$ we have on the one hand

$$g^{(q+1)*}(^{(1)}\omega^i) \equiv {}^{(1)}\omega^i \pmod{{}^{(0)}\omega^1, \ldots, {}^{(1)}\omega^r},$$

and on the other, since $\mathcal{P}^{(q+1)}$ is to be preserved,

$$g^{(q+1)*}(^{(1)}\omega^i) = \text{a linear combination of the left hand sides}$$
$$\text{in the pfaffian equations constituting } \mathcal{P}^{(q+1)}.$$

These two requirements are compatible only if

$$g^{(q+1)*}(^{(1)}\omega^i) = {}^{(1)}\omega^i \quad \text{for } i = 1, \ldots, p_1.$$

Therefore the $^{(1)}\omega^i$ are *invariant one-forms*.

Continuing in this manner we can create $r + (r - h) + p_1 + \cdots + p_q$ invariant one-forms

$$^{(0)}\omega^1, \ldots, {}^{(0)}\omega^r, \, d\hat{x}^{h+1}, \ldots, d\hat{x}^r, \, {}^{(1)}\omega^1, \ldots, {}^{(1)}\omega^{p_1}, \ldots, \, {}^{(q)}\omega^1, \ldots, {}^{(q)}\omega^{p_q},$$

where the last $^{(q)}\omega^i$—but only these—depend on the extra parameters u^1, $\ldots, u^{p_{q+1}}$.

Notations. Let $n := r + (r - h) + p_1 + \cdots + p_q = \dim S^q(G)$, denote the local coordinates $x^1, \ldots, x^r, \hat{x}^{h+1}, \ldots, \hat{x}^r, {}^{(1)}t^1, \ldots, {}^{(1)}t^{p_1}, \ldots, {}^{(q)}t^1, \ldots,$ $^{(q)}t^{p_q}$ for $S^q(G)$ by $x^1, x^2, \ldots, x^n$, denote the invariant one-forms $^{(0)}\omega^1$, $\ldots, {}^{(0)}\omega^r, \ldots, d\hat{x}^{h+1}, \ldots, d\hat{x}^r, {}^{(1)}\omega^1, \ldots, {}^{(1)}\omega^{p_1}, \ldots, {}^{(q)}\omega^1, \ldots, {}^{(q)}\omega^{p_q}$ by $\omega^1, \ldots, \omega^n$, and set $p := p_{q+1}$.

Then the one-forms

$$\omega^i = \sum_{j=1}^n a_j^i(x^1, \ldots, x^n; u^1, \ldots, u^p) \, dx^j \qquad \text{for } i = 1, \ldots, n,$$

make up a parametrized coframe on the n-dimensional $S^q(G)$, with the parameters $u^1, \ldots, u^p$ playing the role of local fibre coordinates for the fibre bundle

$$S^{q+1}(G)$$
$$\downarrow$$
$$S^q(G).$$

The parameters $u^1, \ldots, u^p$ are often called *auxiliary variables*.

Definition. The Lie pseudogroup $\mathfrak{G}^{(q)}$ on $S^q(G)$ consists of all local diffeomorphisms that preserve the invariants $I^1, \ldots, I^h$, and have prolongations to $S^{q+1}(G)$ leaving the one-forms $\omega^1, \ldots, \omega^n$ invariant.

Lemma 14.1.3. *With the defining PDE system $S^q(G)$ of G being involutive there is a one-to-one correspondence*

$$G \longleftrightarrow \mathfrak{G}^{(q)},$$

where the right hand arrow is a prolongation to $S^q(G)$, and the left hand one is induced from the source mapping $\alpha : J^q(\mathbb{C}^r_x, \mathbb{C}^r_{\hat{x}}) \longrightarrow \mathbb{C}^r_x$.

Proof. $\longrightarrow$: G is first prolonged to a pseudogroup of contact transformations on $J^q(\mathbb{C}^r_x, \mathbb{C}^r_{\hat{x}})$, which is then restricted to a pseudogroup on $S^q(G)$. The ω^i have been constructed so as to be invariant under the latter pseudogroup. Because

$\omega^i = \sum_{j=1}^{n} a^i_j(x; u)\, dx^j$, $i = 1, \ldots, n$, are n linearly independent one-forms and $x^1, \ldots, x^n$ make up a system of local coordinates for $S^q(G)$,

the invariance implies that $x^1, \ldots, x^n$ are mapped to functions of $x^1, \ldots, x^n$—so that each $g \in G$ induces a local diffeomorphism on $S^q(G)$, which belongs to $\mathfrak{G}^{(q)}$.

$\longleftarrow$: The first r invariant one-forms are given by

$$\omega^i = \sum_{j=1}^{r} a^i_j(x^1, \ldots, x^r, \hat{x}^{h+1}, \ldots, \hat{x}^r, {}^{(1)}t^1, \ldots, {}^{(1)}t^{p_1})\, dx^j \quad \text{for } i = 1, \ldots, r,$$

where $x^1, \ldots, x^r$ are coordinates for $\mathbb{C}^r_x$. Because these r one-forms are invariant, each $g^{(q)} \in \mathfrak{G}^{(q)}$ restricts to a source mapping

$$g = \alpha_*(g^{(q)}) : \mathbb{C}^r_x \longrightarrow \mathbb{C}^r_x.$$

It is easily seen that this g has but one prolongation leaving $\omega^{r+1}, \ldots, \omega^n$ invariant—namely the $g^{(q)} \in \mathfrak{G}^{(q)}$ that we started from. Because

$$j^q_{g(x)}(g^{-1}) = j^q_{g(x)}(\mathrm{id} \circ g^{-1}) = g^q(j^q_x(\mathrm{id})) \in S^q(G),$$

it follows that $g^{-1} \in G$—and thus also $g \in G$. $\square$

In view of this result it may be appropriate to make the concept of *prolongation of a Lie pseudogroup* more precise.

Definition. Let $\tilde{\mathfrak{G}}$ be a Lie pseudogroup acting on $\mathbb{C}^n_x \times \mathbb{C}^m_y$ with each $\tilde{g} \in \tilde{\mathfrak{G}}$ being of the form

$$\tilde{g} : \mathbb{C}^n_x \times \mathbb{C}^m_y \to \mathbb{C}^n_x \times \mathbb{C}^m_y$$
$$(x, y) \mapsto (g(x), h(x, y)),$$

so that $\tilde{g}$ has a well-defined projection to a local diffeomorphism

$$g : \mathbb{C}^n_x \to \mathbb{C}^n_x$$
$$x \mapsto g(x).$$

Suppose that these projections constitute a Lie pseudogroup $\mathfrak{G}$ on $\mathbb{C}^n_x$. Then

$\mathfrak{G}$ is the projection of $\tilde{\mathfrak{G}}$ to $\mathbb{C}^n_x$, and $\tilde{\mathfrak{G}}$ is a prolongation of $\mathfrak{G}$ to $\mathbb{C}^n_x \times \mathbb{C}^m_y$.

If there corresponds to each $g \in \mathfrak{G}$ *exactly one* $\tilde{g} \in \tilde{\mathfrak{G}}$, $\tilde{\mathfrak{G}}$ is said to be a *one-to-one prolongation of* $\mathfrak{G}$.

However, this concept is often too restricted. A more general one can be explained by means of the theory in section **3.3**: there we have seen how to measure the *size of a Lie pseudogroup* by the *number* s_p of *arbitrary functions in p variables* on which the general local solution of the defining PDE system depends. Using this we now say that $\tilde{\mathfrak{G}}$ is an *isomorphic prolongation* of $\mathfrak{G}$ if $\mathfrak{G}$ is of the same size as $\tilde{\mathfrak{G}}$.

Example. Define the Lie pseudogroup G acting on $\mathbb{C}^3$ by

$$(x, y, z) \mapsto (X(x, y, z), Y(x, y, z), Z(x, y, z)) := (x + a, y + f(x), z + f'(x)),$$

where $a \in \mathbb{C}$ is an arbitrary parameter, and f is an arbitrary function of one variable. This is a Lie pseudogroup with the defining PDE system

$$\begin{cases} \partial X/\partial x = 1, & \partial X/\partial y = 0, & \partial X/\partial z = 0, \\ \partial Y/\partial y = 1, & \partial Y/\partial z = 0, \\ \partial Z/\partial y = 0, & \partial Z/\partial z = 1, \\ \partial(Y - y)/\partial x = Z - z. \end{cases}$$

The two projections

$$\mathbb{C}^3_{xyz}$$

$$\pi_1 \swarrow \qquad \searrow \pi_2$$

$$\mathbb{C}^2_{xy} \qquad\qquad \mathbb{C}^2_{xz}$$

induce the Lie pseudogroups

$$G_1 := \pi_{1\bullet}(G) : (x, y) \mapsto (x + a, y + f(x)) \quad \text{acting on } \mathbb{C}^2_{xy},$$

and

$$G_2 := \pi_{2\bullet}(G) : (x, z) \mapsto (x + a, z + f'(x)) \quad \text{acting on } \mathbb{C}^2_{xz}.$$

Then G is a one-to-one prolongation of G_1, but *not* of G_2. However, with G and G_2 both depending on one arbitrary function in one variable, they are of the same size, and thus G is an isomorphic prolongation of G_2.
    Let us also define

$$G_0 : (x, y, z) \mapsto (x, y + b, z) \quad \text{acting on } \mathbb{C}^3_{xyz},$$

where $b \in \mathbb{C}$ is arbitrary. Then we have the short exact sequences

$$\text{Id} \to G \overset{\pi_{1\bullet}}{\to} G_1 \to \text{Id} \quad \text{and} \quad \text{Id} \to G_0 \to G \overset{\pi_{2\bullet}}{\to} G_2 \to \text{Id},$$

where $\text{Id} = \{\text{id}\}$ denotes the group consisting of the identity transformatiom only. The first shows that $G \cong G_1$, and the second that G_0 is the kernel of the epimorphism $\pi_{2\bullet}$, and as such is an invariant subgroup of

G. Being isomorphic to the group Tr of translations in one variable, G_0 is certainly nontrivial. Because G_1 and G_2 both are of the form

$$(x, t) \mapsto (x + a, t + \text{arbitrary function of } x),$$

they are manifestly isomorphic to each other—and hence also to G. But then we arrive at the paradoxical consequence that

$$\text{Tr} \cong G_0 = \ker\{G \overset{\pi_2}{\to} G_2\} \cong \text{Id}.$$

Thus the concepts invariant subgroup, quotient group and simple group are not evident for general Lie pseudogroups. But here G_0 *depends on one arbitrary constant*, while all three of G, G_1 and G_2 depend on one arbitrary function of one variable.

So from the point of view of section **3.3**, G_0 is to be regarded as a trivial subgroup of G—and then all pieces fit together.

We are now in a position to formulate a first version of **Cartan's first fundamental theorem for Lie pseudogroups**; a second version later on will make the dependence on the auxiliary variables $u^1, \ldots, u^p$ more explicit.

Theorem 14.1.1. *Let G be a Lie pseudogroup acting on $\mathbb{C}^r_x$ with the involutive defining PDE system $S^q(G) \subset J^q(\mathbb{C}^r_x, \mathbb{C}^r_{\hat{x}})$. Supposing that $\dim S^q(G) = n$ and that $\dim S^{q+1}(G) = n + p$, we identify $S^q(G)$ with $\mathbb{C}^n_x$ and $S^{q+1}(G)$ with $\mathbb{C}^n_x \times \mathbb{C}^p_u$ locally. Then G has a one-to-one prolongation to a Lie pseudogroup $\mathfrak{G}^{(q)}$ acting on $S^q(G)$, with $\mathfrak{G}^{(q)}$ being characterized by the following two properties.*

  (i) *$\mathfrak{G}^{(q)}$ leaves certain functions $I^1(x), \ldots, I^h(x)$ invariant.*
  (ii) *There are n linearly independent one-forms*

$$\omega^i(x; u) = \sum_{j=1}^{n} a^i_j(x; u) \, dx^j, \quad j = 1, \ldots, n,$$

  on $\mathbb{C}^n_x \times \mathbb{C}^p_u$ such that the elements of $\mathfrak{G}^{(q)}$ have prolongations to $\mathbb{C}^n_x \times \mathbb{C}^p_u$ leaving the ω^i invariant.   □

Definition. A Lie pseudogroup is said to be given in *Cartan form* if it is defined by such invariance properties.

Thus any Lie pseudogroup has a one-to-one prolongation to a Lie pseudogroup in Cartan form.

Remark. $\mathfrak{G}^{(q)}$ can be simplified somewhat by setting $\hat{x}^{h+1}, \ldots, \hat{x}^r$ equal to suitable constants and the $d\hat{x}^{h+1}, \ldots, d\hat{x}^r$ equal to zero—because it is immediately seen that the resulting Lie pseudogroup will still be a one-to-one prolongation of G.

Let $\mathfrak{G}$ be a Lie pseudogroup in Cartan form with the invariant functions $I^1(x),\ldots,I^h(x)$ and the invariant one-forms $\omega^i = \sum_{j=1}^{n} a_j^i(x;u)\,dx^j$ for $i = 1,\ldots,n$ on $\mathbb{C}_x^n \times \mathbb{C}_u^p$. The local diffeomorphisms in $\mathfrak{G}$ are then obtained as follows.

Let $\mathbb{C}_{\hat{x}}^n \times \mathbb{C}_{\hat{u}}^p$ be a copy of $\mathbb{C}_x^n \times \mathbb{C}_u^p$, and consider $g \in \mathfrak{G}$ as a local diffeomorphism $g: \mathbb{C}_x^n \times \mathbb{C}_u^p \longrightarrow \mathbb{C}_{\hat{x}}^n \times \mathbb{C}_{\hat{u}}^p$. Then the $g \in \mathfrak{G}$ are solutions of the system

$$\begin{cases} I^a(\hat{x}) = I^a(x) & \text{for } a = 1,\ldots,n, \\ \omega^i(\hat{x};\hat{u}) = \omega^i(x;u) & \text{for } i = 1,\ldots,n. \end{cases} \qquad (*)$$

This is really the defining PDE system of $\mathfrak{G}$, although written as a *first order system*, which moreover is *completely symmetric with respect to the source and target variables*.

Since $\sum_{j=1}^{n} a_j^i(\hat{x};\hat{u})\,d\hat{x}^j = \sum_{j=1}^{n} a_j^i(x;u)\,dx^j$ for $i = 1,\ldots,n$ it follows that the $\hat{x}^i$ are functions of $x^1,\ldots,x^n$ only, so that the solutions of (*) take the form

$$(x,u) \longmapsto (g(x), h(x,u)),$$

with the wanted elements of $\mathfrak{G}$ being obtained by projection to the first factor: $x \longmapsto g(x)$.

Because $\mathfrak{G}$ is defined by leaving certain functions and one-forms invariant, it is immediately checked that $\mathfrak{G}$ really is a pseudogroup:

(i) id $\in \mathfrak{G}$, since $\hat{x} = x$, $\hat{u} = u$ is a solution of (*).

(ii) If $g_1, g_2 \in \mathfrak{G}$ and $g_2 \circ g_1$ is defined, $(g_2 \circ g_1)^{\bullet}(\omega^i) = g_1^{\bullet}(g_2^{\bullet}(\omega^i)) = g_1^{\bullet}(\omega^i) = \omega^i$, and analogously for the I^a; therefore $g_2 \circ g_1 \in \mathfrak{G}$ too.

(iii) If $g \in \mathfrak{G}$, $\omega^i = \text{id}^{\bullet}(\omega^i) = (g \circ g^{-1})^{\bullet}(\omega^i) = (g^{-1})^{\bullet}(g^{\bullet}(\omega^i)) = (g^{-1})^{\bullet}(\omega^i)$, i.e., $(g^{-1})^{\bullet}(\omega^i) = \omega^i$, and analogously for the I^a; hence $g^{-1} \in \mathfrak{G}$ too.

14.2 The second fundamental theorem

Let $\mathfrak{G}$ be a Lie pseudogroup on $\mathbb{C}_x^n$ in Cartan form with the invariant functions $I^1(x),\ldots,I^h(x)$ and the invariant one-forms

$$\omega^i(x;u) = \sum_{j=1}^{n} a_j^i(x^1,\ldots,x^n;u^1,\ldots,u^p)\,dx^j, \qquad i = 1,\ldots,n.$$

Then the elements of $\mathfrak{G}$ are projections of local solutions of the pfaffian system $\mathcal{P}$:

$$\begin{cases} \phi^a(x;\hat{x}) := I^a(x) - I^a(\hat{x}) = 0 & \text{for } a = 1,\ldots,h, \\ \theta^i(x,u;\hat{x},\hat{u}) := \omega^i(x;u) - \omega^i(\hat{x};\hat{u}) = 0 & \text{for } i = 1,\ldots,n. \end{cases}$$

This system is constructed from an involutive PDE system, and is therefore itself involutive. In order to solve $\mathcal{P}$ by means of the pfaffian version of Cartan's local existence theorem we have to consider the differentials $d\theta^i$, or—what amounts to the same—the $d\omega^i$.

We know from before that $\det(a^i_j) \neq 0$, and hence the dx^j can conversely be written as linear combinations of the ω^i. Therefore

$$d\omega^i = \sum_{j=1}^{n} da^i_j(x^1,\ldots,x^n;u^1,\ldots,u^p) \wedge dx^j$$

$$= \sum_{j=1}^{n} \left(\sum_{k=1}^{n} \frac{\partial a^i_j}{\partial x^k} dx^k + \sum_{l=1}^{p} \frac{\partial a^i_j}{\partial u^l} du^l \right) \wedge dx^j$$

$$= \sum_{j=1}^{n} \omega^j \wedge \pi^i_j$$

with the π^i_j being certain linear combinations of the du^l and the ω^k.

An element of $\mathfrak{G}$ is a local diffeomorphism g of $\mathbb{C}^n_x$ that preserves the I^a and has a prolongation to a local diffeomorphism $g^{(1)}$ of $\mathbb{C}^n_x \times \mathbb{C}^p_u$ which leaves the ω^i invariant. Identifying g and $g^{(1)}$ we get

$$d\omega^i = g^*(d\omega^i) = \sum_{j=1}^{n} g^*(\omega^j) \wedge g^*(\pi^i_j) = \sum_{j=1}^{n} \omega^j \wedge g^*(\pi^i_j),$$

so that $0 = \sum_{j=1}^{n} \omega^j(g^*(\pi^i_j) - \pi^i_j)$, and hence

$$g^*(\pi^i_j) \equiv \pi^i_j \pmod{\omega^1,\ldots,\omega^n}.$$

There are p of the π^i_j which are linearly independent modulo $\omega^1,\ldots,\omega^n$— i.e., depend in an essential way on $du^1,\ldots,du^p$. With $\pi^1,\ldots,\pi^p$ denoting these, each π^i_j is a linear combination of $\pi^1,\ldots,\pi^p$ and $\omega^1,\ldots,\omega^n$. Thus

$$d\omega^i = \frac{1}{2} \sum_{j,k=1}^{n} c^i_{jk}(x;u)\,\omega^j \wedge \omega^k + \sum_{j=1}^{n}\sum_{l=1}^{p} a^i_{jl}(x;u)\,\omega^j \wedge \pi^l \quad \text{for } i = 1,\ldots,n,$$

where $c^i_{jk} + c^i_{kj} = 0$. In particular

$$d\omega^i \equiv \sum_{j=1}^{n} \sum_{l=1}^{p} a^i_{jl}\, \omega^j \wedge \pi^l \quad (\text{mod } \omega^i \wedge \omega^j \text{ for } 1 \le i < j \le n).$$

Applying g^* to this, we get

$$d\omega^i \equiv \sum_{j=1}^{n} \sum_{l=1}^{p} g^*(a^i_{jl})\omega^j \wedge \pi^l \quad (\text{mod } \omega^i \wedge \omega^j),$$

showing that $g^*(a^i_{jl}) = a^i_{jl}$ for any $g \in \mathfrak{G}$.

Conclusion. *The coefficients a^i_{jl} are **invariants** of $\mathfrak{G}$—that is, functions of $I^1(x),\ldots,I^h(x)$ only.*

Replacing π^l by $\bar{\pi}^l := \pi^l + \sum_{k=1}^{n} b^l_k(x;u)\,\omega^k$ we still have $(\bar{\pi}^1,\ldots,\bar{\pi}^p) \equiv (du^1,\ldots,du^p)$ $(\text{mod } \omega^1,\ldots,\omega^n)$. Under such a change

$$d\omega^i \mapsto \frac{1}{2}\sum_{j,k=1}^{n}\left(c^i_{jk} - 2\sum_{l=1}^{p} a^i_{jl}b^l_k\right)\omega^j \wedge \omega^k + \sum_{j=1}^{n}\sum_{l=1}^{p} a^i_{jl}\,\omega^j \wedge \bar{\pi}^l.$$

In order to simplify the expression for $d\omega^i$ we choose the coefficients b^l_k such that *as many of the coefficients in front of the $\omega^j \wedge \omega^k$ as possible get killed*. This might be done in several different ways, so the resulting c^i_{jk} are not necessarily uniquely determined—but they turn out to be *invariants*.

Lemma 14.2.1. *If we choose the π^l such that a minimal number of the coefficients in front of $\omega^j \wedge \omega^k$ are different from zero, the resulting c^i_{jk} are **invariants** of $\mathfrak{G}$, i.e., functions of $I^1(x),\ldots,I^h(x)$ only.*

Proof. Let $g \in \mathfrak{G}$ be arbitrary. Because the pull-back of the zero function certainly is the zero function,

$$g^*(c^i_{jk}) = c^i_{jk} \quad \text{whenever} \quad c^i_{jk} = 0.$$

Suppose now that there are a $g_1 \in \mathfrak{G}$ and a nonzero $c^{i_0}_{j_0k_0}$ such that $g_1^*(c^{i_0}_{j_0k_0}) \neq c^{i_0}_{j_0k_0}$. Because g_1^* preserves the ω^i and maps π^l to $g_1^*(\pi^l) = \pi^l + \sum_{k=1}^{n} d^l_k\,\omega^k$ for certain functions $d^l_k(x;u)$, g_1^* maps $d\omega^i$ to

$$d\omega^i = \frac{1}{2}\sum_{j,k=1}^{n}\left(g_1^*(c^i_{jk}) - 2\sum_{l=1}^{p} a^i_{jl}d^l_k\right)\omega^j \wedge \omega^k + \sum_{j=1}^{n}\sum_{l=1}^{p} a^i_{jl}\,\omega^j \wedge \pi^l.$$

Therefore

$$g_1^*(c_{jk}^i) = c_{jk}^i + 2\sum_{l=1}^{p} a_{jl}^i d_k^l,$$

where $\sum_{l=1}^{p} a_{jl}^i d_k^l = 0$ whenever $c_{jk}^i = 0$, but $\sum_{l=1}^{p} a_{j_0l}^{i_0} d_{k_0}^l \neq 0$. But then $c_{j_0k_0}^{i_0}$ gets killed if π^l is replaced by

$$\pi^l + \frac{c_{j_0k_0}^{i_0}}{2\sum_{m=1}^{p} a_{j_0m}^{i_0} d_{k_0}^m} \sum_{k=1}^{n} d_k^l \, \omega^k,$$

while the c_{jk}^i that already were zero are not affected. And so the number of nonzero c_{jk}^i was not minimal after all.   □

We have thereby arrived at the **second fundamental theorem for Lie pseudogroups in Cartan form**.

Theorem 14.2.1. *Let $\mathfrak{G}$ be a Lie pseudogroup in Cartan form with the invariant one-forms $\omega^1, \ldots, \omega^n$. The one-forms $\pi^1, \ldots, \pi^p$ involving the differentials of the auxiliary variables $u^1, \ldots, u^p$ may be chosen such that*

$$d\omega^i = \frac{1}{2}\sum_{j,k=1}^{n} c_{jk}^i \, \omega^j \wedge \omega^k + \sum_{j=1}^{n}\sum_{l=1}^{p} a_{jl}^i \, \omega^j \wedge \pi^l,$$

*where the coefficients c_{jk}^i, a_{jl}^i are **invariants** of $\mathfrak{G}$, and $c_{jk}^i + c_{kj}^i = 0$.*   □

If $\mathfrak{G}$ is a finite dimensional Lie parameter group there are no auxiliary variables—and hence no π^l—and no nonconstant invariants. Thus the c_{jk}^i are *constants*, and

$$d\omega^i = \frac{1}{2}\sum_{j,k=1}^{n} c_{jk}^i \, \omega^j \wedge \omega^k.$$

The 1-graphs of the local diffeomorphisms in $\mathfrak{G}$ are integral manifolds of the pfaffian system

$$\begin{cases} \phi^a(x;\hat{x}) := I^a(x) - I^a(\hat{x}) = 0 & \text{for } a = 1,\ldots,h, \\ \theta^i(x,u;\hat{x},\hat{u}) := \omega^i(x;u) - \omega^i(\hat{x},\hat{u}) = 0 & \text{for } i = 1,\ldots,n, \end{cases} \quad (\mathcal{P})$$

on which $\omega^1(x;u) \wedge \cdots \wedge \omega^n(x;u) \neq 0$—or equivalently $\omega^1(\hat{x};\hat{u}) \wedge \cdots \wedge \omega^n(\hat{x};\hat{u}) \neq 0$.

Let us differentiate $\mathcal{P}$. The equations $d\phi^a = 0$ are consequences of the equations $\theta^i = 0$, while the differentials of the θ^i take the form

$$d\theta^i \equiv \sum_{j=1}^{n} \sum_{l=1}^{p} a_{jl}^i \, \omega^j \wedge (\pi^l - \hat{\pi}^l) \quad (\text{mod } \mathcal{P}).$$

If we set $\varpi^l := \pi^l - \hat{\pi}^l$ this goes over into

$$d\theta^i \equiv \sum_{j=1}^{n} \sum_{l=1}^{p} a_{jl}^i \, \omega^j \wedge \varpi^l \quad (\text{mod } \mathcal{P}),$$

which means that $\mathcal{P}$ is of the simple form that is presupposed in Cartan's test for involutivity (see section **4.1**). And this makes it very easy to determine the Cartan characters, which in their turn decide how general the general solution is, as explained in section **3.3**.

It also follows that the involutivity of $\mathcal{P}$ implies that the structure $\{a_{jl}^i\}$ is involutive in the sense of the remark following the involutivity test in section **4.1**.

The second fundamental theorem has an evident converse.

Theorem 14.2.2. *Let $I^1(x), \ldots, I^h(x)$ be functions on $\mathbb{C}_x^n$, let*

$$\omega^i = \sum_{j=1}^{n} a_j^i(x; u) \, dx^j, \qquad i = 1, \ldots, n,$$

be n linearly independent one-forms in $x^1, \ldots, x^n$ and certain auxiliary variables $u^1, \ldots, u^p$, and suppose that

$$d\omega^i = \frac{1}{2} \sum_{j,k=1}^{n} c_{jk}^i(I) \, \omega^j \wedge \omega^k + \sum_{j=1}^{n} \sum_{l=1}^{p} a_{jl}^i(I) \, \omega^j \wedge \pi^l \quad \text{for } i = 1, \ldots, n,$$

$$(*)$$

where the coefficients c_{jk}^i and a_{jl}^i are functions of $I^1, \ldots, I^h$, and where $(\pi^1, \ldots, \pi^p) \equiv (du^1, \ldots, du^p) \pmod{\omega^1, \ldots, \omega^n}$ as modules of one-forms over the ring of functions on $\mathbb{C}_x^n \times \mathbb{C}_u^p$.

Assume moreover that the dI^a are linear combinations of the ω^i with coefficients depending on $I^1, \ldots, I^h$, and that the structure $\{a_{jl}^i\}$ is involutive. Then there is a Lie pseudogroup $\mathfrak{G}$ on $\mathbb{C}_x^n$ having $I^1, \ldots, I^h$ as invariants, and $(*)$ as structure equations.

Proof. The elements of $\mathfrak{G}$ are obtained by solving the system

$$\begin{cases} I^a(x) - I^a(\hat{x}) = 0, & a = 1, \ldots, h, \\ \omega^i(x; u) - \omega^i(\hat{x}, \hat{u}) = 0, & i = 1, \ldots, n \end{cases}$$

by means of Cartan's local existence theorem. The resulting solution set $\mathfrak{G}$ is a Lie pseudogroup by the remark at the end of the preceding section, and it clearly satisfies the requirements of the theorem. □

Let us now consider some examples. The first is our old favourite from chapter **8**.

Example 1. Let G be the affine group in one variable:

$$G: \mathbb{C}_x \to \mathbb{C}_{\hat{x}}$$
$$x \mapsto \hat{x} := ax + b, \quad \text{with } a \neq 0.$$

The defining ODE for G is $d^2\hat{x}/dx^2 = 0$. If we use $x, \hat{x}, p \leftrightarrow d\hat{x}/dx$ and $q \leftrightarrow d^2\hat{x}/dx^2$ as coordinates for $J^2(\mathbb{C}_x, \mathbb{C}_{\hat{x}})$, the defining equation $S^2(G)$ is given by $q = 0$. The restriction of

$$^2Ct^{1,1}: \begin{cases} d\hat{x} = p\,dx, \\ dp = q\,dx \end{cases} \quad \text{to} \quad S^2(G) \quad \text{is therefore} \quad \mathcal{P}: \begin{cases} d\hat{x} = p\,dx, \\ dp = 0. \end{cases}$$

According to the general recipe the first invariant one-form is to be taken as

$$\omega^1 := p\,dx.$$

Differentiating this, we find

$$d\omega^1 = dp \wedge dx = p\,dx \wedge \left(-\frac{dp}{p}\right) = \omega^1 \wedge \left(-\frac{dp}{p}\right).$$

Hence

$$\omega^2 := -\frac{dp}{p}$$

is the second invariant one-form. Moreover the structure equations are

$$d\omega^1 = \omega^1 \wedge \omega^2, \quad d\omega^2 = 0.$$

Next define the Lie pseudogroup $\mathfrak{G}$ on $\mathbb{C}^2_{xp}$ by the property of leaving the one-forms $\omega^1 = p\,dx$ and $\omega^2 = -dp/p$ invariant. If we regard the elements of $\mathfrak{G}$ as mappings $(x, p) \mapsto (\hat{x}, \hat{p}) = (X(x, p), P(x, p))$, this means that the functions $X(x, p)$ and $P(x, p)$ are to satisfy the pfaffian equations

$$P\,dX = p\,dx \quad \text{and} \quad \frac{dP}{P} = \frac{dp}{p}.$$

The first shows that X is a function of x only, which satisfies

$$P\,X'(x) = p, \quad \text{so that} \quad P(x, p) = \frac{p}{X'(x)}.$$

The second says that

$$\frac{dp}{p} = \frac{dP}{P}, \text{ with } \frac{dP}{P} = \frac{X'(x)}{p} \frac{X'(x)\,dp - pX''(x)\,dx}{(X'(x))^2} = \frac{dp}{p} - \frac{X''(x)}{X'(x)}\,dx,$$

whence $X''(x) = 0$. Therefore $X(x) = ax + b$, where a and b are arbitrary parameters, and $\mathfrak{G}$ is given by

$$(x, p) \mapsto \left(ax + b, \frac{p}{a}\right).$$

$\mathfrak{G}$ admits the projection $x \mapsto ax + b$ to $\mathbb{C}_x$—that is, the affine group G which we started from.

Conversely each mapping $x \mapsto ax + b$ in G has a *unique* prolongation to a mapping $(x, p) \mapsto (ax + b, p/a)$ belonging to $\mathfrak{G}$. Hence $\mathfrak{G}$ is a *one-to-one prolongation* of G, and this makes it possible to identify G and $\mathfrak{G}$.

Example 2. Define the 1-parameter group G acting on $\mathbb{C}^2_{xy}$ by

$$(x, y) \mapsto (X(x, y), Y(x, y)) := (x + ay, y),$$

with a being an arbitrary parameter. Then the defining PDE system of G consists of one zeroth order equation, $Y = y$, and the following first order ones:

$$\begin{cases} \partial X/\partial x = 1, & \partial X/\partial y = (X - x)/y, \\ \partial Y/\partial x = 0, & \partial Y/\partial y = 1. \end{cases}$$

The latter can equivalently be written in pfaffian form as

$$\begin{cases} dX = dx + \frac{X-x}{y}\,dy, \\ dY = dy. \end{cases}$$

If we replace X in the right hand side by 0, this gives the invariant one-forms

$$\omega^1 = dx - \frac{x}{y}\,dy \quad \text{and} \quad \omega^2 = dy.$$

With

$$d\omega^1 = -\frac{dx \wedge dy}{y} = -\frac{1}{y}\left(dx - \frac{x}{y}\,dy\right) \wedge dy = -\frac{1}{y}\,\omega^1 \wedge \omega^2,$$

the structure equations are

$$d\omega^1 = -\frac{1}{y}\,\omega^1 \wedge \omega^2 \quad \text{and} \quad d\omega^2 = 0;$$

note that the coefficient $-1/y$ in front of $\omega^1 \wedge \omega^2$ really is an invariant.

Let us conversely define the Lie pseudogroup $\mathfrak{G}$ on $\mathbb{C}^2_{xy}$ by the property of preserving the function y and the one-forms ω^1 and ω^2. That is, $(x, y) \mapsto (X(x, y), Y(x, y))$ belongs to $\mathfrak{G}$ if and only if

$$
\begin{cases}
Y = y, \\
dX - \frac{X}{Y} \, dY = dx - \frac{x}{y} \, dy, \\
dY = dy.
\end{cases}
$$

Then

$$
d(X - x) = \frac{X}{Y} \, dY - \frac{x}{y} \, dy = \frac{X - x}{y} \, dy,
$$

whence $X - x$ is a function of y only: $X - x = f(y)$, where f satisfies

$$
f'(y) = \frac{f(y)}{y}, \quad \text{so that} \quad f(y) = a \, y
$$

with a being an arbitrary parameter. Consequently

$$
X = x + a y \quad \text{and} \quad Y = y,
$$

and we have recovered the group we started from.

Example 3. Define the infinite dimensional Lie pseudogroup G acting on $\mathbb{C}^2_{xy}$ by

$$
(x, y) \mapsto (X(x, y), Y(x, y)) := (x + f(y), y),
$$

where f is an arbitrary function. Then the defining PDE system is

$$
\begin{cases}
Y = y, \\
\partial X / \partial x = 1,
\end{cases}
$$

or equivalently

$$
Y = y \quad \text{and} \quad dX = dx + u \, dy,
$$

with u being an auxiliary parameter. The invariant one-forms of G are

$$
\omega^1 = dy \quad \text{and} \quad \omega^2 = dx + u \, dy,
$$

and they satisfy the structure equations

$$
d\omega^1 = 0 \quad \text{and} \quad d\omega^2 = du \wedge dy = \omega^1 \wedge \pi,
$$

where $\pi = -du$.

Let us conversely define the Lie pseudogroup $\mathfrak{G}$ on $\mathbb{C}^3_{xyu}$ by leaving the function y and the one-forms ω^1 and ω^2 invariant. Then the general element $(x, y, u) \mapsto (X(x, y, u), Y(x, y, u), U(x, y, u))$ of $\mathfrak{G}$ is defined by

$$Y = y \quad \text{and} \quad dX + U\,dY = dx + u\,dy.$$

Hence $d(X - x) = (u - U)\,dy$, showing that $X - x = f(y)$, where $f(y)$ satisfies $f'(y) = u - U$. Thus $\mathfrak{G}$ is given by

$$(x, y, u) \mapsto (x + f(y), y, u - f'(y)),$$

with $f(y)$ being arbitrary. $\mathfrak{G}$ admits the projection

$$(x, y) \mapsto (x + f(y), y) \quad \text{to} \quad \mathbb{C}^2_{xy},$$

which is the group G we started from. Moreover $\mathfrak{G}$ is clearly a one-to-one prolongation of G, and therefore we may identify G and $\mathfrak{G}$.

14.3 The third fundamental theorem

Let us for the sake of simplicity choose the invariants $I^1(x), \ldots, I^h(x)$ as the last h coordinates for $\mathbb{C}^n_x$—that is, setting $v := n - h$ we let

$$x^{v+a} := I^a \quad \text{for } a = 1, \ldots, h.$$

In connection with this we also assume that

$$\omega^{v+a} = dx^{v+a} = dI^a \quad \text{for } a = 1, \ldots, h.$$

Being invariants, the coefficients c^i_{jk} and a^i_{jl} in the structure equations are functions of $x^{v+1}, \ldots, x^n$ only, and therefore

$$dc^i_{jk} = \sum_{a=1}^{h} \frac{\partial c^i_{jk}}{\partial x^{v+a}} dx^{v+a} = \sum_{r=1}^{n} \frac{\partial c^i_{jk}}{\partial x^r} \omega^r,$$

$$da^i_{jl} = \sum_{r=1}^{n} \frac{\partial a^i_{jl}}{\partial x^r} \omega^r.$$

The purpose of the third fundamental theorem is to characterize the coefficients appearing in the structure equations

$$d\omega^i = \frac{1}{2} \sum_{j,k=1}^{n} c^i_{jk}\, \omega^j \wedge \omega^k + \sum_{j=1}^{n} \sum_{l=1}^{p} a^i_{jl}\, \omega^j \wedge \pi^l.$$

Necessary conditions are found by differentiating these:

$$0 = \frac{1}{2} \sum_{j,k} (dc_{jk}^i \wedge \omega^j \wedge \omega^k + c_{jk}^i \, d\omega^j \wedge \omega^k - c_{jk}^i \, \omega^j \wedge d\omega^k)$$

$$+ \sum_{j,l} (da_{jl}^i \wedge \omega^j \wedge \pi^l + a_{jl}^i \, d\omega^j \wedge \pi^l - a_{jl}^i \, \omega^j \wedge d\pi^l)$$

$$= - \sum_{j,q} a_{jq}^i \, \omega^j \wedge d\pi^q + \sum_{j,t} \sum_{(l,k)} (a_{jl}^t a_{tk}^i - a_{jk}^t a_{tl}^i) \, \omega^j \wedge \pi^l \wedge \pi^k$$

$$+ \sum_l \sum_{(j,r)} \left(\sum_t (c_{jr}^t a_{tl}^i + c_{tj}^i a_{rl}^t - c_{tr}^i a_{jl}^t) + \frac{\partial a_{rl}^i}{\partial x^j} - \frac{\partial a_{jl}^i}{\partial x^r} \right) \omega^j \wedge \omega^r \wedge \pi^l$$

$$+ \sum_{(j,r,s)} \left(\sum_t (c_{jr}^t c_{ts}^i + c_{rs}^t c_{tj}^i + c_{sj}^t c_{tr}^i) + \frac{\partial c_{jr}^i}{\partial x^s} - \frac{\partial a_{js}^i}{\partial x^r} + \frac{\partial c_{rs}^i}{\partial x^j} \right) \omega^j \wedge \omega^r \wedge \omega^s$$

where $\sum_{(\ldots)}$ denotes summation over the different permutations of $(\ldots)$.

The structure equations tell us how the $d\omega^i$ look, but we do not know the corresponding formulas for the $d\pi^q$. But as two-forms on $\mathbb{C}_x^n \times \mathbb{C}_u^p$ they can surely be written as

$$d\pi^q = \sum_{(l,k)} \gamma_{lk}^q(x,u) \, \pi^l \wedge \pi^k + \sum_{r,l} \delta_{rl}^q(x,u) \, \omega^r \wedge \pi^l + \sum_{(r,s)} \epsilon_{rs}^q(x,u) \, \omega^r \wedge \omega^s$$

with certain coefficients γ_{lk}^q, δ_{rl}^q and ϵ_{rs}^q, which at the outset are unknown. By the above these unknowns have to satisfy the linear systems

$$\begin{cases} \sum_{t=1}^n (a_{jl}^t a_{tk}^i - a_{jk}^t a_{tl}^i) = \sum_{q=1}^p a_{jq}^i \gamma_{lk}^q \\ \text{for } i,j = 1,\ldots,n \text{ and } k,l = 1,\ldots,p, \end{cases} \tag{1}$$

$$\begin{cases} \sum_{t=1}^n (c_{jr}^t a_{tl}^i + c_{tj}^i a_{rl}^t - c_{tr}^i a_{jl}^t) + \partial a_{rl}^i / \partial x^j - \partial a_{jl}^i / \partial x^r \\ = \sum_{q=1}^p (a_{jq}^i \delta_{rl}^q - a_{rq}^i \delta_{jl}^q) \quad \text{for } i,j,r = 1,\ldots,n \text{ and } l = 1,\ldots,p, \end{cases} \tag{2}$$

$$\begin{cases} \sum_{t=1}^n (c_{jr}^t c_{ts}^i + c_{rs}^t c_{tj}^i + c_{sj}^t c_{tr}^i) + \partial c_{jr}^i / \partial x^s - \partial c_{js}^i / \partial x^r + \partial c_{rs}^i / \partial x^j \\ = \sum_{q=1}^p (a_{jq}^i \epsilon_{rs}^q - a_{rq}^i \epsilon_{js}^q + a_{sq}^i \epsilon_{rj}^q) \\ \text{for } i,j,r,s = 1,\ldots,n. \end{cases} \tag{3}$$

Recall how the π^l are constructed: from the definition of the invariant one-forms ω^i we have

$$d\omega^i \equiv \sum_{j=1}^n \omega^j \wedge \pi_j^i \pmod{\omega^j \wedge \omega^k, \quad 1 \le j < k \le n},$$

and from the π_j^i appearing here we pick out a linearly independent set $\{\pi^1, \ldots, \pi^p\}$; setting $\pi_j^i = \sum_{l=1}^p a_{jl}^i \pi^l$ we then deduce that

$$d\omega^i \equiv \sum_{j=1}^n \sum_{l=1}^p a_{jl}^i \, \omega^j \wedge \pi^l \quad (\mathrm{mod} \ \omega^j \wedge \omega^k).$$

So by the definition of $\pi^1, \ldots, \pi^p$ it is possible to solve the equations

$$\pi_j^i = \sum_{l=1}^p a_{jl}^i \, \pi^l \quad \text{for } i, j = 1, \ldots, n$$

with respect to the π^l. And thus it is also possible to solve the $\frac{1}{2} p^2(p-1)$ coefficients γ_{lk}^q from (1) so as to express them by means of the a_{jl}^i. In particular it follows that the γ_{lk}^q are *invariants of $\mathfrak{G}$*: $\gamma_{lk}^q = \gamma_{lk}^q(x^{\nu+1}, \ldots, x^n)$.

In order to obtain a useful interpretation of (1) we define the vector fields

$$E_l := \sum_{i,j=1}^n a_{jl}^i(x^{\nu+1}, \ldots, x^n) y^j \frac{\partial}{\partial y^i}, \qquad l = 1, \ldots, p$$

on $\mathbb{C}_y^n$, with coefficients belonging to the ring of invariants for $\mathfrak{G}$. Then

$$
\begin{aligned}
[E_l, E_k] &= \left[\sum_{i,j} a_{jl}^i y^j \frac{\partial}{\partial y^i}, \sum_{r,s} a_{sk}^r y^s \frac{\partial}{\partial y^r} \right] \\
&= \sum_{j,r} \left(\sum_i (a_{jl}^i a_{ik}^r - a_{jk}^i a_{il}^r) \right) y^j \frac{\partial}{\partial y^r} \\
&= \sum_{j,r} \sum_q a_{jq}^r \gamma_{lk}^q y^j \frac{\partial}{\partial y^r} = \sum_q \gamma_{lk}^q \left(\sum_{j,r} a_{jq}^r y^j \frac{\partial}{\partial y^r} \right) \\
&= \sum_{q=1}^p \gamma_{lk}^q(x^{\nu+1}, \ldots, x^n) \, E_q,
\end{aligned}
$$

so that the module generated by the vector fields $E_1, \ldots, E_p$ over the ring of invariant functions forms a *Lie algebra over the invariants*, with γ_{lk}^q as *structure functions*.

Having solved γ_{lk}^q from (1) we have only to solve δ_{rl}^q from (2) and ϵ_{rs}^q from (3), which requires that *the linear systems (2) and (3) for the unknowns δ_{rl}^q and ϵ_{rs}^q are compatible*.

Note that if no auxiliary variables u are present, all the a_{jl}^i vanish, and then (1) and (2) are empty. Moreover, if there are no nonconstant invariants, (3) reduces to

$$\sum_{t=1}^{n}(c_{jr}^{t}c_{ts}^{i} + c_{rs}^{t}c_{tj}^{i} + c_{sj}^{t}c_{tr}^{i}) = 0,$$

which is the third fundamental theorem for finite dimensional Lie parameter groups.

The **third fundamental theorem for Lie pseudogroups in Cartan form** can then be stated as follows.

Theorem 14.3.1. *The coefficients c_{jk}^{i} and a_{jl}^{i} in the structure equations*

$$d\omega^{i} = \frac{1}{2}\sum_{j,k=1}^{n} c_{jk}^{i}\,\omega^{j}\wedge\omega^{k} + \sum_{j=1}^{n}\sum_{l=1}^{p} a_{jl}^{i}\,\omega^{j}\wedge\pi^{l}$$

are invariants of $\mathfrak{G}$ which satisfy the following conditions:

  (i) *$c_{jk}^{i} + c_{kj}^{i} = 0$,*
 (ii) *the coefficients $\{a_{jl}^{i}\}$ form an involutive system,*
(iii) *the vector fields $E_{l} = \sum_{i,j=1}^{n} a_{jl}^{i}y^{j}\,(\partial/\partial y^{i})$ for $l = 1,\ldots,p$ form a Lie algebra over the ring of invariants of $\mathfrak{G}$,*
 (iv) *the linear systems (2) and (3) for the unknowns δ_{rl}^{q} and ϵ_{rs}^{q} respectively are compatible.* $\qquad\square$

Assume conversely that c_{jk}^{i} and a_{jl}^{i} are functions of $x^{\nu+1},\ldots,x^{n}$ which satisfy the four conditions above. Then it can be shown that there are one-forms $\omega^{1},\ldots,\omega^{n}$ and $\pi^{1},\ldots,\pi^{n}$ on $\mathbb{C}_{x,u}^{n+p}$ such that

$$d\omega^{i} = \frac{1}{2}\sum_{j,k=1}^{n} c_{jk}^{i}\,\omega^{j}\wedge\omega^{k} + \sum_{j=1}^{n}\sum_{l=1}^{p} a_{jl}^{i}\,\omega^{j}\wedge\pi^{l} \quad \text{for } i = 1,\ldots,n.$$

This being a statement about *two-forms*, the proof—given in [Cartan 1937b] for instance—requires Kähler's generalization of Cartan's local existence theorem to arbitrary differential ideals (and not just pfaffian systems).

Having found the one-forms ω^{i} one then solves the corresponding pfaffian system

$$\begin{cases} x^{\nu+a} - \hat{x}^{\nu+a} = 0, & a = 1,\ldots,h, \\ \omega^{i}(x;u) - \omega^{i}(\hat{x};\hat{u}) = 0, & i = 1,\ldots,n, \end{cases}$$

as in the converse of the second fundamental theorem, and thereby obtains a Lie pseudogroup with the above structure equations.

So given c^i_{jk} and a^i_{jl} satisfying the requirements in the third fundamental theorem, there is a Lie pseudogroup $\mathfrak{G}$ having these as coefficients in its structure equations.

14.4 The stability group and its Lie algebra

By the first fundamental theorem any Lie pseudogroup G has a one-to-one prolongation to a Lie pseudogroup $\mathfrak{G}$ in Cartan form acting on the defining PDE system $S^q(G)$ of G. Moreover the defining pfaffian system of $\mathfrak{G}$ is equivalent to a *first order* PDE system $S^1(\mathfrak{G}) \subset J^1(\mathbb{C}^n_x, \mathbb{C}^n_{\hat{x}})$:

$$\begin{cases} I^a(\hat{x}) = I^a(x) & \text{for } a = 1,\ldots,h, \\ p^i_j = a^i_j(x, \hat{x}, u) & \text{for } i, j = 1,\ldots,n, \end{cases}$$

where $u^1,\ldots,u^p$ are the parametric jet coordinates for $S^1(\mathfrak{G})$. With $d\hat{x}^i = \sum_{j=1}^n p^i_j \, dx^j$ this gives

$$\begin{pmatrix} d\hat{x}^1 \\ \vdots \\ d\hat{x}^n \end{pmatrix} = \begin{pmatrix} a^1_1(x, \hat{x}, u) & \cdots & a^1_n(x, \hat{x}, u) \\ \vdots & \ddots & \vdots \\ a^n_1(x, \hat{x}, u) & \cdots & a^n_n(x, \hat{x}, u) \end{pmatrix} \begin{pmatrix} dx^1 \\ \vdots \\ dx^n \end{pmatrix},$$

or shorter

$$d\hat{x} = A(x, \hat{x}, u) \, dx.$$

Thus the local diffeomorphism $\hat{x} = g(x)$ belongs to $\mathfrak{G}$ if and only if

$$\begin{cases} I^a(g(x)) = I^a(x) & \text{for } a = 1,\ldots,h, \\ \text{and} \\ dg(x) = A(x, g(x), u) \, dx & \text{for a suitable } u. \end{cases}$$

Let us denote the p-parameter family of matrices $A(x, \hat{x}, u)$ by $\mathfrak{A}(\mathfrak{G})$. What does it mean for $\mathfrak{A}(\mathfrak{G})$ that $\mathfrak{G}$ is a Lie pseudogroup? Firstly $g \in \mathfrak{G} \iff g^{-1} \in \mathfrak{G}$, showing that

$$d\hat{x} = A(x, \hat{x}, u) \, dx \iff dx = A(\hat{x}, x, v) \, d\hat{x},$$

for a suitable parameter v, and so $d\hat{x} = A^{-1}(\hat{x}, x, v) \, dx$. But then

$$d\hat{x} = A(x, \hat{x}, u) \, dx = A^{-1}(\hat{x}, x, v) \, dx = A(\hat{x}, x, w) \, dx$$

for some parameter w. Being true for all x and $\hat{x}$, the equality $A(x, \hat{x}, u) =$

$A(\hat{x}, x, w)$ shows that A depends on x and $\hat{x}$ merely through the invariants $I^1, \ldots, I^h$:

$$A = A(I, u).$$

Secondly, $g_1, g_2 \in \mathfrak{G}$ and $g_2 \circ g_1$ is defined $\Longrightarrow g_2 \circ g_1 \in \mathfrak{G}$ too. From this it follows that

$$A(I, u), A(I, v) \in \mathfrak{A}(\mathfrak{G}) \Longrightarrow \exists \phi(u, v) \text{ such that}$$
$$A(I, u)A(I, v) = A(I, \phi(u, v)).$$

Conclusion. $\mathfrak{A}(\mathfrak{G})$ *is a **matrix group** depending on the invariants of $\mathfrak{G}$ and the auxiliary variables $u^1, \ldots, u^p$.*

So starting from the coframe $dx = (dx^1, \ldots, dx^n)^T$ for $\mathbb{C}_x^n$, $\mathfrak{G}$ induces a p-parameter family $A(I, u) dx$ of new coframes. And one passes from one of these to another by means of an element of the matrix group $\mathfrak{A}(\mathfrak{G})$.

Of course there is nothing particular with the coframe dx—it would have been possible to start from any coframe.

Let us for instance consider the coframes given by the invariant one-forms $\omega^i(x; u)$ of $\mathfrak{G}$: $\omega(x; u) = (\omega^1(x; u), \ldots, \omega^n(x; u))^T$ defines a coframe for $\mathbb{C}_x^n$ for each $u \in \mathbb{C}^p$.

Set $e := (0, \ldots, 0) \in \mathbb{C}_u^p$. Because both $\omega(x; u)$ and $\omega(x; e) = (\omega^1(x; e), \ldots, \omega^n(x; e))^T$ are coframes for $\mathbb{C}_x^n$, there are functions $m_j^i(x, u)$ such that

$$\omega(x; u) = \begin{pmatrix} \omega^1(x; u) \\ \vdots \\ \omega^n(x; u) \end{pmatrix} = \begin{pmatrix} m_1^1(x, u) & \cdots & m_n^1(x, u) \\ \vdots & \ddots & \vdots \\ m_1^n(x, u) & \cdots & m_n^n(x, u) \end{pmatrix} \begin{pmatrix} \omega^1(x; e) \\ \vdots \\ \omega^n(x; e) \end{pmatrix}$$
$$=: M(x, u)\omega(x; e).$$

Suppose now that $(\hat{x}; \hat{u})$ and $(x; u)$ are connected through $\mathfrak{G}$—i.e., there is some $g \subset \mathfrak{G}$ with prolongation $g^{(1)}$ such that

$$(\hat{x}; \hat{u}) = (g(x); g^{(1)}(x; u)) \Longleftrightarrow (x; u) = (g^{-1}(\hat{x}), (g^{-1})^{(1)}(\hat{x}; \hat{u})).$$

Then by the invariance property of the ω^i,

$$\omega(\hat{x}; \hat{u}) = \omega(x; u) = M(x, u)\omega(x; e).$$

Setting $\hat{u} := e$ here, $u = (g^{-1})^{(1)}(\hat{x}; e) = (g^{-1})^{(1)}(g(x); e) =: u_g$, and

$$\omega(\hat{x}; e) = M(x, u_g)\omega(x; e) \Longleftrightarrow \omega(x; e) = M^{-1}(x, u_g)\omega(\hat{x}; e).$$

So $g \in \mathfrak{G}$ maps the coframe $\omega(x; e)$ to the coframe $M^{-1}(x, u_g)\omega(\hat{x}; e)$.

Because $g \in \mathfrak{G} \Longleftrightarrow g^{-1} \in \mathfrak{G}$, we also have

$$\omega(x;e) = M(\hat{x}, u_{g^{-1}})\omega(\hat{x};e),$$

and therefore $M^{-1}(x, u_g) = M(\hat{x}, u_{g^{-1}})$. For this to be true for all x and $\hat{x}$, the first arguments in M^{-1} and M can only depend on the invariants $I^1, \ldots, I^h$ of $\mathfrak{G}$:

$$M = M(I, u) \quad \text{and} \quad M^{-1}(I, u_g) = M(I, u_{g^{-1}}).$$

Conclusion. *The local diffeomorphism* $g : \mathbb{C}_x^n \xrightarrow{\cong} \mathbb{C}_x^n$ *belongs to* $\mathfrak{G}$ *if and only if* g *preserves the invariants of* $\mathfrak{G}$, *and*

$$g^*\omega(x;e) = M(I, u_g)\,\omega(x;e)$$

for some parameter u_g. *In particular, for* $g = \mathrm{id}$ *we get*

$$M(I, u_{\mathrm{id}}) = \text{the identity matrix}.$$

If $g_1, g_2 \in \mathfrak{G}$ and $g_2 \circ g_1$ is defined, the latter belongs to $\mathfrak{G}$ too, and therefore

$$\begin{aligned}
M(I, u_{g_2 \circ g_1})\omega(x;e) = (g_2 \circ g_1)^*\omega(x;e) &= g_1^*(g_2^*\omega(x;e)) \\
&= g_1^*(M(I, u_{g_2})\omega(x;e)) = M(I, u_{g_2})g_1^*\omega(x;e) \\
&= M(I, u_{g_2})M(I, u_{g_1})\omega(x;e),
\end{aligned}$$

so that

$$M(I, u_{g_2 \circ g_1}) = M(I, u_{g_2})M(I, u_{g_1}).$$

Thereby we have obtained the following result.

Lemma 14.4.1. *The p-parameter family of matrices* $M(I, u)$ *actually forms a matrix group.* $\qquad\square$

This matrix group is called the *stability group* in [Cartan 1937b], and the *adjoint group* in [Cartan 1905]. We stick to the first-mentioned name, and denote it by $\mathfrak{St}(\mathfrak{G})$.

Because

$$M(I, u), M(I, v) \in \mathfrak{St}(\mathfrak{G}) \Longrightarrow M(I, u)M(I, v) \in \mathfrak{St}(\mathfrak{G}) \text{ too,}$$

there is a function $\phi : \mathbb{C}^p \times \mathbb{C}^p \to \mathbb{C}^p$ such that

$$M(I, u)M(I, v) = M(I, \phi(u, v)).$$

By the reasoning in section **8.1**, $\mathfrak{St}(\mathfrak{G})$ induces in this way a Lie parameter group $\mathfrak{G}_{\text{aux}}$ acting on the space $\mathbb{C}^p_u$ of auxiliary variables. And $\mathfrak{St}(\mathfrak{G})$ is conversely considered as a *realization* of $\mathfrak{G}_{\text{aux}}$ in the form of a matrix group acting on the p-parameter family $\omega(x;u) = (\omega^1(x;u),\ldots,\omega^n(x;u))^T$ of coframes for $\mathbb{C}^n_x$:

$$\omega(x;u) = M(I,u)\,\omega(x;e) \quad \text{and} \quad \omega(x;u) = M(I,\phi(u,v^{-1}))\,\omega(x;v)$$

for $u,v \in \mathbb{C}^p$ and $M \in \mathfrak{St}(\mathfrak{G})$.

From section **8.7** we know that $\mathfrak{G}_{\text{aux}}$ also induces a *Lie algebra of Killing vector fields* on the p-parameter family $\omega(x;u)$ of coframes. This Lie algebra is called the *stability algebra*, and is denoted by $\mathfrak{st}(\mathfrak{G})$. It is constructed in the following way.

Let $U = \sum_{i=1}^p c_i\,(\partial/\partial u^i)$ be a vector at $e = (0,\ldots,0) \in \mathbb{C}^p_u$. Extending U to a vector field near e with constant coefficients c_i, it induces an infinitesimal transformation on the coframe $\omega(x;u) = (\omega^1(x;u),\ldots,\omega^n(x;u))^T$ by means of the *Lie derivative*:

$$\mathcal{L}_U(\omega^i) = U\rfloor d\omega^i + d(\omega^i(U)).$$

Now
$$\omega^i = \sum_{j=1}^n a^i_j(x,u)\,dx^j \quad \text{and} \quad d\omega^i = \tfrac{1}{2}\sum_{j,k=1}^n c^i_{jk}\,\omega^i \wedge \omega^k + \sum_{j=1}^n \sum_{l=1}^p a^i_{jl}\,\omega^j \wedge \pi^l,$$

and therefore

$$\mathcal{L}_U(\omega^i) = U\rfloor\left(\frac{1}{2}\sum_{j,k} c^i_{jk}\,\omega^j \wedge \omega^k + \sum_{j,l} a^i_{jl}\,\omega^j \wedge \pi^l\right) + d(0)$$

$$= -\sum_{j,l} a^i_{jl}\,\omega^j\,\pi^l(U) = \sum_{l=1}^p t^l \sum_{j=1}^n a^i_{jl}\,\omega^j$$

with $t^l := -\pi^l(U)$ for $l = 1,\ldots,p$. Recalling that

$$\pi^l \equiv \sum_{m=1}^p b^l_m\,du^m \quad (\text{mod } \omega^1,\ldots,\omega^n) \quad \text{for} \quad l = 1,\ldots,p,$$

where b^l_m are invariants of $\mathfrak{G}$ and $\det(b^l_m) \neq 0$, we see that the t^l are parameters which depend on the invariants.

So the infinitesimal action induced on the coframe $\omega(x;u)$ is given by

$$\omega(x;u) = \begin{pmatrix} \omega^1(x;u) \\ \vdots \\ \omega^n(x;u) \end{pmatrix} \mapsto \begin{pmatrix} \sum_{l=1}^{p} t^l a_{1l}^1 & \cdots & \sum_{l=1}^{p} t^l a_{nl}^1 \\ \vdots & \ddots & \vdots \\ \sum_{l=1}^{p} t^l a_{1l}^n & \cdots & \sum_{l=1}^{p} t^l a_{nl}^n \end{pmatrix} \begin{pmatrix} \omega^1(x;u) \\ \vdots \\ \omega^n(x;u) \end{pmatrix}$$

$$=: \delta(t)\,\omega(x;u).$$

To obtain a suggestive interpretation of this we define $\partial/\partial\omega^i$ acting on $\omega^1,\ldots,\omega^n$ by

$$\frac{\partial}{\partial\omega^i}(\omega^j) = \begin{cases} 1 & \text{if } j = i, \\ 0 & \text{if } j \neq i, \end{cases}$$

and set

$$E_l := \sum_{i,j=1}^{n} a_{jl}^i \omega^j \frac{\partial}{\partial\omega^i}.$$

Then the infinitesimal action takes the form

$$\omega^i \mapsto \sum_{l=1}^{p} t^l \sum_{j=1}^{n} a_{jl}^i \omega^j = \sum_{l=1}^{p} t^l E_l \omega^i = \delta(t)\,\omega^i$$

with $\delta(t) := \sum_{l=1}^{p} t^l E_l$. Note that

$$[E_k, E_l] = \left[\sum_{i,j=1}^{n} a_{jk}^i \omega^j \frac{\partial}{\partial\omega^i}, \sum_{s,t=1}^{n} a_{tl}^s \omega^t \frac{\partial}{\partial\omega^s} \right]$$

$$= \sum_{i,j,s=1}^{n} (a_{jk}^i a_{il}^s - a_{ik}^s a_{jl}^i)\omega^j \frac{\partial}{\partial\omega^s},$$

which according to the third fundamental theorem, in the preceding section, reduces to

$$[E_k, E_l] = \sum_{j=1}^{p} \gamma_{kl}^j E_j.$$

Thus $E_1,\ldots,E_p$ do generate a Lie algebra over the ring of invariants of $\mathfrak{G}$—and this is the stability algebra $\mathfrak{st}(\mathfrak{G})$ that we have been looking for.

Theorem 14.4.1. *There is a Lie parameter group $\mathfrak{G}_{\text{aux}}$ acting on the space $\mathbb{C}_u^p$ of auxiliary variables which induces the stability group $\mathfrak{St}(\mathfrak{G})$ acting on the p-parameter family $\omega(x;u)$ of coframes on $\mathbb{C}_x^n$, and the stability algebra $\mathfrak{st}(\mathfrak{G}) = \text{span}_{\{\text{invariants of } \mathfrak{G}\}}\{E_1,\ldots,E_p\}$ acting on the same family.* □

Remark. Having found $\mathfrak{st}(\mathfrak{G})$ as described above, we can use the results on local transformation groups in section **8.7** in order to find $\mathfrak{G}_{aux}$ and its realization $\mathfrak{St}(\mathfrak{G})$, as an alternative to defining the latter directly.

As noticed earlier the stability group also characterizes the Lie pseudogroup $\mathfrak{G}$, and hence yields a **second version of the first fundamental theorem**.

Theorem 14.4.2. *Let G be a Lie pseudogroup with the invariants $I^1,\dots,I^h$ and the defining PDE system $S^q(G)$. G has a one-to-one prolongation to a Lie pseudogroup $\mathfrak{G}$ in Cartan form acting on $S^q(G)$, and being defined as follows.*

Let $n := \dim S^q(G)$, and let $\omega^1(x;u),\dots,\omega^n(x;u)$ be the invariant one-forms of the first version. Then $\omega(x;u) = (\omega^1(x;u),\dots,\omega^n(x;u))^T$ constitutes a local coframe for $\mathbb{C}_x^n$ for each $u \in \mathbb{C}_u^p$. With $e := (0,\dots,0) \in \mathbb{C}_u^p$ there is a matrix group $\mathfrak{St}(\mathfrak{G}) = \{M(I,u)\}$ depending on the invariants $I^1,\dots,I^h$ and the auxiliary variables $u^1,\dots,u^p$, called the **stability group**, *such that*

$$\omega(x;u) = M(I,u)\omega(x;e) \quad \text{for each } u.$$

In terms of this set-up,

a local diffeomorphism $g : S^q(G) \xrightarrow{\cong} S^q(G)$ belongs to $\mathfrak{G}$ if and only if g preserves the invariants, and

$$g^*\omega(x;e) = M(I,u)\omega(x;e)$$

for a suitable parameter u.   □

The earlier version of this theorem required g to have a prolongation $g^{(1)}$ to $S^q(G)\times$ (space of auxiliary variables), leaving the one-forms $\omega^1(x;u),\dots,\omega^n(x;u)$ invariant. In this new version there is no need to prolong g; the auxiliary variables $u^1,\dots,u^p$ crop up in the matrices $M(I,u) \in \mathfrak{St}(\mathfrak{G})$ instead.

Finite dimensional Lie parameter groups have no auxiliary variables u, and their stability groups are therefore trivial. So the requirement above reduces to

$$g^*\omega(x) = \omega(x)$$

with $\omega(x) = (\omega^1(x),\dots,\omega^n(x))^T$.

Recall moreover that Lie parameter groups induced from Lie transformation groups automatically are *transitive*—that is, admit no nonconstant invariants. In view of this it is natural to ask: is it possible to get rid of the invariants also in the infinite dimensional case?

The key for answering this question is provided by the set of one-forms

$$\sum_{j=1}^{n} a_{jl}^{i} \, \omega^{j} \quad \text{for } i = 1, \ldots, n \text{ and } l = 1, \ldots, p$$

appearing in the structure equations

$$d\omega^{i} = \frac{1}{2} \sum_{j,k=1}^{n} c_{jk}^{i} \, \omega^{j} \wedge \omega^{k} + \sum_{l=1}^{p} \left(\sum_{j=1}^{n} a_{jl}^{i} \, \omega^{j} \right) \wedge \pi^{l},$$

as well as in the vector fields generating the stability algebra:

$$E_{l} = \sum_{i=1}^{n} \left(\sum_{j=1}^{n} a_{jl}^{i} \, \omega^{j} \right) \frac{\partial}{\partial \omega^{i}}.$$

Definition. The module of one-forms generated by

$$\sum_{j=1}^{n} a_{jl}^{i} \, \omega^{j} \quad \text{for } i = 1, \ldots, n \text{ and } l = 1, \ldots, p$$

over the ring of invariants of $\mathfrak{G}$ is called the *systatic system* of $\mathfrak{G}$, and is denoted by $\mathrm{Sys}(\mathfrak{G})$.

Theorem 14.4.3. $\mathrm{Sys}(\mathfrak{G})$ *is complete.*

Proof. Let $q := \dim \mathrm{Sys}(\mathfrak{G})$. By replacing the one-forms ω^{i} by suitable linear combinations with coefficients that are invariants of $\mathfrak{G}$, it may be assumed that

$$\sum_{j=1}^{n} a_{jl}^{i} \, \omega^{j} = 0 \text{ for } i = 1, \ldots, n \text{ and } l = 1, \ldots, p \iff \omega^{1} = \cdots = \omega^{q} = 0.$$

Then $a_{jl}^{i} = 0$ for $j > q$. Because

$$d\omega^{i} = \frac{1}{2} \sum_{j,k=1}^{n} c_{jk}^{i} \, \omega^{j} \wedge \omega^{k} + \sum_{j=1}^{q} \sum_{l=1}^{p} a_{jl}^{i} \, \omega^{j} \wedge \pi^{l}$$

it therefore suffices to show that $c_{jk}^{i} = 0$ for $i = 1, \ldots, q$ and $j, k > q$. Differentiating $d\omega^{i}$ and considering the coefficients in front of $\omega^{h} \wedge \omega^{k} \wedge \pi^{l}$ for $h, k > q$ we deduce that

$$\sum_{j=1}^{q} a_{jl}^{i} \, c_{hk}^{j} = 0.$$

But since $\sum_{j=1}^{q} a_{jl}^i \omega^j = 0 \Longleftrightarrow \omega^1 = \cdots = \omega^q = 0$, this implies that

$$c_{hk}^1 = \cdots = c_{hk}^q = 0 \qquad \text{for } h, k > q$$

—and we are done. □

With each $\sum_{j=1}^{n} a_{jl}^i \omega^j$ being a linear combination of $dx^1, \ldots, dx^n$, the completeness means that there are functions $f^1(x), \ldots, f^q(x)$ on $\mathbb{C}^n_x$ such that

Sys$(\mathfrak{G}) = (df^1, \ldots, df^q)$ as a module over the ring of invariants.

But then Sys$(\mathfrak{G})$ *really is a module of one-forms living on* $\mathbb{C}^n_x$.

Set $v := n - h$, and assume that the invariants $I^1(x), \ldots, I^h(x)$ have been used as the last h coordinates for $\mathbb{C}^n_x$, i.e.,

$$x^{v+a} = I^a \qquad \text{for } a = 1, \ldots, h.$$

Suppose furthermore that

$$\omega^{v+a} = dx^{v+a} = dI^a \qquad \text{for } a = 1, \ldots, h.$$

Then form linear combinations of the one-forms $\sum_{j=1}^{n} a_{jl}^i \omega^j$ over the invariants, and extract as many as possible which *only depend on the invariants and their differentials*, i.e., are of the form

$$\phi^j = \sum_{l=1}^{h} f_{v+l}^j (x^{v+1}, \ldots, x^n)\, \omega^{v+l}.$$

Assume that m linearly independent ϕ^j can be found in this way. Letting

$$\pi: \qquad \mathbb{C}^n \to \mathbb{C}^h$$
$$(x^1, \ldots, x^n) \mapsto (x^{v+1}, \ldots, x^n)$$

be the natural projection, we then set

$\pi_*(\text{Sys}(\mathfrak{G})) := (\phi^1, \ldots, \phi^m)$, as a module over the ring of invariants.

The completeness of Sys$(\mathfrak{G})$ means in particular that

$$d\phi^j \equiv 0 \pmod{\text{Sys}(\mathfrak{G})} \qquad \text{for } j = 1, \ldots, m.$$

But since $d\phi^j$ only involves $x^{v+1}, \ldots, x^n$ and their differentials, this forces

$$d\phi^j \equiv 0 \pmod{\phi^1, \ldots, \phi^m} \qquad \text{for } j = 1, \ldots, m.$$

Therefore

$\pi_*(\text{Sys}(\mathfrak{G}))$ *is a complete pfaffian system living on the space of invariants of* $\mathfrak{G}$.

Definition. The first integrals of $\pi_*(\mathrm{Sys}(\mathfrak{G}))$ are called *essential invariants*, while the other invariants are said to be *inessential*.

In the next section we will see that it is possible to *get rid of the inessential invariants*—but not of the essential ones.

For finite dimensional Lie parameter groups $\mathfrak{G}$ the a^i_{jl} vanish, so that $\mathrm{Sys}(\mathfrak{G}) = \emptyset$—and hence *all* invariants are inessential in this case.

Example. Let us consider example **2** from section **14.2**, that is, the 1-parameter group G on $\mathbb{C}^2$ defined by

$$(x, y) \mapsto (X(x, y), Y(x, y)) := (x + ay, y)$$

with $a \in \mathbb{C}$ being an arbitrary parameter. We know that the invariant one-forms are given by

$$\omega^1 = dx - \frac{x}{y}\,dy, \quad \omega^2 = dy,$$

and that they satisfy the structure equations

$$d\omega^1 = -\frac{1}{y}\,\omega^1 \wedge \omega^2, \quad d\omega^2 = 0.$$

In particular there are no a^i_{jl}, so $\mathrm{Sys}(\mathfrak{G}) = \emptyset$ and the invariant y is inessential.

There is a general recipe for getting rid of inessential invariants, which will be explained in the next section. Here it amounts to starting from the expression $X = x + ay$, and making the replacements

$$X \mapsto x, \quad x \mapsto x_0 = \text{constant}, \quad a \mapsto \xi = \text{a new variable},$$

so as to end up with $x = x_0 + \xi y$. Choosing $x_0 = 0$ for instance we have $\xi = xy^{-1}$. Using ξ and y as coordinates for $\mathbb{C}^2$ (outside of $\{y = 0\}$), the action of G is given by

$$(\xi, y) = \left(\frac{x}{y}, y\right) \mapsto \left(\frac{x + ay}{y}, y\right) = (\xi + a, y).$$

Hence

$$G \cong \mathrm{Tr} \oplus \mathrm{Id} : \mathbb{C}_\xi \times \mathbb{C}_y \to \mathbb{C}_\xi \times \mathbb{C}_y$$
$$(\xi, y) \mapsto (\xi + a, y),$$

where Tr is the translation group in one variable. Consequently G may be identified with Tr.

14.5 Getting rid of inessential invariants

Let $\mathfrak{G}$ be a Lie pseudogroup in Cartan form acting on $\mathbb{C}_x^n$ with the h invariants $x^{\nu+1},\ldots,x^n$ and the invariant one-forms $\omega^1,\ldots,\omega^\nu,\omega^{\nu+1} = dx^{\nu+1},\ldots,\omega^n = dx^n$. Suppose that the projection $\pi_*(\mathrm{Sys}(\mathfrak{G}))$ of the systatic system to the space $\mathbb{C}_{x^{\nu+1}\ldots x^n}^h$ of invariants admits a functionally independent first integrals $y^1,\ldots,y^a$; these invariants are said to be *essential*. Then we replace $x^{\nu+1},\ldots,x^n$ by $y^1,\ldots,y^a$ and $b := h - a$ functionally independent *inessential* invariants $z^1,\ldots,z^b$.

Theorem 14.5.1. *It is possible to replace $x^1,\ldots,x^\nu$ by new local coordinates $\xi^1,\ldots,\xi^\nu$ such that the action of $\mathfrak{G}$ is given by*

$$\mathfrak{G} : \mathbb{C}_\xi^\nu \times \mathbb{C}_y^a \times \mathbb{C}_z^b \to \mathbb{C}_\xi^\nu \times \mathbb{C}_y^a \times \mathbb{C}_z^b$$
$$(\xi, y, z) \mapsto (g(\xi, y), y, z).$$

Defining $\tilde{\mathfrak{G}}$ by

$$\tilde{\mathfrak{G}} : \mathbb{C}_\xi^\nu \times \mathbb{C}_y^a \to \mathbb{C}_\xi^\nu \times \mathbb{C}_y^a$$
$$(\xi, y) \mapsto (g(\xi, y), y),$$

we deduce that

$$\mathfrak{G} = \tilde{\mathfrak{G}} \oplus \mathrm{Id} : \mathbb{C}_{\xi,y}^{\nu+a} \times \mathbb{C}_z^b \to \mathbb{C}_{\xi,y}^{\nu+a} \times \mathbb{C}_z^b.$$

*In this situation $\mathfrak{G}$ and $\tilde{\mathfrak{G}}$ are regarded as **isomorphic**.*

*If there are no essential invariants at all (i.e., $a = 0$), this means in particular that $\mathfrak{G}$ is isomorphic to a **transitive** Lie pseudogroup.*

Proof. The defining pfaffian system for $\mathfrak{G}$ is

$$\mathcal{P} : \begin{cases} \hat{y}^j - y^j = 0 & \text{for } j = 1,\ldots,a, \\ \hat{z}^k - z^k = 0 & \text{for } k = 1,\ldots,b, \\ \theta^i := \hat{\omega}^i - \omega^i = 0 & \text{for } i = 1,\ldots,n, \end{cases}$$

with

$$d\omega^i = \frac{1}{2} \sum_{j,k=1}^n c_{jk}^i \, \omega^i \wedge \omega^k + \sum_{l=1}^p \left(\sum_{j=1}^n a_{jl}^i \, \omega^j \right) \wedge \pi^l,$$

and analogously for $d\hat{\omega}^i$—so that

$$d\theta^i \equiv \sum_{l=1}^p \left(\sum_{j=1}^n a_{jl}^i \, \omega^j \right) \wedge (\hat{\pi}^l - \pi^l) \pmod{\mathcal{P}}.$$

The complete systatic system

$$\text{Sys}(\mathfrak{G}): \quad \sum_{j=1}^{n} a^i_{jl}\, \omega^j \quad \text{for } i = 1,\ldots,n \text{ and } l = 1,\ldots,p$$

has a well-defined projection $\pi_*(\text{Sys}(\mathfrak{G}))$ to the space of invariants, with $y^1,\ldots,y^a$ as functionally independent first integrals. Let us choose the local coordinates x^i such that the y^j together with $x^1,\ldots,x^q$ form a fundamental system of first integrals for the whole of $\text{Sys}(\mathfrak{G})$. Then the integral manifolds of $\text{Sys}(\mathfrak{G})$ are given by

$$x^i = \text{constant for } i = 1,\ldots,q \quad \text{and} \quad y^j = \text{constant for } j = 1,\ldots,a.$$

Restricting $\mathcal{P}$ to such an integral manifold—i.e., fixing $x^1,\ldots,x^q, y^1,\ldots,y^a$ to $x^1_0,\ldots,x^q_0, y^1_0,\ldots,y^a_0$, say—it follows that $d\theta^i \equiv 0 \pmod{\text{the restricted } \mathcal{P}}$. Therefore the Frobenius theorem applies and gives the integral manifolds

$$\begin{cases} G^i(x,\hat{x},y,z) = A^i & \text{for } i = 1,\ldots,v, \\ \hat{y}^j = y^j & \text{for } j = 1,\ldots,a \quad \text{and} \quad \hat{z}^k = z^k \quad \text{for } k = 1,\ldots,b, \end{cases}$$

where $x = (x^1,\ldots,x^v)$ and $\hat{x} = (\hat{x}^1,\ldots,\hat{x}^v)$. The A^i appearing here are constant on each integral manifold of $\text{Sys}(\mathfrak{G})$. If we do not restrict ourselves to such integral manifolds, the A^i are instead to be regarded as functions of $x^1,\ldots,x^q, y^1,\ldots,y^a$. By adopting the latter point of view and solving the above equations with respect to $\hat{x}^1,\ldots,\hat{x}^v$, the general element of $\mathfrak{G}$ gets expressed in the form

$$\begin{cases} \hat{x}^i = F^i(x,y,z,A^1(x^1,\ldots,x^q;y^1,\ldots,y^a),\ldots,A^v(x^1,\ldots,x^q;y^1,\ldots,y^a)) \\ \quad \text{for } i = 1,\ldots,v, \\ \hat{y}^j = y^j \quad \text{for } j = 1,\ldots,a \quad \text{and} \quad \hat{z}^k = z^k \quad \text{for } k = 1,\ldots,b. \end{cases}$$

For a particular $g \in \mathfrak{G}$—with $\hat{x} = g(x,y,z)$ say—the A^i are determinate functions:

$$A^i_g = G^i(x,g(x,y,z),y,z) \quad \text{for } i = 1,\ldots,v,$$

which however *only depend on* $x^1,\ldots,x^q, y^1,\ldots y^a$—and this is the crucial point.

Now comes the trick (which was used in the example at the end of the preceding section): fix a value x_0 of $x = (x^1,\ldots,x^v)$, and introduce new variables $\xi^1,\ldots,\xi^v$ by means of the equations

$$x^i = F^i(x_0,y,z,\xi^1,\ldots,\xi^v) \quad \text{for } i = 1,\ldots,v,$$

or equivalently $\xi^i = G^i(x_0,x,y,z)$ for $i = 1,\ldots,v$.

In the original coordinates we have $\mathbb{C}^n = \mathbb{C}^v_x \times \mathbb{C}^a_y \times \mathbb{C}^b_z$; replacing $x^1, \ldots, x^v$ by $\xi^1, \ldots, \xi^v$ we instead get $\mathbb{C}^n = \mathbb{C}^v_\xi \times \mathbb{C}^a_y \times \mathbb{C}^b_z$.

Let $g \in \mathfrak{G}$ be defined near the point (x_0, y, z). The old coordinates of the image point $g(x_0, y, z)$ are

$$
\begin{cases}
x^i = F^i(x_0, y, z, A^1_g(x^1_0, \ldots, x^q_0; y^1, \ldots, y^a), \ldots, A^v_g(x^1_0, \ldots, x^q_0; y^1, \ldots, y^a)) \\
\quad \text{for } i = 1, \ldots, v, \\
y^j = y^j \quad \text{for } j = 1, \ldots, a \quad \text{and} \quad z^k = z^k \quad \text{for } k = 1, \ldots, b.
\end{cases}
$$

Comparing this with the definition of the ξ^i we deduce that the ξ-coordinates of $g(x_0, y, z)$ are

$$
\xi^i = A^i_g(x^1_0, \ldots, x^q_0; y^1, \ldots, y^a)
$$

—which thus are *independent of* $z^1, \ldots, z^b$.

Next let (ξ, y, z) be an arbitrary point (expressed in the new coordinates) near (x_0, y, z) (using the old coordinates), and suppose that $h \in \mathfrak{G}$ is defined at (ξ, y, z) by

$$
h(\xi, y, z) = (h^\xi(\xi, y, z), y, z).
$$

Since $\mathfrak{G}$ is transitive with respect to the x- (or equivalently the ξ-) variables, there is a $g_0 \in \mathfrak{G}$ such that $(\xi, y, z) = g_0(x_0, y, z)$. Then

$$
h \circ g_0(x_0, y, z) = h(\xi, y, z) = (h^\xi(\xi, y, z), y, z)
$$

with $h \circ g_0 \in \mathfrak{G}$—which implies that the ξ-coordinates of the image point do not depend on z, so that h^ξ is *a function of* ξ *and* y *only*.

So using the ξ-variables the general element of $\mathfrak{G}$ acquires the form

$$
g : \mathbb{C}^v_\xi \times \mathbb{C}^a_y \times \mathbb{C}^b_z \rightarrow \mathbb{C}^v_\xi \times \mathbb{C}^a_y \times \mathbb{C}^b_z
$$
$$
(\xi, y, z) \mapsto (g^\xi(\xi, y), y, z).
$$

To obtain the $\tilde{\mathfrak{G}}$ announced in the theorem we now set

$$
\tilde{\mathfrak{G}} \ni \tilde{g} : \mathbb{C}^v_\xi \times \mathbb{C}^a_y \rightarrow \mathbb{C}^v_\xi \times \mathbb{C}^a_y
$$
$$
(\xi, y) \mapsto (g^\xi(\xi, y), y).
$$

$\square$

Here are two examples from [Cartan 1937b].

Example 1. Let $\mathfrak{G}$ act on $\mathbb{C}^2$ by

$$
(x, y) \mapsto (x + f(y), y),
$$

where f is an arbitrary function of one variable. Then $\mathfrak{G}$ induces the transitive group $x \mapsto x + f(y)$ on each fibre of

$$\mathbb{C}_x \times \mathbb{C}_y$$
$$\downarrow$$
$$\mathbb{C}_y,$$

and it is intuitively clear that these fibre actions depend in an essential way on the invariant y.

In order to see this from the present point of view, recall from section **14.2** that $\mathfrak{G}$ is defined by the invariant one-forms

$$\omega^1 = dx + u\,dy, \quad \omega^2 = dy,$$

which satisfy the structure equations

$$d\omega^1 = \omega^2 \wedge \pi \quad \text{(with } \pi = -du\text{)}, \quad d\omega^2 = 0.$$

Therefore Sys($\mathfrak{G}$) is given by

$$\omega^2 = 0, \text{ or equivalently } dy = 0,$$

whence y really is an *essential invariant*—which it is not possible to get rid of.

So in contrast to the finite dimensional case there are infinite dimensional Lie pseudogroups which are not isomorphic to transitive ones.

Example 2. Let G act on $\mathbb{C}^3$ by

$$(x, y, z) \mapsto (x + ay + bz, y, z),$$

where $a, b \in \mathbb{C}$ are arbitrary parameters. Then the defining PDE system of G is of the second order:

zeroth order equations $\quad \hat{y} = y, \ \hat{z} = z$;

first order equations $\quad \dfrac{\partial \hat{x}}{\partial x} = 1, \ \dfrac{\partial \hat{x}}{\partial z} = \dfrac{1}{z}\left(\hat{x} - x - y\dfrac{\partial \hat{x}}{\partial y}\right)$;

second order equations $\quad$ all second order derivatives vanish.

Only considering the equations of order ≤ 1 we obtain a Lie pseudogroup $\mathfrak{H}$ in Cartan form, containing G as a subgroup. The defining pfaffian system of $\mathfrak{H}$ is

$$\begin{cases} \hat{y} = y, \ \hat{z} = z, \\ d\hat{x} = dx + u\,dy + z^{-1}(\hat{x} - x - uy)\,dz, \end{cases}$$

where $u = \partial \hat{x} / \partial y$ is an auxiliary variable. Setting $\hat{x} = 0$ we obtain the invariant one-form

$$\omega^1 = dx - \frac{x}{z} dz + u \left(dy - \frac{y}{z} dz \right).$$

Besides this we have the two invariant one-forms $\omega^2 = dy$ and $\omega^3 = dz$, which result from the invariants y and z. Because

$$d\omega^1 = -\frac{1}{z} dx \wedge dz + du \wedge \left(dy - \frac{y}{z} dz \right) - \frac{u}{z} dy \wedge dz$$

$$= -\frac{1}{z} \omega^1 \wedge \omega^3 + du \wedge \left(\omega^2 - \frac{y}{z} \omega^3 \right),$$

the structure equations of $\mathfrak{H}$ are

$$d\omega^1 = -\frac{1}{z} \omega^1 \wedge \omega^3 + \left(\omega^2 - \frac{y}{z} \omega^3 \right) \wedge \pi \quad \text{and} \quad d\omega^2 = d\omega^3 = 0,$$

with $\pi = -du$. Hence the systatic system of $\mathfrak{H}$ is given by

$$\omega^2 - \frac{y}{z} \omega^3 = 0, \quad \text{i.e.,} \quad 0 = dy - \frac{y}{z} dz = z \, d \left(\frac{y}{z} \right).$$

Consequently y/z is an *essential invariant* for $\mathfrak{H}$.

Introducing y/z as a new invariant instead of y outside of $\{z = 0\}$, $\mathfrak{H}$ is defined by leaving

$$\frac{y}{z}, \ z \quad \text{and} \quad \omega^1 = dx - x \frac{dz}{z} + uz \, d \left(\frac{y}{z} \right)$$

invariant. The restriction of $\mathfrak{H}$ to a manifold $\{y/z = \text{constant}\}$ is defined by leaving

$$z \quad \text{and} \quad dx - x \frac{dz}{z}$$

invariant, that is,

$$\hat{z} = z \quad \text{and} \quad d\hat{x} - \hat{x} \frac{d\hat{z}}{z} = dx - x \frac{dz}{z}.$$

Thus

$$d(\hat{x} - x) = (\hat{x} - x) \frac{dz}{z} \quad \text{or} \quad \frac{d(\hat{x} - x)}{\hat{x} - x} = \frac{dz}{z},$$

showing that

$$\hat{x} - x = Az$$

for some constant A. Letting y/z vary we instead get

$$\hat{x} = x + A \left(\frac{y}{z} \right) z.$$

With z being an *inessental invariant* for $\mathfrak{H}$, $\mathfrak{H}$ is isomorphic to the Lie pseudogroup $\tilde{\mathfrak{H}}$ obtained by setting $z = 1$ (for instance)—

$$\tilde{\mathfrak{H}}: \quad (x, y) \mapsto (x + A(y), y)$$

—so $\tilde{\mathfrak{H}}$ is the Lie pseudogroup of the preceding example.

In accordance with the general recipe we obtain the new variable ξ from $\hat{x} = x + Az$ by the replacements

$$x \mapsto x_0 \quad (= 0 \text{ for instance}), \quad \hat{x} \mapsto x, \quad \text{and } A \mapsto \xi,$$

giving

$$x = \xi z \quad \text{or} \quad \xi = \frac{x}{z}.$$

Using $\xi = x/z$, $\eta := y/z$ and z as local coordinates for $\mathbb{C}^3$ outside of $\{z = 0\}$, the action of the original group G is given by

$$(\xi, \eta, z) = \left(\frac{x}{z}, \frac{y}{z}, z\right) \mapsto \left(\frac{x + ay + bz}{z}, \frac{y}{z}, z\right) = (\xi + a\eta + b, \eta, z).$$

That is,

$$G = \tilde{G} \oplus \mathrm{Id} : \mathbb{C}^2_{\xi\eta} \times \mathbb{C}_z \to \mathbb{C}^2_{\xi\eta} \times \mathbb{C}_z$$
$$(\xi, \eta, z) \mapsto (\xi + a\eta + b, \eta, z).$$

Thus G is isomorphic to $\tilde{G} : (\xi, \eta) \mapsto (\xi + a\eta + b, \eta)$, having one invariant fewer than G.

14.6 Normal prolongations

Let

$$S: \quad F^a(x^i; z^j; p^j_{i_1}; p^j_{i_1 i_2}; \dots) = 0, \quad a = 1, \dots, A, \quad 1 \le i, i_1, i_2, \dots \le n$$
$$\text{and} \quad 1 \le j \le m,$$

be a PDE system. Recall that its first prolongation $S^{(1)}$ is defined by

$$S^{(1)}: \quad F^a = 0, \quad \frac{dF^a}{dx^i} = 0 \quad \text{for } a = 1, \dots, A \text{ and } i = 1, \dots, n,$$

with

$$\frac{d}{dx^i} = \frac{\partial}{\partial x^i} + \sum_{j=1}^{m} \left(p^j_i \frac{\partial}{\partial z^j} + \sum_{i_1=1}^{n} p^j_{ii_1} \frac{\partial}{\partial p^j_{i_1}} + \sum_{i_1,i_2=1}^{n} p^j_{ii_1 i_2} \frac{\partial}{\partial p^j_{i_1 i_2}} + \cdots \right).$$

(We could also have obtained a *partial prolongation* by only adding *some* of the extra equations $dF^a/dx^i = 0$, but that type of prolongation is not considered here.)

Higher order prolongations are obtained successively:

$$S^{(k+1)} = (S^{(k)})^{(1)} \quad \text{for } k = 1, 2, \ldots.$$

And it is clear that all these systems have the same solution sets.

All of this applies in particular to the defining PDE systems of Lie pseudogroups.

So let G be a Lie pseudogroup with the involutive defining PDE system $S^q(G) \subset J^q(\mathbb{C}^r_x, \mathbb{C}^r_{\hat{x}})$, and let $S^{q+k}(G) \subset J^{q+k}(\mathbb{C}^r_x, \mathbb{C}^r_{\hat{x}})$ be prolongations of $S^q(G)$ for $k = 1, 2, \ldots$. The first fundamental theorem shows that it is possible to define a one-to-one prolongation $\mathfrak{G}$ of G in Cartan form from

$$
\begin{array}{ccc}
S^{q+1}(G) & & \mathbb{C}^n_x \times \mathbb{C}^p_u \\
\downarrow & \cong & \downarrow \\
S^q(G) & & \mathbb{C}^n_x.
\end{array}
$$

More precisely, $\mathfrak{G}$ consists of all local diffeomorphisms on $S^q(G) \cong \mathbb{C}^n_x$ which

(i) leave h functions $I^1, \ldots, I^h$ on $\mathbb{C}^n_x$ invariant, and
(ii) have prolongations acting on $S^{q+1}(G) \cong \mathbb{C}^{n+p}_{x,u}$ that leave n one-forms $\omega^i = \sum_{j=1}^n a^i_j(x, u)\, dx^j$, $i = 1, \ldots, n$, invariant, where $\{\omega^1, \ldots, \omega^n\}$ constitutes a local coframe for $\mathbb{C}^n_x$ for each u.

The invariant one-forms satisfy the structure equations

$$d\omega^i = \frac{1}{2} \sum_{j,k=1}^n c^i_{jk}\, \omega^j \wedge \omega^k + \sum_{j=1}^n \sum_{l=1}^p a^i_{jl}\, \omega^j \wedge \pi^l \quad \text{for } i = 1, \ldots, n,$$

where $\pi^1, \ldots, \pi^p$ are one-forms which together with $\omega^1, \ldots, \omega^n$ form a local coframe for $\mathbb{C}^{n+p}_{x,u}$.

But we could just as well have constructed one-to-one prolongations of G in Cartan form by starting from *any* of the fibre bundles $S^{q+k+1}(G) \longrightarrow S^{q+k}(G)$ for $k = 1, 2, \ldots$.

For instance, if

$$
\begin{array}{ccc}
S^{q+2}(G) & & \mathbb{C}^{n+p}_{x,u} \times \mathbb{C}^q_v \\
\downarrow & \cong & \downarrow \\
S^{q+1}(G) & & \mathbb{C}^{n+p}_{x,u},
\end{array}
$$

the corresponding Lie pseudogroup $\mathfrak{G}^{(1)}$ in Cartan form will consist of those local diffeomorphisms on $S^{q+1}(G) \cong \mathbb{C}^{n+p}_{x,u}$ which

(i) leave the functions $I^1, \ldots, I^h$ on $\mathbb{C}^n_x$ invariant,

(ii) leave the one-forms $\omega^1, \ldots, \omega^n$ on $\mathbb{C}^{n+p}_{x,u}$ invariant, and

(iii) have prolongations acting on $\mathcal{S}^{q+2}(G) \cong \mathbb{C}^{n+p+q}_{x,u,v}$ that leave certain one-forms $\omega^{n+1}, \ldots, \omega^{n+p}$ defined there invariant, and for which moreover $\{\omega^1, \ldots, \omega^n, \omega^{n+1}, \ldots, \omega^{n+p}\}$ is a local coframe for $\mathbb{C}^{n+p}_{x,u}$ for each fixed v.

The fibre bundle $\mathcal{S}^{q+3}(G) \longrightarrow \mathcal{S}^{q+2}(G)$ analogously yields $\mathfrak{G}^{(2)}$, and so on.

From an intrinsic point of view it is, however, preferable to derive the prolonged Cartan forms directly from the defining pfaffian system of $\mathfrak{G}$.

The local diffeomorphisms $x \mapsto \hat{x}$ of $\mathfrak{G}$ are solutions of the pfaffian system

$$\mathcal{P}: \begin{cases} I^a(\hat{x}) - I^a(x) = 0, & a = 1, \ldots, h, \\ \theta^i := \omega^i(\hat{x}, \hat{u}) - \omega^i(x, u) = 0, & i = 1, \ldots, n, \end{cases}$$

where

$$d\theta^i \equiv \sum_{j=1}^{n} \sum_{l=1}^{p} a^i_{jl}(I)\, \omega^j \wedge (\hat{\pi}^l - \pi^l) \pmod{\mathcal{P}}.$$

Considering $\hat{\pi}^l - \pi^l$ as unknowns in the system

$$\sum_{j,l} a^i_{jl}(I)\, \omega^j \wedge (\hat{\pi}^l - \pi^l) = 0, \quad i = 1, \ldots, n, \tag{*}$$

we infer that

$$\hat{\pi}^l - \pi^l = \text{ a linear combination of } \omega^1, \ldots, \omega^n.$$

From Cartan's test for involutivity in section **4.1** it follows that $q := s_1 + 2s_2 + \cdots + ns_n$ of the coefficients in this linear combination are parametric, while the other ones are expressed linearly in these, with $s_1, \ldots, s_n$ being the Cartan characters of $\mathcal{P}$. Hence the most general solution of (*) can be written as

$$\hat{\pi}^l - \pi^l = \sum_{j=1}^{n} \sum_{k=1}^{q} b^l_{jk} v^k \omega^j, l = 1 \ldots, p,$$

where $v^1, \ldots, v^q$ are arbitrary parameters, while the b^l_{jk} are determinate functions on $\mathbb{C}^{n+p}_{x,u}$.

Let us now consider the first prolongation $\mathfrak{G}^{(1)}$ of $\mathfrak{G}$. Its elements are local diffeomorphisms $(x, u) \mapsto (\hat{x}, \hat{u})$ which preserve $I^1, \ldots, I^h$ and $\omega^1, \ldots, \omega^n$, and furthermore induce the most general pull-backs of $\hat{\pi}^l$ that are compatible with the invariance of $I^1,\dots,I^h$ and $\omega^1,\dots,\omega^n$. Because of the structure equations

$$d\omega^i = \frac{1}{2} \sum_{j,k} c^i_{jk}(I)\,\omega^j \wedge \omega^k + \sum_{j,l} a^i_{jl}(I)\,\omega^j \wedge \pi^l,$$

the latter means that

$$\sum_{j,l} a^i_{jl}(I)\,\omega^j \wedge (\hat{\pi}^l - \pi^l) = 0 \quad \text{for } i = 1,\dots,n,$$

—which by above is equivalent to

$$\hat{\pi}^l = \pi^l + \sum_{j=1}^{n} \sum_{k=1}^{q} b^l_{jk} v^k\,\omega^j \quad \text{for } l = 1,\dots,p,$$

with $q := s_1 + 2s_2 + \cdots + ns_n$ new variables $v^1,\dots,v^q$. The latter are regarded as *auxiliary variables for* $\mathfrak{G}^{(1)}$.

Conclusion. *The one-forms*

$$\omega^{n+l} := \pi^l + \sum_{j=1}^{n} \sum_{k=1}^{q} b^l_{jk} v^k\,\omega^j, \qquad l = 1,\dots,p,$$

on $\mathbb{C}^{n+p+q}_{x,u,v}$ *are the most general prolongations of the* π^l *which leave the structure equations of* $\mathfrak{G}$ *invariant. Note moreover that* $\{\omega^1,\dots,\omega^n,\omega^{n+1},\dots$ $\omega^{n+p}\}$ *is a local coframe for* $\mathbb{C}^{n+p}_{x,u}$ *for each* $v \in \mathbb{C}^q$.

Now $d\pi^l$ and db^l_{jk} are forms on $\mathbb{C}^{n+p}_{x,u}$, and can therefore be expressed by means of $\omega^1,\dots,\omega^{n+p}$. Consequently it is possible to write $d\omega^{n+l}$ as

$$d\omega^{n+l} = \frac{1}{2} \sum_{j,k=1}^{n+p} \tilde{c}^{n+l}_{jk}\,\omega^j \wedge \omega^k + \sum_{j=1}^{n} \sum_{k=1}^{q} b^l_{jk}\,\omega^j \wedge (-dv^k),$$

with $\tilde{c}^{n+l}_{jk} = -\tilde{c}^{n+l}_{kj}$. As in connection with the second fundamental theorem we then replace $-dv^k$ by

$$\chi^k := -dv^k + \text{linear combination of } \omega^1,\dots,\omega^{n+p}$$

in such a way that the second sum goes over into $\sum_j \sum_k b^l_{jk}\omega^j \wedge \chi^k$, while killing as many as possible of the coefficients in front of $\omega^j \wedge \omega^k$.

Denoting the remaining nonzero ones by c_{jk}^{n+l} we finally have

$$d\omega^{n+l} = \frac{1}{2} \sum_{j,k=1}^{n+p} c_{jk}^{n+l} \omega^j \wedge \omega^k + \sum_{j=1}^{n} \sum_{k=1}^{q} b_{jk}^l \omega^j \wedge \chi^k \qquad \text{for } l = 1,\dots,p.$$

But then it should be clear how to define $\mathfrak{G}^{(1)}$.

Definition. $\mathfrak{G}^{(1)}$ consists of all local diffeomorphisms on $\mathbb{C}_{x,u}^{n+p}$ that

   (i) leave the functions $I^1,\dots,I^h$ on $\mathbb{C}_x^n$ invariant,
   (ii) leave the one-forms $\omega^1,\dots,\omega^n$ on $\mathbb{C}_{x,u}^{n+p}$ invariant, and
   (iii) have prolongations to $\mathbb{C}_{x,u,v}^{n+p+q}$ leaving the one-forms $\omega^{n+1},\dots,\omega^{n+p}$ invariant.

The structure equations of $\mathfrak{G}^{(1)}$ are

$$\begin{cases} d\omega^i = \frac{1}{2} \sum_{j,k=1}^{n} c_{jk}^i \omega^j \wedge \omega^k + \sum_{j=1}^{n} \sum_{l=1}^{p} a_{jl}^i \omega^j \wedge \omega^{n+l} & \text{for } i = 1,\dots,n, \\ d\omega^{n+l} = \frac{1}{2} \sum_{j,k=1}^{n+p} c_{jk}^{n+l} \omega^j \wedge \omega^k + \sum_{j=1}^{n} \sum_{k=1}^{q} b_{jk}^l \omega^j \wedge \chi^k & \text{for } l = 1,\dots,p, \end{cases}$$

where—as in the second fundamental theorem for $\mathfrak{G}$—the coefficients are *invariants*, i.e., functions of $I^1,\dots,I^h$.

Definition. $\mathfrak{G}^{(1)}$ is called the *normal prolongation* of $\mathfrak{G}$. The second normal prolongation $\mathfrak{G}^{(2)}$ is obtained as the normal prolongation of $\mathfrak{G}^{(1)}$, and so on.

Remark. To a finite dimensional Lie transformation group G is associated a *unique* Lie parameter group $\mathfrak{G}$. This uniqueness is lost in the infinite dimensional case: to an infinite dimensional G there is associated a *whole chain* $\mathfrak{G}$, $\mathfrak{G}^{(1)}$, $\mathfrak{G}^{(2)}$, ... *of Lie pseudogroups in Cartan form*, with $\mathfrak{G}^{(k+1)}$ being the normal prolongation of $\mathfrak{G}^{(k)}$.

Example. Let us consider the general group in one variable—i.e., the Lie pseudogroup G consisting of *all* local diffeomorphisms

$$\mathbb{C}_x \to \mathbb{C}_{\hat{x}}$$
$$x \mapsto f(x), \qquad \text{with } f'(x) \neq 0.$$

The defining system of G is simply '0=0'. Therefore $S^1(G) = J^1(\mathbb{C}_x, \mathbb{C}_{\hat{x}})$, with the contact equation $d\hat{x} - p\,dx = 0$. This gives the invariant one-form

$$\omega^1 := d\hat{x} = p\,dx,$$

where p is an auxiliary variable. Then

$$d\omega^1 = dp \wedge dx = p\,dx \wedge \left(-\frac{dp}{p} \right) = \omega^1 \wedge \left(-\frac{dp}{p} \right),$$

so that ω^1 satisfies the structure equation

$$d\omega^1 = \omega^1 \wedge \pi \quad \text{with} \quad \pi := -\frac{dp}{p}.$$

Thus the corresponding Lie pseudogroup $\mathfrak{G}$ in Cartan form is given by leaving ω^1 invariant.

Check: We have that $\hat{p} \, d\hat{x} = p \, dx \Longrightarrow \hat{x}$ is a function of x: $\hat{x} = f(x)$, and

$$\hat{p} f'(x) = p, \text{ i.e., } \hat{p} = \frac{p}{f'(x)}.$$

Here $x \mapsto f(x)$ defines $\mathfrak{G}$ (so in particular $\mathfrak{G} = G$ in this case), while $\mathfrak{G}^{(1)}$ is defined by $(x, p) \mapsto (f(x), p/f'(x))$.

In order to determine the invariant one-forms of $\mathfrak{G}^{(1)}$ we next look for the most general one-form ω^2 satisfying $\omega^1 \wedge \omega^2 = \omega^1 \wedge \pi$, or $p \, dx \wedge (\omega^2 - \pi) = 0$—that is

$$\omega^2 = \pi + q \, dx = -\frac{dp}{p} + q \, dx,$$

where q is a new auxiliary variable. Then

$$d\omega^2 = dq \wedge dx = p \, dx \wedge \left(-\frac{dq}{p} \right) = \omega^1 \wedge \chi \quad \text{with} \quad \chi := -\frac{dq}{p}.$$

Hence $\mathfrak{G}^{(1)}$ is defined by the invariant one-forms

$$\omega^1 = p \, dx \quad \text{and} \quad \omega^2 = -\frac{dp}{p} + q \, dx,$$

satisfying the structure equations

$$d\omega^1 = \omega^1 \wedge \omega^2 \quad \text{and} \quad d\omega^2 = \omega^1 \wedge \chi.$$

Check: The invariance of ω^1 gives $(x, p) \mapsto (f(x), p/f'(x))$, and then the invariance of ω^2 amounts to

$$-\frac{d(p/f'(x))}{p/f'(x)} + \hat{q} \, f'(x) \, dx = -\frac{dp}{p} + q \, dx,$$

whence

$$\hat{q} = \frac{q}{f'(x)} - \frac{f''(x)}{(f'(x))^2}.$$

Consequently the next prolongation $\mathfrak{G}^{(2)}$ acts on $\mathbb{C}^3_{xpq}$ by

$$(x, p, q) \mapsto \left(f(x), \frac{p}{f(x)}, \frac{q}{f'(x)} - \frac{f''(x)}{(f'(x))^2} \right).$$

Let us finally determine the invariant one-forms of $\mathfrak{G}^{(2)}$. The most general one-form ω^3 satisfying

$$\omega^1 \wedge \omega^3 = \omega^1 \wedge \chi \quad \text{or} \quad p\,dx \wedge \left(\omega^3 + \frac{dq}{p}\right) = 0$$

is

$$\omega^3 = -\frac{dq}{p} + r\,dx,$$

where r is one more auxiliary variable. Then

$$d\omega^3 = \frac{dp \wedge dq}{p^2} + dr \wedge dx$$

$$= \left(-\frac{dp}{p} + q\,dx\right) \wedge \left(-\frac{dq}{p} + r\,dx\right) + p\,dx \wedge \left(-\frac{r\,dp}{p^2} + \frac{q\,dq}{p^2} - \frac{dr}{p}\right)$$

$$= \omega^2 \wedge \omega^3 + \omega^1 \wedge \psi \quad \text{with} \quad \psi := -\frac{r\,dp}{p^2} + \frac{q\,dq}{p^2} - \frac{dr}{p}.$$

Thus $\mathfrak{G}^{(2)}$ is defined by the three invariant one-forms

$$\omega^1 = p\,dx, \quad \omega^2 = -\frac{dp}{p} + q\,dx, \quad \omega^3 = -\frac{dq}{p} + r\,dx,$$

satisfying the structure equations

$$d\omega^1 = \omega^1 \wedge \omega^2, \quad d\omega^2 = \omega^1 \wedge \omega^3 \quad \text{and} \quad d\omega^3 = \omega^2 \wedge \omega^3 + \omega^1 \wedge \psi.$$

Remark. Note that the systatic systems of $\mathfrak{G}$, $\mathfrak{G}^{(1)}$ and $\mathfrak{G}^{(2)}$ are all given by $\omega^1 = 0$.

This is no coincidence—for it can be proved that

$$\mathrm{Sys}(\mathfrak{G}) = \mathrm{Sys}(\mathfrak{G}^{(k)}) \quad \text{for } k = 1, 2, 3, \ldots$$

for *any* $\mathfrak{G}$. A corollary of this is that the normal prolongations of a Lie pseudogroup in Cartan form all have the same essential invariants.

With the structure equations of $\mathfrak{G}$ being given by

$$d\omega^i = \sum_{j<k} c^i_{jk}\,\omega^j \wedge \omega^k + \sum_{j,l} a^i_{jl}\,\omega^j \wedge \pi^l \quad \text{for } i = 1, \ldots, n,$$

the stability algebra $\mathfrak{st}(\mathfrak{G})$ is generated by the vector fields

$$E_l = \sum_{i,j=1}^{n} a^i_{jl}\,\omega^j \frac{\partial}{\partial \omega^i} \quad \text{for } l = 1, \ldots, p.$$

The general element $\delta = \delta(t) = \sum_{l=1}^{p} t^l E_l$ of $\mathfrak{st}(\mathfrak{G})$ thus acts on the coframe $\{\omega^1, \ldots, \omega^n\}$ by

$$\omega^i \mapsto \delta\omega^i = \sum_{j,l} a_{jl}^i t^l \omega^j \quad \text{for } i = 1, \ldots, n.$$

In the prolongation process $\pi^1, \ldots, \pi^p$ are replaced by $\omega^{n+1}, \ldots, \omega^{n+p}$ in such a way that the structure equations remain invariant. The prolonged stability algebra is then to act on the local coframe $\{\omega^1, \ldots, \omega^n, \omega^{n+1}, \ldots, \omega^{n+p}\}$.

The following theorem defines such a prolongation in a manner which will be convenient for later applications. In order to simplify the proof somewhat we assume that $\mathfrak{G}$ *is transitive*—so that the coefficients c_{jk}^i and a_{jl}^i are *constants*.

Theorem 14.6.1. *The general element δ of the stability algebra can be prolonged to act on the coframe $\{\omega^1, \ldots, \omega^{n+p}\}$ in such a way that δ commutes with d:*

$$d\delta\omega^i = \delta d\omega^i \quad \text{for } i = 1, \ldots, n.$$

Proof. We have that δ acts on ω^i by $\delta\omega^i = \sum_{j,l} a_{jl}^i t^l \omega^j$, and therefore

$$d\delta\omega^i = \sum_{j,l} a_{jl}^i \, dt^l \wedge \omega^j + \sum_{j,l} a_{jl}^i t^l \, d\omega^j.$$

On the other hand, letting δ act on the structure equations above we get

$$\delta d\omega^i = \sum_{j<k} c_{jk}^i (\delta\omega^j \wedge \omega^k + \omega^j \wedge \delta\omega^k) + \sum_{j,l} a_{jl}^i (\delta\omega^j \wedge \omega^{n+l} + \omega^j \wedge \delta\omega^{n+l}),$$

where $\delta\omega^{n+l}$ is to be defined so that $0 = d\delta\omega^i - \delta d\omega^i$:

$$0 = -\sum_{j,l} a_{jl}^i \omega^j \wedge (dt^l + \delta\omega^{n+l})$$

$$+ \sum_{j,l,m} \sum_s (a_{sl}^i a_{jm}^s - a_{sm}^i a_{jl}^s) t^l \omega^j \wedge \omega^{n+m}$$

$$+ \sum_{j<k} \sum_{l,s} (c_{jk}^s a_{sl}^i + c_{sj}^i a_{kl}^s - c_{sk}^i a_{jl}^s) t^l \omega^j \wedge \omega^k.$$

Now the third fundamental theorem, in section **14.3**, says that

$$\sum_{s=1}^{n} (a_{jl}^s a_{sm}^i - a_{jm}^s a_{sl}^i) = \sum_{q=1}^{p} a_{jq}^i \gamma_{lm}^q,$$

where $\{\gamma^q_{lm}\}$ are the structure constants of the stability algebra. Inserting this into the second sum above we get

$$\sum_{j,l} a^i_{jl}\omega^j \wedge \left(\delta\omega^{n+l} + dt^l + \sum_{a,b} \gamma^l_{ab} t^a \omega^{n+b} \right)$$

$$= \sum_{j<k} \sum_{m,s} (c^s_{jk} a^i_{sm} + c^i_{sj} a^s_{km} - c^s_{sk} a^s_{jm}) t^m \omega^j \wedge \omega^k.$$

From this the $\delta\omega^{n+l}$ can be solved as

$$\delta\omega^{n+l} = -dt^l - \sum_{a,b} \gamma^l_{ab} t^a \omega^{n+b} - \sum_{k,m} z^l_{km} t^m \omega^k,$$

provided that it is possible to find z^l_{km} satisfying

$$\sum_{l=1}^{p} (a^i_{jl} z^l_{km} - a^i_{kl} z^l_{jm}) = \sum_{s=1}^{n} (c^s_{jk} a^i_{sm} + c^i_{sj} a^s_{km} - c^i_{sk} a^s_{jm})$$

—and that is exactly what can be done, by comparison with the linear system (2) in the third fundamental theorem, proved in section **14.3**. □

Let us rephrase this in a slightly different way.

Corollary 14.6.1. *The prolonged stability algebra acting on the coframe* $\{\omega^1,\dots,\omega^{n+p}\}$ *is the most general prolongation which preserves the structure equations*

$$d\omega^i = \frac{1}{2} \sum_{j,k} c^i_{jk} \omega^j \wedge \omega^k + \sum_{j=1}^{n} \sum_{l=1}^{p} a^i_{jl} \omega^j \wedge \omega^{n+l}.$$

□

We conclude this short review of Cartan's theory of Lie pseudogroups by discussing the question of *isomorphism of Lie pseudogroups*—although this will not be used in the following.

If G_1 and G_2 are finite dimensional Lie transformation groups (possibly acting on manifolds of different dimensions), they are said to be locally isomorphic if their corresponding Lie parameter groups $\mathfrak{G}_1$ and $\mathfrak{G}_2$ are— in a rather obvious sense.

The general case is more difficult, since there no longer are any parameter groups to compare. However, the sequence of normal prolongations of a Cartan form provide a weak substitute for the parameter group.

Consider first two Lie pseudogroups G_1 and G_2 acting on $\mathbb{C}^n$—let us say that G_1 acts on $\mathbb{C}_x^n$, while G_2 acts on $\mathbb{C}_y^n$. Then G_1 and G_2 are *locally equivalent* if there are local diffeomorphisms

$$\phi : \mathbb{C}_x^n \xrightarrow{\cong} \mathbb{C}_y^n$$

such that

$$\phi^{-1} G_2 \phi := \{\phi^{-1} \circ g \circ \phi \mid g \in G_2\} = G_1,$$

i.e., if G_1 and G_2 happen to be expressed by means of different local coordinates, but really are identical from an intrinsic point of view. In the following we identify locally equivalent Lie pseudogroups.

The general definition of isomorphism was proposed simultaneously (and independently) by Cartan and Vessiot in 1902:

G_1 and G_2 are locally isomorphic if they admit isomorphic prolongations H_1 and H_2, which in their turn are locally equivalent:

$$
\begin{array}{ccc}
H_1 & \cong & H_2 \\
\cong \downarrow & & \downarrow \cong \\
G_1 & & G_2.
\end{array}
$$

Or identifying equivalent Lie pseudogroups we need only find a Lie pseudogroup H which is simultaneously an isomorphic prolongation of G_1 and of G_2:

$$
\begin{array}{ccc}
 & H & \\
\cong \swarrow & & \searrow \cong \\
G_1 & & G_2.
\end{array}
$$

The trouble with this definition is that it is not evident that it is *transitive*:

$$G_1 \cong G_2 \text{ and } G_2 \cong G_3 \implies G_1 \cong G_3?$$

In other words:

$$\exists\, H_{12} \text{ and } H_{23} \text{ such that}$$

$$
\begin{array}{ccccccc}
 & H_{12} & & & & H_{23} & \\
\cong \swarrow & & \searrow \cong & \text{and} & \cong \swarrow & & \searrow \cong \\
G_1 & & G_2 & & G_2 & & G_3
\end{array}
$$

$$
\xRightarrow{?} \exists\, H_{13} \text{ such that }
\begin{array}{ccc}
 & H_{13} & \\
\cong \swarrow & & \searrow \cong \\
G_1 & & G_3
\end{array} ??
$$

In the section called "Détermination des groupes dont G est le prolongement holoédrique" of [Cartan 1905], quotient groups and normal subgroups of a given Lie pseudogroup are investigated by means of invariant foliations. The following is a spin-off effect of that study.

> Let $\mathfrak{G}$ and $\mathfrak{H}$ be two Lie pseudogroups in Cartan form such that $\mathfrak{H}$ is an isomorphic prolongation of $\mathfrak{G}$. Assuming that the invariants of $\mathfrak{H}$ can be expressed by means of the local coordinates of the manifold that $\mathfrak{G}$ acts on, there is a *normal prolongation* $\mathfrak{G}^{(m)}$ of $\mathfrak{G}$ which in its turn is an isomorphic prolongation of $\mathfrak{H}$:

$$
\begin{array}{ccc}
\mathfrak{G}^{(m)} & & \\
& \searrow \cong & \\
\cong \downarrow & & \mathfrak{H} \\
& \nearrow \cong & \\
\mathfrak{G} & &
\end{array}
$$

> If $\mathfrak{H}$ admits some invariants which are not acted upon by $\mathfrak{G}$, $\mathfrak{G}$ is extended by acting trivially on these, and then one instead gets

$$
\begin{array}{ccc}
\mathfrak{G}^{(m)} \oplus \mathrm{Id} & & \\
& \searrow \cong & \\
\cong \downarrow & & \mathfrak{H} \\
& \nearrow \cong & \\
\mathfrak{G} \oplus \mathrm{Id} & &
\end{array}
$$

Expressing this intuitively we can say that the *normal prolongations of $\mathfrak{G}$ are universal* in the sense that any other isomorphic prolongation can be subsumed under a normal one.

Now it is easily seen that the concept of isomorphism is transitive. For let $\mathfrak{G}_1$, $\mathfrak{G}_2$ and $\mathfrak{G}_3$ be Lie pseudogroups in Cartan form such that $\mathfrak{G}_1$ is isomorphic to $\mathfrak{G}_2$, and $\mathfrak{G}_2$ is isomorphic to $\mathfrak{G}_3$:

$$
\begin{array}{ccccccc}
 & \mathfrak{H}_{12} & & & & \mathfrak{H}_{23} & \\
\cong \nearrow & & \searrow \cong & \text{and} & \cong \nearrow & & \searrow \cong \\
\mathfrak{G}_1 & & \mathfrak{G}_2 & & \mathfrak{G}_2 & & \mathfrak{G}_3.
\end{array}
$$

Then there are integers m_{12} and m_{23} such that

$$\mathfrak{G}_2^{(m_{12})} \oplus \mathrm{Id} \text{ is an isomorphic prolongation of } \mathfrak{H}_{12},$$

and

$$\mathfrak{G}_2^{(m_{23})} \oplus \mathrm{Id} \text{ is an isomorphic prolongation of } \mathfrak{H}_{23}.$$

But $\mathfrak{G}_2^{(k)}$ is a one-to-one prolongation of $\mathfrak{G}_2^{(j)}$ whenever $k \geq j$, so setting

$m := \max\{m_{12}, m_{23}\}$ we find that $\mathfrak{G}_2^{(m)} \oplus \mathrm{Id}$ is an isomorphic prolongation of both $\mathfrak{H}_{12}$ and $\mathfrak{H}_{23}$:

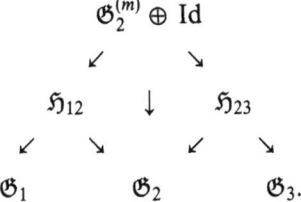

Hence $\mathfrak{G}_1$ and $\mathfrak{G}_3$ are really isomorphic.

15

The equivalence problem

In chapters **5** and **11** we have seen that any second order PDE $\mathcal{S} \subset J^2(\mathbb{C}^2_{xy}, \mathbb{C}_z)$ admits *two Monge characteristic vector field systems*. $\mathcal{S}$ is said to be *hyperbolic* if these are different, and *parabolic* if they coincide. Vessiot's study of those hyperbolic PDEs for which each Monge system admits at least two functionally independent first integrals led to *a remarkable classification*, presented in chapters **12** and **13**.

The corresponding classification of those parabolic PDEs for which the double Monge system admits at least two functionally independent first integrals is the subject of Cartan's 'five variable paper' [Cartan 1910].

While Vessiot's classification is based upon a straightforward stepwise simplification of the Monge systems, Cartan instead starts by solving a very general *equivalence problem*, which he then applies to the special case of parabolic PDEs.

Cartan's ideas for solving the equivalence problem were first announced in the *Comptes Rendus* note [Cartan 1902]. In order to make this comprehensible he then went on with his general theory of Lie pseudogroups in [Cartan 1904] and [Cartan 1905], whereupon the first chapter of [Cartan 1908] is devoted to a full explanation of the equivalence problem. And this is then applied to the classification of parabolic PDEs in [Cartan 1910].

This chapter is a survey of Cartan's solution of the equivalence problem, mainly following [Cartan 1937a]. For more modern references see e.g. [Gardner 1989], [Morimoto 1993], and [Olver 1995].

A different and very interesting approach to the equivalence problem has been developed by Pommaret, taking [Vessiot 1903] as point of departure—see e.g. [Pommaret 1978] and [Pommaret 1983].

The *general equivalence problem* amounts to the following. Let **M** and

421

$\hat{\mathbf{M}}$ be manifolds of the same dimension which are provided with the local structures S and $\hat{S}$ respectively. Then $(\mathbf{M}, S)$ and $(\hat{\mathbf{M}}, \hat{S})$ are *locally equivalent* if there are local diffeomorphisms $\phi : \mathbf{M} \xrightarrow{\cong} \hat{\mathbf{M}}$ inducing local isomorphisms $S \cong \hat{S}$.

> **Question**: How to decide whether $(\mathbf{M}, S)$ and $(\hat{\mathbf{M}}, \hat{S})$ are locally equivalent or not?

The main trouble consists in the fact that the concept 'local structure' is so vague that it is hard to imagine any general method at all.

Cartan's idea is to first consider the *symmetry group* $\mathrm{sym}(\mathbf{M}, S)$ consisting of all local diffeomorphisms of $\mathbf{M}$ preserving S. Under rather general circumstances this is a *Lie pseudogroup*, which by the first fundamental theorem has a one-to-one prolongation to a *Lie pseudogroup* $\mathfrak{G}(\mathbf{M}, S)$ *in Cartan form*. The second version of this theorem specifies $\mathfrak{G}(\mathbf{M}, S)$ as a transformation group acting on some $\mathbb{C}_x^n$ by means of

- a number of invariant functions $I^1, \ldots, I^h$,
- a local coframe $\{\omega^1(x), \ldots, \omega^n(x)\}$,
- a matrix group $\mathfrak{M}$ depending on the invariants and on p auxiliary variables $u^1, \ldots, u^p$,

in the following way:

$\phi : \mathbb{C}_x^n \xrightarrow{\cong} \mathbb{C}_x^n$ belongs to $\mathfrak{G}(\mathbf{M}, S)$ if and only if

$$\phi^*(I^a) = I^a \quad \text{for } a = 1, \ldots, h,$$

and

$$\phi^* \begin{pmatrix} \omega^1 \\ \vdots \\ \omega^n \end{pmatrix} = M_\phi \begin{pmatrix} \omega^1 \\ \vdots \\ \omega^n \end{pmatrix} \quad \text{for some } M_\phi \in \mathfrak{M}.$$

Thus Lie pseudogroups in Cartan form have a *very specific description*— in contrast to local structures.

If $(\mathbf{M}, S)$ and $(\hat{\mathbf{M}}, \hat{S})$ are locally equivalent, certainly also $\mathfrak{G}(\mathbf{M}, S)$ and $\mathfrak{G}(\hat{\mathbf{M}}, \hat{S})$ are. And this observation gives rise to the equivalence problem of Cartan:

Let $\mathbb{C}_x^n$ be provided with a Cartan structure S_C consisting of

- certain functions $I^1(x), \ldots, I^h(x)$,
- a local coframe $\{\omega^1(x), \ldots, \omega^n(x)\}$,

- a matrix group $\mathfrak{M}$ depending on $I^1,\ldots,I^h$ and on p parameters $u^1,\ldots,u^p$,

and let $\mathbb{C}_{\hat{x}}^n$ be given an analogous Cartan structure $\hat{S}_C$, with the same integers h and p. Then $(\mathbb{C}_x^n, S_C)$ and $(\mathbb{C}_{\hat{x}}^n, \hat{S}_C)$ are *locally equivalent* if there are local diffeomorphisms $\phi: \mathbb{C}_x^n \xrightarrow{\cong} \mathbb{C}_{\hat{x}}^n$ such that

$$\phi^*(I^a) = \hat{I}^a \quad \text{for } a = 1,\ldots,h,$$

and

$$\phi^* \begin{pmatrix} \hat{\omega}^1 \\ \vdots \\ \hat{\omega}^n \end{pmatrix} = M_\phi \begin{pmatrix} \omega^1 \\ \vdots \\ \omega^n \end{pmatrix} \quad \text{for some } M_\phi \in \mathfrak{M}.$$

In order to obtain a more symmetric formulation of the problem of how to decide whether $(\mathbb{C}_x^n, S)$ and $(\mathbb{C}_{\hat{x}}^n, \hat{S}_C)$ are locally equivalent we let the general element of $\mathfrak{M}$ act on the coframe $\{\omega^1(x),\ldots,\omega^n(x)\}$, and set

$$\omega(x;u) \equiv \begin{pmatrix} \omega^1(x;u) \\ \vdots \\ \omega^n(x;u) \end{pmatrix} := M(I,u) \begin{pmatrix} \omega^1(x) \\ \vdots \\ \omega^n(x) \end{pmatrix} \quad \text{for } u \in \mathbb{C}^p,$$

thereby obtaining a p-parameter family of local coframes $(\omega(x;u))$ for $\mathbb{C}_x^n$—and analogously for $\mathbb{C}_{\hat{x}}^n$.

Then $\phi: \mathbb{C}_x^n \xrightarrow{\cong} \mathbb{C}_{\hat{x}}^n$ is a local equivalence if $\phi^*(\hat{I}^a) = I^a$ for $a = 1,\ldots,h$ and if ϕ has a prolongation $\phi^{(1)}: \mathbb{C}_x^n \times \mathbb{C}_u^p \longrightarrow \mathbb{C}_{\hat{x}}^n \times \mathbb{C}_{\hat{u}}^p$ such that

$$\phi^{(1)*}\hat{\omega}(\hat{x};\hat{u}) = \omega(x;u).$$

With $\hat{\omega}(\hat{x};\hat{u})$ being a local coframe of $\mathbb{C}_{\hat{x}}^n$ for each $\hat{u}$,

$$\phi^{(1)*}(d\hat{x}^i) = \phi^{(1)*}(\text{linear combination of } \hat{\omega}^1(\hat{x};\hat{u}),\ldots,\hat{\omega}^n(\hat{x};\hat{u}))$$
$$= \text{linear combination of } \omega^1(x;u),\ldots,\omega^n(x;u)$$
$$= \text{linear combination of } dx^1,\ldots,dx^n,$$

so that $\phi^{(1)}$ is of the form $(x,u) \mapsto (\phi(x), \psi(x,u))$. And then the projection $x \mapsto \phi(x)$ to $\mathbb{C}_x^n \xrightarrow{\cong} \mathbb{C}_{\hat{x}}^n$ is the local equivalence that we really are looking for.

Remark. Note that the p-parameter family of coframes $\omega(x;u)$ is not entirely determined by the matrix group $\mathfrak{M}(I,u)$ and the coframe $\omega(x;u)$—for we can always multiply $(\omega^1(x;u),\ldots,\omega^n(x;u))^T$ on the left by an arbitrary matrix in $\mathfrak{M}$.

In the above manner the original equivalence problem of Cartan has been transformed into the *lifted equivalence problem*:

determine $\phi^{(1)} : \mathbb{C}^{n+p}_{x,u} \longrightarrow \mathbb{C}^{n+p}_{\hat{x},\hat{u}}$ such that

$$\phi^{(1)*}(\hat{I}^a) = I^a \quad \text{for } a = 1,\dots,h \quad \text{and} \quad \phi^{(1)*}\hat{\omega}(\hat{x};\hat{u}) = \omega(x;u).$$

Suppose now that we have been able to solve this equivalence problem, and hence know that $\mathfrak{G}(\mathbf{M},S) \cong \mathfrak{G}(\hat{\mathbf{M}},\hat{S})$. Does it then follow that $(\mathbf{M},S)$ and $(\hat{\mathbf{M}},\hat{S})$ are also locally equivalent? In general this is of course not true, but from a moral point of view we could expect the answer to be yes *provided that the symmetry groups involved are big enough*—at least we have to require that there are no inessential invariants. (For more about the interesting connection between geometrical structures and their symmetry groups, see e.g. [Banyaga 1997].)

In practice it is often complicated to determine $\mathfrak{G}(\mathbf{M},S)$ from $(\mathbf{M},S)$ anyway. Therefore one usually tries a more direct approach:

find local coframes for $\mathbf{M}$ and $\hat{\mathbf{M}}$ which are adapted to the local structures S and $\hat{S}$, and then use these coframes for showing that the equivalence problem for $(\mathbf{M},S)$ and $(\hat{\mathbf{M}},\hat{S})$ can be expressed directly in Cartan form!

Now it turns out to be an *empirical fact* that *surprisingly many* equivalence problems indeed can be reduced to Cartan form in this way—see section **15.3** for two examples, and the standard reference [Gardner 1989] for a lot more. But it is not so easy to figure out the reasons behind Cartan's version of the equivalence problem without a prior knowledge of his theory of Lie pseudogroups.

Having accepted that the seemingly so particular equivalence problem of Cartan does have a great general interest, it remains to see how it is to be solved. There turn out to be two aspects of this: a *general approach*, and the *approach by invariants*.

The general approach. Since the lifted equivalence problem gives rise to the pfaffian system

$$\mathcal{P}: \quad \begin{cases} \hat{I}^a(\hat{x}) - I^a(x) = 0 & \text{for } a = 1,\dots,h, \\ \theta^i := \hat{\omega}^i(\hat{x};\hat{u}) - \omega^i(x;u) = 0 & \text{for } i = 1,\dots,n, \end{cases}$$

it is tailor made for Cartan's local existence theorem.

If $\mathcal{P}$ is not involutive, the prolongation theorem of Cartan and Janet

states that a finite number of prolongations will lead either to an inconsistent system, or to an involutive system $\tilde{\mathcal{P}}$—and in the latter case the
local existence theorem applies if all data are real analytic.

Not wanting to assume analyticity one can but hope that the final
system $\tilde{\mathcal{P}}$ is *complete*, and thus can be solved by means of the Frobenius
theorem.

In each prolongation a new set of auxiliary variables is introduced, and
the integral manifolds eventually found are really higher order graphs
of the local equivalences sought for; the latter are then obtained by
projection to the x- and $\hat{x}$-variables. So the auxiliary variables are just a
means for finding the wanted solutions—once they have done their duty,
they are dispensed with.

The approach by invariants. Given two structured manifolds $(\mathbf{M}, S)$ and
$(\hat{\mathbf{M}}, \hat{S})$ which are suspected to be locally equivalent, one tries to find as
many invariants as possible for each of these, and then equality of the
corresponding invariants yields a necessary—and in the most favourable
cases also sufficient—condition for local equivalence.

An invariant belonging to $(\mathbf{M}, S)$ is a function on $\mathbf{M}$ which is invariant
under the symmetry group $\mathfrak{G}(\mathbf{M}, S)$. With $I^1(x), \ldots, I^h(x)$ being invariants,
new ones can be constructed by expressing the dI^a in the given coframe:

$$dI^a = \sum_{j=1}^{n} I^a_{.j} \, \omega^j$$

—for since dI^a and the ω^j are invariant, the $I^a_{.j}$ must also be. The latter
are called *covariant derivatives* of I^a (with respect to the given coframe).

Introducing the dual vector fields $\partial/\partial\omega^k$ by

$$\omega^j(\partial/\partial\omega^k) = \begin{cases} 1 & \text{if } j = k, \\ 0 & \text{if } j \neq k \end{cases} \qquad \text{for } j, k = 1, \ldots, n,$$

we have

$$\frac{\partial}{\partial\omega^k} I^a = dI^a\left(\frac{\partial}{\partial\omega^k}\right) = \sum_{j=1}^{n} I^a_{.j} \omega^j\left(\frac{\partial}{\partial\omega^k}\right) = I^a_{.k},$$

so that

$$I^a_{.k} = \frac{\partial I^a}{\partial\omega^k} \qquad \text{for } k = 1, \ldots, n.$$

Conclusion. *If I is an invariant of $(\mathbf{M}, S)$, the covariant derivatives $\partial/\partial\omega^k(I)$
also are, for $k = 1, \ldots, n$.*

Notation. For some obscure reason Lie and Cartan call the $\partial/\partial\omega^k$ 'linear differential parameters'. Perhaps the name *invariance preserving vector fields* would be more appropriate.

So starting from h invariants $I^1,\ldots,I^h$, we can create an infinite number of new ones by means of the $\partial/\partial\omega^k$:

$$I^a_{.i_1} = \left(\partial/\partial\omega^{i_1}\right)I^a, \quad I^a_{.i_1 i_2} = \left(\partial/\partial\omega^{i_2}\right)I^a_{.i_1},\ldots.$$

Suppose that of all these only $I^1,\ldots,I^h$ are functionally independent. Then the others are functions of these:

$$I^a_{.i_1\ldots i_k} = J^a_{i_1\ldots i_k}(I^1,\ldots,I^h).$$

Making the same calculations for $(\hat{M},\hat{S})$ we obtain a corresponding chain of invariants there:

$$\hat{I}^a_{.i_1\ldots i_k} = \hat{J}^a_{i_1\ldots i_k}(\hat{I}^1,\ldots,\hat{I}^h).$$

Then a *necessary condition* for the local equivalence of (M,S) and $(\hat{M},\hat{S})$ is that

$$J^a_{i_1\ldots i_k}(t^1,\ldots,t^h) \text{ and } \hat{J}^a_{i_1\ldots i_k}(t^1,\ldots,t^h) \text{ are the same functions!}$$

And this leads to the important question: *is it possible to find so many invariants that this condition turns out to be sufficient as well?*

15.1 The simplest case: e-structures

Let us start with the simplest case imaginable: there are no auxiliary variables, the matrix group is reduced to the identity matrix, and there are no invariant functions at the outset. In the modern literature this is called the 'theory of e-structures', with e denoting the identity matrix.

So we are given a coframe $\omega(x) = \{\omega^1(x),\ldots,\omega^n(x)\}$ on $\mathbb{C}^n_x$, and a coframe $\hat{\omega}(\hat{x}) = \{\hat{\omega}^1(\hat{x}),\ldots,\hat{\omega}^n(\hat{x})\}$ on $\mathbb{C}^n_{\hat{x}}$, and are looking for local diffeomorphisms $\phi\colon \mathbb{C}^n_x \overset{\cong}{\longrightarrow} \mathbb{C}^n_{\hat{x}}$ satisfying $\phi^*\hat{\omega}^i = \omega^i$ for $i = 1,\ldots,n$.

The graphs of such diffeomorphisms are integral manifolds on which $\omega^1\wedge\cdots\wedge\omega^n \neq 0$ (or equivalently $\hat{\omega}^1\wedge\cdots\wedge\hat{\omega}^n \neq 0$) of the pfaffian system

$$\mathcal{P}: \quad \theta^i := \hat{\omega}^i(\hat{x}) - \omega^i(x) = 0, \quad i = 1,\ldots,n,$$

defined on $\mathbb{C}^n_x \times \mathbb{C}^n_{\hat{x}}$. Then also

$$0 = d\theta^i = d\hat{\omega}^i - d\omega^i \quad \text{for } i = 1,\ldots,n,$$

where the $d\omega^i$ can be written as

$$d\omega^i = \frac{1}{2} \sum_{j,k=1}^n c_{jk}^i(x)\,\omega^j \wedge \omega^k \quad \text{with } c_{jk}^i = -c_{kj}^i,$$

since $\{\omega^1,\ldots,\omega^n\}$ is a coframe for $\mathbb{C}_x^n$—and analogously for the $d\hat{\omega}^i$.
From the fact that

$$\hat{\omega}^i = \omega^i \quad \text{and} \quad \sum_{j<k} \hat{c}_{jk}^i(\hat{x})\,\hat{\omega}^j \wedge \hat{\omega}^k = \sum_{j<k} c_{jk}^i(x)\,\omega^j \wedge \omega^k$$

on the integral manifolds, it follows that

$$\hat{c}_{jk}^i(\hat{x}) = c_{jk}^i(x) \quad \text{for } i,j,k = 1,\ldots,n$$

there. Differentiating these equations, we get

$$0 = d\hat{c}_{jk}^i(\hat{x}) - dc_{jk}^i(x) = \sum_{l=1}^n \hat{c}_{jk.l}^i(\hat{x})\,\hat{\omega}^l - \sum_{l=1}^n c_{jk.l}^i(x)\,\omega^l,$$

and therefore also

$$\hat{c}_{jk.l}^i(\hat{x}) = c_{jk.l}^i(x) \quad \text{for } i,j,k,l = 1,\ldots,n$$

on the integral manifolds. Continuing in this manner we find an infinite number of invariants:

$$\hat{c}_{jk.l_1,\ldots l_a}^i(\hat{x}) = c_{jk.l_1\ldots l_a}^i(x) \quad \text{for } a = 0,1,2,\ldots.$$

Assume now that precisely p members of the set $\{c_{jk.l_1\ldots l_a}^i(x)\}_{a=0}^\infty$ are functionally independent (locally, as always). Then in case of local equivalence also exactly p of the $\{\hat{c}_{jk.l_1\ldots l_a}^i(\hat{x})\}_{a=0}^\infty$ are to be functionally independent.

First case: $p = 0$. In this case the structure functions c_{jk}^i are *constants*, and a *necessary* condition for local equivalence is that

$$c_{jk}^i = \hat{c}_{jk}^i \quad \text{for } i,j,k = 1,\ldots,n.$$

This is also *sufficient*, since it implies that

$$d\theta^i = \frac{1}{2} \sum_{j,k=1}^n (\hat{c}_{jk}^i\,\hat{\omega}^i - c_{jk}^i\omega^i) \equiv 0 \pmod{\mathcal{P}},$$

showing the pfaffian system $\mathcal{P}$ to be *complete*. But then the Frobenius theorem yields an n-parameter family of local equivalences.

Note that the symmetry group of the structure on $\mathbb{C}_x^n$ which is given by $\omega^1(x), \ldots, \omega^n(x)$ is an n-dimensional Lie parameter group (and analogously on $\hat{C}^n_{\hat{x}}$)—in fact this is more or less Cartan's version of the first fundamental theorem for Lie parameter groups.

Second case: $p = n$. In this case we can select n functionally independent functions $I^1(x), \ldots, I^n(x)$ among the $c^i_{jk.l_1 \ldots l_a}(x)$. Making the same choice among the $\hat{c}^i_{jk.l_1 \ldots l_a}(\hat{x})$ we get functions $\hat{I}^1(\hat{x}), \ldots, \hat{I}^n(\hat{x})$ such that

$$\hat{I}^i(\hat{x}) = I^i(x) \quad \text{for } i = 1, \ldots, n$$

on the graph of the wanted equivalence. These equations do define a local diffeomorphism $x \mapsto \hat{x}$, but this is not necessarily a local equivalence. Indeed, by covariant differentiation

$$\sum_{j=1}^n \hat{I}^i_{.j}(\hat{x}) \, \hat{\omega}^j = \sum_{j=1}^n I^i_{.j}(x) \, \omega^j \quad \text{for } i = 1, \ldots, n,$$

from which

$$\hat{I}^i_{.j}(\hat{x}) = I^i_{.j}(x) \quad \text{for } i, j = 1, \ldots, n.$$

Theorem 15.1.1. *Necessary and sufficient conditions for local equivalence are that the equations $\hat{I}^i_{.j}(\hat{x}) = I^i_{.j}(x)$ for $i, j = 1, \ldots, n$ are consequences of $\hat{I}^i(\hat{x}) = I^i(x)$ for $i = 1, \ldots, n$.*

Proof. Necessity is clear. As to sufficiency, note first that

$$\begin{pmatrix} dI^1 \\ \vdots \\ dI^n \end{pmatrix} = \begin{pmatrix} I^1_{.1} & \cdots & I^1_{.n} \\ \vdots & \ddots & \vdots \\ I^n_{.1} & \cdots & I^n_{.n} \end{pmatrix} \begin{pmatrix} \omega^1 \\ \vdots \\ \omega^n \end{pmatrix}, \quad \text{or} \quad dI = (I^i_{.j}) \, \omega.$$

Because both dI and ω are coframes, the matrix $(I^i_{.j})$ is invertible, and hence

$$\omega = (I^i_{.j})^{-1} \, dI.$$

Arguing in the same way on $\hat{C}^n_{\hat{x}}$ and using the assumptions we then conclude that

$$\hat{\omega} = (\hat{I}^i_{.j})^{-1} \, d\hat{I} = (I^i_{.j})^{-1} \, dI = \omega.$$

$\square$

The general case: $0 < p < n$. Here we pick out precisely p functionally independent functions $I^1(x), \ldots, I^p(x)$ from among the $c^i_{jk.l_1 \ldots l_a}(x)$

for $a = 0, 1, 2, \dots$. Let us assume that these p functions occur among $\{c^i_{jk.l_1 \dots l_a}(\hat{x})\}^m_{a=0}$, but not among $\{c^i_{jk.l_1 \dots l_a}(x)\}^{m-1}_{a=0}$.

Then we make the analogous choice among the $\hat{c}^i_{jk.l_1 \dots l_a}(\hat{x})$.

Theorem 15.1.2. *Necessary and sufficient conditions for local equivalence are that all the equations*

$$\hat{c}^i_{jk.l_1 \dots l_a}(\hat{x}) = c^i_{jk.l_1 \dots l_a}(x) \quad \text{for } a = 0, 1, \dots, m+1$$

are consequences of the equations $\hat{I}^i(\hat{x}) = I^i(x)$ for $i = 1, \dots, p$. Or put somewhat differently: expressing $c^i_{jk.l_1 \dots l_a}$ and $\hat{c}^i_{jk.l_1 \dots l_a}$ as functions of $I^1, \dots, I^p$ and $\hat{I}^1, \dots, \hat{I}^p$ respectively, these functions have to coincide.

Proof. To prove sufficiency we consider the pfaffian system

$$Q : \begin{cases} \hat{\omega}^i(\hat{x}) - \omega^i(x) = 0, \\ \hat{c}^i_{jk}(\hat{x}) - c^i_{jk}(x) = 0, \\ \dots \\ \hat{c}^i_{jk.l_1 \dots l_m}(\hat{x}) - c^i_{jk.l_1 \dots l_m}(x) = 0. \end{cases}$$

Then our assumptions imply that modulo Q,

$$d\hat{\omega}^i(\hat{x}) - d\omega^i(x) = \frac{1}{2} \sum_{j,k=1}^n (\hat{c}^i_{jk} \, \hat{\omega}^j \wedge \hat{\omega}^k - c^i_{jk} \, \omega^j \wedge \omega^k) \equiv 0,$$

$$d\hat{c}^i_{jk} - dc^i_{jk} = \sum_{l=1}^n (\hat{c}^i_{jk.l} \, \hat{\omega}^l - c^i_{jk.l} \, \omega^l) \equiv 0,$$

$$\dots$$

$$d\hat{c}^i_{jk.l_1 \dots l_m} - dc^i_{jk.l_1 \dots l_m} = \sum_{l_{m+1}}^n (\hat{c}_{jk.l_1 \dots l_{m+1}} \, \hat{\omega}^{l_{m+1}} - c_{jk.l_1 \dots l_{m+1}} \, \omega^{l_{m+1}}) \equiv 0,$$

whence Q is a complete system. Applying the Frobenius theorem to Q, we then see that there is an $(n-p)$-parameter family of local equivalences. $\square$

15.2 The general equivalence problem of Cartan

Let us now tackle the general lifted equivalence problem—which means that on $\mathbb{C}^n_x$ there are given

- certain invariants $I^1(x), \dots, I^h(x)$, and
- a family of local coframes $\omega(x; u) = \{\omega^1(x; u), \dots, \omega^n(x; u)\}$, where $\omega^i(x; u) = \sum_{j=1}^n a^i_j(x^1, \dots, x^n, u^1, \dots, u^p) \, dx^j$ for $i = 1, \dots, n$,

and that we have the corresponding data on $\mathbb{C}^n_{\hat{x}}$. Then we are looking for local diffeomorphisms $\phi : \mathbb{C}^n_x \xrightarrow{\cong} \mathbb{C}^n_{\hat{x}}$ having prolongations

$$\phi^{(1)} : \mathbb{C}^{n+p}_{x,u} \to \mathbb{C}^{n+p}_{\hat{x},\hat{u}}$$

$$(x,u) \mapsto (\phi(x), \psi(x,u))$$

such that

$$\phi^* \hat{I}^a(\hat{x}) = I^a(x) \text{ for } a = 1,\ldots,h, \quad \text{and} \quad \phi^{(1)*}\hat{\omega}(\hat{x};\hat{u}) = \omega(x;u).$$

The graphs of the wanted functions $\phi^{(1)}$ are n-dimensional integral manifolds on which $\omega^1 \wedge \cdots \wedge \omega^n \neq 0$ (or equivalently $\hat{\omega}^1 \wedge \cdots \wedge \hat{\omega}^n \neq 0$) of the pfaffian system

$$\mathcal{P}: \begin{cases} \hat{I}^a(\hat{x}) - I^a(x) = 0 & \text{for } a = 1,\ldots,h, \\ \theta^i := \hat{\omega}^i(\hat{x};\hat{u}) - \omega^i(x;u) = 0 & \text{for } i = 1,\ldots,n. \end{cases}$$

Note that the equations

$$0 = \hat{\omega}^i(\hat{x};\hat{u}) - \omega^i(x;u) = \sum_{j=1}^n \hat{a}^i_j(\hat{x},\hat{u}) \, d\hat{x}^j - \sum_{j=1}^n a^i_j(x,u) \, dx^j \text{ for } i = 1,\ldots,n$$

imply that the $\hat{x}^j$ are functions of $x^1,\ldots,x^n$ only—so that $\phi^{(1)}$ does have the form $(x,u) \mapsto (\phi(x), \psi(x,u))$. And the local equivalences we really want to find are then given by $x \mapsto \phi(x)$.

From $\mathcal{P}$ it follows that $0 = d\theta^i = d\hat{\omega}^i(\hat{x};\hat{u}) - d\omega^i(x;u)$, with

$$d\omega^i(x;u) = \sum_{j=1}^n \left(\sum_{k=1}^n \frac{\partial a^i_j}{\partial x^k} \, dx^k + \sum_{l=1}^p \frac{\partial a^i_j}{\partial u^l} \, du^l \right) \wedge dx^j$$

(and analogously for $d\hat{\omega}^i$). Since the dx^i may be expressed as linear combinations of the $\omega^i(x;u)$, the $d\omega^i$ can be written as

$$d\omega^i = \sum_{j=1}^n \omega^i \wedge \varpi^i_j,$$

where the one-forms ϖ^i_j are certain linear combinations of $du^1,\ldots,du^p$ and $\omega^1,\ldots,\omega^n$. Let us select a maximal number of the ϖ^i_j which are linearly independent modulo $\omega^1,\ldots,\omega^n$—say $\varpi^1,\ldots,\varpi^q$. Then there are functions $a^i_{jl}(x,u)$ such that

$$\varpi^i_j \equiv \sum_{l=1}^q a^i_{jl}(x,u)\varpi^l \pmod{\omega^1,\ldots,\omega^n} \quad \text{for } i,j = 1,\ldots,n.$$

Treating the $d\hat{\omega}^i$ in the same way, we also have

$$d\hat{\omega}^i = \sum_{j=1}^{n} \hat{\omega}^j \wedge \hat{\varpi}^i_j \quad \text{with} \quad \hat{\varpi}^i_j \equiv \sum_{l=1}^{q} \hat{a}^i_{jl}(\hat{x}, \hat{u})\,\hat{\omega}^l \quad (\text{mod } \hat{\omega}^1, \ldots, \hat{\omega}^n).$$

Because $\hat{\omega}^i = \omega^i$ and $d\hat{\omega}^i = d\omega^i$ on the wanted integral manifolds, it follows that on these

$$0 = \sum_{j=1}^{n} \omega^j \wedge (\hat{\varpi}^i_j - \varpi^i_j), \quad \text{whence} \quad \hat{\varpi}^i_j \equiv \varpi^i_j \quad (\text{mod } \omega^1 \ldots, \omega^n).$$

But then also

$$\hat{a}^i_{jl}(\hat{x}, \hat{u}) = a^i_{jl}(x, u) \quad \text{for } i, j = 1, \ldots, n \text{ and } l = 1, \ldots, q,$$

on the integral manifolds of $\mathcal{P}$, so that these coefficients are *invariants* of our problem.

Suppose next that at least one of the a^i_{jl} really depends on the auxiliary variables $u^1, \ldots, u^p$—then the corresponding $\hat{a}^i_{jl}$ will analogously depend on $\hat{u}^1, \ldots, \hat{u}^p$. Let us set

$$a^i_{jl}(x, u) = \hat{a}^i_{jl}(\hat{x}, \hat{u}) = \text{ a suitable constant,}$$

and use the implicit function theorem to express one of the u^l (and the corresponding $\hat{u}^l$) as a function of the other variables. By continuing in this way the number of auxiliary variables is decreased until eventually the a^i_{jl} only depend on the x-variables (and the $\hat{a}^i_{jl}$ only on the $\hat{x}$-variables).

Remark. The choice of constant here seems to be rather ambiguous. But the point is that in the end we will project the $\phi^{(1)} : (x, u) \mapsto (\phi(x), \psi(x, u))$ found to $x \mapsto \phi(x)$, and then the possible ambiguity will disappear.

Assuming that $I(x)$ is an invariant (for instance one of the $a^i_{jl}(x)$), covariant differentiation gives

$$dI(x) = \sum_{j=1}^{n} I_{,j}(x, u)\,\omega^j(x; u),$$

where the $I_{,j}$ are new invariants. If some of these do depend on the u-variables, we can exploit them in order to reduce the number of auxiliary variables further.

The best that can happen is that *all* auxiliary variables vanish—for then no ϖ^l remain, the given matrix group consists of the identity matrix only, and the *p*-parameter family of coframes is replaced by a *fixed* coframe. That is, we are reduced to an *e*-structure.

Otherwise we denote the number of remaining auxiliary variables by *p*, and suppose that there are *q* one-forms ϖ^l that are linearly independent modulo $\omega^1, \ldots, \omega^n$. Then

$$d\omega^i = \sum_{j=1}^{n} \omega^j \wedge \varpi^i_j \equiv \sum_{j=1}^{n} \sum_{l=1}^{q} a^i_{jl}(x)\, \omega^j \wedge \varpi^l \quad (\text{mod } \omega^1, \ldots, \omega^n)$$

for $i = 1, \ldots, n$. Now the ϖ^l are at the outset only determined up to a linear combination of the ω^i, and these equations do not specify them precisely either. Let us therefore set

$$\pi^l := \varpi^l - \sum_{k=1}^{n} v^l_k\, \omega^k \quad \text{for } l = 1, \ldots, q,$$

introducing new auxiliary variables v^l_k, so that the $d\omega^i$ acquire the form

$$d\omega^i = \frac{1}{2} \sum_{j,k=1}^{n} C^i_{jk}\, \omega^j \wedge \omega^k + \sum_{j=1}^{n} \sum_{l=1}^{q} a^i_{jl}\, \omega^j \wedge \pi^l, \quad \text{with } C^i_{jk} = -C^i_{kj}.$$

As in connection with the first fundamental theorem for Lie pseudogroups we then *choose the v^l_k such that as many as possible of the C^i_{jk} vanish*. Let us denote the remaining C^i_{jk} by c^i_{jk}, and the resulting π^l by ω^{n+l}, for $l = 1, \ldots, q$, so as to get

$$d\omega^i = \frac{1}{2} \sum_{j,k=1}^{n} c^i_{jk}\, \omega^j \wedge \omega^k + \sum_{j=1}^{n} \sum_{l=1}^{q} a^i_{jl}\, \omega^j \wedge \omega^{n+l}.$$

Making the corresponding calculations on $\mathbb{C}^{n+p}_{\hat{x},\hat{u}}$ we analogously get

$$d\hat{\omega}^i - \frac{1}{2} \sum_{j,k=1}^{n} \hat{c}^i_{jk}\, \hat{\omega}^i \wedge \hat{\omega}^k + \sum_{j=1}^{n} \sum_{l=1}^{q} \hat{a}^i_{jl}\, \hat{\omega}^j \wedge \hat{\omega}^{n+l}.$$

Then from $\hat{\omega}^i \equiv \omega^i$ (mod $\mathcal{P}$) and $\hat{a}^i_{jl} = a^i_{jl}$ on the integral manifolds of $\mathcal{P}$, it follows that the remaining nonzero c^i_{jk} and $\hat{c}^i_{jk}$ coincide on these integral manifolds:

$$\hat{c}^i_{jk}(\hat{x}) = c^i_{jk}(x) \quad \text{for } i, j, k = 1, \ldots, n,$$

so these coefficients are also invariants of our problem. If some of them depend on the auxiliary variables u^l, the number of the latter can be reduced further.

In the end it may therefore be assumed that all invariants (including the a_{jl}^i, the c_{jk}^i, and their covariant derivatives) are functions of $x^1, \ldots, x^n$ only.

If $\phi : x \mapsto \hat{x}$ is to be a local equivalence, we must have

$$\hat{I}(\hat{x}) = I(x) \quad \text{for } all \text{ invariants.}$$

This being in particular true for the c_{jk}^i and a_{jl}^i, it follows that

$$d\theta^i = d\hat{\omega}^i - d\omega^i \equiv \sum_{j=1}^n \sum_{l=1}^q a_{jl}^i \, \omega^j \wedge (\hat{\omega}^{n+l} - \omega^{n+l}) \pmod{\mathcal{P}}, \quad i = 1, \ldots, n.$$

If the coefficient system $\{a_{jl}^i\}$ is *involutive* (see the remark after Cartan's test for involutivity in section **4.1**) and *all data are analytic*, Cartan's local existence theorem will then provide the wanted solutions.

If $\{a_{jl}^i\}$ is *not involutive*, $\mathcal{P}$ is prolonged to

$$\mathcal{P}^{(1)} : \quad \begin{cases} \hat{I}(\hat{x}) - I(x) = 0 & \text{for all invariants,} \\ \hat{\omega}^i(\hat{x}; \hat{u}) - \omega^i(x; u) = 0 & \text{for } i = 1, \ldots, n, \\ \hat{\omega}^{n+l}(\hat{x}; \hat{u}; \hat{v}) - \omega^{n+l}(x; u; v) = 0 & \text{for } l = 1, \ldots, q, \end{cases}$$

where the ω^{n+l} in general depend on the new auxiliary variables v_k^l (namely those which remain after the process of killing a maximal number of the C_{jk}^i). Then we treat $\mathcal{P}^{(1)}$ in the same way as $\mathcal{P}$.

Continuing this prolongation process we know by the prolongation theorem of Cartan and Janet that we end up either with an inconsistent system, or with a system which is involutive—and then Cartan's local existence theorem can be used in the analytic context.

Not wanting to assume analyticity we keep our fingers crossed that the following favourable situation will occur. After having made m prolongations so as to arrive at $\mathcal{P}^{(m)}$, all the new auxiliary variables introduced when passing from $\mathcal{P}^{(m)}$ to the next prolongation $\mathcal{P}^{(m+1)}$ are annihilated by setting invariants equal to constants. Then we are reduced to an e-structure—in the original variables and the auxiliary variables introduced when prolonging $\mathcal{P}$ to $\mathcal{P}^{(m)}$—which is solved by means of the Frobenius theorem.

15.3 Examples

We consider two of the examples given in [Cartan 1937a]. For more we also refer to [Gardner 1989] and [Olver 1995].

Example 1. According to the local rectification lemma for vector fields in section **2.1**, any nonzero vector field X on $\mathbb{R}^n_x$ can locally be brought to the simple form $\partial/\partial t^1$ on $\mathbb{R}^n_t$ by a suitable local diffeomorphism $\phi : \mathbb{R}^n_x \overset{\cong}{\longrightarrow} \mathbb{R}^n_t$: $\phi_* X = \partial/\partial t^1$. So in particular *all nonzero vector fields on $\mathbb{R}^n$ are locally equivalent under the Lie pseudogroup consisting of all local diffeomorphisms of $\mathbb{R}^n$.*

But what about local equivalence under smaller Lie pseudogroups?

Here we will consider *local equivalence of nonzero vector fields on the Euclidean plane under the Lie pseudogroup of conformal mappings.*

More precisely: given nonzero vector fields X on $\mathbb{R}^2_{xy}$ and $\hat{X}$ on $\mathbb{R}^2_{\hat{x}\hat{y}}$ respectively, does there exist a conformal mapping $\phi : \mathbb{R}^2_{xy} \longrightarrow \mathbb{R}^2_{\hat{x}\hat{y}}$ such that $\phi_* X$ is a multiple of $\hat{X}$?

With the lengths of the vector fields being unimportant here, we normalize X and $\hat{X}$ to have the length 1. Then they can be written as

$$X = \cos\theta\,\partial/\partial x + \sin\theta\,\partial/\partial y, \quad \hat{X} = \cos\hat{\theta}\,\partial/\partial\hat{x} + \sin\hat{\theta}\,\partial/\partial\hat{y},$$

where θ is a function of x, y, and $\hat{\theta}$ is a function of $\hat{x}, \hat{y}$. Let us also define the corresponding orthogonal vector fields

$$X^{\text{orth}} = -\sin\theta\,\partial/\partial x + \cos\theta\,\partial/\partial y \text{ and } \hat{X}^{\text{orth}} = -\sin\hat{\theta}\,\partial/\partial\hat{x} + \cos\hat{\theta}\,\partial/\partial\hat{y},$$

so that $\{X, X^{\text{orth}}\}$ and $\{\hat{X}, \hat{X}^{\text{orth}}\}$ form frames on $\mathbb{R}^2_{xy}$ and $\mathbb{R}^2_{\hat{x}\hat{y}}$ respectively. Being conformal, the wanted ϕ preserves angles, and therefore $\phi_* X^{\text{orth}}$ has to be a multiple of $\hat{X}^{\text{orth}}$.

However, in the equivalence problem of Cartan it is preferable to work with coframes. Let us therefore introduce the dual one-forms α^1 and α^2 of X and X^{orth} respectively:

$$\alpha^1 := \cos\theta\,dx + \sin\theta\,dy, \quad \alpha^2 := -\sin\theta\,dx + \cos\theta\,dy,$$

and define $\hat{\alpha}^1, \hat{\alpha}^2$ analogously. Our requirements are then that ϕ pulls back the pfaffian equations $\hat{\alpha}^i = 0$ to $\alpha^i = 0$, or equivalently

$$\phi^* \hat{\alpha}^i = u_i \alpha^i, \quad i = 1, 2,$$

for suitable functions u_1 and u_2. Following the general theory we therefore set

$$\omega^i := u_i \alpha^i \quad \text{and} \quad \hat{\omega}^i = \hat{u}_i \hat{\alpha}^i \quad \text{for } i = 1, 2,$$

where $u_i, \hat{u}_i$ are considered as new variables, and ask for a prolongation

$\phi^{(1)} : \mathbb{R}^2_{xy} \times \mathbb{R}_{u_1 u_2} \longrightarrow \mathbb{R}^2_{\hat{x}\hat{y}} \times \mathbb{R}^2_{\hat{u}_1 \hat{u}_2}$ of ϕ satisfying

$$\phi^{(1)*} \begin{pmatrix} \hat{\omega}^1 \\ \hat{\omega}^2 \end{pmatrix} = \begin{pmatrix} \omega^1 \\ \omega^2 \end{pmatrix}. \tag{*}$$

Because of the conformality ϕ must satisfy

$$\phi^*(d\hat{x}^2 + d\hat{y}^2) = \lambda^2(x, y)(dx^2 + dy^2)$$

for some nonvanishing function λ. For this to be a consequence of (*) we have to require that

$$u_1 = u_2 = u, \quad \hat{u}_1 = \hat{u}_2 = \hat{u};$$

indeed, in that case

$$\hat{u}^2 \left((d\hat{x})^2 + (d\hat{y})^2 \right) = \hat{u}^2 \left((\hat{\alpha}^1)^2 + (\hat{\alpha}^2)^2 \right)$$

$$\xrightarrow{\phi^{(1)*}} u^2 \left((\alpha^1)^2 + (\alpha^2)^2 \right) = u^2(dx^2 + dy^2).$$

Conclusion. *On* $\mathbb{R}^2_{xy}$ *we have*

- *the coframe* $\alpha = (\cos\theta\, dx + \sin\theta dy, -\sin\theta dx + \cos\theta dy)^T$, *and*
- *the matrix group*

$$\mathfrak{M}(u) := \begin{pmatrix} u & 0 \\ 0 & u \end{pmatrix} = u \begin{pmatrix} 1 & 0 \\ 0 & 1 \end{pmatrix},$$

and these together define the 1-parameter family of coframes

$$(\omega^1, \omega^2)^T = (u(\cos\theta\, dx + \sin\theta\, dy), u(-\sin\theta\, dx + \cos\theta\, dy))^T \text{ on } \mathbb{R}^2_{xy}.$$

Having defined the corresponding objects on $\mathbb{R}^2_{\hat{x}\hat{y}}$ we then look for mappings $\phi^{(1)} : \mathbb{R}^3_{xyu} \longrightarrow \mathbb{R}^3_{\hat{x}\hat{y}\hat{u}}$ that satisfy

$$\phi^{(1)*} \begin{pmatrix} \hat{\omega}^1 \\ \hat{\omega}^2 \end{pmatrix} = \begin{pmatrix} \omega^1 \\ \omega^2 \end{pmatrix}.$$

According to the general method we start by differentiating ω^1 and ω^2. Noting that $\omega^1 \wedge \omega^2 = u^2\, dx \wedge dy$,

$$d\omega^1 = du \wedge (\cos\theta\, dx + \sin\theta\, dy) + u \left(\sin\theta\, \frac{\partial\theta}{\partial y} + \cos\theta\, \frac{\partial\theta}{\partial x} \right) dx \wedge dy$$

$$= \omega^1 \wedge \left(-\frac{du}{u} \right) + \frac{1}{u} \left(\cos\theta\, \frac{\partial\theta}{\partial x} + \sin\theta\, \frac{\partial\theta}{\partial y} \right) \omega^1 \wedge \omega^2,$$

and similarly

$$d\omega^2 = \omega^2 \wedge \left(-\frac{du}{u}\right) + \frac{1}{u}\left(-\sin\theta\,\frac{\partial\theta}{\partial x} + \cos\theta\,\frac{\partial\theta}{\partial y}\right)\omega^1 \wedge \omega^2.$$

Still following the general recipe we then introduce the one-form

$$\pi := -\frac{du}{u} - v_1\,\omega^1 - v_2\,\omega^2,$$

where v_1 and v_2 should be chosen so as to get rid of $\omega^1 \wedge \omega^2$ in the expressions for $d\omega^1$ and $d\omega^2$—if possible. Now

$$d\omega^1 = \omega^1 \wedge \pi + u^{-1}\left(uv_2 + \cos\theta\,\frac{\partial\theta}{\partial x} + \sin\theta\,\frac{\partial\theta}{\partial y}\right)\omega^1 \wedge \omega^2,$$

and

$$d\omega^2 = \omega^2 \wedge \pi + u^{-1}\left(-uv_1 - \sin\theta\,\frac{\partial\theta}{\partial x} + \cos\theta\,\frac{\partial\theta}{\partial y}\right)\omega^1 \wedge \omega^2,$$

so there is but one choice for v_1 and v_2,

$$v_1 = -u^{-1}\left(\sin\theta\,\frac{\partial\theta}{\partial x} - \cos\theta\,\frac{\partial\theta}{\partial y}\right), \text{ and } v_2 = -u^{-1}\left(\cos\theta\,\frac{\partial\theta}{\partial x} + \sin\theta\,\frac{\partial\theta}{\partial y}\right),$$

making π go over into

$$\omega^3 := -\frac{du}{u} - \frac{\partial\theta}{\partial y}\,dx + \frac{\partial\theta}{\partial x}\,dy.$$

Then

$$d\omega^3 = \left(\frac{\partial^2\theta}{\partial x^2} + \frac{\partial^2\theta}{\partial y^2}\right)dx \wedge dy = \frac{\Delta_{xy}\theta}{u^2}\,\omega^1 \wedge \omega^2,$$

with $\Delta_{xy} := \partial^2/\partial x^2 + \partial^2/\partial y^2$.

Thus the new auxiliary variables v_1, v_2 have disappeared from the scene, and we have been led to the coframe $\{\omega^1, \omega^2, \omega^3\}$ for $\{(x, y, u) \in \mathbb{R}^3 \mid u \neq 0\}$. The ω^i appearing here satisfy the structure equations

$$d\omega^1 = \omega^1 \wedge \omega^3,$$
$$d\omega^2 = \omega^2 \wedge \omega^3,$$
$$d\omega^3 = \frac{\Delta_{xy}\theta}{u^2}\,\omega^1 \wedge \omega^2.$$

By symmetry we also obtain the corresponding data in the variables $\hat{x}, \hat{y}$, and $\hat{u}$.

These structure equations disclose *one invariant* :

$$I(x, y, u) := \frac{\Delta_{xy}\theta(x, y)}{u^2}\text{—and the corresponding } \hat{I}(\hat{x}, \hat{y}, \hat{u}).$$

Such an invariant is used in order to reduce the number of auxiliary variables. Here we have to distinguish between two cases:

- $\Delta_{xy}\theta = 0$—with no reduction being possible;
- $\Delta_{xy}\theta \neq 0$—with $\partial I/\partial u \neq 0$.

Special case: $\Delta_{xy}\theta = 0$. Then necessarily also $\Delta_{\hat{x}\hat{y}}\hat{\theta} = 0$, so that θ and $\hat{\theta}$ are *harmonic functions*.

Following the general method we prolong the original pfaffian system

$$\mathcal{P}: \quad \theta^i := \hat{\omega}^i - \omega^i \quad \text{for } i = 1, 2$$

to

$$\mathcal{P}^{(1)}: \quad \begin{cases} \theta^1 := \hat{\omega}^1 - \omega^1, \\ \theta^2 := \hat{\omega}^2 - \omega^2, \\ \theta^3 := \hat{\omega}^3 - \omega^3 \end{cases} \quad \text{on } \mathbb{R}^3_{xyu} \times \mathbb{R}^3_{\hat{x}\hat{y}\hat{u}}.$$

The structure equations

$$d\omega^1 = \omega^1 \wedge \omega^3, \quad d\omega^2 = \omega^2 \wedge \omega^3, \quad d\omega^3 = 0,$$
$$d\hat{\omega}^1 = \hat{\omega}^1 \wedge \hat{\omega}^3, \quad d\hat{\omega}^2 = \hat{\omega}^2 \wedge \hat{\omega}^3, \quad d\hat{\omega}^3 = 0,$$

show that $\mathcal{P}^{(1)}$ is complete, i.e., $d\theta^i \equiv 0 \pmod{\mathcal{P}^{(1)}}$. So the Frobenius theorem applies and gives a 3-parameter family of local equivalences.

General case: $\Delta_{xy}\theta \neq 0$. Here it is possible to get rid of the auxiliary variable u by setting the invariant I equal to a suitable constant. Let us for instance set

$$I(x, y, u) = \frac{\Delta_{xy}\theta(x, y)}{u^2} := \begin{cases} 1 & \text{if } \Delta_{xy}\theta > 0, \\ -1 & \text{if } \Delta_{xy}\theta < 0, \end{cases}$$

from which u gets expressed as

$$u = \sqrt{|\Delta_{xy}\theta|}.$$

Then

$$\omega^1 = \sqrt{|\Delta_{xy}\theta|}\,(\cos\theta\,dx + \sin\theta\,dy),$$
$$\omega^2 = \sqrt{|\Delta_{xy}\theta|}\,(-\sin\theta\,dx + \cos\theta\,dy),$$

while ω^3 is reduced to a linear combination of dx and dy, or equivalently of ω^1 and ω^2:

$$\omega^3 := a(x, y)\,\omega^1 + b(x, y)\,\omega^2,$$

and analogously for $\hat{\omega}^1$, $\hat{\omega}^2$ and $\hat{\omega}^3$.

From $\hat{\omega}^i = \omega^i$ for $i = 1, 2, 3$ we then conclude that

$$\hat{a}(\hat{x}, \hat{y}) = a(x, y) \quad \text{and} \quad \hat{b}(\hat{x}, \hat{y}) = b(x, y),$$

thereby obtaining two more invariants. These give rise to three subcases:

- a and b are functionally independent—in which case there is at most one local equivalence, given by solving the system

$$\hat{a}(\hat{x}, \hat{y}) = a(x, y), \quad \hat{b}(\hat{x}, \hat{y}) = b(x, y)$$

  with respect to $\hat{x}$ and $\hat{y}$;
- a and b are functionally dependent, but not both constant—then there is at most a 1-parameter family of local equivalences;
- a and b are constants.

The last case cannot possibly occur, however, since $\omega^3 = a\omega^1 + b\omega^2$ combined with the structure equations

$$\begin{cases} d\omega^1 = \omega^1 \wedge \omega^3 = b\,\omega^1 \wedge \omega^2, \\ d\omega^2 = \omega^2 \wedge \omega^3 = -a\,\omega^1 \wedge \omega^2 \end{cases}$$

reveal that

$$\begin{aligned} d\omega^3 &= da \wedge \omega^1 + db \wedge \omega^2 + a\,d\omega^1 + b\,d\omega^2 \\ &= (-a_{.2} + b_{.1} + ab - ba)\,\omega^1 \wedge \omega^2 = (b_{.1} - a_{.2})\,\omega^1 \wedge \omega^2. \end{aligned}$$

But we had chosen u such that $d\omega^3 = \pm\omega^1 \wedge \omega^2$, and hence $b_{.1} - a_{.2} = \pm 1$. As a consequence, a and b cannot both be constants.

Example 2. Suppose that $\mathbb{R}^2_{xy}$ is provided with the Riemannian metric ds^2, and $\mathbb{R}^2_{\hat{x}\hat{y}}$ with the Riemannian metric $d\hat{s}^2$. Determine local equivalences $\phi: \mathbb{R}^2_{xy} \xrightarrow{\cong} \mathbb{R}^2_{\hat{x}\hat{y}}$ satisfying $\phi^*(d\hat{s}^2) = ds^2$.

It is possible to define ds^2 by means of two one-forms α^1 and α^2 living on $\mathbb{R}^2_{xy}$:

$$ds^2 = (\alpha^1)^2 + (\alpha^2)^2.$$

These forms are, however, not uniquely determined by ds^2, since

$$\omega^1 := \cos\theta\,\alpha^1 + \sin\theta\,\alpha^2, \quad \omega^2 := -\sin\theta\,\alpha^1 + \cos\theta\,\alpha^2$$

work just as well for an arbitrary parameter θ. Therefore we arrive at the following data on $\mathbb{R}^2_{xy}$:

- a local coframe $(\alpha^1, \alpha^2)^T$,
- the matrix group of orthogonal transformations

$$\mathfrak{M}(\theta) = \begin{pmatrix} \cos\theta & \sin\theta \\ -\sin\theta & \cos\theta \end{pmatrix}, \quad \theta \in \mathbb{R}.$$

And these together define the 1-parameter family of local coframes

$$\begin{pmatrix} \omega^1 \\ \omega^2 \end{pmatrix} = \mathfrak{M}(\theta) \begin{pmatrix} \alpha^1 \\ \alpha^2 \end{pmatrix} = \begin{pmatrix} \cos\theta\,\alpha^1 + \sin\theta\,\alpha^2 \\ -\sin\theta\,\alpha^1 + \cos\theta\,\alpha^2 \end{pmatrix} \quad \text{for } \mathbb{R}^2_{xy}.$$

By symmetry we also have the analogous objects on $\mathbb{R}^2_{\hat{x}\hat{y}}$.

In order to solve the equivalence problem we firstly have to differentiate the ω^i:

$$d\omega^1 = d\theta \wedge (-\sin\theta\,\alpha^1 + \cos\theta\,\alpha^2) + \cos\theta\,d\alpha^1 + \sin\theta\,d\alpha^2,$$

where $d\alpha^1$ and $d\alpha^2$ are certain multiples of $dx \wedge dy$, or equivalently of $\omega^1 \wedge \omega^2$—whence $d\omega^1$ may be expressed as

$$d\omega^1 = -\omega^2 \wedge d\theta + a(x, y, \theta)\,\omega^1 \wedge \omega^2.$$

Similarly

$$d\omega^2 = \omega^1 \wedge d\theta + b(x, y, \theta)\,\omega^1 \wedge \omega^2.$$

In accordance with the general theory we then set

$$\pi := d\theta - v_1\,\omega^1 - v_2\,\omega^2$$

so as to obtain

$$d\omega^1 = -\omega^2 \wedge \pi + (v_1 + a)\,\omega^1 \wedge \omega^2 \quad \text{and} \quad d\omega^2 = \omega^1 \wedge \pi + (v_2 + b)\,\omega^1 \wedge \omega^2.$$

This shows that we can get rid of $\omega^1 \wedge \omega^2$ by choosing

$$v_1 := -a \quad \text{and} \quad v_2 := -b.$$

This choice makes π go over into $\omega^3 := d\theta + a\,\omega^1 + b\,\omega^2$, and simplifies the $d\omega^i$ to

$$\begin{cases} d\omega^1 = -\omega^2 \wedge \omega^3, \\ d\omega^2 = \omega^1 \wedge \omega^3. \end{cases}$$

Differentiating the latter, we get

$$\begin{cases} 0 = d^2\omega^1 = -d\omega^2 \wedge \omega^3 + \omega^2 \wedge d\omega^3 = \omega^2 \wedge d\omega^3, \\ 0 = d^2\omega^2 = d\omega^1 \wedge \omega^3 - \omega^1 \wedge d\omega^3 = -\omega^1 \wedge d\omega^3. \end{cases}$$

But this can be true only if $d\omega^3$ is a multiple of $\omega^1 \wedge \omega^2$, i.e.,

$$d\omega^3 = -K(x, y, \theta)\, \omega^1 \wedge \omega^2.$$

The function K appearing here is then an *invariant*. Actually it does not depend on θ, since

$$0 = d^2\omega^3 = -dK \wedge \omega^1 \wedge \omega^2 - K\, d\omega^1 \wedge \omega^2 + K\, \omega^1 \wedge d\omega^2 = -dK \wedge \omega^1 \wedge \omega^2,$$

showing that dK is a linear combination of ω^1 and ω^2, or equivalently of dx and dy.

So by means of the equivalence problem we have found an invariant $K(x, y)$ attached to the Riemannian metric ds^2—and this K is nothing but the *Gaussian curvature*.

If K is a *constant*, the corresponding $\hat{K}$ must assume the same constant value. The local equivalences are found by solving the pfaffian system

$$\mathcal{P}^{(1)}: \quad \begin{cases} \hat{\omega}^1 - \omega^1 = 0, \\ \hat{\omega}^2 - \omega^2 = 0, \qquad \text{on } \mathbb{R}^3_{xy\theta} \times \mathbb{R}^3_{\hat{x}\hat{y}\hat{\theta}}, \\ \hat{\omega}^3 - \omega^3 = 0 \end{cases}$$

which by the structure equations is *complete*. Thus the Frobenius theorem yields a 3-parameter family of local equivalences.

If $K(x, y)$ is *not constant*, we get new invariants by means of covariant differentiation:

$$dK(x, y) = K_{.1}(x, y, \theta)\, \omega^1(x, y, \theta) + K_{.2}(x, y, \theta)\, \omega^2(x, y, \theta).$$

If for instance $\partial K_{.1}/\partial\theta \neq 0$, it is possible to get rid of the auxiliary variable θ by solving the equation $K_{.1}(x, y, \theta) = 0$ with respect to θ.

Let us now specialize to the Euclidean metric $ds^2 = dx^2 + dy^2$, i.e., $\alpha^1 = dx$ and $\alpha^2 = dy$. Then

$$\omega^1 = \cos\theta\, dx + \sin\theta\, dy, \qquad \omega^2 = -\sin\theta\, dx + \cos\theta\, dy,$$

from which we obtain

$$d\omega^1 = d\theta \wedge \omega^2 \quad \text{and} \quad d\omega^2 = -d\theta \wedge \omega^1.$$

Identifying this with the expressions above we see that $\omega^3 = d\theta$ here.

Moreover the ω^i—which make up a coframe $\{\omega^1, \omega^2, \omega^3\}$ for $\mathbb{R}^3_{xy\theta}$—satisfy the structure equations

$$d\omega^1 = -\omega^2 \wedge \omega^3, \quad d\omega^2 = \omega^1 \wedge \omega^3 \quad \text{and} \quad d\omega^3 = 0.$$

So by Cartan's version of the first fundamental theorem for Lie parameter groups the ω^i define a 3-dimensional Lie group—namely the group of *Euclidean motions in the plane*.

Let us now restrict x and y to a curve

$$C : \mathbb{R}_t \to \mathbb{R}^2_{xy}$$
$$t \mapsto (\xi(t), \eta(t)).$$

Then

$$\omega^1 \mapsto (\cos\theta\, \xi'(t) + \sin\theta\, \eta'(t))\, dt = \omega^1{}_{|C},$$
$$\omega^2 \mapsto (-\sin\theta\, \xi'(t) + \cos\theta\, \eta'(t))\, dt = \omega^2{}_{|C},$$

while $\omega^3 = d\theta$ is not affected. With ω^1 and ω^2 being invariant under the group of Euclidean motions, the quotient

$$\frac{\omega^2{}_{|C}}{\omega^1{}_{|C}} = \frac{-\sin\theta\, \xi'(t) + \cos\theta\, \eta'(t)}{\cos\theta\, \xi'(t) + \sin\theta\, \eta'(t)}$$

is also an invariant. We get rid of the auxiliary variable θ by setting this equal to 0:

$$\tan\theta = \frac{\eta'(t)}{\xi'(t)}, \quad \text{or} \quad \theta = \arctan\left(\frac{\eta'(t)}{\xi'(t)}\right).$$

With this choice of θ, $\omega^1{}_{|C}$ goes over into

$$\omega^1{}_{|C} = \sqrt{(\xi'(t))^2 + (\eta'(t))^2}\, dt, \quad \text{which is the arclength } ds.$$

Another invariant is found by forming the quotient of the invariant one-forms ω^3 and $\omega^1{}_{|C}$:

$$\frac{\omega^3}{\omega^1{}_{|C}} = \frac{\xi'(t)\eta''(t) - \eta'(t)\xi''(t)}{((\xi'(t))^2 + (\eta'(t))^2)^{3/2}}, \quad \text{which is the curvature of the curve } C.$$

Remark. We have introduced and employed invariants as a means for finding equivalences. But this last example illustrates another point of view: whenever a local structure is given, it is of interest to determine all invariants attached to this, that is all functions which are invariant under the symmetry group of the structure—and often enough Cartan's procedure for solving the equivalence problem provides a very convenient way for finding these invariants.

16

Parabolic PDEs and associated PDE systems

Chapters **12** and **13** were devoted to the study of those hyperbolic second order PDEs in one dependent and two independent variables for which each Monge system admits at least two functionally independent first integrals. In this and the following chapter we consider the corresponding problem for *parabolic second order PDEs*.

In this case also the theory started with a paper of Goursat—[Goursat 1895]—in bad need of clarification. Satisfactory such was presented in the 'five variable paper' [Cartan 1910], which in conformity with Vessiot's theory of hyperbolic PDEs points out the crucial role played by Lie groups when trying to understand the underlying reasons for the results obtained.

An arbitrary second order PDE S_1 in one dependent and two independent variables can be regarded as a 7-dimensional submanifold of the 8-dimensional jet bundle $J^2(\mathbb{C}^2, \mathbb{C})$, provided with the restriction of the contact pfaffian system ${}^2Ct^{2,1}$ of $J^2(\mathbb{C}^2, \mathbb{C})$ to S_1. In this way our PDE is identified with a 3-dimensional pfaffian system $\mathcal{P}_1$ on a 7-dimensional manifold.

We restrict the study to *parabolic PDEs for which the Monge system admits at least two functionally independent first integrals*. A useful observation in section **16.1** is that the 2-dimensional Monge system is *complete* in this case, and hence in fact admits five functionally independent first integrals. Morover such a parabolic PDE may be identified with a *2-dimensional pfaffian system $Q = (\omega^1, \omega^2)$ on a 6-dimensional manifold*, and—the decisive point—this Q admits a Cauchy characteristic vector field. Because of the latter we are then *reduced to finding integral curves of a 2-dimensional pfaffian system in five variables*, whence we only need to solve ODE systems.

The Lie structure of $\mathcal{Q}$ is

$$\begin{cases} d\omega^1 \equiv \pi^3 \wedge \pi^4, \\ d\omega^2 \equiv \pi^3 \wedge \pi^5 \end{cases} \quad (\mathrm{mod}\ \omega^1, \omega^2).$$

From this it is inferred that the integral curves of $\mathcal{Q}$ that we are looking for are also integral curves of the anti-derived system $'\mathcal{Q} = (\omega^1, \omega^2, \pi^3)$—i.e., a *3-dimensional pfaffian system in five variables*.

In the special case when $'\mathcal{Q}$ is complete, the corresponding PDE is of *Monge–Ampère type*. With the theory of Monge–Ampère equations being fairly well-known, we concentrate on the *general case when $'\mathcal{Q}$ is not complete* in the following.

In a rather round-about way the study of parabolic PDEs has thus been reduced to that of *3- and 2-dimensional pfaffian systems in five variables*. To gain a general perspective on the latter, section **16.2** is devoted to a structural classification of such pfaffian systems.

Let S_2 be a system of two PDEs in one dependent and two independent variables. Then S_2 can be identified with a 3-dimensional pfaffian system $\mathcal{P}_2$ on a 6-dimensional manifold. If $\mathcal{P}_2$ happens to admit a Cauchy characteristic vector field, this can be used in order to reduce $\mathcal{P}_2$ to a *3-dimensional pfaffian system in five variables*—a case which is studied in section **16.3**.

There is one special and easy case, namely when S_2 is quasi-linear. Disregarding this the Lie structure of $\mathcal{P}_2 = (\omega^1, \omega^2, \omega^3)$ has the form

$$\begin{cases} d\omega^1 \equiv 0, \\ d\omega^2 \equiv 0, \\ d\omega^3 \equiv \omega^4 \wedge \omega^5 \end{cases} \quad (\mathrm{mod}\ \omega^1, \omega^2, \omega^3),$$

showing that the derived system $\mathcal{P}_2'$ equals (ω^1, ω^2), that is, a *2-dimensional pfaffian system in five variables*. As such it is associated to a parabolic PDE S_1.

In this way we get a *correspondence between general parabolic PDEs with complete Monge systems, and general systems of two PDEs admitting a Cauchy characteristic vector field.*

In the next chapter we will then consider the local classification of these objects, using Cartan's solution of the equivalence problem.

16.1 Parabolic PDEs for which the Monge system admits at least two first integrals

With the coordinates of the 8-dimensional jet bundle $J^2(\mathbb{C}^2_{xy}, \mathbb{C}_z)$ being denoted by x, y, z, p, q, r, s and t, the contact pfaffian system ${}^2Ct^{2,1}$ of the latter is generated by the one-forms

$$\begin{cases} \theta^1 := dz - p\,dx - q\,dy, \\ \theta^2 := dp - r\,dx - s\,dy, \\ \theta^3 := dq - s\,dx - t\,dy. \end{cases}$$

Noting that

$$d\theta^1 = dx \wedge dp + dy \wedge dq = dx \wedge (\theta^2 + r\,dx + s\,dy) + dy \wedge (\theta^3 + s\,dx + t\,dy)$$
$$= dx \wedge \theta^2 + dy \wedge \theta^3 \equiv 0 \pmod{\theta^1, \theta^2, \theta^3},$$

we have that the Lie structure of ${}^2Ct^{2,1}$ is given by

$$\begin{cases} d\theta^1 \equiv 0 \pmod{\theta^1, \theta^2, \theta^3}, \\ d\theta^2 = dx \wedge dr + dy \wedge ds, \\ d\theta^3 = dx \wedge ds + dy \wedge dt. \end{cases}$$

Let $\mathcal{S}_1$ be a second order PDE in one dependent and two independent variables. Being a hypersurface of $J^2(\mathbb{C}^2_{xy}, \mathbb{C}_z)$, $\mathcal{S}_1$ may be written in the parametrized form

$$\mathcal{S}_1: \quad \begin{cases} r = \rho(x, y, z, p, q, \lambda, \mu), \\ s = \sigma(x, y, z, p, q, \lambda, \mu), \\ t = \tau(x, y, z, p, q, \lambda, \mu), \end{cases}$$

so that x, y, z, p, q, λ and μ can be used as local coordinates on $\mathcal{S}_1$. The pfaffian system $\mathcal{P}_1$ associated to $\mathcal{S}_1$ is the restriction of ${}^2Ct^{2,1}$ to $\mathcal{S}_1$. Setting $\varpi^i := \theta^i|_{\mathcal{S}_1}$ for $i = 1, 2, 3$, and introducing one-forms $\varpi^4, \ldots, \varpi^7$ such that $\{\varpi^1, \ldots, \varpi^7\}$ constitutes a local coframe for $\mathcal{S}_1$, $\mathcal{P}_1$ equals $(\varpi^1, \varpi^2, \varpi^3)$ and has a Lie structure of the form

$$\begin{cases} d\varpi^1 \equiv 0, \\ d\varpi^2 \equiv \sum_{4 \le i < j \le 7} a^2_{ij}\, \varpi^i \wedge \varpi^j, \qquad (\mathrm{mod}\ \varpi^1, \varpi^2, \varpi^3). \\ d\varpi^3 \equiv \sum_{4 \le i < j \le 7} a^3_{ij}\, \varpi^i \wedge \varpi^j \end{cases}$$

In order to discover the Monge systems of $\mathcal{P}_1$ we first simplify this structure by a suitable linear transformation

$$
\begin{pmatrix} \varpi^1 \\ \varpi^2 \\ \vdots \\ \varpi^7 \end{pmatrix} \longmapsto \begin{pmatrix} 1 & 0 & 0 & 0 & \cdots & 0 \\ 0 & * & * & 0 & \cdots & 0 \\ 0 & * & * & * & \cdots & * \\ & & \cdots\cdots\cdots\cdots \\ 0 & * & * & * & \cdots & * \end{pmatrix} \begin{pmatrix} \varpi^1 \\ \varpi^2 \\ \vdots \\ \varpi^7 \end{pmatrix} =: \begin{pmatrix} \omega^1 \\ \omega^2 \\ \vdots \\ \omega^7 \end{pmatrix}
$$

changing at least one of the complicated right hand sides above into a *single wedge product*. In particular we want to find functions u and v such that $\omega := u\varpi^2 + v\varpi^3$ satisfies

$$
d\omega \equiv u\,d\varpi^2 + v\,d\varpi^3 = (ua_{45}^2 + va_{45}^3)\,\varpi^4 \wedge \varpi^5 + \cdots
$$
$$
\equiv \text{a single wedge product—say } \alpha \wedge \beta \quad (\mathrm{mod}\ \mathcal{P}_1).
$$

Then

$$
d\omega \wedge d\omega \equiv \alpha \wedge \beta \wedge \alpha \wedge \beta = 0 \quad (\mathrm{mod}\ \mathcal{P}_1).
$$

Now $d\omega \wedge d\omega$ is a multiple of $\varpi^4 \wedge \varpi^5 \wedge \varpi^6 \wedge \varpi^7$ (mod $\mathcal{P}_1$), and setting this multiple equal to 0 we get

$$
u^2 \cdot (a_{45}^2 a_{67}^2 - a_{46}^2 a_{57}^2 + a_{47}^2 a_{56}^2)
$$
$$
+ uv \cdot (a_{45}^2 a_{67}^3 + a_{45}^3 a_{67}^2 - a_{46}^2 a_{57}^3 - a_{46}^3 a_{57}^2 + a_{47}^2 a_{56}^3 + a_{47}^3 a_{56}^2)
$$
$$
+ v^2 \cdot (a_{45}^3 a_{67}^3 - a_{46}^3 a_{57}^3 + a_{47}^3 a_{56}^3) = 0.
$$

The second order equation for u and v thus obtained is called the *characteristic equation of $\mathcal{P}_1$*.

Let us for each $(x, y, z, p, q, \lambda, \mu) \in \mathcal{S}_1$ consider $(u(x, y, z, p, q, \lambda, \mu), v(x, y, z, p, q, \lambda, \mu))$ as homogeneous coordinates of the Riemann sphere $\mathbb{P}$. Then the roots of the characteristic equation will be sections of the product bundle

$$
\mathcal{S}_1 \times \mathbb{P}
$$
$$
\downarrow
$$
$$
\mathcal{S}_1.
$$

As usual there are two cases:

- $\mathcal{S}_1$ is *hyperbolic* if the roots are different,
- $\mathcal{S}_1$ is *parabolic* if the roots coincide.

The hyperbolic case. Here there are two different linear combinations of

ϖ^2 and ϖ^3 having the property that their differentials are single wedge products modulo $\mathcal{P}_1$.

The characteristic equation can be reduced to $uv = 0$, with the roots $(0, 1)$ and $(1, 0)$, by means of a linear transformation of $\varpi^1, \ldots, \varpi^7$ making the coefficients a_{ij}^2 and a_{ij}^3 go over into

$$a_{45}^2 = a_{67}^3 = 1, \quad \text{and the other } a_{ij}^2, a_{ij}^3 \text{ being equal to zero.}$$

Consequently $\mathcal{P}_1$ can be written as $(\omega^1, \omega^2, \omega^3)$ with the structure

$$\begin{cases} d\omega^1 \equiv 0, \\ d\omega^2 \equiv \omega^4 \wedge \omega^5, \qquad (\text{mod } \omega^1, \omega^2, \omega^3). \\ d\omega^3 \equiv \omega^6 \wedge \omega^7 \end{cases}$$

To see the connection with Monge systems we recall how to construct 2-dimensional involutions of $\mathcal{P}_1$: first take an arbitrary nonzero vector field $X \in \mathcal{P}_1^\perp$, and then determine all $Y \in \mathcal{P}_1^\perp$ in involution with X, i.e., satisfying

$$X \rfloor d\omega^i(Y) = 0 \quad \text{for } i = 1, 2, 3.$$

Since $d\omega^1 \equiv 0$ (mod $\mathcal{P}_1$) this in general yields *two independent conditions* for Y. But it reduces to *only one* if

$$\text{either } X \rfloor d\omega^2 \equiv 0 \quad (\text{mod } \mathcal{P}_1) \quad \text{or} \quad X \rfloor d\omega^3 \equiv 0 \quad (\text{mod } \mathcal{P}_1),$$

in which case X is said to be *singular*.

Hence our $\mathcal{P}_1$ admits *two* singular vector field systems:

$$(\omega^1, \omega^2, \omega^3, \omega^4, \omega^5)^\perp = (\partial/\partial\omega^6, \partial/\partial\omega^7)$$

and

$$(\omega^1, \omega^2, \omega^3, \omega^6, \omega^7)^\perp = (\partial/\partial\omega^4, \partial/\partial\omega^5),$$

and these are the two *Monge systems* of the hyperbolic case that we are so familiar with.

The parabolic case. In this case the characteristic equation can be reduced to $v^2 = 0$, with the double root $(1, 0)$. Then there is only one ω having the property that $d\omega$ is a single wedge product. If we call it ω^2, there are $\omega^3, \omega^4, \ldots$ such that $\mathcal{P}_1 = (\omega^1, \omega^2, \omega^3)$ and

$$\begin{cases} d\omega^1 \equiv 0, \\ d\omega^2 \equiv \omega^4 \wedge \omega^5 \end{cases} \qquad (\text{mod } \omega^1, \omega^2, \omega^3).$$

The precise expression for $d\omega^3$ will not be important in the following, but with a little effort we can find a linear transformation of $\varpi^1, \ldots, \varpi^7$ making the coefficients a_{ij}^2 and a_{ij}^3 go over into

$$a_{45}^2 = a_{46}^3 = a_{57}^3 = 1, \quad \text{and the other } a_{ij}^2, a_{ij}^3 \text{ all vanishing,}$$

and then $\mathcal{P}_1$ is endowed with the Lie structure

$$\begin{cases} d\omega^1 \equiv 0, \\ d\omega^2 \equiv \omega^4 \wedge \omega^5, \\ d\omega^3 \equiv \omega^4 \wedge \omega^6 + \omega^5 \wedge \omega^7 \end{cases} \quad (\text{mod } \omega^1, \omega^2, \omega^3).$$

Because $X \lrcorner d\omega^3 \not\equiv 0 \pmod{\mathcal{P}_1}$ for any nonzero $X \in \mathcal{P}_1^\perp$, there is only one Monge characteristic vector field system in this case:

$$\mathcal{M}(\mathcal{P}_1)^\perp = (\omega^1, \omega^2, \omega^3, \omega^4, \omega^5)^\perp = (\partial/\partial\omega^6, \partial/\partial\omega^7),$$

where $\mathcal{M}(\mathcal{P}_1)$ denotes the Monge characteristic pfaffian system of $\mathcal{P}_1$.

Remembering the considerations in section **3.4** we also note that $\mathcal{P}_1$ does not admit any Cauchy characteristic vector field; in fact

$$\mathcal{C}(\mathcal{P}_1)^\perp = (\omega^1, \omega^2, \omega^3, \omega^4, \omega^5, \omega^6, \omega^7)^\perp = \{0\}.$$

In the earlier study of hyperbolic PDEs it was assumed that each Monge system admits at least two first integrals; let us make an analogous supposition here.

Assumption. *The Monge system $\mathcal{M}(\mathcal{P}_1)$ admits at least two functionally independent first integrals. That is, there are functions f_1 and f_2 such that $df_1 \wedge df_2 \neq 0$ and $df_1, df_2 \in \mathcal{M}(\mathcal{P}_1)$.*

Lemma 16.1.1. *If $\mathcal{M}(\mathcal{P}_1)$ admits two functionally independent first integrals, it is in fact complete—and thus admits five functionally independent first integrals.*

Proof. Note that $d\omega^1$, $d\omega^2$ and $d\omega^3$ are all nonzero, and also that the one-forms ω^4, ω^5 appearing in the structure equations

$$\begin{cases} d\omega^1 \equiv 0, \\ d\omega^2 \equiv \omega^4 \wedge \omega^5, \\ d\omega^3 \equiv \omega^4 \wedge \omega^6 + \omega^5 \wedge \omega^7 \end{cases} \quad (\text{mod } \omega^1, \omega^2, \omega^3)$$

are defined only up to linear combinations of ω^1, ω^2 and ω^3. Adding suitable such linear combinations it may therefore be assumed that the two exact one-forms in $(\omega^1, \ldots, \omega^5)$ are given by ω^4 and ω^5—i.e., $d\omega^4 = d\omega^5 = 0$. But then the structure equations immediately show that

$$d\omega^i \equiv 0 \quad (\text{mod } \omega^1, \ldots, \omega^5) \quad \text{for } i = 1, \ldots, 5,$$

and hence $\mathcal{M}(\mathcal{P}_1) = (\omega^1, \ldots, \omega^5)$ is indeed complete. $\qquad \square$

The main idea in the following is to drop the troublesome ω^3 and consider the 2-dimensional pfaffian system $\mathcal{Q} := (\omega^1, \omega^2)$ instead of $\mathcal{P}_1 = (\omega^1, \omega^2, \omega^3)$.

The structure equations of $\mathcal{P}_1$ imply that the Lie structure of $\mathcal{Q}$ can be written as

$$\begin{cases} d\omega^1 \equiv \omega^3 \wedge (a_4\,\omega^4 + a_5\,\omega^5 + a_6\,\omega^6 + a_7\,\omega^7), \\ d\omega^2 \equiv \omega^4 \wedge \omega^5 + \omega^3 \wedge (b_4\,\omega^4 + b_5\,\omega^5 + b_6\,\omega^6 + b_7\,\omega^7) \end{cases} \quad (\text{mod } \omega^1, \omega^2).$$

Recalling that $d\omega^4 \equiv d\omega^5 \equiv 0$ (mod $\omega^1, \ldots, \omega^5$) and expressing $d^2\omega^1$, $d^2\omega^2$ in the basis $\{\omega^i \wedge \omega^j \wedge \omega^k\}_{1 \le i < j < k \le 7}$ we obtain

$$\begin{cases} 0 = d^2\omega^1 = \cdots + a_7\,\omega^4 \wedge \omega^6 \wedge \omega^7 - a_6\,\omega^5 \wedge \omega^6 \wedge \omega^7 + \cdots, \\ 0 = d^2\omega^2 = \cdots + b_7\,\omega^4 \wedge \omega^6 \wedge \omega^7 - b_6\,\omega^5 \wedge \omega^6 \wedge \omega^7 + \cdots, \end{cases}$$

and therefore $a_7 = a_6 = b_7 = b_6 = 0$. But then $\mathcal{Q}$ has the structure

$$\begin{cases} d\omega^1 \equiv a_4\,\omega^3 \wedge \omega^4 + a_5\,\omega^3 \wedge \omega^5, \\ d\omega^2 \equiv b_4\,\omega^3 \wedge \omega^4 + b_5\,\omega^3 \wedge \omega^5 + \omega^4 \wedge \omega^5 \end{cases} \quad (\text{mod } \omega^1, \omega^2).$$

In particular

$$C(\mathcal{Q}) = (\omega^1, \omega^2, \omega^3, \omega^4, \omega^5),$$

so that *the Cauchy characteristic system of $\mathcal{Q}$ coincides with the Monge characteristic system $\mathcal{M}(\mathcal{P}_1)$ of the original $\mathcal{P}_1$.*

Letting $x^1, \ldots, x^5$ be five functionally independent first integrals of $C(\mathcal{Q}) = \mathcal{M}(\mathcal{P}_1)$, the theory in section **3.4** shows that it is possible to reduce $\mathcal{Q}$ to a *pfaffian system $\mathcal{Q}_{\text{red}}$ in the five variables $x^1, \ldots, x^5$.* And in order to find 2-dimensional integral manifolds of $\mathcal{Q}$ it then *suffices to find integral curves of $\mathcal{Q}_{\text{red}}$ in these five variables*—a task which can be performed by solving ODE systems.

Let us now use the information obtained in order to determine the explicit form of those PDEs that do have a complete Monge system.

The one-form ω^1 in $\mathcal{Q}$ is just $dz - p\,dx - q\,dy$, while ω^2 is a certain linear combination of $dp - r\,dx - s\,dy$ and $dq - s\,dx - t\,dy$—say

$$\omega^2 = dp + u\,dq + v\,dx + w\,dy,$$

where u, v, w are functions of the local coordinates $x, y, z, p, q, \lambda, \mu$ of S_1. Then modulo ω^1, ω^2,

$$d\omega^1 = dx \wedge dp + dy \wedge dq \equiv -u\,dx \wedge dq - w\,dx \wedge dy + dy \wedge dq,$$
$$d\omega^2 = du \wedge dq + dv \wedge dx + dw \wedge dy$$
$$\equiv d\lambda \wedge \left(\frac{\partial u}{\partial \lambda} dq + \frac{\partial v}{\partial \lambda} dx + \frac{\partial w}{\partial \lambda} dy \right) + d\mu \wedge \left(\frac{\partial u}{\partial \mu} dq + \frac{\partial v}{\partial \mu} dx + \frac{\partial w}{\partial \mu} dy \right)$$
$$+ c_{xq}\,dx \wedge dq + c_{xy}\,dx \wedge dy + c_{yq}\,dy \wedge dq,$$

where

$$c_{xq} = \frac{\partial u}{\partial x} + p\frac{\partial u}{\partial z} - v\frac{\partial u}{\partial p} - \frac{\partial v}{\partial q} + u\frac{\partial v}{\partial p},$$
$$c_{xy} = \frac{\partial w}{\partial x} + p\frac{\partial w}{\partial z} - u\frac{\partial w}{\partial p} - \frac{\partial v}{\partial y} - q\frac{\partial v}{\partial z} + w\frac{\partial v}{\partial p},$$
$$c_{yq} = \frac{\partial u}{\partial y} + q\frac{\partial u}{\partial z} - w\frac{\partial u}{\partial p} - \frac{\partial w}{\partial q} + u\frac{\partial w}{\partial p}.$$

Knowing that

$$\begin{cases} d\omega^1 \equiv a_4\,\omega^3 \wedge \omega^4 + a_5\,\omega^3 \wedge \omega^5, \\ d\omega^2 \equiv b_4\,\omega^3 \wedge \omega^4 + b_5\,\omega^3 \wedge \omega^5 + \omega^4 \wedge \omega^5 \end{cases} \quad (\text{mod } \omega^1, \omega^2),$$

we have

$$d\omega^1 \wedge d\omega^2 \equiv 0 \quad \text{and} \quad d\omega^2 \wedge d\omega^2 \equiv 0 \quad (\text{mod } \omega^1, \omega^2).$$

Forming $d\omega^1 \wedge d\omega^2$ and considering the coefficients in front of $dx \wedge dy \wedge d\lambda \wedge dq$ and $dx \wedge dy \wedge d\mu \wedge dq$ we get

$$u\frac{\partial w}{\partial \lambda} - w\frac{\partial u}{\partial \lambda} + \frac{\partial v}{\partial \lambda} = 0 \quad \text{and} \quad u\frac{\partial w}{\partial \mu} - w\frac{\partial u}{\partial \mu} + \frac{\partial v}{\partial \mu} = 0. \qquad (*)$$

Considering the coefficients of $d\lambda \wedge d\mu \wedge dx \wedge dy$ and $d\lambda \wedge d\mu \wedge dy \wedge dq$ in $d\omega^2 \wedge d\omega^2 \equiv 0$, we next find that

$$\frac{\partial u/\partial \mu}{\partial u/\partial \lambda} = \frac{\partial v/\partial \mu}{\partial v/\partial \lambda} = \frac{\partial w/\partial \mu}{\partial w/\partial \lambda}.$$

With the common value being denoted by $\kappa = \kappa(x, y, z, p, q, \lambda, \mu)$, this means that the vector field $\partial/\partial\mu - \kappa\,\partial/\partial\lambda$ kills u, v and w. By the local rectification lemma for vector fields it is possible to replace λ, μ by new local coordinates $\bar{\lambda}$, $\bar{\mu}$ such that this vector field is simplified to $\partial/\partial\bar{\mu}$. Forgetting the bars we conclude that

> u, v, w *do not depend on* μ, *and hence we are reduced to a problem in the* **six** *variables* x, y, z, p, q, λ *only.*

Taking account of the coefficient of $dx \wedge dy \wedge d\lambda \wedge dq$ in $d\omega^2 \wedge d\omega^2$ we also deduce that

$$c_{xq} \frac{\partial w}{\partial \lambda} - c_{xy} \frac{\partial u}{\partial \lambda} - c_{yq} \frac{\partial v}{\partial \lambda} = 0.$$

In the generic case at least one of the functions u, v, w does depend on λ. Assuming e.g. that $\partial u/\partial \lambda \neq 0$, we can write

$$c_{xy} = \frac{\partial w/\partial \lambda}{\partial u/\partial \lambda} c_{xq} - \frac{\partial v/\partial \lambda}{\partial u/\partial \lambda} c_{yq}. \qquad (**)$$

Inserting this into the expression for $d\omega^2$, setting

$$\pi^3 := \frac{\partial u}{\partial \lambda} dq + \frac{\partial v}{\partial \lambda} dx + \frac{\partial w}{\partial \lambda} dy$$

and remembering that μ has disappeared, we find

$$d\omega^2 \equiv d\lambda \wedge \pi^3 + c_{xq} \left(dx \wedge dq + \frac{\partial w/\partial \lambda}{\partial u/\partial \lambda} dx \wedge dy \right)$$

$$+ c_{yq} \left(dy \wedge dq - \frac{\partial v/\partial \lambda}{\partial u/\partial \lambda} dx \wedge dy \right) \pmod{\omega^1, \omega^2}$$

$$= d\lambda \wedge \pi^3 + \frac{1}{\partial u/\partial \lambda} (c_{xq} \, dx + c_{yq} \, dy) \wedge \pi^3$$

$$= \pi^3 \wedge \left(-d\lambda - \frac{1}{\partial u/\partial \lambda} (c_{xq} \, dx + c_{yq} \, dy) \right).$$

Analogously

$$d\omega^1 \equiv (-u \, dx + dy) \wedge dq - w \, dx \wedge dy \pmod{\omega^1, \omega^2}$$

$$= \frac{-u \, dx + dy}{\partial u/\partial \lambda} \wedge \left(\pi^3 - \frac{\partial v}{\partial \lambda} dx - \frac{\partial w}{\partial \lambda} dy \right) - w \, dx \wedge dy$$

$$= \frac{-u \, dx + dy}{\partial u/\partial \lambda} \wedge \pi^3 + \frac{1}{\partial u/\partial \lambda} \left(u \frac{\partial w}{\partial \lambda} + \frac{\partial v}{\partial \lambda} - w \frac{\partial u}{\partial \lambda} \right) dx \wedge dy$$

$$= \pi^3 \wedge \frac{u \, dx - dy}{\partial u/\partial \lambda} \qquad \text{because of } (*).$$

Conclusion. *With*

$$\omega^1 = dz - p \, dx - q \, dy,$$

$$\omega^2 = dp + u \, dq + v \, dx + w \, dy,$$

$$\pi^3 = \frac{\partial u}{\partial \lambda} dq + \frac{\partial v}{\partial \lambda} dx + \frac{\partial w}{\partial \lambda} dy \qquad \left(\textit{where } \frac{\partial u}{\partial \lambda} \neq 0 \right),$$

there are further one-forms

$$\pi^4 := \frac{u\,dx - dy}{\partial u/\partial \lambda},$$

$$\pi^5 := -d\lambda - \frac{c_{xq}\,dx + c_{yq}\,dy}{\partial u/\partial \lambda}$$

such that

$$\begin{cases} d\omega^1 \equiv \pi^3 \wedge \pi^4, \\ d\omega^2 \equiv \pi^3 \wedge \pi^5 \end{cases} \quad (\mathrm{mod}\ \omega^1, \omega^2).$$

It follows in particular that the Cauchy characteristic vector field system of $\mathcal{Q} = (\omega^1, \omega^2)$ is given by

$$C(\mathcal{Q})^\perp = (\omega^1, \omega^2, \pi^3, \pi^4, \pi^5)^\perp,$$

so that $\mathcal{Q}$ admits one Cauchy characteristic vector field on $\mathbb{C}^6_{x,y,z,p,q,\lambda}$. Using this, $\mathcal{Q}$ can be reduced to a 2-dimensional pfaffian system in five variables.

In order to simplify things a bit we may use the assumption $\partial u/\partial \lambda \neq 0$ to replace the local coordinate λ by u. Or more brutally: identify u with the independent variable λ. Then $c_{xq} = -\partial v/\partial q + \lambda\,\partial v/\partial p$, $c_{yq} = -\partial w/\partial q + \lambda\,\partial w/\partial p$, and (*) goes over into

$$\frac{\partial v}{\partial \lambda} = w - \lambda\frac{\partial w}{\partial \lambda}.$$

If we regard this as a first order PDE for the unknowns v and w, w can firstly be chosen arbitrarily, whereupon v is obtained by a quadrature. Let us for instance set $w = \partial \psi/\partial \lambda$ for an arbitrary function $\psi(x, y, z, p, q, \lambda)$. Then

$$\frac{\partial v}{\partial \lambda} = \frac{\partial \psi}{\partial \lambda} - \lambda\frac{\partial^2 \psi}{\partial \lambda^2} = \frac{\partial}{\partial \lambda}\left(2\psi - \lambda\frac{\partial \psi}{\partial \lambda}\right),$$

so that

$$v = 2\psi - \lambda\frac{\partial \psi}{\partial \lambda}.$$

However, not only (*), but also (**) must be satisfied. This yields a complicated second order PDE for ψ:

$$2\frac{\partial \psi}{\partial y} + 2q\frac{\partial \psi}{\partial z} - 2\left(\frac{\partial \psi}{\partial \lambda} - \lambda\frac{\partial^2 \psi}{\partial \lambda^2}\right)\frac{\partial \psi}{\partial p} - 2\frac{\partial^2 \psi}{\partial \lambda^2}\frac{\partial \psi}{\partial q}$$

$$= \frac{\partial^2 \psi}{\partial x \partial \lambda} + \lambda\frac{\partial^2 \psi}{\partial y \partial \lambda} + (p + \lambda q)\frac{\partial^2 \psi}{\partial z \partial \lambda} - \left(2\psi - \lambda\frac{\partial \psi}{\partial \lambda}\right)\frac{\partial^2 \psi}{\partial p \partial \lambda} - \frac{\partial \psi}{\partial \lambda}\frac{\partial^2 \psi}{\partial q \partial \lambda}.$$

$$(\dagger)$$

Let us now summarize what we have done thus far!

Theorem 16.1.1. *Any parabolic PDE $\mathcal{S}_1 \subset J^2(\mathbb{C}^2_{xy}, \mathbb{C}_z)$ is equivalent to a 3-dimensional pfaffian system $\mathcal{P}_1 = (\omega^1, \omega^2, \omega^3)$ in seven variables, with $\mathcal{P}_1$ having the Lie structure*

$$\begin{cases} d\omega^1 \equiv 0, \\ d\omega^2 \equiv \omega^4 \wedge \omega^5, \\ d\omega^3 \equiv \omega^4 \wedge \omega^6 + \omega^4 \wedge \omega^7 \end{cases} \qquad (\mathrm{mod}\ \omega^1, \omega^2, \omega^3).$$

The Monge system $\mathcal{M}(\mathcal{P}_1) = (\omega^1, \omega^2, \omega^3, \omega^4, \omega^5)$ is complete if and only if ω^1 and ω^2 can be written as

$$\begin{cases} \omega^1 = dz - p\,dx - q\,dy, \\ \omega^2 = dp + \lambda\,dq + (2\psi - \lambda\,\partial\psi/\partial\lambda)\,dx + \partial\psi/\partial\lambda\,dy \end{cases}$$

in the six variables x, y, z, p, q, λ, where the function $\psi(x, y, z, p, q, \lambda)$ satisfies the second order PDE (†).

   Moreover $\mathcal{Q} = (\omega^1, \omega^2)$ has the Lie structure

$$\begin{cases} d\omega^1 \equiv \pi^3 \wedge \pi^4, \\ d\omega^2 \equiv \pi^3 \wedge \pi^5 \end{cases} \qquad (\mathrm{mod}\ \omega^1, \omega^2)$$

with

$$\pi^3 = dq + \left(\frac{\partial\psi}{\partial\lambda} - \lambda\frac{\partial^2\psi}{\partial\lambda^2} \right) dx + \frac{\partial^2\psi}{\partial\lambda^2}\,dy,$$

$$\pi^4 = \lambda\,dx - dy,$$

$$\pi^5 = -d\lambda + \left(2\frac{\partial\psi}{\partial q} - \lambda\frac{\partial^2\psi}{\partial q\partial\lambda} - 2\lambda\frac{\partial\psi}{\partial p} + \lambda^2\frac{\partial^2\psi}{\partial p\partial\lambda} \right) dx$$

$$+ \left(\frac{\partial^2\psi}{\partial q\partial\lambda} - \lambda\frac{\partial^2\psi}{\partial p\partial\lambda} \right) dy.$$

The structure equations reveal that the Cauchy characteristic system of $\mathcal{Q}$ is

$$C(\mathcal{Q}) = (\omega^1, \omega^2, \pi^3, \pi^4, \pi^5),$$

whence $\mathcal{Q}$ admits the Cauchy characteristic vector field C defined by $\omega^1 = \omega^2 = \pi^3 = \pi^4 = \pi^5 = 0$. Note in particular that $dy = \lambda\,dx$ on C.

   Because of this Cauchy characteristic vector field it is possible to reduce $\mathcal{Q}$ to a 2-dimensional pfaffian system in **five** *variables.*   □

Urgent question: Why pay attention to $Q = (\omega^1, \omega^2)$? In order to understand this we consider the problem of determining 2-dimensional integral
manifolds of Q on which $dx \wedge dy \neq 0$. On such an integral manifold

$$0 = \omega^1 = dz - p\,dx - q\,dy \Longleftrightarrow p = \frac{\partial z}{\partial x}, \quad q = \frac{\partial z}{\partial y},$$

and also

$$0 = \omega^2 = dp + \lambda\,dq + \left(2\psi - \lambda\frac{\partial\psi}{\partial\lambda}\right)dx + \frac{\partial\psi}{\partial\lambda}\,dy,$$

from which

$$\frac{\partial^2 z}{\partial x^2} + \lambda\frac{\partial^2 z}{\partial x \partial y} + 2\psi - \lambda\frac{\partial\psi}{\partial\lambda} = 0 \quad \text{and} \quad \frac{\partial^2 z}{\partial x \partial y} + \lambda\frac{\partial^2 z}{\partial y^2} + \frac{\partial\psi}{\partial\lambda} = 0,$$

or equivalently

$$\begin{cases} r + \lambda s + 2\psi - \lambda\,\partial\psi/\partial\lambda = 0, \\ s + \lambda t + \partial\psi/\partial\lambda = 0. \end{cases}$$

Eliminating λ we get a second order PDE $S_1^* \subset J^2(\mathbb{C}^2_{xy}, \mathbb{C}_z)$ with
the property that all 2-dimensional integral manifolds of Q on which
$dx \wedge dy \neq 0$ are 2-graphs of solutions of S_1^*, and vice versa.

But the point of departure for the considerations here was just a
second order PDE $S_1 \subset J^2(\mathbb{C}^2_{xy}, \mathbb{C}_z)$—namely a parabolic PDE with
complete Monge system. This PDE is equivalent to the pfaffian system
$\mathcal{P}_1 = (\omega^1, \omega^2, \omega^3)$ in the sense that the set of local solutions of S_1 coincides
with the set of local integral manifolds of $\mathcal{P}_1$ on which $dx \wedge dy \neq 0$. The
latter are clearly integral manifolds of $Q = (\omega^1, \omega^2)$ as well, and hence 2-
graphs of solutions of S_1^*. Letting sol(S) denote the set of local solutions
of a PDE system S, we thus see that

$$\mathrm{sol}\,(S_1) \subseteq \mathrm{sol}\,(S_1^*).$$

Assumption: S_1 *is locally solvable—that is, through each point of* $S_1 \subset$
$J^2(\mathbb{C}^2_{xy}, \mathbb{C}_z)$ *there passes a 2-graph of a local solution.*

Then

$$S_1 = \bigcup_{\mathrm{sol}(S_1)} \{\text{2-graphs of local solutions of } S_1\}$$

$$\subseteq \bigcup_{\mathrm{sol}(S_1^*)} \{\text{2-graphs of local solutions of } S_1^*\} \subseteq S_1^*.$$

so that $S_1 \subseteq S_1^*$, regarded as subsets of $J^2(\mathbb{C}^2_{xy}, \mathbb{C}_z)$. But S_1 and S_1^* are both hypersurfaces of this jet bundle. So if S_1^* is *irreducible* in a suitable sense, this inclusion must in fact be an equality: $S_1 = S_1^*$.

Theorem 16.1.2. *If we assume the parabolic PDE S_1 to be locally solvable and the PDE S_1^* to be irreducible, the problem of solving S_1 is equivalent to that of finding local integral manifolds of the pfaffian system $\mathcal{Q}$ on which $dx \wedge dy \neq 0$—and this explains the importance of $\mathcal{Q}$.* □

Remark. A point $s_1 \in S_1 \subset J^2(\mathbb{C}^2_{xy}, \mathbb{C}_z)$ represents a certain second order Taylor polynomial. S_1 is said to be *formally solvable* if for each $s_1 \in S_1$ the associated Taylor polynomial can be *prolonged to a formal power series solution* of S_1.

Now *local solvability* in this theorem can be replaced by the weaker *formal solvability*. For to each $s_1 \in S_1$ there corresponds a formal power series solution σ_1 of S_1, which automatically also is a formal solution of S_1^*. Considering σ_1 from the latter aspect and forgetting all terms of order greater than 2 we get a second order Taylor polynomial which represents a point $s_1^* \in S_1^*$. Thus $S_1 \subseteq S_1^*$. If S_1^* is irreducible it then follows that $S_1 = S_1^*$.

So there remains to integrate $\mathcal{Q}$. Regarding this as a pfaffian system on $\mathbb{C}^6_{xyzpq\lambda}$, it admits the Cauchy characteristic vector field C defined by $\omega^1 = \omega^2 = \pi^3 = \pi^4 = \pi^5 = 0$. Using this it suffices to take any vector field $X \in \mathcal{Q}^\perp$ which is not a multiple of C, and then let integral curves of C sweep through each point of an integral curve of X in order to generate a 2-dimensional integral manifold of $\mathcal{Q}$. And hence we are reduced to *solving ODE systems only*.

But the task of integrating $\mathcal{Q}$ can also be seen from quite a different point of view.

Let $\mathcal{P}$ be a general noncomplete pfaffian system. Integrating $\mathcal{P}$ is equivalent to finding *minimal complete* pfaffian systems $\mathcal{R} \supset \mathcal{P}$, where completeness means that $d\mathcal{R} \equiv 0 \pmod{\mathcal{R}}$. As a first step towards finding such $\mathcal{R}$ it is natural to look for minimal $^{(1)}\mathcal{P} \supset \mathcal{P}$ with the property $d\mathcal{P} \equiv 0 \pmod{^{(1)}\mathcal{P}}$. In general there are several different $^{(1)}\mathcal{P}$ of minimal dimension satisfying this.

Definition: If there is a *unique* $^{(1)}\mathcal{P}$ of minimal dimension satisfying $d\mathcal{P} \equiv 0 \pmod{^{(1)}\mathcal{P}}$, this $^{(1)}\mathcal{P}$ is denoted by $'\mathcal{P}$, and is called the *anti-derived system* of $\mathcal{P}$.

The next step consists in finding a minimal $^{(2)}\mathcal{P} \supset {}^{(1)}\mathcal{P}$ such that $d^{(1)}\mathcal{P} \equiv \pmod{^{(2)}\mathcal{P}}$—and then one continues in this way until finding a

$^{(m+1)}\mathcal{P}$ equal to its predecessor $^{(m)}\mathcal{P}$, in which case $d^{(m)}\mathcal{P} \equiv 0 \pmod{^{(m)}\mathcal{P}}$, so that $^{(m)}\mathcal{P}$ is complete.

Let us now specialize this to the pfaffian system $\mathcal{Q} = (\omega^1, \omega^2)$. The Lie structure

$$\begin{cases} d\omega^1 \equiv \pi^3 \wedge \pi^4, \\ d\omega^2 \equiv \pi^3 \wedge \pi^5 \end{cases} \pmod{\omega^1, \omega^2}$$

does exhibit a unique minimal pfaffian system $'\mathcal{Q} \supset \mathcal{Q}$ with the property $d\mathcal{Q} \equiv 0 \pmod{'\mathcal{Q}}$—namely $'\mathcal{Q} := (\omega^1, \omega^2, \pi^3)$.

Note that $d\mathcal{Q} \equiv 0 \pmod{(\omega^1, \omega^2, \pi^4, \pi^5)}$ too. However, the Cauchy characteristic vector field C satisfies $\pi^4 = 0$, that is, $dy = \lambda\, dx$. So if we want to find a vector field $X \in \mathcal{Q}^{\perp}$ which together with C generates 2-dimensional integral manifolds of $\mathcal{Q}$ on which $dx \wedge dy \neq 0$, we must require $\pi^4(X)$ to be nonzero. And therefore $(\omega^1, \omega^2, \pi^4, \pi^5)$ cannot be used for finding such integral manifolds.

Theorem 16.1.3. *It is possible to construct two-dimensional integral manifolds of $\mathcal{Q}$ by letting integral curves of the Cauchy characteristic vector field C sweep through all the points of integral curves of vector fields $X \in \mathcal{Q}^{\perp}$, transverse to C. In order to obtain integral manifolds on which $dx \wedge dy \neq 0$, the integral curves in question must be integral curves of $'\mathcal{Q} = (\omega^1, \omega^2, \pi^3)$, on which $\pi^4 \neq 0$.*

And thus the integration of $\mathcal{Q}$ is equivalent to that of $'\mathcal{Q}$! □

Let us recall the explicit expressions for ω^1, ω^2 and π^3:

$$\omega^1 = dz - p\, dx - q\, dy,$$

$$\omega^2 = dp + \lambda\, dq + \left(2\psi - \lambda\frac{\partial\psi}{\partial\lambda}\right) dx + \frac{\partial\psi}{\partial\lambda}\, dy,$$

$$\pi^3 = dq + \left(\frac{\partial\psi}{\partial\lambda} - \lambda\frac{\partial^2\psi}{\partial\lambda^2}\right) dx + \frac{\partial^2\psi}{\partial\lambda^2}\, dy.$$

Since

$$\omega^2 - \lambda\pi^3 = dp + \left(2\psi - 2\lambda\frac{\partial\psi}{\partial\lambda} + \lambda^2\frac{\partial^2\psi}{\partial\lambda^2}\right) dx + \left(\frac{\partial\psi}{\partial\lambda} - \lambda\frac{\partial^2\psi}{\partial\lambda^2}\right) dy,$$

$'Q$ is also generated by the three one-forms

$$\omega^1 = dz - p\,dx - q\,dy,$$

$$\pi^2 = dp - \left(-2\psi + 2\lambda\frac{\partial\psi}{\partial\lambda} - \lambda^2\frac{\partial^2\psi}{\partial\lambda^2}\right)dx - \left(-\frac{\partial\psi}{\partial\lambda} + \lambda\frac{\partial^2\psi}{\partial\lambda^2}\right)dy,$$

$$\pi^3 = dq - \left(-\frac{\partial\psi}{\partial\lambda} + \lambda\frac{\partial^2\psi}{\partial\lambda^2}\right)dx - \left(-\frac{\partial^2\psi}{\partial\lambda^2}\right)dy.$$

But the system generated by these clearly coincides with the restriction of the contact pfaffian system of $J^2(\mathbb{C}^2_{xy}, \mathbb{C}_z)$ to the 6-dimensional submanifold S_2 defined by eliminating λ from the three equations

$$r = -2\psi + 2\lambda\frac{\partial\psi}{\partial\lambda} - \lambda^2\frac{\partial^2\psi}{\partial\lambda^2},$$

$$s = -\frac{\partial\psi}{\partial\lambda} + \lambda\frac{\partial^2\psi}{\partial\lambda^2},$$

$$t = -\frac{\partial^2\psi}{\partial\lambda^2}$$

—i.e., $'Q$ is the pfaffian system associated to the PDE system S_2.

Since it is possible to consider Q as a pfaffian system in five variables only, the same is true for the intrinsically defined $'Q$. So regarding the latter as living on the 6-dimensional S_2, we deduce that $'Q$ admits a Cauchy characterisitic vector field there.

Summary. Under certain general assumptions we have the following equivalences:

> solving the parabolic PDE $S_1 \subset J^2(\mathbb{C}^2_{xy}, \mathbb{C}_z)$, which has a complete Monge system $\Longleftrightarrow$ integrating $P_1 = (\omega^1, \omega^2, \omega^3) \Longleftrightarrow$ integrating $Q = (\omega^1, \omega^2) \Longleftrightarrow$ integrating $'Q = (\omega^1, \omega^2, \omega^3) \Longleftrightarrow$ solving the system $S_2 \subset J^2(\mathbb{C}^2_{xy}, \mathbb{C}_z)$, admitting a Cauchy characteristic vector field.

As regards $'Q = (\omega^1, \omega^2, \omega^3)$ there are two different cases to consider.

Special case: $'Q$ is complete. Then all vector fields in $('Q)^\perp$ are of course Cauchy characteristic. Note conversely that if we can find just one more Cauchy characteristic vector field of $('Q)^\perp$ besides the one already given, $'Q$ can be reduced to a 3-dimensional pfaffian system in four variables, and as such it is automatically complete.

Keeping track of $d\lambda$ in the expressions for $d\pi^2$ and $d\pi^3$ we have

$$d\pi^2 = \lambda^2\frac{\partial^3\psi}{\partial\lambda^3}\,d\lambda \wedge dx - \lambda\frac{\partial^3\psi}{\partial\lambda^3}\,d\lambda \wedge dy + \cdots,$$

$$d\pi^3 = -\lambda\frac{\partial^3\psi}{\partial\lambda^3}\,d\lambda \wedge dx + \frac{\partial^3\psi}{\partial\lambda^3}\,d\lambda \wedge dy + \cdots,$$

where $\cdots$ denotes terms not involving $d\lambda$. Consequently

$$d\lambda \text{ does not occur in } d\omega^1, d\pi^2, d\pi^3 \iff \frac{\partial^3 \psi}{\partial \lambda^3} = 0.$$

So if ψ, regarded as a function of λ, is a polynomial of degree ≤ 2, then $\partial/\partial\lambda$ is Cauchy characteristic for $'Q$, and thus the latter is complete. Assuming this to be the case we set

$$\psi(x, y, z, p, q, \lambda) := \frac{1}{2}C\lambda^2 + B\lambda + \frac{1}{2}A,$$

where A, B and C are functions of x, y, z, p, q making ψ satisfy the complicated PDE ($\dagger$).

The parabolic PDE $\mathcal{S}_1$ corresponding to Q is obtained by eliminating λ from the system

$$\begin{cases} r + \lambda s + 2\psi - \lambda\,\partial\psi/\partial\lambda = 0, \\ s + \lambda t + \partial\psi/\partial\lambda = 0, \end{cases}$$

which here reduces to

$$\begin{cases} r + \lambda s + B\lambda + A = 0, \\ s + \lambda t + C\lambda + B = 0. \end{cases}$$

Thus

$$\lambda = -\frac{r+A}{s+B} = -\frac{s+B}{t+C},$$

and so $(r + A)(t + C) - (s + B)^2 = 0$, *which is a PDE of Monge–Ampère type.*

Conversely it is not hard to see that if $\mathcal{S}_1$ is of this form, the corresponding pfaffian system $'Q$ is indeed complete.

Theorem 16.1.4. $'Q$ *is complete* $\iff$ *the associated parabolic PDE is given by* $(r + A)(t + C) - (s + B)^2 = 0$, *where A, B and C are certain functions of x, y, z, p, q.* $\square$

Regarding $'Q = (\omega^1, \omega^2, \pi^3)$ as a 3-dimensional complete pfaffian system in five variables $x^1, \ldots, x^5$, we can choose the latter such that

$$'Q = (dx^1, dx^2, dx^3).$$

Then the generators ω^1 and ω^2 of Q can be written as

$$\omega^1 = \sum_{i=1}^{3} a_i^1(x^1, \ldots, x^5)\,dx^i \quad \text{and} \quad \omega^2 = \sum_{i=1}^{3} a_i^2(x^1, \ldots, x^5)\,dx^i.$$

By using Gaussian elimination and changing indices if necessary these can be replaced by

$$dx^2 - a(x^1, \ldots, x^5)\, dx^1 \quad \text{and} \quad dx^3 - b(x^1, \ldots, x^5)\, dx^1.$$

In the generic case $dx^1 \wedge dx^2 \wedge dx^3 \wedge da \wedge db \ne 0$, which makes it possible to introduce a and b as new local coordinates instead of x^4 and x^5—i.e., we may assume that $a = x^4$ and $b = x^5$. The pfaffian equations to be solved are then

$$\begin{cases} dx^2 - x^4\, dx^1 = 0, \\ dx^3 - x^5\, dx^1 = 0. \end{cases}$$

The general integral curve of Q on which $dx^1 \ne 0$ is accordingly given by

$$x^2 = f(x^1), \quad x^4 = f'(x^1); \quad x^3 = g(x^1), \quad x^5 = g'(x^1),$$

where f and g are arbitrary functions of one variable.

General case: $'Q$ *is not complete.* In this case the Lie structure of $'Q = (\omega^1, \omega^2, \pi^3)$ can be simplified to

$$\begin{cases} d\omega^1 \equiv 0, \\ d\omega^2 \equiv 0, \qquad (\mathrm{mod}\ \omega^1, \omega^2, \pi^3). \\ d\pi^3 \equiv \pi^4 \wedge \pi^5 \end{cases}$$

Note in particular that

$$('Q)' = \{\theta \in {}'Q \mid d\theta \equiv 0 \pmod{'Q}\} = (\omega^1, \omega^2) = Q,$$

so Q is recovered from $'Q$ by derivation.

This is the interesting case, to be discussed further in the next chapter.

The parabolic PDEs associated to pfaffian systems of this kind are called *Goursat equations* by Cartan—so as to honour the initiator of the study of parabolic (as well as hyperbolic) PDEs.

16.2 Pfaffian systems of three and two dimensions in five variables

The study of parabolic PDEs in the last section led to the consideration of 3- and 2-dimensional pfaffian systems in five variables. Here we make a *structural classification* of such systems—that is, we identify systems with the same Lie structure, and determine a simple canonical form for each equivalence class.

Let us start with a 3-dimensional pfaffian system $\mathcal{P}$ on $\mathbb{C}^5$. We choose one-forms $\omega^1,\ldots,\omega^5$ such that $\mathcal{P} = (\omega^1, \omega^2, \omega^3)$, and $\{\omega^1,\ldots,\omega^5\}$ is a local coframe for $\mathbb{C}^5$.

The very simplest case occurs if $\mathcal{P}$ is *complete*. According to the Frobenius theorem it is then possible to find local coordinates $x^1,\ldots,x^5$ for $\mathbb{C}^5$ such that $\mathcal{P} = (dx^1, dx^2, dx^3)$.

The next simplest case is that when $\dim \mathcal{P}^{(i)} = \dim \mathcal{P}^{(i-1)} - 1$ for $i = 1, 2, \ldots$, until one eventually arrives at a complete system. This has been treated in section **6.7**—although from a dual vector field point of view. The main result of that section is:

Let $\mathcal{V}$ be a 2-dimensional vector field system on $\mathbb{C}^n$ with $C(\mathcal{V}) = \{0\}$, $\dim \mathcal{V}^{(i)} = \dim \mathcal{V}^{(i-1)} + 1$ for $i = 1, \ldots, l+1$, and $\mathcal{V}^{(l+1)}$ being complete. Then there are local coordinates $x, x^0, x^1, \ldots, x^{n-2}$ for $\mathbb{C}^n$ such that

$$\mathcal{V} = \left(\frac{\partial}{\partial x} + x^1 \frac{\partial}{\partial x^0} + \cdots + x^{l+1} \frac{\partial}{\partial x^l}, \frac{\partial}{\partial x^{l+1}} \right).$$

For a one-form $\theta = a\, dx + a_0\, dx^0 + \cdots + a_{n-2}\, dx^{n-2}$ belonging to the dual pfaffian system $\mathcal{P} = \mathcal{V}^{\perp}$, $a_{l+1} = 0$ and $a + a_0 x^1 + \cdots + a_l x^{l+1} = 0$. Thus

$$\theta = a_0(dx^0 - x^1\, dx) + a_1(dx^1 - x^2\, dx) + \cdots + a_l(dx^l - x^{l+1}\, dx)$$
$$+ a_{l+2}\, dx^{l+2} + \cdots + a_{n-2}\, dx^{n-2},$$

so that

$$\mathcal{P} = (dx^0 - x^1\, dx, dx^1 - x^2\, dx, \ldots, dx^l - x^{l+1}\, dx, dx^{l+2}, \ldots, dx^{n-2}).$$

Note that

$$\begin{cases} d(dx^0 - x^1\, dx) = dx \wedge dx^1 \equiv dx \wedge x^2\, dx = 0, \\ \cdots \qquad\qquad\qquad\qquad\qquad\qquad\qquad\qquad\text{(mod } \mathcal{P}\text{)}, \\ d(dx^{l-1} - x^l\, dx) = dx \wedge dx^l \equiv dx \wedge x^{l+1}\, dx = 0 \end{cases}$$

while $d(dx^l - x^{l+1}\, dx) \not\equiv 0 \pmod{\mathcal{P}}$. Therefore

$$\mathcal{P}' = (dx^0 - x^1\, dx, \ldots, dx^{l-1} - x^l\, dx, dx^{l+2}, \ldots, dx^{n-2}).$$

Analogously

$$\mathcal{P}'' = (dx^0 - x^1\, dx, \ldots, dx^{l-2} - x^{l-1}\, dx, dx^{l+2}, \ldots, dx^{n-2}),$$

and continuing in this way one finally obtains the complete

$$\mathcal{P}^{(l+1)} = (dx^{l+2}, \ldots, dx^{n-2}).$$

Here we are interested in what happens when $n = 5$, $\dim \mathcal{P} = 3$ and $l = 0, 1$ or 2:

- $l = 0 \iff \dim \mathcal{P}' = 2$ and $\mathcal{P}'$ is complete; then $\mathcal{P} = (dx^0 - x^1 \, dx,$ $dx^2, dx^3)$;
- $l = 1 \iff \dim \mathcal{P}' = 2$, $\dim \mathcal{P}'' = 1$ and $\mathcal{P}''$ is complete; then $\mathcal{P} = (dx^0 - x^1 \, dx, dx^1 - x^2 \, dx, dx^3)$;
- $l = 2 \iff \dim \mathcal{P}' = 2$, $\dim \mathcal{P}'' = 1$ and $\dim \mathcal{P}''' = 0$; then $\mathcal{P} = (dx^0 - x^1 \, dx, dx^1 - x^2 \, dx, dx^2 - x^3 \, dx)$.

Let us now return to a general 3-dimensional pfaffian system $\mathcal{P} = (\omega^1, \omega^2, \omega^3)$ on $\mathbb{C}^5$. Its Lie structure can be written as

$$\begin{cases} d\omega^1 \equiv A\,\omega^4 \wedge \omega^5, \\ d\omega^2 \equiv B\,\omega^4 \wedge \omega^5, \qquad (\mathrm{mod}\ \omega^1, \omega^2, \omega^3). \\ d\omega^3 \equiv C\,\omega^4 \wedge \omega^5 \end{cases}$$

1. Suppose first that $A = B = C = 0$. Then $\mathcal{P}$ is complete, and can consequently be brought to the form $\mathcal{P} = (dx^1, dx^2, dx^3)$.

2. Next assume that at least one of the coefficients is nonzero—for instance, $C \neq 0$. If we replace $\omega^1, \omega^2, \omega^3$ by

$$\bar{\omega}^1 := \omega^1 - \frac{A}{C}\,\omega^3, \quad \bar{\omega}^2 := \omega^2 - \frac{B}{C}\,\omega^3, \quad \bar{\omega}^3 := \frac{1}{C}\,\omega^3,$$

and forget the bars afterwards, the structure is simplified to

$$\begin{cases} d\omega^1 \equiv 0, \\ d\omega^2 \equiv 0, \qquad (\mathrm{mod}\ \omega^1, \omega^2, \omega^3). \\ d\omega^3 \equiv \omega^4 \wedge \omega^5 \end{cases}$$

Therefore $\mathcal{P}' = (\omega^1, \omega^2)$ with $\dim \mathcal{P}' = 2 = \dim \mathcal{P} - 1$.

Let us then pay attention to $\mathcal{P}'$. Its Lie structure

$$\begin{cases} d\omega^1 \equiv u\,\omega^3 \wedge \omega^4 + b\,\omega^3 \wedge \omega^5, \\ d\omega^2 \equiv a'\,\omega^3 \wedge \omega^4 + b'\,\omega^3 \wedge \omega^5 \end{cases} \qquad (\mathrm{mod}\ \omega^1, \omega^2)$$

gives rise to three subcases:

a. $a = b = a' = b' = 0$. Then $\mathcal{P}' = (\omega^1, \omega^2)$ is complete. So this is the case $l = 0$ above, and hence there are local coordinates x, x^0, x^1, x^2, x^3 for $\mathbb{C}^5$ such that

$$\mathcal{P} = (dx^0 - x^1 \, dx, dx^2, dx^3).$$

b. In this case a, b, a', b' are not all zero, but $ab' - a'b = 0$. Then by a linear transformation of ω^1 and ω^2 it can be arranged that $a = b = a' = 0$, $b' = 1$, whence the Lie structure of $\mathcal{P}'$ takes the form

$$\begin{cases} d\omega^1 \equiv 0, \\ d\omega^2 \equiv \omega^3 \wedge \omega^4 \end{cases} \quad (\text{mod } \omega^1, \omega^2).$$

Thus $\mathcal{P}'' = (\omega^1)$, and so $\dim \mathcal{P}'' = 1 = \dim \mathcal{P}' - 1$.

If $\mathcal{P}''$ is complete we have the case $l = 1$ above: there are local coordinates x, x^0, x^1, x^2, x^3 such that

$$\mathcal{P} = (dx^0 - x^1 \, dx, dx^1 - x^2 \, dx, dx^3).$$

If $\mathcal{P}''$ is not complete, necessarily $\mathcal{P}''' = \{0\}$, and this corresponds to the case $l = 2$ above:

$$\mathcal{P} = (dx^0 - x^1 \, dx, dx^1 - x^2 \, dx, dx^2 - x^3 \, dx).$$

c. $ab' - ba' \neq 0$. If we replace the ω^i by suitable linear combinations we may assume that $a = b' = 1$ and $b = a' = 0$, so that the structure of $\mathcal{P}'$ is given by

$$\begin{cases} d\omega^1 \equiv \omega^3 \wedge \omega^4, \\ d\omega^2 \equiv \omega^3 \wedge \omega^5 \end{cases} \quad (\text{mod } \omega^1, \omega^2).$$

Then $\mathcal{P}'' = \{0\}$—i.e., $\dim \mathcal{P}'' = \dim \mathcal{P}' - 2$. Hence the theory in section **6.7** does not apply to this case.

Conclusion. *The only case not covered by section **6.7** (and therefore the most interesting) is when $\mathcal{P} = (\omega^1, \omega^2)$ has the structure*

$$\begin{cases} d\omega^1 \equiv \omega^3 \wedge \omega^4 \quad (\text{mod } \omega^1, \omega^2), \\ d\omega^2 \equiv \omega^3 \wedge \omega^5 \quad (\text{mod } \omega^1, \omega^2), \\ d\omega^3 \equiv \omega^4 \wedge \omega^5 \quad (\text{mod } \omega^1, \omega^2, \omega^3). \end{cases}$$

In this case $\mathcal{P}' = (\omega^1, \omega^2)$, while $\mathcal{P}'' = \{0\}$.

Let us now consider a 2-dimensional pfaffian system $\mathcal{Q} = (\omega^1, \omega^2)$ on $\mathbb{C}^5$. Its Lie structure can be written as

$$\begin{cases} d\omega^1 \equiv A \, \omega^3 \wedge \omega^4 + B \, \omega^3 \wedge \omega^5 + C \, \omega^4 \wedge \omega^5, \\ d\omega^2 \equiv A' \, \omega^3 \wedge \omega^4 + B' \, \omega^3 \wedge \omega^5 + C' \, \omega^4 \wedge \omega^5 \end{cases} \quad (\text{mod } \omega^1, \omega^2)$$

for certain functions A, B, C, A', B', C' on $\mathbb{C}^5$.

There are three main cases to consider.

1. Suppose that each coefficient is zero. Then Q is *complete*, and can therefore be brought to the form $Q = (dx^1, dx^2)$.

2. Suppose that at least one coefficient is nonvanishing, but that $AB' - BA' = AC' - CA' = BC' - CB' = 0$. In this case it is possible to replace the ω^i by suitable linear combinations such that the coefficients are transformed into $A = B = C = 0$ and $A' = 1$. Then $d\omega^1 \equiv 0$ (mod ω^1, ω^2), while $d\omega^2 \equiv \omega^3 \wedge \omega^4 + B'\,\omega^3 \wedge \omega^5 + C'\,\omega^4 \wedge \omega^5 = (\omega^3 - C'\,\omega^5) \wedge (\omega^4 + B'\,\omega^5)$ (mod ω^1, ω^2). So modifying ω^3 and ω^4, the structure can be simplified to

$$\begin{cases} d\omega^1 \equiv 0, \\ d\omega^2 \equiv \omega^3 \wedge \omega^4 \end{cases} \qquad (\text{mod } \omega^1, \omega^2).$$

As a consequence $C(Q) = (\omega^1, \omega^2, \omega^3, \omega^4)$, showing that Q *admits one Cauchy characteristic vector field*. Therefore Q may be considered as a 2-dimensional pfaffian system in four variables, with $\dim Q^\perp = 2$—and once again it is possible to apply the results from section **6.7**.

Clearly $Q' = (\omega^1)$, so that $\dim Q' = 1 = \dim Q - 1$. There are two subcases:

- if Q' is *complete*, the canonical form for Q is

$$Q = (dx^0 - x^1\, dx, dx^1);$$

- if Q' is *not complete*, necessarily $Q'' = \{0\}$, and then $\dim Q'' = 0 = \dim Q' - 1$; hence the canonical form for Q is

$$Q = (dx^0 - x^1\, dx, dx^1 - x^2\, dx).$$

3. Suppose finally that at least one of the determinants $AB' - BA'$, $AC' - CA'$ and $BC' - CB'$ is nonzero—for instance that $AB' - BA' \neq 0$. Then ω^3, ω^4 and ω^5 may be chosen such that $A = B' = 1$, while the other coefficients vanish. Thus the Lie structure of Q simplifies to

$$\begin{cases} d\omega^1 \equiv \omega^3 \wedge \omega^4, \\ d\omega^2 \equiv \omega^3 \wedge \omega^5 \end{cases} \qquad (\text{mod } \omega^1, \omega^2).$$

Note that Q has a well-defined anti-derived system in this case—namely $'Q = (\omega^1, \omega^2, \omega^3)$.

Again there are two subcases.

- $'Q$ *is complete*: By the last section Q has the canonical form

$$Q = (dx^0 - x^1\,dx, dx^2 - x^3\,dx) \quad \text{with} \quad 'Q = (dx^0, dx^2, dx).$$

- $'Q$ *is not complete*: Then $d\omega^3 \equiv D\,\omega^4 \wedge \omega^5$ (mod $'Q$) for a nonzero D—which moreover can be replaced by 1 by a suitable choice of ω^4 and ω^5. Hence Q has the structure

$$\begin{cases} d\omega^1 \equiv \omega^3 \wedge \omega^4 \pmod{\omega^1, \omega^2}, \\ d\omega^2 \equiv \omega^3 \wedge \omega^5 \pmod{\omega^1, \omega^2}, \\ d\omega^3 \equiv \omega^4 \wedge \omega^5 \pmod{\omega^1, \omega^2, \omega^3} \end{cases}$$

—and this is the general case.

Important remark. The Lie structure of the *general* 3-dimensional pfaffian system $\mathcal{P} = (\omega^1, \omega^2, \omega^3)$ in five variables is

$$\begin{cases} d\omega^1 \equiv 0, \\ d\omega^2 \equiv 0, \\ d\omega^3 \equiv \omega^4 \wedge \omega^5 \end{cases} \quad (\text{mod } \omega^1, \omega^2, \omega^3),$$

with $\mathcal{P}' = (\omega^1, \omega^2)$ having the structure

$$\begin{cases} d\omega^1 \equiv \omega^3 \wedge \omega^4, \\ d\omega^2 \equiv \omega^3 \wedge \omega^5 \end{cases} \quad (\text{mod } \omega^1, \omega^2).$$

The Lie structure of the *general* 2-dimensional pfaffian system $Q = (\omega^1, \omega^2)$ in five variables is

$$\begin{cases} d\omega^1 \equiv \omega^3 \wedge \omega^4, \\ d\omega^2 \equiv \omega^3 \wedge \omega^5 \end{cases} \quad (\text{mod } \omega^1, \omega^2)$$

with $'Q = (\omega^1, \omega^2, \omega^3)$ having the structure

$$\begin{cases} d\omega^1 \equiv 0, \\ d\omega^2 \equiv 0, \\ d\omega^3 \equiv \omega^4 \wedge \omega^5 \end{cases} \quad (\text{mod } \omega^1, \omega^2, \omega^3).$$

Hence there is a one-to-one correspondence between general 3-

dimensional pfaffian systems and general 2-dimensional pfaffian systems in five variables, given by derivation and anti-derivation:

$$\text{3-dimensional systems} \longleftrightarrow \text{2-dimensional systems}$$

$$\mathcal{P} \quad \longrightarrow \quad \mathcal{P}'$$

$$'\mathcal{Q} \quad \longleftarrow \quad \mathcal{Q}.$$

16.3 Systems of two PDEs having a Cauchy characteristic vector field

When studying parabolic PDEs in section **16.1** we were led to consider *systems of two PDEs, admitting a Cauchy characteristic vector field*—let us therefore take a closer look at such systems here.

Assuming $\mathcal{S}_2 \subset J^2(\mathbb{C}^2_{xy}, \mathbb{C}_z)$ to be given by

$$\mathcal{S}_2: \quad \begin{cases} r = \rho(x, y, z, p, q, t), \\ s = \sigma(x, y, z, p, q, t), \end{cases}$$

the corresponding pfaffian system $\mathcal{P}_2$ in the six variables x, y, z, p, q, t is generated by the one-forms

$$\begin{cases} \omega^1 := dz - p\, dx - q\, dy, \\ \omega^2 := dp - \rho\, dx - \sigma\, dy, \\ \omega^3 := dq - \sigma\, dx - t\, dy. \end{cases}$$

Then $d\omega^1 \equiv 0 \pmod{\omega^1, \omega^2, \omega^3}$, $d\omega^2 = dx \wedge dp + dy \wedge d\sigma$ and $d\omega^3 = dx \wedge d\sigma + dy \wedge dt$, where for instance

$$d\rho = \frac{\partial\rho}{\partial x}\, dx + \frac{\partial\rho}{\partial y}\, dy + \frac{\partial\rho}{\partial z}\, dz + \frac{\partial\rho}{\partial p}\, dp + \frac{\partial\rho}{\partial q}\, dq + \frac{\partial\rho}{\partial t}\, dt$$

$$\equiv \left(\frac{\partial\rho}{\partial x} + p\frac{\partial\rho}{\partial z} + \rho\frac{\partial\rho}{\partial p} + \sigma\frac{\partial\rho}{\partial q} \right) dx$$

$$+ \left(\frac{\partial\rho}{\partial y} + q\frac{\partial\rho}{\partial z} + \sigma\frac{\partial\rho}{\partial p} + t\frac{\partial\rho}{\partial q} \right) dy + \frac{\partial\rho}{\partial t}\, dt$$

$$= \frac{d\rho}{dx}\, dx + \frac{d\rho}{dy}\, dy + \frac{\partial\rho}{\partial t}\, dt \quad (\text{mod } \omega^1, \omega^2, \omega^3),$$

with

$$\frac{d}{dx} := \frac{\partial}{\partial x} + p\frac{\partial}{\partial z} + \rho\frac{\partial}{\partial p} + \sigma\frac{\partial}{\partial q} \quad \text{and} \quad \frac{d}{dy} := \frac{\partial}{\partial y} + q\frac{\partial}{\partial z} + \sigma\frac{\partial}{\partial p} + t\frac{\partial}{\partial q}.$$

Calculating $d\sigma$ in the same way we get

$$d\omega^2 \equiv dx \wedge \left(\frac{\partial\rho}{\partial t}\, dt + \frac{d\rho}{dy}\, dy \right) + dy \wedge \left(\frac{\partial\sigma}{\partial t}\, dt + \frac{d\sigma}{dx}\, dx \right).$$

Treating $d\omega^3$ analogously, the Lie structure of $\mathcal{P}_2$ is found to be

$$d\omega^1 \equiv 0,$$

$$d\omega^2 \equiv \frac{\partial\rho}{\partial t}\, dx \wedge dt + \frac{\partial\sigma}{\partial t}\, dy \wedge dt + \left(\frac{d\rho}{dy} - \frac{d\sigma}{dx}\right) dx \wedge dy,$$

$$d\omega^3 \equiv \frac{\partial\sigma}{\partial t}\, dx \wedge dt + dy \wedge dt + \frac{d\sigma}{dy}\, dx \wedge dy,$$

modulo ω^1, ω^2 and ω^3. We require $\mathcal{P}_2$ to admit a Cauchy characteristic vector field—so that its reduction $(\mathcal{P}_2)_{\text{red}}$ depends on five variables only. Hence there must be a coframe $\{\omega^1,\ldots,\omega^6\}$ for $\mathbb{C}^6$ simplifying the Lie structure to

$$\begin{cases} d\omega^1 \equiv 0, \\ d\omega^2 \equiv A\,\omega^4 \wedge \omega^5, \qquad (\mathrm{mod}\ \omega^1,\omega^2,\omega^3), \\ d\omega^3 \equiv B\,\omega^4 \wedge \omega^5 \end{cases}$$

with the Cauchy characteristic vector field being defined by $\omega^1 = \omega^2 = \omega^3 = \omega^4 = \omega^5 = 0$.

A necessary and sufficient condition for this to happen is that $d\omega^2$ and $d\omega^3$ are linearly dependent modulo $(\omega^1, \omega^2, \omega^3)$, i.e.,

$$\frac{\partial\rho/\partial t}{\partial\sigma/\partial t} = \frac{\partial\sigma/\partial t}{1} = \frac{d\rho/dy - d\sigma/dx}{d\sigma/dy},$$

or

$$\frac{\partial\rho}{\partial t} = \left(\frac{\partial\sigma}{\partial t}\right)^2 \quad \text{and} \quad \frac{d\rho}{dy} - \frac{d\sigma}{dx} = \frac{\partial\sigma}{\partial t}\frac{d\sigma}{dy}.$$

Then $d\omega^2 \equiv (\partial\sigma/\partial t)\,d\omega^3$, where

$$d\omega^3 \equiv \frac{\partial\sigma}{\partial t}\, dx \wedge dt + dy \wedge dt + \frac{d\sigma}{dy}\, dx \wedge dy \quad (\mathrm{mod}\ \omega^1,\omega^2,\omega^3)$$

$$= \left(dy + \frac{\partial\sigma}{\partial t}\, dx\right) \wedge \left(dt - \frac{d\sigma}{dy}\, dx\right).$$

So setting

$$\omega^4 := dy + \frac{\partial\sigma}{\partial t}\, dx \quad \text{and} \quad \omega^5 := dt - \frac{d\sigma}{dy}\, dx$$

we get

$$d\omega^2 \equiv \frac{\partial\sigma}{\partial t}\, \omega^4 \wedge \omega^5 \quad \text{and} \quad d\omega^3 \equiv \omega^4 \wedge \omega^5 \quad (\mathrm{mod}\ \omega^1,\omega^2,\omega^3).$$

In this way the Lie structure of $\mathcal{P}_2$ is simplified to

$$\begin{cases} d\omega^1 \equiv 0, \\ d\omega^2 \equiv (\partial\sigma/\partial t)\,\omega^4 \wedge \omega^5, & (\mathrm{mod}\ \omega^1,\omega^2,\omega^3), \\ d\omega^3 \equiv \omega^4 \wedge \omega^5 \end{cases}$$

and the Cauchy characteristic vector field is given by

$$\begin{aligned} C := &\frac{\partial}{\partial x} - \frac{\partial\sigma}{\partial t}\frac{\partial}{\partial y} + \left(p - q\frac{\partial\sigma}{\partial t}\right)\frac{\partial}{\partial z} \\ &+ \left(\rho - \sigma\frac{\partial\sigma}{\partial t}\right)\frac{\partial}{\partial p} + \left(\sigma - t\frac{\partial\sigma}{\partial t}\right)\frac{\partial}{\partial q} + \frac{d\sigma}{dy}\frac{\partial}{\partial t}. \end{aligned}$$

Because

$$d\left(\omega^2 - \frac{\partial\sigma}{\partial t}\,\omega^3\right) \equiv 0 \quad (\mathrm{mod}\ \omega^1,\omega^2,\omega^3),$$

the structure is further simplified if ω^2 is replaced by $\omega^2 - (\partial\sigma/\partial t)\,\omega^3$. Accordingly the definite choice of our ω^i is

$$\begin{cases} \omega^1 = dz - p\,dx - q\,dy, \\ \omega^2 = dp - \rho\,dx - \sigma\,dy - (\partial\sigma/\partial t)(dq - \sigma\,dx - t\,dy), \\ \omega^3 = dq - \sigma\,dx - t\,dy, \\ \omega^4 = dy + (\partial\sigma/\partial t)\,dx, \\ \omega^5 = dt - (d\sigma/dy)\,dx, \end{cases}$$

with $\mathcal{P}_2 = (\omega^1,\omega^2,\omega^3)$ having the structure

$$\begin{cases} d\omega^1 \equiv 0, \\ d\omega^2 \equiv 0, & (\mathrm{mod}\ \omega^1,\omega^2,\omega^3). \\ d\omega^3 \equiv \omega^4 \wedge \omega^5 \end{cases}$$

From this we obtain the derived system

$$\mathcal{P}'_2 = (\omega^1,\omega^2)$$

—and the next task is to determine the Lie structure of $\mathcal{P}'_2$.

Now $\mathcal{P}'_2$ cannot possibly be complete, since $d\omega^1 = dx \wedge dp + dy \wedge dq$ and $dy \wedge dq \not\equiv 0$ $(\mathrm{mod}\ \omega^1,\omega^2)$. This leaves two possibilities for $d\omega^1$ and $d\omega^2$:

- either they are linearly dependent $(\mathrm{mod}\ \omega^1,\omega^2)$,
- or they are not.

Because $d\omega^1$ does not contain dt, it is a multiple of $\omega^3 \wedge \omega^4$ (mod ω^1, ω^2). So in the first case also $d\omega^2$ is free from dt, and hence is a multiple of $\omega^3 \wedge \omega^4$ (mod ω^1, ω^2). Let us therefore collect those terms in $d\omega^2$ which do contain dt:

$$d\omega^2 = dt \wedge \left(-\frac{\partial\rho}{\partial t} dx - \frac{\partial\sigma}{\partial t} dy - \frac{\partial^2\sigma}{\partial t^2} (dq - \sigma\, dx - t\, dy) \right)$$

$$+ \left(\frac{\partial\sigma}{\partial t} \right)^2 dt \wedge dx + \frac{\partial\sigma}{\partial t} dt \wedge dy + \cdots$$

$$= -\frac{\partial^2\sigma}{\partial t^2} dt \wedge (dq - \sigma\, dx - t\, dy) + \cdots \qquad \text{since } \frac{\partial\rho}{\partial t} = \left(\frac{\partial\sigma}{\partial t} \right)^2.$$

Consequently

$$d\omega^1 \text{ and } d\omega^2 \text{ are linearly dependent} \quad (\text{mod } \omega^1, \omega^2) \iff \frac{\partial^2\sigma}{\partial t^2} = 0.$$

Special case: $\partial^2\sigma/\partial t^2 = 0$. Then σ can be expressed as

$$\sigma = a(x, y, z, p, q)t + b(x, y, z, p, q),$$

whereupon

$$\frac{\partial\rho}{\partial t} = \left(\frac{\partial\sigma}{\partial t} \right)^2 = a^2(x, y, z, p, q),$$

implying that

$$\rho = a^2(x, y, z, p, q)t + \text{ a function of } x, y, z, p, q$$
$$= a(x, y, z, p, q)\sigma + c(x, y, z, p, q),$$

so that the system S_2 is *quasi-linear* in this case.

Not containing dt, $d\omega^1$ and $d\omega^2$ do not involve $d\omega^5 = dt - (d\sigma/dy)\, dy$ either, and therefore the Lie structure of P'_2 can be written as

$$\begin{cases} d\omega^1 \equiv A\, \omega^3 \wedge \omega^4, \\ d\omega^2 \equiv B\, \omega^3 \wedge \omega^4 \end{cases} \quad (\text{mod } \omega^1, \omega^2).$$

But then P'_2 admits the Cauchy charateristic vector field defined by $\omega^1 = \omega^2 = \omega^3 = \omega^4 = 0$, and can thus be reduced to a *2-dimensional pfaffian system in four variables*. From the preceding section we know that the latter are divided into two equivalence classes:

- either $P'_2 = (dx^0 - x^1\, dx, dx^2)$,
- or $P'_2 = (dx^0 - x^1\, dx, dx^1 - x^2\, dx)$.

The first of these is characterized by the presence of a nonconstant first integral—namely x^2.

Let us therefore investigate whether $\mathcal{P}'_2 = (\omega^1, \omega^2)$ admits a first integral, i.e., if there is a linear combination $\omega^2 + u\,\omega^1$ which is *exact*—and since we are only interested in local properties, this is equivalent to being *closed*. Because

$$d(\omega^2 + u\,\omega^1) \equiv d\omega^2 + u\,d\omega^1 = d\omega^2 + u\,(dx \wedge dp + dy \wedge dq) \pmod{\omega^1, \omega^2},$$

a *necessary* condition for $d(\omega^2 + u\,\omega^1)$ to vanish is that $dy \wedge dq$ disappears from this expression. With $\sigma = at + b$ and $\rho = a\sigma + c = a^2 t + ab + c$,

$$\omega^2 = dp - \rho\,dx - \sigma\,dy - \frac{\partial \sigma}{\partial t}(dq - \sigma dx - t\,dy)$$

$$= dp - a\,dq - c\,dx - b\,dy.$$

Therefore

$$d\omega^2 = dq \wedge da + dx \wedge dc + dy \wedge db$$

$$= dq \wedge \left(\frac{\partial a}{\partial x} dx + \frac{\partial a}{\partial y} dy + \frac{\partial a}{\partial z} dz + \frac{\partial a}{\partial p} dp \right) + dx \wedge dc$$

$$+ dy \wedge \left(\frac{\partial b}{\partial x} dx + \frac{\partial b}{\partial z} dz + \frac{\partial b}{\partial p} dp + \frac{\partial b}{\partial q} dq \right)$$

$$\equiv dq \wedge \left(\cdots \frac{\partial a}{\partial y} dy + q \frac{\partial a}{\partial z} dy + b \frac{\partial a}{\partial p} dy + \cdots \right)$$

$$+ dy \wedge \left(\cdots + a \frac{\partial b}{\partial y} dq + \frac{\partial b}{\partial q} dq + \cdots \right) \pmod{\omega^1, \omega^2}$$

$$= dy \wedge dq \left(-\frac{\partial a}{\partial y} - q \frac{\partial a}{\partial z} - b \frac{\partial a}{\partial p} + a \frac{\partial b}{\partial p} + \frac{\partial b}{\partial q} \right) + \cdots,$$

and so there is but one choice for u:

$$u = \frac{\partial a}{\partial y} + q \frac{\partial a}{\partial z} + b \frac{\partial a}{\partial p} - a \frac{\partial h}{\partial p} - \frac{\partial h}{\partial q}.$$

a. *If* $d(\omega^2 + u\,\omega^1) = 0$, there is a locally defined function x^3 such that $\omega^2 + u\,\omega^1 = dx^3$. The canonical form for the 3-dimensional pfaffian system $\mathcal{P}_2 = (\omega^1, \omega^2, \omega^3)$ is in this case

$$\mathcal{P}_2 = (dx^0 - x^1\,dx, dx^1 - x^2\,dx, dx^3),$$

with the derived system $\mathcal{P}'_2 = (dx^0 - x^1\,dx, dx^3)$.

The general integral curve of $\mathcal{P}_2$ on which $dx \neq 0$ is given by

$$x^0 = f(x), \quad x^1 = f'(x), \quad x^2 = f''(x), \quad x^3 = \text{constant},$$

where f is an arbitrary function in one variable.

b. *If* $d(\omega^2 + u\,\omega^1) \neq 0$, the canonical form for $\mathcal{P}_2 = (\omega^1, \omega^2, \omega^3)$ is

$$\mathcal{P}_2 = (dx^0 - x^1\,dx,\, dx^1 - x^2\,dx,\, dx^2 - x^3\,dx)$$

with the derived system $\mathcal{P}_2' = (dx^0 - x^1\,dx,\, dx^1 - x^2\,dx)$.

The general integral curve of $\mathcal{P}_2$ on which $dx \neq 0$ is given by

$$x^0 = f(x),\ \ x^1 = f'(x),\ \ x^2 = f''(x),\ \ x^3 = f'''(x),$$

with f being an arbitrary function in one variable.

General case: $\partial^2 \sigma / \partial t^2 \neq 0$. Here $d\omega^1$ and $d\omega^2$ are linearly independent (mod ω^1, ω^2). The results of the preceding section then imply that $\mathcal{P}_2 = (\omega^1, \omega^2, \omega^3)$ has the following structure:

$$\begin{cases} d\omega^1 \equiv \omega^3 \wedge \omega^4 \pmod{\omega^1, \omega^2}, \\ d\omega^2 \equiv \omega^3 \wedge \omega^5 \pmod{\omega^1, \omega^2}, \\ d\omega^3 \equiv \omega^4 \wedge \omega^5 \pmod{\omega^1, \omega^2, \omega^3}. \end{cases}$$

So this is the interesting case!

Summary. The problems of solving a *general* parabolic PDE with complete Monge system and solving a *general* system of two PDEs, admitting a Cauchy characteristic vector field, both reduce to the problem of finding integral curves of a 3-dimensional pfaffian system $\mathcal{P} = (\omega^1, \omega^2, \omega^3)$ in five variables, having the structure

$$\begin{cases} d\omega^1 \equiv \omega^3 \wedge \omega^4 \pmod{\omega^1, \omega^2}, \\ d\omega^2 \equiv \omega^3 \wedge \omega^5 \pmod{\omega^1, \omega^2}, \\ d\omega^3 \equiv \omega^4 \wedge \omega^5 \pmod{\omega^1, \omega^2, \omega^3}. \end{cases}$$

Finding such integral curves is quite easy from a general point of view: just take any vector field $X = a\,\partial/\partial\omega^4 + b\,\partial/\partial\omega^5 \in \mathcal{P}^\perp$, and determine its integral curves by solving ODE systems.

But the main point of Cartan's five variable paper is that we *should try to avoid the general local existence theorem for solutions of ODE systems*.

To do so we first declare that two 3-dimensional pfaffian systems $\mathcal{P}$ and $\hat{\mathcal{P}}$ in five variables with the above structure are *locally equivalent* if there are local equivalences $\phi : \mathbb{C}^5_x \xrightarrow{\cong} \mathbb{C}^5_{\hat{x}}$ such that $\phi^*(\hat{\mathcal{P}}) = \mathcal{P}$ and $(\phi^{-1})^*(\mathcal{P}) = \hat{\mathcal{P}}$. Then we solve the equivalence problem for such systems, and determine canonical forms for the different equivalence classes. Finally we sincerely hope that it is possible to

- transform the pfaffian systems to their canonical forms, and then
- solve these canonical forms,

without using the general existence theorem for local solutions of ODE systems.

17

The equivalence problem for general 3-dimensional pfaffian systems in five variables

Let $\mathcal{P} = (\omega^1, \omega^2, \omega^3)$ be a pfaffian system on $\mathbb{C}_x^5$ which is *general* in the sense that its structure can be written as

$$\begin{cases} d\omega^1 \equiv \omega^3 \wedge \omega^4 \pmod{\omega^1, \omega^2}, \\ d\omega^2 \equiv \omega^3 \wedge \omega^5 \pmod{\omega^1, \omega^2}, \\ d\omega^3 \equiv \omega^4 \wedge \omega^5 \pmod{\omega^1, \omega^2, \omega^3}, \end{cases}$$

and let $\hat{\mathcal{P}} = (\hat{\omega}^1, \hat{\omega}^2, \hat{\omega}^3)$ be an analogous pfaffian system on $\mathbb{C}_{\hat{x}}^5$. In this chapter we will study *local equivalences of $\mathcal{P}$ and $\hat{\mathcal{P}}$*, i.e., local diffeomorphisms $\phi : \mathbb{C}_x^5 \xrightarrow{\cong} \mathbb{C}_{\hat{x}}^5$ satisfying

$$\phi^*(\hat{\mathcal{P}}) = \mathcal{P} \quad \text{and} \quad (\phi^{-1})^*(\mathcal{P}) = \hat{\mathcal{P}}.$$

Note that if ϕ is one such local equivalence, the most general is of the form $\phi \circ \sigma$ or $\hat{\sigma} \circ \phi$, where σ and $\hat{\sigma}$ are general elements of the symmetry groups of $\mathcal{P}$ and $\hat{\mathcal{P}}$ respectively. Because of this the symmetry group sym($\mathcal{P}$) (or group of self-equivalences) of $\mathcal{P}$ will play a dominant role in the following.

The matrix group associated to this equivalence problem should really depend on twelve parameters, but the one presented in the five variable paper [Cartan 1910] depends only on ten. Accepting the latter point of view it is then possible to make a final reduction to a *matrix group depending on seven parameters*.

Hence the lifted equivalence problem involves seven auxiliary variables. When prolonging this two new auxiliary variables emerge, so the prolonged problem depends on $5 + 7 + 2 = 14$ variables.

From an abstract point of view the solution of the equivalence problem as presented in chapter **15** is straightforward and even elegant—but in this concrete case the formulas get terribly complicated already after one prolongation, and it seems hopeless to try to follow the general solution scheme. Fortunately the action of the stability algebra on the coefficients in the structure equations gives rise to a certain *homogeneous polynomial* $\mathcal{F}$ *of degree four in two variables*, and from this a preliminary *rough classification* is deduced.

To wit, if the two variables are interpreted as homogeneous coordinates of the Riemann sphere $\mathbb{P}$, there are five different possibilities for the roots of $\mathcal{F}$ on $\mathbb{P}$:

1. $\mathcal{F}$ has four simple roots.
2. $\mathcal{F}$ vanishes identically.
3. $\mathcal{F}$ has a root of multiplicity 4.
4. $\mathcal{F}$ has one triple and one simple root.
5. $\mathcal{F}$ has two double roots.

If $\mathcal{P}$ and $\hat{\mathcal{P}}$ are locally equivalent, the associated polynomials $\mathcal{F}$ and $\hat{\mathcal{F}}$ necessarily have the same types of roots. In this way one gets five *rough equivalence classes*, and it suffices to solve the equivalence problem within each one of these.

1. In this case—which is the general one—it is possible to get rid of all the nine auxiliary variables, and we are reduced to an *e*-structure in five variables.

Then dim sym($\mathcal{P}$) = 5−(number of functionally independent invariants).

2. This gives rise to an *e*-structure in 14 variables, with *no invariants*—and therefore there is only one local equivalence class. The 14-dimensional sym($\mathcal{P}$) turns out to coincide with the exceptional simple Lie group G_2.

The canonical form for the associated PDE system $\mathcal{S}_2$ is

$$\mathcal{S}_2^{\mathrm{can}}: \quad s = \frac{1}{2}t^2, \ r = \frac{1}{3}t^3.$$

The reduction of a general $\mathcal{S}_2$ in the equivalence class to this simple form is performed by solving a Lie equation associated to G_2 (recall that we discussed Lie equations in chapter **10**).

If a particular solution of $\mathcal{S}_2^{\mathrm{can}}$ is known, the general solution is obtained by solving a Riccati equation and by quadratures.

For the remaining three rough equivalence classes Cartan is content with finding *e-structures in at least six variables*. The results are as follows.

3. Here we are reduced to an *e*-structure in seven variables, and sym($\mathcal{P}$) is either a 7-dimensional or a 6-dimensional Lie group.

Examples of PDE systems $\mathcal{S}_2$ belonging to this rough equivalence class are

$$s = t^m \text{ and } r = \frac{m^2}{2m-1} t^{2m-1} \quad \text{for } m \neq 2, -1, \frac{1}{3}, \frac{2}{3}.$$

4. In this case there are *e*-structures only in less than six variables.

5. Here we are reduced to an *e*-structure in six variables. As regards sym($\mathcal{P}$) there are three subcases:

(i) sym($\mathcal{P}$) is a direct product of two 3-dimensional simple Lie groups,
(ii) sym($\mathcal{P}$) is a direct product of a 3-dimensional simple Lie group and a 3-dimensional solvable Lie group,
(iii) sym($\mathcal{P}$) is isomorphic to the group of Euclidean motions in $\mathbb{C}^3$.

A somewhat surprising consequence is that all symmetry groups appearing here are *finite dimensional*—and hence our study of Lie pseudogroups seems rather pointless. But in fact we do need the more general infinite dimensional frame in order to arrive at the results above.

Remark. The main purpose of this monograph is to point out the importance of

- rewriting PDE systems as vector field or pfaffian systems,
- deriving the Lie structures of the latter,
- using the structure equations for finding singular vector fields—in particular, Cauchy and Monge characteristic vector fields—and then utilizing these in order to construct integral manifolds.

However, in following this route a quite different item has repeatedly cropped up without originally having been asked for—namely Lie groups.

Conversely it would of course have been possible to use groups as a starting point instead: given a pfaffian system $\mathcal{P}$ one deduces its symmetry group sym($\mathcal{P}$), and then derives integration methods from the properties of sym($\mathcal{P}$). And this is more or less what is done in the very active field of modern mathematics called 'group analysis of differential equations' (see e.g. [Olver 1986] and [Ibragimow 1994–96]).

Now when using Cartan's method for solving equivalence problems for pfaffian systems one automatically obtains symmetry groups of these systems. It is therefore natural to use group theoretic methods in order to simplify the integration process as much as possible. But because of the vast literature on this subject, there is no need to discuss that matter here.

We also refrain from explaining standard facts from Lie group theory such as semi-simplicty, solvability, Cartan's criteria for these, and so on—again since information concerning such matters is easily available in the modern literature.

17.1 The naturally associated homogeneous polynomial $\mathcal{F}$ of degree four in two variables

Let $\{\omega^1,\ldots,\omega^5\}$ be a local coframe for $\mathbb{C}^5_x$, let $\mathcal{P} = (\omega^1, \omega^2, \omega^3)$ be a 3-dimensional pfaffian system, and let $\hat{\mathcal{P}} = (\hat{\omega}^1, \hat{\omega}^2, \hat{\omega}^3)$ be an analogous system on $\mathbb{C}^5_{\hat{x}}$. Then $\phi : \mathbb{C}^5_x \xrightarrow{\cong} \mathbb{C}^5_{\hat{x}}$ is a local equivalence of $\mathcal{P}$ and $\hat{\mathcal{P}}$ if and only if

$$
\phi^* \begin{pmatrix} \hat{\omega}^1 \\ \hat{\omega}^2 \\ \hat{\omega}^3 \\ \hat{\omega}^4 \\ \hat{\omega}^5 \end{pmatrix} = \begin{pmatrix} u_1^1 & u_2^1 & u_3^1 & 0 & 0 \\ u_1^2 & u_2^2 & u_3^2 & 0 & 0 \\ u_1^3 & u_2^3 & u_3^3 & 0 & 0 \\ u_1^4 & u_2^4 & u_3^4 & \alpha & \beta \\ u_1^5 & u_2^5 & u_3^5 & \gamma & \delta \end{pmatrix} \begin{pmatrix} \omega^1 \\ \omega^2 \\ \omega^3 \\ \omega^4 \\ \omega^5 \end{pmatrix}, \quad \text{or } \phi^*(\hat{\omega}) = M_{19}\, \omega,
$$

with a matrix involving 19 parameters. To get the lifted equivalence problem we then define the new one-forms

$$
\bar{\omega}^1 := \sum_{i=1}^{3} u_i^1 \omega^i, \quad \bar{\omega}^2 := \sum_{i=1}^{3} u_i^2 \omega^i, \quad \bar{\omega}^3 := \sum_{i=1}^{3} u_i^3 \omega^i,
$$

$$
\bar{\omega}^4 := \sum_{i=1}^{3} u_i^4 \omega^i + \alpha \omega^4 + \beta \omega^5, \quad \bar{\omega}^5 := \sum_{i=1}^{3} u_i^5 \omega^i + \gamma \omega^4 + \delta \omega^5
$$

on $\mathbb{C}^5_x \times \mathbb{C}^{19}$.

Next we require $\mathcal{P}$ and $\hat{\mathcal{P}}$ to have the structures of *general* 3-dimensional pfaffian systems in five variables, i.e.,

$$
\begin{cases} d\omega^1 \equiv \omega^3 \wedge \omega^4 \pmod{\omega^1, \omega^2}, \\ d\omega^2 \equiv \omega^3 \wedge \omega^5 \pmod{\omega^1, \omega^2}, \\ d\omega^3 \equiv \omega^4 \wedge \omega^5 \pmod{\omega^1, \omega^2, \omega^3}, \end{cases}
$$

and similarly for $\hat{\mathcal{P}}$. In order for ϕ to respect this structure, we must for instance have

$$
d(\phi^* \hat{\omega}^3) = \phi^* d\hat{\omega}^3 \equiv \phi^* \omega^4 \wedge \phi^* \omega^5, \quad \text{or} \quad d\bar{\omega}^3 \equiv \bar{\omega}^4 \wedge \bar{\omega}^5.
$$

Since

$$
\begin{cases} \bar{\omega}^4 \equiv \alpha \omega^4 + \beta \omega^5, \\ \bar{\omega}^5 \equiv \gamma \omega^4 + \delta \omega^5 \end{cases} \pmod{\omega^1, \omega^2, \omega^3},
$$

this implies that

$$d\bar{\omega}^3 \equiv (\alpha\delta - \beta\gamma)\,\omega^4 \wedge \omega^5 \equiv (\alpha\beta - \gamma\delta)\,d\omega^3 \quad (\mathrm{mod}\ \omega^1, \omega^2, \omega^3).$$

Therefore

$$\bar{\omega}^3 = u_1^3\,\omega^1 + u_2^3\,\omega^2 + \Delta\,\omega^3,$$

where $\Delta := \alpha\delta - \beta\gamma$. In the same way

$$\begin{aligned}
d\bar{\omega}^1 &\equiv \bar{\omega}^3 \wedge \bar{\omega}^4 \equiv \Delta\omega^3 \wedge (u_3^4\,\omega^3 + \alpha\,\omega^4 + \beta\,\omega^5) \\
&\equiv \Delta(\alpha\,d\omega^1 + \beta\,d\omega^2) \quad (\mathrm{mod}\ \omega^1, \omega^2),
\end{aligned}$$

which shows that

$$\bar{\omega}^1 = \Delta\alpha\,\omega^1 + \Delta\beta\omega^2.$$

With analogous calculations revealing that $\bar{\omega}^2 = \Delta\gamma\,\omega^1 + \Delta\delta\,\omega^2$, we end up with a 12-parameter matrix group $\mathfrak{M}_{12}$:

$$\begin{pmatrix} \bar{\omega}^1 \\ \bar{\omega}^2 \\ \bar{\omega}^3 \\ \bar{\omega}^4 \\ \bar{\omega}^5 \end{pmatrix} = \begin{pmatrix} \Delta\alpha & \Delta\beta & 0 & 0 & 0 \\ \Delta\gamma & \Delta\delta & 0 & 0 & 0 \\ u_1^3 & u_2^3 & \Delta & 0 & 0 \\ u_1^4 & u_2^4 & u_3^4 & \alpha & \beta \\ u_1^5 & u_2^5 & u_3^5 & \gamma & \delta \end{pmatrix} \begin{pmatrix} \omega^1 \\ \omega^2 \\ \omega^3 \\ \omega^4 \\ \omega^5 \end{pmatrix}, \quad \text{or} \quad \bar{\omega} = \mathsf{M}_{12}\,\omega.$$

Thus the $\bar{\omega}^i$ live on $\mathbb{C}_x^5 \times \mathbb{C}^{12}$—so there are 12 auxiliary variables.

The next task is to derive the structure equations for these $\bar{\omega}^i$. In principle this is straightforward: differentiate the above expressions for $\bar{\omega}^i$, write the $d\omega^j$ as

$$\begin{cases}
d\omega^1 = \omega^1 \wedge \left(\sum_{k=2}^5 c_{1k}^1\,\omega^k\right) + \omega^2 \wedge \left(\sum_{k=3}^5 c_{2k}^1\,\omega^k\right) + \omega^3 \wedge \omega^4, \\
\ldots \\
d\omega^5 = \omega^1 \wedge \left(\sum_{k=2}^5 c_{1k}^5\,\omega^k\right) + \omega^2 \wedge \left(\sum_{k=3}^5 c_{2k}^5\,\omega^k\right) \\
\qquad\quad + \omega^3 \wedge \left(\sum_{k=4}^5 c_{3k}^5\,\omega^k\right) + \omega^4 \wedge \omega^5,
\end{cases}$$

and finally express the ω^j as linear combinations of the $\bar{\omega}^i$ by $\omega = \mathsf{M}_{12}^{-1}\,\bar{\omega}$.

The trouble is that it turns out to be extremely tricky to rediscover the formulas indicated by Cartan—forgetting the bars, he claims that the wanted structure equations are given by

$$
\begin{cases}
d\omega^1 = \omega^1 \wedge (2\pi^1 + \pi^4) + \omega^2 \wedge \pi^2 + \omega^3 \wedge \omega^4, \\
d\omega^2 = \omega^1 \wedge \pi^3 + \omega^2 \wedge (\pi^1 + 2\pi^4) + \omega^3 \wedge \omega^5, \\
d\omega^3 = \omega^1 \wedge \pi^5 + \omega^2 \wedge \pi^6 + \omega^3 \wedge (\pi^1 + \pi^4) + \omega^4 \wedge \omega^5, \\
d\omega^4 = \omega^1 \wedge \pi^7 + \omega^2 \wedge \pi^8 + \tfrac{4}{3}\omega^3 \wedge \pi^6 + \omega^4 \wedge \pi^1 + \omega^5 \wedge \pi^2, \\
d\omega^5 = \omega^1 \wedge \pi^9 + \omega^2 \wedge \pi^{10} - \tfrac{4}{3}\omega^3 \wedge \pi^5 + \omega^4 \wedge \pi^3 + \omega^5 \wedge \pi^4,
\end{cases}
$$

where each π^i is a differential of an auxiliary variable plus a suitable linear combination of the ω^i. With 12 auxiliary variables there thus should be 12 one-forms π^i—but there are only 10, which feels a bit awkward.

Fortunately this obscurity has been cleared up in [Hsiao 1980], which contains page after page of very explicit calculations.

Fact. In order to obtain the structure equations of Cartan it is necessary to assume that

$$
u_3^4 = \frac{4}{3\Delta}(\alpha u_2^3 - \beta u_1^3) \quad \text{and} \quad u_3^5 = \frac{4}{3\Delta}(\gamma u_2^3 - \delta u_1^3),
$$

in which case our matrix group depends on 10 parameters only: $\mathfrak{M} = \mathfrak{M}_{10}$.

This means that Cartan considers not the *maximal* symmetry group, but a somewhat smaller subgroup, and hence the corresponding equivalence relation might be somewhat more restricted than originally intended.

But however that may be, in the following we stick to Cartan's classification.

There is a simplification of the structure equations which is explained in detail in the third section of [Hsiao 1980]: using some tricks it may be assumed that $\pi^8 = \pi^9 = 0$ and $\pi^{10} = \pi^7$. Hence the definite structure equations for the lifted equivalence problem are

$$
\begin{cases}
d\omega^1 = \omega^1 \wedge (2\pi^1 + \pi^4) + \omega^2 \wedge \pi^2 + \omega^3 \wedge \omega^4, \\
d\omega^2 = \omega^1 \wedge \pi^3 + \omega^2 \wedge (\pi^1 + 2\pi^4) + \omega^3 \wedge \omega^5, \\
d\omega^3 = \omega^1 \wedge \pi^5 + \omega^2 \wedge \pi^6 + \omega^3 \wedge (\pi^1 + \pi^4) + \omega^4 \wedge \omega^5, \qquad (1) \\
d\omega^4 = \omega^1 \wedge \pi^7 + \tfrac{4}{3}\omega^3 \wedge \pi^6 + \omega^4 \wedge \pi^1 + \omega^5 \wedge \pi^2, \\
d\omega^5 = \omega^2 \wedge \pi^7 - \tfrac{4}{3}\omega^3 \wedge \pi^5 + \omega^4 \wedge \pi^3 + \omega^5 \wedge \pi^4.
\end{cases}
$$

Since there only are seven π^i here, the corresponding matrix group should depend on just seven parameters. And in fact the matrix group yielding these structure equations is given by

$$
\begin{pmatrix}
\Delta\alpha & \Delta\beta & 0 & 0 & 0 \\[4pt]
\Delta\gamma & \Delta\delta & 0 & 0 & 0 \\[4pt]
u_1 & u_2 & \Delta & 0 & 0 \\[4pt]
\dfrac{\alpha u_3}{\Delta} + \dfrac{2u_1(\alpha u_2 - \beta u_1)}{3\Delta^2} & \dfrac{\beta u_3}{\Delta} + \dfrac{2u_2(\alpha u_2 - \beta u_1)}{3\Delta^2} & \dfrac{4(\alpha u_2 - \beta u_1)}{3\Delta} & \alpha & \beta \\[10pt]
\dfrac{\gamma u_3}{\Delta} + \dfrac{2u_1(\gamma u_2 - \delta u_1)}{3\Delta^2} & \dfrac{\delta u_3}{\Delta} + \dfrac{2u_2(\gamma u_2 - \delta u_1)}{3\Delta^2} & \dfrac{4(\gamma u_2 - \delta u_1)}{3\Delta} & \gamma & \delta
\end{pmatrix}
$$

in the seven auxiliary variables $\alpha, \beta, \gamma, \delta, u_1, u_2, u_3$.

So the original 12-parameter matrix group has in two steps been reduced to one in seven parameters:

$$
\mathfrak{M}_{12} \xrightarrow{\text{mistake(?)}} \mathfrak{M}_{10} \xrightarrow[\text{reduction}]{\text{honest}} \mathfrak{M}_7.
$$

The most general linear replacements $\pi^i \mapsto \pi^i +$ (linear combination of the ω^i) which leave (1) invariant are

$$
\pi^1 \mapsto \pi^1 + v_1\,\omega^1, \quad \pi^2 \mapsto \pi^2 + v_2\,\omega^1, \quad \pi^3 \mapsto \pi^3 + v_1\,\omega^2,
$$
$$
\pi^4 \mapsto \pi^4 + v_2\,\omega^2, \quad \pi^5 \mapsto \pi^5 + v_1\,\omega^3, \quad \pi^6 \mapsto \pi^6 + v_2\,\omega^3,
$$
$$
\pi^7 \mapsto \pi^7 + v_1\,\omega^4 + v_2\,\omega^5,
$$

where v_1 and v_2 are arbitrary parameters. In accordance with the general theory of section **15.2** we therefore replace the π^i by the corresponding right hand sides, and consider v_1 and v_2 as two more auxiliary variables.

Thus we now have $7+2 = 9$ auxiliary variables, and also 12 one-forms $\omega^1, \ldots, \omega^5, \pi^1, \ldots, \pi^7$ living on $\mathbb{C}_x^5 \times \mathbb{C}_{\text{aux}}^9 = \mathbb{C}^{14}$.

Recall that the matrix group $\mathfrak{M}_7$ induces a Lie algebra on $\mathbb{C}^7_{\alpha\beta\gamma\delta u_1 u_2 u_3}$, which in turn induces the stability algebra $\text{span}_{\mathbb{C}}\{E_1, \ldots, E_7\}$ acting on $\omega^1, \ldots, \omega^5$. If we let $\delta := \sum_{i=1}^7 t^i E_i$ be the general element of this algebra, the theory in section **14.4** applied to (1) shows that in our case

$$
\begin{cases}
\delta\omega^1 = (2t^1 + t^4)\,\omega^1 + t^2\,\omega^2, \\[4pt]
\delta\omega^2 = t^3\,\omega^1 + (t^1 + 2t^4)\,\omega^2, \\[4pt]
\delta\omega^3 = t^5\,\omega^1 + t^6\,\omega^2 + (t^1 + t^4)\,\omega^3, \\[4pt]
\delta\omega^4 = t^7\,\omega^1 + \tfrac{4}{3}t^6\,\omega^3 + t^1\,\omega^4 + t^2\,\omega^5, \\[4pt]
\delta\omega^5 = t^7\,\omega^2 - \tfrac{4}{3}t^5\,\omega^3 + t^3\,\omega^4 + t^4\,\omega^5.
\end{cases}
\tag{2}
$$

As explained in section **14.6** this δ has a unique prolongation acting on $\omega^1,\ldots,\omega^5,\pi^1,\ldots,\pi^7$ such that $d\circ\delta = \delta\circ d$. Because of the new auxiliary variables v_1 and v_2 the prolonged stability algebra is 9-dimensional, $\mathrm{span}\,_{\mathbb{C}}\{E_1,\ldots,E_9\}$, and is induced by a 9-dimensional matrix group $\mathfrak{M}_9$.

Setting $\delta(1) = d(2)$ we infer that the prolonged δ acts on the π^j by

$$
\begin{cases}
\delta\pi^1 = & -dt^1 + t^2\,\pi^3 - t^3\,\pi^2 + \tfrac{1}{3}t^7\,\omega^3 - \tfrac{2}{3}t^5\,\omega^4 + \tfrac{1}{3}t^6\,\omega^5 + t^8\,\omega^1,\\[2pt]
\delta\pi^2 = & -dt^2 + (t^1 - t^4)\,\pi^2 - t^2\,(\pi^1 - \pi^4) - t^6\,\omega^4 + t^9\,\omega^1,\\[2pt]
\delta\pi^3 = & -dt^3 + (t^4 - t^1)\,\pi^3 - t^3\,(\pi^4 - \pi^1) - t^5\,\omega^5 + t^8\,\omega^2,\\[2pt]
\delta\pi^4 = & -dt^4 + t^3\,\pi^2 - t^2\,\pi^3 + \tfrac{1}{3}t^7\,\omega^3 + \tfrac{1}{3}t^5\,\omega^4 - \tfrac{2}{3}t^6\,\omega^5 + t^9\,\omega^2,\\[2pt]
\delta\pi^5 = & -dt^5 + t^5\,\pi^1 - t^1\,\pi^5 + t^6\,\pi^3 - t^3\,\pi^6 - t^7\,\omega^5 + t^8\,\omega^3,\\[2pt]
\delta\pi^6 = & -dt^6 + t^5\,\pi^2 - t^2\,\pi^5 + t^6\,\pi^4 - t^4\,\pi^6 + t^7\,\omega^4 + t^9\,\omega^3,\\[2pt]
\delta\pi^7 = & -dt^7 + t^7\,(\pi^1 + \pi^4) - (t^1 + t^4)\,\pi^7 + \tfrac{4}{3}t^6\,\pi^5 - \tfrac{4}{3}t^5\,\pi^6\\[2pt]
& +t^8\,\omega^4 + t^9\,\omega^5,
\end{cases}
\tag{3}
$$

where t^8,t^9 are the coefficients of the new vector fields E_8, E_9.

The involutivity criterion applied to (1) shows that the pfaffian system $(\omega^1,\ldots,\omega^5)$ is not involutive, and therefore we have to make a prolongation (as explained in section **14.6**) by adding $\pi^1,\ldots,\pi^7$—involving the two new auxiliary variables v_1 and v_2. To derive the structure equations for the prolonged system $(\omega^1,\ldots,\omega^5,\pi^1,\ldots,\pi^7)$ we differentiate (1) so as to get

$$
0 = d^2\omega^i = \cdots \qquad \text{for } i = 1,\ldots,5,
$$

where the right hand sides contain the wanted $d\pi^1,\ldots,d\pi^7$. Considering the latter as unknowns we then seek the *most general solutions*, which from a computational point of view is quite complicated—as the similar case concerning the third fundamental theorem in section **14.3**. The result is as follows, with

$$
\chi^k := dv_k + \text{ a suitable linear combination of } \omega^1,\ldots,\omega^5,\pi^1,\ldots,\pi^7:
$$

$$
\begin{cases}
d\pi^1 = & \pi^3 \wedge \pi^2 + \tfrac{1}{3}\omega^3 \wedge \pi^7 - \tfrac{2}{3}\omega^4 \wedge \pi^5 + \tfrac{1}{3}\omega^5 \wedge \pi^6 + \omega^1 \wedge \chi^1 \\
& +2B_2\,\omega^1 \wedge \omega^3 + B_3\,\omega^2 \wedge \omega^3 + 2A_2\,\omega^1 \wedge \omega^4 + 2A_3\,\omega^1 \wedge \omega^5 \\
& +A_3\,\omega^2 \wedge \omega^4 + A_4\,\omega^2 \wedge \omega^5, \\[6pt]

d\pi^2 = & \pi^2 \wedge (\pi^1 - \pi^4) - \omega^4 \wedge \pi^6 + \omega^1 \wedge \chi^2 + B_4\,\omega^2 \wedge \omega^3 \\
& +A_4\,\omega^2 \wedge \omega^4 + A_5\,\omega^2 \wedge \omega^5, \\[6pt]

d\pi^3 = & \pi^3 \wedge (\pi^4 - \pi^1) - \omega^5 \wedge \pi^5 + \omega^2 \wedge \chi^1 - B_1\,\omega^1 \wedge \omega^3 \\
& -A_1\,\omega^1 \wedge \omega^4 - A_2\,\omega^1 \wedge \omega^5, \\[6pt]

d\pi^4 = & \pi^2 \wedge \pi^3 + \tfrac{1}{3}\omega^3 \wedge \pi^7 + \tfrac{1}{3}\omega^4 \wedge \pi^5 - \tfrac{2}{3}\omega^5 \wedge \pi^6 + \omega^2 \wedge \chi^2 \\
& -B_2\,\omega^1 \wedge \omega^3 - 2B_3\,\omega^2 \wedge \omega^3 - A_2\,\omega^1 \wedge \omega^4 - A_3\,\omega^1 \wedge \omega^5 \\
& -2A_3\,\omega^2 \wedge \omega^4 - 2A_4\,\omega^2 \wedge \omega^5, \\[6pt]

d\pi^5 = & \pi^1 \wedge \pi^5 + \pi^3 \wedge \pi^6 - \omega^5 \wedge \pi^7 + \omega^3 \wedge \chi^1 + \tfrac{9}{32}D_1\,\omega^1 \wedge \omega^2 \\
& +\tfrac{9}{8}C_1\,\omega^1 \wedge \omega^3 + \tfrac{9}{8}C_2\,\omega^2 \wedge \omega^3 + A_2\,\omega^3 \wedge \omega^4 + A_3\,\omega^3 \wedge \omega^5 \\
& +\tfrac{3}{4}B_1\,\omega^1 \wedge \omega^4 + \tfrac{3}{4}B_2\,(\omega^1 \wedge \omega^5 + \omega^2 \wedge \omega^4) + \tfrac{3}{4}B_3\,\omega^2 \wedge \omega^5, \\[6pt]

d\pi^6 = & \pi^2 \wedge \pi^5 + \pi^4 \wedge \pi^6 + \omega^4 \wedge \pi^7 + \omega^3 \wedge \chi^2 + \tfrac{9}{32}D_2\,\omega^1 \wedge \omega^2 \\
& +\tfrac{9}{8}C_2\,\omega^1 \wedge \omega^3 + \tfrac{9}{8}C_3\,\omega^2 \wedge \omega^3 - A_3\,\omega^3 \wedge \omega^4 - A_4\,\omega^3 \wedge \omega^5 \\
& +\tfrac{3}{4}B_2\,\omega^1 \wedge \omega^4 + \tfrac{3}{4}B_3\,(\omega^1 \wedge \omega^5 + \omega^2 \wedge \omega^4) + \tfrac{3}{4}B_4\omega^2 \wedge \omega^5, \\[6pt]

d\pi^7 = & \tfrac{4}{3}\pi^5 \wedge \pi^6 + (\pi^1 + \pi^4) \wedge \pi^7 + \omega^4 \wedge \chi^1 + \omega^5 \wedge \chi^2 \\
& +\tfrac{9}{64}E\,\omega^1 \wedge \omega^2 - \tfrac{3}{8}D_1\,\omega^1 \wedge \omega^3 - \tfrac{3}{8}D_2\,\omega^2 \wedge \omega^3 + 2A_3\,\omega^4 \wedge \omega^5 \\
& -B_2\,\omega^3 \wedge \omega^4 + B_3\,\omega^3 \wedge \omega^5.
\end{cases}
$$

$$(4)$$

The coefficients A, B, C, D, E appearing here are functions of the five original variables and the nine auxiliary variables. Now the matrix group $\mathfrak{M}_9$ induces a Lie algebra of vector fields acting on the space of nine auxiliary variables, and so it also acts on A, B, C, D, E in a way that makes the formula $d \circ \delta = \delta \circ d$ valid.

Setting $\delta(4) = d(3)$ we get a linear system for δA, δB, δC, δD and δE

with the following solutions:

$$\begin{cases}
\delta A_1 = -4t^1 A_1 - 4t^3 A_2, \\
\delta A_2 = -t^2 A_1 - (3t^1 + t^4)A_2 - 3t^3 A_3, \\
\delta A_3 = -2t^2 A_2 - 2(t^1 + t^4)A_3 - 2t^3 A_4, \\
\delta A_4 = -3t^2 A_3 - (t^1 + 3t^4)A_4 - t^3 A_5, \\
\delta A_5 = -4t^2 A_4 - 4t^4 A_5, \\
\delta B_1 = -(4t^1 + t^4)B_1 - 3t^3 B_2 - \frac{4}{3}t^6 A_1 + \frac{4}{3}t^5 A_2, \\
\delta B_2 = -t^2 B_1 - (3t^1 + 2t^4)B_2 - 2t^3 B_3 - \frac{4}{3}t^6 A_2 + \frac{4}{3}t^5 A_3, \\
\delta B_3 = -2t^2 B_2 - (2t^1 + 3t^4)B_2 - t^3 B_4 - \frac{4}{3}t^6 A_3 + \frac{4}{3}t^5 A_4, \\
\delta B_4 = -3t^2 B_3 - (t^1 + 4t^4)B_4 - \frac{4}{3}t^6 A_4 + \frac{4}{3}t^5 A_5, \\
\delta C_1 = -2(2t^1 + t^4)C_1 - 2t^3 C_2 - \frac{8}{3}t^6 B_1 + \frac{8}{3}t^5 B_2, \\
\delta C_2 = -t^2 C_1 - 3(t^1 + t^4)C_2 - t^3 C_3 - \frac{8}{3}t^6 B_2 + \frac{8}{3}t^5 B_3, \\
\delta C_3 = -2t^2 C_2 - 2(t^1 + 2t^4)C_3 - \frac{8}{3}t^6 B_3 + \frac{8}{3}t^5 B_4, \\
\delta D_1 = -(4t^1 + 3t^4)D_1 - t^3 D_2 - 4t^6 C_1 + 4t^5 C_2, \\
\delta D_2 = -t^2 D_1 - (3t^1 + 4t^4)D_2 - 4t^6 C_2 + 4t^5 C_3, \\
\delta E = -4(t^1 + t^4)E + \frac{16}{3}t^6 D_1 - \frac{16}{3}t^5 D_2.
\end{cases} \qquad (5)$$

Let us check the last formula for instance—using the derivation operator $\partial/\partial(\omega^i \wedge \omega^j)$, defined as follows.

Notation. Let Ω be an m-form which is expressed in the basis

$$\{\omega^{i_1} \wedge \cdots \wedge \omega^{i_m}\}_{1 \le i_1 < \cdots < i_m \le n}.$$

On fixing two indices $i < j$, Ω can be written as

$$\Omega = \omega^i \wedge \omega^j \wedge (\text{something}) + \{\text{terms not involving } \omega^i \wedge \omega^j\}.$$

Then this (something) is denoted by $\partial\Omega/\partial(\omega^i \wedge \omega^j)$. Or equivalently

$$\frac{\partial}{\partial(\omega^i \wedge \omega^j)} = \frac{\partial}{\partial \omega^j} \lrcorner \left(\frac{\partial}{\partial \omega^i} \lrcorner \right).$$

Letting δ act on (4) and using (2) we get

$$\frac{\partial}{\partial(\omega^1 \wedge \omega^2)} \delta d\pi^7 = \frac{9}{64} \delta E + \frac{9}{64} E(2t^1 + t^4) + \frac{9}{64} E(t^1 + 2t^4)$$
$$- \frac{3}{8} t^6 D_1 + \frac{3}{8} t^5 D_2.$$

On the other hand, letting d act on (3) and using (4), we have

$$\frac{\partial}{\partial(\omega^1 \wedge \omega^2)} d\delta\pi^7 = -(t^1 + t^4)\frac{9}{64}E + \frac{4}{3}t^6 \cdot \frac{9}{32}D_1 - \frac{4}{3}t^5 \cdot \frac{9}{32}D_2.$$

Because $\delta \circ d = d \circ \delta$,

$$\frac{9}{64}\delta E + \frac{9}{64}(3t^1 + 3t^4)E - \frac{3}{8}t^6 D_1 + \frac{3}{8}t^6 D_2$$

$$= -\frac{9}{64}(t^1 + t^4)E + \frac{3}{8}t^6 D_1 - \frac{3}{8}t^5 D_2,$$

and so

$$\delta E = -4(t^1 + t^4)E + \frac{16}{3}t^6 D_1 - \frac{16}{3}t^5 D_2,$$

i.e., the last formula in (5).

Formula (5) shows that δ induces a matrix algebra acting on the coefficients A, B, C, D, E, and in particular a matrix subalgebra $\mathfrak{A}_4$ acting on $A_1, \ldots, A_5$:

$$\begin{pmatrix} A_1 \\ A_2 \\ A_3 \\ A_4 \\ A_5 \end{pmatrix} \overset{\delta}{\mapsto} \begin{pmatrix} -4t^1 & -4t^3 & 0 & 0 & 0 \\ -t^2 & -(3t^1 + t^4) & -3t^3 & 0 & 0 \\ 0 & -2t^2 & -2(t^1 + t^4) & -2t^3 & 0 \\ 0 & 0 & -3t^2 & -(t^1 + 3t^4) & -t^3 \\ 0 & 0 & 0 & -4t^2 & -4t^4 \end{pmatrix} \begin{pmatrix} A_1 \\ A_2 \\ A_3 \\ A_4 \\ A_5 \end{pmatrix}.$$

Question: Is there some natural realization of this 4-parameter matrix algebra $\mathfrak{A}_4$?

Cartan's answer might seem a bit far-fetched today, but it was not so when he wrote his five variable paper—in fact it is most natural from the point of view of the invariant theory so popular at that time.

Recall that δ originally has arisen from the matrix group $\mathfrak{M}_7$, which in particular transforms ω^4 and ω^5 as

$$\begin{pmatrix} \omega^4 \\ \omega^5 \end{pmatrix} \mapsto \begin{pmatrix} \alpha & \beta \\ \gamma & \delta \end{pmatrix} \begin{pmatrix} \omega^4 \\ \omega^5 \end{pmatrix} \quad (\mathrm{mod}\ \omega^1, \omega^2, \omega^3),$$

i.e., by a *bilinear transformation*.

If x and y are homogeneous coordinates for the Riemann sphere $\mathbb{P}$, the group of bilinear transformations

$$\begin{cases} x \mapsto \alpha x + \beta y, \\ y \mapsto \gamma x + \delta y \end{cases} \quad \text{acting on } \mathbb{P}$$

naturally induces an action on the coefficients of any homogeneous polynomial in x and y. In this case we want there to be five coefficients

$A_1, \ldots, A_5$, and therefore must have a homogeneous polynomial of *degree* 4:

$$\mathcal{F}(x, y) = A_1 x^4 + 4A_2 x^3 y + 6A_3 x^2 y^2 + 4A_4 xy^3 + A_5 y^4.$$

This gets transformed into

$$\bar{\mathcal{F}} = A_1(\alpha x + \beta y)^4 + 4A_2(\alpha x + \beta y)^3(\gamma x + \delta y) + \cdots$$
$$= \bar{A}_1 x^4 + 4\bar{A}_2 x^3 y + 6\bar{A}_3 x^2 y^2 + 4\bar{A}_4 xy^3 + \bar{A}_5 y^4,$$

inducing a transformation $(A_1, \ldots, A_5)^T \mapsto (\bar{A}_1, \ldots, \bar{A}_5)^T$, where the $\bar{A}_i$ are linear combinations of $A_1, \ldots, A_5$ with coefficients depending on α, β, γ and δ.

What we are actually interested in are the corresponding infinitesimal transformations. To obtain these, let $\mathbf{v} = (v^1, v^2, v^3, v^4)$ be a vector at the identity element $e = (1, 0, 0, 1)$ in the space of parameters $\alpha, \beta, \gamma, \delta$, and let $t \mapsto (\alpha(t), \beta(t), \gamma(t), \delta(t))$ be a curve through e having the tangent vector $\mathbf{v}$ there: $(\alpha(0), \beta(0), \gamma(0), \delta(0)) = (1, 0, 0, 1)$, and $(\dot{\alpha}(0), \dot{\beta}(0), \dot{\gamma}(0), \dot{\delta}(0)) = (v^1, v^2, v^3, v^4)$. Along this curve the $\bar{A}_i$ are functions of t, and the infinitesimal transformation corresponding to $\mathbf{v}$ is given by differentiation with respect to this variable:

$$\delta A_i := \frac{d\bar{A}_i}{dt}\Big|_{t=0}.$$

For instance, $\bar{A}_1 = A_1 \alpha^4 + 4A_2 \alpha^3 \gamma + 6A_3 \alpha^2 \gamma^2 + 4A_4 \alpha\gamma^3 + A_5 \gamma^4$ implies that

$$\delta A_1 = \frac{d\bar{A}_1}{dt}\Big|_{t=0} = 4A_1\dot{\alpha}(0) + 4A_2\dot{\gamma}(0) = 4v^1 A_1 + 4v^3 A_2.$$

Calculating $\delta A_2, \ldots, \delta A_5$ in the same way and setting $t^i := -v^i$ for $i = 1, \ldots, 4$ we see that the induced infinitesimal transformations indeed coincide with those of $\mathfrak{A}_4$.

Remark. The minus sign here is due to the fact that we really should have let the *inverse* of the bilinear transformation act on $(x, y)^T$.

By using symmetric products instead of wedge products the one-forms ω^4, ω^5 can be identified with 'usual' variables x, y, and in this way we obtain the homogeneous polynomial

$$\mathcal{F}(\omega^4, \omega^5) = A_1(\omega^4)^4 + 4A_2(\omega^4)^3\omega^5 + 6A_3(\omega^4)^2(\omega^5)^2$$
$$+ 4A_4\omega^4(\omega^5)^3 + A_5(\omega^5)^4$$

naturally associated to our equivalence problem.

This polynomial is not in itself an invariant, because we can always let an element of $\mathfrak{M}_7$ act on $(\omega^1,\ldots,\omega^5)^T$. It follows in particular that the group of bilinear transformations acts on $(\omega^4,\omega^5)^T$ modulo $\omega^1,\omega^2,\omega^3$. If we identify ω^4 and ω^5 with homogeneous coordinates of $\mathbb{P}$ (*sic*!), this means that the group of bilinear transformations can be used to move the zeros of $\mathcal{F}$ on $\mathbb{P}$ around.

All the calculations for $\mathcal{P} = (\omega^1,\omega^2,\omega^3)$ are also performed on the corresponding pfaffian system $\hat{P} = (\hat{\omega}^1,\hat{\omega}^2,\hat{\omega}^3)$, and thus we in particular obtain a homogeneous polynomial $\hat{\mathcal{F}}(\hat{\omega}^4,\hat{\omega}^5)$ associated to $\hat{P}$. Now *if* $\mathcal{P}$ and $\hat{P}$ are locally equivalent, and if $\mathcal{F}$ has four simple roots on $\mathbb{P}$ (say), then necessarily also $\hat{\mathcal{F}}$ has four simple roots—for the bilinear transformations cannot make two different roots coincide.

In this way $\mathcal{F}$ gives rise to *five rough equivalence classes*:

- $\mathcal{F}$ has four simple roots (which is the general case),
- $\mathcal{F}$ vanishes identically,
- $\mathcal{F}$ has a root of multiplicity 4,
- $\mathcal{F}$ has one triple and one simple root,
- $\mathcal{F}$ has two double roots.

Let us return to δ acting on the coefficients A,B,C,D,E appearing in the structure equations (4):

$$\begin{pmatrix} A \\ B \\ C \\ D \\ E \end{pmatrix} \overset{\delta}{\longmapsto} \begin{pmatrix} * & 0 & 0 & 0 & 0 \\ * & * & 0 & 0 & 0 \\ 0 & * & * & 0 & 0 \\ 0 & 0 & * & * & 0 \\ 0 & 0 & 0 & * & * \end{pmatrix} \begin{pmatrix} A \\ B \\ C \\ D \\ E \end{pmatrix}.$$

The action of $\mathfrak{M}_7$ on $(\omega^3,\omega^4,\omega^5)^T$ is given by

$$\begin{pmatrix} \omega^3 \\ \omega^4 \\ \omega^5 \end{pmatrix} \longmapsto \begin{pmatrix} \Delta & 0 & 0 \\ \lambda & \alpha & \beta \\ \mu & \gamma & \delta \end{pmatrix} \begin{pmatrix} \omega^3 \\ \omega^4 \\ \omega^5 \end{pmatrix} \quad (\mathrm{mod}\ \omega^1,\omega^2).$$

Letting this 6-parameter group act (preferably through its inverse) on the homogeneous polynomial

$$\begin{aligned}
\mathcal{G}(\omega^3,\omega^4,\omega^5) :=& A_1\,(\omega^4)^4 + 4A_2\,(\omega^4)^3\omega^5 + 6A_3\,(\omega^4)^2(\omega^5)^2 \\
&+ 4A_4\,\omega^4(\omega^5)^3 + A_5\,(\omega^5)^4 \\
&+ 4\left(B_1\,(\omega^4)^3 + 3B_2\,(\omega^4)^2\omega^5 + 3B_3\,\omega^4(\omega^5)^2 + B_4\,(\omega^5)^3\right)\omega^3 \\
&+ 6\left(C_1\,(\omega^4)^2 + 2C_2\,\omega^4\omega^5 + C_3\,(\omega^5)^2\right)(\omega^3)^2 \\
&+ 4\left(D_1\,\omega^4 + D_2\,\omega^5\right)(\omega^3)^3 \\
&+ E\,(\omega^3)^4,
\end{aligned}$$

we see that the corresponding infinitesimal action on the coefficients A, B, C, D, E is identical to that given by δ above. So also $\mathcal{G}(\omega^3, \omega^4, \omega^5)$ is intrinsically associated to our equivalence problem.—Note by the way that $\mathcal{G}(0, \omega^4, \omega^5) = \mathcal{F}(\omega^4, \omega^5)$.

Obviously the structure equations (1) and (4) are so complicated that it would be hopeless to use the general solution scheme of chapter **15** with these as starting point. What does happen, however, is that if we place ourselves within each of the rough equivalence classes above, the equivalence problem simplifies to become more or less manageable.

17.2 $\mathcal{F}$ has four simple roots

If we somewhat recklessly consider ω^4 and ω^5 as homogeneous co-ordinates of the Riemann sphere $\mathbb{P}$, $\mathcal{F}(\omega^4, \omega^5)$ may be regarded as a polynomial in the complex variable $z := \omega^4/\omega^5$:

$$(\omega^5)^{-4} \mathcal{F}(\omega^4, \omega^5) = \mathcal{F}(\frac{\omega^4}{\omega^5}, 1) = \mathcal{F}(z, 1) =: F(z).$$

In this section it is assumed that $F(z)$ has *four different roots on* $\mathbb{P}$. As remarked earlier the matrix group $\mathfrak{M}_7$ induces an action of the group of bilinear transformations on $\mathbb{P}$, and this may be used to move the roots around.

If the roots of $F(z)$ are z_1, z_2, z_3, z_4, we can for instance move them to $0, \infty, \zeta, \zeta^{-1}$ respectively, for a certain ζ, by means of the bilinear transformation

$$z \mapsto c \frac{z - z_1}{z - z_2}, \quad \text{with } c \text{ satisfying} \quad c \frac{z_3 - z_1}{z_3 - z_2} \cdot c \frac{z_4 - z_1}{z_4 - z_2} = 1.$$

Thus up to a nonzero factor

$$F(z) = 4z(z - \zeta)(z - \zeta^{-1}) = 4z^3 + 6Az^2 + 4z \quad \text{with} \quad A = -\frac{2}{3}(\zeta + \zeta^{-1}).$$

Conclusion. *The associated homogeneous polynomial*

$$\mathcal{F}(\omega^4, \omega^5) = A_1 (\omega^4)^4 + 4A_2 (\omega^4)^3 \omega^5 + 6A_3 (\omega^4)^2 (\omega^5)^2$$
$$+ 4A_4 \omega^4 (\omega^5)^3 + A_5 (\omega^5)^4$$

can in this case be reduced to

$$\mathcal{F}(\omega^4, \omega^5) = 4(\omega^4)^3 \omega^5 + 6A(\omega^4)^2 (\omega^5)^2 + 4\omega^4 (\omega^5)^3,$$

—so that $A_1 = 0$, $A_2 = 1$, $A_3 = A$, $A_4 = 1$ and $A_5 = 0$.

But then the formulas

$$\begin{cases} \delta A_1 = -4t^1 A_1 - 4t^3 A_2, \\ \delta A_2 = -t^2 A_1 - (3t^1 + t^4)A_2 - 3t^3 A_3, \\ \delta A_4 = -3t^2 A_3 - (t^1 + 3t^4)A_4 - t^3 A_5, \\ \delta A_5 = -4t^2 A_4 - 4t^4 A_5 \end{cases}$$

occurring in (5) of the preceding section imply that $t^1 = t^2 = t^3 = t^4 = 0$. Hence the stability algebra $\text{span}_\mathbb{C}\{E_1, \ldots, E_9\}$ is reduced to five dimensions only, $\text{span}_\mathbb{C}\{E_5, \ldots, E_9\}$, and therefore π^1, π^2, π^3 and π^4 are *linear combinations of* $\omega^1, \ldots, \omega^5$.

Let us consider the one-form $\pi^1 + \pi^4$ for instance. By (3) in the preceding section,

$$\delta(\pi^1 + \pi^4) = \frac{2}{3}t^7\omega^3 - \frac{1}{3}t^5\omega^4 - \frac{1}{3}t^6\omega^5 + t^8\omega^1 + t^9\omega^2.$$

Writing $\pi^1 + \pi^4 = \sum_{i=1}^5 a_i \omega^i$ we also have

$$\delta(\pi^1 + \pi^4) = \sum_{i=1}^5 (\delta a_i \, \omega^i + a_i \, \delta\omega^i)$$
$$= \omega^1(\delta a_1 + a_3 t^5 + a_4 t^7) + \omega^2(\delta a_2 + a_3 t^6 + a_5 t^7) + \cdots$$

by formula (2) of the preceding section. Identifying these two expressions for $\delta(\pi^1 + \pi^4)$, we get

$$\begin{cases} \delta a_1 = t^8 - t^5 a_3 - t^7 a_4, \\ \delta a_2 = t^9 - t^6 a_3 - t^7 a_5, \\ \delta a_3 = \frac{2}{3}t^7 - \frac{4}{3}t^6 a_4 + \frac{4}{3}t^5 a_5, \\ \delta a_4 = -\frac{1}{3}t^5, \\ \delta a_5 = -\frac{1}{3}t^6. \end{cases}$$

The last equation shows that the 1-parameter group generated by E_6 can be used to kill a_5, the next but last that the 1-parameter group generated by E_5 can be used to kill a_4—and continuing in this manner we see that by letting the stability group act suitably it can be arranged that

$$a_5 = a_4 = a_3 = a_2 = a_1 = 0\text{—that is, } \pi^1 + \pi^4 = 0.$$

Having fixed the a_i in this way the equations above conversely imply that

$$t^5 = t^6 = t^7 = t^8 = t^9 = 0,$$

so that the stability algebra has been reduced to $\{0\}$, and the stability group to the identity—whence we have *got rid of all auxiliary variables.*

Conclusion. *The two requirements*

- $\mathcal{F}(\omega^4, \omega^5) = 4(\omega^4)^3 \omega^5 + 6A(\omega^4)^2(\omega^5)^2 + 4\omega^4(\omega^5)^3,$
- $\pi^1 + \pi^4 = 0$

are compatible with the action of the stability group $\mathfrak{M}_9$, and reduce the latter to consisting of the identity transformation only.

Hence we are left with an e-structure in five variables, and as explained in section **15.1** there then remains to determine the possible invariants.

In the prolonged problem not only the ω^i but also the π^j are invariant one-forms. Knowing that each π^j is a linear combination of $\omega^1, \ldots, \omega^5$:

$$\pi^j = \sum_{i=1}^{5} \alpha_i^j \, \omega^i \quad \text{for } j = 1, \ldots, 7,$$

we deduce that the α_i^j are invariants of our problem.

Because $\pi^4 = -\pi^1$ we in this way get $(7-1) \cdot 5 = 30$ invariants. By applying the structure equations (4) of the preceding section to $d\pi^4 + d\pi^1 = 0$ it is moreover inferred that

$$\alpha_4^7 = -\frac{1}{2}\alpha_3^5, \quad \alpha_5^7 = -\frac{1}{2}\alpha_3^6 \quad \text{and} \quad \alpha_4^6 = \alpha_5^5,$$

and thus there remain *27 invariants* α_i^j—of which at most 5 can be functionally independent on $\mathbb{C}_x^5$.

Cartan calls these 27 invariants *fundamental*—because *any* invariant is a function of these and their covariant derivatives.

For instance, with the π^j being linear combinations of the ω^i, the coefficients $A, B_1, B_2, B_3, B_4, C_1, C_2, C_3, D_1, D_2, E$ in the structure equations are all invariants, and thus it should be possible to express them by means of the α_j^i and their covariant derivatives.

To see why this is so we write $d\pi^1$ in two ways:

$$d\pi^1 = \sum_{i=1}^{5} (\alpha_i^1 \, d\omega^i + d\alpha_i^1 \wedge \omega^i),$$

and also, by the structure equations,

$$d\pi^1 = \pi^3 \wedge \pi^2 + \frac{1}{3}\omega^3 \wedge \pi^7 - \frac{2}{3}\omega^1 \wedge \pi^5 + \frac{1}{3}\omega^5 \wedge \pi^6 + \omega^1 \wedge \chi^1$$
$$+ 2B_2\omega^1 \wedge \omega^3 + B_3\omega^2 \wedge \omega^3 + 2\omega^1 \wedge \omega^4 + 2A\omega^1 \wedge \omega^5$$
$$+ A\omega^2 \wedge \omega^4 + \omega^2 \wedge \omega^5,$$

then plug in $d\alpha_i^1 = \sum_{k=1}^{5} \alpha_{i,k}^1 \omega^k$, $\pi^j = \sum_{i=1}^{5} \alpha_i^j \omega^i$, and so obtain 15 equations by identifying the coefficients of $\omega^i \wedge \omega^j$ for $1 \le i < j \le 5$.

Doing the same for $d\pi^2, \ldots, d\pi^7$, all the coefficients $A, B_1, B_2, \ldots, E$ can be expressed in terms of the α_i^j and $\alpha_{i,k}^j$—and besides we also get relations between the latter.

Having obtained the 27 fundamental invariants, the number of functionally independent invariants—0, 1, 2, 3, 4 or 5—is determined, and then the methods of of section **15.1** are used for solving the equivalence problem.

17.3 *F* vanishes identically

Here we study the case when *all* coefficients A, B, C, D, E in the stucture equations for the $d\pi^j$ vanish—so that the formulas (1) and (4) from section **17.1** simplify to

$$
\begin{cases}
d\omega^1 = \omega^1 \wedge (2\pi^1 + \pi^4) + \omega^2 \wedge \pi^2 + \omega^3 \wedge \omega^4, \\
d\omega^2 = \omega^1 \wedge \pi^3 + \omega^2 \wedge (\pi^1 + 2\pi^4) + \omega^3 \wedge \omega^5, \\
d\omega^3 = \omega^1 \wedge \pi^5 + \omega^2 \wedge \pi^6 + \omega^3 \wedge (\pi^1 + \pi^4) + \omega^4 \wedge \omega^5, \\
d\omega^4 = \omega^1 \wedge \pi^7 + \tfrac{4}{3}\omega^3 \wedge \pi^6 + \omega^4 \wedge \pi^1 + \omega^5 \wedge \pi^2, \\
d\omega^5 = \omega^2 \wedge \pi^7 - \tfrac{4}{3}\omega^3 \wedge \pi^5 + \omega^4 \wedge \pi^3 + \omega^5 \wedge \pi^4, \\
d\pi^1 = \pi^3 \wedge \pi^2 + \tfrac{1}{3}\omega^3 \wedge \pi^7 - \tfrac{2}{3}\omega^4 \wedge \pi^5 + \tfrac{1}{3}\omega^5 \wedge \pi^6 + \omega^1 \wedge \chi^1, \\
d\pi^2 = \pi^2 \wedge (\pi^1 - \pi^4) - \omega^4 \wedge \pi^6 + \omega^1 \wedge \chi^2, \\
d\pi^3 = \pi^3 \wedge (\pi^4 - \pi^1) - \omega^5 \wedge \pi^5 + \omega^2 \wedge \chi^1, \\
d\pi^4 = \pi^2 \wedge \pi^3 + \tfrac{1}{3}\omega^3 \wedge \pi^7 + \tfrac{1}{3}\omega^4 \wedge \pi^5 - \tfrac{2}{3}\omega^5 \wedge \pi^6 + \omega^2 \wedge \chi^2, \\
d\pi^5 = \pi^1 \wedge \pi^5 + \pi^3 \wedge \pi^6 - \omega^5 \wedge \pi^7 + \omega^3 \wedge \chi^1, \\
d\pi^6 = \pi^2 \wedge \pi^5 + \pi^4 \wedge \pi^6 + \omega^4 \wedge \pi^7 + \omega^3 \wedge \chi^2, \\
d\pi^7 = \tfrac{4}{3}\pi^5 \wedge \pi^6 + (\pi^1 + \pi^4) \wedge \pi^7 + \omega^4 \wedge \chi^1 + \omega^5 \wedge \chi^2.
\end{cases}
\tag{*}
$$

According to these the pfaffian system $\mathcal{P}^{(1)} = (\omega^1, \ldots, \omega^5, \pi^1, \ldots, \pi^7)$ is *not involutive*, and hence we have to make a normal prolongation—as described in section **14.6**.

There are two extra one-forms χ^1 and χ^2 in (*), and first of all we look for the *most general replacements*

$$\chi^i \mapsto \chi^i + \text{ linear combination of } \omega^1, \ldots, \omega^5, \pi^1, \ldots, \pi^7$$

which leave (*) invariant. But by considering $d\pi^1$, $d\pi^3$ and $d\pi^2$, $d\pi^4$ we see that only $\chi^i \mapsto \chi^i$ is allowed—and hence there are *no new auxiliary variables* for the prolongation $\mathcal{P}^{(2)} = (\omega^1, \ldots, \omega^5, \pi^1, \ldots, \pi^7, \chi^1, \chi^2)$.

With $\mathcal{P}^{(2)}$ being a 14-dimensional pfaffian system in 14 variables, it is *automatically complete*.

In order to have the structure equations for $\mathcal{P}^{(2)}$ we need to know $d\chi^1$ and $d\chi^2$, and these are found by differentiating (*). The result is

$$\begin{cases} d\chi^1 = \pi^5 \wedge \pi^7 + (2\pi^1 + \pi^4) \wedge \chi^1 + \pi^3 \wedge \chi^2, \\ d\chi^2 = \pi^6 \wedge \pi^7 + \pi^2 \wedge \chi^1 + (\pi^1 + 2\pi^4) \wedge \chi^2. \end{cases} \tag{**}$$

Formulas (*) and (**) together define the structure of a 14-dimensional Lie parameter group—which turns out to be nothing but the *exceptional simple Lie group* G_2 (some hints explaining this can be found in [Gardner 1989], page 107).

Because (*) and (**) do not contain any nonconstant coefficients, there are no invariants in this case. In particular,

there is only one local equivalence class!

To find a simple canonical form we look for the simplest one-forms satisfying (*) and (**). These are found by setting $\pi^1 = \cdots = \pi^7 = \chi^1 = \chi^2 = 0$, and then choosing $\omega^1, \ldots, \omega^5$ such that

$$\begin{cases} d\omega^1 = \omega^3 \wedge \omega^4, \\ d\omega^2 = \omega^3 \wedge \omega^5, \\ d\omega^3 = \omega^4 \wedge \omega^5, \\ d\omega^4 = 0, \\ d\omega^5 = 0. \end{cases}$$

The latter can be realized by putting $\omega^4 = dx^4$, $\omega^5 = dx^5$, whereupon

$$d\omega^3 = dx^4 \wedge dx^5 = \frac{1}{2} d(x^4 \, dx^5 - x^5 \, dx^4) \implies \omega^3 = dx^3 + \frac{1}{2} x^4 \, dx^5 - \frac{1}{2} x^5 dx^4,$$

$$d\omega^2 = dx^3 \wedge dx^5 - \frac{1}{2} x^5 \, dx^4 \wedge dx^5 = d\left(x^3 - \frac{1}{2} x^4 x^5\right) \wedge dx^5$$

$$\implies \omega^2 = dx^2 + \left(x^3 - \frac{1}{2} x^4 x^5\right) dx^5, \quad \text{and}$$

$$d\omega^1 = dx^3 \wedge dx^4 + \frac{1}{2} x^4 \, dx^5 \wedge dx^4 = d\left(x^3 + \frac{1}{2} x^4 x^5\right) \wedge dx^4$$

$$\implies \omega^1 = dx^1 + \left(x^3 + \frac{1}{2} x^4 x^5\right) dx^4.$$

Thus

$$\begin{cases} \omega^1 = dx^1 + (x^3 + \frac{1}{2}x^4x^5)\,dx^4, \\ \omega^2 = dx^2 + (x^3 - \frac{1}{2}x^4x^5)\,dx^5, \\ \omega^3 = dx^3 + \frac{1}{2}x^4\,dx^5 - \frac{1}{2}x^5\,dx^4, \\ \omega^4 = dx^4, \\ \omega^5 = dx^5. \end{cases}$$

Now we were originally looking for integral curves of $\mathcal{P} = (\omega^1, \omega^2, \omega^3)$. If we set $x^4 := s =$ (an arbitrary parameter), and assume ds to be nonzero on the wanted integral curves, the equation $\omega^1 = 0$ shows that

$$x^1 = f(s) \quad \text{and} \quad x^3 + \frac{1}{2}sx^5 = -f'(s),$$

where f is an arbitrary function of one variable. Consequently

$$dx^3 + \frac{1}{2}s\,dx^5 + \frac{1}{2}x^5\,ds = -f''(s)\,ds,$$

which combined with $\omega^3 = 0$ implies that

$$x^5\,ds = -f''(s)\,ds, \quad \text{i.e.,} \quad x^5 = -f''(s).$$

But then

$$x^3 = -\frac{1}{2}sx^5 - f'(s) = -f'(s) + \frac{1}{2}sf''(s).$$

Finally we obtain x^2 by inserting these results into $\omega^2 = 0$, and conclude that the general integral curve of $\mathcal{P}$ on which $ds \neq 0$ is given by

$$\begin{cases} x^1 = f(s), \\ x^2 = -f''(s)\,(f'(s) - \frac{1}{2}sf''(s)) + \frac{1}{2}\int f''^2(s)\,ds, \\ x^3 = -f'(s) + \frac{1}{2}sf''(s), \\ x^4 = s, \\ x^5 = -f''(s). \end{cases}$$

In section **16.3** we saw how to associate a pfaffian system $\mathcal{P} = (\omega^1, \omega^2, \omega^3)$ in five variables to a PDE system $\mathcal{S}_2$ admitting a Cauchy characteristic vector field. To pass in the opposite direction from $\mathcal{P}$ to $\mathcal{S}_2$ we introduce a new variable u and use the methods in chapter **6** to put the pfaffian equation $\omega^1 + u\omega^2 = 0$ in six variables into the canonical form $dz - p\,dx - q\,dy = 0$. The new local coordinates accomplishing this are

$$\begin{cases} x := u, \\ y := x^4 + ux^5, \\ z := x^1 + ux^2 - \tfrac{1}{2}ux^4(x^5)^2 - \tfrac{1}{6}u^2(x^5)^3, \\ p := x^2 + x^3x^5 + \tfrac{1}{6}u(x^5)^3, \\ q := -x^3 - \tfrac{1}{2}x^4x^5 - \tfrac{1}{2}u(x^5)^2, \\ x^5 := x^5. \end{cases}$$

Expressing $\omega^1 + u\omega^2$, ω^2 and ω^3 in these, we have

$$\begin{cases} \omega^1 + u\omega^2 & = dz - p\,dx - q\,dy, \\ \omega^2 & = dp + x^5\,dq - \tfrac{1}{6}(x^5)^3\,dx + \tfrac{1}{2}(x^5)^2\,dy, \\ \omega^3 & = -dq + \tfrac{1}{2}(x^5)^2\,dx - x^5\,dy, \end{cases}$$

or

$$\begin{cases} \omega^1 + u\omega^2 & = dz - p\,dx - q\,dy, \\ \omega^2 + x^5\,\omega^3 & = dp + \tfrac{1}{3}(x^5)^3\,dx - \tfrac{1}{2}(x^5)^2\,dy, \\ -\omega^3 & = dq - \tfrac{1}{2}(x^5)^2\,dx + x^5\,dy. \end{cases}$$

Comparing this with the contact pfaffian system generated by

$$\begin{cases} dz - p\,dx - q\,dy, \\ dp - r\,dx - s\,dy, \\ dq - s\,dx - t\,dy \end{cases}$$

we see that $t = -x^5$, and then that $r = \tfrac{1}{3}t^3$, $s = \tfrac{1}{2}t^2$, so that S_2 is given by

$$r = \frac{1}{3}t^3 \quad \text{and} \quad s = \frac{1}{2}t^2.$$

From section **16.1** we know that the corresponding parabolic PDE S_1 is obtained by eliminating λ from the system

$$\begin{cases} r + \lambda s + 2\psi - \lambda\,\partial\psi/\partial\lambda = 0, \\ s + \lambda t + \partial\psi/\partial\lambda = 0, \end{cases}$$

where λ and ψ can be read off from the expression for ω^2:

$$\omega^2 = dp + \lambda\,dq + \left(2\psi - \lambda\frac{\partial\psi}{\partial\lambda} \right) dx + \frac{\partial\psi}{\partial\lambda}\,dy.$$

So here

$$\lambda = x^5, \quad 2\psi - \lambda \frac{\partial \psi}{\partial \lambda} = -\frac{1}{6}(x^5)^3 \quad \text{and} \quad \frac{\partial \psi}{\partial \lambda} = \frac{1}{2}(x^5)^2,$$

whence $\mathcal{S}_1$ is the PDE obtained by eliminating λ from

$$\begin{cases} r + \lambda s - \frac{1}{6}\lambda^3 = 0, \\ s + \lambda t + \frac{1}{2}\lambda^2 = 0. \end{cases}$$

Finally we wish to relate the case considered in this section to the homogeneous polynomial $\mathcal{F}(\omega^4, \omega^5)$.

Lemma 17.3.1. *All the coefficients A, B, C, D, E in the structure equations vanish $\iff$ all A vanish $\iff$ $\mathcal{F}$ vanishes identically.*

Proof. We must show that if all A vanish, the remaining coefficients B, C, D, E also do.

The formulas (**) for $d\chi^1$ and $d\chi^2$ are valid if all coefficients vanish. Otherwise certain two-forms Ω^1 and Ω^2 have to be added:

$$\begin{cases} d\chi^1 = \pi^5 \wedge \pi^7 + (2\pi^1 + \pi^4) \wedge \chi^1 + \pi^3 \wedge \chi^2 + \Omega^1, \\ d\chi^2 = \pi^6 \wedge \pi^7 + \pi^2 \wedge \chi^1 + (\pi^1 + 2\pi^4) \wedge \chi^2 + \Omega^2. \end{cases}$$

The general idea now is to differentiate the formulas (4) in section **17.1** for $d\pi^i$, plug in the expressions for $d\omega^i$ from (1), $d\pi^j$ from (4) and the above for $d\chi^k$, then express the three-forms $0 = d^2\pi^i$ for $i = 1, \ldots, 7$ in the basis provided by the one-forms $\omega^1, \ldots, \omega^5, \pi^1, \ldots, \pi^7, \chi^1, \chi^2$, and finally set all coefficients equal to zero.

For instance, considering the coefficient in front of $\omega^1 \wedge \omega^4 \wedge \omega^5$ in $d\pi^1$ and the coefficient in front of $\omega^2 \wedge \omega^4 \wedge \omega^5$ in $d\pi^3$, we find that

$$-\frac{1}{2}B_2 - \frac{1}{4}B_2 - \frac{\partial\Omega^1}{\partial(\omega^4 \wedge \omega^5)} - 2B_2 = 0 \quad \text{and} \quad \frac{3}{4}B_2 - \frac{\partial\Omega^1}{\partial(\omega^4 \wedge \omega^5)} = 0,$$

from which $B_2 = 0$. And considering the coefficient in front of $\omega^2 \wedge \omega^4 \wedge \omega^5$ in $d\pi^2$ we see that $-\frac{3}{4}B_4 - B_4 = 0$, whence also $B_4 = 0$.

Continuing in this way and putting in a lot of effort we finally obtain the desired result: $B_1 = B_2 = B_3 = B_4 = C_1 = C_2 = C_3 = D_1 = D_2 = E = 0$. $\qquad\square$

Remark. In order to avoid to many complications when treating the remaining rough equivalence classes, only local equivalence classes connected to *e*-structures in at least six variables will be considered in the following.

17.4 $\mathcal{F}$ has a root of multiplicity 4

If $\mathcal{F}$ has a root of multiplicity 4, we may suppose this to be situated at ∞, and so

$$\mathcal{F}(\omega^4, \omega^5) = (\omega^5)^4.$$

Then the coefficients A_i in the general expression for $\mathcal{F}$ take the values $A_1 = A_2 = A_3 = A_4 = 0$ and $A_5 = 1$. From

$$\begin{cases} \delta A_4 = -3t^2 A_3 - (t^1 + 3t^4)A_4 - t^3 A_5 = 0 \Longleftrightarrow 0 = -t^3 \cdot 1, \\ \delta A_5 = -4t^2 A_4 - 4t^4 A_5 \Longleftrightarrow 0 = -4t^4 \cdot 1 \end{cases}$$

it moreover follows that $t^3 = t^4 = 0$. Hence π^3 and π^4 are linear combinations of $\omega^1, \ldots, \omega^5$.

Lemma 17.4.1. $B_1 = B_2 = B_3 = C_1 = C_2 = D_1 = 0$, *so that only* $A_5 = 1$, B_2, C_3, D_2 *and* E *are left nonzero (thus far).*

Proof. To see why $B_1 = 0$ we differentiate the identity

$$d\pi^3 = \pi^3 \wedge (\pi^4 - \pi^1) - \omega^5 \wedge \pi^5 + \omega^2 \wedge \chi^1 - B_1 \omega^1 \wedge \omega^3,$$

and use the structure formulas in section **17.1** together with $d\chi^1 = \pi^5 \wedge \pi^7 + (2\pi^1 + \pi^4) \wedge \chi^1 + \pi^3 \wedge \chi^2 + \Omega^1$. The only way to obtain the wedge product $\omega^1 \wedge \omega^4 \wedge \omega^5$ is through

$$B_1 \omega^1 \wedge d\omega^3 = B_1 \omega^1 \wedge \omega^4 \wedge \omega^5 + \cdots,$$

showing that $B_1 = 0$. And the only way to get $\omega^2 \wedge \omega^4 \wedge \omega^5$ is by

$$\omega^5 \wedge d\pi^5 - \omega^2 \wedge d\chi^1 = \left(\frac{3}{4} B_2 - \frac{\partial \Omega^1}{\partial(\omega^4 \wedge \omega^5)} \right) \omega^2 \wedge \omega^4 \wedge \omega^5 + \cdots,$$

whence

$$\frac{3}{4} B_2 = \frac{\partial \Omega^1}{\partial(\omega^4 \wedge \omega^5)}.$$

Differentiating

$$d\pi^1 = \pi^3 \wedge \pi^2 + \frac{1}{3} \omega^3 \wedge \pi^7 - \frac{2}{3} \omega^4 \wedge \pi^5 + \frac{1}{3} \omega^5 \wedge \pi^6$$
$$+ \omega^1 \wedge \chi^1 + 2B_2 \omega^1 \wedge \omega^3 + B_3 \omega^2 \wedge \omega^3$$

and looking for terms containing $\omega^1 \wedge \omega^4 \wedge \omega^5$ one similarly finds

$$\frac{2}{3} \omega^4 \wedge d\pi^5 - \frac{1}{3} \omega^5 \wedge d\pi^6 - \omega^1 \wedge d\chi^1 - 2B_2 \omega^1 \wedge d\omega^3$$
$$= \left(-\frac{2}{3} \cdot \frac{3}{4} B_2 - \frac{1}{3} \cdot \frac{3}{4} B_2 - \frac{\partial \Omega^1}{\partial(\omega^4 \wedge \omega^5)} - 2B_2 \right) \omega^1 \wedge \omega^4 \wedge \omega^5 + \cdots,$$

so that

$$-\frac{11}{4}B_2 = \frac{\partial\Omega^1}{\partial(\omega^4 \wedge \omega^5)}.$$

Combining the two expressions found for B_2, we see that $B_2 = 0$.

With $B_1 = 0$ the only way to obtain $\omega^1 \wedge \omega^3 \wedge \omega^5$ in $d^2\pi^3$ is through

$$\omega^5 \wedge d\pi^5 = \frac{9}{8}C_1 \, \omega^1 \wedge \omega^3 \wedge \omega^5 + \cdots,$$

and hence $C_1 = 0$.

By studying $d^2\pi^3$, $d^2\pi^1$ and $d^2\pi^5$ analogously it is inferred that also $B_3 = C_2 = D_1 = 0$. □

Using $t^3 = t^4 = 0$ and the lemma we see from (5) in section **17.1** that

$$\delta B_4 = -t^1 B_4 + \frac{4}{3}t^5 \quad \text{and} \quad \delta C_3 = -2t^1 C_3 + \frac{8}{3}t^5 B_4.$$

By the first equality $\delta(B_4^2) = 2B_4 \, \delta B_4 = -2t^1 B_4^2 + \frac{8}{3}t^5 B_4$, and so

$$\delta(C_3 - B_4^2) = -2t^1(C_3 - B_4^2).$$

This leaves two alternatives to be investigated further:

$$C_3 - B_4^2 \neq 0 \quad \text{or} \quad C_3 - B_4^2 = 0.$$

By employing the same type of arguments as elsewhere in this chapter and doing a lot of calculations it can be seen that

if $C_3 - B_4^2 \neq 0$ it is possible to get rid of all auxiliary variables, thereby obtaining an e-structure in five variables.

Let us therefore concentrate on the second alternative: $C_3 - B_4^2 = 0$.

From $\delta B_4 = -t^1 B_4 + \frac{4}{3}t^5$ it is inferred that the 1-parameter subgroup of the stability group generated by E_5 can be used to kill B_4—and thereby also C_3. Having fixed B_4 to 0 we then deduce that $t^5 = 0$.

Lemma 17.4.2. $D_2 = E = 0$ too, so that the only remaining nonzero coefficient is $A_5 = 1$.

Proof. To see this we differentiate the expressions

$$\begin{cases} d\pi^2 = \pi^2 \wedge (\pi^1 - \pi^4) - \omega^4 \wedge \pi^6 + \omega^1 \wedge \chi^2 + \omega^2 \wedge \omega^5, \\ d\pi^6 = \pi^2 \wedge \pi^5 + \pi^4 \wedge \pi^6 + \omega^4 \wedge \pi^7 + \omega^3 \wedge \chi^2 + \frac{9}{32}D_2 \, \omega^1 \wedge \omega^2 \end{cases}$$

from (4) in section **17.1** and use $d\chi^2 = \pi^6 \wedge \pi^7 + \pi^2 \wedge \chi^1 + (\pi^1 + 2\pi^4) \wedge \chi^2 + \Omega^2$ in order to deduce that

$$0 = \frac{\partial (d^2 \pi^2)}{\partial (\omega^1 \wedge \omega^2 \wedge \omega^4)} = \frac{\partial d\pi^6}{\partial (\omega^1 \wedge \omega^2)} - \frac{\partial \Omega^2}{\partial (\omega^2 \wedge \omega^4)} + \frac{\partial d\omega^5}{\partial (\omega^1 \wedge \omega^4)}$$

$$= \frac{9}{32} D_2 - \frac{\partial \Omega^2}{\partial (\omega^2 \wedge \omega^4)} - \frac{\partial \pi^3}{\partial \omega^1},$$

and

$$0 = \frac{\partial (d^2 \pi^6)}{\partial (\omega^2 \wedge \omega^3 \wedge \omega^4)} = -\frac{\partial d\pi^7}{\partial (\omega^2 \wedge \omega^3)} + \frac{\partial \Omega^2}{\partial (\omega^2 \wedge \omega^4)} + \frac{9}{32} D_2 \frac{\partial d\omega^1}{\partial (\omega^3 \wedge \omega^4)}$$

$$= \frac{3}{8} D_2 + \frac{\partial \Omega^2}{\partial (\omega^2 \wedge \omega^4)} + \frac{9}{32} D_2.$$

Adding these equations, we get

$$\frac{15}{16} D_2 - \frac{\partial \pi^3}{\partial \omega^1} = 0.$$

If we consider

$$0 = \frac{\partial (d^2 \pi^1)}{\partial (\omega^1 \wedge \omega^2 \wedge \omega^5)} \quad \text{and} \quad 0 = \frac{\partial (d^2 \pi^5)}{\partial (\omega^2 \wedge \omega^3 \wedge \omega^5)}$$

in the same way it analogously follows that

$$\frac{15}{32} D_2 + \frac{\partial \pi^3}{\partial \omega^1} = 0.$$

Thus $D_2 = 0$. Finally

$$d\pi^5 = \pi^1 \wedge \pi^5 + \pi^3 \wedge \pi^6 - \omega^5 \wedge \pi^7 + \omega^3 \wedge \chi^1$$

shows that

$$0 = \frac{\partial (d^2 \pi^5)}{\partial (\omega^1 \wedge \omega^2 \wedge \omega^5)} = \frac{\partial d\pi^7}{\partial (\omega^1 \wedge \omega^2)} = \frac{9}{64} E, \quad \text{i.e.,} \quad E = 0.$$

$\square$

From $t^3 = t^4 = t^5 = 0$ it follows that π^3, π^4 and π^5 are linear combinations of $\omega^1, \dots, \omega^5$:

$$\pi^j = \sum_{k=1}^{5} a_k^j \omega^k \quad \text{for } j = 3, 4, 5.$$

If we insert these expressions into $d^2\pi^j = 0$ for $j = 1,\ldots,7$, a boring calculation leads to

$$\begin{cases} \pi^3 = a_1\,\omega^2 + a_2\,\omega^5, \\ \pi^4 = a_3\,\omega^1 + a_4\,\omega^2 + a_5\,\omega^3 + \tfrac{1}{4}a_2\,\omega^4 + a_6\,\omega^5, \\ \pi^5 = a_7\,\omega^2 + (a_1 - 4a_3)\,\omega^3 - 3a_5\,\omega^5, \end{cases}$$

with just seven coefficients a_k. The π^j are invariant one-forms for the prolonged equivalence problem, and therefore all the a_k are also invariants of this problem.

Lemma 17.4.3. *Suppose that I is an invariant of the prolonged equivalence problem with $\partial dI/\partial\omega^4 \neq 0$. Then it is possible to get rid of all auxiliary variables, so that we are left with an e-structure in five variables.*

Proof. Writing

$$dI \equiv \sum_{k=1}^{5} I_{k}\,\omega^k \quad (\mathrm{mod}\ \pi^1, \pi^2, \pi^6, \pi^7)$$

and using the formula $\delta\omega^4 = t^7\omega^1 + \tfrac{4}{3}t^6\omega^3 + t^1\omega^4 + t^2\omega^5$ we see that the subgroup of the stability group generated by E_1, E_2, E_6, E_7 can be used to make $I_{.1}, I_{.3}, I_{.5}$ vanish and $I_{.4}$ become equal to 1. Then the coframe $\{\omega^1,\ldots,\omega^5\}$ has been fixed in such a way that $dI \equiv \omega^4 + I_{.2}\,\omega^2$, and the stability group has been reduced to Id. $\qquad\square$

In view of this lemma we assume that any invariant I appearing in the following satisfies $\partial dI/\partial\omega^4 = 0$ (note that the case we are studying is becoming more and more special).

Now a_2 is such an invariant, and therefore

$$\frac{\partial(d\pi^3)}{\partial(\omega^4 \wedge \omega^5)} = a_2\frac{\partial(d\omega^5)}{\partial(\omega^4 \wedge \omega^5)} = a_2\left(a_2 - \frac{1}{4}a_2\right),$$

since $d\omega^5 = \cdots + \omega^4 \wedge \pi^3 + \omega^5 \wedge \pi^4$. But on the other hand we know from (4) in section **17.1** that $\partial(d\pi^3)/\partial(\omega^4 \wedge \omega^5) = 0$. Hence $a_2 = 0$, and so π^3 simplifies to $\pi^3 = a_1\,\omega^2$.

Since $t^3 = t^4 = t^5 = 0$ the action of δ described in section **17.1** here takes the form

$$\begin{cases} \delta\omega^2 = t^1\,\omega^2, \\ \delta\omega^5 = t^7\,\omega^2, \\ \delta\pi^3 = -t^1\,\pi^3 + t^8\,\omega^2, \\ \delta\pi^4 = -t^2\,\pi^3 + t^9\,\omega^2 + \frac{1}{3}t^7\,\omega^3 - \frac{2}{3}t^6\,\omega^5, \\ \delta\pi^5 = -t^1\,\pi^5 + t^6\,\pi^3 + t^8\,\omega^3 - t^7\,\omega^5. \end{cases}$$

In particular

$$\delta\pi^3 = \delta a_1\,\omega^2 + a_1\,\delta\omega^2 = \delta a_1\,\omega^2 + a_1 t^1\,\omega^2$$
$$= -t^1 a_1\,\omega^2 + t^8\,\omega^2, \quad \text{i.e.,} \quad \delta a_1 = t^8 - 2t^1 a_1.$$

Consequently the 1-parameter subgroup generated by E_8 may be used to kill a_1—whereby also π^3 vanishes. Having fixed a_1 to 0 we deduce that $t^8 = 0$.

What remains in $0 = d\pi^3 = \pi^3 \wedge (\pi^4 - \pi^1) - \omega^5 \wedge \pi^5 + \omega^2 \wedge \chi^1$ after this reduction is

$$0 = -\omega^5 \wedge (a_7\,\omega^2 - 4a_3\,\omega^3) + \omega^2 \wedge \chi^1 = \omega^2 \wedge (\chi^1 + a_7\,\omega^5) - 4a_3\,\omega^3 \wedge \omega^5.$$

Hence $a_3 = 0$, and $\chi^1 = -a_7\,\omega^5 + \lambda\,\omega^2$ for some function λ.

Applying δ to $\pi^5 = a_7\,\omega^2 - 3a_5\,\omega^5$ we next get

$$-t^1(a_7\,\omega^2 - 3a_5\,\omega^5) + t^8\,\omega^3 - t^7\,\omega^5 = \delta a_7\,\omega^2 + a_7 t^1\,\omega^2 - 3\delta a_5\,\omega^5 - 3a_5 t^7\,\omega^2;$$

equating the coefficients in front of ω^5, we have

$$3a_5 t^1 - t^7 = -3\delta a_5, \quad \text{i.e.,} \quad \delta a_5 = \frac{1}{3}t^7 - t^1 a_5.$$

Using E_7 we may thus suppose that $a_5 = 0$, whereupon $t^7 = 0$ and $\pi^5 = a_7\,\omega^2$. Then

$$\frac{\partial(d\pi^5)}{\partial(\omega^3 \wedge \omega^5)} = \frac{\partial}{\partial(\omega^3 \wedge \omega^5)}(a_7\,d\omega^2 + da_7 \wedge \omega^2) = a_7,$$

while

$$\frac{\partial(d\pi^5)}{\partial(\omega^3 \wedge \omega^5)} = A_3 = 0, \quad \text{according to (4) in section **17.1**.}$$

Thus $a_7 = 0$, which then gives $\pi^5 = 0$ and $\chi^1 = \lambda\,\omega^2$.

Applying the same kind of reasoning to π^3 it is found that

$$\delta a_4 = t^8 - t^1 a_4 \quad \text{and} \quad \delta a_6 = -\frac{2}{3}t^6.$$

Therefore the stability group can be employed to make a_4 and a_6 vanish. Having fixed a_4 and a_6 to 0 we then deduce that $t^7 = t^6 = 0$.

With all of $a_1, \ldots, a_7$ killed,

$$\pi^3 = \pi^4 = \pi^5 = 0.$$

But then

$$\begin{cases} 0 = d\pi^5 = -\omega^5 \wedge \pi^7 + \omega^3 \wedge \lambda \, \omega^2, \\ 0 = d\pi^4 = \frac{1}{3} \omega^3 \wedge \pi^7 - \frac{2}{3} \omega^5 \wedge \pi^6 + \omega^2 \wedge \chi^2, \end{cases}$$

implying that

$$\lambda = 0\text{---and hence also } \chi^1 = 0, \ \pi^7 = b_1 \, \omega^5\text{---}$$

and that π^6, χ^2 are certain linear combinations of ω^2, ω^3 and ω^5—which can be written as

$$\pi^6 = b_2 \, \omega^2 - \frac{1}{2} b_1 \, \omega^3 + b_3 \, \omega^5 \quad \text{and} \quad \chi^2 = b_4 \, \omega^2 - \frac{2}{3} b_2 \, \omega^5.$$

Here the b_i are invariants, so that by assumption $\partial d b_i / \partial \omega^4 = 0$.

Differentiating π^6 we see that

$$\frac{\partial \, d\pi^6}{\partial(\omega^4 \wedge \omega^5)} = -\frac{1}{2} b_1;$$

comparing this with (4) in section **17.1** we deduce that $b_1 = 0$—and hence also $\pi^7 = 0$. Next

$$0 = d\pi^7 = \ldots + \omega^5 \wedge \chi^2 = \cdots + \omega^5 \wedge \left(b_4 \, \omega^2 - \frac{2}{3} b_2 \, \omega^5 \right) = \cdots - b_4 \, \omega^2 \wedge \omega^5,$$

and therefore $b_4 = 0$. So now

$$\pi^6 = b_2 \, \omega^2 + b_3 \, \omega^5 \quad \text{and} \quad \chi^2 = -\frac{2}{3} b_2 \, \omega^5.$$

Differentiating π^6 and comparing with (4) in section **17.1**, we find

$$\frac{\partial \, d\pi^6}{\partial(\omega^3 \wedge \omega^5)} = \frac{\partial \, db_3}{\partial \omega^3} = \frac{\partial \chi^2}{\partial \omega^5} = -\frac{2}{3} b_2,$$

whence db_3 takes the form

$$db_3 = -\frac{2}{3} b_2 \, \omega^3 + c \, \omega^5 + \cdots.$$

Applying d to this, we have

$$0 = \frac{\partial \, d^2 b_3}{\partial(\omega^4 \wedge \omega^5)} = -\frac{2}{3} b_2 + \frac{\partial dc}{\partial \omega^4}.$$

But c is an invariant, and hence

$$0 = \frac{\partial dc}{\partial \omega^4} = \frac{2}{3}b_2, \quad \text{i.e.,} \quad b_2 = 0.$$

Thus $b_1 = b_2 = b_4 = 0$, and from this it is realized that $\pi^7 = \chi^2 = 0$.

Conclusion. *Exploiting the stability group in a suitable way we may assume that*

$$\pi^3 = \pi^4 = \pi^5 = \pi^7 = \chi^1 = \chi^2 = 0 \quad and \quad \pi^6 = I\,\omega^5,$$

with I being an invariant. After we have achieved this the stability algebra is 2-dimensional: $\operatorname{span}_{\mathbb{C}}\{E_1, E_2\}$.

If we now add π^1 and π^2 to $(\omega^1, \ldots, \omega^5)$ and reinterpret the two remaining auxiliary variables as original ones, we get a 7-dimensional pfaffian system $(\omega^1, \ldots, \omega^5, \pi^1, \pi^2)$ in $5+2 = 7$ variables—so that $\{\omega^1, \ldots, \omega^4, \pi^1, \pi^2\}$ defines a local coframe.

Conclusion. *The set* $\{\omega^1, \ldots, \omega^5, \pi^1, \pi^2\}$ *constitutes an e-structure in seven variables.*

The structure equations for this are

$$\begin{cases}
d\omega^1 = 2\omega^1 \wedge \pi^1 + \omega^2 \wedge \pi^2 + \omega^3 \wedge \omega^4, \\
d\omega^2 = \omega^2 \wedge \pi^1 + \omega^3 \wedge \omega^5, \\
d\omega^3 = I\,\omega^2 \wedge \omega^5 + \omega^3 \wedge \pi^1 + \omega^4 \wedge \omega^5, \\
d\omega^4 = \frac{4}{3}I\,\omega^3 \wedge \omega^5 + \omega^4 \wedge \pi^1 + \omega^5 \wedge \pi^2, \\
d\omega^5 = 0, \\
d\pi^1 = 0, \\
d\pi^2 = \pi^2 \wedge \pi^1 - I\,\omega^4 \wedge \omega^5 + \omega^2 \wedge \omega^5.
\end{cases} \qquad (*)$$

These give rise to two different cases.

1. *The invariant I is a constant.* A necessary and sufficient condition for local equivalence in this case is that I and the corresponding invariant $\hat{I}$ have the same values.

Formula (*) provides the *structure constants* for the symmetry group $\operatorname{sym}(\mathcal{P})$.

Having obtained a simple canonical form for $\{\omega^1, \ldots, \omega^5, \pi^1, \pi^2\}$, one also gets a simple expression for the pfaffian system $\mathcal{P} = (\omega^1, \omega^2, \omega^3)$ that was the starting point for our study.

From $\mathcal{P}$ it is finally possible to read off canonical forms for the parabolic PDE $\mathcal{S}_1$ and the PDE system $\mathcal{S}_2$ in analogy with what was done in the preceding section.

2. *The invariant I is not a constant.* By differentiating $d\omega^3$, $d\omega^4$ and $d\pi^2$ it is readily found that dI is a multiple of ω^5:

$$dI = J\,\omega^5.$$

I is a *fundamental invariant* in the sense that all other invariants (including J) are functions of I.

The two pfaffian systems $\mathcal{P}$ and $\hat{\mathcal{P}}$ are locally equivalent if and only if $J = f(I)$ and $\hat{J} = f(\hat{I})$ with the same function f.

Let us now consider sym $(\mathcal{P})$. With $d\omega^5 = 0$ it may be assumed that ω^5 is the differential of a function x^5: $\omega^5 = dx^5$. In that case I is a function of x^5, and hence we might as well regard x^5 as a fundamental invariant. Since there are no auxiliary variables, the invariant x^5 is inessential (see section **14.5**). Therefore we obtain an isomorphic Lie group by setting x^5 equal to a suitable constant—for instance 0. Then $\omega^5 = 0$, $I = 0$, and (*) simplifies to

$$\begin{cases} d\omega^1 = 2\omega^1 \wedge \pi^1 + \omega^2 \wedge \pi^2 + \omega^3 \wedge \omega^4, \\ d\omega^2 = \omega^2 \wedge \pi^1, \\ d\omega^3 = \omega^3 \wedge \pi^1, \\ d\omega^4 = \omega^4 \wedge \pi^1, \\ d\pi^1 = 0, \\ d\pi^2 = \pi^2 \wedge \pi^1 \end{cases}$$

—which are the structure equations of a 6-dimensional Lie parameter group.

17.5 F has one triple and one simple root

If we suppose that $\mathcal{F}$ has a triple root at ∞ and a simple root at 0, $\mathcal{F}$ can be written as

$$\mathcal{F}(\omega^4, \omega^5) = 4\,\omega^4(\omega^5)^3.$$

Thus $A_4 = 1$, while the other A_i vanish.

Since we are only interested in e-structures in more than five variables, this case turns out to be a disappointment: there are no such.

17.6 $\mathcal{F}$ has two double roots

Assuming $\mathcal{F}$ to have two double roots, these may be moved to 0 and ∞ respectively by a bilinear transformation, and then $\mathcal{F}$ can be written as

$$\mathcal{F}(\omega^4, \omega^5) = 6\,(\omega^4)^2(\omega^5)^2,$$

so that $A_3 = 1$, and the other A_i vanish. The formulas

$$\begin{cases} \delta A_2 = -t^2 A_1 - (3t^1 + t^4)A_2 - 3t^3 A_3, \\ \delta A_3 = -2t^2 A_2 - 2(t^1 + t^4)A_3 - 2t^3 A_4, \\ \delta A_4 = -3t^2 A_3 - (t^1 + 3t^4)A_4 - t^3 A_5 \end{cases}$$

from (5) in section **17.1** imply in this case that

$$t^3 = t^1 + t^4 = t^2 = 0.$$

Thus π^2, π^3 and $\pi^1 + \pi^4$ are linear combinations of $\omega^1, \ldots, \omega^5$. With $\pi^1 + \pi^4 = \sum_{i=1}^{5} a_i\,\omega^i$ we have on the one hand

$$\delta(\pi^1 + \pi^4) = \sum_{i=1}^{5}(\delta a_i\,\omega^i + a_i\,\delta\omega^i),$$

and on the other, from (4) in section **17.1**,

$$\delta(\pi^1 + \pi^4) = \frac{2}{3}t^7\,\omega^3 - \frac{1}{3}t^5\omega^4 - \frac{1}{3}t^6\,\omega^5 + t^8\,\omega^1 + t^9\,\omega^2.$$

Comparing these expressions we see—just as in section **17.2**—that $E_5, \ldots, E_9$ can be used to kill all the a_i, whereupon $\pi^1 + \pi^4$ vanishes. Once we have achieved this, $t^5 = t^6 = t^7 = t^8 = t^9 = 0$, and then *just one auxiliary variable remains*.

Only being interested in e-structures involving more than the original five variables we have to keep this auxiliary variable alive, and thus necessarily $t^1 = -t^4 \neq 0$.

With $t^i = 0$ for $i = 2, 3, 5, 6, 7, 8, 9$ and $t^4 = -t^1$, δ acts on the ω^i and π^j by

$$\begin{cases} \delta\omega^1 = t^1\,\omega^1, & \delta\omega^2 = -t^1\,\omega^2, & \delta\omega^3 = 0, & \delta\omega^4 = t^1\omega^4, & \delta\omega^5 = -t^1\,\omega^5, \\ \delta\pi^2 = 2t^1\,\pi^2, & \delta\pi^3 = -2t^1\,\pi^3, & \delta\pi^5 = -t^1\pi^5, & \delta\pi^6 = t^1\,\pi^6, & \delta\pi^7 = 0. \end{cases}$$

Because $t^1 + t^4 = t^2 = t^3 = t^5 = t^6 = t^7 = t^8 = t^9 = 0$, the one-forms $\pi^1 + \pi^4$, π^2, π^3, π^5, π^6, π^7, π^8 and π^9 are all linear combinations of ω^1, $\ldots$, ω^5 with coefficients that are invariants of the prolonged equivalence problem. Now if any of these coefficients does depend on the remaining auxiliary variable, the latter can be got rid of by setting the corresponding coefficient equal to a suitable constant. Not wanting this to happen, we require that

all the coefficients are functions of the original variables only, and are thus killed by δ.

By the above the ω^i are 'eigenforms' of δ with the eigenvalues t^1, $-t^1$, 0, t^1 and $-t^1$ respectively.

Also the π^j are 'eigenforms' of δ, and so must be *linear combinations of those ω^i that belong to the same eigenvalues*. In this way it follows that

$$\pi^2 = \pi^3 = 0 \qquad\qquad \text{(because } \pi^2, \pi^3 \text{ have the eigenvalue } 2t^1\text{)},$$
$$\pi^5 = a\,\omega^2 + b\,\omega^5 \qquad \text{(because } \pi^5 \text{ has the eigenvalue } -t^1\text{)},$$
$$\pi^6 = a'\,\omega^1 + b'\,\omega^4 \qquad \text{(because } \pi^6 \text{ has the eigenvalue } t^1\text{)},$$
$$\pi^7 = c\,\omega^3 \qquad\qquad \text{(because } \pi^7 \text{ has the eigenvalue } 0\text{)}.$$

From (4) in section **17.1** we then see for instance that

$$0 = d\pi^3 = -\omega^5 \wedge (a\,\omega^2 + b\,\omega^5) + \omega^2 \wedge \chi^1 - B_1\,\omega^1 \wedge \omega^2$$
$$= \omega^2 \wedge (\chi^1 + a\,\omega^5) - B_1\,\omega^1 \wedge \omega^2.$$

Consequently $B_1 = 0$, and $\chi^1 = -a\,\omega^5 + e\,\omega^2$ for some function e.
Considering $0 = d\pi^2$ in the same way, we have

$$B_4 = 0 \quad \text{and} \quad \chi^2 = -a'\,\omega^4 + f\,\omega^1 \quad \text{for some } f.$$

Finally, from $0 = d\pi^1 + d\pi^4$ it is inferred that

$$B_2 = B_3 = 0, \quad f = e, \quad b' = b, \quad \text{and} \quad a = -a' = 3/4.$$

Hence

$$\begin{cases} \pi^5 = \tfrac{3}{4}\,\omega^2 + b\,\omega^5, \\ \pi^6 = -\tfrac{3}{4}\,\omega^1 + b\,\omega^4, \\ \chi^1 = -\tfrac{3}{4}\,\omega^5 + e\,\omega^2, \\ \chi^2 = \tfrac{3}{4}\,\omega^4 + e\,\omega^1, \end{cases}$$

where b and e are invariants.

Lemma 17.6.1. *If there is to be one auxiliary variable, the coefficients b, c and e appearing in the expressions for π^5, π^6, π^7, χ^1 and χ^2 are* **constants**.

Proof. If I is an invariant in the five original variables,

$$dI = \sum_{i=1}^{5} I_{,i}\,\omega^i \quad \text{and} \quad 0 = \sum_{i=1}^{5} (\delta I_{,i}\,\omega^i + I_{,i}\,\delta\omega^i).$$

Suppose that $I_{,1} \neq 0$. Because $\delta\omega^1 = t^1\,\omega^1$, E_1 can be used to fix the value of $I_{,1}$—for instance to 1—whereupon the auxiliary variable vanishes.

The same reasoning works for $I_{,2}$, $I_{,4}$ and $I_{,5}$ as well, but *not* for $I_{,3}$, since $\delta\omega^3 = 0$. Hence the only possibility for dI is $dI = J\,\omega^3$ for some function J. But then

$$0 = d^2 I = dJ \wedge \omega^3 + J\,d\omega^3 = dJ \wedge \omega^3 + J\,(\cdots + \omega^4 \wedge \omega^5),$$

and so

$$0 = \frac{\partial(d^2 I)}{\partial(\omega^4 \wedge \omega^5)} = J, \quad \text{i.e.,} \quad dI = 0.$$

$\square$

Now $d\pi^5$, $d\pi^6$ and $d\pi^7$ can be calculated in two ways: either as

$$d\pi^5 = \frac{3}{4}\,d\omega^2 + b\,d\omega^5 = \cdots, \quad d\pi^6 = -\frac{3}{4}\,d\omega^1 + b\,d\omega^4 = \cdots,$$

$$d\pi^7 = c\,d\omega^3 = \cdots,$$

or by using (4) in section **17.1**. A comparison shows that

$$C_1 = C_3 = D_1 = D_2 = 0, \quad C_2 = \frac{16}{9}b, \quad E = -\frac{128}{9}b^2, \quad e = b\left(\frac{1}{2} + \frac{4}{3}b^2\right).$$

Of the π^i, only $\pi^1 = -\pi^4$ contains the auxiliary variable. On adjoining the latter to the five original variables, the prolonged equivalence problem is defined on a 6-dimensional manifold, admitting $\{\omega^1, \ldots, \omega^5, \pi^1\}$ as a local coframe.

Thus $\{\omega^1, \ldots, \omega^5, \pi^1\}$ defines an e-structure in six variables!

Inserting the results above into the expressions for $d\omega^1$, …, $d\omega^5$, $d\pi^1$ given in section **17.1**, the structure equations are found to be

$$\begin{cases} d\omega^1 = \omega^1 \wedge \pi^1 + \omega^3 \wedge \omega^4, \\ d\omega^2 = \omega^2 \wedge \pi^1 + \omega^3 \wedge \omega^5, \\ d\omega^3 = \frac{3}{2}\,\omega^1 \wedge \omega^2 + b\,\omega^1 \wedge \omega^5 + b\,\omega^2 \wedge \omega^4 + \omega^4 \wedge \omega^5, \\ d\omega^4 = \left(\frac{3}{2} - \frac{4}{3}b^2\right)\omega^1 \wedge \omega^3 + \frac{4}{3}b\,\omega^3 \wedge \omega^4 + \omega^4 \wedge \pi^1, \\ d\omega^5 = \left(\frac{3}{2} - \frac{4}{3}b^2\right)\omega^2 \wedge \omega^3 - \frac{4}{3}b\,\omega^3 \wedge \omega^5 - \omega^5 \wedge \pi^1, \\ d\pi^1 = b\left(\frac{1}{2} + \frac{4}{3}b^2\right)\omega^1 \wedge \omega^2 + \frac{3}{2}\,\omega^1 \wedge \omega^5 + \frac{3}{2}\,\omega^2 \wedge \omega^4 - b\,\omega^4 \wedge \omega^5, \end{cases}$$

where b is an arbitrary parameter. By the converse of the second fundamental theorem, in section **8.5**, this structure defines a 6-dimensional Lie parameter group for each b.

This Lie group turns out to be decomposable for generic values of b. As we do not want to get involved in any general theory of decomposition of Lie groups, let us just state the following **fact**.

Let ρ and σ be the roots of the second order equation

$$x^2 + \frac{4}{3}bx + \frac{3}{2} - \frac{4}{3}b^2 = 0.$$

If $b \neq \pm\sqrt{27/32}$, these roots are different, and then a new coframe $\{\Omega^1, \Omega^2, \Omega^3, \Pi^1, \Pi^2, \Pi^3\}$ may be defined by

$$
\begin{aligned}
\Omega^1 &:= \omega^4 + \rho\,\omega^1, & \Pi^1 &:= \omega^4 + \sigma\,\omega^1, \\
\Omega^2 &:= \omega^5 - \rho\,\omega^2, & \Pi^2 &:= \omega^5 - \sigma\,\omega^2, \\
\Omega^3 &:= \pi^1 + \sigma\,\omega^3, & \Pi^3 &:= \pi^1 + \rho\,\omega^3.
\end{aligned}
$$

Using this, the structure gets simplified to

$$
\begin{cases}
d\Omega^1 = \Omega^1 \wedge \Omega^3, \\
d\Omega^2 = -\Omega^2 \wedge \Omega^3, \\
d\Omega^3 = \frac{3\rho+7\sigma}{4}\,\Omega^1 \wedge \Omega^2,
\end{cases}
\qquad
\begin{cases}
d\Pi^1 = \Pi^1 \wedge \Pi^3, \\
d\Pi^2 = -\Pi^2 \wedge \Pi^3, \\
d\Pi^3 = \frac{3\sigma+7\rho}{4}\,\Pi^1 \wedge \Pi^2.
\end{cases}
$$

There are two different cases to consider.

1. If $(3\rho + 7\sigma)(3\sigma + 7\rho) \neq 0$, the symmetry group $\mathrm{sym}(\mathcal{P})$ that we are looking for is the direct product of two simple 3-dimensional Lie groups.

2. If $(3\rho + 7\sigma)(3\sigma + 7\rho) = 0$, $\mathrm{sym}(\mathcal{P})$ is the direct product of two 3-dimensional Lie groups, with one being simple and the other solvable.

Finally, if $b = \pm\sqrt{27/32}$ the two roots ρ and σ coincide, and then we instead introduce a new coframe by

$$
\begin{aligned}
\Omega^1 &:= \omega^4 + \frac{8}{3}b\,\omega^1, & \Pi^1 &:= \omega^4 - \frac{2}{3}b\,\omega^1, \\
\Omega^2 &:= \omega^5 - \frac{8}{3}b\,\omega^2, & \Pi^2 &:= \omega^5 + \frac{2}{3}b\,\omega^2, \\
\Omega^3 &:= \omega^3, & \Pi^3 &:= \pi^1 - \frac{2}{3}b\,\omega^3.
\end{aligned}
$$

The corresponding structure equations are

$$
\begin{cases}
d\Omega^1 = \Omega^1 \wedge \Pi^3 + \frac{10}{3}b\,\Omega^3 \wedge \Pi^1, \\
d\Omega^2 = -\Omega^2 \wedge \Pi^3 - \frac{10}{3}b\,\Omega^3 \wedge \Pi^2, \\
d\Omega^3 = \frac{1}{2}\,\Omega^1 \wedge \Pi^2 - \frac{1}{2}\,\Omega^2 \wedge \Pi^1, \\
d\Pi^1 = \Pi^1 \wedge \Pi^3, \\
d\Pi^2 = -\Pi^2 \wedge \Pi^3, \\
d\Pi^3 = -\frac{5}{3}b\,\Pi^1 \wedge \Pi^2.
\end{cases}
$$

Now this is reminiscent of the structure belonging to the group of Euclidean motions in $\mathbb{C}^3$—that is,

$$\begin{cases} d\omega^i = \sum_{k=1}^{3} \omega^k \wedge \omega^i_k & \text{for } i = 1, 2, 3, \\ d\omega^i_j = \sum_{k=1}^{3} \omega^i_k \wedge \omega^k_j & \text{for } i, j = 1, 2, 3, \end{cases}$$

where $\{\omega^1, \omega^2, \omega^3\}$ is a coframe for $\mathbb{C}^3$ and $\omega^i_j = -\omega^j_i$ are connection forms. And in fact

sym $(\mathcal{P})$ *is in this case isomorphic to the group of Euclidean motions in* $\mathbb{C}^3$.

18

Involutive second order PDE systems in one dependent and three independent variables, solved by the method of Monge

Cartan's local existence theorem for solutions of involutive PDE systems is remarkable because of its great generality; one disadvantage however is that it is based upon the Cauchy–Kowalewski theorem, which requires analyticity.

The main aim of this monograph is to avoid power series arguments by exploiting the possible *singular vector field systems and their first integrals*. And we have seen that this works quite well for second order PDE systems in one dependent and *two* independent variables—at least if there are sufficiently many such first integrals.

Unfortunately there seems to be a problem when generalizing these results: as seen in section **5.2** the generic second order PDE in one dependent and more than two independent variables *does not admit any singular vector field*! So the methods used thus far appear to have a rather limited applicability.

In order to gain a more general perspective on these matters, we will in this last chapter study *involutive second order PDE systems in one dependent and three independent variables*, following [Cartan 1911].

To be more precise we are going to look at involutive PDE systems $\mathcal{S}_h \subset J^2(\mathbb{C}^3_x, \mathbb{C}_z)$, given by

$$\mathcal{S}_h: \quad F_k(x^1, x^2, x^3; z; p_1, p_2, p_3; p_{11}, p_{12}, p_{13}, p_{22}, p_{23}, p_{33}) = 0, \quad k = 1, \ldots, h,$$
$$\text{where} \quad dF_1 \wedge \cdots \wedge dF_h \neq 0,$$

and having the property that $\mathcal{S}_h$ maps *onto* $J^1(\mathbb{C}^3_x, \mathbb{C}_z)$ under the canonical projection $J^2(\mathbb{C}^3_x, \mathbb{C}_z) \longrightarrow J^1(\mathbb{C}^3_x, \mathbb{C}_z)$.

There are two extreme cases: $\mathcal{S}_0 = J^2(\mathbb{C}^3_x, \mathbb{C}_z)$, and

$$\mathcal{S}_6: \quad p_{ij} = f_{ij}(x^1, x^2, x^3; z; p_1, p_2, p_3) \quad \text{for } i, j = 1, 2, 3.$$

505

Let us consider S_0 at first. The corresponding pfaffian system is

$$^2Ct^{3,1}: \quad \begin{cases} \theta^0 = dz - \sum_{j=1}^3 p_j \, dx^j, \\ \theta^i = dp_i - \sum_{j=1}^3 p_{ij} \, dx^j \quad \text{for } i = 1,2,3. \end{cases}$$

The general 3-dimensional integral manifold on which $dx^1 \wedge dx^2 \wedge dx^3 \neq 0$ is given by

$$\begin{cases} z = f(x^1, x^2, x^3), \\ p_i = \partial f / \partial x^i, \qquad \text{for } i, j = 1, 2, 3, \\ p_{ij} = \partial^2 f / \partial x^i \partial x^j \end{cases}$$

where f is an arbitrary function of three variables—that is, the integral manifolds are 2-graphs of functions on $\mathbb{C}_x^3$.

The Lie structure of $^2Ct^{3,1}$ is

$$\begin{cases} d\theta^0 \equiv 0, \\ d\theta^i \equiv \sum_{j=1}^3 dx^j \wedge dp_{ij} \end{cases} \qquad (\mathrm{mod}\ {}^2Ct^{3,1}).$$

In the next section we will see that $^2Ct^{3,1}$ is *involutive* (as expected), and has the Cartan characters $s_1 = 3$, $s_2 = 2$ and $s_3 = 1$. Hence the general solution will indeed depend on one arbitrary function of three variables (see section **3.3**).

Let us then consider the other extreme case: S_6, given by $p_{ij} = f_{ij}(x; z; p)$ for $i, j = 1, 2, 3$. From these defining equations it follows that x^1, x^2, x^3, z, p_1, p_2, p_3 can be used as local coordinates on S_6, which therefore—as a pure manifold, disregarding its structure—locally can be identified with $J^1(\mathbb{C}_x^3, \mathbb{C}_z)$.

The corresponding pfaffian system $\mathcal{P}(S_6) = {}^2Ct^{3,1}|_{S_6}$ is generated by the one-forms

$$\begin{cases} \theta^0 = dz - \sum_{j=1}^3 p_j \, dx^j, \\ \theta^i = dp_i - \sum_{j=1}^3 f_{ij}(x; z; p) \, dx^j \quad \text{for } i = 1, 2, 3. \end{cases}$$

Here $d\theta^0 \equiv 0 \ (\mathrm{mod}\ \mathcal{P}(S_6))$, while

$$d\theta^i = \sum_{j=1}^3 dx^j \wedge df_{ij} \equiv \sum_{j=1}^3 dx^j \wedge \left(\sum_{k=1}^3 \frac{df_{ij}}{dx^k} \, dx^k \right) \qquad (\mathrm{mod}\ \mathcal{P}(S_6))$$

$$= \sum_{1 \leq j < k \leq 3} \left(\frac{df_{ij}}{dx^k} - \frac{df_{ik}}{dx^j} \right) dx^j \wedge dx^k,$$

where

$$\frac{d}{dx^k} = \frac{\partial}{\partial x^k} + p_k \frac{\partial}{\partial z} + \sum_{j=1}^{3} f_{jk} \frac{\partial}{\partial p_j}.$$

By assumption $\mathcal{P}(S_6)$ is involutive with respect to the independence condition $dx^1 \wedge dx^2 \wedge dx^3 \neq 0$, and therefore admits an involution (X_1, X_2, X_3), with

$$X_k = \frac{\partial}{\partial x^k} + a_k \frac{\partial}{\partial z} + \sum_{j=1}^{3} b_{kj} \frac{\partial}{\partial p_j} \in \mathcal{P}^\perp \quad \text{for } k = 1, 2, 3,$$

where a_k and b_{kj} are certain functions on S_6. The involutivity means that $d\theta^i(X_k, X_l) = 0$ for $i, k, l = 1, 2, 3$. In our case

$$d\theta^i(X_k, X_l) = \frac{df_{ik}}{dx^l} - \frac{df_{il}}{dx^k} \quad \text{for } k < l,$$

so that

$$S_6 \text{ is involutive} \iff \frac{df_{ik}}{dx^l} = \frac{df_{il}}{dx^k} \quad \text{for } i, k, l = 1, 2, 3.$$

Conclusion. *An involutive S_6 has a trivial Lie structure:*

$$d\theta^i \equiv 0 \pmod{\mathcal{P}(S_6)} \quad \text{for } i = 0, 1, 2, 3.$$

*Therefore $\mathcal{P}(S_6)$ **is complete**, and hence can be integrated by means of the Frobenius theorem.*

Consequently $\mathcal{P}(S_6)$ defines a 4-parameter foliation of the 7-dimensional S_6 with 3-dimensional leaves.

The general idea in the following is that by imposing the condition of involutivity on $S_1, \ldots, S_5$, it should be rather straightforward to make a

• structural classification of these S_h,

and then we use the Lie structures found in order to read off

• the possible singular vector field systems.

We will later see that the following gratifying result comes out of this procedure:

> all involutive second order PDE systems in one dependent and three independent variables consisting of **at least two** PDEs do admit singular vector field systems!

So the next problem is how to utilize the singular vector fields in the best possible way.

There is one case which is crystal clear: when a pfaffian system $\mathcal{P}$ admits Cauchy characteristic vector fields. Recalling that a vector field $C \in \mathcal{P}^{\perp}$ is Cauchy charateristic if and only if $C \rfloor d\mathcal{P} \equiv 0 \pmod{\mathcal{P}}$, we make the following definition in the general case.

Definition. Let $\operatorname{sing}(\mathcal{P}^{\perp})$ be a singular subsystem of the vector field system $\mathcal{P}^{\perp}$, with the dual pfaffian system $\operatorname{sing}(\mathcal{P})$ $(\supset \mathcal{P})$. Then

$$\mathcal{P}_{\text{sing}} := \{\theta \in \mathcal{P} \mid \operatorname{sing}(\mathcal{P}^{\perp}) \rfloor d\theta \equiv 0 \pmod{\mathcal{P}}\}.$$

Thus $\operatorname{sing}(\mathcal{P}^{\perp}) \rfloor d\mathcal{P}_{\text{sing}} \equiv 0 \pmod{\mathcal{P}}$.

Example from section 16.1. Let $\mathcal{P} = (\theta^1, \theta^2, \theta^3)$ be a pfaffian system on a 7-dimensional manifold, with $\mathcal{P}$ having the Lie structure

$$\begin{cases} d\theta^1 \equiv 0, \\ d\theta^2 \equiv \pi^1 \wedge \pi^2, \qquad\qquad (\text{mod } \mathcal{P}). \\ d\theta^3 \equiv \pi^1 \wedge \pi^3 + \pi^2 \wedge \pi^4 \end{cases}$$

Setting $\operatorname{sing}(\mathcal{P}) := (\theta^1, \theta^2, \theta^3, \pi^1, \pi^2)$, we see that the dual vector field system $\operatorname{sing}(\mathcal{P}^{\perp}) = (\partial/\partial\pi^3, \partial/\partial\pi^4)$ is a singular subsystem of $\mathcal{P}^{\perp}$—since $\operatorname{sing}(\mathcal{P}^{\perp}) \rfloor d\theta^2 \equiv 0 \pmod{\mathcal{P}}$. The corresponding $\mathcal{P}_{\text{sing}}$ is

$$\mathcal{P}_{\text{sing}} = \{\theta \in \mathcal{P} \mid \operatorname{sing}(\mathcal{P}^{\perp}) \rfloor d\theta \equiv 0 \pmod{\mathcal{P}}\} = (\theta^1, \theta^2).$$

A remarkable fact proved in section **16.1** is that under fairly general circumstances

$$\text{integrating } \mathcal{P} \Longleftrightarrow \text{integrating } \mathcal{P}_{\text{sing}}.$$

That is, if $\mathcal{N}$ is an integral manifold of $\mathcal{P}_{\text{sing}}$—so that $\theta^1|_{\mathcal{N}} = \theta^2|_{\mathcal{N}} = 0$— then it *automatically follows* that $\theta^3|_{\mathcal{N}} = 0$ too!

Moreover the integration of $\mathcal{P}_{\text{sing}}$ is made easy by the fact that $\mathcal{P}_{\text{sing}}$ admits a 2-dimensional Cauchy characteristic vector field system—in fact

$$C(\mathcal{P}_{\text{sing}}) = (\theta^1, \theta^2, \theta^3, \pi^1, \pi^2) = \operatorname{sing}(\mathcal{P}).$$

Therefore the integration of $\mathcal{P}_{\text{sing}}$ (and hence also of $\mathcal{P}$) is reduced to five variables only.

Because any Cauchy characteristic vector field system is complete, the favourable case $\operatorname{sing}(\mathcal{P}) = C(\mathcal{P}_{\text{sing}})$ can occur only if $\operatorname{sing}(\mathcal{P})$ is complete—which is not always true. However, the study of hyperbolic

PDEs in chapter **12** indicates that there is a legitimate hope for the integration problem if sing $(\mathcal{P})$ *admits sufficiently many first integrals.*

Generalizing from this we can now announce the main principle used in [Cartan 1911], which Cartan calls the *method of Monge*.

The method of Monge. *Suppose that* $\mathcal{P}^{\perp}$ *admits a singular subsystem* sing $(\mathcal{P}^{\perp})$ *having 'enough' first integrals. Then*

- *integrating* $\mathcal{P} \iff$ *integrating* $\mathcal{P}_{\text{sing}}$,
- *it is easier to integrate* $\mathcal{P}_{\text{sing}}$ *than to integrate* $\mathcal{P}$ *directly.*

And in particular it is in this way possible to find the integral manifolds of $\mathcal{P}$ *without using the Cauchy–Kowalewski theorem.*

This general form of the Monge method is neither stated, nor proved, in [Cartan 1911]. But—as will be seen in the following—it is applied by Cartan to a lot of special cases, and given a separate proof in each event. So the overall impression is that the method of Monge really is a powerful principle for utilizing singular vector field systems—and definitely is a *great challenge for future research in PDE theory.*

Let us finally mention another method which sometimes can be used in order to simplify the integration problem for a pfaffian system $\mathcal{P} = (\theta^1, \ldots, \theta^s)$.

According to chapter **6** it is possible to write the single pfaffian equation $\theta^1 = 0$ as $dz - \sum_{j=1}^{g} p_j \, dx^j = 0$ for a certain integer g. Then for some function a, $\theta^1 = a\left(dz - \sum_{j=1}^{g} p_j \, dx^j\right)$, so that

$$d\theta^1 = \frac{da}{a} \wedge \theta^1 + a \, d\theta^1 \equiv a \sum_{j=1}^{g} dx^j \wedge dp_j \pmod{\theta^1}.$$

Hence

$$(d\theta^1)^{\wedge g} := \overbrace{d\theta^1 \wedge \cdots \wedge d\theta^1}^{g \text{ times}} \not\equiv 0 \pmod{\theta^1},$$

while

$$(d\theta^1)^{\wedge(g+1)} = \overbrace{d\theta^1 \wedge \cdots \wedge d\theta^1}^{g+1 \text{ times}} \equiv 0 \pmod{\theta^1}.$$

Comparing this with section **3.5**, we see that

$$g = \text{ the genus of } d\theta^1 \pmod{\theta^1}.$$

The general g-dimensional integral manifold of the pfaffian equation $\theta^1 = 0$ on which $dx^1 \wedge \cdots \wedge dx^g \neq 0$ has the very simple form

$$\mathcal{N}_f: \quad z = f(x^1, \ldots, x^g), \quad p_i = \frac{\partial f}{\partial x^i}(x^1, \ldots, x^g) \quad \text{for } i = 1, \ldots, g,$$

where f is an *arbitrary* function of g variables.

It is now tempting to restrict $\mathcal{P}$ to each of these integral manifolds,

$$\mathcal{P}_f := \mathcal{P}_{|\mathcal{N}_f} = (\theta^2_{|\mathcal{N}_f}, \ldots, \theta^s_{|\mathcal{N}_f}) \quad \text{for each } f(x^1, \ldots, x^g),$$

and then integrate each of the $(s-1)$-dimensional pfaffian systems $\mathcal{P}_f$.

The problem with this approach is that requiring $\mathcal{P}_f$ to be involutive will in general *impose conditions on f*—so that f is not arbitrary after all.

However, supposing s_p to be the last nonzero integer in the list s_1, ..., s_p, s_{p+1}, ... of Cartan characters, we know from section **3.3** that at least the general formal integral manifold of $\mathcal{P}$ depends on s_p *arbitrary functions of p variables*. Therefore, if $g = p$ (or even better: $g < p$), there is some hope for our f to stay arbitrary also when we require $\mathcal{P}_f$ to be involutive.

Conclusion. *If the genus of $d\theta^i$ (mod θ^i) is sufficiently small it may be possible to restrict $\mathcal{P}$ to the integral manifolds $\mathcal{N}_f$ of $\theta^i = 0$, with the function f being arbitrary, and with each $\mathcal{P}_{|\mathcal{N}_f}$ being involutive.*

Special case. If $d\theta^1 \equiv 0$ (mod θ^1), the Frobenius theorem implies that

$$\theta^1 = 0 \Longleftrightarrow f^1 = \text{constant} \quad \text{for a certain function } f^1.$$

Then it clearly suffices to integrate the restrictions of $\mathcal{P} = (\theta^1, \ldots, \theta^s)$ to the level surfaces of f^1.

18.1 Preliminaries

In this section we make some preparations in order to facilitate the structural classification that follows in the next section.

The contact pfaffian system $^2Ct^{3,1}$ of $J^2(\mathbb{C}^3_x, \mathbb{C}_z)$ is generated by the one-forms

$$\theta^0 = dz - \sum_{j=1}^{3} p_j \, dx^j \quad \text{and} \quad \theta^i = dp_i - \sum_{j=1}^{3} p_{ij} \, dx^j \quad \text{for } i = 1, 2, 3.$$

Setting $\omega^i := dx^i$ and $\pi^i_j := dp_{ij} = dp_{ji} = \pi^j_i$ for $i, j = 1, 2, 3$,

$$\{\omega^1, \omega^2, \omega^3, \theta^0, \theta^1, \theta^2, \theta^3, \pi^1_1, \pi^1_2, \pi^1_3, \pi^2_2, \pi^2_3, \pi^3_3\}$$

constitutes a local coframe for $J^2(\mathbb{C}^3_x, \mathbb{C}_z)$. Expressing the structure of $^2Ct^{3,1}$ in this coframe we get

$$d\theta^0 = \sum_{j=1}^{3} \omega^j \wedge dp_j = \sum_{j=1}^{3} \omega^j \wedge \theta^j + \sum_{i,j=1}^{3} p_{ij}\, dx^i \wedge dx^j$$

$$= \sum_{j=1}^{3} \omega^j \wedge \theta^j$$

(with the last equality arising because $p_{ij} = p_{ji}$ and $dx^i \wedge dx^j = -dx^j \wedge dx^i$), and

$$d\theta^i = \sum_{j=1}^{3} \omega^j \wedge \pi^i_j \quad \text{for } i = 1, 2, 3.$$

Now $\{\theta^0, \theta^1, \theta^2, \theta^3\}$ is one possible basis for $^2Ct^{3,1}$. If the functions a^i_j, living on $J^2(\mathbb{C}^3_x, \mathbb{C}_z)$, satisfy $\det(a^i_j) \neq 0$, another basis is given by

$$\begin{pmatrix} \bar{\theta}^0 \\ \bar{\theta}^1 \\ \bar{\theta}^2 \\ \bar{\theta}^3 \end{pmatrix} = \begin{pmatrix} 1 & 0 & 0 & 0 \\ 0 & a^1_1 & a^1_2 & a^1_3 \\ 0 & a^2_1 & a^2_2 & a^2_3 \\ 0 & a^3_1 & a^3_2 & a^3_3 \end{pmatrix} \begin{pmatrix} \theta^0 \\ \theta^1 \\ \theta^2 \\ \theta^3 \end{pmatrix}.$$

The transformations defined in this way are the most general that preserve θ^0 as well as the pfaffian system $(\theta^1, \theta^2, \theta^3)$. With

$$A := \begin{pmatrix} a^1_1 & a^1_2 & a^1_3 \\ a^2_1 & a^2_2 & a^2_3 \\ a^3_1 & a^3_2 & a^3_3 \end{pmatrix}, \quad \text{the matrix above is written as } \begin{pmatrix} 1 & 0 \\ 0 & A \end{pmatrix}.$$

The family of such matrices clearly form a group, which is denoted by $\mathfrak{A}$.

Claim. There is a prolongation $\mathfrak{A}^{(1)}$ of $\mathfrak{A}$ acting on the coframe $\{\omega^1, \omega^2, \omega^3, \theta^0, \theta^1, \theta^2, \theta^3, \pi^1_1, \pi^1_2, \pi^1_3, \pi^2_2, \pi^2_3, \pi^3_3\}$ in such a way as to preserve the Lie structure of $^2Ct^{3,1}$.

To see this, set

$$\theta := \begin{pmatrix} \theta^1 \\ \theta^2 \\ \theta^3 \end{pmatrix} \quad \text{and} \quad \bar{\theta} := \begin{pmatrix} \bar{\theta}^1 \\ \bar{\theta}^2 \\ \bar{\theta}^3 \end{pmatrix},$$

so that $\bar{\theta} = A\,\theta$ and $\theta = A^{-1}\,\bar{\theta}$. With $A^{-1} = (\alpha^i_j)$,

$$d\bar{\theta}^0 = d\theta^0 = \sum_{i=1}^{3} \omega^i \wedge \theta^i = \sum_{i=1}^{3} \omega^i \wedge \left(\sum_{j=1}^{3} \alpha^i_j\, \bar{\theta}^j \right) = \sum_{j=1}^{3} \bar{\omega}^j \wedge \bar{\theta}^j,$$

where $\bar{\omega}^j := \sum_{i=1}^{3} \alpha^i_j\, \omega^i$. Introducing

$$\omega := \begin{pmatrix} \omega^1 \\ \omega^2 \\ \omega^3 \end{pmatrix} \quad \text{and} \quad \bar{\omega} := \begin{pmatrix} \bar{\omega}^1 \\ \bar{\omega}^2 \\ \bar{\omega}^3 \end{pmatrix},$$

we get

$$\bar{\omega} = \left(A^{-1} \right)^T \omega = \left(A^T \right)^{-1} \omega \quad \text{and} \quad \omega = A^T\,\bar{\omega}.$$

Consequently

$$d\bar{\theta}^i \equiv \sum_{j=1}^{3} a^i_j\, d\theta^j = \sum_{j=1}^{3} a^i_j \left(\sum_{k=1}^{3} \omega^k \wedge \pi^j_k \right) \quad (\mathrm{mod}\ \bar{\theta}^1, \bar{\theta}^2, \bar{\theta}^3)$$

$$= \sum_{j=1}^{3} a^i_j \left(\sum_{k=1}^{3} \left(\sum_{l=1}^{3} a^l_k\, \bar{\omega}^l \right) \wedge \pi^j_k \right) = \sum_{l=1}^{3} \bar{\omega}^l \wedge \left(\sum_{j,k=1}^{3} a^i_j a^l_k\, \pi^j_k \right)$$

$$= \sum_{j=1}^{3} \bar{\omega}^j \wedge \bar{\pi}^i_j, \quad \text{where} \quad \bar{\pi}^i_j := \sum_{k,l=1}^{3} a^i_k a^j_l\, \pi^k_l.$$

So if we let $\mathfrak{A}^{(1)}$ act on the ω^i and π^i_j by

$$\omega^i \mapsto \sum_{j=1}^{3} \alpha^j_i\, \omega^j \quad \text{and} \quad \pi^i_j \mapsto \sum_{k,l}^{3} a^i_k a^j_l\, \pi^k_l,$$

the equations $d\theta^i = \sum_{j=1}^{3} \omega^j \wedge \pi^i_j$ for $i = 1, 2, 3$ go over into the congruences

$$d\bar{\theta}^i \equiv \sum_{j=1}^{3} \bar{\omega}^j \wedge \bar{\pi}^i_j \quad (\mathrm{mod}\ \bar{\theta}^0, \ldots, \bar{\theta}^3), \quad i = 1, 2, 3.$$

Note moreover that the π^i_j transform as the coefficients of a quadratic form. For if

$$x^i = \sum_{k=1}^{n} a^i_k\, \bar{x}^k \quad \text{for } i = 1, \ldots, n,$$

the quadratic form $\sum_{i,j=1}^{n} c_{ij} x^i x^j$ gets expressed as follows in the $\bar{x}^i$:

$$\sum_{i,j=1}^{n} c_{ij} x^i x^j = \sum_{i,j=1}^{n} c_{ij} \left(\sum_{k=1}^{n} a_k^i \bar{x}^k \right) \left(\sum_{l=1}^{n} a_l^j \bar{x}^l \right)$$

$$= \sum_{i,j=1}^{n} \bar{c}_{ij} \bar{x}^i \bar{x}^j, \quad \text{with} \quad \bar{c}_{ij} := \sum_{k,l=1}^{n} a_i^k a_j^l c_{kl}.$$

Theorem 18.1.1. $\mathfrak{A}^{(1)}$ *prolongs* $\mathfrak{A}$, *maps linear combinations of the* ω^i *to linear combinations of the* ω^i, *and transforms the structure equations*

$$\begin{cases} d\theta^0 = \sum_{j=1}^{3} \omega^j \wedge \theta^j, \\ d\theta^i = \sum_{j=1}^{3} \omega^j \wedge \pi_j^i & \text{for } i = 1, 2, 3 \end{cases}$$

of $^2 C t^{3,1} = (\theta^0, \theta^1, \theta^2, \theta^3)$ *into*

$$\begin{cases} d\bar{\theta}^0 = \sum_{j=1}^{3} \bar{\omega}^j \wedge \bar{\theta}^j, \\ d\bar{\theta}^i \equiv \sum_{j=1}^{3} \bar{\omega}^j \wedge \bar{\pi}_j^i \pmod{\bar{\theta}^1, \bar{\theta}^2, \bar{\theta}^3} & \text{for } i = 1, 2, 3. \end{cases}$$

Moreover the π_j^i *are transformed as the coefficients of a quadratic form.* $\square$

Later on we will use this group $\mathfrak{A}^{(1)}$ in order to kill as many $\bar{\pi}_j^i$ as possible, thereby making the structure maximally simple.

Let us now consider $(S; \mathcal{P})$, where S is an h-codimensional submanifold of $J^2(\mathbb{C}_x^3, \mathbb{C}_z)$ such that $\omega^1 \wedge \omega^2 \wedge \omega^3 \wedge \theta^0 \wedge \theta^1 \wedge \theta^2 \wedge \theta^3|_S \neq 0$, and $\mathcal{P} = {}^2 C t^{3,1}|_S$.

A local coframe for S is given by the restrictions of $\omega^1, \omega^2, \omega^3, \theta^0, \theta^1, \theta^2, \theta^3$ to S together with $m := \dim S - 7 = \dim J^2(\mathbb{C}_x^3, \mathbb{C}_z) - h - 7 = 6 - h$ complementary one-forms $\pi^1, \ldots, \pi^m$.

The structure equations for $^2 C t^{3,1} = (\theta^0, \theta^1, \theta^2, \theta^3)$ are

$$\begin{cases} d\theta^0 = \sum_{j=1}^{3} \omega^j \wedge \theta^j, \\ d\theta^i = \sum_{j=1}^{3} \omega^j \wedge \pi_j^i & \text{for } i = 1, 2, 3. \end{cases}$$

If we write ω^i and θ^i instead of the more precise $\omega^i|_S$ and $\theta^i|_S$, the inherited structure equations for $\mathcal{P}$ are of the form

$$\begin{cases} d\theta^0 = \sum_{j=1}^{3} \omega^j \wedge \theta^j, \\ d\theta^i \equiv \sum_{j=1}^{3} \sum_{k=1}^{m} c_{jk}^i \, \omega^j \wedge \pi^k \pmod{\mathcal{P}}, & i = 1, 2, 3, \end{cases}$$

for certain structure functions c_{jk}^i defined on S.

When we are looking for integral manifolds on which $\omega^1 \wedge \omega^2 \wedge \omega^3 \neq 0$, the first two reduced characters of $\mathcal{P}$ are given by

$$\sigma_1 = \max_{u^i} \dim \left(\left(\sum_{i=1}^{3} u^i \frac{\partial}{\partial \omega^i} \right) \rfloor d\theta^j \quad (\text{mod } \mathcal{P}) \right),$$

$$\sigma_1 + \sigma_2 = \max_{u^i, v^i} \dim \left(\left(\sum_{i=1}^{3} u^i \frac{\partial}{\partial \omega^i} \right) \rfloor d\theta^j, \left(\sum_{i=1}^{3} v^i \frac{\partial}{\partial \omega^i} \right) \rfloor d\theta^j \quad (\text{mod } \mathcal{P}) \right)$$

where $j = 1, 2, 3$, and u^i, v^i are arbitrary functions on S (as explained in section **4.1**). Note that $\sigma_1 \leq 3$ and $\sigma_1 + \sigma_2 \leq m$.

The general 3-dimensional involution $\mathcal{I}_3$ on which $\omega^1 \wedge \omega^2 \wedge \omega^3 \neq 0$ is generated by vector fields of the form

$$X_i = \frac{\partial}{\partial \omega^i} + \sum_{l=1}^{m} a_i^l \frac{\partial}{\partial \pi^l}, \quad i = 1, 2, 3,$$

with suitable functions a_i^l. The dual pfaffian system $\mathcal{I}_3^{\perp}$ is then generated by $\theta^0, \theta^1, \theta^2, \theta^3$ and the one-forms

$$\psi^l := \pi^l - \sum_{k=1}^{3} a_k^l \omega^k \quad \text{for } l = 1, \ldots, m.$$

The condition for $\mathcal{I}_3$ to be an involution is that the $d\theta^i$ vanish modulo $\theta^0, \ldots, \theta^3, \psi^1, \ldots, \psi^m$, i.e.,

$$0 = \sum_{j=1}^{3} \sum_{l=1}^{m} c_{jl}^i \omega^j \wedge \sum_{k=1}^{3} a_k^l \omega^k = \sum_{1 \leq j < k \leq 3} \left(\sum_{l=1}^{m} (c_{jl}^i a_k^l - c_{kl}^i a_j^l) \right) \omega^j \wedge \omega^k,$$

or

$$\sum_{l=1}^{m} (c_{jl}^i a_k^l - c_{kl}^i a_j^l) = 0 \quad \text{for } i, j, k = 1, 2, 3.$$

This system of linear equations for the a_k^l leaves a certain number—say d—of them arbitrary, while the others are expressed in terms of these d. By our standard conventions we assume d to have the same value at each point of S.

If $\omega^1, \omega^2, \omega^3, \pi^1, \ldots, \pi^m$ all appear in the expressions $d\theta^i \pmod{\mathcal{P}}$, $\mathcal{P}$ admits no Cauchy characteristic vector field. In that case Cartan's involutivity criterion from section **4.1** says that

$\mathcal{P}$ is involutive with respect to $\omega^1 \wedge \omega^2 \wedge \omega^3 \iff d = 2\sigma_1 + \sigma_2 \iff$ the number of arbitrary coefficients a_i^l in $\mathcal{I}_3$ equals $3m - d = 18 - 3h - 2\sigma_1 - \sigma_2$

In order to exemplify this we consider the case $h = 0$—that is, $(\mathcal{S}; \mathcal{P}) = (J^2(\mathbb{C}_x^3, \mathbb{C}_z); {}^2Ct^{3,1})$. With the complementary one-forms $\pi_j^i = dp_{ij} = \pi_i^j$ for $i, j = 1, 2, 3$, the structure equations for $\mathcal{P} = (\theta^0, \theta^1, \theta^2, \theta^3)$ are

$$d\theta^0 = \sum_{j=1}^{3} \omega^j \wedge \theta^j \quad \text{and} \quad d\theta^i = \sum_{j=1}^{3} \omega^j \wedge \pi_j^i \quad \text{for } i = 1, 2, 3.$$

Hence $(\partial/\partial\omega^1)\rfloor d\theta^i = \pi_1^i$ for $i = 1, 2, 3$, giving π_1^1, π_1^2 and π_1^3, so that $\sigma_1 = 3 = $ the largest possible number. Next $(\partial/\partial\omega^2)\rfloor d\theta^i = \pi_2^i$, giving the two new one-forms π_2^2 and π_2^3. Since no linear combination of the ω^i can give any more, this means that $\sigma_2 = 2$. Finally the only remaining one-form π_3^3 is obtained from $(\partial/\partial\omega^3)\rfloor d\theta^3$ (which incidentally means that $\sigma_3 = 1$). Consequently

$$18 - 3h - 2\sigma_1 - \sigma_2 = 18 - 3 \cdot 0 - 2 \cdot 3 - 2 = 10.$$

This is to be compared with the number of arbitrary coefficients in $\mathcal{I}_3$, which is defined by the pfaffian equations

$$\theta^i = 0 \quad \text{and} \quad \pi_j^i - \sum_{k=1}^{3} a_{jk}^i \omega^k = 0.$$

Here the a_{jk}^i are subject to the conditions

$$a_{jk}^i = a_{ik}^j \quad \text{and} \quad 0 = \sum_{j=1}^{3} \omega^j \wedge \left(\sum_{k=1}^{3} a_{jk}^i \omega^k \right) = \sum_{j<k} (a_{jk}^i - a_{kj}^i) \omega^j \wedge \omega^k.$$

Therefore the a_{jk}^i are *symmetric with respect to all three indices*, but are otherwise arbitrary. And so there are as many independent a_{jk}^i as there are 3-tuples (i, j, k) with $1 \le i \le j \le k \le 3$—that is, 10.

Thus Cartan's criterion says that $(J^2(\mathbb{C}_x^3, \mathbb{C}_z); {}^2Ct^{3,1})$ indeed is involutive with respect to $\omega^1 \wedge \omega^2 \wedge \omega^3 = dx^1 \wedge dx^2 \wedge dx^3$.

18.2 Structural classification

This section is devoted to a structural classification of $(\mathcal{S}_h; \mathcal{P}_h)$ for $h = 1, \ldots, 5$, where $\mathcal{S}_h$ is an h-codimensional submanifold of $J^2(\mathbb{C}_x^3, \mathbb{C}_z)$, and $\mathcal{P}_h$ is the corresponding pfaffian system. Admittedly this classification is a bit boring—as most classifications are—but our main interest lies in the fact that it will allow us to determine the possible *singular subsystems*, which in their turn make it possible to apply the *method of Monge*.

$\boxed{h = 1}$ The restrictions of ω^i, θ^i and π_j^i to $\mathcal{S}_1$ are related by *one* linear equation, which we want to simplify as much as possible by means of the group $\mathfrak{A}^{(1)}$.

The structure equations

$$d\theta^i \equiv \sum_{j=1}^{3} \omega^j \wedge \pi_j^i \quad (\text{mod } \theta^0, \ldots, \theta^3), \quad i = 1, 2, 3,$$

obviously remain invariant if the π_j^i are replaced by $\pi_j^i + \sum_{k=1}^{3} a_{jk}^i \omega^k + \sum_{l=0}^{3} b_{jl}^i \theta^l$, where the a_{jk}^i and b_{jl}^i are functions which are arbitrary except for the condition that the a_{jk}^i are symmetric with respect to all three indices. So this is a permissible change of basis.

First we utilize the b_{jl}^i to reduce the given linear equation to the form

$$\sum A_i^j \pi_j^i + \sum B_i \omega^i = 0. \tag{*}$$

And then we replace

$$\begin{pmatrix} \theta^1 \\ \theta^2 \\ \theta^3 \end{pmatrix} \quad \text{by} \quad \begin{pmatrix} a_1^1 & a_2^1 & a_3^1 \\ a_1^2 & a_2^2 & a_3^2 \\ a_1^3 & a_2^3 & a_3^3 \end{pmatrix} \begin{pmatrix} \theta^1 \\ \theta^2 \\ \theta^3 \end{pmatrix},$$

where (a_j^i) is a nonsingular matrix. As explained in the preceding section this transformation can be prolonged to act on the ω^i and π_j^i so as to preserve the structure. Since the π_j^i then will transform as the coefficients of a quadratic form, (*) may be reduced to one of the following three equations:

- $\pi_1^1 + \pi_2^2 + \pi_3^3 + \sum_{i=1}^{3} B_i \omega^i = 0$,
- $\pi_2^1 + \sum_{i=1}^{3} B_i \omega^i = 0 \quad (\text{or } \pi_1^1 - \pi_2^2 + \sum_{i=1}^{3} B_i \omega^i = 0)$,
- $\pi_1^1 + \sum_{i=1}^{3} B_i \omega^i = 0$.

Finally, using the possibility of adding linear combinations of the ω^i to the π_j^i, we can afterwards get rid of the terms $\sum_{i=1}^{3} B_i \omega^i$ as well.

Theorem 18.2.1. *A single PDE gives rise to one linear relation between the one-forms ω^i, θ^i and π_j^i, which can be reduced to one of the following canonical forms:*

(a) $\pi_1^1 + \pi_2^2 + \pi_3^3 = 0$—*the general case,*
(b) $\pi_2^1 = 0$—*the hyperbolic case,*
(c) $\pi_1^1 = 0$—*the parabolic case.*

The corresponding pfaffian systems are involutive with respect to $\omega^1 \wedge \omega^2 \wedge \omega^3$.

Proof. It remains to prove the involutivity. The general 3-dimensional involution of $^2Ct^{3,1}$ is defined by $\theta^0 = \cdots = \theta^3 = 0$ and

$$\pi_j^i = \sum_{k=1}^{3} a_{jk}^i \omega^k \quad (\text{mod } \theta^0, \ldots, \theta^3) \quad \text{for } i, j = 1, 2, 3,$$

where the functions a_{jk}^i are arbitrary except for being symmetric with respect to all indices. Hence there are 10 arbitrary coefficients. Then the equations **(a)**, **(b)** and **(c)** impose the further conditions

$$\sum_{i=1}^{3} a_{ik}^i = 0, \quad a_{2k}^1 = 0 \quad \text{and} \quad a_{1k}^1 = 0 \quad \text{for } k = 1, 2, 3$$

respectively, so in each case there remain 7 arbitrary coefficients.

In case **(a)** the structure equations can be written as

$$\begin{cases} d\theta^0 \equiv 0, \\ d\theta^1 \equiv -\omega^1 \wedge (\pi_2^2 + \pi_3^3) + \omega^2 \wedge \pi_2^1 + \omega^3 \wedge \pi_3^1, \\ d\theta^2 \equiv \sum_{i=1}^{3} \omega^i \wedge \pi_i^2, \\ d\theta^3 \equiv \sum_{i=1}^{3} \omega^i \wedge \pi_i^3 \end{cases} \quad (\text{mod } \theta^0, \ldots, \theta^3).$$

Then $(\partial/\partial\omega^3)\rfloor d\theta^j \equiv \pi_3^j$ for $j = 1, 2, 3$, showing that $\sigma_1 = 3$. Next we have $(\partial/\partial\omega^2)\rfloor d\theta^j \equiv \pi_2^j$ for $j = 1, 2, 3$, giving the new one-forms π_2^1 and π_2^2, so that $\sigma_2 = 2$. Therefore

$$18 - 3h - 2\sigma_1 - \sigma_2 = 18 - 3 \cdot 1 - 2 \cdot 3 - 2 = 7,$$

which by Cartan's criterion proves involutivity.

Quite analogously $\sigma_1 = 3$ and $\sigma_2 = 2$ also in the hyperbolic and parabolic cases, so these are involutive too. □

Let us make some preliminary remarks before entering into the cases $h = 2$ and 3.

Restricting ω^i, θ^i and π_j^i to S_h gives rise to h linear relations. By replacing π_j^i by $\pi_j^i +$ (a suitable linear combination of the θ^i), these may be reduced to

$$\sum_{i,j=1}^{3} A_i^{jk} \pi_j^i + \sum_{l=1}^{3} A_l^k \omega^l \quad \text{for } k = 1, \ldots, h. \quad (**)$$

Let us now use the assumption that *the corresponding pfaffian system* $\mathcal{P}_h$ *is involutive with respect to* $\omega^1 \wedge \omega^2 \wedge \omega^3$. That is, there is *at least one* involution of the form

$$\theta^i = 0 \quad \text{and} \quad \pi^i_j - \sum_{l=1}^{3} \alpha^i_{jl} \omega^l = 0,$$

where the α^i_{jl} are symmetric with respect to all indices. But then

$$\sum_{i,j=1}^{3} A^{jk}_i \alpha^i_{jl} + A^k_l = 0 \quad \text{for } l = 1, 2, 3 \text{ and } k = 1, \dots, h,$$

so that (**) is equivalent to

$$\sum_{i,j=1}^{3} A^{jk}_i \left(\pi^i_j - \sum_{l=1}^{3} \alpha^i_{jl} \omega^l \right) = 0 \quad \text{for } k = 1, \dots, h.$$

Setting $\tilde{\pi}^i_j := \pi^i_j - \sum_{l=1}^{3} \alpha^i_{jl} \omega^l$ and forgetting the tildes afterwards, the h relations are finally reduced to

$$\sum_{i,j=1}^{3} A^{jk}_i \pi^i_j = 0 \quad \text{for } k = 1, \dots, h. \tag{***}$$

When we are looking for the *general involution* $\mathcal{I}_3$, defined by

$$\theta^i = 0 \quad \text{and} \quad \pi^i_j - \sum_{l=1}^{3} a^i_{jl} \omega^l = 0,$$

(***) induces the $3h$ equations

$$\sum_{i,j=1}^{3} A^{jk}_i a^i_{jl} = 0 \quad \text{for } l = 1, 2, 3 \text{ and } k = 1, \dots, h,$$

and the question is how many of the a^i_{jl} are left arbitrary by these.

As noted earlier we always have $\sigma_1 \leq 3$ and $\sigma_1 + \sigma_2 \leq 6 - h$, so that $2\sigma_1 + \sigma_2 \leq 9 - h$. Consequently

> the number of arbitrary coefficients in the general involution
> $= 18 - 3h - 2\sigma_1 - \sigma_2 \geq 18 - 3h - 9 + h = 9 - 2h$,

so that the 10 coefficients a^i_{jl} are related by at most $10 - (9 - 2h) = 2h + 1$ independent linear equations. Expressed differently,

> among the $3h$ equations $\sum_{i,j=1}^{3} A^{jk}_i a^i_{jl} = 0$ there are at most $2h + 1$ which are linearly independent.

Because π^i_j and a^i_{jl} are symmetric with respect to their indices, it makes sense to identify them with elements of the symmetric algebras in two and three variables respectively—i.e.,

$$\pi^i_j \longleftrightarrow v^i v^j \quad \text{and} \quad a^i_{jl} \longleftrightarrow v^i v^j v^l,$$

where v^1, v^2 and v^3 are independent variables. Then the result above can be rephrased in the following way.

Lemma 18.2.1. *Defining the quadratic forms*

$$F^k = \sum_{i,j=1}^{3} A^{jk}_i v^i v^j \quad \text{for } k = 1, \ldots, h,$$

the $3h$ cubic forms $v^i F^k$, where $i = 1, 2, 3$ and $k = 1, \ldots, h$, are related by at least $3h - (2h+1) = h - 1$ linear equations. ☐

Let us now check what happens when $h = 2$ and 3.

$\boxed{h = 2}$ In this case we are given two quadratic forms F^1 and F^2 in three variables v^1, v^2, v^3 such that there is at least *one* linear relation between the six cubic forms

$$v^1 F^1, \quad v^2 F^1, \quad v^3 F^1, \quad v^1 F^2, \quad v^2 F^2 \quad \text{and} \quad v^3 F^2.$$

Consequently there are linear forms l^1 and l^2 in the v^i with the property that

$$l^1 F^1 + l^2 F^2 = 0.$$

The l^i must be linearly independent, because otherwise the $h = 2$ relations between the π^i_j would not be independent. By applying the group $\mathfrak{A}^{(1)}$ there are induced linear transformations of the v^i, and these make it possible to assume that $l^1 = v^2$ and $l^2 = -v^1$, so that $v^2 F^1 - v^1 F^2 = 0$. But then F^1 and F^2 can be written as

$$F^1 = v^1 \cdot \sum_{i=1}^{3} A_i v^i, \quad F^2 = v^2 \cdot \sum_{i=1}^{3} A_i v^i$$

for certain A_i. Supposing that $A_3 \neq 0$, $\sum A_i v^i$ can be replaced by v^3, whence

- $F^1 = v^1 v^3$ and $F^2 = v^2 v^3$.

Otherwise the F^i can be reduced to

- $F^1 = (v^1)^2$ and $F^2 = v^1 v^2$.

With $F^k = \sum A_i^{jk} v^i v^j$ and $\sum A_i^{jk} \pi_j^i = 0$, the corresponding relations between the π_j^i are

$$\text{(a) } \pi_3^1 = \pi_3^2 = 0 \quad \text{and} \quad \text{(b) } \pi_1^1 = \pi_2^1 = 0.$$

Formula **(a)** yields the structure equations

$$\begin{cases} d\theta^0 \equiv 0, \\ d\theta^1 \equiv \omega^1 \wedge \pi_1^1 + \omega^2 \wedge \pi_2^1, \\ d\theta^2 \equiv \omega^1 \wedge \pi_1^2 + \omega^2 \wedge \pi_2^2, \\ d\theta^3 \equiv \qquad\qquad \omega^3 \wedge \pi_3^3 \end{cases} \quad (\bmod\ \theta^0, \ldots, \theta^3).$$

Applying $X := \partial/\partial\omega^1 + \partial/\partial\omega^3$ to the $d\theta^i$, we get

$$X \rfloor d\theta^1 \equiv \pi_1^1, \quad X \rfloor d\theta^2 \equiv \pi_1^2 \quad \text{and} \quad X \rfloor d\theta^3 \equiv \pi_3^3,$$

so that $\sigma_1 = 3$. Then $(\partial/\partial\omega^2) \rfloor d\theta^2$ gives the remaining π_2^2, whence $\sigma_2 = 1$. Consequently

$$18 - 3h - 2\sigma_1 - \sigma_2 = 18 - 3 \cdot 2 - 2 \cdot 3 - 1 = 5.$$

On the other hand, the equations $a_{3l}^1 = a_{3l}^2 = 0$ for $l = 1, 2, 3$ leave 5 of the 10 a_{jk}^i arbitrary—and thus Cartan's criterion shows that we have involutivity.

In case **(b)** we have the structure equations

$$\begin{cases} d\theta^0 \equiv 0, \\ d\theta^1 \equiv \qquad\qquad \omega^3 \wedge \pi_3^1, \\ d\theta^2 \equiv \qquad \omega^2 \wedge \pi_2^2 + \omega^3 \wedge \pi_3^2, \\ d\theta^3 \equiv \omega^1 \wedge \pi_1^3 + \omega^2 \wedge \pi_2^3 + \omega^3 \wedge \pi_3^3 \end{cases} \quad (\bmod\ \theta^0, \ldots, \theta^3).$$

Here $(\partial/\partial\omega^3) \rfloor d\theta^i$ gives π_3^i for $i = 1, 2, 3$, and then the missing π_2^2 is obtained from $(\partial/\partial\omega^2) \rfloor d\theta^2$. So $\sigma_1 = 3$ and $\sigma_1 = 1$, and we get involutivity also in this case.

Theorem 18.2.2. *When $h = 2$ there are two different equivalence classes of involutive systems, represented by*

(a) $\pi_3^1 = \pi_3^2 = 0$,
(b) $\pi_1^1 = \pi_2^1 = 0$ □

$\boxed{h = 3}$ Here we are led to consider three quadratic forms F^i and six linear forms l^i, m^i $(i = 1, 2, 3)$, which satisfy

$$\sum_{i=1}^{3} l^i F^i = 0 \quad \text{and} \quad \sum_{i=1}^{3} m^i F^i = 0.$$

In such a situation it is natural to form a linear combination $\sum_{i=1}^{3}(z_1 l^i + z_2 m^i)F^i = 0$, making one of the coefficients $z_1 l^i + z_2 m^i$ vanish. Or more generally: Choose z_1, z_2 such that the three linear forms $z_1 l^i + z_2 m^i$ become linearly dependent. On regarding z_1 and z_2 as homogeneous coordinates for the Riemann sphere $\mathbb{P}$ (once again!) and setting $z := z_2/z_1$, z is thus to be determined in such a way that the three linear forms $l^i + zm^i$ are dependent. This is achieved by setting a 3×3 determinant equal to 0, so we get an equation with three roots on $\mathbb{P}$.

$\mathfrak{A}^{(1)}$ induces linear transformations of the v^i, which in their turn induce dual linear transformations of z_1 and z_2. Using the latter we may assume that one of the roots is $z = 0$. Then l^1, l^2 and l^3 are linearly dependent, and hence it can be arranged that

$$l^1 = v^1, \quad l^2 = v^2 \quad \text{and} \quad l^3 = 0.$$

With

$$m^1 = av^1 + bv^2 + cv^3, \quad m^2 = a'v^1 + b'v^2 + c'v^3, \quad m^3 = a''v^1 + b''v^2 + c''v^3,$$

the equation that z is to satisfy is

$$0 = \det \begin{pmatrix} 1 + az & bz & cz \\ a'z & 1 + b'z & c'z \\ a''z & b''z & c''z \end{pmatrix} = z \cdot \det \begin{pmatrix} 1 + az & bz & cz \\ a'z & 1 + b'z & c'z \\ a'' & b'' & c'' \end{pmatrix},$$

which we write as $P(z) = 0$.

In analogy with what we did in chapter **17**, there are four different cases to consider. Since these are very similar, we only treat the first one in detail.

1. $P(z) = 0$ *has three simple roots on* $\mathbb{P}$. Applying the linear transformations induced from $\mathfrak{A}^{(1)}$, we may assume that the roots are 0, 1, and ∞. Thus our equation is $z(1 - z) = 0$ (or in homogeneous coordinates: $z_2(z_1 - z_2)z_1 = 0$). Here c'' is the coefficient in front of z, and so must be nonzero in this case. Hence it is possible to substitute $c''v^3 + b''v^2 + a''v^1$ for v^3, so that $a'' = b'' = 0$ and $c'' = 1$. After replacing F^3 by $F^3 + cF^1 + c'F^2$ it may further be assumed that $c = c' = 0$, which leaves us with

$$0 = (ab' - ba')z^3 + (a + b')z^2 + z = -z^2 + z.$$

This shows that $ab' - ba' = 0$, so we can find a linear transformation of the v^i making $b = b' = 0$, and then $a = -1$.

As a result of all this,

$$m^1 = -v^1, \quad m^2 = a'v^1 \quad \text{and} \quad m^3 = v^3,$$

so that the F^i satisfy

$$v^1 F^1 + v^2 F^2 = 0, \quad -v^1 F^1 + a'v^1 F^2 + v^3 F^3 = 0.$$

Or equivalently,

$$\begin{cases} v^1(F^1 - a'F^2) + (v^2 + a'v^1)F^2 = 0, \\ -v^1(F^1 - a'F^2) + v^3 F^3 = 0. \end{cases}$$

By making the substitutions

$$F^1 - a'F^2 \mapsto F^1, \quad v^2 + a'v^1 \mapsto v^2,$$

these equations can be written as

$$v^1 F^1 = -v^2 F^2 = v^3 F^3, \quad \text{or} \quad \frac{F^1}{v^2 v^3} = -\frac{F^2}{v^1 v^3} = \frac{F^3}{v^1 v^2} = c \text{ (say)}.$$

So the end result result is that

$$F^1 = c v^2 v^3, \quad F^2 = -c v^1 v^3, \quad F^3 = c v^1 v^2.$$

Recalling that $F^k = \sum A_i^{jk} v^i v^j$ and that the π_j^i are related by $\sum A_i^{jk} \pi_j^i = 0$, we have

$$\pi_3^2 = \pi_3^1 = \pi_2^1 = 0.$$

The corresponding structure equations are

$$\begin{cases} d\theta^0 \equiv 0, \\ d\theta^1 \equiv \omega^1 \wedge \pi_1^1, \\ d\theta^2 \equiv \qquad \omega^2 \wedge \pi_2^2, \\ d\theta^3 \equiv \qquad\qquad \omega^3 \wedge \pi_3^3 \end{cases} \qquad (\text{mod } \theta^0, \ldots, \theta^3).$$

Since $(\partial/\partial\omega^1 + \partial/\partial\omega^2 + \partial/\partial\omega^3)\rfloor d\theta^i$ gives π_1^1, π_2^2 and π_3^3, we see that $\sigma_1 = 3$ and $\sigma_2 = 0$. Thus $18 - 3h - 2\sigma_1 - \sigma_2 = 18 - 3 \cdot 3 - 2 \cdot 3 = 3$.

On the other hand, $a_{3l}^2 = a_{3l}^1 = a_{2l}^1 = 0$ for $l = 1, 2, 3$, so that only the three coefficients a_{11}^1, a_{22}^2 and a_{33}^3 are nonzero—which proves involutivity.

2. $P(z) = 0$ *has one double and one simple root.* If we suppose the double root to be situated at ∞ and the simple root at 0, the equation is $z = 0$ (or in homogeneous coordinates: $z_1^2 z_2 = 0$). Reasoning as in the preceding case, we can reduce the relations between the π_j^i to

$$\pi_3^2 = \pi_3^1 = \pi_1^1 = 0.$$

The corresponding structure equations are

$$
\begin{cases}
d\theta^0 \equiv 0, \\
d\theta^1 \equiv \qquad\qquad \omega^2 \wedge \pi_2^1, \\
d\theta^3 \equiv \omega^1 \wedge \pi_1^2 + \omega^2 \wedge \pi_2^2, \\
d\theta^3 \equiv \qquad\qquad\qquad \omega^3 \wedge \pi_3^3
\end{cases}
\qquad (\text{mod } \theta^0, \ldots, \theta^3).
$$

Here $(\partial/\partial\omega^2 + \partial/\partial\omega^3) \rfloor d\theta^i$ gives π_2^1, π_2^2 and π_3^3, whence $\sigma_1 = 3$, $\sigma_2 = 0$ and $18 - 3h - 2\sigma_1 - \sigma_2 = 18 - 3 \cdot 3 - 2 \cdot 3 = 3$.

The conditions on the a_{jl}^i are $a_{3l}^2 = a_{3l}^1 = a_{1l}^1 = 0$ for $l = 1, 2, 3$, leaving the three coefficients a_{22}^1, a_{22}^2 and a_{33}^3 nonzero. So we get an involutive system in this case too.

3. $P(z) = 0$ *has the triple root* $z = 0$. The corresponding equation then necessarily is $z^3 = 0$. Reasoning as before we obtain the following reduced equations for the π_j^i:

$$\pi_1^1 = \pi_2^1 = \pi_3^1 - \pi_2^2 = 0,$$

with the corresponding structure equations

$$
\begin{cases}
d\theta^0 \equiv 0, \\
d\theta^1 \equiv \qquad\qquad\qquad \omega^3 \wedge \pi_3^1, \\
d\theta^2 \equiv \qquad \omega^2 \wedge \pi_3^1 + \omega^3 \wedge \pi_3^2, \\
d\theta^3 \equiv \omega^1 \wedge \pi_1^3 + \omega^2 \wedge \pi_2^3 + \omega^3 \wedge \pi_3^3
\end{cases}
\qquad (\text{mod } \theta^0, \ldots, \theta^3).
$$

In this case $(\partial/\partial\omega^3) \rfloor d\theta^i$ gives π_3^1, π_3^2 and π_3^3, so again $\sigma_1 = 3$, $\sigma_2 = 0$, and $18 - 3h - 2\sigma_1 - \sigma_2 = 3$.

The 10 coefficients a_{jl}^i are related by $a_{1l}^1 = a_{2l}^1 = a_{3l}^1 - a_{2l}^2 = 0$ for $l = 1, 2, 3$, i.e., $a_{11}^1 = a_{12}^1 = a_{13}^1 = a_{22}^1 = a_{23}^1 = a_{22}^2 = a_{33}^1 - a_{23}^2 = 0$. So there remain 3 arbitrary a_{jl}^i—proving involutivity.

4. *Finally* $P(z)$ *might vanish identically, i.e.,*

$$
\det \begin{pmatrix}
1 + az & bz & cz \\
a'z & 1 + b'z & c'z \\
a'' & b'' & c''
\end{pmatrix} \equiv 0.
$$

By applying suitable linear transformations, the matrices satisfying this can be reduced

$$\text{either to} \quad \begin{pmatrix} 1 & 0 & 0 \\ 0 & 1 & z \\ 1 & 0 & 0 \end{pmatrix}, \quad \text{or to} \quad \begin{pmatrix} 1 & z & 0 \\ 0 & 1 & 0 \\ 1 & 0 & 0 \end{pmatrix}.$$

In the *first case*, $m^1 = 0$, $m^2 = v^3$, and $m^3 = v^1$, which means that F^1 and F^2 are related by $v^1 F^1 + v^2 F^2 = 0$ and $v^3 F^2 + v^1 F^3 = 0$, or equivalently

$$\frac{F^1}{v^1 v^2} = -\frac{F^2}{(v^1)^2} = \frac{F^3}{v^1 v^3} = c \text{ (say)}.$$

But $F^1 = c v^1 v^2$, $F^2 = -c (v^1)^2$ and $F^3 = c v^1 v^3$ imply that the π^i_j are related by

$$\pi^1_2 = \pi^1_1 = \pi^1_3 = 0.$$

Hence the structure equations are given by

$$\begin{cases} d\theta^0 \equiv 0, \\ d\theta^1 \equiv 0, \\ d\theta^2 \equiv \omega^2 \wedge \pi^2_2 + \omega^3 \wedge \pi^2_3, \\ d\theta^3 \equiv \omega^2 \wedge \pi^3_2 + \omega^3 \wedge \pi^3_3 \end{cases} \quad (\text{mod } \theta^0, \dots, \theta^3).$$

Now $(\partial/\partial\omega^2)\rfloor d\theta^i$ gives π^2_2 and π^3_2, whereupon $(\partial/\partial\omega^3)\rfloor d\theta^3$ yields the remaining π^3_3. Therefore $\sigma_1 = 2$, $\sigma_2 = 1$ and $18 - 3h - 2\sigma_1 - \sigma_2 = 18 - 3 \cdot 3 - 2 \cdot 2 - 1 = 4$.

On the other hand, the equations $a^1_{2l} = a^1_{1l} = a^1_{3l} = 0$ for $l = 1, 2, 3$ leave four of the coefficients nonzero, which again proves involutivity.

In the *second case*, $m^1 = v^2$, $m^2 = 0$, and $m^3 = v^1$, giving $v^1 F^1 + v^2 F^2 = 0$ and $v^2 F^1 + v^1 F^3 = 0$, or

$$\frac{F^1}{v^1 v^2} = -\frac{F^2}{(v^1)^2} = -\frac{F^3}{(v^2)^2}.$$

Thus $\pi^1_2 = \pi^1_1 = \pi^2_2 = 0$, with the corresponding structure equations

$$\begin{cases} d\theta^0 \equiv 0, \\ d\theta^1 \equiv & \omega^3 \wedge \pi^1_3, \\ d\theta^2 \equiv & \omega^3 \wedge \pi^2_3, \\ d\theta^3 \equiv \omega^1 \wedge \pi^3_1 + \omega^2 \wedge \pi^3_2 + \omega^3 \wedge \pi^3_3 \end{cases} \quad (\text{mod } \theta^0, \dots, \theta^3).$$

Now $(\partial/\partial\omega^3)\rfloor d\theta^i$ gives all three of π_3^1, π_3^2, π_3^3, whence $\sigma_1 = 3$, $\sigma_2 = 0$ and $18 - 3h - 2\sigma_1 - \sigma_2 = 18 - 3\cdot3 - 2\cdot3 = 3$.

The 10 coefficients a^i_{jl} are related by $a^1_{2l} = a^1_{1l} = a^2_{2l} = 0$ for $l = 1, 2, 3$, leaving 3 nonzero ones—so this case is also involutive.

Theorem 18.2.3. When $h = 3$ there are five different equivalence classes of involutive systems, represented by

(a) $\pi_2^1 = \pi_3^1 = \pi_3^2 = 0$,

(b) $\pi_1^1 = \pi_3^1 = \pi_3^2 = 0$,

(c) $\pi_1^1 = \pi_2^1 = \pi_3^1 - \pi_2^2 = 0$,

(d) $\pi_1^1 = \pi_2^1 = \pi_2^2 = 0$,

(e) $\pi_1^1 = \pi_2^1 = \pi_3^1 = 0$. □

$\boxed{h = 4}$ Because there are only two complementary one-forms in this case—say π^1 and π^2—it is easiest to base the considerations on these, and try out the few possible combinations that they may give rise to. It turns out that by choosing them suitably and replacing θ^1, θ^2 and θ^3 by appropriate linear combinations ψ^1, ψ^2 and ψ^3, the structure equations of any involutive system can in this case be reduced to one of the following forms, where all congruences are taken modulo $(\theta^0, \psi^1, \psi^2, \psi^3) = (\theta^0, \theta^1, \theta^2, \theta^3)$:

$$\begin{cases} d\theta^0 \equiv 0, \\ d\psi^1 \equiv \omega^1 \wedge \pi^1, \\ d\psi^2 \equiv \omega^2 \wedge \pi^2, \\ d\psi^3 \equiv 0 \end{cases} \qquad \begin{cases} d\theta^0 \equiv 0, \\ d\psi^1 \equiv \omega^1 \wedge \pi^1, \\ d\psi^2 \equiv \omega^1 \wedge \pi^2 + \omega^2 \wedge \pi^1, \\ d\psi^3 \equiv 0 \end{cases} \quad \text{and}$$

$$\begin{cases} d\theta^0 \equiv 0, \\ d\psi^1 \equiv \omega^1 \wedge \pi^1 + \omega^2 \wedge \pi^2, \\ d\psi^2 \equiv 0, \\ d\psi^3 \equiv 0. \end{cases}$$

If $\theta^i = a^i\,\psi^1 + b^i\,\psi^2 + c^i\,\psi^3$ for certain functions a^i, b^i and c^i defined on $\mathcal{S}_4$,

$$d\theta^i \equiv a^i\,d\psi^1 + b^i\,d\psi^2 + c^i\,d\psi^3 \quad (\text{mod } \theta^0, \ldots, \theta^3).$$

In the first case this gives

$$\begin{cases} d\theta^1 \equiv a^1\,\omega^1 \wedge \pi^1 + b^1\,\omega^2 \wedge \pi^2, \\ d\theta^2 \equiv a^2\,\omega^1 \wedge \pi^1 + b^2\,\omega^2 \wedge \pi^2, \qquad (\mathrm{mod}\ \theta^0,\dots,\theta^3), \\ d\theta^3 \equiv a^3\,\omega^1 \wedge \pi^1 + b^3\,\omega^2 \wedge \pi^2 \end{cases}$$

whence

$$\pi_1^1 = a^1\,\pi^1, \quad \pi_2^1 = b^1\,\pi^2 = a^2\pi^1 \implies b^1 = a^2 = 0 \implies \pi_2^1 = 0,$$
$$\pi_3^1 = 0, \quad \pi_2^2 = b^2\,\pi^2, \quad \pi_3^2 = 0, \quad \pi_3^3 = 0.$$

Consequently the structure equations are

$$\begin{cases} d\theta^0 \equiv 0, \\ d\theta^1 \equiv \omega^1 \wedge \pi_1^1, \\ d\theta^2 \equiv \qquad\quad \omega^2 \wedge \pi_2^2, \qquad (\mathrm{mod}\ \theta^0,\dots,\theta^3), \\ d\theta^3 \equiv 0 \end{cases}$$

with $\sigma_1 = 2$ and $\sigma_2 = 0$.

In the second case

$$\begin{cases} d\theta^1 \equiv a^1\,\omega^1 \wedge \pi^1 + b^1\,(\omega^1 \wedge \pi^2 + \omega^2 \wedge \pi^1), \\ d\theta^2 \equiv a^2\,\omega^1 \wedge \pi^1 + b^2\,(\omega^1 \wedge \pi^2 + \omega^2 \wedge \pi^1), \qquad (\mathrm{mod}\ \theta^0,\dots,\theta^3), \\ d\theta^3 \equiv a^3\,\omega^1 \wedge \pi^1 + b^3\,(\omega^1 \wedge \pi^2 + \omega^2 \wedge \pi^1) \end{cases}$$

implying that

$$\pi_1^1 = a^1\,\pi^1 + b^1\,\pi^2, \quad \pi_2^1 = b^1\,\pi^1 = a^2\pi^1 + b^2\,\pi^2 \implies b^2 = 0 \text{ and } b^1 = a^2,$$
$$\pi_3^1 = 0, \quad \pi_2^2 = b^2\,\pi^1 = 0, \quad \pi_3^2 = 0, \quad \pi_3^3 = 0.$$

That is, all π_j^i except π_1^1 and π_2^1 vanish, and the corresponding structure equations are

$$\begin{cases} d\theta^\upsilon \equiv 0, \\ d\theta^1 \equiv \omega^1 \wedge \pi_1^1 + \omega^2 \wedge \pi_2^1, \\ d\theta^2 \equiv \omega^1 \wedge \pi_1^2, \qquad (\mathrm{mod}\ \theta^0,\dots,\theta^3), \\ d\theta^3 \equiv 0 \end{cases}$$

with $\sigma_1 = 2$ and $\sigma_2 = 0$.

In the third case the same type of reasoning shows that *all* π_j^i vanish— but then h equals 6 instead of 4.

Theorem 18.2.4. *When $h = 4$ there are two different equivalence classes of involutive systems, represented by*

(a) $\pi_2^1 = \pi_3^1 = \pi_3^2 = \pi_3^3 = 0$,
(b) $\pi_3^1 = \pi_2^2 = \pi_3^2 = \pi_3^3 = 0$.   □

$\boxed{h = 5}$ In this case there is just one complementary one-form π. On assuming involutivity this can be chosen such that

$$d\theta^i \equiv a^i \omega^1 \wedge \pi \quad (\text{mod } \theta^0, \dots, \theta^3), \quad i = 1, 2, 3,$$

for suitable functions a^i. Then

$$\pi_1^1 = a^1 \pi, \quad \text{while} \quad \pi_2^1 = \pi_3^1 = \pi_2^2 = \pi_3^2 = \pi_3^3 = 0,$$

giving the structure equations

$$\begin{cases} d\theta^0 \equiv 0, \\ d\theta^1 \equiv \omega^1 \wedge \pi_1^1, \\ d\theta^2 \equiv 0, \\ d\theta^3 \equiv 0 \end{cases} \quad (\text{mod } \theta^0, \dots, \theta^3).$$

Clearly $\sigma_1 = 1$ and $\sigma_2 = 0$.

Theorem 18.2.5. *When $h = 5$ there is only one equivalence class of involutive systems, represented by $\pi_2^1 = \pi_3^1 = \pi_2^2 = \pi_3^2 = \pi_3^3 = 0$.*   □

18.3 A single PDE

In the following we will go through the list of canonical Lie structures found in the preceding section, and for each structure

- *determine the associated singular vector field systems,*

and

- *investigate to what extent the latter can be used for finding integral manifolds.*

We will in particular be interested in the *method of Monge*: how general is it?, how about a precise formulation?, how to prove it?—and so on.

The account given certainly follows [Cartan 1911], but it should be remarked that the latter work contains a lot of interesting material not covered here.

We begin by considering a *single PDE*—so that $h = 1$. Since this is the most popular case in the literature, it is tempting to believe that it also is the easiest one. However, at least from the point of view taken here, it rather turns out to be *the worst*. The best case instead occurs when $h = 2$, as will be seen in the next section (by the way, is there any good underlying reason for this?).

Remark. [Parsons 1960] is a very good reference for second order PDEs in one dependent and three independent variables, treated by means of Monge characteristics. But with Cartan's work being closer to our intentions, we stick to that.

When $h = 1$ there are three canonical forms to investigate:

- *the general case* $\pi_1^1 + \pi_2^2 + \pi_3^3 = 0$,
- *the hyperbolic case* $\pi_2^1 = 0$,
- *the parabolic case* $\pi_1^1 = 0$.

The general case. This admits *no singular vector fields*, as already pointed out in secton **5.2**, so this case falls outside of our study.

However, in the following we will see that *of all cases listed in the preceding section, this is the only one lacking singular vector fields*. And this fact makes the method of Monge very interesting—while at the same time being bad luck for Cartan's local existence theorem, where singular integral manifolds have to be handled by means of the complicated prolongation theorem of Cartan–Kuranishi.

The hyperbolic case. Here we consider the pfaffian system $\mathcal{P} = (\theta^0, \theta^1, \theta^2, \theta^3)$ with the structure equations

$$d\theta^0 \equiv \sum_{i=1}^{3} \omega^i \wedge \theta^i \quad (\mathrm{mod}\ \theta^0)$$

and

$$\begin{cases} d\theta^1 \equiv \omega^1 \wedge \pi_1^1 \qquad\qquad + \omega^3 \wedge \pi_3^1, \\ d\theta^2 \equiv \qquad\qquad \omega^2 \wedge \pi_2^2 + \omega^3 \wedge \pi_3^2, \qquad (\mathrm{mod}\ \mathcal{P}). \\ d\theta^3 \equiv \omega^1 \wedge \pi_1^3 + \omega^2 \wedge \pi_2^3 + \omega^3 \wedge \pi_3^3 \end{cases}$$

There are two ways of obtaining singular systems: firstly the dual vector field system of

$$\mathrm{sing}^1(\mathcal{P}) := (\theta^0, \theta^1, \theta^2, \theta^3, \omega^1, \omega^3, \pi_1^1, \pi_3^1)$$

kills not only $d\theta^0$, but also $d\theta^1$ (mod $\mathcal{P}$), and secondly the dual of

$$\mathrm{sing}^2(\mathcal{P}) := (\theta^0, \theta^1, \theta^2, \theta^3, \omega^2, \omega^3, \pi_2^2, \pi_3^2)$$

will kill $d\theta^0$ and $d\theta^2$ (mod $\mathcal{P}$). Note that

$$\dim \operatorname{sing}^i(\mathcal{P})^\perp = \dim \mathcal{S}_1 - \dim \operatorname{sing}^i(\mathcal{P}) = (13-1) - 8 = 4 \quad \text{for } i = 1, 2,$$

which is in accordance with the result in section **5.2**, when there are three independent variables.

It was also seen in section **5.2** that each integral manifold of $\mathcal{P}$ on which $\omega^1 \wedge \omega^2 \wedge \omega^3 \neq 0$ admits one vector field $\partial/\partial\omega^2 + \cdots \in \operatorname{sing}^1(\mathcal{P})^\perp$, and another vector field $\partial/\partial\omega^1 + \cdots \in \operatorname{sing}^2(\mathcal{P})^\perp$, so that the $\operatorname{sing}^i(\mathcal{P}^\perp)$ are Monge systems yielding 1-dimensional Monge characteristics.

Unfortunately it turns out to be difficult to work with the $\operatorname{sing}^i(\mathcal{P})$ directly, but one has better luck when restricting the attention to the one-forms which come from ${}^1Ct^{3,1}$ only—that is, we consider

$$\operatorname{sing}^1_0(\mathcal{P}) := (\theta^0, \theta^1, \theta^2, \theta^3, \omega^1, \omega^3) \quad \text{and} \quad \operatorname{sing}^2_0(\mathcal{P}) := (\theta^0, \theta^1, \theta^2, \theta^3, \omega^2, \omega^3).$$

In analogy with the results obtained for two independent variables we suspect that there might be solution methods if at least one of these systems admits *sufficiently many first integrals*.

One way to determine the first integrals of a general pfaffian system $\mathcal{Q}$ is to form the chain of its derivatives:

$$\mathcal{Q} \supseteq \mathcal{Q}' := \{\theta \in \mathcal{Q} \mid d\theta \equiv 0 \pmod{\mathcal{Q}}\} \supseteq \mathcal{Q}'' := (\mathcal{Q}')' \supseteq \mathcal{Q}''' \supseteq \cdots,$$

which for dimension reasons must stabilize after a finite number of steps:

$$\mathcal{Q} \supset \mathcal{Q}' \supset \cdots \supset \mathcal{Q}^{(k)} = \mathcal{Q}^{(k+1)} = \cdots.$$

Then $\bar{\mathcal{Q}} := \mathcal{Q}^{(k)}$ is complete, and thus the Frobenius theorem shows that there are $m := \dim \bar{\mathcal{Q}}$ functionally independent functions $f^1, \ldots, f^m$ such that $\bar{\mathcal{Q}} = (df^1, \ldots, df^m)$. And these f^i form a fundamental set of first integals for $\mathcal{Q}$.

Lemma 18.3.1. *The singular system* $\operatorname{sing}^1_0(\mathcal{P})$ *admits at most three functionally independent first integrals.*

Proof. Any $\theta \in \operatorname{sing}^1_0(\mathcal{P}) = (\theta^0, \theta^1, \theta^2, \theta^3, \omega^1, \omega^3)$ is of the form $\theta = \sum_{i=0}^3 a_i \theta^i + b_1 \omega^1 + b_3 \omega^3$, where the a_i and b_i are functions on $\mathcal{S}$. Because

$$d\theta^0 \equiv d\theta^1 \equiv 0, \quad d\theta^2 \equiv \omega^2 \wedge \pi^2_2 \quad \text{and} \quad d\theta^3 \equiv \omega^2 \wedge \pi^3_2 \pmod{\operatorname{sing}^1_0(\mathcal{P})},$$

we then have $d\theta \equiv a_2 d\theta^2 + a_3 d\theta^3 + b_1 d\omega^1 + b_3 d\omega^3 \pmod{\operatorname{sing}^1_0(\mathcal{P})}$, or

$$d\theta \equiv b_1 d(\omega^1 + a^1_2 \theta^2 + a^1_3 \theta^3) + b_3 d(\omega^3 + a^3_2 \theta^2 + a^3_3 \theta^3) \pmod{\operatorname{sing}^1_0(\mathcal{P})}$$

for certain functions a_j^i. In the most favourable case the a_j^i can be chosen such that

$$d(\omega^1 + a_2^1\,\theta^2 + a_3^1\,\theta^3) \equiv d(\omega^3 + a_2^3\,\theta^2 + a_3^3\,\theta^3) \equiv 0 \pmod{\mathrm{sing}_0^1(\mathcal{P})},$$

and then

$$\mathrm{sing}_0^1(\mathcal{P})' = (\theta^0, \theta^1, \omega^1 + a_2^1\,\theta^2 + a_3^1\,\theta^3, \omega^3 + a_2^3\,\theta^2 + a_3^3\,\theta^3).$$

Now

$$d\theta^0 \equiv \omega^2 \wedge \theta^2 \pmod{\theta^0, \theta^1, \omega^1, \omega^3} \implies d\theta^0 \not\equiv 0 \pmod{\mathrm{sing}_0^1(\mathcal{P})'},$$

and therefore $\theta^0 \notin \mathrm{sing}_0^1(\mathcal{P})''$. Accordingly $\dim \mathrm{sing}_0^1(\mathcal{P})'' \le 3$, whence also $\dim \overline{\mathrm{sing}_0^1(\mathcal{P})} \le 3$. $\qquad\qquad\square$

If $\dim \overline{\mathrm{sing}_0^1(\mathcal{P})} = 3$, we might as well suppose that

$$\overline{\mathrm{sing}_0^1(\mathcal{P})} = (\theta^1, \omega^1, \omega^3)$$

after having made the substitutions $\omega^1 + a_2^1\,\theta^2 + a_3^1\,\theta^3 \mapsto \omega^1$, $\omega^3 + a_2^3\,\theta^2 + a_3^3\,\theta^3 \mapsto \omega^3$, and perhaps also $\theta^1 + f\,\theta^0 \mapsto \theta^1$ for some function f. Then in particular

$$d\theta^1 \equiv 0 \pmod{\theta^1, \omega^1, \omega^3}.$$

This case occurs for instance if the congruence

$$d\theta^1 \equiv \omega^1 \wedge \pi_1^1 + \omega^3 \wedge \pi_3^1$$

in the structure equations for $\mathcal{P}$ is true modulo (θ^1), and not merely modulo $(\mathcal{P})$.

Theorem 18.3.1. *Suppose that $\mathrm{sing}_0^1(\mathcal{P})$ admits three functionally independent first integrals, and more precisely that $d\theta^1 = \omega^1 \wedge \pi_1^1 + \omega^3 \wedge \pi_3^1$ $\pmod{\theta^1}$. Then the given hyperbolic PDE in three independent variables can be reduced to a PDE in two independent variables only.*

Proof. By the above it may be supposed that $\overline{\mathrm{sing}_0^1(\mathcal{P})} = (\theta^1, \omega^1, \omega^3)$. Then there are functions u, v, w such that

$$(\theta^1, \omega^1, \omega^3) = \overline{\mathrm{sing}_0^1(\mathcal{P})} = (du, dv, dw),$$

where for instance

$$\frac{\partial \theta^1}{\partial dw} \ne 0 \quad \text{and} \quad \frac{\partial(\omega^1 \wedge \omega^3)}{\partial(du \wedge dv)} \ne 0.$$

Then the pfaffian equation $\theta^1 = 0$ can equivalently be written as

$$dw - g\,du - h\,dv = 0. \tag{1}$$

Let us suppose that $du \wedge dv \wedge dw \wedge dg \wedge dh \neq 0$ on S. The integral manifolds of (1) on which $du \wedge dv \neq 0$ (or $\omega^1 \wedge \omega^2 \neq 0$) are given by

$$\mathcal{N}_f: \quad w = f(u), \quad g = \partial f/\partial u, \quad h = \partial f/\partial v,$$

with f being an arbitrary function of two variables. Because

- $\theta^1 = 0$ on $\mathcal{N}_f$,
- $\omega^1 \wedge \omega^3 \neq 0$ on $\mathcal{N}_f$,
- $d\theta^1 \equiv \omega^1 \wedge \pi_1^1 + \omega^3 \wedge \pi_3^1 \pmod{\theta^1}$,

also $\pi_1^1 = \pi_3^1 = 0$ on $\mathcal{N}_f$. If we let $\mathcal{P}_f := \mathcal{P}_{|\mathcal{N}_f}$, and write θ^i instead of $\theta^i_{|\mathcal{N}_f}$, it accordingly follows that $\mathcal{P}_f = (\theta^0, \theta^2, \theta^3)$ has the structure

$$\begin{cases} d\theta^0 \equiv \omega^2 \wedge \theta^2 + \omega^3 \wedge \theta^3 & \pmod{\theta^0}, \\ d\theta^2 \equiv \omega^2 \wedge \pi_2^2 + \omega^3 \wedge \pi_3^2 & \pmod{\mathcal{P}_f}, \\ d\theta^3 \equiv \omega^2 \wedge \pi_2^3 + \omega^3 \wedge \pi_3^3 & \pmod{\mathcal{P}_f}. \end{cases} \tag{2}$$

Thus the Cauchy characteristic system of $\mathcal{P}_f$ is

$$\mathcal{C}(\mathcal{P}_f) = (\theta^0, \theta^2, \theta^3, \omega^2, \omega^3, \pi_2^2, \pi_3^2, \pi_3^3),$$

which has the dimension 8, while $\dim \mathcal{N}_f = \dim S - 3 = 9$. Therefore $\mathcal{P}_f$ admits one Cauchy characteristic vector field on $\mathcal{N}_f$.

The structure of the reduced pfaffian system $(\mathcal{P}_f)_{\text{red}}$ is given by (2), which also represents the structure of the pfaffian system associated to a second order PDE in one dependent and *two* independent variables. Consequently it might be possible to integrate $(\mathcal{P}_f)_{\text{red}}$ by means of the methods described in earlier chapters, whereupon 3-dimensional integral manifolds of $\mathcal{P}_f$ are obtained by using the Cauchy characteristic vector field. $\qquad \square$

The parabolic case. Here $\mathcal{P} = (\theta^0, \theta^1, \theta^2, \theta^3)$ with the structure equations

$$d\theta^0 \equiv \sum_{i=1}^{3} \omega^i \wedge \theta^i \pmod{\theta^0}$$

and

$$\begin{cases} d\theta^1 \equiv \qquad\qquad \omega^2 \wedge \pi_2^1 + \omega^3 \wedge \pi_3^1, \\ d\theta^2 \equiv \omega^1 \wedge \pi_1^2 + \omega^2 \wedge \pi_2^2 + \omega^3 \wedge \pi_3^2, \qquad \pmod{\mathcal{P}}. \\ d\theta^3 \equiv \omega^1 \wedge \pi_1^3 + \omega^2 \wedge \pi_2^3 + \omega^3 \wedge \pi_3^3 \end{cases}$$

Just as in section **5.2**, it is seen that $\mathcal{P}$ admits one Monge characteristic vector field, giving 1-dimensional Monge characteristics—namely the dual of

$$\text{sing}\,(\mathcal{P}) := (\theta^0, \theta^1, \theta^2, \theta^3, \omega^2, \omega^3, \pi^1_2, \pi^1_3),$$

which besides $d\theta^0$ also kills $d\theta^1$ (mod $\mathcal{P}$). In analogy with the hyperbolic case we want the simplest of the congruences above—$d\theta^1 \equiv \omega^2 \wedge \pi^1_2 + \omega^3 \wedge \pi^1_3$ (mod $\mathcal{P}$)—to be valid even modulo θ^1, perhaps after we have modified ω^2, ω^3 and θ^1 somewhat. With $d\theta^1 \equiv \omega^2 \wedge \pi^1_2 + \omega^3 \wedge \pi^1_3$ (mod θ^1), the Cauchy characteristic pfaffian system of (θ^1) is

$$\mathcal{C}(\theta^1) = (\theta^1, \omega^2, \omega^3, \pi^1_2, \pi^1_3),$$

and hence

$$d\theta^1 \equiv d\omega^2 \equiv d\omega^3 \equiv d\pi^1_2 \equiv d\pi^1_3 \equiv 0 \quad (\text{mod } \theta^1, \omega^2, \omega^3, \pi^1_2, \pi^1_3).$$

Combining this with the structure equations above we find that

$$d\theta^0 \equiv d\theta^1 \equiv d\theta^2 \equiv d\theta^3 \equiv d\omega^2 \equiv d\omega^3 \equiv d\pi^1_2 \equiv d\pi^1_3 \equiv 0 \quad (\text{mod sing}\,(\mathcal{P})),$$

so that sing $(\mathcal{P})$ *is complete*—which corresponds to the 2-variable case considered in chapter **16**.

The method of Monge now indicates that we ought to consider $\mathcal{P}_{\text{sing}} := (\theta^0, \theta^1)$.

Theorem 18.3.2. *Suppose that* sing $(\mathcal{P})$ *is complete. Then under fairly general conditions the integation of* $\mathcal{P} = (\theta^0, \theta^1, \theta^2, \theta^3)$ *is equivalent to integrating* $\mathcal{P}_{\text{sing}} = (\theta^0, \theta^1)$, *with the latter having the Lie structure*

$$\begin{cases} d\theta^0 \equiv \omega^2 \wedge \theta^2 + \omega^3 \wedge \theta^3, \\ d\theta^1 = \omega^2 \wedge \pi^1_2 + \omega^3 \wedge \pi^1_3 \end{cases} \quad (\text{mod } \mathcal{P}_{\text{sing}})$$

Proof. The structure equations of $\mathcal{P}$ show that there are functions a_i, b_i and c such that the following hold modulo $(\mathcal{P}_{\text{sing}})$:

$$d\theta^0 \equiv \omega^2 \wedge \theta^2 + \omega^3 \wedge \theta^3,$$
$$d\theta^1 \equiv \omega^2 \wedge \pi^1_2 + \omega^3 \wedge \pi^1_3 + \theta^2 \wedge (a_1\,\omega^1 + a_2\,\omega^2 + a_3\,\omega^3$$
$$+ a_4\,\pi^1_2 + a_5\,\pi^1_3 + a_6\,\pi^2_2 + a_7\,\pi^2_3 + a_8\,\pi^3_3)$$
$$+ \theta^3 \wedge (b_1\,\omega^1 + b_2\,\omega^2 + b_3\,\omega^3 + b_4\,\pi^1_2 + b_5\,\pi^1_3$$
$$+ b_6\,\pi^2_2 + b_7\,\pi^2_3 + b_8\,\pi^3_3) + c\,\theta^2 \wedge \theta^3.$$

Our task is to show that all the coefficients a_i, b_i and c can be assumed to be 0. Because

$$d\theta^1 \equiv \omega^2 \wedge (\pi_2^1 - a_2\,\theta^2 - b_2\,\theta^3) + \omega^3 \wedge (\pi_3^1 - a_3\,\theta^2 - b_3\,\theta^3) + \cdots$$

it is natural to replace $\pi_2^1 - a_2\,\theta^2 - b_2\,\theta^3$ by π_2^1 and $\pi_3^1 - a_3\,\theta^2 - b_3\,\theta^3$ by π_3^1, so that we may suppose that $a_2 = b_2 = a_3 = b_3 = 0$. Then

$$d\theta^1 \equiv -\pi_2^1 \wedge (\omega^2 + a_4\,\theta^2 + b_4\,\theta^3) - \pi_3^1 \wedge (\omega^3 + a_5\,\theta^2 + b_5\,\theta^3) + \cdots.$$

Replacing $\omega^2 + a_4\,\theta^2 + b_4\,\theta^3$ by ω^2 and $\omega^3 + b_4\,\theta^2 + b_5\,\theta^3$ by ω^3 does not alter the congruence

$$d\theta^0 \equiv \omega^2 \wedge \theta^2 + \omega^3 \wedge \theta^3 \quad (\text{mod } \mathcal{P}_{\text{sing}}),$$

while $d\theta^1$ becomes simplified in the sense that a_4, b_4 and b_5 vanish; besides, a_5 is changed a bit. Thus modulo $\mathcal{P}_{\text{sing}}$ we are now left with

$$d\theta^0 \equiv \omega^2 \wedge \theta^2 + \omega^3 \wedge \theta^3,$$
$$d\theta^1 \equiv \omega^2 \wedge \pi_2^1 + \omega^3 \wedge \pi_3^1 + \theta^2 \wedge (a_1\,\omega^1 + a_5\,\pi_3^1 + a_6\,\pi_2^2 + a_7\,\pi_3^2 + a_8\,\pi_3^3)$$
$$+ \theta^3 \wedge (b_1\,\omega^1 + b_6\,\pi_2^2 + b_7\,\pi_3^2 + b_8\,\pi_3^3) + c\,\theta^2 \wedge \theta^3.$$

Let us next investigate the consequences of the vanishing of $d^2\theta^1$. First, the terms in $d^2\theta^1$ containing $\omega^1 \wedge \omega^2 \wedge \pi_2^2$ can only come from $-\omega^2 \wedge d\pi_2^1$ and $a_1\,d\theta^2 \wedge \omega^1$. But π_2^1 belongs to the complete system $\text{sing}(\mathcal{P})$, and therefore $d\pi_2^1$ cannot contain $\omega^1 \wedge \pi_2^2$. Hence only $a_1\,d\theta^2 \wedge \omega^1 = a_1\,\omega^2 \wedge \pi_2^2 \wedge \omega^1 + \cdots$ remains, showing that $a_1 = 0$. By considering the terms in $d^2\theta^1$ containing $\omega^1 \wedge \omega^2 \wedge \pi_3^2$ it is similarly seen that $b_1 = 0$.

The only terms in $d^2\theta^1$ which might contain $\omega^1 \wedge \pi_2^1 \wedge \pi_2^2$ are $d\omega^2 \wedge \pi_2^1$ and $a_6\,d\theta^2 \wedge \pi_2^2$. Since $\omega^2 \in \mathcal{C}(\theta^1)$, $d\omega^2 \equiv 0 \ (\text{mod } \theta^1, \omega^2, \omega^3, \pi_2^1, \pi_3^1)$, and hence $d\omega^2$ cannot be a multiple of $\omega^1 \wedge \pi_2^2$, while $d\theta^2$ does contain $\omega^1 \wedge \pi_2^2$. Thus $a_6 = 0$.

By considering $\omega^1 \wedge \pi_2^1 \wedge \pi_3^2$, $\omega^1 \wedge \pi_2^1 \wedge \pi_3^3$, $\omega^1 \wedge \pi_3^1 \wedge \pi_2^2$, $\omega^1 \wedge \pi_3^1 \wedge \pi_3^2$ and $\omega^1 \wedge \pi_3^1 \wedge \pi_3^3$ in the same way, it is seen that also $a_7 = a_8 = b_6 = b_7 = b_8 = 0$.

Then only a_5 and c are still different from 0, so that

$$d\theta^1 \equiv \omega^2 \wedge \pi_2^1 + \omega^3 \wedge \pi_3^1 + a_5\,\theta^2 \wedge \pi_3^1 + c\,\theta^2 \wedge \theta^3 \quad (\text{mod } \mathcal{P}_{\text{sing}}).$$

The terms in $d^2\theta^1$ that may contain $\omega^1 \wedge \pi_2^1 \wedge \pi_3^1$ are $d\omega^2 \wedge \pi_2^1 + d\omega^3 \wedge \pi_3^1 + a_5\,d\theta^2 \wedge \pi_3^1$. Inserting

$$d\omega^2 = \alpha\,\omega^1 \wedge \pi_3^1 + \cdots \quad \text{and} \quad d\omega^3 = \beta\,\omega^1 \wedge \pi_2^1 + \cdots$$

into this we see that $-\alpha + \beta + a_5 = 0$.

Differentiating $d\theta^0 \equiv \omega^1 \wedge \theta^1 + \omega^2 \wedge \theta^2 + \omega^3 \wedge \theta^3 \pmod{\theta^0}$ and considering the terms involving $\omega^1 \wedge \theta^2 \wedge \pi_3^1$, we get

$$0 = -a_5\,\omega^1 \wedge \theta^2 \wedge \pi_3^1 + \alpha\,\omega^1 \wedge \pi_3^1 \wedge \theta^2 = -(a_5 + \alpha)\,\omega^1 \wedge \theta^2 \wedge \pi_3^1,$$

whence $\alpha = -a_5$.

Looking at the terms in $d^2\theta^0$ containing $\omega^1 \wedge \theta^3 \wedge \pi_2^1$ we similarly see that $\beta = 0$. Consequently $a_5 = \alpha = -\alpha$, and therefore $a_5 = \alpha = 0$, leaving

$$d\theta^1 \equiv \omega^2 \wedge \pi_2^1 + \omega^3 \wedge \pi_3^1 + c\,\theta^2 \wedge \theta^3 \pmod{\mathcal{P}_{\mathrm{sing}}}.$$

The terms in $d^2\theta^1$ which may contain $\omega^1 \wedge \theta^3 \wedge \pi_2^1$ are

$$d\omega^2 \wedge \pi_2^1 + c\,\omega^1 \wedge \pi_1^2 \wedge \theta^3 = (d\omega^2 - c\,\omega^1 \wedge \theta^3) \wedge \pi_2^1,$$

showing that $d\omega^2 = c\,\omega^1 \wedge \theta^3 + \cdots$.

If we consider the terms in $d^2\theta^1$ containing $\omega^1 \wedge \theta^2 \wedge \pi_3^1$ it analogously follows that $d\omega^3 = -c\,\omega^1 \wedge \theta^2 + \cdots$.

Finally, inserting the expressions above for $d\theta^1$, $d\omega^2$, $d\omega^3$ into $0 = d^2\theta^0$, and only keeping the terms involving $\omega^1 \wedge \theta^2 \wedge \theta^3$, we obtain

$$0 = -c\,\omega^1 \wedge \theta^2 \wedge \theta^3 + c\,\omega^1 \wedge \theta^3 \wedge \theta^2 - c\,\omega^1 \wedge \theta^2 \wedge \theta^3 = -3c\,\omega^1 \wedge \theta^2 \wedge \theta^3,$$

forcing c to vanish.

So the wanted Lie structure of $\mathcal{P}_{\mathrm{sing}}$ is given by

$$\begin{cases} d\theta^0 \equiv \omega^2 \wedge \theta^2 + \omega^3 \wedge \theta^3, \\ d\theta^1 \equiv \omega^2 \wedge \pi_2^1 + \omega^3 \wedge \pi_3^1 \end{cases} \pmod{\mathcal{P}_{\mathrm{sing}}},$$

which thus is *much simpler than expected*.

From these structure equations it is readily seen that

$$\mathcal{C}(\mathcal{P}_{\mathrm{sing}}) = (\theta^0, \theta^1, \theta^2, \theta^3, \omega^2, \omega^3, \pi_2^1, \pi_3^1), \quad \text{that is,} \quad \mathcal{C}(\mathcal{P}_{\mathrm{sing}}) = \mathrm{sing}\,(\mathcal{P}).$$

Because

$$(\omega^1, \omega^2, \omega^3, \theta^0, \theta^1, \theta^2, \theta^3) = (dx^1, dx^2, dx^3, dz, dp_1, dp_2, dp_3),$$

the eight first integrals of $\mathcal{C}(\mathcal{P}_{\mathrm{sing}})$ may be expressed by means of x^1, x^2, x^3, z, p_1, p_2, p_3 and two more variables v_2 and v_3. The latter may for instance be chosen such that the pfaffian equation $\theta^1 = 0$ is equivalent to

$$dp_1 - v_2\,dx^2 - v_3\,dx^3 - a^1\,dx^1 - a^2\,dp_2 - a^3\,dp_3 = 0,$$

where the a^i are functions of x^1, x^2, x^3, z, p_1, p_2, p_3, v_2, v_3. Expressed in these variables, $\theta^0 = 0 \iff dz - \sum_{i=1}^3 p_i\,dx^i = 0$. Differentiating this,

$\sum_{i=1}^{3} dp_i \wedge dx^i = 0$, and therefore $dp_i = \sum_{j=1}^{3} p_{ij} dx^j$ for $i = 1, 2, 3$, where $p_{ij} = p_{ji}$ for all i and j. Using these expressions for dp_i, $\theta^1 = 0$ is equivalent to

$$\begin{cases} p_{11} - a^1 - a^2 p_{12} - a^3 p_{13} = 0, \\ p_{12} - v_2 - a^2 p_{22} - a^3 p_{23} = 0, \\ p_{13} - v_3 - a^2 p_{23} - a^3 p_{33} = 0. \end{cases}$$

If we eliminate v_2 and v_3 from this system, there remains just one equation—say

$$F(x^1, x^2, x^3, z, p_1, p_2, p_3, p_{11}, p_{12}, p_{13}, p_{22}, p_{23}, p_{33}) = 0,$$

which defines a PDE $S^* \subset J^2(\mathbb{C}^3_x, \mathbb{C}_z)$. The solutions of S^* correspond to those 3-dimensional integral manifolds of $\mathcal{P}_{\text{sing}}$ on which $dx^1 \wedge dx^2 \wedge dx^3 \neq 0$.

Since the 2-graphs of solutions of the original parabolic PDE $S \subset J^2(\mathbb{C}^3_x, \mathbb{C}_z)$ are integrals of $\mathcal{P}$, they are integrals of $\mathcal{P}_{\text{sing}}$ too, and consequently also solutions of the PDE S^*. With S being supposed to be formally solvable, this implies that $S \subseteq S^*$ (as explained in the remark after theorem **16.1.2**). But both S and S^* are of codimension 1 in $J^2(\mathbb{C}^3_x, \mathbb{C}_z)$, and therefore, if S^* is irreducible in a suitable sense, they must be equal. So finally

$$\text{integrating } \mathcal{P} \iff \text{integrating } \mathcal{P}_{\text{sing}}.$$

$\square$

Let us now integrate the pfaffian system $\mathcal{P}_{\text{sing}}$ on $\mathbb{C}^9$, where the latter is equipped with the local coordinates $x^1, x^2, x^3, z, p_1, p_2, p_3, v_2$ and v_3. Because $\dim \mathcal{C}(\mathcal{P}_{\text{sing}}) = 8$, there is one Cauchy characteristic vector field, namely $\partial/\partial\omega^1 = \partial/\partial x^1 + \cdots$. According to the proof of the local rectification lemma for vector fields it is possible to introduce new coordinates $X^2, X^3, Z, P_1, P_2, P_3, V_2, V_3$ satisfying

$$X^2(0, x^2, \ldots, v_3) = x^2, \ldots, V_3(0, x^2, \ldots, v_3) = v_3,$$

and having the property that the Cauchy characteristic vector field $\partial/\partial\omega^1$ goes over into $\partial/\partial x^1$ when we use $x^1, X^2, \ldots, V_3$ as new local coordinates. On changing to these coordinates, $\mathcal{P}_{\text{sing}}$ is transformed into

$$(\mathcal{P}_{\text{sing}})_{\text{red}} : \quad \begin{cases} dZ - P_2 \, dX^2 - P_3 \, dX^3 = 0, \\ dP_1 - V_2 \, dX^2 - V_3 \, dX^3 - A^2 \, dP_2 - A^3 \, dP_3 = 0, \end{cases}$$

living on $\mathbb{C}^8$, with A^2, A^3 being functions of $X^2, \ldots, V_3$. The 2-dimensional integral manifolds of $(\mathcal{P}_{\text{sing}})_{\text{red}}$ on which $dX^2 \wedge dX^3 \neq 0$ are obtained as follows.

Setting $Z := f(X^2, X^3)$, $P_1 := g(X^2, X^3)$ with arbitrary functions f and g, we have $P_2 = \partial f / \partial X^2$ and $P_3 = \partial f / \partial X^3$, while V_2 and V_3 are obtained by means of the implicit function theorem from the system

$$\frac{\partial g}{\partial X^2} - V_2 - A^2 \frac{\partial^2 f}{(\partial X^2)^2} - A^3 \frac{\partial^2 f}{\partial X^2 \partial X^3} = 0,$$

$$\frac{\partial g}{\partial X^3} - V_3 - A^2 \frac{\partial^2 f}{\partial X^2 \partial X^3} - A^3 \frac{\partial^2 f}{(\partial X^3)^2} = 0.$$

Let $\mathcal{N}_{\text{red}}$ denote the general 2-dimensional integral manifold of $(\mathcal{P}_{\text{sing}})_{\text{red}}$ on $\mathbb{C}^8_{X,Z,P,V}$ obtained in this way. Then the general 3-dimensional integral manifold of $\mathcal{P}_{\text{sing}}$ on $\mathbb{C}^9 = \mathbb{C}^8_{X,Z,P,V} \times \mathbb{C}^1_{x^1}$ is given by $\mathcal{N}_{\text{red}} \times \mathbb{C}^1_{x^1}$.

We see in particular that the general solution of the parabolic PDE S depends on two arbitrary functions of two variables, which is in accordance with the fact that $s_2 = 2$ and $s_3 = 0$.

Conclusion. *Integrating* $\mathcal{P}$ $\iff$ *integrating* $\mathcal{P}_{\text{sing}}$, *and the latter can be done in an elementary way (that is, without using the Cauchy–Kowalewski theorem).*

18.4 Two PDEs

Here we will use the method of Monge to prove that *all involutive systems of **two** second order PDEs in one dependent and three independent variables may be solved by reductions to ODE systems*. Which indeed is surprising—at least when compared with the poor results above for a single PDE.

As seen in section **18.2** there are two canonical forms to consider:

(a) $\pi_3^1 = \pi_3^2 = 0$,
(b) $\pi_1^1 = \pi_2^1 = 0$.

Type (a). In this case $\mathcal{P} = (\theta^0, \theta^1, \theta^2, \theta^3)$, with the structure equations

$$d\theta^0 \equiv \sum_{i=1}^{3} \omega^i \wedge \theta^i \pmod{\theta^0}$$

and

$$\begin{cases} d\theta^1 \equiv \omega^1 \wedge \pi_1^1 + \omega^2 \wedge \pi_2^1, \\ d\theta^2 \equiv \omega^1 \wedge \pi_1^2 + \omega^2 \wedge \pi_2^2, \qquad\qquad (\text{mod } \mathcal{P}). \\ d\theta^3 \equiv \qquad\qquad\qquad \omega^3 \wedge \pi_3^3 \end{cases}$$

Clearly the dual vector field system of $(\theta^0, \theta^1, \theta^2, \theta^3, \omega^1, \omega^2, \pi_1^1, \pi_2^1)$ besides $d\theta^0$ will also kill $d\theta^1$ (mod $\mathcal{P}$), and hence is singular. Note however that on each integral manifold $\mathcal{N}$ of $\mathcal{P}$ on which $\omega^1 \wedge \omega^2 \wedge \omega^3 \neq 0$,

$$0 = d\theta^1 = \omega^1 \wedge \pi_1^1 + \omega^2 \wedge \pi_2^1 \quad \text{and} \quad 0 = d\theta^2 = \omega^1 \wedge \pi_1^2 + \omega^2 \wedge \pi_2^2,$$

so that π_1^1, π_2^1 and π_2^2 are linear combinations of ω^1 and ω^2 on $\mathcal{N}$. Then if $\omega^1 = \omega^2 = 0$, also $\pi_1^1 = \pi_2^1 = \pi_2^2 = 0$ on $\mathcal{N}$. Therefore it seems more natural to consider the bigger singular pfaffian system

$$\text{sing}^1(\mathcal{P}) := (\theta^0, \theta^1, \theta^2, \theta^3, \omega^1, \omega^2, \pi_1^1, \pi_2^1, \pi_2^2),$$

the dual of which besides $d\theta^0$ also kills both $d\theta^1$ and $d\theta^2$ (mod $\mathcal{P}$).

Another singular system is

$$\text{sing}^2(\mathcal{P}) := (\theta^0, \theta^1, \theta^2, \theta^3, \omega^3, \pi_3^3),$$

whose dual kills both $d\theta^0$ and $d\theta^3$ (mod $\mathcal{P}$).

Recall that a singular subsystem $\text{sing}(\mathcal{P}^\perp)$ of $\mathcal{P}^\perp$ (or dually: $\text{sing}(\mathcal{P}) \supset \mathcal{P}$) is *Monge characteristic* if

for any complete subsystem $\mathcal{K}$ of $\mathcal{P}^\perp$, $\mathcal{K} \cap \text{sing}(\mathcal{P}^\perp)$ is complete as well.

In that case the integral manifolds of $\mathcal{K}$ are foliated by *Monge characteristics*, where the latter are the integral manifolds of $\mathcal{K} \cap \text{sing}(\mathcal{P}^\perp)$.

Remark. By abuse of language, a *Monge system* denotes both a Monge characteristic vector field system, and its dual pfaffian system.

Theorem 18.4.1. $\text{sing}^1(\mathcal{P})$ *and* $\text{sing}^2(\mathcal{P})$ *are Monge systems having Monge characteristics of dimension 1 and 2 respectively.*

Proof. We give the proof for $\text{sing}^2(\mathcal{P})$ only. To simplify the notation we denote the one-forms π_1^1, π_2^1, π_2^2 and π_3^3 alternatively by π^1, π^2, π^3 and π^4 here.

Any complete 3-dimensional subsystem $\mathcal{K}$ of $\mathcal{P}^\perp$ on which $\omega^1 \wedge \omega^2 \wedge \omega^3 \neq 0$ is generated by vector fields of the form

$$X_i = \frac{\partial}{\partial \omega^i} + \sum_{j=1}^4 a_i^j \frac{\partial}{\partial \pi^j} \quad \text{for } i = 1, 2, 3.$$

The dual complete pfaffian system $Q := K^{\perp}$ is then given by

$$Q = (\theta^0, \theta^1, \theta^2, \theta^3, \psi^1, \psi^2, \psi^3, \psi^4),$$

where $\psi^k := \pi^k - \sum_{i=1}^3 a_i^k \omega^i$ for $k = 1, 2, 3, 4$. Therefore

$$(K \cap \text{sing}^2(P)^{\perp})^{\perp} = (Q, \text{sing}^2(P)) = (\theta^0, \theta^1, \theta^2, \theta^3, \psi^1, \psi^2, \psi^3, \psi^4, \omega^3, \pi_3^3).$$

Since Q is complete, $d\theta^3 \equiv 0 \pmod{Q}$, implying that $\omega^3 \wedge \pi_3^3 \equiv 0 \pmod{Q}$. Hence π_3^3 is a multiple of $\omega^3 \pmod{Q}$, and can therefore be deleted from the list of generators of $(Q, \text{sing}^2(P))$.

Because of the completeness of Q it then suffices to prove that $d\omega^3 \equiv 0 \pmod{Q, \text{sing}^2(P)}$.

The structure equations of P reveal that

$$\begin{aligned} C(P) &= (\theta^0, \theta^1, \theta^2, \theta^3, \omega^1, \omega^2, \omega^3, \pi_1^1, \pi_2^1, \pi_2^2, \pi_3^2) \\ &= (\theta^0, \theta^1, \theta^2, \theta^3, \omega^1, \omega^2, \omega^3, \psi^1, \psi^2, \psi^3, \psi^4), \end{aligned}$$

the generators of which constitute a local coframe for the PDE system S (so that there is no Cauchy characteristic vector field). Consequently

$$d\omega^3 \equiv a\,\omega^1 \wedge \omega^2 \pmod{Q, \text{sing}^2(P)}$$

for some function a, and we have to prove that

$$a = \frac{\partial d\omega^3}{\partial(\omega^1 \wedge \omega^2)} = 0.$$

This follows when we differentiate $d\theta^3 \equiv \omega^3 \wedge \pi_3^3 \pmod{P}$:

$$0 = d^2\theta^3 \equiv d\omega^3 \wedge \pi_3^3 - \omega^3 \wedge d\pi_3^3 \pmod{P, dP},$$

from which $a\,\omega^1 \wedge \omega^2 \wedge \pi_3^3 + \cdots = 0$—forcing a to vanish. $\square$

Conclusion. *Each 3-dimensional integral manifold of P on which $\omega^1 \wedge \omega^2 \wedge \omega^3 \neq 0$ is foliated by 1-dimensional Monge characteristics from $\text{sing}^1(P)$, generated by $\partial/\partial\omega^3 + \cdots$, and by 2-dimensional Monge characteristics from $\text{sing}^2(P)$, generated by $\partial/\partial\omega^1 + \cdots$ and $\partial/\partial\omega^2 + \cdots$.*

The presence of these Monge systems makes it conceivable that the Monge method can be used in order to find integral manifolds. Applying this method with respect to $\text{sing}^2(P)$, we are to consider

$$P_{\text{sing}} := (\theta^0, \theta^3).$$

Theorem 18.4.2. *It is possible to modify the one-forms θ^3, ω^3 and π_3^3 in such a way that the structure equations of $\mathcal{P}_{\text{sing}}$ are simplified to*

$$\begin{cases} d\theta^0 \equiv \omega^1 \wedge \theta^1 + \omega^2 \wedge \theta^2, \\ d\theta^3 \equiv \omega^3 \wedge \pi_3^3. \end{cases} \quad (\text{mod } \mathcal{P}_{\text{sing}}).$$

Proof. The congruence $d\theta^3 \equiv \omega^3 \wedge \pi_3^3 \ (\text{mod } \mathcal{P})$ implies that

$$\begin{aligned} d\theta^3 &\equiv \omega^3 \wedge \pi_3^3 + \theta^1 \wedge (a_1 \, \omega^1 + a_2 \, \omega^2 + a_3 \, \omega^3 + a_4 \, \pi_1^1 + a_5 \, \pi_2^1 \\ &\quad + a_6 \, \pi_2^2 + a_7 \, \pi_3^3) \\ &\quad + \theta^2 \wedge (b_1 \, \omega^1 + b_2 \, \omega^2 + b_3 \, \omega^3 \\ &\quad + b_4 \, \pi_1^1 + b_5 \, \pi_2^1 + b_6 \, \pi_2^2 + b_7 \, \pi_3^3) + c \, \theta^1 \wedge \theta^2 \quad (\text{mod } \mathcal{P}_{\text{sing}}) \end{aligned}$$

for suitable functions a_i, b_i and c. By setting $\tilde{\omega}^3 := \omega^3 + a_7 \, \theta^1 + b_7 \, \theta^2$ and $\tilde{c} := c - a_3 b_7 - b_3 a_7$, $d\theta^3$ is transformed into

$$\begin{aligned} d\theta^3 &\equiv \tilde{\omega}^3 \wedge \pi_3^3 + \theta^1 \wedge (a_1 \, \omega^1 + a_2 \, \omega^2 + a_3 \, \tilde{\omega}^3 + a_4 \, \pi_1^1 + a_5 \, \pi_2^1 + a_6 \, \pi_2^2) \\ &\quad + \theta^2 \wedge (b_1 \, \omega^1 + b_2 \, \omega^2 + b_3 \, \tilde{\omega}^3 + b_4 \, \pi_1^1 + b_5 \, \pi_2^1 + b_6 \, \pi_2^2) + \tilde{c} \, \theta^1 \wedge \theta^2 \\ &\equiv \tilde{\omega}^3 \wedge (\pi_3^3 - a_3 \, \theta^1 - b_3 \, \theta^2) + \theta^1 \wedge (a_1 \, \omega^1 + a_2 \, \omega^2 + a_4 \, \pi_1^1 + a_5 \, \pi_2^1 + a_6 \, \pi_2^2) \\ &\quad + \theta^2 \wedge (b_1 \, \omega^1 + b_2 \, \omega^2 + b_4 \, \pi_1^1 + b_5 \, \pi_2^1 + b_6 \, \pi_2^2) + \tilde{c} \, \theta^1 \wedge \theta^2 \quad (\text{mod } \mathcal{P}_{\text{sing}}). \end{aligned}$$

By introducing $\tilde{\pi}_3^3 := \pi_3^3 - a_3 \, \theta^1 - b_3 \, \theta^2$ and forgetting the tildes afterwards, $d\theta^3$ is further simplified to

$$\begin{aligned} d\theta^3 &\equiv \omega^3 \wedge \pi_3^3 + \theta^1 \wedge (a_1 \, \omega^1 + a_2 \, \omega^2 + a_4 \, \pi_1^1 + a_5 \, \pi_2^1 + a_6 \, \pi_2^2) \\ &\quad + \theta^2 \wedge (b_1 \, \omega^1 + b_2 \, \omega^2 + b_4 \, \pi_1^1 + b_5 \, \pi_2^1 + b_6 \, \pi_2^2) + c \, \theta^1 \wedge \theta^2 \quad (\text{mod } \mathcal{P}_{\text{sing}}). \end{aligned}$$

The next idea is to study the implications of the vanishing of $d^2\theta^3$. For instance, $\omega^2 \wedge \pi_1^1 \wedge \pi_2^1$ can appear in $d^2\theta^3$ only through

$$a_4 \, d\theta^1 \wedge \pi_1^1 = a_4 \, \omega^2 \wedge \pi_2^1 \wedge \pi_1^1 + \cdots,$$

and hence $a_4 = 0$. If we consider $\omega^2 \wedge \pi_1^1 \wedge \pi_2^2$ and $\omega^2 \wedge \pi_2^1 \wedge \pi_2^2$ in the same way it follows that $b_4 = 0$ and $a_6 - b_5 = 0$.

Looking at the coefficients of $\omega^1 \wedge \pi_1^1 \wedge \pi_2^1$, $\omega^1 \wedge \pi_1^1 \wedge \pi_2^2$ and $\omega^1 \wedge \pi_2^1 \wedge \pi_2^2$ we similarly find that $a_5 - b_4 = 0$, $a_6 = 0$ and $b_6 = 0$.

Finally, consideration of $\omega^1 \wedge \omega^2 \wedge \pi_1^1$, $\omega^1 \wedge \omega^2 \wedge \pi_2^1$ and $\omega^1 \wedge \omega^2 \wedge \pi_2^2$ shows that $a_2 = 0$, $a_1 - b_2 = 0$ and $b_1 = 0$ respectively, so that

$$d\theta^3 \equiv \omega^3 \wedge \pi_3^3 + a \, (\theta^1 \wedge \omega^1 + \theta^2 \wedge \omega^2) + c \, \theta^1 \wedge \theta^2 \quad (\text{mod } \mathcal{P}_{\text{sing}}),$$

where $a := a_1 = b_2$. Or

$$d(\theta^3 - a \, \theta^0) \equiv d\theta^3 - a \, d\theta^0 \equiv \omega^3 \wedge \pi_3^3 + c \, \theta^1 \wedge \theta^2 \quad (\text{mod } \mathcal{P}_{\text{sing}}).$$

So replacing $\theta^3 - a\,\theta^0$ by θ^3, we have

$$d\theta^3 \equiv \omega^3 \wedge \pi_3^3 + c\,\theta^1 \wedge \theta^2 \quad (\mathrm{mod}\ \mathcal{P}_{\mathrm{sing}}).$$

If we differentiate this, the coefficient of $\omega^2 \wedge \theta^1 \wedge \pi_2^2$ in $d^2\theta^3 = 0$ will be equal to $-c$, and therefore $c = 0$. Hence the structure of $\mathcal{P}_{\mathrm{sing}}$ is at last reduced to

$$\begin{cases} d\theta^0 \equiv \omega^1 \wedge \theta^1 + \omega^2 \wedge \theta^2, \\ d\theta^3 \equiv \omega^3 \wedge \pi_3^3 \end{cases} \quad (\mathrm{mod}\ \mathcal{P}_{\mathrm{sing}}).$$

$\square$

These structure equations reveal that

$$\mathcal{C}(\mathcal{P}_{\mathrm{sing}}) = (\theta^0, \theta^1, \theta^2, \theta^3, \omega^1, \omega^2, \omega^3, \pi_3^3) \supsetneq (\theta^0, \theta^1, \theta^2, \theta^3, \omega^3, \pi_3^3) = \mathrm{sing}^2(\mathcal{P}).$$

Because

$$(\omega^1, \omega^2, \omega^3, \theta^0, \theta^1, \theta^2, \theta^3) = (dx^1, dx^2, dx^3, dz, dp_1, dp_2, dp_3)$$

it is thus possible to find a function v such that

$$\mathcal{C}(\mathcal{P}_{\mathrm{sing}}) = (dx^1, dx^2, dx^3, dz, dp_1, dp_2, dp_3, dv).$$

Therefore the one-forms θ^0 and θ^3 in $\mathcal{P}_{\mathrm{sing}} = (\theta^0, \theta^3)$ may be modified so as to live on $J^1(\mathbb{C}_x^3, \mathbb{C}_z) \times \mathbb{C}_v \cong \mathbb{C}^8$ without changing the structure equations. Then $(\mathcal{P}_{\mathrm{sing}})_{\mathrm{red}} = (\theta^0, \theta^3)$, with the right hand side being defined on $\mathbb{C}^8$.

Theorem 18.4.3. *Solving the original PDE system* $\mathcal{S} \subset J^2(\mathbb{C}_x^3, \mathbb{C}_z)$ *is under fairly general conditions equivalent to determining those 3-dimensional integral manifolds of* $(\mathcal{P}_{\mathrm{sing}})_{\mathrm{red}}$ *on which* $dx^1 \wedge dx^2 \wedge dx^3 \neq 0$.

Proof. In fact, when θ^0 and θ^3 are regarded as one-forms on $J^1(\mathbb{C}_x^3, \mathbb{C}_z) \times \mathbb{C}_v$, the system $\theta^0 = \theta^3 = 0$ can be written as

$$\begin{cases} dz - \sum_{i=1}^3 p_i\,dx^i = 0, \\ dp_3 + a_1\,dx^1 + a_2\,dx^2 + a_3\,dx^3 + b^1\,dp_1 + b^2\,dp_2 = 0, \end{cases}$$

where the a_i and b^j are functions of x^1, x^2, x^3, z, p_1, p_2, p_3 and v. Differentiation of the first equation shows that $dp_i = \sum_{j=1}^3 p_{ij}\,dx^j$, where the coefficients satisfy $p_{ij} = p_{ji}$, but are otherwise arbitrary. Inserting this into the second equation, and regarding dx^1, dx^2 and dx^3 as independent on the integral manifolds we are looking for, we get

$$\begin{cases} p_{13} + a_1 + b^1 p_{11} + b^2 p_{12} = 0, \\ p_{23} + a_2 + b^1 p_{12} + b^2 p_{22} = 0, \\ p_{33} + a_3 + b^1 p_{13} + b^2 p_{23} = 0. \end{cases}$$

When we eliminate v between these equations there remain two which define a submanifold $\mathcal{S}^*$ of $J^2(\mathbb{C}^3_x, \mathbb{C}_z)$ of codimension 2. With $\mathcal{S}$ being formally solvable this implies that $\mathcal{S} \subseteq \mathcal{S}^*$, and so, if $\mathcal{S}^*$ is irreducible, $\mathcal{S} = \mathcal{S}^*$ for dimension reasons. $\qquad\square$

Let us next investigate how to integrate $(\mathcal{P}_{\text{sing}})_{\text{red}} = (\theta^0, \theta^3)$ on $J^1(\mathbb{C}^3_x, \mathbb{C}_z)$ $\times \mathbb{C}_v \cong \mathbb{C}^8$. Because $d\theta^3 \equiv \omega^3 \wedge \pi^3_3 \pmod{\theta^0, \theta^3}$ there are functions c_i on $\mathbb{C}^8$ such that modulo θ^3,

$$d\theta^3 \equiv \omega^3 \wedge \pi^3_3 + \theta^0 \wedge (c_1 \omega^1 + c_2 \omega^2 + c_3 \omega^3 + c_4 \theta^1 + c_5 \theta^2 + c_6 \pi^3_3).$$

By differentiating this and considering the coefficients of $\omega^1 \wedge \omega^2 \wedge \theta^2$, $\omega^1 \wedge \omega^2 \wedge \theta^1$, $\omega^2 \wedge \theta^1 \wedge \theta^2$ and $\omega^1 \wedge \theta^1 \wedge \theta^2$ respectively, it is found that $c_1 = c_2 = c_4 = c_5 = 0$, and hence actually

$$\begin{aligned} d\theta^3 &\equiv \omega^3 \wedge \pi^3_3 + c_3 \, \theta^0 \wedge \omega^3 + c_6 \, \theta^0 \wedge \pi^3_3 \\ &\equiv (\omega^3 + c_6 \, \theta^0) \wedge (\pi^3_3 - c_3 \, \theta^0) \pmod{\theta^3}. \end{aligned}$$

Replacing $\omega^3 + c_6 \, \theta^0$ by ω^3 and $\pi^3_3 - c_3 \, \theta^0$ by π^3_3 we finally obtain the congruence

$$d\theta^3 \equiv \omega^3 \wedge \pi^3_3 \pmod{\theta^3}.$$

But then the genus of $d\theta^3 \pmod{\theta^3}$ equals 1, and so the classification of pfaffian equations in chapter **6** shows that there are functions U, V, W such that

$$\theta^3 = 0 \Longleftrightarrow dV - W \, dU = 0.$$

The integral manifolds on which $dU \neq 0$ are thus given by

$$V = F(U) \quad \text{and} \quad W = F'(U),$$

where $F(U)$ is an arbitrary function. Expressing everything in the local coordinates $x^1, x^2, x^3, z, p_1, p_2, p_3, v$, and eliminating v between these two equations, we get a relation

$$R_F(x^1, x^2, x^3, z, p_1, p_2, p_3) = 0,$$

defining a hypersurface $\mathcal{S}_F \subset J^1(\mathbb{C}^3_x, \mathbb{C}_z)$. Then for each F there remains to solve the pfaffian equation $\theta^0 = 0$ on $\mathcal{S}_F$, which according to chapter **6** can be accomplished by solving ordinary differential equations. Because of the last step the general solution will depend on one arbitrary function of two variables.

Type (b). In this case $\pi_1^1 = \pi_2^1 = 0$, and accordingly the structure equations of $\mathcal{P} = (\theta^0, \theta^1, \theta^2, \theta^3)$ are given by

$$d\theta^0 \equiv \sum_{i=1}^{3} \omega^i \wedge \theta^i \pmod{\theta^0}$$

and

$$\begin{cases} d\theta^1 \equiv & \omega^3 \wedge \pi_3^1, \\ d\theta^2 \equiv & \omega^2 \wedge \pi_2^2 + \omega^3 \wedge \pi_3^2, \\ d\theta^3 \equiv \omega^1 \wedge \pi_1^3 + \omega^2 \wedge \pi_2^3 + \omega^3 \wedge \pi_3^3 \end{cases} \pmod{\mathcal{P}}.$$

There is always one Monge system giving 1-dimensional characteristics, namely $\mathrm{sing}^1(\mathcal{P}) := (\theta^0, \theta^1, \theta^2, \theta^3, \omega^2, \omega^3, \pi_3^1, \pi_2^2, \pi_3^2)$, the dual of which besides $d\theta^0$ also kills $d\theta^1$ and $d\theta^2$ (mod $\mathcal{P}$). Then in analogy with type **(a)** it can be seen that $\mathrm{sing}^2(\mathcal{P}) := (\theta^0, \theta^1, \theta^2, \theta^3, \omega^3, \pi_3^1)$—whose dual kills $d\theta^0$ and $d\theta^1$ (mod $\mathcal{P}$)—is a Monge system with 2-dimensional characteristics, provided that

$$\frac{\partial d\omega^3}{\partial(\omega^1 \wedge \omega^2)} = 0.$$

Let us examine $d\theta^1$ in order to see when this is true. Since $d\theta^1 \equiv \omega^3 \wedge \pi_3^1$ (mod $\mathcal{P}$) there are functions a_i, b_i and c such that

$$d\theta^1 \equiv \omega^3 \wedge \pi_3^1 + \theta^2 \wedge (a_1 \omega^1 + a_2 \omega^2 + a_3 \omega^3 + a_4 \pi_3^1 + a_5 \pi_2^2 + a_6 \pi_3^2 + a_7 \pi_3^3)$$
$$+ \theta^3 \wedge (b_1 \omega^1 + b_2 \omega^2 + b_3 \omega^3 + b_4 \pi_3^1 + b_5 \pi_2^2 + b_6 \pi_3^2 + b_7 \pi_3^3)$$
$$+ c \theta^2 \wedge \theta^3 \pmod{\theta^0, \theta^1}$$

Lemma 18.4.1. $\mathrm{sing}^2(\mathcal{P})$ *is a Monge system if and only if* $b_2 = 0$.

Proof. Straightforward modifications of ω^3, π_3^1 and c make it possible to assume that $a_3 = b_3 = a_4 = b_4 = 0$. Differentiating $d\theta^1$ and considering the coefficients of $\omega^1 \wedge \omega^2 \wedge \pi_3^2$, $\omega^1 \wedge \omega^2 \wedge \pi_3^2$, $\omega^2 \wedge \pi_3^2 \wedge \pi_3^3$ and $\omega^2 \wedge \pi_3^2 \wedge \pi_3^3$ it is further seen that $a_1 = b_1 = a_7 = b_7 = 0$. Because the coefficient of $\omega^1 \wedge \pi_3^1 \wedge \pi_3^2$ in $d^2\theta^1$ is to vanish,

$$-\frac{\partial d\omega^3}{\partial(\omega^1 \wedge \pi_3^2)} + b_6 = 0.$$

On the other hand, differentiating $d\theta^0$ and considering the coefficient of $\omega^1 \wedge \theta^3 \wedge \pi_3^2$ we see that

$$-b_6 - \frac{\partial d\omega^3}{\partial(\omega^1 \wedge \pi_3^2)} = 0$$

—whence $b_6 = 0$. In the same way it can be seen that $b_5 = 0$. Then, considering the coefficient of $\omega^2 \wedge \pi_2^2 \wedge \pi_3^2$ in $d^2\theta^1$, we finally realize that $a_6 = 0$.

All this leaves us with

$$d\theta^1 \equiv \omega^3 \wedge \pi_3^1 + \theta^2 \wedge (a_2 \, \omega^2 + a_5 \, \pi_2^2) \qquad (*)$$
$$+ b_2 \, \theta^3 \wedge \omega^2 + c \, \theta^2 \wedge \theta^3 \quad (\mathrm{mod} \ \theta^0, \theta^1).$$

Accordingly the coefficient of $\omega^1 \wedge \omega^2 \wedge \pi_3^1$ in $d^2\theta^1$ is

$$0 = \frac{\partial d\omega^3}{\partial(\omega^1 \wedge \omega^2)} - b_2.$$

$\square$

Differentiating $d\theta^0$ and looking at the coefficients of $\omega^1 \wedge \omega^2 \wedge \theta^3$ and $\omega^1 \wedge \theta^2 \wedge \pi_2^2$ we find that

$$2b_2 + \frac{\partial d\theta^2}{\partial(\omega^1 \wedge \theta^3)} = 0 \quad \text{and} \quad a_5 + \frac{\partial d\omega^2}{\partial(\omega^1 \wedge \pi_2^2)} = 0.$$

By regarding the coefficient of $\omega^1 \wedge \theta^3 \wedge \pi_2^2$ in $d^2\theta^1$ and using the equations above it is then seen that $-2b_2a_5 + b_2a_5 = 0$—that is, $b_2a_5 = 0$.

As a result there are three different subcases to consider:

(1) $b_2 = 0, \ a_5 \neq 0$,
(2) $b_2 \neq 0, \ a_5 = 0$,
(3) $b_2 = a_5 = 0$.

Subcase (1). Let us apply the method of Monge with respect to $\mathrm{sing}^1(\mathcal{P})$ here, and thus consider $\mathcal{P}_{\mathrm{sing}} := (\theta^0, \theta^1, \theta^2)$. By assumption the congruence (*) in this case reduces to

$$d\theta^1 \equiv \omega^3 \wedge \pi_3^1 + \theta^2 \wedge (a\,\omega^2 + b\,\pi_2^2 + c\,\theta^3) \quad (\mathrm{mod} \ \theta^0, \theta^1),$$

where $b \neq 0$. The latter fact makes it possible to replace $b\,\pi_2^2 + a\,\omega^2 + c\,\theta^3$ by π_2^2, so that

$$d\theta^1 \equiv \omega^3 \wedge \pi_3^1 + \theta^2 \wedge \pi_2^2 \quad (\mathrm{mod} \ \theta^0, \theta^1).$$

Let us next investigate the differential of θ^2. We have

$$d\theta^2 \equiv \omega^2 \wedge \pi_2^2 + \omega^3 \wedge \pi_3^2 + \theta^3 \wedge (c_1\,\omega^1 + c_2\,\omega^2 + c_3\,\omega^3$$
$$+ c_4\,\pi_3^1 + c_5\,\pi_2^2 + c_6\,\pi_3^2 + c_7\,\pi_3^3) \quad (\text{mod } \mathcal{P}_{\text{sing}})$$

for suitable functions c_i. By replacing $\pi_3^2 - c_3\,\theta^3$ by π_3^2 it may be assumed that $c_3 = 0$. Then differentiating $d\theta^1$ and considering the coefficients of $\omega^1 \wedge \theta^3 \wedge \pi_2^2$, $\omega^2 \wedge \theta^3 \wedge \pi_2^2$, $\theta^3 \wedge \pi_2^2 \wedge \pi_3^2$ and $\theta^3 \wedge \pi_2^2 \wedge \pi_3^3$ respectively, we see that $c_1 = c_2 = c_6 = c_7 = 0$. Differentiating $d\theta^2$ and looking at the coefficient of $\omega^1 \wedge \pi_3^1 \wedge \pi_2^2$, we find

$$\frac{\partial d\omega^2}{\partial(\omega^1 \wedge \pi_3^1)} + c_5 = 0.$$

Finally, differentiating $d\theta^0$ and considering the coefficient of $\omega^1 \wedge \theta^2 \wedge \pi_3^1$ we deduce that

$$\frac{\partial d\omega^2}{\partial(\omega^1 \wedge \pi_3^1)} = 0,$$

and hence $c_5 = 0$. This leaves us with

$$d\theta^2 \equiv \omega^2 \wedge \pi_2^2 + \omega^3 \wedge \pi_3^2 + c_4\,\theta^3 \wedge \pi_3^1 \quad (\text{mod } \mathcal{P}_{\text{sing}}).$$

If we write c instead of c_4, the Lie structure of $\mathcal{P}_{\text{sing}} = (\theta^0, \theta^1, \theta^2)$ is then given by

$$\begin{cases} d\theta^0 \equiv & \omega^3 \wedge \theta^3, \\ d\theta^1 \equiv & \omega^3 \wedge \pi_3^1, \qquad\qquad (\text{mod } \mathcal{P}_{\text{sing}}) \\ d\theta^2 \equiv \omega^2 \wedge \pi_2^2 + & \omega^3 \wedge \pi_3^2 + c\,\theta^3 \wedge \pi_3^1 \end{cases}$$

From this we read off the Cauchy characteristic system:

$$\mathcal{C}(\mathcal{P}_{\text{sing}}) = (\theta^0, \theta^1, \theta^2, \theta^3, \omega^2, \omega^3, \pi_3^1, \pi_2^2, \pi_3^2)$$

—which coincides with $\text{sing}^1(\mathcal{P})$. In particular it follows that $\text{sing}^1(\mathcal{P})$ is *complete*.

With $\dim \mathcal{C}(\mathcal{P}_{\text{sing}}) = 9$, $(\mathcal{P}_{\text{sing}})_{\text{red}}$ lives on $\mathbb{C}^9$. But since

$$(\omega^1, \omega^2, \omega^3, \theta^0, \theta^1, \theta^2, \theta^3) = (dx^1, dx^2, dx^3, dz, dp_1, dp_2, dp_3)$$

it is possible to make a less drastic reduction by using the jet coordinates $x^1, x^2, x^3, z, p_1, p_2, p_3$ of $J^1(\mathbb{C}^3_x, \mathbb{C}_z)$ together with three more functions v_1, v_2, v_3 as local coordinates on a space $J^1(\mathbb{C}^3_x, \mathbb{C}_z) \times \mathbb{C}^3_v$ to which $\mathcal{P}_{\text{sing}}$ can be reduced.

By a slight abuse of language we denote this partially reduced system on $J^1(\mathbb{C}^3_x, \mathbb{C}_z) \times \mathbb{C}^3_v$ by $\mathcal{P}_{\text{sing}}$ too.

Theorem 18.4.4. *Solving the original PDE system $S \subset J^2(\mathbb{C}_x^3, \mathbb{C}_z)$ is under fairly general conditions equivalent to integrating $\mathcal{P}_{\text{sing}}$ on $J^1(\mathbb{C}_x^3, \mathbb{C}_z) \times \mathbb{C}_v^3$.*

Proof. The pfaffian system $\theta^0 = \theta^1 = \theta^2 = 0$ on $J^1(\mathbb{C}_x^3, \mathbb{C}_z) \times \mathbb{C}_v^3$ may be expressed as

$$
\begin{cases}
dz - \sum_{i=1}^3 p_i \, dx^i = 0, \\
dp_1 + \sum_{i=1}^3 a_i \, dx^i + a_4 \, dp_3 \equiv 0 \quad (\text{mod } \mathcal{P}_{\text{sing}}), \\
dp_2 + \sum_{i=1}^3 b_i \, dx^i + b_4 \, dp_3 \equiv 0 \quad (\text{mod } \mathcal{P}_{\text{sing}}),
\end{cases} \tag{1}
$$

where a_i and b_j are functions of $x^1, x^2, x^3, z, p_1, p_2, p_3, v_1, v_2, v_3$.

Now $d\theta^0 \equiv \omega^3 \wedge \theta^3 \pmod{\mathcal{P}_{\text{sing}}} \Longrightarrow d\theta^0 \wedge d\theta^0 \equiv 0 \pmod{\mathcal{P}_{\text{sing}}}$. But

$$
\begin{aligned}
d\theta^0 = \sum_{i=1}^3 dx^i \wedge dp_i \equiv & -dx^1 \wedge (a_2 \, dx^2 + a_3 \, dx^3 + a_4 \, dp_3) \\
& - dx^2 \wedge (b_1 \, dx^1 + b_3 \, dx^3 + b_4 \, dp_3) \\
& + dx^3 \wedge dp_3 \quad (\text{mod } \mathcal{P}_{\text{sing}}),
\end{aligned}
$$

and therefore the coefficient of $dx^1 \wedge dx^2 \wedge dx^3 \wedge dp_3$ in $d\theta^0 \wedge d\theta^0$ $(\text{mod } \mathcal{P}_{\text{sing}})$ equals $-a_3 b_4 + a_4 b_3 - a_2 + b_1$, so that

$$
-a_3 b_4 + a_4 b_3 - a_2 + b_1 = 0. \tag{2}
$$

The first line in (1) shows as usual that $dp_i = \sum_{j=1}^3 p_{ij} \, dx^j$, where the p_{ij} satisfy $p_{ij} = p_{ji}$, but are arbitrary otherwise. Inserting this into the following two, and regarding dx^1, dx^2, dx^3 as independent, we obtain the six equations

$$
\begin{aligned}
p_{11} + a_1 + a_4 p_{13} = 0, \quad & p_{12} + a_2 + a_4 p_{23} = 0, \quad && p_{13} + a_3 + a_4 p_{33} = 0, \\
p_{12} + b_1 + b_4 p_{13} = 0, \quad & p_{22} + b_2 + b_4 p_{23} = 0, \quad && p_{23} + b_3 + b_4 p_{33} = 0.
\end{aligned}
$$

However, of these only five are independent. In fact, the fourth minus the second gives $b_1 - a_2 + b_4 p_{13} - a_4 p_{23} = 0$, which, using the third and the sixth, is reduced to

$$
b_1 - a_2 - b_4(a_3 + a_4 p_{33}) + a_4(b_3 + b_4 p_{33}) = b_1 - a_2 - b_4 a_3 + a_4 b_3 = 0,
$$

which according to (2) is identically satisfied.

Elimination of v_1, v_2 and v_3 by means of the remaining five equations leaves us with a PDE system $S^* \subset J^2(\mathbb{C}_x^3, \mathbb{C}_z)$ of codimension 2, such that all solutions of the original PDE system $S \subset J^2(\mathbb{C}_x^3, \mathbb{C}_z)$ are solutions of S^* too. Assuming the latter to be irreducible, the wanted result follows for dimension reasons. $\qquad \square$

So there remains to integrate $\mathcal{P}_{\text{sing}} = (\theta^0, \theta^1, \theta^2)$ on $J^1(\mathbb{C}^3_x, \mathbb{C}_z) \times \mathbb{C}^3_v$. With $\mathcal{P}_{\text{sing}}$ admitting the Cauchy characteristic vector field $C = \partial/\partial\omega^1 = \partial/\partial x^1 + \cdots$, the local rectification lemma for vector fields allows us to replace $x^2, x^3, z, p_1, p_2, p_3, v_1, v_2, v_3$ by new local coordinates X^2, X^3, Z, P_1, P_2, P_3, V_1, V_2, V_3 such that

$$X^2(0, x^2, \ldots, v_3) = x^2, \ldots, V_3(0, x^2, \ldots, v_3) = v_3,$$

and $C = \partial/\partial x^1$ when $X^2, \ldots, V_3$ are fixed. In these new coordinates the equations $\theta^0 = \theta^1 = \theta^2 = 0$ will take the form

$$\begin{cases} dZ - P_2 \, dX^2 - P_3 \, dX^3 = 0, \\ dP_1 + A_2 \, dX^2 + A_3 \, dX^3 + A_4 \, dP_3 = 0, \\ dP_2 + B_2 \, dX^2 + B_3 \, dX^3 + B_4 \, dP_3 = 0, \end{cases}$$

where the A_i and B_i are functions of $X^2, X^3, \ldots, V_3$. If we look for integral manifolds on which $dX^2 \wedge dX^3 \neq 0$, the first equation shows that

$$Z = F(X^2, X^3), \quad P_2 = \partial F/\partial X^2 \quad \text{and} \quad P_3 = \partial F/\partial X^3$$

for an arbitrary function F of two variables. Fixing F and inserting these expressions for Z, P_2 and P_3 in the second and third equations above we get the system

$$\begin{cases} \partial P_1/\partial X^2 + A_2 + A_4 \, \partial^2 F/\partial X^2 \partial X^3 = 0, \\ \partial P_1/\partial X^3 + A_3 + A_4 \, \partial^2 F/(\partial X^3)^2 = 0, \\ \partial^2 F/(\partial X^2)^2 + B_2 + B_4 \, \partial^2 F/\partial X^2 \partial X^3 = 0, \\ \partial^2 F/\partial X^2 \partial X^3 + B_3 + B_4 \, \partial^2 F/(\partial X^3)^2 = 0. \end{cases}$$

Eliminating V_1, V_2 and V_3, there results one first order PDE in the dependent variable P_1 and the two independent variables X^2 and X^3. Having determined the solutions of this as described in chapter 6, we then obtain the 3-dimensional integral manifolds of $\mathcal{P}_{\text{sing}}$ by means of the Cauchy characteristic vector field C. So in the end we are reduced to solving ordinary differential equations.

The general solution of S found in this way depends on one arbitrary function of two variables, which is in accordance with the fact that the characters of $\mathcal{P}$ are $s_1 = 3$, $s_2 = 1$ and $s_3 = 0$.

Subcase (2). Let us this time apply the Monge method with respect to $\text{sing}^2(\mathcal{P}) = (\theta^0, \theta^1, \theta^2, \theta^3, \omega^3, \pi_3^1)$, which means that we are to consider $\mathcal{P}_{\text{sing}} := (\theta^0, \theta^1)$.

By assumption

$$d\theta^1 \equiv \omega^3 \wedge \pi^1_3 + \omega^2 \wedge (a\,\theta^2 + b\,\theta^3) + c\,\theta^2 \wedge \theta^3 \quad (\text{mod } \mathcal{P}_{\text{sing}}),$$

with $b \neq 0$. Rewriting this as $d\theta^1 \equiv \omega^3 \wedge \pi^1_3 + (\omega^2 + cb^{-1}\theta^2) \wedge (b\,\theta^3 + a\,\theta^2)$, replacing $\omega^2 + cb^{-1}\theta^2$ by ω^2, and $b\,\theta^3 + a\,\theta^2$ by θ^3, we end up with

$$d\theta^1 \equiv \omega^3 \wedge \pi^1_3 + \omega^2 \wedge \theta^3 \quad (\text{mod } \mathcal{P}_{\text{sing}}).$$

Thus the Lie structure of $\mathcal{P}_{\text{sing}}$ is

$$\begin{cases} d\theta^0 \equiv \omega^2 \wedge \theta^2 + \omega^3 \wedge \theta^3, \\ d\theta^1 \equiv \omega^2 \wedge \theta^3 + \omega^3 \wedge \pi^1_3 \end{cases} \quad (\text{mod } \mathcal{P}_{\text{sing}}),$$

from which the Cauchy characteristic system is seen to be

$$\mathcal{C}(\mathcal{P}_{\text{sing}}) = (\theta^0, \theta^1, \theta^2, \theta^3, \omega^2, \omega^3, \pi^1_3).$$

Therefore $\mathcal{P}_{\text{sing}}$ can be reduced so as to live on a 7-dimensional space. Or we may alternatively regard $\mathcal{P}_{\text{sing}}$ as being defined on $J^1(\mathbb{C}^3_x, \mathbb{C}_c) \times \mathbb{C}_v$ for a suitable function v, and admitting the Cauchy characteristic vector field $C = \partial/\partial\omega^1 = \partial/\partial x^1 + \cdots$.

By using the latter approach the pfaffain system $\theta^0 = \theta^1 = 0$ can be written as

$$\begin{cases} dz - \sum_{i=1}^3 p_i\,dx^i = 0, \\ dp_1 + a_1\,dx^1 + a_2\,dx^2 + a_3\,dx^3 + b^2\,dp_2 + b^3\,dp_3 = 0, \end{cases}$$

where the a_i and b^i are functions of $x^1, x^2, x^3, z, p_1, p_2, p_3$ and v. In the ususal way this yields the system

$$\begin{cases} p_{11} + a_1 + b^2\,p_{12} + b^3\,p_{13} = 0, \\ p_{12} + a_2 + b^2\,p_{22} + b^3\,p_{23} = 0, \\ p_{13} + a_3 + b^2\,p_{23} + b^3\,p_{33} = 0. \end{cases}$$

Elimination of v gives an $\mathcal{S}^* \subset J^2(\mathbb{C}^3_x, \mathbb{C}_z)$ of codimension 2. If this is irreducible one infers as before that solving the original PDE system $\mathcal{S}$ is equivalent to solving $\mathcal{S}^*$. So we have arrived at the following result.

Theorem 18.4.5. *Solving the originally given PDE system $\mathcal{S} \subset J^2(\mathbb{C}^3_x, \mathbb{C}_z)$ is under fairly general conditions equivalent to integrating $\mathcal{P}_{\text{sing}} = (\theta^0, \theta^1)$ on $J^1(\mathbb{C}^3_x, \mathbb{C}_z) \times \mathbb{C}_v$.* $\qquad \square$

In order to integrate $\mathcal{P}_{\text{sing}}$ we replace $x^2, x^3, z, p_1, p_2, p_3, v$ by new coordinates $X^2, X^3, Z, P_1, P_2, P_3, V$ satisfying

$$X^2(0, x^2, \ldots, v) = x^2, \quad \ldots, \quad V(0, x^2, \ldots, v) = v,$$

and making the Cauchy characteristic vector field C go over into $\partial/\partial x^1$. Then the system $\theta^0 = \theta^1 = 0$ is transformed into

$$\begin{cases} dZ - P_2 \, dX^2 - P_3 \, dX^3 = 0, \\ dP_1 + A_2 \, dX^2 + A_3 \, dX^3 + B^2 \, dP_2 + B^3 \, dP_3 = 0, \end{cases}$$

where the A_i and B^j are functions of $X^2, \ldots, V$. As we want integral manifolds on which $dX^2 \wedge dX^3 \neq 0$, the first equation yields

$$Z = F(X^2, X^3), \quad P_2 = \partial F/\partial X^2 \quad \text{and} \quad P_3 = \partial F/\partial X^3$$

with an arbitrary function F of two variables. Then the second equation implies that

$$\begin{cases} \partial P_1/\partial X^2 + A_2 + B^2 \, \partial^2 F/(\partial X^2)^2 + B^3 \, \partial^2 F/\partial X^2 \partial X^3 = 0, \\ \partial P_1/\partial X^3 + A_3 + B^2 \, \partial^2 F/\partial X^2 \partial X^3 + B^3 \, \partial^2 F/(\partial X^3)^2 = 0. \end{cases}$$

Elimination of V gives a first order PDE in the dependent variable P_1 and the independent variables X^2, X^3. Having solved this, the wanted 3-dimensional integral manifolds of $\mathcal{P}_{\text{sing}}$ are then constructed by means of the Cauchy characteristic vector field C.

So again the general solution depends on one arbitrary function of two variables, and is obtained by solving ordinary differential equations.

Subcase (3). We use the singular subsystem $\text{sing}^2(\mathcal{P})$, just as in the preceding subcase, and thus consider $\mathcal{P}_{\text{sing}} = (\theta^0, \theta^1)$ one more time. Here

$$d\theta^1 \equiv \omega^3 \wedge \pi_3^1 + a\,\omega^2 \wedge \theta^2 + b\,\theta^2 \wedge \theta^3 \quad (\text{mod } \mathcal{P}_{\text{sing}}),$$

which can be rewritten as

$$\begin{aligned} d(\theta^1 - a\,\theta^0) &\equiv \omega^3 \wedge \pi_3^1 - a\,\omega^3 \wedge \theta^3 + b\,\theta^2 \wedge \theta^3 \\ &= \omega^3 \wedge (\pi_3^1 - a\,\theta^3) + b\,\theta^2 \wedge \theta^3 \quad (\text{mod } \mathcal{P}_{\text{sing}}). \end{aligned}$$

Replacing $\theta^1 - a\,\theta^0$ by θ^1 and $\pi_3^1 - a\,\theta^3$ by π_3^1, we instead get

$$d\theta^1 \equiv \omega^3 \wedge \pi_3^1 + b\,\theta^2 \wedge \theta^3 \quad (\text{mod } \mathcal{P}_{\text{sing}}).$$

By differentiating this and checking the coefficient of $\omega^2 \wedge \theta^3 \wedge \pi_2^2$ it is found that $b = 0$. Therefore the Lie structure of $\mathcal{P}_{\text{sing}} = (\theta^0, \theta^1)$ is

$$\begin{cases} d\theta^0 \equiv \omega^2 \wedge \theta^2 + \ \omega^3 \wedge \theta^3, \\ d\theta^1 \equiv \qquad\qquad\ \omega^3 \wedge \pi_3^1 \end{cases} \quad (\text{mod } \mathcal{P}_{\text{sing}}),$$

implying that the Cauchy characteristic sytem is

$$\mathcal{C}(\mathcal{P}_{\text{sing}}) = (\theta^0, \theta^1, \theta^2, \theta^3, \omega^2, \omega^3, \pi_3^1).$$

We can then argue as in the preceding case, and will obtain analogous results. In particular the general solution depends on one arbitrary function of two variables, and is found by solving ordinary equations.

Conclusion. *Using **the method of Monge**, all involutive systems of two second order PDEs in one dependent and three independent variables can be solved by reductions to ODE systems.*

18.5 Three PDEs

There are five different classes of involutive systems of three second order PDEs in one dependent and three independent variables, having the following canonical representations:

(a) $\pi_2^1 = \pi_3^1 = \pi_3^2 = 0$,
(b) $\pi_1^1 = \pi_3^1 = \pi_3^2 = 0$,
(c) $\pi_1^1 = \pi_2^1 = \pi_3^1 - \pi_2^2 = 0$,
(d) $\pi_1^1 = \pi_2^1 = \pi_2^2 = 0$,
(e) $\pi_1^1 = \pi_2^1 = \pi_3^1 = 0$.

And we want to investigate whether these might be solvable by means of the Monge method.

Type (a). Here $\mathcal{P} = (\theta^0, \theta^1, \theta^2, \theta^3)$ with the structure equations

$$d\theta^0 \equiv \sum_{i=1}^{3} \omega^i \wedge \theta^i \quad (\text{mod } \theta^0)$$

and

$$\begin{cases} d\theta^1 \equiv \omega^1 \wedge \pi_1^1, \\ d\theta^2 \equiv \quad\quad \omega^2 \wedge \pi_2^2, \\ d\theta^3 \equiv \quad\quad\quad\quad \omega^3 \wedge \pi_3^3 \end{cases} \quad (\text{mod } \mathcal{P}).$$

Lemma 18.5.1. *There are three Monge systems with 2-dimensional Monge characteristics, namely* $\text{sing}^i(\mathcal{P}) := (\theta^0, \theta^1, \theta^2, \theta^3, \omega^i, \pi_i^i)$ *for* $i = 1, 2$ *and* 3.

Proof. $\text{sing}^i(\mathcal{P})$ is certainly singular since its dual vector field system besides $d\theta^0$ also kills $d\theta^i$ (mod $\mathcal{P}$).

Let us consider $\text{sing}^1(\mathcal{P})$. Any complete pfaffian system $\mathcal{K} \supset \mathcal{P}$ with $\omega^1 \wedge \omega^2 \wedge \omega^3 \neq 0$ (mod $\mathcal{K}$) is of the form

$$\mathcal{K} = (\theta^0, \theta^1, \theta^2, \theta^3, \psi^1, \psi^2, \psi^3),$$

where $\psi^i = \pi_i^i - \sum_{k=1}^{3} a_k^i \omega^k$ for certain functions a_k^i. From $d\theta^1 \equiv 0$ (mod $\mathcal{K}$) it follows that also $\omega^1 \wedge \pi_1^1 \equiv 0$ (mod $\mathcal{K}$), and therefore

$$(\mathcal{K}, \mathrm{sing}^1(\mathcal{P})) = (\theta^0, \theta^1, \theta^2, \theta^3, \omega^1, \psi^1, \psi^2, \psi^3).$$

Because $\mathcal{K}$ and $\mathcal{C}(\mathcal{P}) = (\theta^0, \theta^1, \theta^2, \theta^3, \omega^1, \omega^2, \omega^3, \psi^1, \psi^2, \psi^3)$ are complete, it thus suffices to show that

$$\frac{\partial d\omega^1}{\partial(\omega^2 \wedge \omega^3)} = 0$$

in order for $(\mathcal{K}, \mathrm{sing}^1(\mathcal{P})$ to be complete as well. Supposing that $d\omega^1 = a\,\omega^2 \wedge \omega^3 + \cdots$ and differentiating $d\theta^1$ we obtain

$$0 \equiv d\omega^1 \wedge \pi_1^1 - \omega^1 \wedge d\pi_1^1 = a\,\omega^2 \wedge \omega^3 \wedge \pi_1^1 + \cdots$$

And thus a indeed vanishes.   □

As in analogous situations encountered before, $\mathrm{sing}^1(\mathcal{P})$ admits at most three functionally independent first integrals—and there is hope for the integration problem when this maximum number is attained. In that case it is possible to modify θ^1, ω^1 and π_1^1 so as to get the congruence

$$d\theta^1 \equiv \omega^1 \wedge \pi_1^1 \quad (\mathrm{mod}\ \theta^1).$$

Thus the genus of $d\theta^1$ (mod θ^1) equals 1, and so the pfaffian equation $\theta^1 = 0$ can be written as

$$dV - W\,dU = 0$$

for suitable functions U, V and W. The general integral manifold of this on which $dU \neq 0$ is given by

$$\mathcal{N}_F: \quad V = F(U), \quad W = F'(U),$$

where F is an arbitrary function of one variable. With θ^1 vanishing on $\mathcal{N}_F$, there remains to solve the restriction to $\mathcal{N}_F$ of the system $\theta^0 = \theta^2 = \theta^3 = 0$. The structure equations of this are

$$d\theta^0 \equiv \omega^2 \wedge \theta^2 + \omega^3 \wedge \theta^3 \quad (\mathrm{mod}\ \theta^0)$$

and

$$\begin{cases} d\theta^2 \equiv \omega^2 \wedge \pi_2^2, \\ d\theta^3 \equiv \qquad \omega^3 \wedge \pi_3^3 \end{cases} \quad (\mathrm{mod}\ \theta^0, \theta^2, \theta^3).$$

On noting that $\partial/\partial\omega^1$ is Cauchy characteristic for $(\theta^0, \theta^2, \theta^3)$, ω^1 can be forgotten for a while, so that we are reduced to finding 2-dimensional integral manifolds on which $\omega^2 \wedge \omega^3 \neq 0$. But just as in a similar situation discussed in section **18.3**, this problem is equivalent to solving a hyperbolic second order PDE in one dependent and two independent variables. Which we may or may not be able do to.

If $\text{sing}^1(\mathcal{P})$ and $\text{sing}^2(\mathcal{P})$ each admit three first integrals, and these six functions are functionally independent, the system $\theta^1 = \theta^2 = 0$ can be written as

$$dV^i - W^i\, dU^i = 0, \quad i = 1, 2,$$

with the integral manifolds

$$V^i = F^i(U^i), \quad W^i = (F^i)'(U^i), \qquad i = 1, 2,$$

for arbitrary 1-variable functions F^1 and F^2. Then there remains to integrate $\theta^0 = \theta^3 = 0$ on such integral manifolds, with (θ^0, θ^3) having the Lie structure

$$\begin{cases} d\theta^0 \equiv 0, \\ d\theta^3 \equiv \omega^3 \wedge \pi_3^3 \end{cases} \quad (\text{mod } \theta^0, \theta^3).$$

Since $\partial/\partial\omega^1$ and $\partial/\partial\omega^2$ are Cauchy characteistic vector fields for (θ^0, θ^3), this problem is reduced to finding integral curves on which $\omega^3 \neq 0$—that is, to solving ordinary differential equations.

Type (b). Here $\mathcal{P} = (\theta^0, \theta^1, \theta^2, \theta^3)$ with the structure equations

$$d\theta^0 \equiv \sum_{i=1}^{3} \omega^i \wedge \theta^i \quad (\text{mod } \theta^0)$$

and

$$\begin{cases} d\theta^1 \equiv \qquad\quad \omega^2 \wedge \pi_2^1, \\ d\theta^2 \equiv \omega^1 \wedge \pi_1^2 + \omega^2 \wedge \pi_2^2, \\ d\theta^3 \equiv \qquad\qquad\qquad\quad \omega^3 \wedge \pi_3^3 \end{cases} \quad (\text{mod } \mathcal{P}).$$

From this it is readily seen that $\text{sing}^1(\mathcal{P}) := (\theta^0, \theta^1, \theta^2, \theta^3, \omega^1, \omega^2, \pi_2^1, \pi_2^2)$ and $\text{sing}^2(\mathcal{P}) := (\theta^0, \theta^1, \theta^2, \theta^3, \omega^2, \omega^3, \pi_2^1, \pi_3^3)$ are Monge systems giving rise to 1-dimensional Monge characteristics.

Lemma 18.5.2. *There are also two Monge systems yielding 2-dimensional characteristics, namely*

$\text{sing}^3(\mathcal{P}) := (\theta^0, \theta^1, \theta^2, \theta^3, \omega^2, \pi_2^1)$ *and* $\text{sing}^4(\mathcal{P}) := (\theta^0, \theta^1, \theta^2, \theta^3, \omega^3, \pi_3^3).$

Proof. The proof that $\text{sing}^4(\mathcal{P})$ is Monge is similar to the corresponding one in type **(a)**, and therefore we concentrate on $\text{sing}^3(\mathcal{P})$.

According to an argument which is familiar by now, the Monge property follows if we can show that

$$\frac{\partial d\omega^2}{\partial(\omega^1 \wedge \omega^3)} = 0.$$

Differentiating $d\theta^1$, considering $\omega^1 \wedge \omega^3 \wedge \pi_2^1$ and using the fact that $d\theta^2 = \omega^1 \wedge \pi_1^2 + \omega^2 \wedge \pi_2^2 + \cdots$, we find that

$$\frac{\partial d\omega^2}{\partial(\omega^1 \wedge \omega^3)} - \frac{\partial d\theta^1}{\partial(\theta^2 \wedge \omega^3)} = 0.$$

Looking at $\omega^2 \wedge \omega^3 \wedge \pi_2^2$ instead, we have

$$-\frac{\partial d\theta^1}{\partial(\theta^2 \wedge \omega^3)} - \frac{\partial d\pi_2^1}{\partial(\omega^3 \wedge \pi_2^2)} = 0,$$

and hence

$$\frac{\partial d\omega^2}{\partial(\omega^1 \wedge \omega^3)} = -\frac{\partial d\pi_2^1}{\partial(\omega^3 \wedge \pi_2^2)}.$$

By checking the coefficient of $\omega^1 \wedge \omega^3 \wedge \pi_2^2$ in $d^2\theta^2 = 0$, it is found that

$$-\frac{\partial d\pi_2^1}{\partial(\omega^3 \wedge \pi_2^2)} + \frac{\partial d\omega^2}{\partial(\omega^1 \wedge \omega^3)} = 0,$$

so that we finally infer that

$$\frac{\partial d\omega^2}{\partial(\omega^1 \wedge \omega^3)} = 0.$$

$\square$

If $\text{sing}^4(\mathcal{P})$ admits three functionally independent first integrals, the equation $\theta^3 = 0$ can be written as

$$dV - W\, dU = 0.$$

With this being solved, we have only to integrate the restriction of the system $\theta^0 = \theta^1 = \theta^2 = 0$ to each of the integral manifolds found. With $\partial/\partial\omega^3$ being Cauchy characteristic for $(\theta^0, \theta^1, \theta^2)$, this problem is reduced to a problem in two independent variables—in fact to a second order PDE in one dependent and two independent variables.

If $\text{sing}^3(\mathcal{P})$ admits three functionally independent first integrals a similar reduction leads to a hyperbolic second order PDE in one dependent and two independent variables.

Conclusion. *If* $\text{sing}^3(\mathcal{P})$ *or* $\text{sing}^4(\mathcal{P})$ *admits a maximal number of first integrals, the problem may be reduced to solving a second order PDE in one dependent and two independent variables—which may or may not be possible.*

Type (c). Here $\mathcal{P} = (\theta^0, \theta^1, \theta^2, \theta^3)$ has the structure equations

$$d\theta^0 \equiv \sum_{i=1}^{3} \omega^i \wedge \theta^i \pmod{\mathcal{P}}$$

and

$$\begin{cases} d\theta^1 \equiv & \omega^3 \wedge \pi_3^1, \\ d\theta^2 \equiv & \omega^2 \wedge \pi_3^1 + \omega^3 \wedge \pi_3^2, \\ d\theta^3 \equiv \omega^1 \wedge \pi_1^3 + \omega^2 \wedge \pi_2^3 + \omega^3 \wedge \pi_3^3 \end{cases} \pmod{\mathcal{P}}.$$

By using the same type of reasoning as before, it is found that $\text{sing}^1(\mathcal{P}) := (\theta^0, \theta^1, \theta^2, \theta^3, \omega^2, \omega^3, \pi_3^1, \pi_3^2)$ is a Monge system with 1-dimensional characteristics, and that $\text{sing}^2(\mathcal{P}) := (\theta^0, \theta^1, \theta^2, \theta^3, \omega^3, \pi_3^1)$ is a Monge system with 2-dimensional characteristics.

The following cases are treated in detail in [Cartan 1911].

(1) $\text{sing}^1(\mathcal{P})$ is complete.
(2) $\text{sing}^2(\mathcal{P})$ admits three functionally independent first integrals. In this case suitable modifications lead to $d\theta^1 \equiv \omega^3 \wedge \pi_3^1 \pmod{\theta^1}$.
(3) $\text{sing}^2(\mathcal{P})$ is complete.

It turns out that **(2)** and **(3)** actually are special cases of **(1)**. The first case can be reduced to a third order PDE in one dependent and two independent variables, the second to a second order PDE in one dependent and two independent variables, and the third one to ordinary differential equations.

We omit the proofs, since these results are not very startling.

Type (d). In this case the structure equations of $\mathcal{P} = (\theta^0, \theta^1, \theta^2, \theta^3)$ are

$$d\theta^0 \equiv \sum_{i=1}^{3} \omega^i \wedge \theta^i \pmod{\theta^0}$$

and

$$\begin{cases} d\theta^1 \equiv & \omega^3 \wedge \pi_3^1, \\ d\theta^2 \equiv & \omega^3 \wedge \pi_3^2, \\ d\theta^3 \equiv \omega^1 \wedge \pi_1^3 + \omega^2 \wedge \pi_2^3 + \omega^3 \wedge \pi_3^3 \end{cases} \pmod{\mathcal{P}}.$$

From these it follows that $\text{sing}(\mathcal{P}) := (\theta^0, \theta^1, \theta^2, \theta^3, \omega^3, \pi_3^1, \pi_3^2)$ is a singular system—since its dual vector field system besides $d\theta^0$ also kills $d\theta^1$ and $d\theta^2$ (mod $\mathcal{P}$). Applying the method of Monge we are therefore to consider

$$\mathcal{P}_{\text{sing}} := (\theta^0, \theta^1, \theta^2).$$

The structure equations of $\mathcal{P}$ show that there are functions a_i and b_i such that the Lie structure of $\mathcal{P}_{\text{sing}}$ can be expressed in the form

$$\begin{cases} d\theta^0 \equiv \omega^3 \wedge \theta^3, \\ d\theta^1 \equiv \omega^3 \wedge \pi_3^1 + \theta^3 \wedge (a_1 \omega^1 + a_2 \omega^2 + a_3 \omega^3 + a_4 \pi_3^1 + a_5 \pi_3^2 + a_6 \pi_3^3), \\ d\theta^2 \equiv \omega^3 \wedge \pi_3^2 + \theta^3 \wedge (b_1 \omega^1 + b_2 \omega^2 + b_3 \omega^3 + b_4 \pi_3^1 + b_5 \pi_3^2 + b_6 \pi_3^3) \end{cases}$$

modulo $\mathcal{P}_{\text{sing}}$. The last two congruences can be rewritten as

$$d\theta^1 \equiv (\omega^3 + a_4 \theta^3) \wedge (\pi_3^1 - a_3 \theta^3) + \theta^3 \wedge (a_1 \omega^1 + a_2 \omega^2 + a_5 \pi_3^2 + a_6 \pi_3^3)$$
$$d\theta^2 \equiv (\omega^3 + a_4 \theta^3) \wedge (\pi_3^2 - b_3 \theta^3) + \theta^3 \wedge (b_1 \omega^1 + b_2 \omega^2 + b_4 \pi_3^1$$
$$+ (b_5 - a_4) \pi_3^2 + b_6 \pi_3^3),$$

and therefore the replacements $\omega^3 + a_4 \theta^3 \mapsto \omega^3$, $\pi_3^1 - a_3 \theta^3 \mapsto \pi_3^1$, $\pi_3^2 - b_3 \theta^3 \mapsto \pi_3^2$ and $b_5 - a_4 \mapsto b_5$ simplify the structure to

$$\begin{cases} d\theta^0 \equiv \omega^3 \wedge \theta^3, \\ d\theta^1 \equiv \omega^3 \wedge \pi_3^1 + \theta^3 \wedge (a_1 \omega^1 + a_2 \omega^2 + a_5 \pi_3^2 + a_6 \pi_3^3), \\ d\theta^2 \equiv \omega^3 \wedge \pi_3^2 + \theta^3 \wedge (b_1 \omega^1 + b_2 \omega^2 + b_4 \pi_3^1 + b_5 \pi_3^2 + b_6 \pi_3^3) \end{cases}$$

modulo $\mathcal{P}_{\text{sing}}$. By differentiating $d\theta^1$ and considering the coefficients of $\omega^2 \wedge \pi_3^2 \wedge \pi_3^3$ and $\omega^1 \wedge \omega^2 \wedge \pi_3^2$ it is found that $a_6 = a_1 = 0$. Differentiating $d\theta^2$ and considering the coefficients of $\omega^1 \wedge \pi_3^1 \wedge \pi_3^3$ and $\omega^1 \wedge \omega^2 \wedge \pi_3^1$ we likewise have $b_6 = b_1 = 0$. Moreover, checking the coefficients of $\omega^1 \wedge \omega^2 \wedge \pi_3^1$, $\omega^1 \wedge \omega^2 \wedge \pi_3^2$ and $\omega^1 \wedge \omega^2 \wedge \theta^3$ in $d^2\theta^1$, $d^2\theta^2$ and $d^2\theta^0$ respectively, we see that

$$\frac{\partial d\omega^3}{\partial(\omega^1 \wedge \omega^2)} - a_2 = 0, \qquad \frac{\partial d\omega^3}{\partial(\omega^1 \wedge \omega^2)} + b_1 = 0, \qquad \frac{\partial d\omega^3}{\partial(\omega^1 \wedge \omega^2)} + a_2 - b_1 = 0,$$

and hence $a_2 = b_1 = 0$. Considering the coefficients of $\omega^1 \wedge \pi_3^1 \wedge \pi_3^2$ and $\omega^1 \wedge \theta^3 \wedge \pi_3^2$ in $d^2\theta^1$ and $d^2\theta^0$ we next see that

$$\frac{\partial d\omega^3}{\partial(\omega^1 \wedge \pi_3^2)} - a_5 = 0 \quad \text{and} \quad \frac{\partial d\omega^3}{\partial(\omega^1 \wedge \pi_3^2)} + a_5 = 0,$$

whence $a_5 = 0$. In a similar way it is realized that $b_4 = 0$ too. Finally, checking the coefficients of $\omega^1 \wedge \pi_3^1 \wedge \pi_3^2$ and $\omega^1 \wedge \theta^3 \wedge \pi_3^1$ in $d^2\theta^2$ and

$d^2\theta^0$ respectively, we find

$$\frac{\partial d\omega^3}{\partial(\omega^1 \wedge \pi_3^1)} + b_5 = 0 \quad \text{and} \quad \frac{\partial d\omega^3}{\partial(\omega^1 \wedge \pi_3^1)} = 0,$$

implying that $b_5 = 0$.

Thereby the Lie structure of $\mathcal{P}_{\text{sing}} = (\theta^0, \theta^1, \theta^2)$ has been simplified to

$$\begin{cases} d\theta^0 \equiv \omega^3 \wedge \theta^3, \\ d\theta^1 \equiv \omega^3 \wedge \pi_3^1, \qquad (\text{mod } \mathcal{P}_{\text{sing}}). \\ d\theta^2 \equiv \omega^3 \wedge \pi_3^2 \end{cases}$$

From this we see that the Cauchy characteristic system is

$$\mathcal{C}(\mathcal{P}_{\text{sing}}) = (\theta^0, \theta^1, \theta^2, \theta^3, \omega^3, \pi_3^1, \pi_3^2),$$

and consequently coincides with the singular system $\text{sing}(\mathcal{P})$ that we started from.

Thus $\mathcal{P}_{\text{sing}}$ can be reduced to a pfaffian system defined on the 7-dimensional space having the first integrals of $\mathcal{P}_{\text{sing}}$ as local coordinates. In this process there are functions a, b and c such that

$$d\omega^3 \equiv a\,\theta^3 \wedge \pi_3^1 + b\,\theta^3 \wedge \pi_3^2 + c\,\pi_3^1 \wedge \pi_3^2 \quad (\text{mod } \theta^0, \theta^1, \theta^2, \omega^3).$$

If we insert this expression into $0 = d^2\theta^0 = d^2\theta^1 = d^2\theta^2$ it readily follows that $a = b = c = 0$, so that actually $d\omega^3 \equiv 0 \pmod{\theta^0, \theta^1, \theta^2, \omega^3}$. Combining this with the structure equations for $\mathcal{P}_{\text{sing}}$, we have

$$d\theta^0 \equiv d\theta^1 \equiv d\theta^2 \equiv d\omega^3 \equiv 0 \pmod{\theta^0, \theta^1, \theta^2, \omega^3}.$$

Hence we have found the following result.

Theorem 18.5.1. *The pfaffian system* $(\theta^0, \theta^1, \theta^2, \omega^3)$ *is complete.* $\qquad\square$

Because of this and the inclusion

$$(\theta^0, \theta^1, \theta^2, \omega^3) \subset (\omega^1, \omega^2, \omega^3, \theta^0, \theta^1, \theta^2, \theta^3) = (dx^1, dx^2, dx^3, dz, dp_1, dp_2, dp_3)$$

it is possible to find local coordinates X^1, X^2, X^3, Z, P_1, P_2 and P_3 for $J^1(\mathbb{C}^3, \mathbb{C})$ such that the pfaffian systems $(\theta^0, \theta^1, \theta^2, \omega^3)$ and (dZ, dP_1, dP_2, dX_3) become equal. In terms of these new coordinates there is a projection

$$J^1(\mathbb{C}^3, \mathbb{C}) \to \mathbb{C}^4$$
$$(X^1, \ldots, P_3) \mapsto (Z, P_1, P_2, X_3).$$

In order to integrate $\mathcal{P} = (\theta^0, \theta^1, \theta^2, \theta^3)$ on $\mathcal{S} \subset J^2(\mathbb{C}^3, \mathbb{C})$ it then suffices to integrate the restriction of $\theta^3 = 0$ to $\pi^{-1}(C)$ for each integral curve

C of $\theta^0 = \theta^1 = \theta^2 = 0$ on $\mathbb{C}^4$ satisfying $\omega^3|_C \neq 0$, where π denotes the projection $S \longrightarrow J^1(\mathbb{C}^3, \mathbb{C}) \longrightarrow \mathbb{C}^4$.

Hence the integration of $\mathcal{P}$ is reduced to solving ordinary differential equations.

Type (e). Here $\mathcal{P} = (\theta^0, \theta^1, \theta^2, \theta^3)$ with the structure

$$d\theta^0 \equiv \sum_{i=1}^{3} \omega^i \wedge \theta^i \pmod{\theta^0}$$

and

$$\begin{cases} d\theta^1 \equiv 0, \\ d\theta^2 \equiv \omega^2 \wedge \pi_2^2 + \omega^3 \wedge \pi_3^2, \\ d\theta^3 \equiv \omega^2 \wedge \pi_2^3 + \omega^3 \wedge \pi_3^3 \end{cases} \pmod{\mathcal{P}}.$$

From this it follows in particular that there are functions a_i, b_i and c such that

$$\begin{aligned} d\theta^1 \equiv\ & \theta^2 \wedge (a_1 \omega^1 + a_2 \omega^2 + a_3 \omega^3 + a_4 \pi_2^2 + a_5 \pi_3^2 + a_6 \pi_3^3) \\ &+ \theta^3 \wedge (b_1 \omega^1 + b_2 \omega^2 + b_3 \omega^3 + b_4 \pi_2^2 + b_5 \pi_3^2 + b_6 \pi_3^3) \\ &+ c\, \theta^2 \wedge \theta^3 \pmod{\theta^0, \theta^1}. \end{aligned}$$

Differentiating this and using the structure equations we obtain modulo $\mathcal{P}$

$$(\omega^2 \wedge \pi_2^2 + \omega^3 \wedge \pi_3^2) \wedge (a_1 \omega^1 + a_2 \omega^2 + a_3 \omega^3 + a_4 \pi_2^2 + a_5 \pi_3^2 + a_6 \pi_3^3)$$
$$+ (\omega^2 \wedge \pi_2^3 + \omega^3 \wedge \pi_3^3) \wedge (b_1 \omega^1 + b_2 \omega^2 + b_3 \omega^3 + b_4 \pi_2^2 + b_5 \pi_3^2 + b_6 \pi_3^3) \equiv 0.$$

But then

$$a_1 = a_3 = a_4 = a_5 = a_6 = b_1 = b_2 = b_4 = b_5 = b_6 = 0 \quad \text{and} \quad a_2 = b_3,$$

and so

$$d\theta^1 \equiv a\,(\omega^2 \wedge \theta^2 + \omega^3 \wedge \theta^3) + c\,\theta^2 \wedge \theta^3 \pmod{\theta^0, \theta^1},$$

with $a := -a_2 = -b_3$. Rewriting this as $d(\theta^1 - a\,\theta^0) \equiv c\,\theta^0 \wedge \theta^3$ $\pmod{\theta^0, \theta^1}$, and replacing $\theta^1 - a\,\theta^0$ by θ^1, we get

$$d\theta^1 \equiv c\,\theta^2 \wedge \theta^3 \pmod{\theta^0, \theta^1}.$$

Differentiating this and considering the coefficient of $\omega^2 \wedge \theta^3 \wedge \pi_2^2$ we deduce that $c = 0$, and so $d\theta^1 \equiv 0 \pmod{\theta^0, \theta^1}$. This means that there is a one-form χ such that

$$d\theta^1 \equiv \theta^0 \wedge \chi \pmod{\theta^1}.$$

Applying d to this, we find

$$0 \equiv (\omega^2 \wedge \theta^2 + \omega^3 \wedge \theta^3) \wedge \chi \pmod{\theta^0, \theta^1},$$

and therefore

$$\chi = a\,\omega^2 + b\,\theta^2 + c\,\omega^3 + e\,\theta^3 \pmod{\theta^0, \theta^1},$$

where the functions a, b, c, e satisfy

$$0 = a\,\omega^2 \wedge \omega^3 \wedge \theta^3 - b\,\omega^3 \wedge \theta^2 \wedge \theta^3 - c\,\omega^2 \wedge \omega^3 \wedge \theta^2 + e\,\omega^2 \wedge \theta^2 \wedge \theta^3.$$

Consequently $a = b = c = e = 0$, implying that $\chi \equiv 0 \pmod{\theta^0, \theta^1}$, and so $d\theta^1 \equiv 0 \pmod{\theta^1}$.

But then the Frobenius theorem says that there is a function f such that $\theta^1 = 0 \Longleftrightarrow df = 0$. Because θ^1 is defined by means of the first order jet coordinates, f is a function of these too, and

$$\theta^1 = 0 \Longleftrightarrow f(x^1, x^2, x^3, z, p_1, p_2, p_3) = c = \text{ constant}.$$

Differentiating this and regarding x^1, x^2, x^3 as independent variables we get a second order PDE system S^*,

$$\frac{\partial f}{\partial x^i} + p_i \frac{\partial f}{\partial z} + \sum_{j=1}^{3} p_{ij} \frac{\partial f}{\partial p_j} = 0 \quad \text{for } i = 1, 2, 3,$$

such that all solutions of the original $S \subset J^2(\mathbb{C}_x^3, \mathbb{C}_z)$ are also solutions of $S^* \subset J^2(\mathbb{C}_x^3, \mathbb{C}_z)$. If S^* is irreducible it follows as usual for dimension reasons that $S = S^*$. Hence solving S is equivalent to solving the 1-parameter family of first order PDEs

$$f(x^1, x^2, x^3, z, p_1, p_2, p_3) = c, \quad \text{for } c \in \mathbb{C},$$

and this task can be reduced to solving ordinary differential equations.

18.6 Four and five PDEs

There are two types of involutive systems of *four* second order PDEs in one dependent and three independent variables, represented by

(a) $\pi_2^1 = \pi_3^1 = \pi_3^2 = \pi_3^3 = 0$

and

(b) $\pi_3^1 = \pi_2^2 = \pi_3^2 = \pi_3^3 = 0$

respectively.

Type (a). Here $\mathcal{P} = (\theta^0, \theta^1, \theta^2, \theta^3)$ with the structure equations

$$d\theta^0 \equiv \sum_{i=1}^{3} \omega^i \wedge \theta^i \quad (\text{mod } \theta^0)$$

and

$$\begin{cases} d\theta^1 \equiv \omega^1 \wedge \pi_1^1, \\ d\theta^2 \equiv \qquad\quad \omega^2 \wedge \pi_2^2, \qquad (\text{mod } \mathcal{P}). \\ d\theta^3 \equiv 0 \end{cases}$$

These show that $\partial/\partial\omega^3$ is a Cauchy characteristic vector field, and therefore the problem is immediately reduced to one in two independent variables only.

It is easily seen that

$$\text{sing}^1(\mathcal{P}) := (\theta^0, \theta^1, \theta^2, \theta^3, \omega^1, \pi_1^1) \quad \text{and} \quad \text{sing}^2(\mathcal{P}) := (\theta^0, \theta^1, \theta^2, \theta^3, \omega^2, \pi_2^2)$$

are Monge systems with 2-dimensional Monge characteristics.

There are two cases where the methods used earlier apply. The first occurs if we by suitable modifications can arrive at

$$d\theta^i \equiv \omega^i \wedge \pi_i^i \quad (\text{mod } \theta^i) \quad \text{for } i = 1 \text{ or } 2,$$

in which case the pfaffian equation $\theta^i = 0$ can be expressed as $dV^i - W^i dU^i = 0$, and the other if it is possible to find π^i such that

$$\begin{cases} d\theta^i \equiv \omega^i \wedge \pi^i, \\ d\theta^3 \equiv 0 \end{cases} \qquad (\text{mod } \theta^i, \theta^3) \quad \text{for } i = 1 \text{ or } 2.$$

Type (b). In this case the structure equations of $\mathcal{P} = (\theta^0, \theta^1, \theta^2, \theta^3)$ are

$$d\theta^0 \equiv \sum_{i=1}^{3} \omega^i \wedge \theta^i \quad (\text{mod } \theta^0)$$

and

$$\begin{cases} d\theta^1 \equiv \omega^1 \wedge \pi_1^1 + \omega^2 \wedge \pi_2^1, \\ d\theta^2 \equiv \omega^1 \wedge \pi_1^2, \qquad\qquad (\text{mod } \mathcal{P}). \\ d\theta^3 \equiv 0 \end{cases}$$

So $\partial/\partial\omega^3$ is Cauchy characteritic also here, and we are reduced to a problem in two independent variables only.

We readily find that

$$\operatorname{sing}(\mathcal{P}) := (\theta^0, \theta^1, \theta^2, \theta^3, \omega^1, \pi_1^2)$$

is a Monge system with 2-dimensional characteristics, the dual of which kills $d\theta^0$, $d\theta^2$ and $d\theta^3$ (mod $\mathcal{P}$).

Let us now *assume that* $\operatorname{sing}(\mathcal{P})$ *is complete!* Then in particular

$$\frac{\partial d\omega^1}{\partial(\omega^2 \wedge \pi_1^1)} = \frac{\partial d\pi_1^2}{\partial(\omega^2 \wedge \pi_1^1)} = 0.$$

Applying the Monge method with respect to $\operatorname{sing}(\mathcal{P})$ we are then to consider the corresponding system

$$\mathcal{P}_{\operatorname{sing}} := (\theta^0, \theta^2, \theta^3).$$

By the above its Lie structure can be written as

$$\begin{cases} d\theta^0 \equiv \omega^1 \wedge \theta^1, \\ d\theta^2 \equiv \omega^1 \wedge \pi_1^2 + \theta^1 \wedge (a_1 \omega^1 + a_2 \omega^2 + a_3 \pi_1^1 + a_4 \pi_1^2), & (\text{mod } \mathcal{P}_{\operatorname{sing}}), \\ d\theta^3 \equiv \qquad\qquad \theta^1 \wedge (b_1 \omega^1 + b_2 \omega^2 + b_3 \pi_1^1 + b_4 \pi_1^2) \end{cases}$$

for certain functions a_i and b_i. By replacing $\omega^1 + a_4 \theta^1$ by ω^1 and then $\pi_2^1 - a_1 \theta^1$ by π_2^1, it may be assumed that $a_4 = a_1 = 0$. By differentiating $d\theta^2$ and considering the coefficients of $\omega^1 \wedge \omega^2 \wedge \pi_1^1$ and $\omega^2 \wedge \pi_1^1 \wedge \pi_1^2$ it is inferred that $a_2 = a_3 = 0$. Treating $d\theta^3$ in the same way we see that also $b_2 = b_3 = 0$. Finally the replacement $\theta^3 - b_1 \theta^0 \mapsto \theta^3$ makes b_1 vanish, and then the Lie structure has been reduced to

$$\begin{cases} d\theta^0 \equiv \omega^1 \wedge \theta^1, \\ d\theta^2 \equiv \omega^1 \wedge \pi_1^2, & (\text{mod } \mathcal{P}_{\operatorname{sing}}). \\ d\theta^3 \equiv b_4 \theta^1 \wedge \pi_1^2 \end{cases}$$

From this we see that

$$\mathcal{C}(\mathcal{P}_{\operatorname{sing}}) = (\theta^0, \theta^1, \theta^2, \theta^3, \omega^1, \pi_1^2) = \operatorname{sing}(\mathcal{P}).$$

Hence $\mathcal{P}_{\operatorname{sing}}$ can be regarded as living on $J^1(\mathbb{C}_x^3, \mathbb{C}_z) \times \mathbb{C}_v$ for a suitable function v, and admitting the two Cauchy characteristic vector fields $\partial/\partial\omega^2$ and $\partial/\partial\omega^3$.

As in similar cases treated earlier it turns out that the original problem is equivalent to that of integrating $\mathcal{P}_{\operatorname{sing}}$ on $J^1(\mathbb{C}_x^3, \mathbb{C}_z) \times \mathbb{C}_v$. Because $\partial/\partial\omega^2$ and $\partial/\partial\omega^3$ are Cauchy characteristic, the latter is reduced to finding integral curves on which $\omega^1 \neq 0$—and this is achieved by solving ordinary differential equations.

Finally, in the case of involutive systems of *five* second order PDEs in one dependent and three independent variables there is but one type to consider: $\mathcal{P} = (\theta^0, \theta^1, \theta^2, \theta^3)$ with the structure

$$\begin{cases} d\theta^0 \equiv 0, \\ d\theta^1 \equiv \omega^1 \wedge \pi_1^1, \\ d\theta^2 \equiv 0, \\ d\theta^3 \equiv 0 \end{cases} \quad (\text{mod } \mathcal{P}).$$

Hence

$$\mathcal{C}(\mathcal{P}) = (\theta^0, \theta^1, \theta^2, \theta^3, \omega^1, \pi_1^1),$$

so that $\mathcal{P}$ can be regarded as a pfaffian system on $J^1(\mathbb{C}^3_x, \mathbb{C}_z) \times \mathbb{C}_v$ for a suitable function v, admitting $\partial/\partial\omega^2$ and $\partial/\partial\omega^3$ as Cauchy characteristic vector fields. It therefore suffices to find integral curves on which $\omega^1 \neq 0$, and this may be done by solving ordinary differential equations.

Conclusion. *We have seen that*

- **all** *types of involutive systems of second order PDEs in one dependent and three independent variables* **except one** *admit singular systems,*
- *for most types which admit singular systems, the latter can by means of the Monge method be utilized so as to bring about a substantial simplification,*
- *in several cases the method of Monge even gives a reduction to ordinary differential equations.*

The most urgent task remaining is to replace the tedious case-by-case verifications of the Monge method by a more conceptual framework.

18.7 How to go further?

The main reason behind the work of Cartan and Vessiot on second order PDEs in one dependent and two independent variables is to clarify Goursat's remarkable classifications of parabolic and hyperbolic PDEs respectively. The key to this is the exploitation of *Monge characteristic systems*, and the surprising result is that the classifications ultimately are explained by *Lie group theory*.

A spin-off effect is that the use of Monge systems and their first integrals often enough makes it possible to find integral manifolds without having to employ the Cauchy–Kowalewski theorem.

In this last chapter we have seen that the idea of exploiting singular systems is quite successful also in dealing with involutive systems of second order PDEs in one dependent and *three* independent variables—although the Monge method used is not very precise, but rather consists of a number of 'variations on a theme'.

The classification of Drach in section **5.1** implies that in order to understand general PDE systems it suffices to master second order PDE systems in one dependent and n independent variables. If one is willing to believe that the case $n = 3$ is typical, the results found indicate that the consideration of singular systems and their first integrals should be a general clue to solving PDE systems without applying the Cauchy–Kowalewski theorem.

This last point became acute in 1957, when Lewy found his celebrated counterexample to the Cauchy–Kowalewski theorem in the C^∞ category. Thereby—and more tragically—Cartan's local existence theorem also failed in the case of C^∞ data, and so the following fundamental question arose.

The Lewy problem. *What are the necessary and sufficient conditions for an involutive C^∞ PDE system to be locally solvable?*

At present the general answer to this seems to be very far away.

However, the work of Cartan and Vessiot described here can—although pre-Lewy—be seen as an attempt at a solution, focusing on singular systems. Indeed, we have seen that the Cartan–Vessiot theory is very rich in results, although there unfortunately have been too many case-by-case verifications.

A rather different approach is found in [Yang 1987]. It is based upon the following facts:

- the chain of Cauchy–Kowalewski systems used for integrating a vector field system $\mathcal{V}$ as described in section **3.2** is not entirely determined by $\mathcal{V}$, but allows a certain flexibility;
- *hyperbolic* Cauchy–Kowalewski systems *are* locally solvable in the C^∞ category.

Therefore $\mathcal{V}$ can be integrated in the C^∞ category provided that the Cauchy–Kowalewski systems appearing in Cartan's solution procedure can be chosen in such a way that they *all are hyperbolic*. Not surprisingly it turns out that the possibility for achieving this is connected with the existence of Monge systems.

A more general theory, containing Yang's as a special case, has been

given by Kakié in a series of papers—see e.g. [Kakié 1987a], [Kakié 1989]
and [Kakié 1990]. The starting point is an *algebraic* definition of Monge
characteristics. But this is a bit foreign to the point of view used here, so
let us therefore indicate the corresponding differential geometric notion.

Let $\mathcal{P} = (\theta^1, \ldots, \theta^s)$ be a pfaffian system on $\mathbb{R}^r$, which is involutive with
respect to the independence condition $dx^1 \wedge \cdots \wedge dx^n$, where $x^1, \ldots, x^n, x^{n+1}$,
$\ldots, x^r$ are local coordinates for $\mathbb{R}^r$. Setting $x := (x^1, \ldots, x^r)$, we consider
a vector field

$$X := \sum_{i=1}^{n} a^i(x) \frac{\partial}{\partial x^i}.$$

For any integral manifold $\mathcal{N}$ of $\mathcal{P}$ on which $dx^1 \wedge \cdots \wedge dx^n \neq 0$, this X has
a unique prolongation to a vector field $X_{\mathcal{N}}$ on $\mathcal{N}$. Then quite obviously
$\theta^i{}_{|\mathcal{N}}(X_{\mathcal{N}}) = 0$ for $i = 1, \ldots, s$. For *certain* X it *may happen* that there are
one-forms which are linearly independent of $\theta^1, \ldots, \theta^s, dx^1, \ldots, dx^n$, and
have the property that their restrictions to any integral manifold $\mathcal{N}$ will
also kill $X_{\mathcal{N}}$. In this case X is said to be *Monge characteristic*, and the
extra one-forms are called *Monge characteristic one-forms*.

Although the definitions are not exactly the same, there is clearly a
connection to the Monge fields of the Cartan–Vessiot theory.

Schematically the algebraic definition of Monge characteristics goes as
follows:

> involutive pfaffian system $\longrightarrow$ involutive symbol $\longrightarrow$ characteristic
> module $\longrightarrow$ irredundant primary decomposition $\longrightarrow$ characteristic
> vectors.

The principal existence result of Kakié is as follows.

*Let $\mathcal{P}$ be a C^∞ pfaffian system which is in involution with respect to an
independence condition $\omega^1 \wedge \cdots \wedge \omega^n$. Suppose that*

> (i) *there is an integer p, $1 \leq p < n$, such that the Cartan characters
> of $\mathcal{P}$ satisfy*
>
> $$s_1 = \cdots = s_p > 0, \quad s_{p+1} = \cdots = s_n = 0,$$

> (ii) *certain conditions on the primary decomposition of the charac-
> teristic module are satisfied, which from a differential geometric
> point of view roughly amount to the existence of 'sufficiently many'
> Monge characteristic one-forms.*

Then $\mathcal{P}$ admits local n-dimensional integral manifolds.

The first (and easiest) part of the proof consists in showing that $\mathcal{P}$

admits p-dimensional integral manifolds; in the second part these are then uniquely extended to n-dimensional integral manifolds.

For the latter part condition 1 can actually be replaced by the weaker $s_p > 0$ and $s_{p+1} = \cdots = s_n = 0$.

It is this type of general theorem which is missing in the Cartan–Vessiot theory. On the other hand, a disadvantage of the Kakié–Yang theory is that it is difficult to find natural examples for which it works. Here is one, though, taken from [Kakié 1987a].

Example. Let p be an integer such that $1 \le p < n$, let $\mathcal{P} = (\theta^1, \ldots, \theta^s)$ be a C^∞ pfaffian system on $\mathbb{R}^{n+s+sp}$, and let us look for n-dimensional integral manifolds satisfying an independence condition $\omega^1 \wedge \cdots \wedge \omega^n \ne 0$.

Suppose moreover that $\mathcal{P}$ has the Lie structure

$$d\theta^i \equiv \sum_{j \in J^i} \omega^j \wedge \pi^i_j \quad (\bmod \ \mathcal{P}) \quad \text{for } i = 1, \ldots, s,$$

where

(i) $J^1, \ldots, J^s \subset \{1, \ldots, n\}$ each contain p integers,
(ii) the ω^j, θ^i and π^i_j form a local coframe for $\mathbb{R}^{n+s+sp}$.

Then $\mathcal{P}$ is involutive with respect to $\omega^1 \wedge \cdots \wedge \omega^n$, its Cartan characters are $s_1 = \cdots = s_p = s$, $s_{p+1} = \cdots = s_n = 0$, and Kakié's theorem applies—so *that $\mathcal{P}$ can be integrated in the C^∞ category.*

Although impressive, the Kakié–Yang theory does not cover all the cases solved by the Cartan–Vessiot theory—as follows already from the first condition on the Cartan characters.

An urgent task (not as overwhelming as the general Lewy problem) is therefore to connect these two theories, and possibly find a more general one, including them as special cases.

A first step in this direction is to gain a thorough understanding of the classical work of Cartan and Vessiot—which is precisely the object of this monograph.

Final remark. There is a story of an American who dreamed for years of an extended trip to Paris, to live on the left bank, and absorb Parisian life as it really is. Shortly before the long-awaited departure he cancelled his trip, confidently declaring that it was no longer necessary.

The reason? He had just completed reading the *Michelin Guide to Paris*.

Do not make the same mistake as regards this book! Its purpose certainly is not to replace reading the original papers of Lie, Cartan and Vessiot, but quite the opposite: to *encourage it, and hopefully make it more fruitful.*

Bibliography

Banyaga 1997 *The Structure of Classical Diffeomorphism Groups*, Kluwer Academic Puplishers, Dordrecht, Netherlands, 1997.

BCG³ 1991 Bryant, Chern, Gardner, Goldschmidt, Griffiths: *Exterior Differential Systems*, Springer Verlag, New York, Berlin, Heidelberg, 1991.

Cartan 1896 Sur la réduction à sa forme canonique de la structure d'un groupe de transformations fini et continu, *Amer. J. Math.*, **18** (1896), 1–61.

Cartan 1901a Sur l'intégration des systèmes d'équations aux différentielles totales, *Ann. Éc. Normale*, **18** (1901), 241–311.

Cartan 1901b Sur l'intégration de certaines systèmes de Pfaff de caractère deux, *Bull. Soc. Math. France.*, **29** (1901), 233–302.

Cartan 1902 Sur l'équivalence des systèmes différentiels, *C.R. Acad. Sci.*, **135** (1902), 781–783.

Cartan 1904 Sur la structure des groupes infinis de transformations, chapitres I–II, *Ann. Éc. Normale*, **21** (1904), 153–206.

Cartan 1905 Sur la structure des groupes infinis de transformations, chapitres III–IV, *Ann. Éc. Normale*, **22** (1905), 219–308.

Cartan 1908 Les sous-groupes des groupes continus de transformations, *Ann. Éc. Normale*, **25** (1908), 57–194.

Cartan 1910 Les systèmes de Pfaff à cinq variables et les équations aux dérivées partielles du second ordre, *Ann. Éc. Normale*, **27** (1910), 109–192.

Cartan 1911 Sur les systèmes en involution d'équations aux dérivées partielles du second ordre à une fonction inconnue de trois variables indépendantes, *Bull. Soc. Math. France*, **39** (1911), 352–443.

Cartan 1914 Sur l'intégration de certaines systèmes d'équations différenti-elles, *C. R. Acad. Sci.*, **158** (1914), 326–328.

Cartan 1915 Sur les transformations de Bäcklund, *Bull. Soc. Math. France*, **43** (1915), 6–24.

Cartan 1937a Les problèmes d'équivalence, *Œuvres Complètes d'Élie Cartan*, Gauthier-Villars 1953 and Springer Verlag 1984, Partie II, 1311–1334.

Cartan 1937b La structure des groupes infinis, *Œuvres Complètes d'Élie Cartan*, Gauthier-Villars 1953 and Springer Verlag 1984, Partie II, 1335–1384.

Cartan 1945 *Les Systèmes Différentielles Extérieurs et Leurs Applications Géométriques*, Hermann, Paris, 1945.

Cartan 1946 L'œuvre scientifique de M. Ernest Vessiot, *Bull. Soc. Math. France*, **75** (1946), 1–8.

Cartan 1953 *Œuvres Complètes d'Élie Cartan*, Gauthier-Villars 1953 and Springer Verlag 1984—which contains the above references for Cartan, except [Cartan 1945] and [Cartan 1946].

Drach 1897 Sur les systèmes complètement orthogonaux dans l'espace à n dimensions et sur la réduction des systèmes différentielles les plus généraux, *C. R. Acad. Sci.* **125** (1897), 598.

Flanders 1989 *Differential Forms with Applications to the Physical Sciences*, Dover Publications, New York, 1989.

Gardner 1989 *The Method of Equivalence and Its Applications*, SIAM, Philadelphia, 1989.

Goursat 1895 Sur une classe d'équations aux dérivées partielles du second ordre, et sur la théorie des intégrales intermédiaires, *Acta Math.*, **19** (1895), 285–340.

Goursat 1896–98 *Leçons sur l'Intégration des Équations aux Dérivées Partielles du Second Ordre*, Hermann, Paris, Tome I 1896, Tome II 1898.

Goursat 1899 Recherches sur quelques équations aux dérivées partielles du second ordre, *Ann. Fac. Sci. Toulouse*, **1** (1899), 31–78 and 439–464.

Goursat 1918 Sur le problème de Bäcklund et les systèmes de deux équations de Pfaff, *Ann. Fac. Sci. Toulouse*, **10** (1918), 65–173.

Goursat 1922 *Leçons sur le Problème de Pfaff*, Hermann, Paris, 1922.

Hermann 1965 É. Cartan's geometric theory of partial differential equations, *Adv. Math.*, **1** (1965), 265–317.

Hsiao 1980 On Cartan–Sternberg's example of the reduction of G-structure II–III, *Tamkang J. Math.*, **11** (1980), 277–312.

Ibragimow 1994–96 *CRC Handbook of Lie Group Analysis of Differential Equations*, Vol. **1** (1994), Vol. **2** (1995), Vol. **3** (1996), CRC Press, Boca Raton, Florida.

Janet 1929 *Leçons sur les Systèmes d'Équations aux Dérivées Partielles*, Gauthier-Villars, Paris, 1929.

Kähler 1934 *Einführung in die Theorie der Systeme von Differentialgleichungen*, Chelsea Publishing Company, New York, 1949.

Kakié 1977 Algebraic structures of characteristics in involutive systems of non-linear partial differential equations, *Publ. Res. Inst. Math. Sci.*, **13** (1977), 107–158.

Kakié 1979 Cauchy's characteristics of involutive systems of non-linear partial differential equations, *Comment. Math. Univ. St Paul*, **28** (1979), 87–92.

Kakié 1985 A fundamental property of Monge characteristics in involutive systems of non-linear partial differential equations and its application, *Math. Ann.*, **273** (1985), 89–114.

Kakié 1987a The Monge characteristic in involutive Pfaffian systems and its application to the Cauchy problem, *Japan. J. Math.*, **13** (1987), 127–162.

Kakié 1987b Symmetrizable hyperbolic systems of non-linear partial differential equations, *Comment. Math. Univ. St Paul*, **36** (1987), 215–226.

Kakié 1989 On the integration of involutive systems of non-linear partial differential equations, *J. Fac. Sci. Univ. Tokyo Sect. IA Math.*, **36**, 537–569.

Kakié 1990 Existence of C^∞ integral manifolds of an involutive Pfaffian system, *Comment. Math. Univ. St Paul*, **39** (1990), 111–125.

Kuranishi 1959 On the local theory of continuous infinite pseudo groups, *Nagoya Math. J.*, **15** (1959), 225–260.

Kuranishi 1961 On the local theory of continuous infinite pseudo groups, *Nagoya Math. J.*, **19** (1961), 55–91.

Kuranishi 1962 *Lectures on Exterior Differential Systems*, Tata Institute of Fundamental Research, Bombay, 1962.

Lie 1885 Allgemeine Untersuchungen über Differentialgleichungen, die eine kontinuierliche endliche Gruppe gestatten, *Math. Ann.*, **25** (1885), 71–151.

Lie–Scheffers 1893 *Vorlesungen über continuierliche Gruppen mit geometrischen und anderen Anwendungen*, Chelsea Publishing Company, New York, 1971.

Lie–Scheffers 1896 *Geometrie der Berührungstransformationen*, Chelsea Publishing Company, New York, 1977.

Matsuda 1967 Cartan–Kuranishi's prolongation theorem of differential systems combined with that of Lagrange and Jacobi, *Publ. RIMS, Kyoto Univ. Ser. A*, **3** (1967), 69–84.

Matsuda 1972 Generalized Pfaff's problem, *J. Fac. Sci. Univ. Tokyo Sect. IA*, **19** (1972), 231–242.

Morimoto 1993 Geometric structures on filtered manifolds, *Hokkaido Math. J.*, **22** (1993), 263–347.

Olver 1986 *Applications of Lie Groups to Differential Equations*, Graduate Texts in Mathematics, Vol. **107**, Springer Verlag, New York, 1986.

Olver 1995 *Equivalence, Invariants and Symmetry*, Cambridge Univ. Press, New York, 1995.

Parsons 1960 *The Extension of Darboux's Method*, Mémorial Sci. Math. CXLII, Gauthier-Villars, Paris, 1960.

Pommaret 1978 *Systems of Partial Differential Equations and Lie Pseudogroups*, Gordon and Breach, London, New York, 1978.

Pommaret 1983 *Differential Galois Theory*, Gordon and Breach, London, New York, 1983.

Pommaret 1994 *Partial Differential Equations and Group Theory*, Kluwer Academic Publishers, Dordrecht, Netherlands, 1994.

Pontryagin 1966 *Topological Groups*, Gordon and Breach, New York, 1966.

Vessiot 1903 Sur la théorie des groupes continues, *Ann. Éc. Normale*, **20** (1903), 411–451.

Vessiot 1924 Sur une théorie nouvelle des problèmes d'integration, *Bull. Soc. Math. France*, **52** (1924), 336–395.

Vessiot 1928 Sur l'intégration des faisceaux de transformations infinitésimales dans le cas où le degré du faisceau étant n, celui du faisceau dérivée est $n + 1$, *Ann. Éc. Normale*, **45** (1928), 189–253.

Vessiot 1930 Sur les intégrales complètes et les caractéristiques des équations aux dérivées partielles du second ordre à une fonction inconnue et n variables indépendantes, *Praktika Acad. Athènes*, **5** (1930), 424–441.

Vessiot 1936 Sur les faisceaux de transformations infinitésimales associées aux équations aux dérivées partielles du second ordre $F(x, y, z, p, q, r, s, t) = 0$, *J. Math. pure appl.*, **15** (1936), 301–320.

Vessiot 1937 Sur une classe de faisceaux complets de degré 2, *Bull. Soc. Math. France*, **45** (1937), 149–167.

Vessiot 1939 Sur les équations aux dérivées partielles du second ordre, $F(x, y, z, p, q, r, s, t) = 0$, intégrable par la méthode de Darboux, *J. Math. pure appl.*, **18** (1939), 1–61.

Vessiot 1942 Sur les équations aux dérivées partielles du second ordre, $F(x, y, z, p, q, r, s, t) = 0$, intégrables par la méthode de Darboux, *J. Math. pure appl.*, **21** (1942), 1–66.

Yang 1987 Involutive hyperbolic differential systems, *Mem. Amer. Math. Soc.*, **68**, No. 370, Providence, Rhode Island (1987).

Index

Lightning Source UK Ltd.
Milton Keynes UK
UKHW022016160921
390718UK00003B/19